AF328572

Springer Series in Statistics

Advisors:
P. Bickel, P. Diggle, S. Fienberg, K. Krickeberg,
I. Olkin, N. Wermuth, S. Zeger

Springer Science+Business Media, LLC

Springer Series in Statistics

Andersen/Borgan/Gill/Keiding: Statistical Models Based on Counting Processes.
Atkinson/Riani: Robust Diagnotstic Regression Analysis.
Berger: Statistical Decision Theory and Bayesian Analysis, 2nd edition.
Borg/Groenen: Modern Multidimensional Scaling: Theory and Applications
Brockwell/Davis: Time Series: Theory and Methods, 2nd edition.
Chan/Tong: Chaos: A Statistical Perspective.
Chen/Shao/Ibrahim: Monte Carlo Methods in Bayesian Computation.
David/Edwards: Annotated Readings in the History of Statistics.
Devroye/Lugosi: Combinatorial Methods in Density Estimation.
Efromovich: Nonparametric Curve Estimation: Methods, Theory, and Applications.
Eggermont/LaRiccia: Maximum Penalized Likelihood Estimation, Volume I:
     Density Estimation.
Fahrmeir/Tutz: Multivariate Statistical Modelling Based on Generalized Linear
     Models, 2nd edition.
Farebrother: Fitting Linear Relationships: A History of the Calculus of Observations
     1750-1900.
Federer: Statistical Design and Analysis for Intercropping Experiments, Volume I:
     Two Crops.
Federer: Statistical Design and Analysis for Intercropping Experiments, Volume II:
     Three or More Crops.
Glaz/Naus/Wallenstein: Scan Statistics.
Good: Permutation Tests: A Practical Guide to Resampling Methods for Testing
     Hypotheses, 2nd edition.
Gouriéroux: ARCH Models and Financial Applications.
Gu: Smoothing Spline ANOVA Models.
Györfi/Kohler/Krzyżak/ Walk: A Distribution-Free Theory of Nonparametric
     Regression.
Haberman: Advanced Statistics, Volume I: Description of Populations.
Hall: The Bootstrap and Edgeworth Expansion.
Härdle: Smoothing Techniques: With Implementation in S.
Harrell: Regression Modeling Strategies: With Applications to Linear Models,
     Logistic Regression, and Survival Analysis
Hart: Nonparametric Smoothing and Lack-of-Fit Tests.
Hastie/Tibshirani/Friedman: The Elements of Statistical Learning: Data Mining,
     Inference, and Prediction
Hedayat/Sloane/Stufken: Orthogonal Arrays: Theory and Applications.
Heyde: Quasi-Likelihood and its Application: A General Approach to Optimal
     Parameter Estimation.
Huet/Bouvier/Gruet/Jolivet: Statistical Tools for Nonlinear Regression: A Practical
     Guide with S-PLUS Examples.
Ibrahim/Chen/Sinha: Bayesian Survival Analysis.
Jolliffe: Principal Component Analysis.
Kolen/Brennan: Test Equating: Methods and Practices.
Kotz/Johnson (Eds.): Breakthroughs in Statistics Volume I.

(continued after index)

László Györfi   Michael Kohler
Adam Krzyżak   Harro Walk

A Distribution-Free Theory of Nonparametric Regression

With 86 Figures

 Springer

László Györfi
Department of Computer Science and
    Information Theory
Budapest University of Technology and
    Economics
1521 Stoczek, U.2.
Budapest
Hungary
gyorfi@inf.bme.hu

Michael Kohler
Fachbereich Mathematik
Universität Stuttgart
Pfaffenwaldring 57
70569 Stuttgart
Germany
kohler@mathematik.uni-stuttgart.de

Adam Krzyżak
Department of Computer Science
Concordia University
1455 De Maisonneuve Boulevard West
Montreal, Quebec, H3G 1M8
Canada
krzyzak@cs.concordia.ca

Harro Walk
Fachbereich Mathematik
Universität Stuttgart
Pfaffenwaldring 57
70569 Stuttgart
Germany
walk@mathematik.uni-stuttgart.de

Library of Congress Cataloging-in-Publication Data
A distribution-free theory of nonparametric regression / László Györfi . . . [et al.].
        p. cm. — (Springer series in statistics)
    Includes bibliographical references and index.

    DOI 10.1007/978-0-387-22442-8
      1. Regression analysis.   2. Nonparametric statistics.   3. Distribution (Probability theory)
    I. Györfi, László.   II. Series.
    QA278.2 .D57 2002
    519.5′36—dc21   2002021151

    Printed on acid-free paper.

9 8 7 6 5 4 3 2 1   SPIN 10866288

Typesetting: Pages created by the authors using a Springer TEX macro package.

www.springer-ny.com
www.springer.com/mycopy

To our families:

Kati, Kati, and Jancsi
Judith, Iris, and Julius
Henryka, Jakub, and Tomasz
Hildegard

Preface

The *regression estimation problem* has a long history. Already in 1632 Galileo Galilei used a procedure which can be interpreted as fitting a linear relationship to contaminated observed data. Such fitting of a line through a cloud of points is the classical linear regression problem. A solution of this problem is provided by the famous principle of least squares, which was discovered independently by A. M. Legendre and C. F. Gauss and published in 1805 and 1809, respectively. The principle of least squares can also be applied to construct *nonparametric* regression estimates, where one does not restrict the class of possible relationships, and will be one of the approaches studied in this book.

Linear regression analysis, based on the concept of a regression function, was introduced by F. Galton in 1889, while a probabilistic approach in the context of multivariate normal distributions was already given by A. Bravais in 1846. The first nonparametric regression estimate of local averaging type was proposed by J. W. Tukey in 1947. The partitioning regression estimate he introduced, by analogy to the classical partitioning (histogram) density estimate, can be regarded as a special least squares estimate.

Some aspects of nonparametric estimation had already appeared in belletristic literature in 1930/31 in *The Man Without Qualities* by Robert Musil (1880-1942) where, in Section 103 (first book), methods of partitioning estimation are described: "... as happens so often in life, you ... find yourself facing a phenomenon about which you can't quite tell whether it is a law or pure chance; that's where things acquire a human interest. Then you translate a series of observations into a series of figures, which you divide into categories to see which numbers lie between this value and that, and the next, and so on .... You then calculate the degree of aberration, the mean deviation, the degree of deviation from some arbitrary value ... the average value ... and so forth, and with the help of all these concepts you study your given phenomenon" (cited from page 531 of the English translation, Alfred A. Knopf Inc., Picador, 1995).

Besides its long history, the problem of regression estimation is of increasing importance today. Stimulated by the tremendous growth of information technology in the past 20 years, there is a growing demand for procedures capable of automatically extracting useful information from massive highly-dimensional databases that companies gather about their customers. One of the fundamental approaches for dealing with this "data-mining problem" is regression estimation. Usually there is little or no *a priori* information about the data, leaving the researcher with no other choice but a nonparametric approach.

This book presents a modern approach to nonparametric regression with random design. The starting point is a prediction problem where minimization of the mean squared error (or L_2 risk) leads to the regression function. If the goal is to construct an estimate of this function which has mean squared prediction error close to the minimum mean squared error, then this goal naturally leads to the L_2 error criterion used throughout this book.

We study almost all known regression estimates, such as classical local averaging estimates including kernel, partitioning, and nearest neighbor estimates, least squares estimates using splines, neural networks and radial basis function networks, penalized least squares estimates, local polynomial kernel estimates, and orthogonal series estimates. The emphasis is on the *distribution-free* properties of the estimates, and thus most consistency results presented in this book are valid for all distributions of the data. When it is impossible to derive distribution-free results, as is the case for rates of convergence, the emphasis is on results which require as few constraints on distributions as possible, on distribution-free inequalities, and on adaptation.

Our aim in writing this book was to produce a self-contained text intended for a wide audience, including graduate students in statistics, mathematics, computer science, and engineering, as well as researchers in these fields. We start off with elementary techniques and gradually introduce more difficult concepts as we move along. Chapters 1–6 require only a basic knowledge of probability. In Chapters 7 and 8 we use exponential inequalities for the sum of independent random variables and for the sum of martingale differences. These inequalities are proven in Appendix A. The remaining part of the book contains somewhat more advanced concepts, such as almost sure convergence together with the real analysis techniques given in Appendix A. The foundations of the least squares and penalized least squares estimates are given in Chapters 9 and 19, respectively.

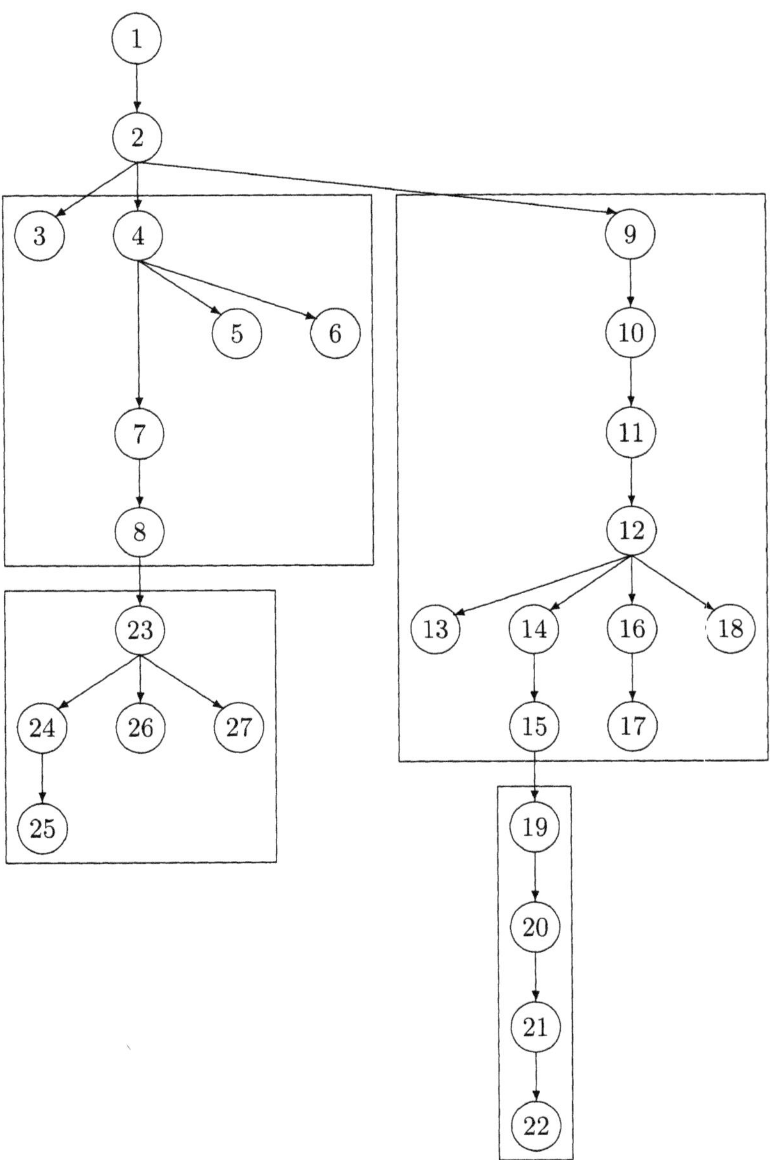

Figure 1. The structure of the book.

The structure of the book is shown in Figure 1. This figure is a precedence tree which could assist an instructor in organizing a course based on this book. It shows the sequence of chapters needed to be covered in order to understand a particular chapter. The focus of the chapters in the upper-left box is on local averaging estimates, in the lower-left box on strong consistency results, in the upper-right box on least squares estimation, and in the lower-right box on penalized least squares.

We would like to acknowledge the contribution of many people who influenced the writing of this book. Luc Devroye, Gábor Lugosi, Eric Regener, and Alexandre Tsybakov made many invaluable suggestions leading to conceptual improvements and better presentation. A number of colleagues and friends have, often without realizing it, contributed to our understanding of nonparametrics. In particular we would like to thank in this respect Paul Algoet, Andrew Barron, Peter Bartlett, Lucien Birgé, Jan Beirlant, Alain Berlinet, Sándor Csibi, Miguel Delgado, Jürgen Dippon, Jerome Friedman, Wlodzimierz Greblicki, Iain Johnstone, Jack Koplowitz, Tamás Linder, Andrew Nobel, Mirek Pawlak, Ewaryst Rafajlowicz, Igor Vajda, Sara van de Geer, Edward van der Meulen, and Sid Yakowitz. András Antos, András György, Michael Hamers, Kinga Máthé, Dániel Nagy, Márta Pintér, Dominik Schäfer and Stefan Winter provided long lists of mistakes and typographical errors. Sándor Győri drew the figures and gave us advice and help on many LaTeX-problems. John Kimmel was helpful, patient and supportive at every stage.

In addition, we gratefully acknowledge the research support of the Budapest University of Technology and Economics, the Hungarian Academy of Sciences (MTA SZTAKI, AKP, and MTA IEKCS), the Hungarian Ministry of Education (FKFP and MÖB), the University of Stuttgart, Deutsche Forschungsgemeinschaft, Stiftung Volkswagenwerk, Deutscher Akademischer Austauschdienst, Alexander von Humboldt Stiftung, Concordia University, Montreal, NSERC Canada, and FCAR Quebec.

Early versions of this text were tried out at a DMV seminar in Oberwolfach, Germany, and in various classes at the Carlos III University of Madrid, the University of Stuttgart, and at the International Centre for Mechanical Sciences in Udine. We would like to thank the students there for useful feedback which improved this book.

László Györfi,   Budapest, Hungary
Michael Kohler,   Stuttgart, Germany
Adam Krzyżak,   Montreal, Canada
Harro Walk,   Stuttgart, Germany

June 6, 2002

Contents

1
Why Is Nonparametric Regression Important?

In the present and following chapters we provide an overview of this book. In this chapter we introduce the problem of regression function estimation and describe important properties of regression estimates. An overview of various approaches to nonparametric regression estimates is provided in the next chapter.

1.1 Regression Analysis and L_2 Risk

In regression analysis one considers a random vector (X, Y), where X is $\mathcal{R}^d$-valued and Y is $\mathcal{R}$-valued, and one is interested how the value of the so-called response variable Y depends on the value of the observation vector X. This means that one wants to find a (measurable) function $f : \mathcal{R}^d \to \mathcal{R}$, such that $f(X)$ is a "good approximation of Y," that is, $f(X)$ should be close to Y in some sense, which is equivalent to making $|f(X) - Y|$ "small." Since X and Y are random vectors, $|f(X) - Y|$ is random as well, therefore it is not clear what "small $|f(X) - Y|$" means. We can resolve this problem by introducing the so-called L_2 *risk* or *mean squared error* of f,

$$\mathbf{E}|f(X) - Y|^2,$$

and requiring it to be as small as possible.

While it seems natural to use the expectation, it is not obvious why one wants to minimize the expectation of the squared distance between $f(X)$

and Y and not, more generally, the L_p risk

$$\mathbf{E}|f(X) - Y|^p$$

for some $p \geq 1$ (especially $p = 1$). There are two reasons for considering the L_2 risk. First, as we will see in the sequel, this simplifies the mathematical treatment of the whole problem. For example, as is shown below, the function which minimizes the L_2 risk can be derived explicitly. Second, and more important, trying to minimize the L_2 risk leads naturally to estimates which can be computed rapidly. This will be described later in detail (see, e.g., Chapter 10).

So we are interested in a (measurable) function $m^* : \mathcal{R}^d \to \mathcal{R}$ such that

$$\mathbf{E}|m^*(X) - Y|^2 = \min_{f:\mathcal{R}^d \to \mathcal{R}} \mathbf{E}|f(X) - Y|^2.$$

Such a function can be obtained explicitly as follows. Let

$$m(x) = \mathbf{E}\{Y|X = x\}$$

be the *regression function*. We will show that the regression function minimizes the L_2 risk. Indeed, for an arbitrary $f : \mathcal{R}^d \to \mathcal{R}$, one has

$$\begin{aligned}
\mathbf{E}|f(X) - Y|^2 &= \mathbf{E}|f(X) - m(X) + m(X) - Y|^2 \\
&= \mathbf{E}|f(X) - m(X)|^2 + \mathbf{E}|m(X) - Y|^2,
\end{aligned}$$

where we have used

$$\begin{aligned}
& \mathbf{E}\left\{(f(X) - m(X))(m(X) - Y)\right\} \\
&= \mathbf{E}\left\{\mathbf{E}\left\{(f(X) - m(X))(m(X) - Y)\big|X\right\}\right\} \\
&= \mathbf{E}\left\{(f(X) - m(X))\mathbf{E}\{m(X) - Y|X\}\right\} \\
&= \mathbf{E}\left\{(f(X) - m(X))(m(X) - m(X))\right\} \\
&= 0.
\end{aligned}$$

Hence,

$$\mathbf{E}|f(X) - Y|^2 = \int_{\mathcal{R}^d} |f(x) - m(x)|^2 \mu(dx) + \mathbf{E}|m(X) - Y|^2, \qquad (1.1)$$

where μ denotes the distribution of X. The first term is called the L_2 error of f. It is always nonnegative and is zero if $f(x) = m(x)$. Therefore, $m^*(x) = m(x)$, i.e., the optimal approximation (with respect to the L_2 risk) of Y by a function of X is given by $m(X)$.

1.2 Regression Function Estimation and L_2 Error

In applications the distribution of (X, Y) (and hence also the regression function) is usually unknown. Therefore it is impossible to predict Y using

$m(X)$. But it is often possible to observe data according to the distribution of (X, Y) and to estimate the regression function from these data.

To be more precise, denote by (X, Y), (X_1, Y_1), $(X_2, Y_2), \ldots$ independent and identically distributed (i.i.d.) random variables with $\mathbf{E} Y^2 < \infty$. Let D_n be the set of *data* defined by

$$D_n = \{(X_1, Y_1), \ldots, (X_n, Y_n)\}.$$

In the regression function estimation problem one wants to use the data D_n in order to construct an estimate $m_n : \mathcal{R}^d \to \mathcal{R}$ of the regression function m. Here $m_n(x) = m_n(x, D_n)$ is a measurable function of x and the data. For simplicity, we will suppress D_n in the notation and write $m_n(x)$ instead of $m_n(x, D_n)$.

In general, estimates will not be equal to the regression function. To compare different estimates, we need an error criterion which measures the difference between the regression function and an arbitrary estimate m_n. In the literature, several distinct error criteria are used: first, the pointwise error,

$$|m_n(x) - m(x)| \text{ for some fixed } x \in \mathcal{R}^d,$$

second, the supremum norm error,

$$\|m_n - m\|_\infty = \sup_{x \in C} |m_n(x) - m(x)| \text{ for some fixed set } C \subseteq \mathcal{R}^d,$$

which is mostly used for $d = 1$ where C is a compact subset of $\mathcal{R}$ and, third, the L_p error,

$$\int_C |m_n(x) - m(x)|^p dx,$$

where the integration is with respect to the Lebesgue measure, C is a fixed subset of $\mathcal{R}^d$, and $p \geq 1$ is arbitrary (often $p = 2$ is used).

One of the key points we would like to make is that the motivation for introducing the regression function leads naturally to an L_2 error criterion for measuring the performance of the regression function estimate. Recall that the main goal was to find a function f such that the L_2 risk $\mathbf{E}|f(X) - Y|^2$ is small. The minimal value of this L_2 risk is $\mathbf{E}|m(X) - Y|^2$, and it is achieved by the regression function m. Similarly to (1.1), one can show that the L_2 risk $\mathbf{E}\{|m_n(X) - Y|^2 | D_n\}$ of an estimate m_n satisfies

$$\mathbf{E}\left\{|m_n(X) - Y|^2 | D_n\right\} = \int_{\mathcal{R}^d} |m_n(x) - m(x)|^2 \mu(dx) + \mathbf{E}|m(X) - Y|^2. \quad (1.2)$$

Thus the L_2 risk of an estimate m_n is close to the optimal value if and only if the L_2 error

$$\int_{\mathcal{R}^d} |m_n(x) - m(x)|^2 \mu(dx) \quad (1.3)$$

is close to zero. Therefore we will use the L_2 error (1.3) in order to measure the quality of an estimate and we will study estimates for which this L_2 error is small.

1.3   Practical Applications

In this section we describe several applications in order to illustrate the practical relevance of regression estimation.

Example 1.1. *Loan management.*

A bank is interested in predicting the return Y on a loan given to a customer. Available to the bank is the profile X of the customer including his credit history, assets, profession, income, age, etc. The predicted return affects the decision as to whether to issue or refuse a loan, as well as the conditions of the loan. For more details refer to Krahl et al. (1998).

Example 1.2. *Profit prediction in marketing.*

A company is interested in boosting its sales by mailing product advertisements to potential customers. Typically 50,000 people are selected. If selection is done randomly, then typically only one or two percent respond to the advertisement. This way, about 49,000 letters are wasted. What is more, many respondents out of this number will choose only discounted offers on which the company loses money, or they will buy a product at the regular price but later return it. The company makes money only on the remaining respondents. Clearly, the company is interested in predicting the profit (or loss) for a potential customer. It is easy to obtain a list of names and addresses of potential customers by choosing them randomly from the telephone book or by buying a list from another company. Furthermore, there are databases available which provide, for each name and address, attributes like sex, age, education, profession, etc., describing the person (or a small group of people to which he belongs). The company is interested in using the vector X of attributes for a particular customer to predict the profit Y.

Example 1.3. *Boston housing values.*

Harrison and Rubinfeld (1978) considered the effect of air pollution concentration on housing values in Boston. The data consisted of 506 samples of median home values Y in a neighborhood with attributes X such as nitrogen oxide concentration, crime rate, average number of rooms, percentage of nonretail businesses, etc. A regression estimate was fitted to the data and it was then used to determine the median value of homes as a function of air pollution measured by nitrogen oxide concentration. For more details refer to Harrison and Rubinfeld (1978) and Breiman et al. (1984).

Example 1.4. *Wheat crop prediction.*

The Ministry for Agriculture of Hungary supported a research project for estimating the total expected crop yield of corn and wheat in order to plan and schedule the export-import of these commodities. They tried to predict the corn and wheat yield per unit area based on measurements of the reflected light spectrum obtained from satellite images taken by the LANDSAT 7 satellite (Asmus et al. (1987)). The satellite computes the integrals of spectral density (the energy of the light) in the following spectrum bands (wavelengths in μm):

(1) $[0.45, 0.52]$ blue;
(2) $[0.52, 0.60]$ green;
(3) $[0.63, 0.69]$ yellow;
(4) $[0.76, 0.90]$ red;
(5) $[1.55, 1.75]$ infrared;
(6) $[2.08, 2.35]$ infrared; and
(7) $[10.40, 12.50]$ infrared.

These are the components of the observation vector X which is used to predict crop yields for corn and wheat. The bands (2), (3), and (4) turned out to be the most relevant for this task.

Example 1.5. *Fat-free weight.*

A variety of health books suggest that the readers assess their health–at least in part–by considering the percentage of fat-free weight. Exact determination of this quantity requires knowledge of the body volume, which is not easily measurable. It can be computed from an underwater weighing: for this, a person has to undress and submerge in water in order to compute the increase of volume. This procedure is very inconvenient and so one wishes to estimate the body fat content Y from indirect measurements X, such as electrical impedance of the skin, weight, height, age, and sex.

Example 1.6. *Survival analysis.*

In survival analysis one is interested in predicting the survival time Y of a patient with a life-threatening disease given a description X of the case, such as type of disease, blood measurements, sex, age, therapy, etc. The result can be used to determine the appropriate therapy for a patient by maximizing the predicted survival time with respect to the therapy (see, e.g., Dippon, Fritz, and Kohler (2002) for an application in connection with breast cancer data). One specific feature in this application is that usually one cannot observe the survival time of a patient. Instead, one gets only the minimum of the survival time and a censoring time together with the information as to whether the survival time is less than the censoring time or not. We deal with regression function estimation from such censored data in Chapter 26.

Most of these applications are concerned with the *prediction* of Y from X. But some of them (see Examples 1.3 and 1.6) also deal with *interpretation* of the dependency of Y on X.

1.4 Application to Pattern Recognition

In pattern recognition, Y takes only finitely many values. For simplicity assume that Y takes two values, say 0 and 1. The aim is to predict the value of Y given the value of X (e.g., to predict whether a patient has a special disease or not, given some measurements of the patient like body temperature, blood pressure, etc.). The goal is to find a function $g^* : \mathcal{R}^d \to \{0, 1\}$ which minimizes the probability of $g^*(X) \neq Y$, i.e., to find a function g^* such that

$$\mathbf{P}\{g^*(X) \neq Y\} = \min_{g:\mathcal{R}^d \to \{0,1\}} \mathbf{P}\{g(X) \neq Y\}, \qquad (1.4)$$

where g^* is called the Bayes decision function, and $\mathbf{P}\{g(X) \neq Y)$ is the probability of misclassification.

The Bayes decision function can be obtained explicitly.

Lemma 1.1.

$$g^*(x) = \begin{cases} 1 & \textit{if} \quad \mathbf{P}\{Y = 1 | X = x\} \geq 1/2, \\ 0 & \textit{if} \quad \mathbf{P}\{Y = 1 | X = x\} < 1/2, \end{cases} \qquad (1.5)$$

is the Bayes decision function, i.e., g^* satisfies (1.4).

PROOF. Let $g : \mathcal{R}^d \to \{0, 1\}$ be an arbitrary (measurable) function. Fix $x \in \mathcal{R}^d$. Then

$$\begin{aligned} \mathbf{P}\{g(X) \neq Y | X = x\} &= 1 - \mathbf{P}\{g(X) = Y | X = x\} \\ &= 1 - \mathbf{P}\{Y = g(x) | X = x\}. \end{aligned}$$

Hence,

$$\mathbf{P}\{g(X) \neq Y | X = x\} - \mathbf{P}\{g^*(X) \neq Y | X = x\}$$
$$= \mathbf{P}\{Y = g^*(x) | X = x\} - \mathbf{P}\{Y = g(x) | X = x\} \geq 0,$$

because

$$\mathbf{P}\{Y = g^*(x) | X = x\} = \max\{\mathbf{P}\{Y = 0 | X = x\}, \mathbf{P}\{Y = 1 | X = x\}\}$$

by the definition of g^*. This proves

$$\mathbf{P}\{g^*(X) \neq Y | X = x\} \leq \mathbf{P}\{g(X) \neq Y | X = x\}$$

for all $x \in \mathcal{R}^d$, which implies

$$\mathbf{P}\{g^*(X) \neq Y\} = \int \mathbf{P}\{g^*(X) \neq Y | X = x\} \mu(dx)$$

$$\leq \int \mathbf{P}\{g(X) \neq Y | X = x\} \mu(dx)$$

$$= \mathbf{P}\{g(X) \neq Y\}.$$

$\square$

$\mathbf{P}\{Y = 1 | X = x\}$ and $\mathbf{P}\{Y = 0 | X = x\}$ are the so-called a posteriori probabilities. Observe that

$$\mathbf{P}\{Y = 1 | X = x\} = \mathbf{E}\{Y | X = x\} = m(x).$$

A natural approach is to estimate the regression function m by an estimate m_n using data $D_n = \{(X_1, Y_1), \ldots, (X_n, Y_n)\}$ and then to use a so-called plug-in estimate

$$g_n(x) = \begin{cases} 1 & \text{if } m_n(x) \geq 1/2, \\ 0 & \text{if } m_n(x) < 1/2, \end{cases} \tag{1.6}$$

to estimate g^*. The next theorem implies that if m_n is close to the real regression function m, then the error probability of decision g_n is near to the error probability of the optimal decision g^*.

Theorem 1.1. *Let $\hat{m} : \mathcal{R}^d \to \mathcal{R}$ be a fixed function and define the plug-in decision $\hat{g}$ by*

$$\hat{g}(x) = \begin{cases} 1 & \text{if } \hat{m}(x) \geq 1/2, \\ 0 & \text{if } \hat{m}(x) < 1/2. \end{cases}$$

Then

$$0 \;\leq\; \mathbf{P}\{\hat{g}(X) \neq Y\} - \mathbf{P}\{g^*(X) \neq Y\}$$

$$\leq\; 2 \int_{\mathcal{R}^d} |\hat{m}(x) - m(x)| \mu(dx)$$

$$\leq\; 2 \left(\int_{\mathcal{R}^d} |\hat{m}(x) - m(x)|^2 \mu(dx) \right)^{\frac{1}{2}}.$$

PROOF. It follows from the proof of Lemma 1.1 that, for arbitrary $x \in \mathcal{R}^d$,

$$\mathbf{P}\{\hat{g}(X) \neq Y | X = x\} - \mathbf{P}\{g^*(X) \neq Y | X = x\}$$

$$= \mathbf{P}\{Y = g^*(x) | X = x\} - \mathbf{P}\{Y = \hat{g}(x) | X = x\}$$

$$= I_{\{g^*(x)=1\}} m(x) + I_{\{g^*(x)=0\}} (1 - m(x))$$

$$\quad - \left(I_{\{\hat{g}(x)=1\}} m(x) + I_{\{\hat{g}(x)=0\}} (1 - m(x)) \right)$$

$$= I_{\{g^*(x)=1\}} m(x) + I_{\{g^*(x)=0\}} (1 - m(x))$$

$$\quad - \left(I_{\{g^*(x)=1\}} \hat{m}(x) + I_{\{g^*(x)=0\}} (1 - \hat{m}(x)) \right)$$

$$\quad + \left(I_{\{g^*(x)=1\}} \hat{m}(x) + I_{\{g^*(x)=0\}} (1 - \hat{m}(x)) \right)$$

$$- \left(I_{\{\hat{g}(x)=1\}}\hat{m}(x) + I_{\{\hat{g}(x)=0\}}(1 - \hat{m}(x)) \right)$$

$$+ \left(I_{\{\hat{g}(x)=1\}}\hat{m}(x) + I_{\{\hat{g}(x)=0\}}(1 - \hat{m}(x)) \right)$$

$$- \left(I_{\{\hat{g}(x)=1\}}m(x) + I_{\{\hat{g}(x)=0\}}(1 - m(x)) \right)$$

$$\leq I_{\{g^*(x)=1\}}(m(x) - \hat{m}(x)) + I_{\{g^*(x)=0\}}(\hat{m}(x) - m(x))$$

$$+ I_{\{\hat{g}(x)=1\}}(\hat{m}(x) - m(x)) + I_{\{\hat{g}(x)=0\}}(m(x) - \hat{m}(x))$$

(because of

$$I_{\{\hat{g}(x)=1\}}\hat{m}(x) + I_{\{\hat{g}(x)=0\}}(1 - \hat{m}(x)) = \max\{\hat{m}(x), 1 - \hat{m}(x)\}$$

by definition of $\hat{g}$)

$$\leq 2|\hat{m}(x) - m(x)|.$$

Hence

$$\begin{aligned}
0 \;\;\leq\;\; & \mathbf{P}\{\hat{g}(X) \neq Y\} - \mathbf{P}\{g^*(X) \neq Y\} \\
=\;\; & \int \left(\mathbf{P}\{\hat{g}(X) \neq Y | X = x\} - \mathbf{P}\{g^*(X) \neq Y | X = x\} \right) \mu(dx) \\
\leq\;\; & 2 \int |\hat{m}(x) - m(x)| \mu(dx).
\end{aligned}$$

The second assertion follows from the Cauchy-Schwarz inequality. $\qquad\square$

In Theorem 1.1, the second inequality in particular is not tight. Therefore pattern recognition is easier than regression estimation (cf. Devroye, Györfi, and Lugosi (1996)).

It follows from Theorem 1.1 that the error probability of the plug-in decision g_n defined above satisfies

$$\begin{aligned}
0 \;\;\leq\;\; & \mathbf{P}\{g_n(X) \neq Y | D_n\} - \mathbf{P}\{g^*(X) \neq Y\} \\
\leq\;\; & 2 \int_{\mathcal{R}^d} |m_n(x) - m(x)| \mu(dx) \\
\leq\;\; & 2 \left(\int_{\mathcal{R}^d} |m_n(x) - m(x)|^2 \mu(dx) \right)^{\frac{1}{2}}.
\end{aligned}$$

Thus estimates m_n with small L_2 error automatically lead to estimates g_n with small misclassification probability. Observe, however, that for (1.6) to be a good approximation of (1.5) it is not important that $m_n(x)$ be close to $m(x)$. Instead it is only important that $m_n(x)$ should be on the same side of the decision boundary as $m(x)$, i.e., that $m_n(x) > \frac{1}{2}$ whenever $m(x) > \frac{1}{2}$ and $m_n(x) < \frac{1}{2}$ whenever $m(x) < \frac{1}{2}$. Nevertheless, one often constructs estimates by minimizing the L_2 risk $\mathbf{E}\{|m_n(X) - Y|^2 | D_n\}$ and using the plug-in rule (1.6), because trying to minimize the L_2 risk leads to estimates which can be computed efficiently.

This can be generalized to the case where Y takes $M \geq 2$ distinct values, without loss of generality (w.l.o.g.) $1, \ldots, M$ (e.g., depending on whether a patient has a special type of disease or no disease). The goal is to find a function $g^* : \mathcal{R}^d \to \{1, \ldots, M\}$ such that

$$\mathbf{P}\{g^*(X) \neq Y\} = \min_{g : \mathcal{R}^d \to \{1, \ldots, M\}} \mathbf{P}\{g(X) \neq Y\}, \tag{1.7}$$

where g^* is called the Bayes decision function. It can be computed using the a posteriori probabilities $\mathbf{P}\{Y = k | X = x\}$ ($k \in \{1, \ldots, M\}$):

$$g^*(x) = \arg \max_{1 \leq k \leq M} \mathbf{P}\{Y = k | X = x\} \tag{1.8}$$

(cf. Problem 1.4).

The a posteriori probabilities are the regression functions

$$\mathbf{P}\{Y = k | X = x\} = \mathbf{E}\{I_{\{Y=k\}} | X = x\} = m^{(k)}(x).$$

Given data $D_n = \{(X_1, Y_1), \ldots, (X_n, Y_n)\}$, estimates $m_n^{(k)}$ of $m^{(k)}$ can be constructed from the data set

$$D_n^{(k)} = \{(X_1, I_{\{Y_1=k\}}), \ldots, (X_n, I_{\{Y_n=k\}})\},$$

and one can use a plug-in estimate

$$g_n(x) = \arg \max_{1 \leq k \leq M} m_n^{(k)}(x) \tag{1.9}$$

to estimate g^*. If the estimates $m_n^{(k)}$ are close to the a posteriori probabilities, then again the error of the plug-in estimate (1.9) is close to the optimal error (cf. Problem 1.5).

1.5 Parametric versus Nonparametric Estimation

The classical approach for estimating a regression function is the so-called parametric regression estimation. Here one assumes that the structure of the regression function is known and depends only on finitely many parameters, and one uses the data to estimate the (unknown) values of these parameters.

The linear regression estimate is an example of such an estimate. In linear regression one assumes that the regression function is a linear combination of the components of $x = (x^{(1)}, \ldots, x^{(d)})^T$, i.e.,

$$m(x^{(1)}, \ldots, x^{(d)}) = a_0 + \sum_{i=1}^{d} a_i x^{(i)} \quad ((x^{(1)}, \ldots, x^{(d)})^T \in \mathcal{R}^d)$$

for some unknown $a_0, \ldots, a_d \in \mathcal{R}$. Then one uses the data to estimate these parameters, e.g., by applying the principle of least squares, where one chooses the coefficients $a_0, \ldots, a_d$ of the linear function such that it best fits the given data:

$$(\hat{a}_0, \ldots, \hat{a}_d) = \arg \min_{a_0, \ldots, a_d \in \mathcal{R}^d} \left\{ \frac{1}{n} \sum_{j=1}^{n} \left| Y_j - a_0 - \sum_{i=1}^{d} a_i X_j^{(i)} \right|^2 \right\}.$$

Here $X_j^{(i)}$ denotes the ith component of X_j and $z = \arg \min_{x \in D} f(x)$ is the abbreviation for $z \in D$ and $f(z) = \min_{x \in D} f(x)$. Finally one defines the estimate by

$$\hat{m}_n(x) = \hat{a}_0 + \sum_{i=1}^{d} \hat{a}_i x^{(i)} \quad ((x^{(1)}, \ldots, x^{(d)})^T \in \mathcal{R}^d).$$

Parametric estimates usually depend only on a few parameters, therefore they are suitable even for small sample sizes n, if the parametric model is appropriately chosen. Furthermore, they are often easy to interpret. For instance in a linear model (when $m(x)$ is a linear function) the absolute value of the coefficient $\hat{a}_i$ indicates how much influence the ith component of X has on the value of Y, and the sign of $\hat{a}_i$ describes the nature of this influence (increasing or decreasing the value of Y).

However, parametric estimates have a big drawback. Regardless of the data, a parametric estimate cannot approximate the regression function better than the best function which has the assumed parametric structure. For example, a linear regression estimate will produce a large error for every sample size if the true underlying regression function is not linear and cannot be well approximated by linear functions.

For univariate X one can often use a plot of the data to choose a proper parametric estimate. But this is not always possible, as we now illustrate using simulated data. These data will be used throughout the book. They consist of $n = 200$ points such that X is standard normal restricted to $[-1, 1]$, i.e., the density of X is proportional to the standard normal density on $[-1, 1]$ and is zero elsewhere. The regression function is piecewise polynomial:

$$m(x) = \begin{cases} (x+2)^2/2 & \text{if} \quad -1 \leq x < -0.5, \\ x/2 + 0.875 & \text{if} \quad -0.5 \leq x < 0, \\ -5(x-0.2)^2 + 1.075 & \text{if} \quad 0 < x \leq 0.5, \\ x + 0.125 & \text{if} \quad 0.5 \leq x < 1. \end{cases}$$

Given X, the conditional distribution of $Y - m(X)$ is normal with mean zero and standard deviation

$$\sigma(X) = 0.2 - 0.1 \cos(2\pi X).$$

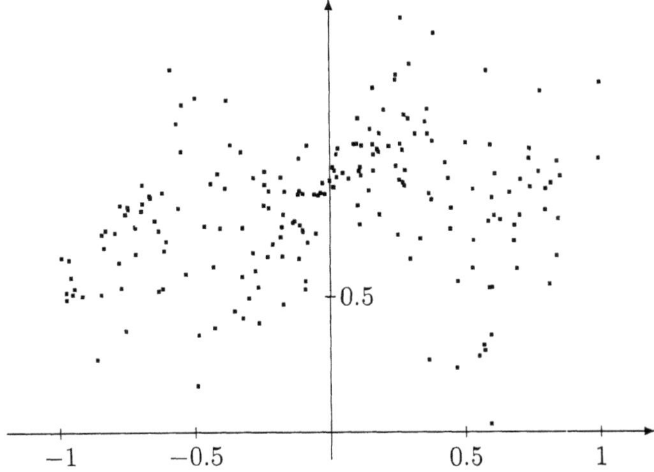

Figure 1.1. Simulated data points.

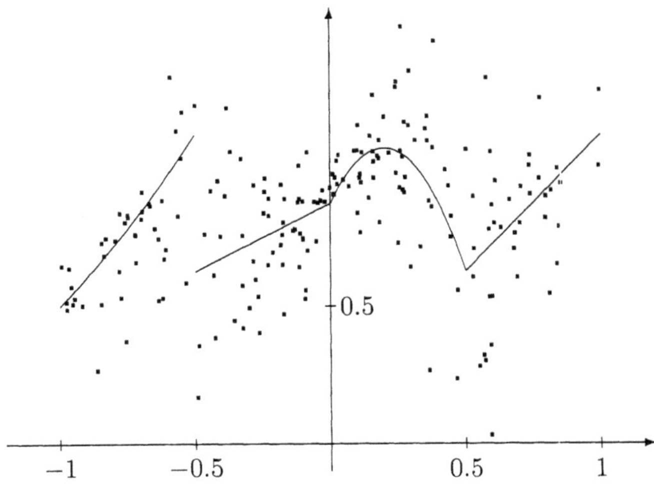

Figure 1.2. Data points and regression function.

Figure 1.1 shows the data points. In this example the human eye is not able to see from the data points what the regression function looks like. In Figure 1.2 the data points are shown together with the regression function.

In Figure 1.3 a linear estimate is constructed for these simulated data. Obviously, a linear function does not approximate the regression function well.

Furthermore, for multivariate X, there is no easy way to visualize the data. Thus, especially for multivariate X, it is not clear how to choose a

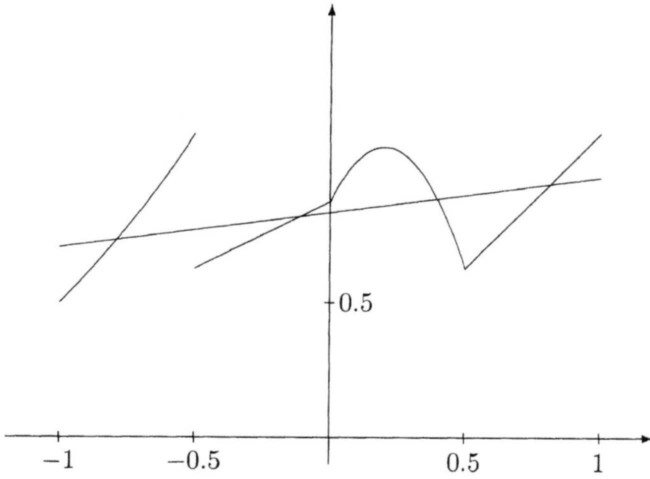

Figure 1.3. Linear regression estimate.

proper form of a parametric estimate, and a wrong form will lead to a bad estimate.

This inflexibility concerning the structure of the regression function is avoided by so-called nonparametric regression estimates. These methods, which do not assume that the regression function can be described by finitely many parameters, are introduced in Chapter 2 and are the main subject of this book.

1.6   Consistency

We will now define the modes of convergence of the regression estimates that we will study in this book.

The first and weakest property an estimate should have is that, as the sample size grows, it should converge to the estimated quantity, i.e., the error of the estimate should converge to zero for a sample size tending to infinity. Estimates which have this property are called consistent.

To measure the error of a regression estimate, we use the L_2 error

$$\int |m_n(x) - m(x)|^2 \mu(dx).$$

The estimate m_n depends on the data D_n, therefore the L_2 error is a random variable. We are interested in the convergence of the expectation of this random variable to zero as well as in the almost sure $(a.s.)$ convergence of this random variable to zero.

Definition 1.1. *A sequence of regression function estimates $\{m_n\}$ is called* **weakly consistent for a certain distribution of** (X, Y), *if*

$$\lim_{n \to \infty} \mathbf{E} \left\{ \int (m_n(x) - m(x))^2 \mu(dx) \right\} = 0.$$

Definition 1.2. *A sequence of regression function estimates $\{m_n\}$ is called* **strongly consistent for a certain distribution of** (X, Y), *if*

$$\lim_{n \to \infty} \int (m_n(x) - m(x))^2 \mu(dx) = 0 \quad \textit{with probability one.}$$

It may be that a regression function estimate is consistent for a certain class of distributions of (X, Y), but not consistent for others. It is clearly desirable to have estimates that are consistent for a large class of distributions. In this monograph we are interested in properties of m_n that are valid for all distributions of (X, Y), that is, in distribution-free or universal properties. The concept of universal consistency is important in nonparametric regression because the mere use of a nonparametric estimate is normally a consequence of the partial or total lack of information about the distribution of (X, Y). Since in many situations we do not have any prior information about the distribution, it is essential to have estimates that perform well for *all* distributions. This very strong requirement of universal goodness is formulated as follows:

Definition 1.3. *A sequence of regression function estimates $\{m_n\}$ is called* **weakly universally consistent** *if it is weakly consistent for all distributions of (X, Y) with $\mathbf{E}\{Y^2\} < \infty$.*

Definition 1.4. *A sequence of regression function estimates $\{m_n\}$ is called* **strongly universally consistent** *if it is strongly consistent for all distributions of (X, Y) with $\mathbf{E}\{Y^2\} < \infty$.*

We will later give many examples of estimates that are weakly and strongly universally consistent.

1.7 Rate of Convergence

If an estimate is universally consistent, then, regardless of the true underlying distribution of (X, Y), the L_2 error of the estimate converges to zero for a sample size tending to infinity. But this says nothing about how fast this happens. Clearly, it is desirable to have estimates for which the L_2 error converges to zero as fast as possible.

To decide about the rate of convergence of an estimate m_n, we will look at the expectation of the L_2 error,

$$\mathbf{E} \int |m_n(x) - m(x)|^2 \mu(dx). \tag{1.10}$$

A natural question to ask is whether there exist estimates for which (1.10) converges to zero at some fixed, nontrivial rate for all distributions of (X, Y). Unfortunately, as we will see in Chapter 3, such estimates do not exist, i.e., for any estimate the rate of convergence may be arbitrarily slow. In order to get nontrivial rates of convergence, one has to restrict the class of distributions, e.g., by imposing some smoothness assumptions on the regression function.

In Chapter 3 we will define classes $\mathcal{F}_p$ of the distributions of (X, Y) where the corresponding regression function satisfies some smoothness condition depending on a parameter p (e.g., m is p times continuously differentiable). We then use the classical minimax approach to define the optimal rate of convergence for such classes $\mathcal{F}_p$. This means that we will try to minimize the maximal value of (1.10) within the class $\mathcal{F}_p$ of the distributions of (X, Y), i.e., we will look at

$$\inf_{\hat{m}_n} \sup_{(X,Y) \in \mathcal{F}_p} \mathbf{E} \int |\hat{m}_n(x) - m(x)|^2 \mu(dx), \qquad (1.11)$$

where the infimum is taken over all estimates $\hat{m}_n$. We are interested in optimal estimates m_n, for which the maximal value of (1.10) within $\mathcal{F}_p$, i.e.,

$$\sup_{(X,Y) \in \mathcal{F}_p} \mathbf{E} \int |m_n(x) - m(x)|^2 \mu(dx), \qquad (1.12)$$

is close to (1.11).

To simplify our analysis, we will only look at the asymptotic behavior of (1.11) and (1.12), i.e., we will determine the rate of convergence of (1.11) to zero for a sample size tending to infinity, and we will construct estimates which achieve (up to some constant factor) the same rate of convergence.

For classes $\mathcal{F}_p$, where m is p times continuously differentiable, the optimal rate of convergence will be $n^{-\frac{2p}{2p+d}}$.

1.8 Adaptation

Often, estimates which achieve the optimal minimax rate of convergence for a given class $\mathcal{F}_{p_0}$ of distributions (where, e.g., m is p_0 times continuously differentiable) require the knowledge of p_0 and are adjusted perfectly to this class of distributions. Therefore they don't achieve the optimal rate of convergence for other classes $\mathcal{F}_p$, $p \neq p_0$.

If one could find out in an application to which classes of distributions the true underlying distribution belongs, then one could choose that class which has the best rate of convergence (which will be the smallest class in the case of nested classes), and could choose an estimate which achieves the optimal minimax rate of convergence within this class. This, however, would require knowledge about the smoothness of the regression function. In applications such knowledge is typically not available and, unfortunately it is not possible to use the data to decide about the smoothness of the regression function (at least, we do not know of any test which can decide how smooth the regression function is, e.g., whether m is continuous or not).

Therefore, instead of looking at each class $\mathcal{F}_p$ of distributions separately, and constructing estimates which are optimal for this class only, one tries to construct estimates which achieve the optimal (or a nearly optimal) minimax rate of convergence simultaneously for many different classes of distributions. Such estimates are called adaptive and will be used throughout this book. Several possibilities for constructing adaptive estimates will be described in the next chapter.

1.9 Fixed versus Random Design Regression

The problem studied in this book is also called regression estimation with random design, which means that the X_i's are random variables. We want to mention that there exists a related problem, called regression estimation with fixed design. This section is about similarities and differences between these two problems.

Regression function estimation with fixed design can be described as follows: one observes values of some function at some fixed (given) points with additive random errors, and wants to recover the true value of the function at these points. More precisely, given data $(x_1, Y_1), \ldots, (x_n, Y_n)$, where $x_1, \ldots, x_n$ are fixed (nonrandom) points in $\mathcal{R}^d$ and

$$Y_i = f(x_i) + \sigma_i \cdot \epsilon_i \quad (i = 1, \ldots, n) \tag{1.13}$$

for some (unknown) function $f : \mathcal{R}^d \to \mathcal{R}$, some $\sigma_1, \ldots, \sigma_n \in \mathcal{R}_+$, and some independent and identically distributed random variables $\epsilon_1, \ldots, \epsilon_n$ with $\mathbf{E}\epsilon_1 = 0$ and $\mathbf{E}\epsilon_1^2 = 1$, one wants to estimate the values of f at the so-called design points $x_1, \ldots, x_n$. Typically, in this problem, one has $d = 1$, sometimes also $d = 2$ (image reconstruction). Often the x_i are equidistant, e.g., in $[0, 1]$, and one assumes that the variance σ_i^2 of the additive error (noise) $\sigma_i \cdot \epsilon_i$ is constant, i.e., $\sigma_1^2 = \cdots = \sigma_n^2 = \sigma^2$.

Clearly, this problem has some similarity with the problem we study in this book. This becomes obvious when one rewrites the data in our model as

$$Y_i = m(X_i) + \epsilon_i, \tag{1.14}$$

where $\epsilon_i = \epsilon_i(X_i) = Y_i - m(X_i)$ satisfies $\mathbf{E}\{\epsilon_i | X_i\} = 0$.

It may seem that fixed design regression is a more general approach than random design and that one can handle random design regression estimation by imposing conditions on the design points and then applying results for fixed design regression. We want to point out that this is not true, because the assumptions in both models are fundamentally different. First, in (1.14) the error ϵ_i depends on X_i, thus it is not independent of X_i and its whole structure (i.e., kind of distribution) can change with X_i. Second, the design points $X_1, \ldots, X_n$ in (1.14) are typically far from being uniformly distributed. And third, while in fixed design regression the design points are typically univariate or at most bivariate, the dimension d of X in random design regression is often much larger than two, which fundamentally changes the problem (cf. Section 2.2).

1.10   Bibliographic Notes

We have made a computer search for nonparametric regression, resulting in 3457 items. It is clear that we cannot cite all of them and we apologize at this point for the many good papers which we didn't cite. In the later chapters we refer only to publications on L_2 theory. Concerning nonparametric regression estimation including pointwise or uniform consistency properties, we refer to the following monographs (with further references therein): Bartlett and Anthony (1999), Bickel et al. (1993), Bosq (1996), Bosq and Lecoutre (1987), Breiman et al. (1984), Collomb (1980), Devroye, Györfi, and Lugosi (1996), Devroye and Lugosi (2001), Efromovich (1999), Eggermont and La Riccia (2001), Eubank (1999), Fan and Gijbels (1995), Härdle (1990), Härdle et al. (1998), Hart (1997), Hastie, Tibshirani and Friedman (2001), Horowitz (1998), Korostelev and Tsybakov (1993), Nadaraya (1989), Prakasa Rao (1983), Simonoff (1996), Thompson and Tapia (1990), Vapnik (1982; 1998), Vapnik and Chervonenkis (1974), Wand and Jones (1995). We refer also to the bibliography of Collomb (1985) and to the survey on fixed design regression of Gasser and Müller (1979).

For parametric methods we refer to Rao (1973), Seber (1977), Draper and Smith (1981) and Farebrother (1988) and the literature cited therein.

Lemma 1.1 and Theorem 1.1 are well-known in the literature. Concerning Theorem 1.1 see, e.g., Van Ryzin (1966), Wolverton and Wagner (1969a), Glick (1973), Csibi (1971), Györfi (1975; 1978), Devroye and Wagner (1976), Devroye (1982b), or Devroye and Györfi (1985).

The concept of (weak) universal consistency goes back to Stone (1977).

Problems and Exercises

PROBLEM 1.1. Show that the regression function also has the following pointwise optimality property:

$$\mathbf{E}\left\{|m(X) - Y|^2 | X = x\right\} = \min_{f} \mathbf{E}\left\{|f(X) - Y|^2 | X = x\right\}$$

for μ-almost all $x \in \mathcal{R}^d$.

PROBLEM 1.2. Prove (1.2).

PROBLEM 1.3. Let (X, Y) be an $\mathcal{R}^d \times \mathcal{R}$-valued random variable with $\mathbf{E}|Y| < \infty$. Determine a function $f^* : \mathcal{R}^d \to \mathcal{R}$ which minimizes the L_1 risk, i.e., which satisfies

$$\mathbf{E}\{|f^*(X) - Y|\} = \min_{f:\mathcal{R}^d \to \mathcal{R}} \mathbf{E}\{|f(X) - Y|\}.$$

PROBLEM 1.4. Prove that the decision rule (1.8) satisfies (1.7).

PROBLEM 1.5. Show that the error probability of the plug-in decision rule (1.9) satisfies

$$\begin{aligned}
0 &\leq \mathbf{P}\{g_n(X) \neq Y | D_n\} - \mathbf{P}\{g^*(X) \neq Y\} \\
&\leq \sum_{k=1}^{M} \int_{\mathcal{R}^d} |m_n^{(k)}(x) - m^{(k)}(x)| \mu(dx) \\
&\leq \sum_{k=1}^{M} \left\{ \int_{\mathcal{R}^d} |m_n^{(k)}(x) - m^{(k)}(x)|^2 \mu(dx) \right\}^{\frac{1}{2}}.
\end{aligned}$$

2
How to Construct Nonparametric Regression Estimates?

In this chapter we give an overview of various ways to define nonparametric regression estimates. In Section 2.1 we introduce four related paradigms for nonparametric regression. For multivariate X there are some special modifications of the resulting estimates due to the so-called "curse of dimensionality." These will be described in Section 2.2. The estimates depend on smoothing parameters. The choice of these parameters is important because of the so-called bias-variance tradeoff, which will be described in Section 2.3. Finally, in Section 2.4, several methods for choosing the smoothing parameters are introduced.

2.1 Four Related Paradigms

In this section we describe four paradigms of nonparametric regression: **local averaging**, **local modeling**, **global modeling** (or **least squares estimation**), and **penalized modeling**.

Recall that the data can be written as

$$Y_i = m(X_i) + \epsilon_i,$$

where $\epsilon_i = Y_i - m(X_i)$ satisfies $\mathbf{E}(\epsilon_i|X_i) = 0$. Thus Y_i can be considered as the sum of the value of the regression function at X_i and some error ϵ_i, where the expected value of the error is zero. This motivates the construction of the estimates by **local averaging**, i.e., estimation of $m(x)$ by the average of those Y_i where X_i is "close" to x. Such an estimate can be written as

$$m_n(x) = \sum_{i=1}^{n} W_{n,i}(x) \cdot Y_i,$$

where the weights $W_{n,i}(x) = W_{n,i}(x, X_1, \ldots, X_n) \in \mathcal{R}$ depend on $X_1, \ldots, X_n$. Usually the weights are nonnegative and $W_{n,i}(x)$ is "small" if X_i is "far" from x.

An example of such an estimate is the *partitioning estimate*. Here one chooses a finite or countably infinite partition $\mathcal{P}_n = \{A_{n,1}, A_{n,2}, \ldots\}$ of $\mathcal{R}^d$ consisting of Borel sets $A_{n,j} \subseteq \mathcal{R}^d$ and defines, for $x \in A_{n,j}$, the estimate by averaging Y_i's with the corresponding X_i's in $A_{n,j}$, i.e..

$$m_n(x) = \frac{\sum_{i=1}^{n} I_{\{X_i \in A_{n,j}\}} Y_i}{\sum_{i=1}^{n} I_{\{X_i \in A_{n,j}\}}} \quad \text{for } x \in A_{n,j}. \tag{2.1}$$

where I_A denotes the indicator function of set A, so

$$W_{n,i}(x) = \frac{I_{\{X_i \in A_{n,j}\}}}{\sum_{l=1}^{n} I_{\{X_l \in A_{n,j}\}}} \quad \text{for } x \in A_{n,j}.$$

Here and in the following we use the convention $\frac{0}{0} = 0$.

The second example of a local averaging estimate is the *Nadaraya–Watson kernel estimate*. Let $K : \mathcal{R}^d \to \mathcal{R}_+$ be a function called the kernel function, and let $h > 0$ be a bandwidth. The kernel estimate is defined by

$$m_n(x) = \frac{\sum_{i=1}^{n} K\left(\frac{x-X_i}{h}\right) Y_i}{\sum_{i=1}^{n} K\left(\frac{x-X_i}{h}\right)}, \tag{2.2}$$

so

$$W_{n,i}(x) = \frac{K\left(\frac{x-X_i}{h}\right)}{\sum_{j=1}^{n} K\left(\frac{x-X_j}{h}\right)}.$$

If one uses the so-called naive kernel (or window kernel) $K(x) = I_{\{\|x\| \leq 1\}}$, then

$$m_n(x) = \frac{\sum_{i=1}^{n} I_{\{\|x-X_i\| \leq h\}} Y_i}{\sum_{i=1}^{n} I_{\{\|x-X_i\| \leq h\}}},$$

i.e., one estimates $m(x)$ by averaging Y_i's such that the distance between X_i and x is not greater than h.

For more general $K : \mathcal{R}^d \to \mathcal{R}_+$ one uses a weighted average of the Y_i, where the weight of Y_i (i.e., the influence of Y_i on the value of the estimate at x) depends on the distance between X_i and x.

Our final example of local averaging estimates is the *k-nearest neighbor* (*k-NN*) *estimate*. Here one determines the k nearest X_i's to x in terms of distance $\|x - X_i\|$ and estimates $m(x)$ by the average of the corresponding Y_i's. More precisely, for $x \in \mathcal{R}^d$, let

$$(X_{(1)}(x), Y_{(1)}(x)), \ldots, (X_{(n)}(x), Y_{(n)}(x))$$

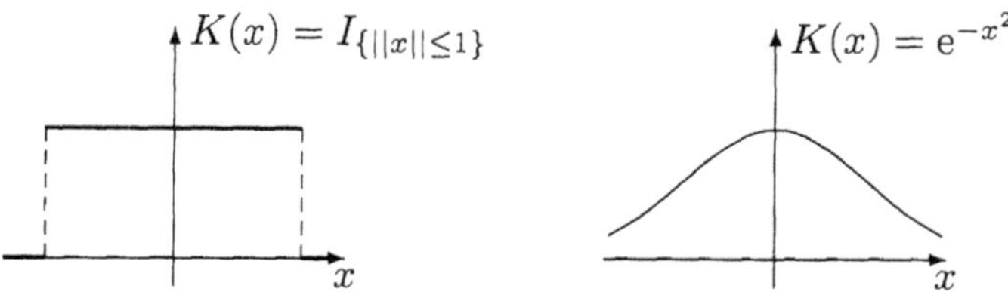

Figure 2.1. Examples of kernels: window (naive) kernel and Gaussian kernel.

Figure 2.2. Nearest neighbors to x.

be a permutation of

$$(X_1, Y_1), \ldots, (X_n, Y_n)$$

such that

$$\|x - X_{(1)}(x)\| \leq \cdots \leq \|x - X_{(n)}(x)\|.$$

The k-NN estimate is defined by

$$m_n(x) = \frac{1}{k} \sum_{i=1}^{k} Y_{(i)}(x). \tag{2.3}$$

Here the weight $W_{ni}(x)$ equals $1/k$ if X_i is among the k nearest neighbors of x, and equals 0 otherwise.

The kernel estimate (2.2) can be considered as locally fitting a constant to the data. In fact, it is easy to see (cf. Problem 2.2) that it satisfies

$$m_n(x) = \arg\min_{c \in \mathcal{R}} \frac{1}{n} \sum_{i=1}^{n} K\left(\frac{x - X_i}{h}\right) (Y_i - c)^2. \tag{2.4}$$

A generalization of this leads to the **local modeling** paradigm: instead of locally fitting a constant to the data, locally fit a more general function, which depends on several parameters. Let $g(\cdot, \{a_k\}_{k=1}^{l}) : \mathcal{R}^d \to \mathcal{R}$ be a function depending on parameters $\{a_k\}_{k=1}^{l}$. For each $x \in \mathcal{R}^d$, choose values of these parameters by a local least squares criterion

$$\{\hat{a}_k(x)\}_{k=1}^{l} = \underset{\{a_k\}_{k=1}^{l}}{\arg\min} \frac{1}{n} \sum_{i=1}^{n} K\left(\frac{x - X_i}{h}\right) \left(Y_i - g\left(X_i, \{a_k\}_{k=1}^{l}\right)\right)^2. \quad (2.5)$$

Here we do not require that the minimum in (2.5) be unique. In case there are several points at which the minimum is attained we use an arbitrary rule (e.g., by flipping a coin) to choose one of these points. Evaluate the function g for these parameters at the point x and use this as an estimate of $m(x)$:

$$m_n(x) = g\left(x, \{\hat{a}_k(x)\}_{k=1}^{l}\right). \quad (2.6)$$

If one chooses $g(x, \{c\}) = c$ $(x \in \mathcal{R}^d)$, then this leads to the Nadaraya–Watson kernel estimate.

The most popular example of a local modeling estimate is the *local polynomial kernel estimate*. Here one locally fits a polynomial to the data. For example, for $d = 1$, X is real-valued and

$$g\left(x, \{a_k\}_{k=1}^{l}\right) = \sum_{k=1}^{l} a_k x^{k-1}$$

is a polynomial of degree $l - 1$ (or less) in x.

A generalization of the partitioning estimate leads to **global modeling** or **least squares estimates**. Let $\mathcal{P}_n = \{A_{n,1}, A_{n,2}, \ldots\}$ be a partition of $\mathcal{R}^d$ and let $\mathcal{F}_n$ be the set of all piecewise constant functions with respect to that partition, i.e.,

$$\mathcal{F}_n = \left\{\sum_{j} a_j I_{A_{n,j}} \ : \ a_j \in \mathcal{R}\right\}. \quad (2.7)$$

Then it is easy to see (cf. Problem 2.3) that the partitioning estimate (2.1) satisfies

$$m_n(\cdot) = \underset{f \in \mathcal{F}_n}{\arg\min} \left\{\frac{1}{n} \sum_{i=1}^{n} |f(X_i) - Y_i|^2\right\}. \quad (2.8)$$

Hence it minimizes the empirical L_2 risk

$$\frac{1}{n} \sum_{i=1}^{n} |f(X_i) - Y_i|^2 \quad (2.9)$$

over $\mathcal{F}_n$. Least squares estimates are defined by minimizing the empirical L_2 risk over a general set of functions $\mathcal{F}_n$ (instead of (2.7)). Observe that it doesn't make sense to minimize (2.9) over all (measurable) functions f, because this may lead to a function which interpolates the data and hence is not a reasonable estimate. Thus one has to restrict the set of functions over

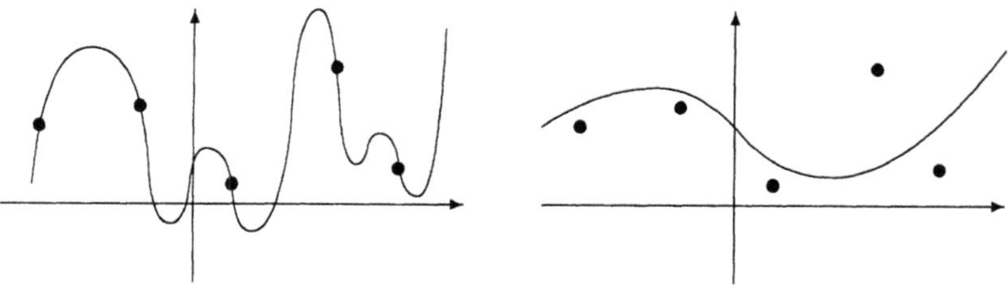

Figure 2.3. The estimate on the right seems to be more reasonable than the estimate on the left, which interpolates the data.

which one minimizes the empirical L_2 risk. Examples of possible choices of the set $\mathcal{F}_n$ are sets of piecewise polynomials with respect to a partition $\mathcal{P}_n$, or sets of smooth piecewise polynomials (splines). The use of spline spaces ensures that the estimate is a smooth function.

Instead of restricting the set of functions over which one minimizes, one can also add a penalty term to the functional to be minimized. Let $J_n(f) \geq 0$ be a penalty term penalizing the "roughness" of a function f. The **penalized modeling** or **penalized least squares estimate** m_n is defined by

$$m_n = \arg\min_f \left\{ \frac{1}{n} \sum_{i=1}^{n} |f(X_i) - Y_i|^2 + J_n(f) \right\}, \qquad (2.10)$$

where one minimizes over all measurable functions f. Again we do not require that the minimum in (2.10) be unique. In the case it is not unique, we randomly select one function which achieves the minimum.

A popular choice for $J_n(f)$ in the case $d = 1$ is

$$J_n(f) = \lambda_n \int |f''(t)|^2 dt, \qquad (2.11)$$

where f'' denotes the second derivative of f and λ_n is some positive constant. We will show in Chapter 20 that for this penalty term the minimum in (2.10) is achieved by a cubic spline with knots at the X_i's, i.e., by a twice differentiable function which is equal to a polynomial of degree 3 (or less) between adjacent values of the X_i's (a so-called smoothing spline). A generalization of (2.11) is

$$J_{n,k}(f) = \lambda_n \int |f^{(k)}(t)|^2 dt,$$

where $f^{(k)}$ denotes the k-th derivative of f. For multivariate X one can use

$$J_{n,k}(f) = \lambda_n \int \sum_{i_1,\dots,i_k \in \{1,\dots,d\}} \left| \frac{\partial^k f}{\partial x_{i_1} \dots \partial x_{i_k}} \right|^2 dx,$$

which leads to the so-called thin plate spline estimates.

2.2 Curse of Dimensionality

If X takes values in a high-dimensional space (i.e., if d is large), estimating the regression function is especially difficult. The reason for this is that in the case of large d it is, in general, not possible to densely pack the space of X with finitely many sample points, even if the sample size n is very large. This fact is often referred to as the "curse of dimensionality." In the sequel we will illustrate this with an example.

Let $X, X_1, \ldots, X_n$ be independent and identically distributed $\mathcal{R}^d$-valued random variables with X uniformly distributed in the hypercube $[0,1]^d$. Denote the expected supremum-norm distance of X to its nearest neighbor in $X_1, \ldots, X_n$ by $d_\infty(d, n)$, i.e., set

$$d_\infty(d, n) = \mathbf{E}\left\{ \min_{i=1,\ldots,n} \|X - X_i\|_\infty \right\}.$$

Here $\|x\|_\infty$ is the supremum norm of a vector $x = (x^{(1)}, \ldots, x^{(d)})^T \in \mathcal{R}^d$ defined by

$$\|x\|_\infty = \max_{l=1,\ldots,d} \left| x^{(l)} \right|.$$

Then

$$
\begin{aligned}
d_\infty(d, n) &= \int_0^\infty \mathbf{P}\left\{ \min_{i=1,\ldots,n} \|X - X_i\|_\infty > t \right\} dt \\
&= \int_0^\infty \left(1 - \mathbf{P}\left\{ \min_{i=1,\ldots,n} \|X - X_i\|_\infty \leq t \right\} \right) dt.
\end{aligned}
$$

The bound

$$
\begin{aligned}
\mathbf{P}\left\{ \min_{i=1,\ldots,n} \|X - X_i\|_\infty \leq t \right\} &\leq n \cdot \mathbf{P}\left\{ \|X - X_1\|_\infty \leq t \right\} \\
&\leq n \cdot (2t)^d
\end{aligned}
$$

implies

$$
\begin{aligned}
d_\infty(d, n) &\geq \int_0^{1/(2n^{1/d})} \left(1 - n \cdot (2t)^d \right) dt \\
&= \left[t - n \cdot 2^d \cdot \frac{t^{d+1}}{d+1} \right]_{t=0}^{1/(2n^{1/d})} \\
&= \frac{1}{2 \cdot n^{1/d}} - \frac{1}{d+1} \cdot \frac{1}{2 \cdot n^{1/d}} \\
&= \frac{d}{2(d+1)} \cdot \frac{1}{n^{1/d}}.
\end{aligned}
$$

Table 2.1 shows values of this lower bound for various values of d and n. As one can see, for dimension $d = 10$ or $d = 20$ this lower bound is not

Table 2.1. Lower bounds for $d_\infty(d, n)$.

	$n = 100$	$n = 1000$	$n = 10,000$	$n = 100,000$
$d_\infty(1, n)$	≥ 0.0025	≥ 0.00025	≥ 0.000025	≥ 0.0000025
$d_\infty(10, n)$	≥ 0.28	≥ 0.22	≥ 0.18	≥ 0.14
$d_\infty(20, n)$	≥ 0.37	≥ 0.34	≥ 0.30	≥ 0.26

close to zero even if the sample size n is very large. So for most values of x one only has data points (X_i, Y_i) available where X_i is not close to x. But at such data points $m(X_i)$ will, in general, not be close to $m(x)$ even for a smooth regression function.

The only way to overcome the curse of dimensionality is to incorporate additional assumptions about the regression function besides the sample. This is implicitly done by nearly all multivariate estimation procedures, including projection pursuit, neural networks, radial basis function networks, trees, etc.

As we will see in Problem 2.4, a similar problem also occurs if one replaces the supremum norm by the Euclidean norm. Of course, the arguments above are no longer valid if the components of X are not independent (e.g., if all components of X are equal and hence all values of X lie on a line in $\mathcal{R}^d$). But in this case they are (roughly speaking) still valid with d replaced by the number of independent components of X (or, more generally, the "intrinsic" dimension of X), which for large d may still be a large number.

2.3 Bias–Variance Tradeoff

Let m_n be an arbitrary estimate. For any $x \in \mathcal{R}^d$ we can write the expected squared error of m_n at x as

$$\mathbf{E}\{|m_n(x) - m(x)|^2\}$$

$$= \mathbf{E}\{|m_n(x) - \mathbf{E}\{m_n(x)\}|^2\} + |\mathbf{E}\{m_n(x)\} - m(x)|^2$$

$$= \mathbf{Var}(m_n(x)) + |bias(m_n(x))|^2.$$

Here $\mathbf{Var}(m_n(x))$ is the variance of the random variable $m_n(x)$ and $bias(m_n(x))$ is the difference between the expectation of $m_n(x)$ and $m(x)$. This also leads to a similar decomposition of the expected L_2 error:

$$\mathbf{E}\left\{\int |m_n(x) - m(x)|^2 \mu(dx)\right\}$$

$$= \int \mathbf{E}\{|m_n(x) - m(x)|^2\}\mu(dx)$$

$$= \int \mathbf{Var}(m_n(x))\mu(dx) + \int |bias(m_n(x))|^2 \mu(dx).$$

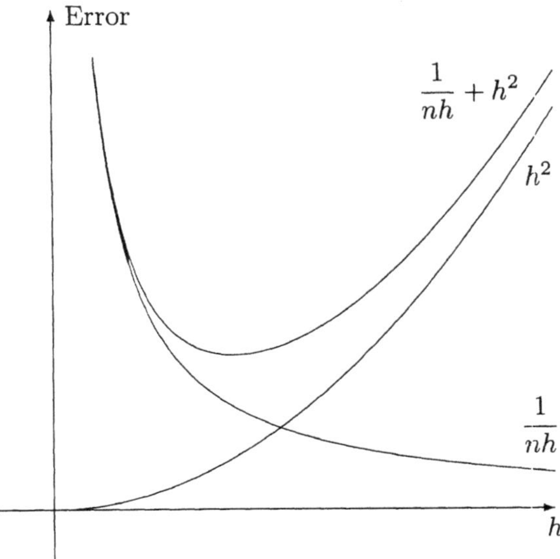

Figure 2.4. Bias–variance tradeoff.

The importance of these decompositions is that the integrated variance and the integrated squared bias depend in opposite ways on the wiggliness of an estimate. If one increases the wiggliness of an estimate, then usually the integrated bias will decrease, but the integrated variance will increase (so-called **bias–variance tradeoff**).

In Figure 2.4 this is illustrated for the kernel estimate, where one has, under some regularity conditions on the underlying distribution and for the naive kernel,

$$\int_{\mathcal{R}^d} \mathbf{Var}(m_n(x))\mu(dx) = c_1 \frac{1}{nh^d} + o\left(\frac{1}{nh^d}\right)$$

and

$$\int_{\mathcal{R}^d} |bias(m_n(x))|^2\mu(dx) = c_2 h^2 + o\left(h^2\right).$$

Here h denotes the bandwidth of the kernel estimate which controls the wiggliness of the estimate, c_1 is some constant depending on the conditional variance $\mathbf{Var}\{Y|X = x\}$, the regression function is assumed to be Lipschitz continuous, and c_2 is some constant depending on the Lipschitz constant.

The value h^* of the bandwidth for which the sum of the integrated variance and the squared bias is minimal depends on c_1 and c_2. Since the underlying distribution, and hence also c_1 and c_2, are unknown in an application, it is important to have methods which choose the bandwidth automatically using only the data D_n. Such methods will be described in the next section.

2.4 Choice of Smoothing Parameters and Adaptation

Recall that we want to construct estimates $m_{n,p}$ such that the L_2 risk

$$\mathbf{E}\{|m_{n,p}(X) - Y|^2 | D_n\} \tag{2.12}$$

is small. Hence the smoothing parameter p of an estimate (e.g., bandwidth $p = h$ of a kernel estimate or number $p = K$ of cells of a partitioning estimate) should be chosen to make (2.12) small.

For a fixed function $f : \mathcal{R}^d \to \mathcal{R}$, the L_2 risk $\mathbf{E}\{|f(X) - Y|^2\}$ can be estimated by the empirical L_2 risk (error on the sample)

$$\frac{1}{n} \sum_{i=1}^{n} |f(X_i) - Y_i|^2. \tag{2.13}$$

The **resubstitution method** also uses this estimate for $m_{n,p}$, i.e., it chooses the smoothing parameter p that minimizes

$$\frac{1}{n} \sum_{i=1}^{n} |m_{n,p}(X_i) - Y_i|^2. \tag{2.14}$$

Usually this leads to overly optimistic estimates of the L_2 risk and is hence not useful. The reason for this behavior is that (2.14) favors estimates which are too well-adapted to the data and are not reasonable for new observations (X, Y).

This problem doesn't occur if one uses a new sample

$$(\bar{X}_1, \bar{Y}_1), \ldots, (\bar{X}_n, \bar{Y}_n)$$

to estimate (2.12), where

$$(X_1, Y_1), \ldots, (X_n, Y_n), (\bar{X}_1, \bar{Y}_1), \ldots, (\bar{X}_n, \bar{Y}_n),$$

are i.i.d., i.e., if one minimizes

$$\frac{1}{n} \sum_{i=1}^{n} |m_{n,p}(\bar{X}_i) - \bar{Y}_i|^2. \tag{2.15}$$

Of course, in the regression function estimation problem one doesn't have an additional sample. But this isn't a real problem, because we can simply **split** the sample D_n into two parts: a learning sample

$$D_{n_1} = \{(X_1, Y_1), \ldots, (X_{n_1}, Y_{n_1})\}$$

which we use to construct estimates

$$m_{n_1,p}(\cdot) = m_{n_1,p}(\cdot, D_{n_1})$$

depending on some parameter p, and a test sample

$$\{(X_{n_1+1}, Y_{n_1+1}), \ldots, (X_n, Y_n)\}$$

which we use to choose the parameter p of the estimate by minimizing

$$\frac{1}{n-n_1} \sum_{i=n_1+1}^{n} |m_{n_1,p}(X_i) - Y_i|^2. \qquad (2.16)$$

In applications one often uses $n_1 = \frac{2}{3}n$ or $n_1 = \frac{n}{2}$.

If n is large, especially if n is so large that it is computationally difficult to construct an estimate m_n using all the data, then this is a very reasonable method (cf. Chapter 7). But it has the drawback that it chooses one estimate from the family $\{m_{n_1,p} : p\}$ of estimates which depend only on $n_1 < n$ of the sample points. To avoid this problem one can take a parameter p^* for which (2.16) is minimal and use it for an estimate m_{n,p^*} which uses the whole sample D_n. But then one introduces some instability into the estimate: if one splits the sample D_n, in a different way, into a learning sample and a test sample, then one might get another parameter $\bar{p}$ for which the error on the test sample is minimal and, hence, one might end up with another estimate $m_{n,\bar{p}}$, which doesn't seem to be reasonable. This can be avoided if one repeats this procedure for all possible splits of the sample and averages (2.16) for all these splits. In general, this is computationally intractable, therefore one averages (2.16) only over some of all the possible splits.

For k-**fold cross-validation** these splits are chosen in a special deterministic way. Let $1 \le k \le n$. For simplicity we assume that $\frac{n}{k}$ is an integer. Divide the data into k groups of equal size $\frac{n}{k}$ and denote the set of data consisting of all groups, except the lth one, by $D_{n,l}$:

$$D_{n,l} = \left\{ (X_1, Y_1), \ldots, (X_{\frac{n}{k}(l-1)}, Y_{\frac{n}{k}(l-1)}), (X_{\frac{n}{k}l+1}, Y_{\frac{n}{k}l+1}), \ldots, (X_n, Y_n) \right\}.$$

For each data set $D_{n,l}$ construct estimates $m_{n-\frac{n}{k},p}(\cdot, D_{n,l})$. Then choose the parameter p such that

$$\frac{1}{k} \sum_{l=1}^{k} \frac{1}{n/k} \sum_{i=\frac{n}{k}(l-1)+1}^{\frac{n}{k}l} \left| m_{n-\frac{n}{k},p}(X_i, D_{n,l}) - Y_i \right|^2 \qquad (2.17)$$

is minimal, and use this parameter p^* for an estimate m_{n,p^*} constructed from the whole sample D_n.

n-fold cross-validation, where $D_{n,l}$ is the whole sample except (X_l, Y_l) and where one minimizes

$$\frac{1}{n} \sum_{l=1}^{n} |m_{n-1,p}(X_l, D_{n,l}) - Y_l|^2,$$

is often referred to as **cross-validation**. Here $m_{n-1,p}(\cdot, D_{n,l})$ is the estimate computed with parameter value p and based upon the whole sample except (X_l, Y_l) (so it is based upon $n-1$ of the n data points) and $m_{n-1,p}(X_l, D_{n,l})$ is the value of this estimate at the point X_l, i.e., at that x-value of the sample which is not used in the construction of the estimate.

There is an important difference between the estimates (2.16) and (2.17) of the L_2 risk (2.12). Formula (2.16) estimates the L_2 risk of the estimate $m_{n_1,p}(\cdot, D_{n_1})$, i.e., the L_2 risk of an estimate constructed with the data D_{n_1}, while (2.17) estimates the average L_2 risk of an estimate constructed with $n - \frac{n}{k}$ of the data points in D_n.

For least squares and penalized least squares estimates we will also study another method called **complexity regularization** (cf. Chapter 12) for choosing the smoothing parameter. The idea there is to derive an upper bound on the L_2 error of the estimate and to choose the parameter such that this upper bound is minimal. The upper bound will be of the form

$$\frac{1}{n} \sum_{i=1}^{n} \left(|m_{n,p}(X_i) - Y_i|^2 - |m(X_i) - Y_i|^2 \right) + pen_n(m_{n,p}), \qquad (2.18)$$

where $pen_n(m_{n,p})$ is a penalty term penalizing the complexity of the estimate. Observe that minimization of (2.18) with respect to p is equivalent to minimization of

$$\frac{1}{n} \sum_{i=1}^{n} |m_{n,p}(X_i) - Y_i|^2 + pen_n(m_{n,p}),$$

and that the latter term depends only on $m_{n,p}$ and the data. If $m_{n,p}$ is defined by minimizing the empirical L_2 risk over some linear vector space $\mathcal{F}_{n,p}$ of functions with dimension K_p, then the penalty will be of the form

$$pen_n(m_{n,p}) = c \cdot \frac{K_p}{n} \quad \text{or} \quad pen_n(m_{n,p}) = c \cdot \log(n) \cdot \frac{K_p}{n}.$$

In contrast to the definition of penalized least squares estimates (cf. (2.10) and (2.11)) the penalty here depends not directly on $m_{n,p}$ but on the class of functions $\mathcal{F}_n$ to which $m_{n,p}$ belongs.

2.5   Bibliographic Notes

The description in Section 2.1 concerning the four paradigms in nonparametric regression is based on Friedman (1991). The partitioning estimate was introduced under the name *regressogram* by Tukey (1947; 1961). The kernel estimate is due to Nadaraya (1964; 1970) and Watson (1964). Nearest neighbor estimates were introduced in pattern recognition by Fix and Hodges (1951) and also used in density estimation and regression estimation by Loftsgaarden and Quesenberry (1965), Royall (1966), Cover and Hart (1967), Cover (1968a), and Stone (1977), respectively.

The principle of least squares, which is behind the global modeling estimates, is much older. It was independently proposed by A. M. Legendre in 1805 and by C. F. Gauss with a publication in 1809 (but of course applied in a parametric setting). Further developments are due to P. S. Laplace

1816, P. L. Chebyshev 1859, F. R. Helmert 1872, J. P. Gram 1879, and T. N. Thiele 1903 (with the aspect of a suitable termination of an orthogonal series expansion). F. Galton's 1889 work has been continued by F. Y. Edgeworth, K. Pearson, G. U. Yule, and R. A. Fisher in the last decade of the nineteenth century and in the first decade of the twentieth century. For historical details we refer to Hald (1998), Farebrother (1999), and Stigler (1999).

The principle of penalized modeling, in particular, smoothing splines, goes back to Whittaker (1923), Schoenberg (1964), and Reinsch (1967); see Wahba (1990) for additional references.

The phrase "curse of dimensionality" is due to Bellman (1961).

The concept of cross-validation in statistics was introduced by Lunts and Brailovsky (1967), Allen (1974) and M. Stone (1974). Complexity regularization was introduced under the name "structural risk minimization" by Vapnik, see Vapnik (1982; 1998) and the references therein. There are many other ways of using the data for choosing a smoothing parameter, see, e.g., Chapter 4 in Fan and Gijbels (1995) and the references therein.

Problems and Exercises

PROBLEM 2.1. Let $z_1, \ldots, z_n \in \mathcal{R}$ and set $\bar{z} = (1/n) \sum_{i=1}^{n} z_i$.
(a) Show, for any $c \in \mathcal{R}$,

$$\frac{1}{n} \sum_{i=1}^{n} |c - z_i|^2 = |c - \bar{z}|^2 + \frac{1}{n} \sum_{i=1}^{n} |\bar{z} - z_i|^2 .$$

(b) Conclude from (a):

$$\frac{1}{n} \sum_{i=1}^{n} |\bar{z} - z_i|^2 = \min_{c \in \mathcal{R}} \frac{1}{n} \sum_{i=1}^{n} |c - z_i|^2 .$$

PROBLEM 2.2. Prove that the Nadaraya–Watson kernel estimate defined by (2.2) satisfies (2.4).

HINT: Let $m_n(x)$ be the Nadaraya–Watson kernel estimate. Show that, for any $c \in \mathcal{R}$,

$$\sum_{i=1}^{n} K\left(\frac{x - X_i}{h}\right) (Y_i - c)^2$$

$$= \sum_{i=1}^{n} K\left(\frac{x - X_i}{h}\right) (Y_i - m_n(x))^2 + \sum_{i=1}^{n} K\left(\frac{x - X_i}{h}\right) (m_n(x) - c)^2 .$$

PROBLEM 2.3. Prove that the partitioning estimate defined by (2.1) satisfies (2.8).

HINT: Let $\mathcal{F}$ be defined by (2.7) and let m_n be the partitioning estimate defined by (2.1). Show by the aid of Problem 2.1 that, for any $f \in \mathcal{F}$,

$$\sum_{i=1}^{n} |f(X_i) - Y_i|^2 = \sum_{i=1}^{n} |f(X_i) - m_n(X_i)|^2 + \sum_{i=1}^{n} |m_n(X_i) - Y_i|^2.$$

PROBLEM 2.4. Let $X, X_1, \ldots, X_n$ be independent and uniformly distributed on $[0,1]^d$. Prove

$$\mathbf{E}\left\{\min_{i=1,\ldots,n} \|X - X_i\|\right\} \geq \frac{d}{d+1} \left(\frac{\Gamma\left(\frac{d}{2}+1\right)^{1/d}}{\sqrt{\pi}}\right) \cdot \frac{1}{n^{1/d}}.$$

HINT: The volume of a ball in $\mathcal{R}^d$ with radius t is given by

$$\frac{\pi^{d/2}}{\Gamma\left(\frac{d}{2}+1\right)} \cdot t^d,$$

where $\Gamma(x) = \int_0^\infty t^{x-1} e^{-t} dt$ $(x > 0)$ satisfies $\Gamma(x+1) = x \cdot \Gamma(x)$, $\Gamma(1) = 1$, and $\Gamma(1/2) = \sqrt{\pi}$. Show that this implies

$$\mathbf{P}\left\{\min_{i=1,\ldots,n} \|X - X_i\| \leq t\right\} \leq n \cdot \frac{\pi^{d/2}}{\Gamma\left(\frac{d}{2}+1\right)} \cdot t^d.$$

3
Lower Bounds

3.1 Slow Rate

Recall that the nonparametric regression problem is formulated as follows: Given the observation X and the training data

$$D_n = \{(X_1, Y_1), \ldots, (X_n, Y_n)\}$$

of independent and identically distributed random variables, estimate the random variable Y by a regression function estimate

$$m_n(X) = m_n(X, D_n).$$

The error criterion is the L_2 error

$$\|m_n - m\|^2 = \int (m_n(x) - m(x))^2 \mu(dx).$$

Obviously, the average L_2 error $\mathbf{E}\|m_n - m\|^2$ is completely determined by the distribution of the pair (X, Y) and the regression function estimator m_n. We shall see in Chapters 4, 5, 6, etc., that there exist universally consistent regression estimates. The next question is whether there are regression function estimates with $\mathbf{E}\|m_n - m\|^2$ tending to 0 with a guaranteed rate of convergence. Disappointingly, such estimates do not exist. As our next theorem indicates, it is impossible to obtain a nontrivial rate of convergence results without imposing strong restrictions on the distribution of (X, Y), because even when the distribution of X is good and $Y = m(X)$, i.e., Y is noiseless, the rate of convergence of any estimate can be arbitrarily slow.

Theorem 3.1. *Let $\{a_n\}$ be a sequence of positive numbers converging to zero. For every sequence of regression estimates, there exists a distribution of (X, Y), such that X is uniformly distributed on $[0,1]$, $Y = m(X)$, m is ± 1 valued, and*

$$\limsup_{n \to \infty} \frac{\mathbf{E}\|m_n - m\|^2}{a_n} \geq 1.$$

PROOF. Without loss of generality we assume that $1/4 \geq a_1 \geq a_2 \geq \cdots > 0$, otherwise replace a_n by $\min\{1/4, a_1, \ldots, a_n\}$. Let $\{p_j\}$ be a probability distribution and let $\mathcal{P} = \{A_j\}$ be a partition of $[0,1]$ such that A_j is an interval of length p_j. We consider regression functions indexed by a parameter

$$c = (c_1, c_2, \ldots),$$

where $c_j \in \{-1, 1\}$. Denote the set of all such parameters by $\mathcal{C}$. For $c = (c_1, c_2, \ldots) \in \mathcal{C}$ define $m^{(c)} : [0,1] \to \{-1, 1\}$ by

$$m^{(c)}(x) = c_j \quad \text{if} \quad x \in A_j,$$

i.e., $m^{(c)}$ is piecewise constant with respect to the partition $\mathcal{P}$ and takes a value c_j on A_j.

Fix a sequence of regression estimates m_n. We will show that there exists a distribution of (X, Y) such that X is uniformly distributed on $[0,1]$, $Y = m^{(c)}(X)$ for some $c \in \mathcal{C}$, and

$$\limsup_{n \to \infty} \frac{\mathbf{E}\int |m_n(x) - m^{(c)}(x)|^2 \mu(dx)}{a_n} \geq 1.$$

Let $\hat{m}_n$ be the projection of m_n on the set of all functions which are piecewise constant with respect to $\mathcal{P}$, i.e., for $x \in A_j$, set

$$\hat{m}_n(x) = \frac{1}{p_j} \int_{A_j} m_n(z)\mu(dz).$$

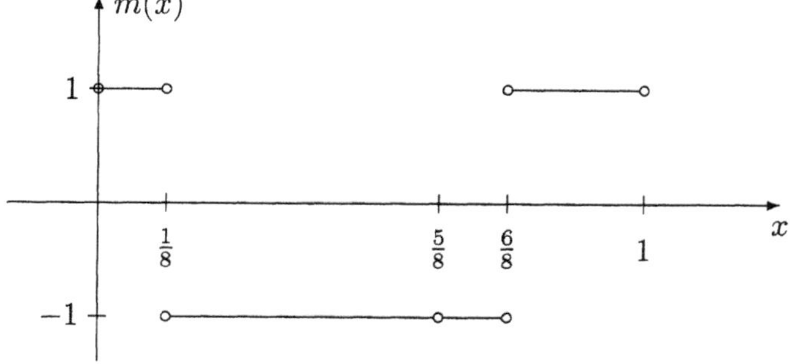

Figure 3.1. Regression function for $p_1 = \frac{1}{8}$, $p_2 = \frac{4}{8}$, $p_3 = \frac{1}{8}$, $p_4 = \frac{2}{8}$, and $c_1 = c_4 = 1$, $c_2 = c_3 = -1$.

Then

$$\int_{A_j} |m_n(x) - m^{(c)}(x)|^2 \mu(dx)$$

$$= \int_{A_j} |m_n(x) - \hat{m}_n(x)|^2 \mu(dx) + \int_{A_j} |\hat{m}_n(x) - m^{(c)}(x)|^2 \mu(dx)$$

$$\geq \int_{A_j} |\hat{m}_n(x) - m^{(c)}(x)|^2 \mu(dx).$$

Set

$$\hat{c}_{nj} = \begin{cases} 1 & \text{if } \int_{A_j} m_n(z)\mu(dz)/p_j \geq 0, \\ -1 & \text{otherwise.} \end{cases}$$

Fix $x \in A_j$. If $\hat{c}_{nj} = 1$ and $c_j = -1$, then $\hat{m}_n(x) \geq 0$ and $m^{(c)}(x) = -1$, which implies

$$|\hat{m}_n(x) - m^{(c)}(x)|^2 \geq 1.$$

If $\hat{c}_{nj} = -1$ and $c_j = 1$, then $\hat{m}_n(x) < 0$ and $m^{(c)}(x) = 1$, which again implies

$$|\hat{m}_n(x) - m^{(c)}(x)|^2 \geq 1.$$

It follows that

$$\int_{A_j} |m_n(x) - m^{(c)}(x)|^2 \mu(dx) \geq \int_{A_j} |\hat{m}_n(x) - m^{(c)}(x)|^2 \mu(dx)$$

$$\geq I_{\{\hat{c}_{nj} \neq c_j\}} \cdot \int_{A_j} 1 \mu(dx)$$

$$= I_{\{\hat{c}_{nj} \neq c_j\}} \cdot p_j$$

$$\geq I_{\{\hat{c}_{nj} \neq c_j\}} \cdot I_{\{\mu_n(A_j)=0\}} \cdot p_j,$$

and so

$$\mathbf{E}\left\{ \int |m_n(x) - m^{(c)}(x)|^2 \mu(dx) \right\} \geq \sum_{j=1}^{\infty} \mathbf{P}\{\hat{c}_{nj} \neq c_j, \mu_n(A_j) = 0\} p_j$$

$$=: R_n(c).$$

Now we randomize c. Let $C_1, C_2, \ldots$ be a sequence of independent and identically distributed random variables independent of $X_1, X_2, \ldots$ which satisfy

$$\mathbf{P}\{C_1 = 1\} = \mathbf{P}\{C_1 = -1\} = \frac{1}{2},$$

and set

$$C = (C_1, C_2, \ldots).$$

On the one hand,

$$R_n(c) \ \leq \ \sum_{j=1}^{\infty} \mathbf{P}\{\mu_n(A_j) = 0\}p_j$$

$$= \ \sum_{j=1}^{\infty}(1 - p_j)^n p_j,$$

and, on the other hand,

$$\mathbf{E}\{R_n(C)\} \ = \ \sum_{j=1}^{\infty} \mathbf{P}\{\hat{c}_{nj} \neq C_j, \mu_n(A_j) = 0\}p_j$$

$$= \ \sum_{j=1}^{\infty} \mathbf{E}\{\mathbf{P}\{\hat{c}_{nj} \neq C_j, \mu_n(A_j) = 0 | X_1, \ldots, X_n\}\}p_j$$

$$= \ \sum_{j=1}^{\infty} \mathbf{E}\{I_{\{\mu_n(A_j)=0\}}\mathbf{P}\{\hat{c}_{nj} \neq C_j | X_1, \ldots, X_n\}\}p_j$$

$$= \ \sum_{j=1}^{\infty} \mathbf{E}\left\{I_{\{\mu_n(A_j)=0\}}\frac{1}{2}\right\} p_j$$

$$= \ \frac{1}{2}\sum_{j=1}^{\infty}(1 - p_j)^n p_j,$$

where we used the fact that for $\mu_n(A_j) = 0$ the random variables $\hat{c}_{nj}$ and C_j are independent given $X_1, \ldots, X_n$. Thus

$$\frac{R_n(c)}{\mathbf{E}R_n(C)} \leq 2.$$

We can apply the Fatou lemma:

$$\mathbf{E}\left\{\limsup_{n\to\infty} \frac{R_n(C)}{\mathbf{E}R_n(C)}\right\} \ \geq \ \limsup_{n\to\infty} \mathbf{E}\left\{\frac{R_n(C)}{\mathbf{E}R_n(C)}\right\} = 1,$$

which implies that there exists $c \in \mathcal{C}$ such that

$$\limsup_{n\to\infty} \frac{R_n(c)}{\mathbf{E}R_n(C)} \geq 1.$$

Summarizing the above results we get that there exists a distribution of (X, Y) such that X is uniformly distributed on $[0, 1]$, $Y = m^{(c)}(X)$ for some $c \in \mathcal{C}$, and

$$\limsup_{n\to\infty} \frac{\mathbf{E}\{\int |m_n(x) - m(x)|^2 \mu(dx)\}}{\frac{1}{2}\sum_{j=1}^{\infty}(1 - p_j)^n p_j} \geq 1.$$

What is left is to show that for a_n satisfying $1/4 \geq a_1 \geq a_2 \geq \cdots > 0$ and $a_n \to 0$, there is a distribution $\{p_j\}$ such that

$$\frac{1}{2}\sum_{j=1}^{\infty}(1-p_j)^n p_j \geq a_n$$

for all n. This is shown by the following lemma:

Lemma 3.1. *Let $\{a_n\}$ be a sequence of positive numbers converging to zero with $1/2 \geq a_1 \geq a_2 \geq \cdots$. Then there is a distribution $\{p_j\}$ such that*

$$\sum_{j=1}^{\infty}(1-p_j)^n p_j \geq a_n$$

for all n.

PROOF. Set

$$p_1 = 1 - 2a_1$$

and choose integers $\{k_n\}$,

$$1 = k_1 < k_2 < \cdots$$

and $p_2, p_3, \ldots$ such that, for $i > k_n$,

$$p_i \leq \frac{1}{2n}$$

and

$$\sum_{i=k_n+1}^{k_{n+1}} p_i = 2(a_n - a_{n+1}).$$

Then

$$\sum_{j=1}^{\infty} p_j = p_1 + \sum_{n=1}^{\infty} 2(a_n - a_{n+1}) = 1$$

and

$$\sum_{j=1}^{\infty}(1-p_j)^n p_j \geq \left(1 - \frac{1}{2n}\right)^n \sum_{p_j \leq 1/2n} p_j$$

$$\geq \frac{1}{2} \sum_{p_j \leq 1/2n} p_j$$

$$\text{(because } (1 - 1/(2n))^n \geq 1/2, \text{ cf. Problem 3.2)}$$

$$\geq \frac{1}{2} \sum_{j=k_n+1}^{\infty} p_j$$

$$= \frac{1}{2} \sum_{i=n}^{\infty} 2(a_i - a_{i+1})$$
$$= a_n.$$

$\square$

Pattern recognition is easier than regression estimation (cf. Theorem 1.1), therefore lower bounds for pattern recognition imply lower bounds for regression estimation, so Theorem 3.1 can be sharpened as follows: Let $\{a_n\}$ be a sequence of positive numbers converging to zero with $1/64 \geq a_1 \geq a_2 \geq \cdots$. For every sequence of regression estimates, there exists a distribution of (X, Y), such that X is uniformly distributed on $[0, 1]$, $Y = m(X)$, and

$$\mathbf{E}\|m_n - m\|^2 \geq a_n$$

for all n.

The proof of this statement applies the combination of Theorem 7.2 and Problem 7.2 in Devroye, Györfi, and Lugosi (1996): Consider the classification problem. Let $\{a'_n\}$ be a sequence of positive numbers converging to zero with $1/16 \geq a'_1 \geq a'_2 \geq \cdots$. For every sequence of classification rules $\{g_n\}$, there exists a distribution of (X, Y), such that X is uniformly distributed on $[0, 1]$, $Y = m(X)$, and

$$\mathbf{P}\{g_n(X) \neq Y\} \geq a'_n$$

for all n. Using this, for an arbitrary regression estimate m_n, introduce the classification rule g_n such that $g_n(X) = 1$ if $m_n(X) \geq 1/2$ and 0 otherwise. Apply the above-mentioned result for $a'_n = 4a_n$. Then

$$
\begin{aligned}
\mathbf{E}\|m_n - m\|^2 &= \mathbf{E}\{(m_n(X) - m(X))^2\} \\
&= \mathbf{E}\{(m_n(X) - Y)^2\} \\
&\geq \mathbf{E}\{(g_n(X) - Y)^2\}/4 \\
&= \mathbf{P}\{g_n(X) \neq Y\}/4 \\
&\geq a'_n/4 = a_n.
\end{aligned}
$$

3.2   Minimax Lower Bounds

Theorem 3.1 shows that universally good regression estimates do not exist even in the case of a nice distribution of X and noiseless Y. Rate of convergence studies for particular estimates must necessarily be accompanied by conditions on (X, Y). Under certain regularity conditions it is possible to obtain upper bounds for the rates of convergence to 0 for $\mathbf{E}\|m_n - m\|^2$ of certain estimates. Then it is natural to ask what the fastest achievable rate is for the given class of distributions.

Let $\mathcal{D}$ be a class of distributions of (X, Y). Given data

$$D_n = \{(X_1, Y_1), \ldots, (X_n, Y_n)\}$$

an arbitrary regression estimate is denoted by m_n.

In the classical minimax approach one tries to minimize the maximal error within a class of distributions. If we use $\mathbf{E}\|m_n - m\|^2$ as error, then this means that one tries to minimize

$$\sup_{(X,Y)\in\mathcal{D}} \mathbf{E}\{(m_n(X) - m(X))^2\}.$$

In the sequel we will derive asymptotic lower bounds of

$$\inf_{m_n} \sup_{(X,Y)\in\mathcal{D}} \mathbf{E}\{(m_n(X) - m(X))^2\}$$

for special classes $\mathcal{D}$ of distributions. Here the infimum is taken over all estimates m_n, i.e., over all measurable functions of the data.

Definition 3.1. *The sequence of positive numbers a_n is called the* **lower minimax rate of convergence for the class** $\mathcal{D}$ *if*

$$\liminf_{n\to\infty} \inf_{m_n} \sup_{(X,Y)\in\mathcal{D}} \frac{\mathbf{E}\{\|m_n - m\|^2\}}{a_n} = C_1 > 0.$$

Definition 3.2. *The sequence of positive numbers a_n is called the* **optimal rate of convergence for the class** $\mathcal{D}$ *if it is a lower minimax rate of convergence and there is an estimate m_n such that*

$$\limsup_{n\to\infty} \sup_{(X,Y)\in\mathcal{D}} \frac{\mathbf{E}\{\|m_n - m\|^2\}}{a_n} = C_0 < \infty.$$

We will derive rate of convergence results for classes of distributions where the regression function satisfies the following smoothness condition:

Definition 3.3. *Let $p = k + \beta$ for some $k \in \mathcal{N}_0$ and $0 < \beta \leq 1$, and let $C > 0$. A function $f : \mathcal{R}^d \to \mathcal{R}$ is called (p, C)-smooth if for every $\alpha = (\alpha_1, \ldots, \alpha_d)$, $\alpha_i \in \mathcal{N}_0$, $\sum_{j=1}^{d} \alpha_j = k$ the partial derivative $\frac{\partial^k f}{\partial x_1^{\alpha_1} \ldots \partial x_d^{\alpha_d}}$ exists and satisfies*

$$\left| \frac{\partial^k f}{\partial x_1^{\alpha_1} \ldots \partial x_d^{\alpha_d}}(x) - \frac{\partial^k f}{\partial x_1^{\alpha_1} \ldots \partial x_d^{\alpha_d}}(z) \right| \leq C \cdot \|x - z\|^\beta \quad (x, z \in \mathcal{R}^d).$$

Let $\mathcal{F}^{(p,C)}$ be the set of all (p, C)-smooth functions $f : \mathcal{R}^d \to \mathcal{R}$.

Clearly, if $\mathcal{D} \subseteq \bar{\mathcal{D}}$ and $\{a_n\}$ is an lower minimax rate of convergence for $\mathcal{D}$, then it is also a lower minimax rate of convergence for $\bar{\mathcal{D}}$. Thus, to determine lower minimax rates of convergence, it might be useful to restrict the class of distributions. It turns out that it suffices to look at classes of distributions where X is uniformly distributed on $[0, 1]^d$ and $Y - m(X)$ has a normal distribution.

Definition 3.4. *Let $\mathcal{D}^{(p,C)}$ be the class of distributions of (X, Y) such that:*

   (i) X *is uniformly distributed on* $[0, 1]^d$;

   (ii) $Y = m(X) + N$, *where X and N are independent and N is standard normal; and*

   (iii) $m \in \mathcal{F}^{(p,C)}$.

In the next theorem we derive a lower minimax rate of convergence for this class of distributions.

Theorem 3.2. *For the class $\mathcal{D}^{(p,C)}$, the sequence*

$$a_n = n^{-\frac{2p}{2p+d}}$$

is a lower minimax rate of convergence. In particular,

$$\liminf_{n \to \infty} \inf_{m_n} \sup_{(X,Y) \in \mathcal{D}^{(p,C)}} \frac{\mathbf{E}\{\|m_n - m\|^2\}}{C^{\frac{2d}{2p+d}} n^{-\frac{2p}{2p+d}}} \geq C_1 > 0$$

for some constant C_1 independent of C.

For bounded X, in the subsequent chapters, we will discuss estimates which achieve the lower minimax rate of convergence in Theorem 3.2 for Lipschitz continuous m (cf. Theorems 4.3, 5.2, and 6.2) and for m from a higher smoothness class (cf. Corollary 11.2, Theorem 14.5, and Corollary 19.1). Therefore this is the optimal rate of convergence for this class.

The proof of Theorem 3.2 applies the following lemma:

Lemma 3.2. *Let u be an l-dimensional real vector, let C be a zero mean random variable taking values in $\{-1, +1\}$, and let N be an l-dimensional standard normal random variable, independent of C. Set*

$$Z = Cu + N.$$

Then the error probability of the Bayes decision for C based on Z is

$$L^* := \min_{g : \mathcal{R}^l \to \mathcal{R}} \mathbf{P}\{g(Z) \neq C\} = \Phi(-\|u\|),$$

where Φ is the standard normal distribution function.

PROOF. Let φ be the density of an l–dimensional standard normal random variable. The conditional density of Z, given $C = 1$, is $\varphi(z - u)$ and, given $C = -1$, is $\varphi(z + u)$. For an arbitrary decision rule $g : \mathcal{R}^l \to \mathcal{R}$ one obtains

$$
\begin{aligned}
&\mathbf{P}\{g(Z) \neq C\} \\
=\;& \mathbf{P}\{C = 1\}\mathbf{P}\{g(Z) \neq C | C = 1\} + \mathbf{P}\{C = -1\}\mathbf{P}\{g(Z) \neq C | C = -1\} \\
=\;& \frac{1}{2}\mathbf{P}\{g(Z) = -1 | C = 1\} + \frac{1}{2}\mathbf{P}\{g(Z) = 1 | C = -1\} \\
=\;& \frac{1}{2}\int I_{\{g(z)=-1\}}\varphi(z - u)\,dz + \frac{1}{2}\int I_{\{g(z)=+1\}}\varphi(z + u)\,dz
\end{aligned}
$$

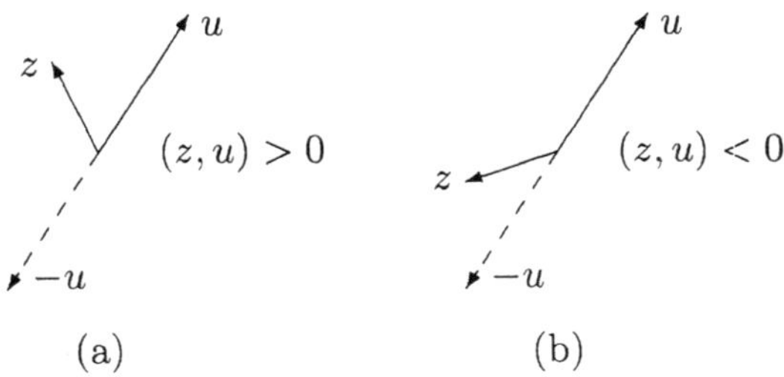

Figure 3.2. (a) z is closer to u than to $-u$, (b) z is closer to $-u$ than to u.

$$= \frac{1}{2} \int \left(I_{\{g(z)=-1\}} \varphi(z - u) + I_{\{g(z)=+1\}} \varphi(z + u) \right) \, dz.$$

The above expression is minimized by the Bayes decision rule

$$g^*(z) = \begin{cases} 1 & \text{if } \varphi(z - u) > \varphi(z + u), \\ -1 & \text{otherwise.} \end{cases}$$

This together with

$$\varphi(z - u) > \varphi(z + u) \Leftrightarrow z \text{ is closer to } u \text{ than to } -u \Leftrightarrow (u, z) > 0$$

proves that the Bayes decision rule is given by

$$g^*(z) = \begin{cases} 1 & \text{if } (u, z) > 0, \\ -1 & \text{if } (u, z) \leq 0, \end{cases}$$

where (u, z) denotes the inner product of u and z. Hence

$$
\begin{aligned}
L^* &= \mathbf{P}\{g^*(Z) \neq C\} \\
&= \mathbf{P}\{g^*(Z) = -1, C = 1\} + \mathbf{P}\{g^*(Z) = 1, C = -1\} \\
&= \mathbf{P}\{(u, Z) \leq 0, C = 1\} + \mathbf{P}\{(u, Z) > 0, C = -1\} \\
&= \mathbf{P}\{\|u\|^2 + (u, N) \leq 0, C = 1\} + \mathbf{P}\{-\|u\|^2 + (u, N) > 0, C = -1\} \\
&\qquad\qquad\qquad\qquad\qquad\qquad\qquad\qquad (\text{because of } Z = Cu + N) \\
&\leq \frac{1}{2}\mathbf{P}\{(u, N) \leq -\|u\|^2\} + \frac{1}{2}\mathbf{P}\{(u, N) > \|u\|^2\}.
\end{aligned}
$$

For $u = 0$, one obtains

$$L^* = \frac{1}{2} \cdot 1 + \frac{1}{2} \cdot 0 = \Phi(-\|u\|).$$

For $u \neq 0$,

$$\frac{(u, N)}{\|u\|} = \left(\frac{u}{\|u\|}, N \right)$$

is a one-dimensional standard normal variable, which implies

$$L^* = \frac{1}{2}\mathbf{P}\left\{\left(\frac{u}{\|u\|}, N\right) \le -\|u\|\right\} + \frac{1}{2}\mathbf{P}\left\{\left(\frac{u}{\|u\|}, N\right) > \|u\|\right\}$$

$$= \Phi(-\|u\|).$$

$\square$

PROOF OF THEOREM 3.2. First we define (depending on n) subclasses of distributions (X, Y) contained in $\mathcal{D}^{(p,C)}$. Set

$$M_n = \left\lceil (C^2 n)^{\frac{1}{2p+d}} \right\rceil.$$

Partition $[0, 1]^d$ by M_n^d cubes $\{A_{n,j}\}$ of side length $1/M_n$ and with centers $\{a_{n,j}\}$. Choose a function $\bar{g} : \mathcal{R}^d \to \mathcal{R}$ such that the support of $\bar{g}$ is a subset of $[-\frac{1}{2}, \frac{1}{2}]^d$, $\int \bar{g}^2(x)\,dx > 0$, and $\bar{g} \in \mathcal{F}^{(p,2^{\beta-1})}$. Define $g : \mathcal{R}^d \to \mathcal{R}$ by $g(x) = C \cdot \bar{g}(x)$. Then:

(I)   the support of g is a subset of $[-\frac{1}{2}, \frac{1}{2}]^d$;
(II) $\int g^2(x)\,dx = C^2 \int \bar{g}^2(x)\,dx$ and $\int \bar{g}^2(x)\,dx > 0$; and
(III) $g \in \mathcal{F}^{(p,C2^{\beta-1})}$.

The class of regression functions is indexed by a vector

$$c_n = (c_{n,1}, \ldots, c_{n,M_n^d})$$

of $+1$ or -1 components, so the "worst regression function" will depend on the sample size n. Denote the set of all such vectors by $\mathcal{C}_n$. For $c_n = (c_{n,1}, \ldots, c_{n,M_n^d}) \in \mathcal{C}_n$ define the function

$$m^{(c_n)}(x) = \sum_{j=1}^{M_n^d} c_{n,j} g_{n,j}(x),$$

where

$$g_{n,j}(x) = M_n^{-p} g(M_n(x - a_{n,j})).$$

Next we show that, because of (III),

$$m^{(c_n)} \in \mathcal{F}^{(p,C)}.$$

Let $\alpha = (\alpha_1, \ldots, \alpha_d)$, $\alpha_i \in \mathcal{N}_0$, and $\sum_{j=1}^d \alpha_j = k$. Set $D^\alpha = \frac{\partial^k}{\partial x_1^{\alpha_1} \ldots \partial x_d^{\alpha_d}}$. If $x, z \in A_{n,i}$ then (I) and (III) imply

$$|D^\alpha m^{(c_n)}(x) - D^\alpha m^{(c_n)}(z)|$$
$$= |c_{n,i}| \cdot |D^\alpha g_{n,i}(x) - D^\alpha g_{n,i}(z)|$$
$$\le C2^{\beta-1} M_n^{-p} M_n^k \|M_n(x - a_{n,i}) - M_n(z - a_{n,i})\|^\beta$$
$$\le C2^{\beta-1} \|x - z\|^\beta$$
$$\le C\|x - z\|^\beta.$$

Now assume that $x \in A_{n,i}$ and $z \in A_{n,j}$ for $i \ne j$. Choose $\bar{x}$, $\bar{z}$ on the line between x and z such that $\bar{x}$ is on the boundary of $A_{n,i}$, $\bar{z}$ is on the boundary of $A_{n,j}$, and $\|x - \bar{x}\| + \|\bar{z} - z\| \leq \|x - z\|$. Then

$$
\begin{aligned}
&|D^\alpha m^{(c_n)}(x) - D^\alpha m^{(c_n)}(z)| \\
={}& |c_{n,i} D^\alpha g_{n,i}(x) - c_{n,j} D^\alpha g_{n,j}(z)| \\
\leq{}& |c_{n,i} D^\alpha g_{n,i}(x)| + |c_{n,j} D^\alpha g_{n,j}(z)| \\
={}& |c_{n,i}| \cdot |D^\alpha g_{n,i}(x) - D^\alpha g_{n,i}(\bar{x})| + |c_{n,j}| \cdot |D^\alpha g_{n,j}(z) - D^\alpha g_{n,j}(\bar{z})| \\
&\qquad \text{(because of } D^\alpha g_{n,i}(\bar{x}) = D^\alpha g_{n,j}(\bar{z}) = 0) \\
\leq{}& C 2^{\beta-1}(\|x - \bar{x}\|^\beta + \|z - \bar{z}\|^\beta) \\
&\qquad \text{(as in the first case)} \\
={}& C 2^\beta \left(\frac{1}{2}\|x - \bar{x}\|^\beta + \frac{1}{2}\|z - \bar{z}\|^\beta \right) \\
\leq{}& C 2^\beta \left(\frac{\|x - \bar{x}\|}{2} + \frac{\|z - \bar{z}\|}{2} \right)^\beta \\
&\qquad \text{(by Jensen's inequality)} \\
\leq{}& C\|x - z\|^\beta.
\end{aligned}
$$

Hence, each distribution of (X, Y) with $Y = m^{(c)}(X) + N$ for some $c \in \mathcal{C}_n$ is contained in $\mathcal{D}^{(p,C)}$, and it suffices to show

$$
\liminf_{n \to \infty} \inf_{m_n} \sup_{(X,Y):Y=m^{(c)}(X)+N,\, c \in \mathcal{C}_n} \frac{M_n^{2p}}{C^2} \mathbf{E}\{\|m_n - m^{(c)}\|^2\} > 0.
$$

Let m_n be an arbitrary estimate. By definition, $\{g_{n,j} : j\}$ is an orthogonal system in L_2, therefore the projection $\hat{m}_n$ of m_n to $\{m^{(c)} : c \in \mathcal{C}_n\}$ is given by

$$
\hat{m}_n(x) = \sum_{j=1}^{M_n^d} \hat{c}_{n,j} g_{n,j}(x),
$$

where

$$
\hat{c}_{n,j} = \frac{\int_{A_{n,j}} m_n(x) g_{n,j}(x)\, dx}{\int_{A_{n,j}} g_{n,j}^2(x)\, dx}.
$$

Let $c \in \mathcal{C}_n$ be arbitrary. Then

$$
\begin{aligned}
\|m_n - m^{(c)}\|^2 &\geq \|\hat{m}_n - m^{(c)}\|^2 \\
&= \sum_{j=1}^{M_n^d} \int_{A_{n,j}} (\hat{c}_{n,j} g_{n,j}(x) - c_{n,j} g_{n,j}(x))^2\, dx \\
&= \sum_{j=1}^{M_n^d} \int_{A_{n,j}} (\hat{c}_{n,j} - c_{n,j})^2 g_{n,j}^2(x)\, dx
\end{aligned}
$$

$$= \int g^2(x)\, dx \sum_{j=1}^{M_n^d} (\hat{c}_{n,j} - c_{n,j})^2 \frac{1}{M_n^{2p+d}}.$$

Let $\tilde{c}_{n,j}$ be 1 if $\hat{c}_{n,j} \geq 0$ and -1 otherwise. Because of

$$|\hat{c}_{n,j} - c_{n,j}| \geq |\tilde{c}_{n,j} - c_{n,j}|/2,$$

we get

$$\|m_n - m^{(c)}\|^2 \;\geq\; \int g^2(x)\, dx \cdot \frac{1}{4} \frac{1}{M_n^{2p+d}} \sum_{j=1}^{M_n^d} (\tilde{c}_{n,j} - c_{n,j})^2$$

$$\geq \int g^2(x)\, dx \cdot \frac{1}{M_n^{2p+d}} \sum_{j=1}^{M_n^d} I_{\{\tilde{c}_{n,j} \neq c_{n,j}\}}$$

$$= \frac{C^2}{M_n^{2p}} \cdot \int \bar{g}^2(x)\, dx \cdot \frac{1}{M_n^d} \sum_{j=1}^{M_n^d} I_{\{\tilde{c}_{n,j} \neq c_{n,j}\}}.$$

Thus for the proof we need

$$\liminf_{n \to \infty} \inf_{\tilde{c}_n} \sup_{c_n} \frac{1}{M_n^d} \sum_{j=1}^{M_n^d} \mathbf{P}\{\tilde{c}_{n,j} \neq c_{n,j}\} > 0.$$

Now we randomize c_n. Let $C_{n,1}, \ldots, C_{n,M_n^d}$ be a sequence of i.i.d. random variables independent of $(X_1, N_1), (X_2, N_2), \ldots$, which satisfy

$$\mathbf{P}\{C_{n,1} = 1\} = \mathbf{P}\{C_{n,1} = -1\} = \frac{1}{2}.$$

Set

$$C_n = (C_{n,1}, \ldots, C_{n,M_n^d}).$$

Then

$$\inf_{\tilde{c}_n} \sup_{c_n} \frac{1}{M_n^d} \sum_{j=1}^{M_n^d} \mathbf{P}\{\tilde{c}_{n,j} \neq c_{n,j}\} \geq \inf_{\tilde{c}_n} \frac{1}{M_n^d} \sum_{j=1}^{M_n^d} \mathbf{P}\{\tilde{c}_{n,j} \neq C_{n,j}\},$$

where $\tilde{c}_{n,j}$ can be interpreted as a decision on $C_{n,j}$ using D_n. Its error probability is minimal for the Bayes decision $\bar{C}_{n,j}$, which is 1 if $\mathbf{P}\{C_{n,j} = 1 | D_n\} \geq \frac{1}{2}$ and -1 otherwise, therefore

$$\inf_{\tilde{c}_n} \frac{1}{M_n^d} \sum_{j=1}^{M_n^d} \mathbf{P}\{\tilde{c}_{n,j} \neq C_{n,j}\} \;\geq\; \frac{1}{M_n^d} \sum_{j=1}^{M_n^d} \mathbf{P}\{\bar{C}_{n,j} \neq C_{n,j}\}$$

$$= \mathbf{P}\{\bar{C}_{n,1} \neq C_{n,1}\}$$

$$= \mathbf{E}\{\mathbf{P}\{\bar{C}_{n,1} \neq C_{n,1} | X_1, \ldots, X_n\}\}.$$

Let $X_{i_1}, \ldots, X_{i_l}$ be those $X_i \in A_{n,1}$. Then

$$(Y_{i_1}, \ldots, Y_{i_l}) = C_{n,1} \cdot (g_{n,1}(X_{i_1}), \ldots, g_{n,1}(X_{i_l})) + (N_{i_1}, \ldots, N_{i_l}),$$

while

$$\{Y_1, \ldots, Y_n\} \setminus \{Y_{i_1}, \ldots, Y_{i_l}\}$$

depends only on $C \setminus \{C_{n,1}\}$ and on X_r's and N_r's with $r \notin \{i_1, \ldots, i_l\}$, and therefore is independent of $C_{n,1}$ given $X_1, \ldots, X_n$. Now conditioning on $X_1, \ldots, X_n$, the error of the conditional Bayes decision for $C_{n,1}$ based on $(Y_1, \ldots, Y_n)$ depends only on $(Y_{i_1}, \ldots, Y_{i_l})$, hence Lemma 3.2 implies

$$\mathbf{P}\{\bar{C}_{n,1} \neq C_{n,1} | X_1, \ldots, X_n\} = \Phi\left(-\sqrt{\sum_{r=1}^{l} g_{n,1}^2(X_{i_r})}\right)$$

$$= \Phi\left(-\sqrt{\sum_{i=1}^{n} g_{n,1}^2(X_i)}\right),$$

where Φ is the standard normal distribution function. The second derivative of $\Phi(-\sqrt{x})$ is positive, therefore $\Phi(-\sqrt{x})$ is convex, so by Jensen's inequality

$$\mathbf{P}\{\bar{C}_{n,1} \neq C_{n,1}\} = \mathbf{E}\left\{\Phi\left(-\sqrt{\sum_{i=1}^{n} g_{n,1}^2(X_i)}\right)\right\}$$

$$\geq \Phi\left(-\sqrt{\mathbf{E}\left\{\sum_{i=1}^{n} g_{n,1}^2(X_i)\right\}}\right)$$

$$= \Phi\left(-\sqrt{n\mathbf{E}\{g_{n,1}^2(X_1)\}}\right)$$

$$= \Phi\left(-\sqrt{nM_n^{-(2p+d)}\int g^2(x)\,dx}\right)$$

$$\geq \Phi\left(-\sqrt{\int \bar{g}^2(x)\,dx}\right) > 0.$$

$\square$

3.3 Individual Lower Bounds

In some sense, the lower bounds in Section 3.2 are not satisfactory. They do not tell us anything about the way the error decreases as the sample size is increased for a given regression problem. These bounds, for each n, give information about the maximal error within the class, but not about the behavior of the error for a single fixed distribution as the sample size n increases. In other words, the "bad" distribution, causing the largest error for an estimator, may be different for each n. For example, the lower bound for the class $\mathcal{D}^{(p,C)}$ does not exclude the possibility that there exists a sequence of estimators $\{m_n\}$ such that for *every* distribution in $\mathcal{D}^{(p,C)}$, the expected error $\mathbf{E}\{\|m_n - m\|^2\}$ decreases at an exponential rate in n.

In this section, we are interested in "individual" minimax lower bounds that describe the behavior of the error for a fixed distribution of (X, Y) as the sample size n grows.

Definition 3.5. *A sequence of positive numbers a_n is called the* **individual lower rate of convergence for a class** $\mathcal{D}$ *of distributions of (X, Y), if*

$$\inf_{\{m_n\}} \sup_{(X,Y)\in\mathcal{D}} \limsup_{n\to\infty} \frac{\mathbf{E}\{\|m_n - m\|^2\}}{a_n} > 0 \, ,$$

where the infimum is taken over all sequences $\{m_n\}$ of the estimates.

In this definition the $\limsup_{n\to\infty}$ can be replaced by $\liminf_{n\to\infty}$, here we consider $\limsup_{n\to\infty}$ for the sake of simplicity.

We will show that for every sequence $\{b_n\}$ tending to zero, $b_n n^{-\frac{2p}{2p+d}}$ is an individual lower rate of convergence of the class $\mathcal{D}^{(p,C)}$. Hence, there exist individual lower rates of these classes, which are arbitrarily close to the optimal lower rates.

Theorem 3.3. *Let $\{b_n\}$ be an arbitrary positive sequence tending to zero. Then the sequence*

$$b_n a_n = b_n n^{-\frac{2p}{2p+d}}$$

is an individual lower rate of convergence for the class $\mathcal{D}^{(p,C)}$.

For the sequence $\{\sqrt{b_n}\}$ Theorem 3.3 implies that for all $\{m_n\}$ there is $(X, Y) \in \mathcal{D}^{(p,C)}$ such that

$$\limsup_{n\to\infty} \frac{\mathbf{E}\{\|m_n - m\|^2\}}{\sqrt{b_n}\, a_n} > 0,$$

thus

$$\limsup_{n\to\infty} \frac{\mathbf{E}\{\|m_n - m\|^2\}}{b_n a_n} = \infty.$$

PROOF OF THEOREM 3.3. The proof is an extension of the proof of Theorem 3.2, but is a little involved. We therefore recommend skipping it during the first reading. First we define a subclass of distributions of (X, Y) contained in $\mathcal{D}^{(p,C)}$. We pack infinitely many disjoint cubes into $[0, 1]^d$ in the following way: For a given probability distribution $\{p_j\}$, let $\{B_j\}$ be a partition of $[0, 1]$ such that B_j is an interval of length p_j. We pack disjoint

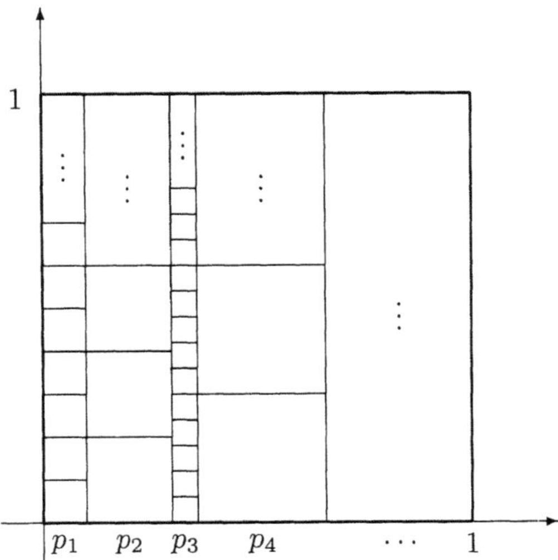

Figure 3.3. Two dimensional partition.

cubes of volume p_j^d into the rectangle

$$B_j \times [0,1]^{d-1}.$$

Denote these cubes by

$$A_{j,1}, \ldots, A_{j,S_j},$$

where

$$S_j = \left\lfloor \frac{1}{p_j} \right\rfloor^{d-1}.$$

Let $a_{j,k}$ be the center of $A_{j,k}$. Choose a function $g : \mathcal{R}^d \to \mathcal{R}$ such that:

(I)   the support of g is a subset of $[-\frac{1}{2}, \frac{1}{2}]^d$;

(II) $\int g^2(x)\,dx > 0$; and

(III) $g \in \mathcal{F}^{(p,C2^{\beta-1})}$.

The class of regression functions is indexed by a vector

$$c = (c_{1,1}, c_{1,2}, \ldots, c_{1,S_1}, c_{2,1}, c_{2,2}, \ldots, c_{2,S_2}, \ldots)$$

of $+1$ or -1 components. Denote the set of all such vectors by $\mathcal{C}$. For $c \in \mathcal{C}$ define the function

$$m^{(c)}(x) = \sum_{j=1}^{\infty} \sum_{k=1}^{S_j} c_{j,k} g_{j,k}(x),$$

where

$$g_{j,k}(x) = p_j^p g(p_j^{-1}(x - a_{j,k})).$$

As in the proof of Theorem 3.2, (III) implies that

$$m^{(c)} \in \mathcal{F}^{(p,C)}.$$

Hence, each distribution of (X, Y) with $Y = m^{(c)}(X) + N$ for some $c \in \mathcal{C}$ is contained in $\mathcal{D}^{(p,C)}$, which implies

$$\inf_{\{m_n\}} \sup_{(X,Y) \in \mathcal{D}^{(p,C)}} \limsup_{n \to \infty} \frac{\mathbf{E}\{\|m_n - m\|^2\}}{b_n a_n}$$

$$\geq \inf_{\{m_n\}} \sup_{(X,Y):Y=m^{(c)}(X)+N, c \in \mathcal{C}} \limsup_{n \to \infty} \frac{\mathbf{E}\{\|m_n - m^{(c)}\|^2\}}{b_n a_n}. \quad (3.1)$$

Let m_n be an arbitrary estimate. By definition, $\{g_{j,k} : j, k\}$ is an orthogonal system in L_2, therefore the projection $\hat{m}_n$ of m_n to $\{m^{(c)} : c \in \mathcal{C}\}$ is given by

$$\hat{m}_n(x) = \sum_{j,k} \hat{c}_{n,j,k} g_{j,k}(x),$$

where

$$\hat{c}_{n,j,k} = \frac{\int_{A_{j,k}} m_n(x) g_{j,k}(x)\, dx}{\int_{A_{j,k}} g_{j,k}^2(x)\, dx}.$$

Let $c \in \mathcal{C}$ be arbitrary. Then

$$\|m_n - m^{(c)}\|^2 \geq \|\hat{m}_n - m^{(c)}\|^2$$

$$= \sum_{j,k} \int_{A_{j,k}} (\hat{c}_{n,j,k} g_{j,k}(x) - c_{j,k} g_{j,k}(x))^2\, dx$$

$$= \sum_{j,k} \int_{A_{j,k}} (\hat{c}_{n,j,k} - c_{j,k})^2 g_{j,k}^2(x)\, dx$$

$$= \int g^2(x)\, dx \sum_{j,k} (\hat{c}_{n,j,k} - c_{j,k})^2 p_j^{2p+d}.$$

Let $\tilde{c}_{n,j,k}$ be 1 if $\hat{c}_{n,j,k} \geq 0$ and -1 otherwise. Because of

$$|\hat{c}_{n,j,k} - c_{j,k}| \geq |\tilde{c}_{n,j,k} - c_{j,k}|/2,$$

we get

$$\|m_n - m^{(c)}\|^2 \geq \int g^2(x)\, dx \frac{1}{4} \sum_{j,k} (\tilde{c}_{n,j,k} - c_{j,k})^2 p_j^{2p+d}$$

$$\geq \int g^2(x)\, dx \sum_{j,k} I_{\{\tilde{c}_{n,j,k} \neq c_{j,k}\}} p_j^{2p+d}.$$

This proves

$$\mathbf{E}\{\|m_n - m^{(c)}\|^2\} \geq \int g^2(x)\, dx \cdot R_n(c), \quad (3.2)$$

where

$$R_n(c) = \sum_{j:np_j^{2p+d}\leq 1} \sum_{k=1}^{S_j} p_j^{2p+d} \cdot \mathbf{P}\{\tilde{c}_{n,j,k} \neq c_{j,k}\}. \tag{3.3}$$

Relations (3.1) and (3.2) imply

$$\inf_{\{m_n\}} \sup_{(X,Y)\in\mathcal{D}^{(p,C)}} \limsup_{n\to\infty} \frac{\mathbf{E}\{\|m_n - m\|^2\}}{b_n a_n}$$

$$\geq \int g^2(x)\,dx \, \inf_{\{m_n\}} \sup_{c\in\mathcal{C}} \limsup_{n\to\infty} \frac{R_n(c)}{b_n a_n}. \tag{3.4}$$

To bound the last term, we fix a sequence $\{m_n\}$ of estimates and choose $c \in \mathcal{C}$ randomly. Let $C_{1,1},\ldots,C_{1,S_1},C_{2,1},\ldots,C_{2,S_2},\ldots$ be a sequence of independent and identically distributed random variables independent of (X_1, N_1), (X_2, N_2), $\ldots$, which satisfy

$$\mathbf{P}\{C_{1,1} = 1\} = \mathbf{P}\{C_{1,1} = -1\} = \frac{1}{2}.$$

Set

$$C = (C_{1,1},\ldots,C_{1,S_1},C_{2,1},\ldots,C_{2,S_2},\ldots).$$

Next we derive a lower bound for

$$\mathbf{E}R_n(C) = \sum_{j:np_j^{2p+d}\leq 1} \sum_{k=1}^{S_j} p_j^{2p+d} \cdot \mathbf{P}\{\tilde{c}_{n,j,k} \neq C_{j,k}\},$$

where $\tilde{c}_{n,j,k}$ can be interpreted as a decision on $C_{j,k}$ using D_n. Its error probability is minimal for the Bayes decision $\bar{C}_{n,j,k}$, which is 1 if $\mathbf{P}\{C_{j,k} = 1|D_n\} \geq \frac{1}{2}$ and -1 otherwise, therefore

$$\mathbf{P}\{\tilde{c}_{n,j,k} \neq C_{j,k}\} \geq \mathbf{P}\{\bar{C}_{n,j,k} \neq C_{j,k}\}.$$

Let $X_{i_1},\ldots,X_{i_l}$ be those $X_i \in A_{j,k}$. Then

$$(Y_{i_1},\ldots,Y_{i_l}) = C_{j,k} \cdot (g_{j,k}(X_{i_1}),\ldots,g_{j,k}(X_{i_l})) + (N_{i_1},\ldots,N_{i_l}),$$

while

$$\{Y_1,\ldots,Y_n\} \setminus \{Y_{i_1},\ldots,Y_{i_l}\}$$

depends only on $C \setminus \{C_{j,k}\}$ and on X_r's and N_r's with $r \notin \{i_1,\ldots,i_l\}$, and therefore is independent of $C_{j,k}$ given $X_1,\ldots,X_n$. Now conditioning on $X_1,\ldots,X_n$, the error of the conditional Bayes decision for $C_{j,k}$ based on $(Y_1,\ldots,Y_n)$ depends only on $(Y_{i_1},\ldots,Y_{i_l})$, hence Lemma 3.2 implies

$$\mathbf{P}\{\bar{C}_{n,j,k} \neq C_{j,k}|X_1,\ldots,X_n\} = \Phi\left(-\sqrt{\sum_{r=1}^{l} g_{j,k}^2(X_{i_r})}\right)$$

$$= \Phi\left(-\sqrt{\sum_{i=1}^{n} g_{j,k}^2(X_i)}\right).$$

Since $\Phi(-\sqrt{x})$ is convex, by Jensen's inequality

$$
\begin{aligned}
\mathbf{P}\{\bar{C}_{n,j,k} \neq C_{j,k}\} &= \mathbf{E}\{\mathbf{P}\{\bar{C}_{n,j,k} \neq C_{j,k}|X_1,\ldots,X_n\}\} \\
&= \mathbf{E}\left\{\Phi\left(-\sqrt{\sum_{i=1}^{n} g_{j,k}^2(X_i)}\right)\right\} \\
&\geq \Phi\left(-\sqrt{\mathbf{E}\left\{\sum_{i=1}^{n} g_{j,k}^2(X_i)\right\}}\right) \\
&= \Phi\left(-\sqrt{n\mathbf{E}\{g_{j,k}^2(X_1)\}}\right) \\
&= \Phi\left(-\sqrt{np_j^{2p+d}\int g^2(x)\,dx}\right)
\end{aligned}
$$

independently of k, thus

$$
\begin{aligned}
\mathbf{E}R_n(C) &\geq \sum_{j:np_j^{2p+d}\leq 1}\sum_{k=1}^{S_j} p_j^{2p+d}\Phi\left(-\sqrt{np_j^{2p+d}\int g^2(x)\,dx}\right) \\
&\geq \Phi\left(-\sqrt{\int g^2(x)\,dx}\right)\sum_{j:np_j^{2p+d}\leq 1} S_j p_j^{2p+d} \\
&\geq K_1 \cdot \sum_{j:np_j^{2p+d}\leq 1} p_j^{2p+1},
\end{aligned}
\tag{3.5}
$$

where

$$K_1 = \Phi\left(-\sqrt{\int g^2(x)\,dx}\right)\left(\frac{1}{2}\right)^{d-1}.$$

Since b_n and a_n tend to zero we can take a subsequence $\{n_t\}_{t\in\mathcal{N}}$ of $\{n\}_{n\in\mathcal{N}}$ with

$$b_{n_t} \leq 2^{-t}$$

and

$$a_{n_t}^{1/2p} \leq 2^{-t}.$$

Define q_t such that

$$\frac{2^{-t}}{q_t} = \left\lceil\frac{2^{-t}}{a_{n_t}^{1/2p}}\right\rceil,$$

and choose $\{p_j\}$ as

$$q_1, \ldots, q_1, q_2, \ldots, q_2, \ldots, q_t, \ldots, q_t, \ldots,$$

where q_t is repeated $2^{-t}/q_t$ times. So because of $a_n = n^{-\frac{2p}{2p+d}}$,

$$\sum_{j:np_j^{2p+d}\leq 1} p_j^{2p+1} = \sum_{t:nq_t^{2p+d}\leq 1} \frac{2^{-t}}{q_t} q_t^{2p+1}$$

$$\geq \sum_{t:nq_t^{2p+d}\leq 1} b_{n_t} q_t^{2p}$$

$$= \sum_{t:\lceil 2^{-t}a_{n_t}^{-1/2p}\rceil \geq 2^{-t}a_n^{-1/2p}} b_{n_t} \left(\frac{2^{-t}}{\left\lceil \frac{2^{-t}}{a_{n_t}^{1/2p}} \right\rceil}\right)^{2p}$$

$$\geq \sum_{t:a_{n_t}\leq a_n} b_{n_t} \left(\frac{2^{-t}}{\frac{2^{-t}}{a_{n_t}^{1/2p}}+1}\right)^{2p}$$

$$= \sum_{t:n_t\geq n} b_{n_t} \left(\frac{a_{n_t}^{1/2p}}{1+2^t a_{n_t}^{1/2p}}\right)^{2p}$$

$$\geq \sum_{t:n_t\geq n} \frac{b_{n_t} a_{n_t}}{2^{2p}}$$

by $a_{n_t}^{1/2p} \leq 2^{-t}$ and, especially, for $n = n_s$ (3.5) implies

$$\mathbf{E}R_{n_s}(C) \geq K_1 \sum_{j:n_s p_j^{2p+d}\leq 1} p_j^{2p+1} \geq \frac{K_1}{2^{2p}} \sum_{t\geq s} b_{n_t} a_{n_t} \geq \frac{K_1}{2^{2p}} b_{n_s} a_{n_s}. \qquad (3.6)$$

Using (3.6) one gets

$$\inf_{\{m_n\}} \sup_{c\in\mathcal{C}} \limsup_{n\to\infty} \frac{R_n(c)}{b_n a_n} \geq \inf_{\{m_n\}} \sup_{c\in\mathcal{C}} \limsup_{s\to\infty} \frac{R_{n_s}(c)}{b_{n_s} a_{n_s}}$$

$$\geq \frac{K_1}{2^{2p}} \inf_{\{m_n\}} \sup_{c\in\mathcal{C}} \limsup_{s\to\infty} \frac{R_{n_s}(c)}{\mathbf{E}R_{n_s}(C)}$$

$$\geq \frac{K_1}{2^{2p}} \inf_{\{m_n\}} \mathbf{E}\left\{\limsup_{s\to\infty} \frac{R_{n_s}(C)}{\mathbf{E}R_{n_s}(C)}\right\}.$$

Because of (3.5) and the fact that, for all $c \in \mathcal{C}$,

$$R_n(c) \leq \sum_{j:np_j^{2p+d}\leq 1} S_j p_j^{2p+d} \leq \sum_{j:np_j^{2p+d}\leq 1} p_j^{2p+1},$$

the sequence $R_{n_s}(C)/\mathbf{E}R_{n_s}(C)$ is uniformly bounded, so we can apply Fatou's lemma to get

$$\inf_{\{m_n\}} \sup_{c \in \mathcal{C}} \limsup_{n \to \infty} \frac{R_n(c)}{b_n a_n} \geq \frac{K_1}{2^{2p}} \inf_{\{m_n\}} \limsup_{s \to \infty} \mathbf{E}\left(\frac{R_{n_s}(C)}{\mathbf{E}R_{n_s}(C)}\right) = \frac{K_1}{2^{2p}} > 0.$$

This together with (3.4) implies the assertion. $\qquad\square$

3.4 Bibliographic Notes

Versions of Theorem 3.1 appeared earlier in the literature. First, Cover (1968b) showed that for any sequence of classification rules, for sequences $\{a_n\}$ converging to zero at arbitrarily slow algebraic rates (i.e., as $1/n^\delta$ for arbitrarily small $\delta > 0$), there exists a distribution such that the error probability $\geq L^* + a_n$ infinitely often. Devroye (1982b) strengthened Cover's result allowing sequences tending to zero arbitrarily slowly. Lemma 3.1 is due to Devroye and Györfi (1985). Theorem 3.2 has been proved by Stone (1982). For related results on the general minimax theory of statistical estimates see Ibragimov and Khasminskii (1980; 1981; 1982), Bretagnolle and Huber (1979), Birgé (1983), and Korostelev and Tsybakov (1993). The concept of individual lower rates has been introduced in Birgé (1986) concerning density estimation. Theorem 3.3 is from Antos, Györfi, and Kohler (2000). It uses the ideas from Antos and Lugosi (1998), who proved the related results in pattern recognition.

Problems and Exercises

PROBLEM 3.1. Prove a version of Theorem 3.1 when m is arbitrarily many times differentiable on $[0, 1)$.

HINT: Let g be an arbitrarily many times differentiable function on $\mathcal{R}$ such that it is zero outside $[0, 1]$. In the proof of Theorem 3.1, let a_j denote the left end point of A_j. Put

$$m(x) = \sum_{j=1}^{\infty} g\left(\frac{x - a_j}{p_j}\right).$$

Then m is arbitrarily many times differentiable on $[0, 1)$. Follow the line of the proof of Theorem 3.1.

PROBLEM 3.2. Show that

$$\left(1 - \frac{1}{2n}\right)^n \geq \frac{1}{2}.$$

HINT: For $n \geq 2$,

$$\left(1 - \frac{1}{2n}\right)^n \geq \left(\frac{1}{1 + 1/(2n-1)}\right)^n \geq \left(\frac{1}{e^{1/(2n-1)}}\right)^n \geq \frac{1}{e^{2/3}} \geq \frac{1}{2}.$$

PROBLEM 3.3. Sometimes rate of convergence results for nonparametric regression estimation are only shown for bounded $|Y|$, so it is reasonable to consider minimax lower bounds for such classes. Let $\mathcal{D}^{*(p,C)}$ be the class of distributions of (X, Y) such that:

(I') X is uniformly distributed on $[0,1]^d$;

(II') $Y \in \{0,1\}$ a.s. and $\mathbf{P}\{Y = 1 | X = x\} = 1 - \mathbf{P}\{Y = 0 | X = x\} = m(x)$
for all $x \in [0,1]^d$; and

(III') $m \in \mathcal{F}^{(p,C)}$ and $m(x) \in [0,1]$ for all $x \in [0,1]^d$.

Prove that for the class $\mathcal{D}^{*(p,C)}$ the sequence

$$a_n = n^{-\frac{2p}{2p+d}}$$

is a lower minimax rate of convergence.

PROBLEM 3.4. Often minimax rates of convergence are defined using tail probabilities instead of expectations. For example, in Stone (1982) $\{a_n\}$ is called the lower rate of convergence for the class $\mathcal{D}$ if

$$\liminf_{n \to \infty} \inf_{m_n} \sup_{(X,Y) \in \mathcal{D}} \mathbf{P}\left\{ \frac{\|m_n - m\|^2}{a_n} \geq c \right\} = C_0 > 0. \tag{3.7}$$

Show that (3.7) implies that $\{a_n\}$ is a lower minimax rate of convergence according to Definition 3.1.

PROBLEM 3.5. Show that (3.7) holds for $\mathcal{D} = \mathcal{D}^{(p,C)}$ and

$$a_n = n^{-\frac{2p}{2p+d}}.$$

HINT: Stone (1982).

PROBLEM 3.6. Show that (3.7) holds for $\mathcal{D} = \mathcal{D}^{*(p,C)}$ (as defined in Problem 3.3) and

$$a_n = n^{-\frac{2p}{2p+d}}.$$

PROBLEM 3.7. Give a definition of individual lower rates of convergence using tail probabilities.

PROBLEM 3.8. Let $\{b_n\}$ be an arbitrary positive sequence tending to zero. Prove that for your definition in Problem 3.7 the sequence

$$b_n a_n = b_n n^{-\frac{2p}{2p+d}}$$

is an individual lower rate of convergence for the class $\mathcal{D}^{(p,C)}$.

4
Partitioning Estimates

4.1   Introduction

Let $\mathcal{P}_n = \{A_{n,1}, A_{n,2}, \ldots\}$ be a partition of $\mathcal{R}^d$ and for each $x \in \mathcal{R}^d$ let $A_n(x)$ denote the cell of $\mathcal{P}_n$ containing x. The partitioning estimate (histogram) of the regression function is defined as

$$m_n(x) = \frac{\sum_{i=1}^n Y_i I_{\{X_i \in A_n(x)\}}}{\sum_{i=1}^n I_{\{X_i \in A_n(x)\}}}$$

with $0/0 = 0$ by definition. This means that the partitioning estimate is a local averaging estimate such for a given x we take the average of those Y_i's for which X_i belongs to the same cell into which x falls.

The simplest version of this estimate is obtained for $d = 1$ and when the cells $A_{n,j}$ are intervals of size $h = h_n$. Figures 4.1 – 4.3 show the estimates for various choices of h for our simulated data introduced in Chapter 1. In the first figure h is too small (undersmoothing, large variance), in the second choice it is about right, while in the third it is too large (oversmoothing, large bias).

For $d > 1$ one can use, e.g., a cubic partition, where the cells $A_{n,j}$ are cubes of volume h_n^d, or a rectangle partition which consists of rectangles $A_{n,j}$ with side lengths $h_{n1}, \ldots, h_{nd}$. For the sake of illustration we generated two-dimensional data when the actual distribution is a correlated normal distribution. The partition in Figure 4.4 is cubic, and the partition in Figure 4.5 is made of rectangles.

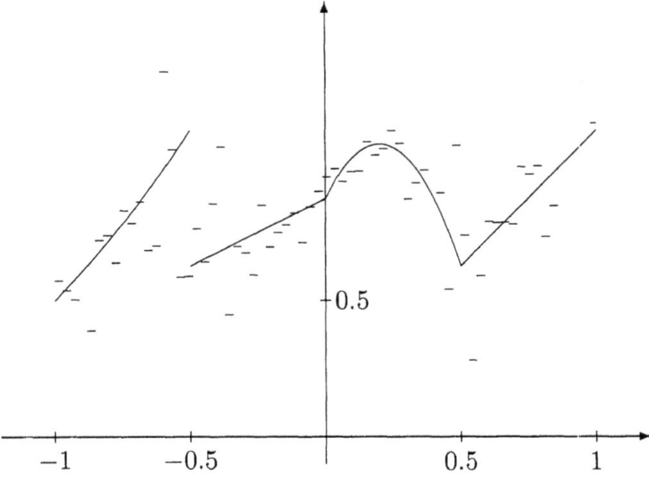

Figure 4.1. Undersmoothing: $h = 0.03$, L_2 error $= 0.062433$.

Cubic and rectangle partitions are particularly attractive from the computational point of view, because the set $A_n(x)$ can be determined for each x in constant time, provided that we use an appropriate data structure. In most cases, partitioning estimates are computationally superior to the other nonparametric estimates, particularly if the search for $A_n(x)$ is organized using binary decision trees (cf. Friedman (1977)).

The partitions may depend on the data. Figure 4.6 shows such a partition, where each cell contains an equal number of points. This par-

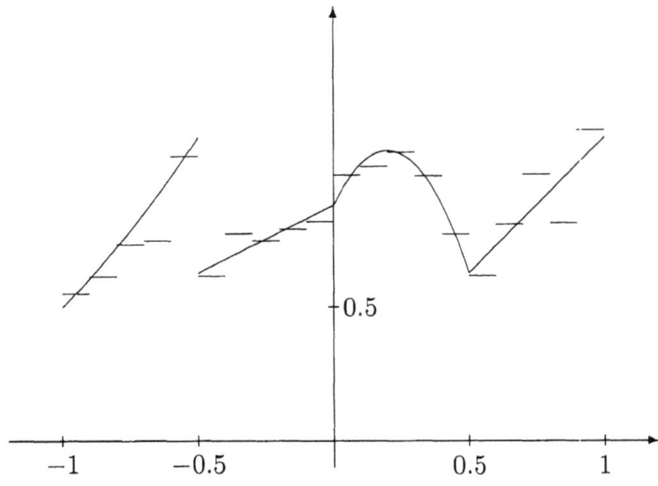

Figure 4.2. Good choice: $h = 0.1$, L_2 error $= 0.003642$.

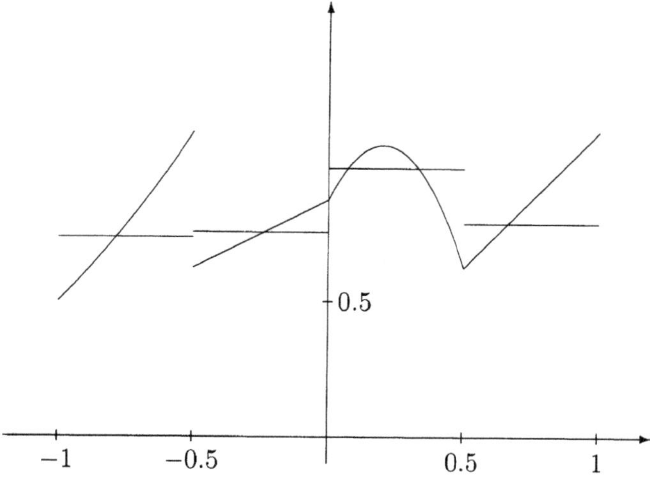

Figure 4.3. Oversmoothing: $h = 0.5$, L_2 error $= 0.013208$.

tition consists of so-called statistically equivalent blocks. Data-dependent partitions are dealt with in Chapter 13.

Another advantage of the partitioning estimate is that it can be represented or compressed very efficiently. Instead of storing all data D_n, one should only know the estimate for each nonempty cell, i.e., for cells $A_{n,j}$ for which $\mu_n(A_{n,j}) > 0$, where μ_n denotes the empirical distribution. The number of nonempty cells is much smaller than n (cf. Problem 4.8).

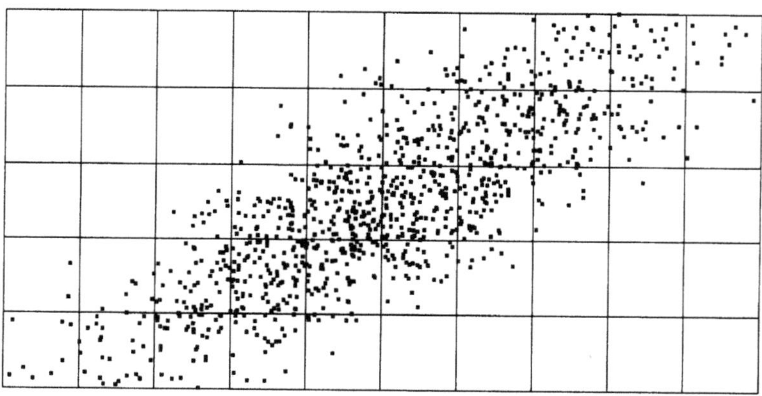

Figure 4.4. Cubic partition.

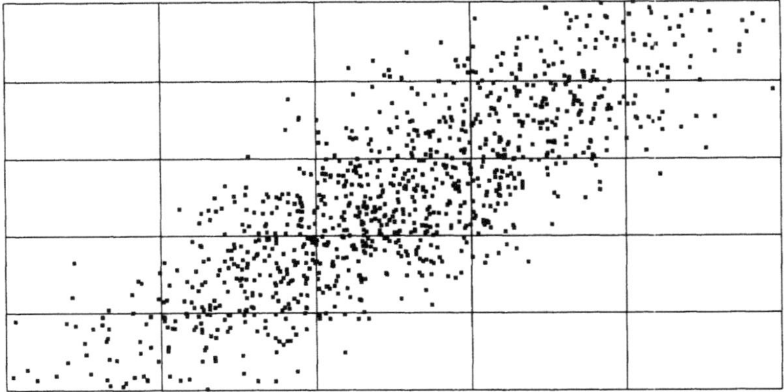

Figure 4.5. Rectangle partition.

4.2 Stone's Theorem

In the next section we will prove the weak universal consistency of partitioning estimates. In the proof we will use Stone's theorem (Theorem 4.1 below) which is a powerful tool for proving weak consistency for local averaging regression function estimates. It will also be applied to prove the weak universal consistency of kernel and nearest neighbor estimates in Chapters 5 and 6.

Local averaging regression function estimates take the form

$$m_n(x) = \sum_{i=1}^{n} W_{ni}(x) \cdot Y_i,$$

where the weights $W_{n,i}(x) = W_{n,i}(x, X_1, \dots, X_n) \in \mathcal{R}$ are depending on $X_1, \dots, X_n$.

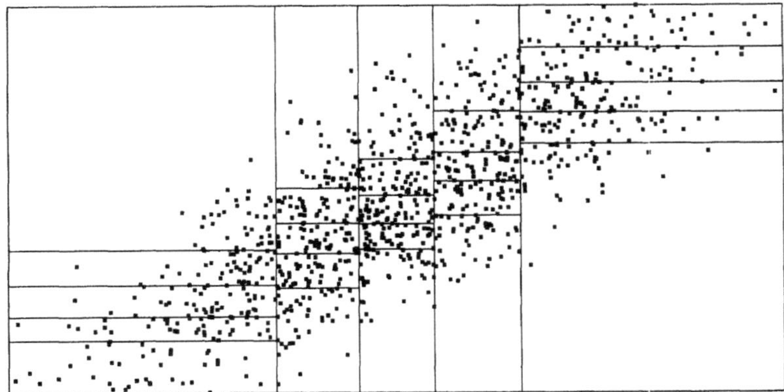

Figure 4.6. Statistically equivalent blocks.

Usually the weights are nonnegative and $W_{n,i}(x)$ is "small" if X_i is "far" from x. The most common examples for weights are the weights for partitioning, kernel, and nearest neighbor estimates (cf. Section 2.1). The next theorem states conditions on the weights which guarantee the weak universal consistency of the local averaging estimates.

Theorem 4.1. (STONE'S THEOREM). *Assume that the following conditions are satisfied for any distribution of X:*

(i)   *There is a constant c such that for every nonnegative measurable function f satisfying $\mathbf{E}f(X) < \infty$ and any n,*

$$\mathbf{E}\left\{\sum_{i=1}^{n}|W_{n,i}(X)|f(X_i)\right\} \leq c\mathbf{E}f(X).$$

(ii)   *There is a $D \geq 1$ such that*

$$\mathbf{P}\left\{\sum_{i=1}^{n}|W_{n,i}(X)| \leq D\right\} = 1,$$

    for all n.

(iii)   *For all $a > 0$,*

$$\lim_{n\to\infty}\mathbf{E}\left\{\sum_{i=1}^{n}|W_{n,i}(X)|I_{\{\|X_i-X\|>a\}}\right\} = 0.$$

(iv)

$$\sum_{i=1}^{n}W_{n,i}(X) \to 1$$

    in probability.

(v)

$$\lim_{n\to\infty}\mathbf{E}\left\{\sum_{i=1}^{n}W_{n,i}(X)^2\right\} = 0.$$

Then the corresponding regression function estimate m_n is weakly universally consistent, i.e.,

$$\lim_{n\to\infty}\mathbf{E}\left\{\int(m_n(x)-m(x))^2\mu(dx)\right\} = 0$$

for all distributions of (X,Y) with $\mathbf{E}Y^2 < \infty$.

For nonnegative weights and noiseless data (i.e., $Y = m(X) \geq 0$) condition (i) says that the mean value of the estimate is bounded above by some constant times the mean value of the regression function. Conditions (ii) and (iv) state that the sum of the weights is bounded and is asymptotically 1. Condition (iii) ensures that the estimate at a point x is asymptotically influenced only by the data close to x. Condition (v) states that asymptotically all weights become small.

PROOF OF THEOREM 4.1. Because of $(a + b + c)^2 \leq 3a^2 + 3b^2 + 3c^2$ we have that

$$
\begin{aligned}
\mathbf{E}\{m_n(X) - m(X)\}^2 \;\leq\; & 3\mathbf{E}\left\{\left(\sum_{i=1}^{n} W_{n,i}(X)(Y_i - m(X_i))\right)^2\right\} \\
& + 3\mathbf{E}\left\{\left(\sum_{i=1}^{n} W_{n,i}(X)(m(X_i) - m(X))\right)^2\right\} \\
& + 3\mathbf{E}\left\{\left(\left(\sum_{i=1}^{n} W_{n,i}(X) - 1\right) m(X)\right)^2\right\} \\
=\; & 3I_n + 3J_n + 3L_n.
\end{aligned}
$$

By the Cauchy–Schwarz inequality, and condition (ii),

$$
\begin{aligned}
J_n \;\leq\; & \mathbf{E}\left\{\left(\sum_{i=1}^{n} \sqrt{|W_{n,i}(X)|}\sqrt{|W_{n,i}(X)|}\,|m(X_i) - m(X)|\right)^2\right\} \\
\leq\; & \mathbf{E}\left\{\left(\sum_{i=1}^{n} |W_{n,i}(X)|\right)\left(\sum_{i=1}^{n} |W_{n,i}(X)|(m(X_i) - m(X))^2\right)\right\} \\
\leq\; & D\mathbf{E}\left\{\sum_{i=1}^{n} |W_{n,i}(X)|(m(X_i) - m(X))^2\right\} \\
=\; & DJ'_n.
\end{aligned}
$$

Because of Theorem A.1, for $\epsilon > 0$ we can choose $\tilde{m}$ bounded and uniformly continuous such that

$$
\mathbf{E}\{(m(X) - \tilde{m}(X))^2\} < \epsilon.
$$

Then

$$
\begin{aligned}
J'_n \;\leq\; & 3\mathbf{E}\left\{\sum_{i=1}^{n} |W_{n,i}(X)|(m(X_i) - \tilde{m}(X_i))^2\right\} \\
& + 3\mathbf{E}\left\{\sum_{i=1}^{n} |W_{n,i}(X)|(\tilde{m}(X_i) - \tilde{m}(X))^2\right\} \\
& + 3\mathbf{E}\left\{\sum_{i=1}^{n} |W_{n,i}(X)|(\tilde{m}(X) - m(X))^2\right\} \\
=\; & 3J_{n1} + 3J_{n2} + 3J_{n3}.
\end{aligned}
$$

For arbitrary $\delta > 0$,

$$
\begin{aligned}
J_{n2} &= \mathbf{E}\left\{\sum_{i=1}^{n}|W_{n,i}(X)|\cdot(\tilde{m}(X_i)-\tilde{m}(X))^2 I_{\{\|X_i-X\|>\delta\}}\right\} \\
&\quad + \mathbf{E}\left\{\sum_{i=1}^{n}|W_{n,i}(X)|\cdot(\tilde{m}(X_i)-\tilde{m}(X))^2 I_{\{\|X_i-X\|\leq\delta\}}\right\} \\
&\leq \mathbf{E}\left\{\sum_{i=1}^{n}|W_{n,i}(X)|\cdot(2\tilde{m}(X_i)^2+2\tilde{m}(X)^2) I_{\{\|X_i-X\|>\delta\}}\right\} \\
&\quad + \mathbf{E}\left\{\sum_{i=1}^{n}|W_{n,i}(X)|\cdot(\tilde{m}(X_i)-\tilde{m}(X))^2 I_{\{\|X_i-X\|\leq\delta\}}\right\} \\
&\leq 4\cdot\sup_{u\in\mathcal{R}^d}|\tilde{m}(u)|^2\cdot\mathbf{E}\left\{\sum_{i=1}^{n}|W_{n,i}(X)|\cdot I_{\{\|X_i-X\|>\delta\}}\right\} \\
&\quad + D\cdot\left(\sup_{u,v\in\mathcal{R}^d\,:\,\|u-v\|\leq\delta}|\tilde{m}(u)-\tilde{m}(v)|\right)^2.
\end{aligned}
$$

By (iii),

$$
\limsup_{n\to\infty} J_{n2} \leq D\cdot\left(\sup_{u,v\in\mathcal{R}^d\,:\,\|u-v\|\leq\delta}|\tilde{m}(u)-\tilde{m}(v)|\right)^2.
$$

Using $\tilde{m}$ uniformly continuous we get, with $\delta\to 0$,

$$
J_{n2}\to 0.
$$

By (ii),

$$
J_{n3} \leq D\mathbf{E}\{(\tilde{m}(X)-m(X))^2\} < D\epsilon,
$$

moreover, by (i),

$$
\limsup_{n\to\infty} J_{n1} \leq c\mathbf{E}\{(\tilde{m}(X)-m(X))^2\} \leq c\epsilon,
$$

so

$$
\limsup_{n\to\infty} J_n' \leq 3c\epsilon + 3D\epsilon.
$$

Put

$$
\sigma^2(x) = \mathbf{E}\{(Y-m(X))^2|X=x\},
$$

then $\mathbf{E}Y^2 < \infty$ implies that $\mathbf{E}\sigma^2(X) < \infty$, and

$$
I_n = \mathbf{E}\left\{\left(\sum_{i=1}^{n}W_{n,i}(X)(Y_i-m(X_i))\right)^2\right\}
$$

$$= \sum_{i=1}^{n}\sum_{j=1}^{n}\mathbf{E}\{W_{n,i}(X)W_{n,j}(X)(Y_i - m(X_i))(Y_j - m(X_j))\}.$$

For $i \neq j$,

$$\begin{aligned}
&\mathbf{E}\{W_{n,i}(X)W_{n,j}(X)(Y_i - m(X_i))(Y_j - m(X_j))\}\\
=\ &\mathbf{E}\{\mathbf{E}\{W_{n,i}(X)W_{n,j}(X)(Y_i - m(X_i))(Y_j - m(X_j))|X_1,\ldots,X_n,Y_i\}\}\\
=\ &\mathbf{E}\{W_{n,i}(X)W_{n,j}(X)(Y_i - m(X_i))\mathbf{E}\{(Y_j - m(X_j))|X_1,\ldots,X_n,Y_i\}\}\\
=\ &\mathbf{E}\{W_{n,i}(X)W_{n,j}(X)(Y_i - m(X_i))(m(X_j) - m(X_j))\}\\
=\ &0,
\end{aligned}$$

hence,

$$\begin{aligned}
I_n &= \mathbf{E}\left\{\sum_{i=1}^{n}W_{n,i}(X)^2(Y_i - m(X_i))^2\right\}\\
&= \mathbf{E}\left\{\sum_{i=1}^{n}W_{n,i}(X)^2\sigma^2(X_i)\right\}.
\end{aligned}$$

If $\sigma^2(x)$ is bounded then (v) implies that $I_n \to 0$. For general $\sigma^2(x)$ and $\epsilon > 0$, Theorem A.1 implies that there exists bounded $\tilde{\sigma}^2(x) \leq L$ such that

$$\mathbf{E}\{|\tilde{\sigma}^2(X) - \sigma^2(X)|\} < \epsilon.$$

Then, by (ii),

$$\begin{aligned}
I_n &\leq \mathbf{E}\left\{\sum_{i=1}^{n}W_{n,i}(X)^2\tilde{\sigma}^2(X_i)\right\} + \mathbf{E}\left\{\sum_{i=1}^{n}W_{n,i}(X)^2|\sigma^2(X_i) - \tilde{\sigma}^2(X_i)|\right\}\\
&\leq L\mathbf{E}\left\{\sum_{i=1}^{n}W_{n,i}(X)^2\right\} + D\mathbf{E}\left\{\sum_{i=1}^{n}|W_{n,i}(X)||\sigma^2(X_i) - \tilde{\sigma}^2(X_i)|\right\},
\end{aligned}$$

therefore, by (i) and (v),

$$\limsup_{n\to\infty} I_n \leq cD\mathbf{E}\{|\tilde{\sigma}^2(X) - \sigma^2(X)|\} < cD\epsilon.$$

Concerning the third term

$$L_n = \mathbf{E}\left\{\left(\left(\sum_{i=1}^{n}W_{n,i}(X) - 1\right)m(X)\right)^2\right\} \to 0$$

by conditions (ii), (iv), and by the dominated convergence theorem. $\quad\square$

From the proof it is clear that under conditions (ii), (iii), (iv), and (v) alone weak consistency holds if the regression function is uniformly continuous and the conditional variance function $\sigma^2(x)$ is bounded. Condition (i) makes the extension possible. For nonnegative weights conditions (i), (iii), and (v) are necessary (see Problems 4.1 – 4.4).

Definition 4.1. *The weights $\{W_{n,i}\}$ are called normal if $\sum_{i=1}^{n} W_{n,i}(x) = 1$. The weights $\{W_{n,i}\}$ are called subprobability weights if they are nonnegative and sum up to ≤ 1. They are called probability weights if they are nonnegative and sum up to 1.*

Obviously for subprobability weights condition (ii) is satisfied, and for probability weights conditions (ii) and (iv) are satisfied.

4.3 Consistency

The purpose of this section is to prove the *weak* universal consistency of the partitioning estimates. This is the first such result that we mention. Later we will prove the same property for other estimates, too. The next theorem provides sufficient conditions for the weak universal consistency of the partitioning estimate. The first condition ensures that the cells of the underlying partition shrink to zero inside a bounded set, so the estimate is local in this sense. The second condition means that the number of cells inside a bounded set is small with respect to n, which implies that with large probability each cell contains many data points.

Theorem 4.2. *If for each sphere S centered at the origin*

$$\lim_{n \to \infty} \max_{j : A_{n,j} \cap S \neq \emptyset} diam(A_{n,j}) = 0 \tag{4.1}$$

and

$$\lim_{n \to \infty} \frac{|\{j : A_{n,j} \cap S \neq \emptyset\}|}{n} = 0 \tag{4.2}$$

then the partitioning regression function estimate is weakly universally consistent.

For cubic partitions,

$$\lim_{n \to \infty} h_n = 0 \quad \text{and} \quad \lim_{n \to \infty} nh_n^d = \infty$$

imply (4.1) and (4.2).

In order to prove Theorem 4.2 we will verify the conditions of Stone's theorem. For this we need the following technical lemma. An integer-valued random variable $B(n,p)$ is said to be binomially distributed with parameters n and $0 \leq p \leq 1$ if

$$\mathbf{P}\{B(n,p) = k\} = \binom{n}{k} p^k (1-p)^{n-k}, \quad k = 0, 1, \ldots, n.$$

Lemma 4.1. *Let the random variable $B(n,p)$ be binomially distributed with parameters n and p. Then:*

(i)

$$\mathbf{E}\left\{\frac{1}{1+B(n,p)}\right\} \leq \frac{1}{(n+1)p},$$

(ii)

$$\mathbf{E}\left\{\frac{1}{B(n,p)}I_{\{B(n,p)>0\}}\right\} \leq \frac{2}{(n+1)p}.$$

PROOF. Part (i) follows from the following simple calculation:

$$\mathbf{E}\left\{\frac{1}{1+B(n,p)}\right\} = \sum_{k=0}^{n}\frac{1}{k+1}\binom{n}{k}p^k(1-p)^{n-k}$$

$$= \frac{1}{(n+1)p}\sum_{k=0}^{n}\binom{n+1}{k+1}p^{k+1}(1-p)^{n-k}$$

$$\leq \frac{1}{(n+1)p}\sum_{k=0}^{n+1}\binom{n+1}{k}p^k(1-p)^{n-k+1}$$

$$= \frac{1}{(n+1)p}(p+(1-p))^{n+1}$$

$$= \frac{1}{(n+1)p}.$$

For (ii) we have

$$\mathbf{E}\left\{\frac{1}{B(n,p)}I_{\{B(n,p)>0\}}\right\} \leq \mathbf{E}\left\{\frac{2}{1+B(n,p)}\right\} \leq \frac{2}{(n+1)p}$$

by (i). $\qquad\qquad\qquad\square$

PROOF OF THEOREM 4.2. The proof proceeds by checking the conditions of Stone's theorem (Theorem 4.1). Note that if $0/0 = 0$ by definition, then

$$W_{n,i}(x) = I_{\{X_i \in A_n(x)\}}\Big/ \sum_{l=1}^{n}I_{\{X_l \in A_n(x)\}}.$$

To verify (i), it suffices to show that there is a constant $c > 0$, such that for any nonnegative function f with $\mathbf{E}f(X) < \infty$,

$$\mathbf{E}\left\{\sum_{i=1}^{n}f(X_i)\frac{I_{\{X_i \in A_n(X)\}}}{\sum_{l=1}^{n}I_{\{X_l \in A_n(X)\}}}\right\} \leq c\mathbf{E}f(X).$$

Observe that

$$\mathbf{E}\left\{\sum_{i=1}^{n} f(X_i)\frac{I_{\{X_i \in A_n(X)\}}}{\sum_{l=1}^{n} I_{\{X_l \in A_n(X)\}}}\right\}$$

$$= \sum_{i=1}^{n}\mathbf{E}\left\{f(X_i)\frac{I_{\{X_i \in A_n(X)\}}}{1 + \sum_{l \neq i} I_{\{X_l \in A_n(X)\}}}\right\}$$

$$= n\mathbf{E}\left\{f(X_1)I_{\{X_1 \in A_n(X)\}}\frac{1}{1 + \sum_{l \neq 1} I_{\{X_l \in A_n(X)\}}}\right\}$$

$$= n\mathbf{E}\left\{\mathbf{E}\left\{f(X_1)I_{\{X_1 \in A_n(X)\}}\frac{1}{1 + \sum_{l=2}^{n} I_{\{X_l \in A_n(X)\}}}\,\middle|\,X, X_1\right\}\right\}$$

$$= n\mathbf{E}\left\{f(X_1)I_{\{X_1 \in A_n(X)\}}\mathbf{E}\left\{\frac{1}{1 + \sum_{l=2}^{n} I_{\{X_l \in A_n(X)\}}}\,\middle|\,X, X_1\right\}\right\}$$

$$= n\mathbf{E}\left\{f(X_1)I_{\{X_1 \in A_n(X)\}}\mathbf{E}\left\{\frac{1}{1 + \sum_{l=2}^{n} I_{\{X_l \in A_n(X)\}}}\,\middle|\,X\right\}\right\}$$

by the independence of the random variables $X, X_1, \ldots, X_n$. Using Lemma 4.1, the expected value above can be bounded by

$$n\mathbf{E}\left\{f(X_1)I_{\{X_1 \in A_n(X)\}}\frac{1}{n\mu(A_n(X))}\right\}$$

$$= \sum_{j}\mathbf{P}\{X \in A_{nj}\}\int_{A_{nj}} f(u)\mu(du)\frac{1}{\mu(A_{nj})}$$

$$= \int_{\mathcal{R}^d} f(u)\mu(du) = \mathbf{E}f(X).$$

Therefore, the condition is satisfied with $c = 1$. The weights are sub-probability weights, so (ii) is satisfied. To see that condition (iii) is satisfied first choose a ball S centered at the origin, and then by condition (4.1) a large n such that for $A_{n,j} \cap S \neq \emptyset$ we have $\operatorname{diam}(A_{n,j}) < a$. Thus $X \in S$ and $\|X_i - X\| > a$ imply $X_i \notin A_n(X)$, therefore

$$I_{\{X \in S\}}\sum_{i=1}^{n} W_{n,i}(X)I_{\{\|X_i - X\| > a\}}$$

$$= I_{\{X \in S\}}\frac{\sum_{i=1}^{n} I_{\{X_i \in A_n(X),\|X - X_i\| > a\}}}{n\mu_n(A_n(X))}$$

$$= I_{\{X \in S\}}\frac{\sum_{i=1}^{n} I_{\{X_i \in A_n(X),X_i \notin A_n(X),\|X - X_i\| > a\}}}{n\mu_n(A_n(X))}$$

$$= 0.$$

Thus

$$\limsup_n \mathbf{E} \sum_{i=1}^n W_{n,i}(X) I_{\{\|X_i - X\| > a\}} \le \mu(S^c).$$

Concerning (iv) note that

$$\mathbf{P}\left\{ \sum_{i=1}^n W_{n,i}(X) \ne 1 \right\}$$

$$= \mathbf{P}\{\mu_n(A_n(X)) = 0\}$$

$$= \sum_j \mathbf{P}\{X \in A_{n,j}, \mu_n(A_{n,j}) = 0\}$$

$$= \sum_j \mu(A_{n,j})(1 - \mu(A_{n,j}))^n$$

$$\le \sum_{j:A_{n,j} \cap S = \emptyset} \mu(A_{n,j}) + \sum_{j:A_{n,j} \cap S \ne \emptyset} \mu(A_{n,j})(1 - \mu(A_{n,j}))^n.$$

Elementary inequalities

$$x(1 - x)^n \le x e^{-nx} \le \frac{1}{en} \qquad (0 \le x \le 1)$$

yield

$$\mathbf{P}\left\{ \sum_{i=1}^n W_{n,i}(X) \ne 1 \right\} \le \mu(S^c) + \frac{1}{en} |\{j : A_{n,j} \cap S \ne \emptyset\}|.$$

The first term on the right-hand side can be made arbitrarily small by the choice of S, while the second term goes to zero by (4.2). To prove that condition (v) holds, observe that

$$\sum_{i=1}^n W_{n,i}(x)^2 = \begin{cases} \dfrac{1}{\sum_{l=1}^n I_{\{X_l \in A_n(x)\}}} & \text{if } \mu_n(A_n(x)) > 0, \\ 0 & \text{if } \mu_n(A_n(x)) = 0. \end{cases}$$

Then we have

$$\mathbf{E}\left\{ \sum_{i=1}^n W_{n,i}(X)^2 \right\}$$

$$\le \mathbf{P}\{X \in S^c\} + \sum_{j:A_{n,j} \cap S \ne \emptyset} \mathbf{E}\left\{ I_{\{X \in A_{n,j}\}} \frac{1}{n\mu_n(A_{n,j})} I_{\{\mu_n(A_{n,j}) > 0\}} \right\}$$

$$\le \mu(S^c) + \sum_{j:A_{n,j} \cap S \ne \emptyset} \mu(A_{n,j}) \frac{2}{n\mu(A_{n,j})}$$

$$\text{(by Lemma 4.1)}$$

$$= \mu(S^c) + \frac{2}{n} |\{j : A_{n,j} \cap S \ne \emptyset\}|.$$

A similar argument to the previous one concludes the proof.   □

4.4   Rate of Convergence

In this section we bound the rate of convergence of $\mathbf{E}\|m_n - m\|^2$ for cubic partitions and regression functions which are Lipschitz continuous.

Theorem 4.3. *For a cubic partition with side length h_n assume that*

$$\mathbf{Var}(Y|X = x) \leq \sigma^2, \ x \in \mathcal{R}^d,$$

$$|m(x) - m(z)| \leq C\|x - z\|, \ x, z \in \mathcal{R}^d, \tag{4.3}$$

and that X has a compact support S. Then

$$\mathbf{E}\|m_n - m\|^2 \leq \hat{c}\frac{\sigma^2 + \sup_{z \in S} |m(z)|^2}{n \cdot h_n^d} + d \cdot C^2 \cdot h_n^2,$$

where $\hat{c}$ depends only on d and on the diameter of S, thus for

$$h_n = c' \left(\frac{\sigma^2 + \sup_{z \in S} |m(z)|^2}{C^2}\right)^{1/(d+2)} n^{-\frac{1}{d+2}}$$

we get

$$\mathbf{E}\|m_n - m\|^2 \leq c'' \left(\sigma^2 + \sup_{z \in S} |m(z)|^2\right)^{2/(d+2)} C^{2d/(d+2)} n^{-2/(d+2)}.$$

PROOF. Set

$$\hat{m}_n(x) = \mathbf{E}\{m_n(x)|X_1, \ldots, X_n\} = \frac{\sum_{i=1}^n m(X_i) I_{\{X_i \in A_n(x)\}}}{n\mu_n(A_n(x))}.$$

Then

$$\mathbf{E}\{(m_n(x) - m(x))^2|X_1, \ldots, X_n\}$$
$$= \ \mathbf{E}\{(m_n(x) - \hat{m}_n(x))^2|X_1, \ldots, X_n\} + (\hat{m}_n(x) - m(x))^2. \tag{4.4}$$

We have

$$\mathbf{E}\{(m_n(x) - \hat{m}_n(x))^2|X_1, \ldots, X_n\}$$
$$= \ \mathbf{E}\left\{\left(\frac{\sum_{i=1}^n (Y_i - m(X_i)) I_{\{X_i \in A_n(x)\}}}{n\mu_n(A_n(x))}\right)^2 \Big| X_1, \ldots, X_n\right\}$$
$$= \ \frac{\sum_{i=1}^n \mathbf{Var}(Y_i|X_i) I_{\{X_i \in A_n(x)\}}}{(n\mu_n(A_n(x)))^2}$$
$$\leq \ \frac{\sigma^2}{n\mu_n(A_n(x))} I_{\{n\mu_n(A_n(x))>0\}}.$$

By Jensen's inequality

$$(\hat{m}_n(x) - m(x))^2 \;=\; \left(\frac{\sum_{i=1}^{n}(m(X_i) - m(x))I_{\{X_i \in A_n(x)\}}}{n\mu_n(A_n(x))}\right)^2 I_{\{n\mu_n(A_n(x))>0\}}$$

$$+ m(x)^2 I_{\{n\mu_n(A_n(x))=0\}}$$

$$\leq\; \frac{\sum_{i=1}^{n}(m(X_i) - m(x))^2 I_{\{X_i \in A_n(x)\}}}{n\mu_n(A_n(x))} I_{\{n\mu_n(A_n(x))>0\}}$$

$$+ m(x)^2 I_{\{n\mu_n(A_n(x))=0\}}$$

$$\leq\; d \cdot C^2 h_n^2 I_{\{n\mu_n(A_n(x))>0\}} + m(x)^2 I_{\{n\mu_n(A_n(x))=0\}}$$

$$\left(\text{by } (4.3) \text{ and } \max_{z \in A_n(x)} \|x - z\| \leq d \cdot h_n^2\right)$$

$$\leq\; d \cdot C^2 h_n^2 + m(x)^2 I_{\{n\mu_n(A_n(x))=0\}}.$$

Without loss of generality assume that S is a cube and the union of $A_{n,1}, \ldots, A_{n,l_n}$ is S. Then

$$l_n \leq \frac{\tilde{c}}{h_n^d}$$

for some constant $\tilde{c}$ proportional to the volume of S and, by Lemma 4.1 and (4.4),

$$\mathbf{E}\left\{\int (m_n(x) - m(x))^2 \mu(dx)\right\}$$

$$=\; \mathbf{E}\left\{\int (m_n(x) - \hat{m}_n(x))^2 \mu(dx)\right\} + \mathbf{E}\left\{\int (\hat{m}_n(x) - m(x))^2 \mu(dx)\right\}$$

$$=\; \sum_{j=1}^{l_n} \mathbf{E}\left\{\int_{A_{n,j}} (m_n(x) - \hat{m}_n(x))^2 \mu(dx)\right\}$$

$$+ \sum_{j=1}^{l_n} \mathbf{E}\left\{\int_{A_{n,j}} (\hat{m}_n(x) - m(x))^2 \mu(dx)\right\}$$

$$\leq\; \sum_{j=1}^{l_n} \mathbf{E}\left\{\frac{\sigma^2 \mu(A_{n,j})}{n\mu_n(A_{n,j})} I_{\{\mu_n(A_{n,j})>0\}}\right\} + dC^2 h_n^2$$

$$+ \sum_{j=1}^{l_n} \mathbf{E}\left\{\int_{A_{n,j}} m(x)^2 \mu(dx) I_{\{\mu_n(A_{n,j})=0\}}\right\}$$

$$\leq\; \sum_{j=1}^{l_n} \frac{2\sigma^2 \mu(A_{n,j})}{n\mu(A_{n,j})} + dC^2 h_n^2 + \sum_{j=1}^{l_n} \int_{A_{n,j}} m(x)^2 \mu(dx) \mathbf{P}\{\mu_n(A_{n,j}) = 0\}$$

$$\leq \quad l_n \frac{2\sigma^2}{n} + dC^2 h_n^2 + \sup_{z \in S}\{m(z)^2\} \sum_{j=1}^{l_n} \mu(A_{n,j})(1-\mu(A_{n,j}))^n$$

$$\leq \quad l_n \frac{2\sigma^2}{n} + dC^2 h_n^2 + l_n \frac{\sup_{z\in S} m(z)^2}{n} \sup_j n\mu(A_{n,j})e^{-n\mu(A_{n,j})}$$

$$\leq \quad l_n \frac{2\sigma^2}{n} + dC^2 h_n^2 + l_n \frac{\sup_{z\in S} m(z)^2 e^{-1}}{n}$$

$$(\text{since } \sup_z ze^{-z} = e^{-1})$$

$$\leq \quad \frac{(2\sigma^2 + \sup_{z\in S} m(z)^2 e^{-1})\tilde{c}}{nh_n^d} + dC^2 h_n^2.$$

$\square$

According to Theorem 4.3 the cubic partition estimate has optimal rate in the class $\mathcal{D}^{(1,C)}$ (cf. Theorem 3.2) only under condition that X has distribution with compact support.

Unfortunately, the partitioning estimate is not optimal for smoother regression functions. For example, let, for $d=1$,

$$m(x) = x,$$

X be uniformly distributed on $[0,1]$ and let

$$Y = m(X) + N,$$

where N is standard normal and X and N are independent. Put

$$\bar{m}_n(x) = \frac{\int_{A_n(x)} m(z)\,dz}{h_n}.$$

Assume that $l_n h_n = 1$ for some integer l_n. Then

$$\int_0^1 (\hat{m}_n(x) - m(x))^2\,dx$$

$$= \int_0^1 (\hat{m}_n(x) - \bar{m}_n(x))^2\,dx + \int_0^1 (\bar{m}_n(x) - m(x))^2\,dx$$

$$\geq \int_0^1 (\bar{m}_n(x) - m(x))^2\,dx$$

$$= \sum_{j=1}^{l_n} \int_{(j-1)h_n}^{jh_n} (\bar{m}_n(x) - m(x))^2\,dx$$

$$= \sum_{j=1}^{l_n} \int_{(j-1)h_n}^{jh_n} ((j-1/2)h_n - x)^2\,dx$$

$$= l_n h_n^3/12$$

$$= h_n^2/12.$$

Then, according to the decomposition (4.4),

$$\mathbf{E}\left\{\int (m_n(x) - m(x))^2 \mu(dx)\right\}$$

$$= \mathbf{E}\left\{\int (m_n(x) - \hat{m}_n(x))^2 \mu(dx)\right\} + \mathbf{E}\left\{\int (\hat{m}_n(x) - m(x))^2 \mu(dx)\right\}$$

$$\geq \sum_{j=1}^{l_n} \mathbf{E}\left\{\frac{\mu(A_{n,j})}{n\mu_n(A_{n,j})} I_{\{\mu_n(A_{n,j})>0\}}\right\} + \int_0^1 (\bar{m}_n(x) - m(x))^2 \, dx$$

$$\geq \sum_{j=1}^{l_n} \frac{\mu(A_{n,j})(1 - (1 - \mu(A_{n,j}))^n)^2}{n\mu(A_{n,j})} + h_n^2/12$$

$$\text{(by Problem 4.10)}$$

$$= \frac{l_n}{n}(1 - (1 - h_n)^n)^2 + h_n^2/12$$

$$= \frac{1}{nh_n}(1 + o(1)) + h_n^2/12,$$

if $nh_n \to \infty$ and $h_n \to 0$ as $n \to \infty$. The above term is minimal for $h_n = c'n^{-\frac{1}{3}}$, hence

$$\mathbf{E}\|m_n - m\|^2 \geq c''n^{-\frac{2}{3}},$$

which is not the optimal rate, because this distribution of (X, Y) is in $\mathcal{D}^{(p,C)}$ for all $p \geq 1$, with the optimal rate of $n^{-\frac{2p}{2p+1}}$ for $d = 1$ (cf. Theorem 3.2).

The optimality of the partitioning estimates can be extended using a local polynomial partitioning estimate where, within each cell, the estimate is a polynomial (cf. Section 11.2).

4.5 Bibliographic Notes

Theorem 4.1 is due to Stone (1977). The partitioning estimate, called a regressogram, was introduced by Tukey (1947; 1961) and studied by Collomb (1977), Bosq and Lecoutre (1987), and Lecoutre (1980). Concerning its consistency, see Devroye and Györfi (1983) and Györfi (1991). In general, the behavior of $\|m_n - m\|^2$ cannot be characterized by the rate of convergence of its expectation $\mathbf{E}\|m_n - m\|^2$. However, here the random variable is close to its expectation in the sense that the ratio of the two is close to 1 with large probability. In this respect, Beirlant and Györfi (1998) proved that under some conditions and for cubic partitions

$$nh_n^{d/2} \left(\|m_n - m\|^2 - \mathbf{E}\|m_n - m\|^2\right)/\sigma_0 \xrightarrow{\mathcal{D}} N(0, 1),$$

which means that

$$\|m_n - m\|^2 - \mathbf{E}\|m_n - m\|^2$$

is of order $\frac{1}{nh_n^{d/2}}$. This should be compared with the rate of convergence of $\mathbf{E}\|m_n - m\|^2$ which is at least $\frac{1}{nh_n^d}$ (cf. Problem 4.9). This implies that

$$\frac{\|m_n - m\|^2}{\mathbf{E}\|m_n - m\|^2} \approx 1,$$

thus the L_2 error is relatively stable. The relative stability holds under more general conditions (cf. Györfi, Schäfer, and Walk (2002)).

Problems and Exercises

PROBLEM 4.1. Assume that there is a constant c such that for every nonnegative measurable function f, satisfying $\mathbf{E}f(X) < \infty$,

$$\limsup_{n \to \infty} \mathbf{E}\left\{ \sum_{i=1}^{n} |W_{n,i}(X)| f(X_i) \right\} \le c\mathbf{E}f(X).$$

Prove that (i) in Theorem 4.1 is satisfied (Proposition 7 in Stone (1977)).
   HINT: Apply an indirect proof.

PROBLEM 4.2. Assume that the weights $\{W_{n,i}\}$ are nonnegative and the estimate is weakly universally consistent. Prove that (i) in Theorem 4.1 is satisfied (Stone (1977)).
   HINT: Apply Problem 4.1 for

$$Y = f(X) = m(X),$$

and show that

$$\lim_{n \to \infty} \left(\mathbf{E}\left\{ \sum_{i=1}^{n} |W_{n,i}(X)| f(X_i) \right\} - \mathbf{E}f(X) \right)^2 = 0.$$

PROBLEM 4.3. Assume that a local averaging estimate is weakly universally consistent. Prove that (v) in Theorem 4.1 is satisfied (Stone (1977)).
   HINT: Consider the case $m = 0$ and $Y \pm 1$ valued.

PROBLEM 4.4. Assume that the weights $\{W_{n,i}\}$ are nonnegative and that the corresponding local averaging estimate is weakly universally consistent. Prove that (iii) in Theorem 4.1 is satisfied (Stone (1977)).
   HINT: For any fixed x_0 and $a > 0$ let f be a nonnegative continuous function which is 0 on $S_{x_0,a/3}$ and is 1 on $S_{x_0,2a/3}^c$. Choose $Y = f(X) = m(X)$, then

$$I_{\{X \in S_{x_0,a/3}\}} \sum_{i=1}^{n} W_{n,i}(X) f(X_i) \ge I_{\{X \in S_{x_0,a/3}\}} \sum_{i=1}^{n} W_{n,i}(X) I_{\{\|X_i - X\| > a\}} \to 0$$

in probability, therefore, for any compact set B,

$$I_{\{X \in B\}} \sum_{i=1}^{n} W_{ni}(X) I_{\{\|X_i - X\| > a\}} \to 0$$

in probability.

PROBLEM 4.5. Noiseless observations. Call the observations noiseless if $Y_i = m(X_i)$, so that it is the problem of function interpolation for a random design. Prove that under the conditions (i) – (iv) of Theorem 4.1 the regression estimate is weakly consistent for noiseless observations.

HINT: Check the proof of Theorem 4.1.

PROBLEM 4.6. Let a rectangle partition consist of rectangles with side lengths $h_{n1}, \ldots, h_{nd}$. Prove weak universal consistency for

$$\lim_{n \to \infty} h_{nj} = 0 \, (j = 1, \ldots, d) \text{ and } \lim_{n \to \infty} n h_{n1} \ldots h_{nd} = \infty.$$

PROBLEM 4.7. Prove the extension of Theorem 4.3 for rectangle partitions:

$$\mathbf{E}\|m_n - m\|^2 \leq \frac{\hat{c}}{n \prod_{j=1}^{d} h_{nj}} + C^2 \sum_{j=1}^{d} h_{nj}^2.$$

PROBLEM 4.8. A cell A is called empty if $\mu_n(A) = 0$. Let M_n be the number of nonempty cells for $\mathcal{P}_n$. Prove that under the condition (4.2), $M_n/n \to 0$ a.s.

HINT: For a sufficiently large sphere S consider $\frac{1}{n} \sum_{i=1}^{n} I_{\{X_i \notin S\}}$ for $n \to \infty$.

PROBLEM 4.9. Prove that, for $\sigma^2(x) \geq c_1 > 0$,

$$\mathbf{E}\|m_n - m\|^2 \geq \frac{c_2}{n h_n^d}$$

with some constant c_2 (cf. Beirlant and Györfi (1998)).

HINT: Take the lower bound of the first term in the right-hand side in the decomposition (4.4).

PROBLEM 4.10. Let the random variable $B(n,p)$ be binomially distributed with parameters n and p. Then

$$\mathbf{E}\left\{ \frac{1}{B(n,p)} I_{\{B(n,p) > 0\}} \right\} \geq \frac{1}{np} (1 - (1-p)^n)^2.$$

HINT: Apply the Jensen inequality.

5
Kernel Estimates

5.1 Introduction

The kernel estimate of a regression function takes the form

$$m_n(x) = \frac{\sum_{i=1}^n Y_i K\left(\frac{x-X_i}{h_n}\right)}{\sum_{i=1}^n K\left(\frac{x-X_i}{h_n}\right)},$$

if the denominator is nonzero, and 0 otherwise. Here the bandwidth $h_n > 0$ depends only on the sample size n, and the function $K : \mathcal{R}^d \to [0, \infty)$ is called a kernel. (See Figure 5.1 for some examples.) Usually $K(x)$ is "large" if $\|x\|$ is "small," therefore the kernel estimate again is a local averaging estimate.

Figures 5.2–5.5 show the kernel estimate for the naive kernel ($K(x) = I_{\{\|x\|\leq 1\}}$) and for the Epanechnikov kernel ($K(x) = (1 - \|x\|^2)_+$) using various choices for h_n for our simulated data introduced in Chapter 1.

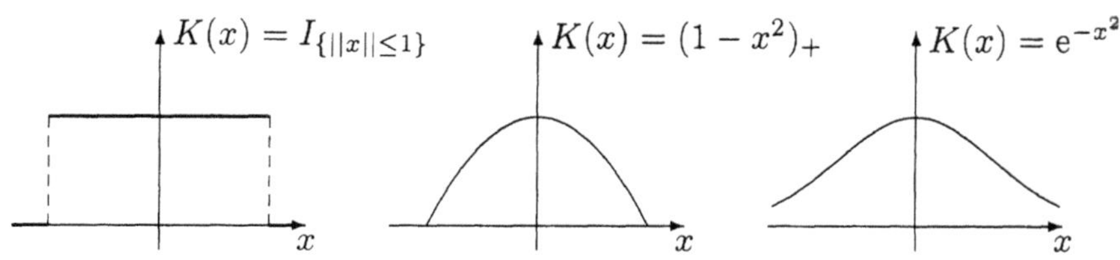

Figure 5.1. Examples for univariate kernels.

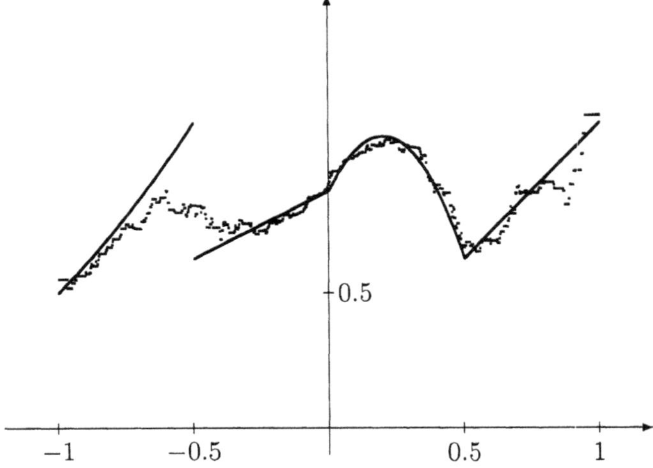

Figure 5.2. Kernel estimate for the naive kernel: $h = 0.1$, L_2 error $= 0.004066$.

Figure 5.6 shows the L_2 error as a function of h.

5.2 Consistency

In this section we use Stone's theorem (Theorem 4.1) in order to prove the weak universal consistency of kernel estimates under general conditions on h and K.

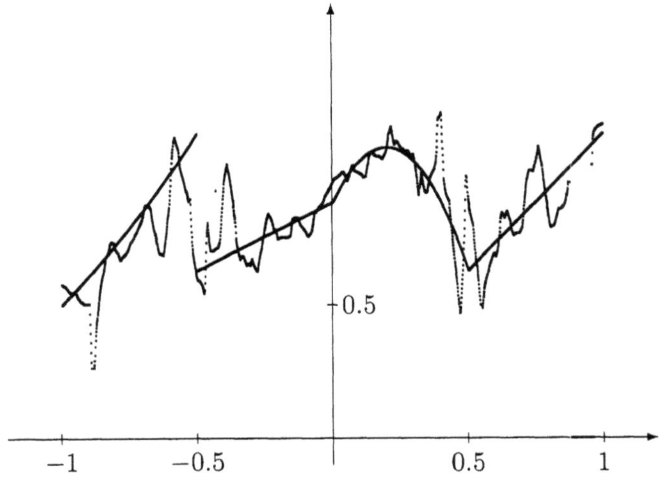

Figure 5.3. Undersmoothing for the Epanechnikov kernel: $h = 0.03$, L_2 error $= 0.031560$.

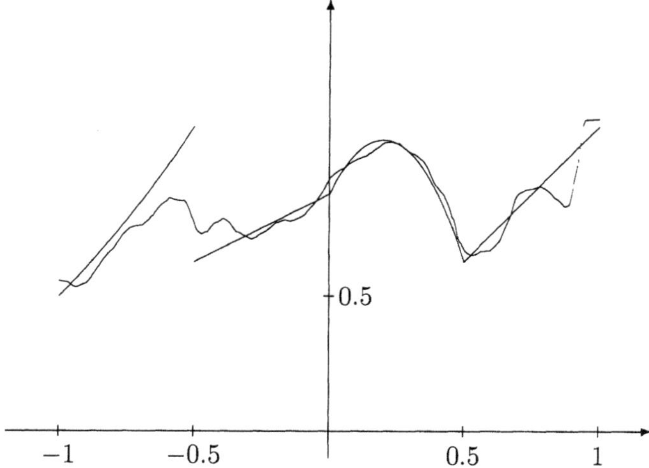

Figure 5.4. Kernel estimate for the Epanechnikov kernel: $h = 0.1$, L_2 error = 0.003608.

Theorem 5.1. *Assume that there are balls $S_{0,r}$ of radius r and balls $S_{0,R}$ of radius R centered at the origin ($0 < r \leq R$), and constant $b > 0$ such that*

$$I_{\{x \in S_{0,R}\}} \geq K(x) \geq bI_{\{x \in S_{0,r}\}}$$

(boxed kernel), and consider the kernel estimate m_n. If $h_n \to 0$ and $nh_n^d \to \infty$, then the kernel estimate is weakly universally consistent.

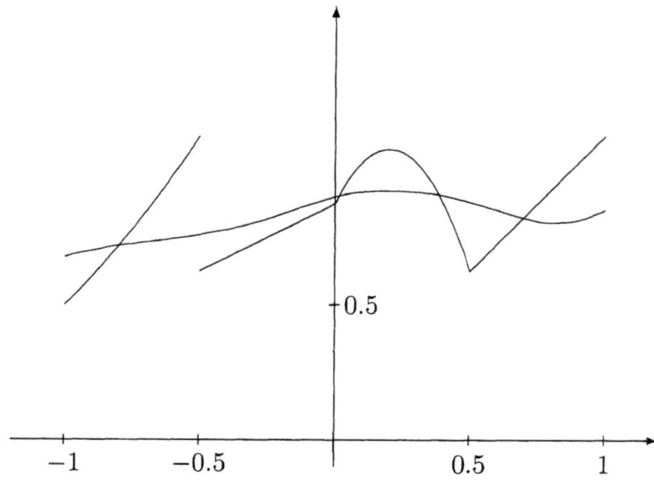

Figure 5.5. Oversmoothing for the Epanechnikov kernel: $h = 0.5$, L_2 error = 0.012551.

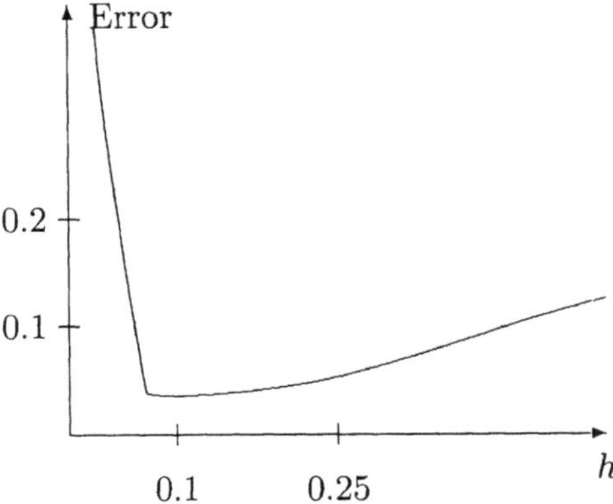

Figure 5.6. The L_2 error for the Epanechnikov kernel as a function of h.

As one can see in Figure 5.7, the weak consistency holds for a bounded kernel with compact support such that it is bounded away from zero at the origin. The bandwidth must converge to zero but not too fast.

PROOF. Put

$$K_h(x) = K(x/h).$$

We check the conditions of Theorem 4.1 for the weights

$$W_{n,i}(x) = \frac{K_h(x - X_i)}{\sum_{j=1}^{n} K_h(x - X_j)}.$$

Condition (i) means that

$$\mathbf{E}\left\{ \frac{\sum_{i=1}^{n} K_h(X - X_i)f(X_i)}{\sum_{j=1}^{n} K_h(X - X_j)} \right\} \le c\mathbf{E}\{f(X)\}$$

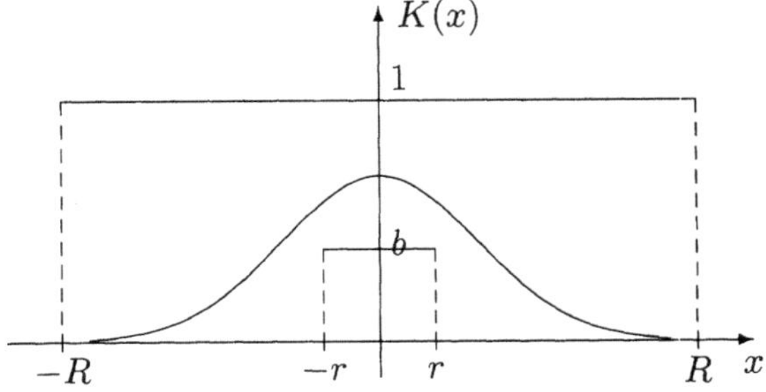

Figure 5.7. Boxed kernel.

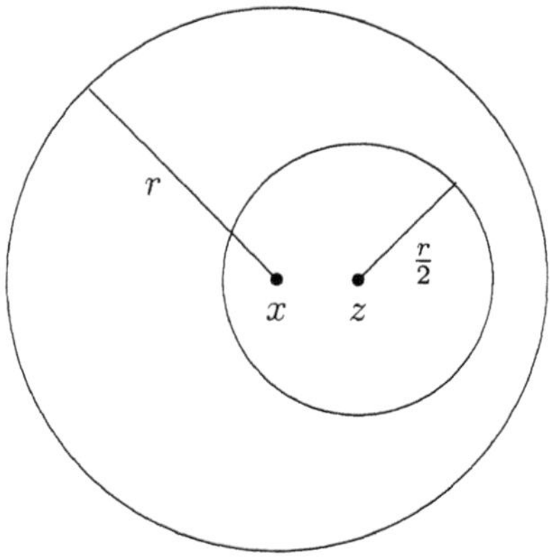

Figure 5.8. If $x \in S_{z,r/2}$, then $S_{z,r/2} \subseteq S_{x,r}$.

with $c > 0$. Because of

$$
\mathbf{E}\left\{ \frac{\sum_{i=1}^{n} K_h(X - X_i)f(X_i)}{\sum_{j=1}^{n} K_h(X - X_j)} \right\}
$$

$$
= \ n\mathbf{E}\left\{ \frac{K_h(X - X_1)f(X_1)}{\sum_{j=1}^{n} K_h(X - X_j)} \right\}
$$

$$
= \ n\mathbf{E}\left\{ \frac{K_h(X - X_1)f(X_1)}{K_h(X - X_1) + \sum_{j=2}^{n} K_h(X - X_j)} \right\}
$$

$$
= \ n \int f(u) \left[\mathbf{E}\left\{ \int \frac{K_h(x - u)}{K_h(x - u) + \sum_{j=2}^{n} K_h(x - X_j)} \mu(dx) \right\} \right] \mu(du)
$$

it suffices to show that, for all u and n,

$$
\mathbf{E}\left\{ \int \frac{K_h(x - u)}{K_h(x - u) + \sum_{j=2}^{n} K_h(x - X_j)} \mu(dx) \right\} \leq \frac{c}{n}.
$$

The compact support of K can be covered by finitely many balls, with translates of $S_{0,r/2}$, where $r > 0$ is the constant appearing in the condition on the kernel K, and with centers x_i, $i = 1, 2, \ldots, M$. Then, for all x and u,

$$
K_h(x - u) \leq \sum_{k=1}^{M} I_{\{x \in u + hx_k + S_{0,rh/2}\}}.
$$

Furthermore, $x \in u + hx_k + S_{0,rh/2}$ implies that

$$
u + hx_k + S_{0,rh/2} \subset x + S_{0,rh}
$$

(cf. Figure 5.8). Now, by these two inequalities,

$$\mathbf{E}\left\{\int \frac{K_h(x-u)}{K_h(x-u)+\sum_{j=2}^n K_h(x-X_j)}\mu(dx)\right\}$$

$$\leq \sum_{k=1}^M \mathbf{E}\left\{\int_{u+hx_k+S_{0,rh/2}} \frac{K_h(x-u)}{K_h(x-u)+\sum_{j=2}^n K_h(x-X_j)}\mu(dx)\right\}$$

$$\leq \sum_{k=1}^M \mathbf{E}\left\{\int_{u+hx_k+S_{0,rh/2}} \frac{1}{1+\sum_{j=2}^n K_h(x-X_j)}\mu(dx)\right\}$$

$$\leq \frac{1}{b}\sum_{k=1}^M \mathbf{E}\left\{\int_{u+hx_k+S_{0,rh/2}} \frac{1}{1+\sum_{j=2}^n I_{\{X_j\in x+S_{0,rh}\}}}\mu(dx)\right\}$$

$$\leq \frac{1}{b}\sum_{k=1}^M \mathbf{E}\left\{\int_{u+hx_k+S_{0,rh/2}} \frac{1}{1+\sum_{j=2}^n I_{\{X_j\in u+hx_k+S_{0,rh/2}\}}}\mu(dx)\right\}$$

$$= \frac{1}{b}\sum_{k=1}^M \mathbf{E}\left\{\frac{\mu(u+hx_k+S_{0,rh/2})}{1+\sum_{j=2}^n I_{\{X_j\in u+hx_k+S_{0,rh/2}\}}}\right\}$$

$$\leq \frac{1}{b}\sum_{k=1}^M \frac{\mu(u+hx_k+S_{0,rh/2})}{n\mu(u+hx_k+S_{0,rh/2})}$$

(by Lemma 4.1)

$$\leq \frac{M}{nb}.$$

The condition (ii) holds since the weights are subprobability weights.
    Concerning (iii) notice that, for $h_n R < a$,

$$\sum_{i=1}^n |W_{n,i}(X)| I_{\{\|X_i-X\|>a\}} = \frac{\sum_{i=1}^n K_{h_n}(X-X_i)I_{\{\|X_i-X\|>a\}}}{\sum_{i=1}^n K_{h_n}(X-X_i)} = 0.$$

In order to show (iv), mention that

$$1-\sum_{i=1}^n W_{n,i}(X) = I_{\{\sum_{i=1}^n K_{h_n}(X-X_i)=0\}},$$

therefore,

$$\mathbf{P}\left\{1\neq\sum_{i=1}^n W_{n,i}(X)\right\} = \mathbf{P}\left\{\sum_{i=1}^n K_{h_n}(X-X_i)=0\right\}$$

$$\leq \mathbf{P}\left\{\sum_{i=1}^n I_{\{X_i\notin S_{X,rh_n}\}}=0\right\}$$

$$= \mathbf{P}\left\{\mu_n(S_{X,rh_n})=0\right\}$$

$$= \int (1 - \mu(S_{x,rh_n}))^n \mu(dx).$$

Choose a sphere S centered at the origin, then

$$\mathbf{P}\left\{ 1 \neq \sum_{i=1}^n W_{n,i}(X) \right\}$$

$$\leq \int_S e^{-n\mu(S_{x,rh_n})} \mu(dx) + \mu(S^c)$$

$$= \int_S n\mu(S_{x,rh_n}) e^{-n\mu(S_{x,rh_n})} \frac{1}{n\mu(S_{x,rh_n})} \mu(dx) + \mu(S^c)$$

$$= \max_u u e^{-u} \int_S \frac{1}{n\mu(S_{x,rh_n})} \mu(dx) + \mu(S^c).$$

By the choice of S, the second term can be small. For the first term we can find $z_1, \ldots, z_{M_n}$ such that the union of $S_{z_1,rh_n/2}, \ldots, S_{z_{M_n},rh_n/2}$ covers S, and

$$M_n \leq \frac{\tilde{c}}{h_n^d}.$$

Then

$$\int_S \frac{1}{n\mu(S_{x,rh_n})} \mu(dx) \;\leq\; \sum_{j=1}^{M_n} \int \frac{I_{\{x \in S_{z_j,rh_n/2}\}}}{n\mu(S_{x,rh_n})} \mu(dx)$$

$$\leq\; \sum_{j=1}^{M_n} \int \frac{I_{\{x \in S_{z_j,rh_n/2}\}}}{n\mu(S_{z_j,rh_n/2})} \mu(dx)$$

$$\leq\; \frac{M_n}{n}$$

$$\leq\; \frac{\tilde{c}}{nh_n^d} \to 0. \tag{5.1}$$

Concerning (v), since $K(x) \leq 1$ we get that, for any $\delta > 0$,

$$\sum_{i=1}^n W_{n,i}(X)^2 \;=\; \frac{\sum_{i=1}^n K_{h_n}(X - X_i)^2}{(\sum_{i=1}^n K_{h_n}(X - X_i))^2}$$

$$\leq\; \frac{\sum_{i=1}^n K_{h_n}(X - X_i)}{(\sum_{i=1}^n K_{h_n}(X - X_i))^2}$$

$$\leq\; \min\left\{ \delta, \frac{1}{\sum_{i=1}^n K_{h_n}(X - X_i)} \right\}$$

$$\leq\; \min\left\{ \delta, \frac{1}{\sum_{i=1}^n b I_{\{X_i \in S_{X,rh_n}\}}} \right\}$$

$$\leq \quad \delta + \frac{1}{\sum_{i=1}^{n} b I_{\{X_i \in S_{X,rh_n}\}}} I\Big\{\sum_{i=1}^{n} I_{\{X_i \in S_{X,rh_n}\}} > 0\Big\},$$

therefore it is enough to show that

$$\mathbf{E}\left\{\frac{1}{\sum_{i=1}^{n} I_{\{X_i \in S_{X,rh_n}\}}} I\Big\{\sum_{i=1}^{n} I_{\{X_i \in S_{X,rh_n}\}} > 0\Big\}\right\} \to 0.$$

Let S be as above, then

$$\mathbf{E}\left\{\frac{1}{\sum_{i=1}^{n} I_{\{X_i \in S_{X,rh_n}\}}} I\Big\{\sum_{i=1}^{n} I_{\{X_i \in S_{X,rh_n}\}} > 0\Big\}\right\}$$

$$\leq \quad \mathbf{E}\left\{\frac{1}{\sum_{i=1}^{n} I_{\{X_i \in S_{X,rh_n}\}}} I\Big\{\sum_{i=1}^{n} I_{\{X_i \in S_{X,rh_n}\}} > 0\Big\} I_{\{X \in S\}}\right\} + \mu(S^c)$$

$$\leq \quad 2\mathbf{E}\left\{\frac{1}{(n+1)\mu(S_{X,h_n})} I_{\{X \in S\}}\right\} + \mu(S^c)$$

(by Lemma 4.1)

$$\to \quad \mu(S^c)$$

as above. $\qquad\qquad\qquad\qquad\qquad\qquad\qquad\qquad\qquad\qquad\qquad\square$

5.3 Rate of Convergence

In this section we bound the rate of convergence of $\mathbf{E}\|m_n - m\|^2$ for a naive kernel and a Lipschitz continuous regression function.

Theorem 5.2. *For a kernel estimate with a naive kernel assume that*

$$\mathbf{Var}(Y|X = x) \leq \sigma^2, \ x \in \mathcal{R}^d,$$

and

$$|m(x) - m(z)| \leq C\|x - z\|, \ x, z \in \mathcal{R}^d,$$

and X has a compact support S^*. Then

$$\mathbf{E}\|m_n - m\|^2 \leq \hat{c}\frac{\sigma^2 + \sup_{z \in S^*} |m(z)|^2}{n \cdot h_n^d} + C^2 h_n^2,$$

where $\hat{c}$ depends only on the diameter of S^* and on d, thus for

$$h_n = c' \left(\frac{\sigma^2 + \sup_{z \in S^*} |m(z)|^2}{C^2}\right)^{1/(d+2)} n^{-\frac{1}{d+2}}$$

we have

$$\mathbf{E}\|m_n - m\|^2 \leq c'' \left(\sigma^2 + \sup_{z \in S^*} |m(z)|^2\right)^{2/(d+2)} C^{2d/(d+2)} n^{-2/(d+2)}.$$

PROOF. We proceed similarly to Theorem 4.3. Put

$$\hat{m}_n(x) = \frac{\sum_{i=1}^n m(X_i)I_{\{X_i \in S_{x,h_n}\}}}{n\mu_n(S_{x,h_n})},$$

then we have the decomposition (4.4). If $B_n(x) = \{n\mu_n(S_{x,h_n}) > 0\}$, then

$$\mathbf{E}\{(m_n(x) - \hat{m}_n(x))^2 | X_1, \ldots, X_n\}$$

$$= \mathbf{E}\left\{\left(\frac{\sum_{i=1}^n (Y_i - m(X_i))I_{\{X_i \in S_{x,h_n}\}}}{n\mu_n(S_{x,h_n})}\right)^2 | X_1, \ldots, X_n\right\}$$

$$= \frac{\sum_{i=1}^n \mathbf{Var}(Y_i|X_i)I_{\{X_i \in S_{x,h_n}\}}}{(n\mu_n(S_{x,h_n}))^2}$$

$$\leq \frac{\sigma^2}{n\mu_n(S_{x,h_n})}I_{B_n(x)}.$$

By Jensen's inequality and the Lipschitz property of m,

$$(\hat{m}_n(x) - m(x))^2$$

$$= \left(\frac{\sum_{i=1}^n (m(X_i) - m(x))I_{\{X_i \in S_{x,h_n}\}}}{n\mu_n(S_{x,h_n})}\right)^2 I_{B_n(x)} + m(x)^2 I_{B_n(x)^c}$$

$$\leq \frac{\sum_{i=1}^n (m(X_i) - m(x))^2 I_{\{X_i \in S_{x,h_n}\}}}{n\mu_n(S_{x,h_n})}I_{B_n(x)} + m(x)^2 I_{B_n(x)^c}$$

$$\leq C^2 h_n^2 I_{B_n(x)} + m(x)^2 I_{B_n(x)^c}$$

$$\leq C^2 h_n^2 + m(x)^2 I_{B_n(x)^c}.$$

Using this, together with Lemma 4.1,

$$\mathbf{E}\left\{\int (m_n(x) - m(x))^2 \mu(dx)\right\}$$

$$= \mathbf{E}\left\{\int (m_n(x) - \hat{m}_n(x))^2 \mu(dx)\right\} + \mathbf{E}\left\{\int (\hat{m}_n(x) - m(x))^2 \mu(dx)\right\}$$

$$\leq \int_{S^*} \mathbf{E}\left\{\frac{\sigma^2}{n\mu_n(S_{x,h_n})}I_{\{\mu_n(S_{x,h_n}) > 0\}}\right\} \mu(dx) + C^2 h_n^2$$

$$+ \int_{S^*} \mathbf{E}\left\{m(x)^2 I_{\{\mu_n(S_{x,h_n}) = 0\}}\right\} \mu(dx)$$

$$\leq \int_{S^*} \frac{2\sigma^2}{n\mu(S_{x,h_n})}\mu(dx) + C^2 h_n^2 + \int_{S^*} m(x)^2 (1 - \mu(S_{x,h_n}))^n \mu(dx)$$

$$\leq \int_{S^*} \frac{2\sigma^2}{n\mu(S_{x,h_n})}\mu(dx) + C^2 h_n^2 + \sup_{z \in S^*} m(z)^2 \int_{S^*} e^{-n\mu(S_{x,h_n})}\mu(dx)$$

$$\leq \ 2\sigma^2 \int_{S^*} \frac{1}{n\mu(S_{x,h_n})}\mu(dx) + C^2 h_n^2$$

$$+ \sup_{z \in S^*} m(z)^2 \max_u ue^{-u} \int_{S^*} \frac{1}{n\mu(S_{x,h_n})}\mu(dx).$$

Now we refer to (5.1) such that there the set S is a sphere containing S^*. Combining these inequalities the proof is complete. $\square$

According to Theorem 5.2, the kernel estimate is of optimum rate for the class $\mathcal{D}^{(1,C)}$ (cf. Definition 3.2 and Theorem 3.2). In Theorem 5.2 the only condition on X is that it has compact support, there is no density assumption.

In contrast to the partitioning estimate, the kernel estimate can "track" the derivative of a differentiable regression function. Using a nonnegative symmetric kernel, in the pointwise theory the kernel regression estimate is of the optimum rate of convergence for the class $\mathcal{D}^{(2,C)}$ (cf. Härdle (1990)). Unfortunately, in the L_2 theory, this is not the case.

In order to show this, consider the following example:
- X is uniform on $[0,1]$;
- $m(x) = x$; and
- $Y = X + N$, where N is standard normal and is independent of X.

This example belongs to $\mathcal{D}^{(p,C)}$ for any $p \geq 1$. In Problem 5.1 we show that for a naive kernel and for $h_n \to 0$ and $nh_n \to \infty$,

$$\mathbf{E}\int_0^1 (m_n(x) - m(x))^2\mu(dx) \geq \frac{1+o(1)}{2}\frac{1}{nh_n} + \frac{1+o(1)}{54}h_n^3, \qquad (5.2)$$

where the lower bound is minimized by $h_n = cn^{-\frac{1}{4}}$, and then

$$\mathbf{E}\int_0^1 (m_n(x) - m(x))^2\mu(dx) \geq c'n^{-\frac{3}{4}},$$

therefore the kernel estimate is not optimal for $\mathcal{D}^{(2,C)}$.

The main point of this example is that because of the end points of the uniform density the squared bias is of order h_n^3 and not h_n^4. From this, one can expect that the kernel regression estimate is optimal for the class $\mathcal{D}^{(1.5,C)}$.

Theorem 5.3. *For a naive kernel the kernel estimate is of an optimal rate of convergence for the class $\mathcal{D}^{(1.5,C)}$.*

PROOF. See Problem 5.2. $\square$

In the pointwise theory, the kernel estimate with a nonnegative kernel can have an optimal rate only for the class $\mathcal{D}^{(p,C)}$ with $p \leq 2$ (cf. Härdle (1990)). If $p > 2$, then that theory suggests a higher-order kernels which can take negative values, too. A kernel is called a higher-order kernel, i.e.,

K is a kernel of order k, if

$$\int K(x)x^j \, dx = 0$$

for $1 \le j \le k - 1$, and

$$0 < \int K(x)x^k \, dx < \infty.$$

So the naive kernel is of order 2. An example for a forth-order kernel is

$$K(x) = \frac{3}{8}(3 - 5x^2)I_{\{|x| \le 1\}}. \tag{5.3}$$

Unfortunately, in the L_2 theory, we lose the consistency if the kernel can take on negative values. For example, let X be uniformly distributed on $[0, 1]$, Y is ± 1 valued with $\mathbf{E}Y = 0$, and X and Y are independent. These conditions imply that $m = 0$. In Problem 5.4 we show, for the kernel defined by (5.3), that

$$\mathbf{E} \int (m_n(x) - m(x))^2 \mu(dx) = \infty. \tag{5.4}$$

5.4   Local Polynomial Kernel Estimates

In the pointwise theory, the optimality of a kernel estimate can be extended using a local polynomial kernel estimate. Similarly to the partitioning estimate, notice that the kernel estimate can be written as a solution of the following minimization problem:

$$m_n(x) = \arg\min_{c} \sum_{i=1}^{n}(Y_i - c)^2 K_{h_n}(x - X_i)$$

(cf. Problem 2.2). To generalize this, choose functions $\phi_0, \ldots, \phi_M$ on $\mathcal{R}^d$, and define the estimate by

$$m_n(x) = \sum_{l=0}^{M} c_l(x)\phi_l(x), \tag{5.5}$$

where

$$(c_0(x), \ldots, c_M(x)) = \arg\min_{(c_0,\ldots,c_M)} \sum_{i=1}^{n}\left(Y_i - \sum_{l=0}^{M} c_l\phi_l(X_i)\right)^2 K_{h_n}(x - X_i). \tag{5.6}$$

The most popular example for estimates of this kind is the local polynomial kernel estimate, where the $\phi_l(x)$'s are monomials of the components of x. For simplicity we consider only $d = 1$. Then $\phi_l(x) = x^l$ $(l = 0, 1, \ldots M)$, and the estimate m_n is defined by locally fitting (via (5.5) and (5.6)) a polynomial to the data. If $M = 0$, then m_n is the standard kernel estimate. If $M = 1$, then m_n is the so-called locally linear kernel estimate.

The locally polynomial estimate has no global consistency properties. Similarly to the piecewise linear partitioning estimate, the locally linear kernel estimate m_n is not weakly universally consistent. This can be shown by the same example: let X be uniformly distributed on $[0, 1]$, Y is ± 1 valued with $\mathbf{E}Y = 0$, and X and Y are independent. These conditions imply that $m = 0$. Then for the naive kernel

$$\mathbf{E} \int (m_n(x) - m(x))^2 \mu(dx) = \infty \tag{5.7}$$

(cf. Problem 5.8).

The main point in the above counterexample for consistency is that, due to interpolation effects, the locally linear kernel estimate can take arbitrarily large values even for bounded data. These interpolation effects occur only with very small probability but, nevertheless, they force the expectation of the L_2 error to be infinity.

This problem can be avoided if one minimizes in (5.6) only over coefficients which are bounded in absolute value by some constant depending on n and converging to infinity.

Theorem 5.4. *Let $M \in \mathcal{N}_0$. For $n \in \mathcal{N}$ choose $\beta_n, h_n > 0$ such that*

$$\beta_n \to \infty, \quad h_n \beta_n \to 0$$

and

$$\frac{n h_n}{\beta_n^2 \log n} \to \infty.$$

Let K be the naive kernel. Define the estimate m_n by

$$m_n(x) = \sum_{l=0}^{M} c_l(x) x^l,$$

where $c_0(x), \ldots, c_M(x) \in [-\beta_n, \beta_n]$ is chosen such that

$$\sum_{i=1}^{n} \left(Y_i - \sum_{l=0}^{M} c_l(x) X_i^l \right)^2 K_{h_n}(x - X_i)$$

$$\leq \min_{c_0(x),\ldots,c_M(x) \in [-\beta_n, \beta_n]} \sum_{i=1}^{n} \left(Y_i - \sum_{l=0}^{M} c_l X_i^l \right)^2 K_{h_n}(x - X_i) + \frac{1}{n}.$$

Then

$$\mathbf{E}\left\{ \int (m_n(x) - m(x))^2 \mu(dx) \right\} \to 0$$

for all distributions of (X, Y) with X bounded a.s. and $\mathbf{E}\{Y^2\} < \infty$.

PROOF. See Kohler (2002b). $\square$

The assumption that X is bounded a.s., can be avoided if one sets the estimate to zero outside an interval which depends on the sample size n and tends to $\mathcal{R}$ for n tending to infinity (cf. Kohler (2002b)).

5.5   Bibliographic Notes

Kernel regression estimates were originally derived from the kernel estimate in density estimation studied by Parzen (1962), Rosenblatt (1956), Akaike (1954), and Cacoullos (1965). They were introduced in regression estimation by Nadaraya (1964; 1970) and Watson (1964). Statistical analysis of the kernel regression function estimate can be found in Nadaraya (1964; 1970), Rejtő and Révész (1973), Devroye and Wagner (1976; 1980a; 1980b), Greblicki (1974; 1978b; 1978a), Krzyżak (1986; 1990), Krzyżak and Pawlak (1984b), Devroye (1978b), Devroye and Krzyżak (1989), and Pawlak (1991), etc. Theorem 5.1 is due to Devroye and Wagner (1980a) and to Spiegelman and Sacks (1980).

Several authors studied the pointwise properties of the kernel estimates, i.e., the pointwise optimality of the locally polynomial kernel estimates under some regularity conditions on m and μ: Stone (1977; 1980), Katkovnik (1979; 1983; 1985), Korostelev and Tsybakov (1993), Cleveland (1979), Härdle (1990), Fan and Gijbels (1992; 1995), Fan (1993), Tsybakov (1986), and Fan, Hu, and Truong (1994). Kernel regression estimate without bandwidth, called Hilbert kernel estimate, was investigated by Devroye, Györfi, and Krzyżak (1998).

The counterexample for the consistency of the local polynomial kernel estimate in Problem 5.8 is due to Devroye (personal communication, 1998). Under regularity conditions on the distribution of X (in particular, for X uniformly distributed on $[0, 1]$), Stone (1982) showed that the L_2 error of the local polynomial kernel estimate converges in probability to zero with the rate $n^{-2p/(2p+1)}$ if the regression function is (p, C)-smooth. So the above-mentioned counterexample is true only for the expected L_2 error. It is an open problem whether the result of Stone (1982) holds without any regularity assumptions on the distribution of X besides boundedness, and whether the expected L_2 error of the estimate in Theorem 5.4 converges to zero with the optimal rate of convergence if the regression function is (p, C)-smooth.

Problems and Exercises

PROBLEM 5.1. Prove (5.2).
  HINT:

Step (a).

$$\mathbf{E}(m_n(x) - m(x))^2 \geq \mathbf{E}(m_n(x) - \mathbf{E}\{m_n(x)|X_1,\ldots X_n\})^2 + (\mathbf{E}\{m_n(x)\} - m(x))^2.$$

Step (b).

$$\mathbf{E}(m_n(x) - \mathbf{E}\{m_n(x)|X_1,\ldots X_n\})^2$$

$$= \frac{1}{n\{\mu([x - h_n, x + h_n])} \mathbf{P}\{\mu_n([x - h_n, x + h_n]) > 0\}^2.$$

Step (c).

$$\mathbf{E}\int_0^1 (m_n(x) - \mathbf{E}\{m_n(x)|X_1,\ldots X_n\})^2 \mu(dx)$$

$$\geq \int_{h_n}^{1-h_n} \frac{1}{2nh_n}(1 - (1 - 2h_n)^n)^2 \, dx$$

$$\geq \frac{1 + o(1)}{2nh_n}.$$

Step (d).

$$(\mathbf{E}\{m_n(x)\} - m(x))^2$$

$$= \left(n\mathbf{E}\left\{X_1 I_{\{|X_1 - x| \leq h_n\}}\right\} \mathbf{E}\left\{\frac{1}{1 + \sum_{i=2}^n I_{\{|X_i - x| \leq h_n\}}}\right\} - x\right)^2.$$

Step (e). Fix $0 \leq x \leq h_n/2$, then

$$\mathbf{E}\left\{X_1 I_{\{|X_1 - x| \leq h_n\}}\right\} = \frac{(x + h_n)^2}{2},$$

Step (f).

$$\mathbf{E}\left\{\frac{1}{1 + \sum_{i=2}^n I_{\{|X_i - x| \leq h_n\}}}\right\} \geq \frac{1}{1 + (n - 1)(x + h_n)}.$$

Step (g). For large nh_n,

$$n\mathbf{E}\left\{X_1 I_{\{|X_1 - x| \leq h_n\}}\right\} \mathbf{E}\left\{\frac{1}{1 + \sum_{i=2}^n I_{\{|X_i - x| \leq h_n\}}}\right\} \geq \frac{x + h_n}{3} \geq x.$$

Step (h). For large nh_n,

$$\int_0^1 (\mathbf{E}\{m_n(x)\} - m(x))^2 \mu(dx) \geq \int_0^{h_n/2} \left(\frac{x + h_n}{3} - x\right)^2 dx$$

$$= \frac{h_n^3}{54}.$$

PROBLEM 5.2. Prove Theorem 5.3.

HINT: For the class $\mathcal{D}^{(1.5,C)}$, m is differentiable and

$$\|m'(x) - m'(z)\| \leq C\|x - z\|^{1/2}.$$

Show that

$$\mathbf{E}\|m_n - m\|^2 \leq \frac{\hat{c}}{nh_n^d} + \tilde{C}^2 h_n^3,$$

thus, for $h_n = c'n^{-\frac{1}{d+3}}$,

$$\mathbf{E}\|m_n - m\|^2 \leq c''n^{-\frac{3}{d+3}}.$$

With respect to the proof of Theorem 5.2 the difference is that we show

$$\mathbf{E}\left\{\int (\hat{m}_n(x) - m(x))^2 I_{\{\mu_n(S_{x,h_n})>0\}}\mu(dx)\right\} \leq C'h_n^3 + C''h_n^2\frac{1}{nh_n^d},$$

i.e.,

$$\mathbf{E}\left\{\int \left(\frac{\sum_{i=1}^n (m(X_i) - m(x))I_{\{X_i \in S_{x,h_n}\}}}{n\mu_n(S_{x,h_n})}\right)^2 \mu(dx)\right\} \leq C'h_n^3 + C''h_n^2\frac{1}{nh_n^d}.$$

Step (a).

$$\mathbf{E}\left\{\int \left(\frac{\sum_{i=1}^n (m(X_i) - m(x))I_{\{X_i \in S_{x,h_n}\}}}{n\mu_n(S_{x,h_n})}\right)^2 \mu(dx)\right\}$$

$$\leq \quad C''h_n^2\frac{1}{nh_n^d} + 2\int \frac{(\int_{S_{x,h_n}}(m(u) - m(x))\mu(du))^2}{\mu(S_{x,h_n})^2}\mu(dx).$$

Step (b). By the mean value theorem, for a convex linear combination u' of x and u,

$$m(u) - m(x) = (m(u')', u - x),$$

therefore,

$$\left(\frac{\int_{S_{x,h_n}}(m(u) - m(x))\mu(du)}{\mu(S_{x,h_n})}\right)^2 \leq 2\left(m(x)', \frac{\int_{S_{x,h_n}}u\mu(du)}{\mu(S_{x,h_n})} - x\right)^2 + 2C^2h_n^3.$$

Step (c). Let B be the set of points of $[0,1]^d$ which are further from the border of $[0,1]^d$ than h_n. Then

$$\int_{[0,1]^d} \left(m(x)', \frac{\int_{S_{x,h_n}}u\mu(du)}{\mu(S_{x,h_n})} - x\right)^2 \mu(dx) \leq 2^d \max_x \|m(x)'\|^2 h_n^3.$$

PROBLEM 5.3. Prove that the kernel defined by (5.3) is of order 4.

PROBLEM 5.4. Prove (5.4).

HINT: Let A be the event that $X_1, X_2 \in [0, 4h]$, $2\sqrt{\frac{3}{5}}h \leq |X_1 - X_2| \leq 2h$, $X_3, \ldots, X_n \in [6h, 1]$, and $Y_1 \neq Y_2$. Then $\mathbf{P}\{A\} > 0$ and

$$\mathbf{E}\int (m_n(x) - m(x))^2\mu(dx) \geq \left(\mathbf{E}\left\{\int |m_n(x)|dx|A\right\}\mathbf{P}\{A\}\right)^2.$$

For the event A and for $x \in [0, 4h]$, the quantity $|m_n(x)|$ is a ratio of $|K((x - X_1)/h) - K((x - X_2)/h)|$ and $|K((x - X_1)/h) + K((x - X_2)/h)|$, such that there are two x's for which the denominator is 0 and the numerator is positive, so the integral of $|m(x)|$ is ∞.

PROBLEM 5.5. The kernel estimate can be generalized to product kernels. Let K_j^* $(j = 1, 2, \ldots, d)$ be kernels defined on $\mathcal{R}$ and $h_{n1}, \ldots, h_{nd}$ bandwidth sequences. Put

$$K_n(x - z) = \prod_{j=1}^{d} K_j^* \left(\frac{x_j - z_j}{h_{nj}} \right).$$

Formulate consistency conditions on $h_{n1}, \ldots, h_{nd}$.

PROBLEM 5.6. Extend Theorem 5.2 to boxed kernels.

PROBLEM 5.7. Extend Theorem 5.2 to product kernels:

$$\mathbf{E}\|m_n - m\|^2 \leq \frac{\hat{c}}{n \prod_{j=1}^{d} h_{nj}} + C^2 \sum_{j=1}^{d} h_{nj}^2.$$

PROBLEM 5.8. Prove (5.7).

HINT:

Step (a). Let A be the event that $X_1, X_2 \in [h/2, 3h/2]$, $X_3, \ldots, X_n \in [5h/2, 1]$, and $Y_1 \neq Y_2$. Then $\mathbf{P}\{A\} > 0$ and

$$\mathbf{E} \int (m_n(x) - m(x))^2 \mu(dx) = \mathbf{E} \int m_n(x)^2 dx \geq \left(\mathbf{E} \left\{ \int |m_n(x)| dx | A \right\} \mathbf{P}\{A\} \right)^2.$$

Step (b). Given A, on $[h/2, 3h/2]$, the locally linear kernel estimate m_n has the form

$$m_n(x) = \frac{\pm 2}{\Delta}(x - c),$$

where $\Delta = |X_1 - X_2|$ and $h/2 \leq c \leq 3h/2$. Then

$$\mathbf{E} \left\{ \int |m_n(x)| dx | A \right\} \geq \mathbf{E} \left\{ \frac{2}{\Delta} \int_{h/2}^{3h/2} |x - h| dx | A \right\}.$$

Step (c).

$$\mathbf{E} \left\{ \frac{1}{\Delta} | A \right\} = \mathbf{E} \left\{ \frac{1}{\Delta} \right\} = \infty.$$

6
k-NN Estimates

6.1 Introduction

We fix $x \in \mathcal{R}^d$, and reorder the data $(X_1, Y_1), \ldots, (X_n, Y_n)$ according to increasing values of $\|X_i - x\|$. The reordered data sequence is denoted by

$$(X_{(1,n)}(x), Y_{(1,n)}(x)), \ldots, (X_{(n,n)}(x), Y_{(n,n)}(x))$$

or by

$$(X_{(1,n)}, Y_{(1,n)}), \ldots, (X_{(n,n)}, Y_{(n,n)})$$

if no confusion is possible. $X_{(k,n)}(x)$ is called the kth nearest neighbor (k-NN) of x.

The k_n-NN regression function estimate is defined by

$$m_n(x) = \frac{1}{k_n} \sum_{i=1}^{k_n} Y_{(i,n)}(x).$$

If X_i and X_j are equidistant from x, i.e., $\|X_i - x\| = \|X_j - x\|$, then we have a tie. There are several rules for tie breaking. For example, X_i might be declared "closer" if $i < j$, i.e., the tie breaking is done by indices. For the sake of simplicity we assume that ties occur with probability 0. In principle, this is an assumption on μ, so the statements are formally not universal, but adding a component to the observation vector X we can automatically satisfy this condition as follows: Let (X, Z) be a random vector, where Z is independent of (X, Y) and uniformly distributed on $[0, 1]$. We also artificially enlarge the data set by introducing $Z_1, Z_2, \ldots, Z_n$, where the

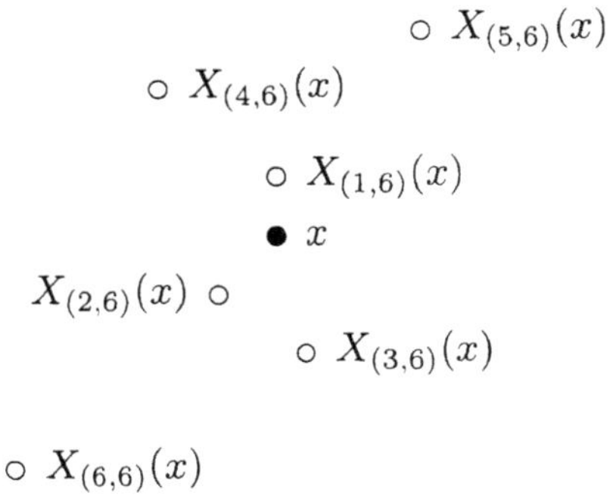

Figure 6.1. Illustration of nearest neighbors.

Z_i's are i.i.d. uniform $[0,1]$ as well. Thus, each (X_i, Z_i) is distributed as (X, Z). Then ties occur with probability 0. In the sequel we shall assume that X has such a component and, therefore, for each x the random variable $\|X - x\|^2$ is absolutely continuous, since it is a sum of two independent random variables such that one of the two is absolutely continuous.

Figures 6.2 – 6.4 show k_n-NN estimates for various choices of k_n for our simulated data introduced in Chapter 1. Figure 6.5 shows the L_2 error as a function of k_n.

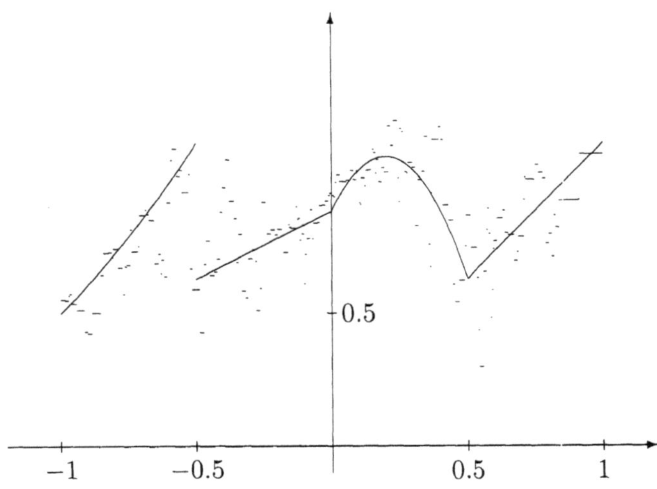

Figure 6.2. Undersmoothing: $k_n = 3$, L_2 error $=0.011703$.

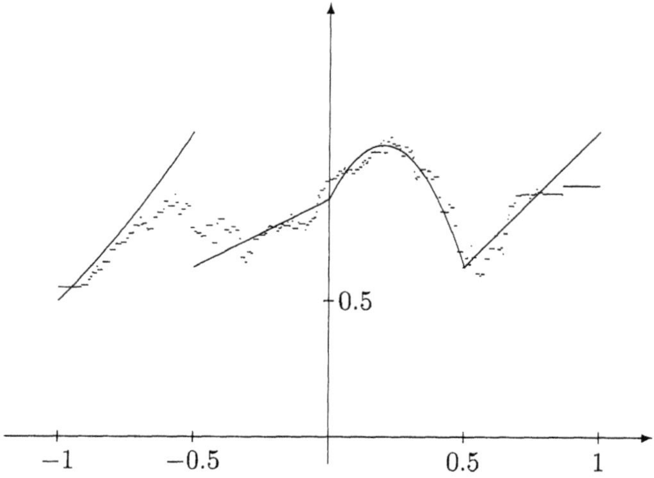

Figure 6.3. Good choice: $k_n = 12$, L_2 error $=0.004247$.

6.2 Consistency

In this section we use Stone's theorem (Theorem 4.1) in order to prove
weak universal consistency of the k-NN estimate. The main result is the
following theorem:

Theorem 6.1. *If $k_n \to \infty$, $k_n/n \to 0$, then the k_n-NN regression function
estimate is weakly consistent for all distributions of (X, Y) where ties occur
with probability zero and $\mathbf{E}Y^2 < \infty$.*

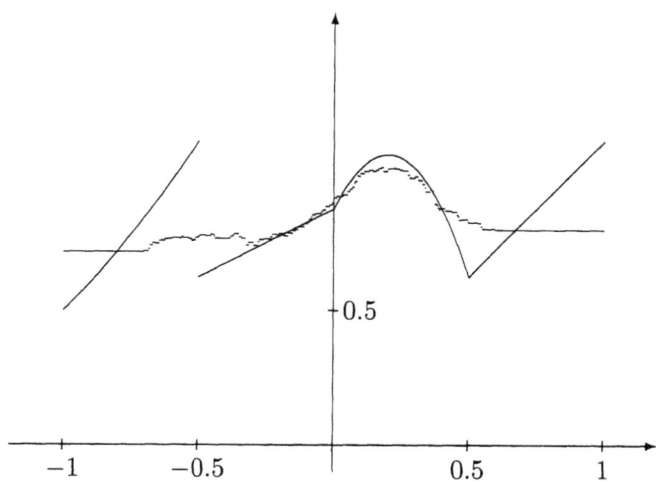

Figure 6.4. Oversmoothing: $k_n = 50$, L_2 error $=0.009931$.

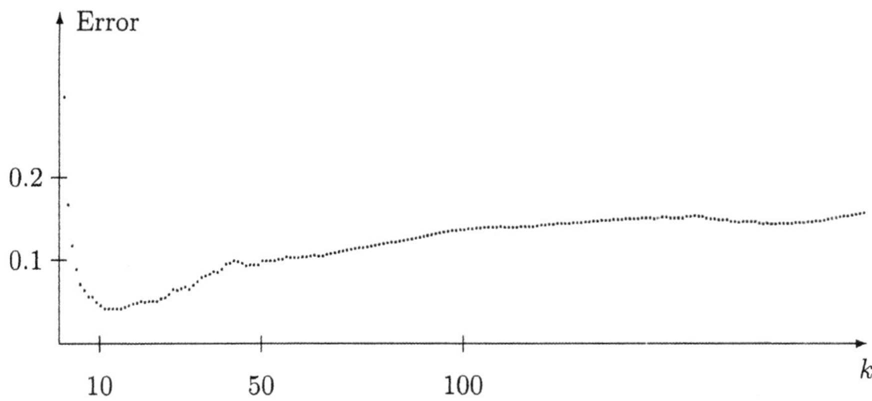

Figure 6.5. L_2 error of the k-NN estimate as a function of k.

According to Theorem 6.1 the number of nearest neighbors (k_n), over which one averages in order to estimate the regression function, should on the one hand converge to infinity but should, on the other hand, be small with respect to the sample size n. To verify the conditions of Stone's theorem we need several lemmas.

We will use Lemma 6.1 to verify condition (iii) of Stone's theorem. Denote the probability measure for X by μ, and let $S_{x,\epsilon}$ be the closed ball centered at x of radius $\epsilon > 0$. The collection of all x with $\mu(S_{x,\epsilon}) > 0$ for all $\epsilon > 0$ is called the support of X or μ. This set plays a key role because of the following property:

Lemma 6.1. *If $x \in \text{support}(\mu)$ and $\lim_{n\to\infty} k_n/n = 0$, then*

$$\|X_{(k_n,n)}(x) - x\| \to 0$$

with probability one.

PROOF. Take $\epsilon > 0$. By definition, $x \in \text{support}(\mu)$ implies that $\mu(S_{x,\epsilon}) > 0$. Observe that

$$\{\|X_{(k_n,n)}(x) - x\| > \epsilon\} = \left\{\frac{1}{n}\sum_{i=1}^{n} I_{\{X_i \in S_{x,\epsilon}\}} < \frac{k_n}{n}\right\}.$$

By the strong law of large numbers,

$$\frac{1}{n}\sum_{i=1}^{n} I_{\{X_i \in S_{x,\epsilon}\}} \to \mu(S_{x,\epsilon}) > 0$$

with probability one, while, by assumption,

$$\frac{k_n}{n} \to 0.$$

Therefore, $\|X_{(k_n,n)}(x) - x\| \to 0$ with probability one. $\square$

The next two lemmas will enable us to establish condition (i) of Stone's theorem.

Lemma 6.2. *Let*

$$B_a(x') = \left\{ x : \mu(S_{x,\|x-x'\|}) \le a \right\}.$$

Then, for all $x' \in \mathcal{R}^d$,

$$\mu(B_a(x')) \le \gamma_d a,$$

where γ_d depends on the dimension d only.

PROOF. Let $C_j \subset \mathcal{R}^d$ be a cone of angle $\pi/3$ and centered at 0. It is a property of cones that if $u, u' \in C_j$ and $\|u\| < \|u'\|$, then $\|u - u'\| < \|u'\|$ (cf. Figure 6.6). Let $C_1, \ldots, C_{\gamma_d}$ be a collection of such cones with different central directions such that their union covers $\mathcal{R}^d$:

$$\bigcup_{j=1}^{\gamma_d} C_j = \mathcal{R}^d.$$

Then

$$\mu(B_a(x')) \le \sum_{i=1}^{\gamma_d} \mu(\{x' + C_i\} \cap B_a(x')).$$

Let $x^* \in \{x' + C_i\} \cap B_a(x')$. Then, by the property of cones mentioned above, we have

$$\mu(\{x' + C_i\} \cap S_{x',\|x'-x^*\|} \cap B_a(x')) \le \mu(S_{x^*,\|x'-x^*\|}) \le a,$$

where we use the fact that $x^* \in B_a(x')$. Since x^* is arbitrary,

$$\mu(\{x' + C_i\} \cap B_a(x')) \le a,$$

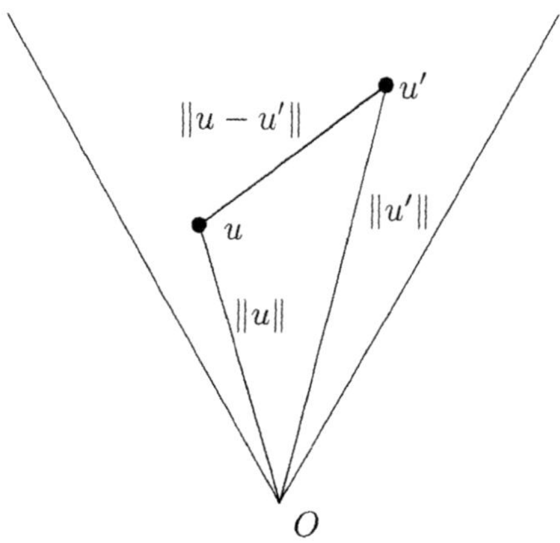

Figure 6.6. The cone property.

which completes the proof of the lemma. $\square$

An immediate consequence of the lemma is that the number of points among $X_1, \ldots, X_n$, such that X is one of their k nearest neighbors, is not more than a constant times k.

Corollary 6.1. *Assume that ties occur with probability zero. Then*

$$\sum_{i=1}^{n} I_{\{X \text{ is among the } k \text{ NNs of } X_i \text{ in } \{X_1,\ldots,X_{i-1},X,X_{i+1},\ldots,X_n\}\}} \leq k\gamma_d$$

a.s.

PROOF. Apply Lemma 6.2 with $a = k/n$ and let μ be the empirical measure μ_n of $X_1, \ldots, X_n$, i.e., for each Borel set $A \subseteq \mathcal{R}^d$, $\mu_n(A) = (1/n)\sum_{i=1}^{n} I_{\{X_i \in A\}}$. Then

$$B_{k/n}(X) = \left\{x \; : \; \mu_n(S_{x,\|x-X\|}) \leq k/n\right\}$$

and

$$X_i \in B_{k/n}(X)$$
$$\Leftrightarrow \quad \mu_n(S_{X_i,\|X_i-X\|}) \leq k/n$$
$$\Leftrightarrow \quad X \text{ is among the } k \text{ NNs of } X_i \text{ in } \{X_1, \ldots, X_{i-1}, X, X_{i+1}, \ldots, X_n\}$$

a.s., where for the second $\Leftrightarrow$ we applied the condition that ties occur with probability zero. This, together with Lemma 6.2, yields

$$\sum_{i=1}^{n} I_{\{X \text{ is among the } k \text{ NNs of } X_i \text{ in } \{X_1,\ldots,X_{i-1},X,X_{i+1},\ldots,X_n\}\}}$$
$$= \quad \sum_{i=1}^{n} I_{\{X_i \in B_{k/n}(X)\}}$$
$$= \quad n \cdot \mu_n(B_{k/n}(X))$$
$$\leq \quad k\gamma_d$$

a.s. $\square$

Lemma 6.3. *Assume that ties occur with probability zero. Then for any integrable function f, any n, and any $k \leq n$,*

$$\sum_{i=1}^{k} \mathbf{E}\left\{|f(X_{(i,n)}(X))|\right\} \leq k\gamma_d \mathbf{E}\{|f(X)|\},$$

where γ_d depends upon the dimension only.

PROOF. If f is a nonnegative function,

$$\sum_{i=1}^{k} \mathbf{E}\left\{f(X_{(i,n)}(X))\right\}$$

$$= \mathbf{E}\left\{\sum_{i=1}^{n} I_{\{X_i \text{ is among the k NNs of } X \text{ in } \{X_1,\dots,X_n\}\}} f(X_i)\right\}$$

$$= \mathbf{E}\left\{f(X) \sum_{i=1}^{n} I_{\{X \text{ is among the k NNs of } X_i \text{ in } \{X_1,\dots,X_{i-1},X,X_{i+1},\dots,X_n\}\}}\right\}$$

(by exchanging X and X_i)

$$\leq \mathbf{E}\{f(X)k\gamma_d\},$$

by Corollary 6.1. This concludes the proof of the lemma. $\qquad\square$

PROOF OF THEOREM 6.1. We proceed by checking the conditions of Stone's weak convergence theorem (Theorem 4.1) under the condition that ties occur with probability zero. The weight $W_{n,i}(X)$ in Theorem 4.1 equals $1/k_n$ if X_i is among the k_n nearest neighbors of X, and equals 0 otherwise, thus the weights are probability weights, and (ii) and (iv) are automatically satisfied. Condition (v) is obvious since $k_n \to \infty$. For condition (iii) observe that, for each $\epsilon > 0$,

$$\mathbf{E}\left\{\sum_{i=1}^{n} W_{n,i}(X)I_{\{\|X_i-X\|>\epsilon\}}\right\}$$

$$= \int \mathbf{E}\left\{\sum_{i=1}^{n} W_{n,i}(x)I_{\{\|X_i-x\|>\epsilon\}}\right\} \mu(dx)$$

$$= \int \mathbf{E}\left\{\frac{1}{k_n} \sum_{i=1}^{k_n} I_{\{\|X_{(i,n)}(x)-x\|>\epsilon\}}\right\} \mu(dx) \to 0$$

holds whenever

$$\int \mathbf{P}\left\{\|X_{(k_n,n)}(x) - x\| > \epsilon\right\} \mu(dx) \to 0, \tag{6.1}$$

where $X_{(k_n,n)}(x)$ denotes the k_nth nearest neighbor of x among $X_1,\dots,X_n$. For $x \in support(\mu)$, $k_n/n \to 0$, together with Lemma 6.1, implies

$$\mathbf{P}\left\{\|X_{(k_n,n)}(x) - x\| > \epsilon\right\} \to 0 \quad (n \to \infty).$$

This together with the dominated convergence theorem implies (6.1). Finally, we consider condition (i). It suffices to show that for any nonnegative measurable function f with $\mathbf{E}\{f(X)\} < \infty$, and any n,

$$\mathbf{E}\left\{\sum_{i=1}^{n} \frac{1}{k_n} I_{\{X_i \text{ is among the } k_n \text{ NNs of } X\}} f(X_i)\right\} \leq c \cdot \mathbf{E}\{f(X)\}$$

for some constant c. But we have shown in Lemma 6.3 that this inequality always holds with $c = \gamma_d$. Thus, condition (i) is verified. $\qquad\square$

6.3 Rate of Convergence

In this section we bound the rate of convergence of $\mathbf{E}\|m_n - m\|^2$ for a k_n-nearest neighbor estimate.

Theorem 6.2. *Assume that X is bounded,*

$$\sigma^2(x) = \mathbf{Var}(Y|X = x) \le \sigma^2 \quad (x \in \mathcal{R}^d)$$

and

$$|m(x) - m(z)| \le C\|x - z\| \quad (x, z \in \mathcal{R}^d).$$

Assume that $d \ge 3$. Let m_n be the k_n-NN estimate. Then

$$\mathbf{E}\|m_n - m\|^2 \le \frac{\sigma^2}{k_n} + c_1 \cdot C^2 \left(\frac{k_n}{n}\right)^{2/d},$$

thus for $k_n = c' \left(\sigma^2/C^2\right)^{d/(2+d)} n^{\frac{2}{d+2}}$,

$$\mathbf{E}\|m_n - m\|^2 \le c'' \sigma^{\frac{4}{d+2}} C^{\frac{2d}{2+d}} n^{-\frac{2}{d+2}}.$$

For the proof of Theorem 6.2 we need the rate of convergence of nearest neighbor distances.

Lemma 6.4. *Assume that X is bounded. If $d \ge 3$, then*

$$\mathbf{E}\{\|X_{(1,n)}(X) - X\|^2\} \le \frac{\tilde{c}}{n^{2/d}}.$$

PROOF. For fixed $\epsilon > 0$,

$$\mathbf{P}\{\|X_{(1,n)}(X) - X\| > \epsilon\} = \mathbf{E}\{(1 - \mu(S_{X,\epsilon}))^n\}.$$

Let $A_1, \ldots, A_{N(\epsilon)}$ be a cubic partition of the bounded support of μ such that the A_j's have diameter ϵ and

$$N(\epsilon) \le \frac{c}{\epsilon^d}.$$

If $x \in A_j$, then $A_j \subset S_{x,\epsilon}$, therefore

$$\mathbf{E}\{(1 - \mu(S_{X,\epsilon}))^n\} = \sum_{j=1}^{N(\epsilon)} \int_{A_j} (1 - \mu(S_{x,\epsilon}))^n \mu(dx)$$

$$\le \sum_{j=1}^{N(\epsilon)} \int_{A_j} (1 - \mu(A_j))^n \mu(dx)$$

$$= \sum_{j=1}^{N(\epsilon)} \mu(A_j)(1 - \mu(A_j))^n.$$

Obviously,

$$\sum_{j=1}^{N(\epsilon)} \mu(A_j)(1 - \mu(A_j))^n \leq \sum_{j=1}^{N(\epsilon)} \max_z z(1 - z)^n$$

$$\leq \sum_{j=1}^{N(\epsilon)} \max_z z e^{-nz}$$

$$= \frac{e^{-1} N(\epsilon)}{n}.$$

If L stands for the diameter of the support of μ, then

$$\mathbf{E}\{\|X_{(1,n)}(X) - X\|^2\} = \int_0^\infty \mathbf{P}\{\|X_{(1,n)}(X) - X\|^2 > \epsilon\}\, d\epsilon$$

$$= \int_0^{L^2} \mathbf{P}\{\|X_{(1,n)}(X) - X\| > \sqrt{\epsilon}\}\, d\epsilon$$

$$\leq \int_0^{L^2} \min\left\{1, \frac{e^{-1} N(\sqrt{\epsilon})}{n}\right\}\, d\epsilon$$

$$\leq \int_0^{L^2} \min\left\{1, \frac{c}{en} \epsilon^{-d/2}\right\}\, d\epsilon$$

$$= \int_0^{(c/(en))^{2/d}} 1\, d\epsilon + \frac{c}{en} \int_{(c/(en))^{2/d}}^{L^2} \epsilon^{-d/2} d\epsilon$$

$$\leq \frac{\tilde{c}}{n^{2/d}}$$

for $d \geq 3$. $\qquad\qquad\qquad\qquad\qquad\qquad\qquad\qquad\qquad\qquad\qquad\qquad\square$

PROOF OF THEOREM 6.2. We have the decomposition

$$\mathbf{E}\{(m_n(x) - m(x))^2\} = \mathbf{E}\{(m_n(x) - \mathbf{E}\{m_n(x)|X_1, \ldots, X_n\})^2\}$$

$$+ \mathbf{E}\{(\mathbf{E}\{m_n(x)|X_1, \ldots, X_n\} - m(x))^2\}$$

$$= I_1(x) + I_2(x).$$

The first term is easier:

$$I_1(x) = \mathbf{E}\left\{\left(\frac{1}{k_n} \sum_{i=1}^{k_n} \left(Y_{(i,n)}(x) - m(X_{(i,n)}(x))\right)\right)^2\right\}$$

$$= \mathbf{E}\left\{\frac{1}{k_n^2} \sum_{i=1}^{k_n} \sigma^2(X_{(i,n)}(x))\right\}$$

$$\leq \frac{\sigma^2}{k_n}.$$

For the second term

$$
\begin{aligned}
I_2(x) &= \mathbf{E}\left\{\left(\frac{1}{k_n}\sum_{i=1}^{k_n}(m(X_{(i,n)}(x)) - m(x))\right)^2\right\} \\
&\leq \mathbf{E}\left\{\left(\frac{1}{k_n}\sum_{i=1}^{k_n}|m(X_{(i,n)}(x)) - m(x)|\right)^2\right\} \\
&\leq \mathbf{E}\left\{\left(\frac{1}{k_n}\sum_{i=1}^{k_n}C\|X_{(i,n)}(x) - x\|\right)^2\right\}.
\end{aligned}
$$

Put $N = k_n\lfloor\frac{n}{k_n}\rfloor$. Split the data $X_1,\ldots,X_n$ into $k_n + 1$ segments such that the first k_n segments have length $\lfloor\frac{n}{k_n}\rfloor$, and let $\tilde{X}_j^x$ be the first nearest neighbor of x from the jth segment. Then $\tilde{X}_1^x, \ldots, \tilde{X}_{k_n}^x$ are k_n different elements of $\{X_1,\ldots,X_n\}$, which implies

$$\sum_{i=1}^{k_n}\|X_{(i,n)}(x) - x\| \leq \sum_{j=1}^{k_n}\|\tilde{X}_j^x - x\|,$$

therefore, by Jensen's inequality,

$$
\begin{aligned}
I_2(x) &\leq C^2\mathbf{E}\left\{\left(\frac{1}{k_n}\sum_{j=1}^{k_n}\|\tilde{X}_j^x - x\|\right)^2\right\} \\
&\leq C^2\frac{1}{k_n}\sum_{j=1}^{k_n}\mathbf{E}\left\{\|\tilde{X}_j^x - x\|^2\right\} \\
&= C^2\mathbf{E}\left\{\|\tilde{X}_1^x - x\|^2\right\} \\
&= C^2\mathbf{E}\left\{\|X_{(1,\lfloor\frac{n}{k_n}\rfloor)}(x) - x\|^2\right\}.
\end{aligned}
$$

Thus, by Lemma 6.4,

$$
\begin{aligned}
\frac{1}{C^2}\left\lfloor\frac{n}{k_n}\right\rfloor^{2/d}\int I_2(x)\mu(dx) &\leq \left\lfloor\frac{n}{k_n}\right\rfloor^{2/d}\mathbf{E}\left\{\|X_{(1,\lfloor\frac{n}{k_n}\rfloor)}(X) - X\|^2\right\} \\
&\leq const.
\end{aligned}
$$

$\square$

For $d \leq 2$ the rate of convergence of Theorem 6.2 holds under additional conditions on μ (cf. Problem 6.7).

According to Theorem 6.2, the nearest neighbor estimate is of optimum rate for the class $\mathcal{D}^{(1,C)}$ (cf. Definition 3.2 and Theorem 3.2). In Theorem

6.2 the only condition on X for $d \geq 3$ is that it has compact support, there is no density assumption.

Similarly to the partitioning estimate, the nearest neighbor estimate cannot "track" the derivative of a differentiable regression function. In the pointwise theory the nearest neighbor regression estimate has the optimum rate of convergence for the class $\mathcal{D}^{(2,C)}$ (cf. Härdle (1990)). Unfortunately, in the L_2 theory, this is not the case.

In order to show this consider the following example:
- X is uniform on $[0,1]$;
- $m(x) = x$; and
- $Y = X + N$, where N is standard normal and is independent of X.

This example belongs to $\mathcal{D}^{(p,C)}$ for any $p \geq 1$. In Problem 6.2 we will see that, for $k_n/n \to 0$ and $k_n \to \infty$,

$$\mathbf{E} \int_0^1 (m_n(x) - m(x))^2 \mu(dx) \geq \frac{1}{k_n} + \frac{1}{24} \left(\frac{k_n}{n+1} \right)^3, \qquad (6.2)$$

where the lower bound is minimized by $k_n = cn^{3/4}$, and thus

$$\mathbf{E} \int_0^1 (m_n(x) - m(x))^2 \mu(dx) \geq c'n^{-\frac{3}{4}},$$

therefore the nearest neighbor estimate is not optimal for $\mathcal{D}^{(2,C)}$.

The main point of this example is that because of the end points of the uniform density the squared bias is of order $\left(\frac{k_n}{n} \right)^3$ and not $\left(\frac{k_n}{n} \right)^4$. From this one may conjecture that the nearest neighbor regression estimate is optimal for the class $\mathcal{D}^{(1.5,C)}$.

6.4   Bibliographic Notes

The consistency of the k_n-nearest neighbor classification, and the corresponding regression and density estimation has been studied by many researchers. See Beck (1979), Bhattacharya and Mack (1987), Bickel and Breiman (1983), Cheng (1995), Collomb (1979; 1980; 1981), Cover (1968a), Cover and Hart (1967), Devroye (1978a; 1981; 1982b), Devroye and Györfi (1985), Devroye et al. (1994), Fix and Hodges (1951; 1952), Guerre (2000) Györfi and Györfi (1975), Mack (1981), Stone (1977), Stute (1984), and Zhao (1987). Theorem 6.1 is due to Stone (1977). Various versions of Lemma 6.2 appeared in Fritz (1974), Stone (1977), Devroye and Györfi (1985). Lemma 6.4 is a special case of the result of Kulkarni and Posner (1995).

Problems and Exercises

PROBLEM 6.1. Prove that for $d \leq 2$ Lemma 6.4 is not distribution-free, i.e., construct a distribution of X for which Lemma 6.4 does not hold.

HINT: Put $d = 1$ and assume a density $f(x) = 3x^2$, then $F(x) = x^3$ and

$$
\begin{aligned}
\mathbf{E}\{\|X_{(1,n)}(X) - X\|^2\} \quad &\geq \quad \int_0^{1/4} \int_0^{\sqrt{\epsilon}} (1 - [F(x + \sqrt{\epsilon}) - F(x - \sqrt{\epsilon})])^n f(x)\, dx\, d\epsilon \\
&\geq \quad \frac{C}{n^{5/3}}.
\end{aligned}
$$

PROBLEM 6.2. Prove (6.2).

HINT:

Step (a).

$$
\begin{aligned}
&\mathbf{E}(m_n(x) - m(x))^2 \\
\geq \quad &\mathbf{E}(m_n(x) - \mathbf{E}\{m_n(x)|X_1,\ldots X_n\})^2 + (\mathbf{E}\{m_n(x)\} - m(x))^2.
\end{aligned}
$$

Step (b).

$$
\mathbf{E}\{(m_n(x) - \mathbf{E}\{m_n(x)|X_1,\ldots X_n\})^2\} = \frac{1}{k_n}.
$$

Step (c). Observe that the function

$$
\mathbf{E}\{m_n(x)|X_1,\ldots,X_n\} = \frac{1}{k_n} \sum_{i=1}^{k_n} X_{(i,n)}(x)
$$

is a monotone increasing function of x, therefore

$$
\mathbf{E}\{m_n(x)|X_1,\ldots,X_n\} \geq \frac{1}{k_n} \sum_{i=1}^{k_n} X_{(i,n)}(0).
$$

Let $X_1^*,\ldots,X_n^*$ be the ordered sample of $X_1,\ldots,X_n$, then $X_{(i,n)}(0) = X_i^*$, and so

$$
\mathbf{E}\{m_n(x)\} \geq \mathbf{E}\left\{\frac{1}{k_n} \sum_{i=1}^{k_n} X_i^*\right\} = \alpha_{k_n}.
$$

Thus

$$
\int_0^1 (\mathbf{E}\{m_n(x)\} - m(x))^2 \mu(dx) \geq \frac{\alpha_{k_n}^3}{3}.
$$

Step (d).

$$
\alpha_{k_n} = \frac{1}{2} \frac{k_n}{n + 1}.
$$

PROBLEM 6.3. Prove that for fixed k the k-NN regression estimate is weakly consistent for noiseless observations.

HINT: See Problem 4.5.

PROBLEM 6.4. Let $m_n(x)$ be the k-NN regression estimate. Prove that, for fixed k,

$$\lim_{n \to \infty} \mathbf{E} \int (m_n(x) - m(x))^2 \mu(dx) = \frac{\mathbf{E}(Y - m(X))^2}{k}$$

for all distributions of (X, Y) with $\mathbf{E}Y^2 < \infty$.

HINT: Use the decomposition

$$m_n(x) = \frac{1}{k} \sum_{i=1}^{k} m(X_{(i,n)}(x)) + \frac{1}{k} \sum_{i=1}^{k} (Y_{(i,n)}(x) - m(X_{(i,n)}(x))).$$

Handle the first term by Problem 6.3. Show that

$$\mathbf{E} \int \left(\frac{1}{k} \sum_{i=1}^{k} (Y_{(i,n)}(x) - m(X_{(i,n)}(x))) \right)^2 \mu(dx) \quad = \quad \frac{1}{k^2} \sum_{i=1}^{k} \mathbf{E}\{\sigma^2(X_{(i,n)}(X))\}$$

$$\to \quad \frac{\mathbf{E}(Y - m(X))^2}{k}.$$

PROBLEM 6.5. Let g_n be the k-NN classification rule for M classes:

$$g_n(x) = \arg\max_{1 \le j \le M} \sum_{i=1}^{k} I_{\{Y_{(i,n)}(x)=j\}}.$$

Show that, for $k_n \to \infty$ and $k_n/n \to 0$,

$$\lim_{n \to \infty} \mathbf{P}\{g_n(X) \ne Y\} = \mathbf{P}\{g^*(X) \ne Y\}$$

for all distributions of (X, Y), where g^* is the Bayes decision rule (Devroye, Györfi, and Lugosi (1996)).

HINT: Apply Problem 1.5 and Theorem 6.1.

PROBLEM 6.6. Let g_n be the 1-NN classification rule. Prove that

$$\lim_{n \to \infty} \mathbf{P}\{g_n(X) \ne Y\} = 1 - \sum_{j=1}^{M} \mathbf{E}\{m^{(j)}(X)^2\}$$

for all distributions of (X, Y), where $m^{(j)}(X) = \mathbf{P}\{Y = j|X\}$ (Cover and Hart (1967), Stone (1977)).

HINT:

Step (a). Show that

$$\mathbf{P}\{g_n(X) \ne Y\} \quad = \quad 1 - \sum_{j=1}^{M} \mathbf{P}\{Y = j, g_n(X) = j\}$$

$$= \quad 1 - \sum_{j=1}^{M} \mathbf{E}\{m^{(j)}(X)m^{(j)}(X_{(1,n)}(X))\}.$$

Step (b). Problem 6.3 implies that

$$\lim_{n \to \infty} \mathbf{E}\{(m^{(j)}(X) - m^{(j)}(X_{(1,n)}(X)))^2\} = 0.$$

PROBLEM 6.7. For $d \le 2$ assume that there exist $\epsilon_0 > 0$, a nonnegative function g such that for all $x \in \mathcal{R}^d$, and $0 < \epsilon \le \epsilon_0$,

$$\mu(S_{x,\epsilon}) > g(x)\epsilon^d \tag{6.3}$$

and

$$\int \frac{1}{g(x)^{2/d}} \mu(dx) < \infty.$$

Prove the rate of convergence given in Theorem 6.2.

HINT: Prove that under the conditions of the problem

$$\mathbf{E}\{\|X_{(1,n)}(X) - X\|^2\} \le \frac{\tilde{c}}{n^{2/d}}.$$

Formula (6.3) implies that for almost all $x \bmod \mu$ and $\epsilon_0 < \epsilon < L$,

$$\mu(S_{x,\epsilon}) \ge \mu(S_{x,\epsilon_0}) \ge g(x)\epsilon_0^d \ge g(x)\left(\frac{\epsilon_0}{L}\right)^d \epsilon^d,$$

hence we can assume w.l.o.g. that (6.3) holds for all $0 < \epsilon < L$. In this case, we get, for fixed $L > \epsilon > 0$,

$$
\begin{aligned}
\mathbf{P}\{\|X_{(1,n)}(X) - X\| > \epsilon\} &= \mathbf{E}\{(1 - \mu(S_{X,\epsilon}))^n\} \\
&\le \mathbf{E}\left\{e^{-n\mu(S_{X,\epsilon})}\right\} \\
&\le \mathbf{E}\left\{e^{-ng(X)\epsilon^d}\right\},
\end{aligned}
$$

therefore,

$$
\begin{aligned}
\mathbf{E}\{\|X_{(1,n)}(X) - X\|^2\} &= \int_0^{L^2} \mathbf{P}\{\|X_{(1,n)}(X) - X\| > \sqrt{\epsilon}\}\, d\epsilon \\
&\le \int_0^{L^2} \mathbf{E}\left\{e^{-ng(X)\epsilon^{d/2}}\right\} d\epsilon \\
&\le \int \int_0^{\infty} e^{-ng(x)\epsilon^{d/2}}\, d\epsilon\, \mu(dx) \\
&= \int \frac{1}{n^{2/d}g(x)^{2/d}} \int_0^{\infty} e^{-z^{d/2}}\, dz\, \mu(dx) \\
&= \frac{\tilde{c}}{n^{2/d}}.
\end{aligned}
$$

7
Splitting the Sample

In the previous chapters the parameters of the estimates with the optimal rate of convergence depend on the unknown distribution of (X, Y), especially on the smoothness of the regression function. In this and in the following chapter we present data-dependent choices of the smoothing parameters. We show that for bounded Y the estimates with parameters chosen in such an adaptive way achieve the optimal rate of convergence.

7.1 Best Random Choice of a Parameter

Let $D_n = \{(X_1, Y_1), \ldots, (X_n, Y_n)\}$ be the sample as before. Assume a finite set $\mathcal{Q}_n$ of parameters such that for every parameter $h \in \mathcal{Q}_n$ there is a regression function estimate $m_n^{(h)}(\cdot) = m_n^{(h)}(\cdot, D_n)$. Let $\hat{h} = \hat{h}(D_n) \in \mathcal{Q}_n$ be such that

$$\int |m_n^{(\hat{h})}(x) - m(x)|^2 \mu(dx) = \min_{h \in \mathcal{Q}_n} \int |m_n^{(h)}(x) - m(x)|^2 \mu(dx),$$

where $\hat{h}$ is called the best random choice of the parameter. Obviously, $\hat{h}$ is not an estimate, it depends on the unknown m and μ.

This best random choice can be approximated by splitting the data. Let $D_{n_l} = \{(X_1, Y_1), \ldots, (X_{n_l}, Y_{n_l})\}$ be the learning (training) data of size n_l and $D_n \backslash D_{n_l}$ the testing data of size n_t ($n = n_l + n_t \geq 2$). For every parameter $h \in \mathcal{Q}_n$ let $m_{n_l}^{(h)}(\cdot) = m_{n_l}^{(h)}(\cdot, D_{n_l})$ be an estimate of m depending only on the learning data D_{n_l} of the sample D_n. Use the testing data to choose a parameter $H = H(D_n) \in \mathcal{Q}_n$:

$$\frac{1}{n_t} \sum_{i=n_l+1}^{n_l+n_t} |m_{n_l}^{(H)}(X_i) - Y_i|^2 = \min_{h \in \mathcal{Q}_n} \frac{1}{n_t} \sum_{i=n_l+1}^{n_l+n_t} |m_{n_l}^{(h)}(X_i) - Y_i|^2. \quad (7.1)$$

Define the estimate by

$$m_n(x) = m_n(x, D_n) = m_{n_l}^{(H)}(x, D_{n_l}). \quad (7.2)$$

We show that H approximates the best random choice $\hat{h}$ in the sense that $\mathbf{E} \int |m_n(x) - m(x)|^2 \mu(dx)$ approximates $\mathbf{E} \int |m_{n_l}^{(\hat{h})}(x) - m(x)|^2 \mu(dx)$.

Theorem 7.1. *Let* $0 < L < \infty$*. Assume*

$$|Y| \leq L \quad a.s. \quad (7.3)$$

and

$$\max_{h \in \mathcal{Q}_n} \|m_{n_l}^{(h)}\|_\infty \leq L \quad a.s. \quad (7.4)$$

Then, for any $\delta > 0$,

$$\mathbf{E} \int |m_n(x) - m(x)|^2 \mu(dx)$$

$$\leq (1 + \delta)\mathbf{E} \int |m_{n_l}^{(\hat{h})}(x) - m(x)|^2 \mu(dx) + c\frac{1 + \log(|\mathcal{Q}_n|)}{n_t}, \quad (7.5)$$

where $\hat{h} = \hat{h}(D_{n_l})$ *and* $c = L^2(16/\delta + 35 + 19\delta)$.

The only assumption on the underlying distribution in Theorem 7.1 is the boundedness of $|Y|$ (cf. (7.3)). It can be applied to any estimate which is bounded in supremum norm by the same bound as the data (cf. (7.4)). We can always truncate an estimate at $\pm L$, which implies that (7.4) holds. If (7.3) holds, then the regression function will be bounded in absolute value by L, too, and hence the L_2 error of the truncated estimate will be less than or equal to the L_2 error of the original estimate, so the truncation has no negative consequence in view of the error of the estimate.

In the next section we will apply this theorem to partitioning, kernel, and nearest neighbor estimates. We will choose $\mathcal{Q}_n$ and n_t such that the second term on the right-hand side of (7.5) is less than the first term. This implies that the expected L_2 error of the estimate is bounded by some constant times the expected L_2 error of an estimate, which is applied to a data set of size n_l (rather than n) and where the parameter is chosen in an optimal way for this data set. Observe that this is not only true asymptotically, but true for each finite sample size.

PROOF OF THEOREM 7.1. An essential tool in the proof will be Bernstein's inequality together with majorization of a variance by some constant times the corresponding expectation. This will yield the denominator n_t in the result instead of $\sqrt{n_t}$ attainable by the use of Hoeffding's inequality (cf. Problem 7.2).

We will use the error decomposition

$$\mathbf{E}\left\{\int |m_n(x) - m(x)|^2 \mu(dx)\Big|D_{n_l}\right\}$$

$$= \mathbf{E}\left\{\int |m_{n_l}^{(H)}(x) - m(x)|^2 \mu(dx)\Big|D_{n_l}\right\}$$

$$= \mathbf{E}\left\{|m_{n_l}^{(H)}(X) - Y|^2\Big|D_{n_l}\right\} - \mathbf{E}|m(X) - Y|^2$$

$$=: \ T_{1,n} + T_{2,n},$$

where

$$T_{1,n} = \mathbf{E}\left\{|m_{n_l}^{(H)}(X) - Y|^2\Big|D_{n_l}\right\} - \mathbf{E}|m(X) - Y|^2 - T_{2,n}$$

and

$$T_{2,n} = (1 + \delta)\frac{1}{n_t}\sum_{i=n_l+1}^{n_l+n_t}\left(|m_{n_l}^{(H)}(X_i) - Y_i|^2 - |m(X_i) - Y_i|^2\right).$$

Because of (7.1),

$$T_{2,n} \le (1 + \delta)\frac{1}{n_t}\sum_{i=n_l+1}^{n_l+n_t}\left(|m_{n_l}^{(\hat{h})}(X_i) - Y_i|^2 - |m(X_i) - Y_i|^2\}\right),$$

hence,

$$\mathbf{E}\{T_{2,n}|D_{n_l}\} \le (1 + \delta)\left(\mathbf{E}\left\{|m_{n_l}^{(\hat{h})}(X) - Y|^2\Big|D_{n_l}\right\} - \mathbf{E}|m(X) - Y|^2\right)$$

$$= (1 + \delta)\int |m_{n_l}^{(\hat{h})}(x) - m(x)|^2 \mu(dx).$$

In the sequel we will show

$$\mathbf{E}\{T_{1,n}|D_{n_l}\} \le c\frac{(1 + \log|\mathcal{Q}_n|)}{n_t}, \tag{7.6}$$

which, together with the inequality above, implies the assertion. Let $s > 0$ be arbitrary. Then

$$\mathbf{P}\{T_{1,n} \ge s|D_{n_l}\}$$

$$= \mathbf{P}\left\{(1 + \delta)\left(\mathbf{E}\{|m_{n_l}^{(H)}(X) - Y|^2|D_{n_l}\} - \mathbf{E}|m(X) - Y|^2\right.\right.$$

$$\left.\left. - \frac{1}{n_t}\sum_{i=n_l+1}^{n}\{|m_{n_l}^{(H)}(X_i) - Y_i|^2 - |m(X_i) - Y_i|^2\}\right)\right.$$

$$\geq s + \delta \left(\mathbf{E}\{|m_{n_l}^{(H)}(X) - Y|^2 | D_{n_l}\} - \mathbf{E}|m(X) - Y|^2 \right) \Big| D_{n_l} \Bigg\}$$

$$\leq \mathbf{P}\Bigg\{ \exists h \in \mathcal{Q}_n : \mathbf{E}\{|m_{n_l}^{(h)}(X) - Y|^2 | D_{n_l}\} - \mathbf{E}|m(X) - Y|^2$$

$$- \frac{1}{n_t} \sum_{i=n_l+1}^{n} \left\{ |m_{n_l}^{(h)}(X_i) - Y_i|^2 - |m(X_i) - Y_i|^2 \right\}$$

$$\geq \frac{1}{1+\delta} \left(s + \delta \mathbf{E}\left\{ |m_{n_l}^{(h)}(X) - Y|^2 - |m(X) - Y|^2 \Big| D_{n_l} \right\} \right) \Big| D_{n_l} \Bigg\}$$

$$\leq |\mathcal{Q}_n| \max_{h \in \mathcal{Q}_n} \mathbf{P}\Bigg\{ \left\{ \mathbf{E}\left\{ |m_{n_l}^{(h)}(X) - Y|^2 | D_{n_l} \right\} - \mathbf{E}|m(X) - Y|^2 \right.$$

$$- \frac{1}{n_t} \sum_{i=n_l+1}^{n} \left\{ |m_{n_l}^{(h)}(X_i) - Y_i|^2 - |m(X_i) - Y_i|^2 \right\} \Bigg\}$$

$$\geq \frac{1}{1+\delta} \left(s + \delta \mathbf{E}\left\{ |m_{n_l}^{(h)}(X) - Y|^2 - |m(X) - Y|^2 \Big| D_{n_l} \right\} \right) \Big| D_{n_l} \Bigg\}.$$

Fix $h \in \mathcal{Q}_n$. Set

$$Z = |m_{n_l}^{(h)}(X) - Y|^2 - |m(X) - Y|^2$$

and

$$Z_i = |m_{n_l}^{(h)}(X_{n_l+i}) - Y_{n_l+i}|^2 - |m(X_{n_l+i}) - Y_{n_l+i}|^2 \quad (i = 1, \ldots, n - n_l).$$

Using Bernstein's inequality (see Lemma A.2) and

$$\begin{aligned}
\sigma^2 \quad &:= \quad \mathbf{Var}\{Z|D_{n_l}\} \\
&\leq \quad \mathbf{E}\{Z^2|D_{n_l}\} \\
&= \quad \mathbf{E}\left\{ \left| (m_{n_l}^{(h)}(X) - Y) - (m(X) - Y) \right|^2 \right. \\
&\qquad\qquad \times \left| (m_{n_l}^{(h)}(X) - Y) + (m(X) - Y) \right|^2 \Big| D_{n_l} \Bigg\} \\
&\leq \quad 16L^2 \int |m_{n_l}^{(h)}(x) - m(x)|^2 \mu(dx) \\
&= \quad 16L^2 \mathbf{E}\{Z|D_{n_l}\}
\end{aligned} \tag{7.7}$$

we get

$$\mathbf{P}\Bigg\{\bigg\{\mathbf{E}\{|m_{n_l}^{(h)}(X) - Y|^2|D_{n_l}\} - \mathbf{E}|m(X) - Y|^2$$

$$-\frac{1}{n_t}\sum_{i=n_l+1}^{n}\{|m_{n_l}^{(h)}(X_i) - Y_i|^2 - |m(X_i) - Y_i|^2\}\bigg\}$$

$$\geq \frac{1}{1+\delta}\left(s + \delta\mathbf{E}\left\{|m_{n_l}^{(h)}(X) - Y|^2 - |m(X) - Y|^2\bigg|D_{n_l}\right\}\right)\bigg|D_{n_l}\Bigg\}$$

$$= \mathbf{P}\left\{\mathbf{E}\{Z|D_{n_l}\} - \frac{1}{n_t}\sum_{i=1}^{n_t}Z_i \geq \frac{1}{1+\delta}\left(s + \delta\cdot\mathbf{E}\{Z|D_{n_l}\}\right)\bigg|D_{n_l}\right\}$$

$$\leq \mathbf{P}\left\{\mathbf{E}\{Z|D_{n_l}\} - \frac{1}{n_t}\sum_{i=1}^{n_t}Z_i \geq \frac{1}{1+\delta}\left(s + \delta\cdot\frac{\sigma^2}{16L^2}\right)\bigg|D_{n_l}\right\}$$

$$\leq \exp\left(-n_t\frac{\frac{1}{(1+\delta)^2}\left(s + \delta\frac{\sigma^2}{16L^2}\right)^2}{2\sigma^2 + \frac{2}{3}\frac{8L^2}{1+\delta}\left(s + \delta\frac{\sigma^2}{16L^2}\right)}\right).$$

Here we don't need the factor 2 before the exponential term because we don't have absolute value inside the probability (cf. proof of Lemma A.2). Next we observe

$$\frac{\frac{1}{(1+\delta)^2}\left(s + \delta\frac{\sigma^2}{16L^2}\right)^2}{2\sigma^2 + \frac{2}{3}\frac{8L^2}{1+\delta}\left(s + \delta\frac{\sigma^2}{16L^2}\right)}$$

$$\geq \frac{s^2 + 2s\delta\frac{\sigma^2}{16L^2}}{\frac{16}{3}L^2(1+\delta)s + \sigma^2\left(2(1+\delta)^2 + \frac{1}{3}\delta(1+\delta)\right)}.$$

An easy but tedious computation (cf. Problem 7.1) shows

$$\frac{s^2 + 2s\delta\frac{\sigma^2}{16L^2}}{\frac{16}{3}L^2(1+\delta)s + \sigma^2\left(2(1+\delta)^2 + \frac{1}{3}\delta(1+\delta)\right)} \geq \frac{s}{c}, \qquad (7.8)$$

where $c = L^2(16/\delta + 35 + 19\delta)$. Using this we get that

$$\mathbf{P}\{T_{1,n} \geq s|D_{n_l}\} \leq |\mathcal{Q}_n|\exp\left(-n_t\frac{s}{c}\right).$$

It follows, for arbitrary $u > 0$,

$$\mathbf{E}\{T_{1,n}|D_{n_l}\} \leq u + \int_u^\infty \mathbf{P}\{T_{1,n} > s|D_{n_l}\}\,ds$$

$$\leq u + \frac{|\mathcal{Q}_n|\,c}{n_t}\exp\left(-\frac{n_t\,u}{c}\right).$$

Setting $u = \frac{c\log(|\mathcal{Q}_n|)}{n_t}$, this implies (7.6), which in turn implies the assertion.

$\square$

7.2 Partitioning, Kernel, and Nearest Neighbor Estimates

In Theorems 4.3, 5.2, and 6.2 we showed that partitioning, kernel, and nearest neighbor estimates are able to achieve the minimax lower bound for the
estimation of (p, C)-smooth regression functions if $p = 1$ and if the parameters are chosen depending on C (the Lipschitz constant of the regression
function). Obviously, the value of C will be unknown in an application,
therefore, one cannot use estimates where the parameters depend on C
in applications. In the sequel we show that, in the case of bounded data,
one can also derive similar bounds for estimates where the parameters are
chosen by splitting the sample.

We start with the kernel estimate. Let $m_n^{(h)}$ be the kernel estimate with
naive kernel and bandwidth h. We choose the finite set $\mathcal{Q}_n$ of bandwidths
such that we can approach the choice of the bandwidth in Theorem 5.2 up
to some factor less than some constant, e.g., up to factor 2. This can be
done, e.g., by setting

$$\mathcal{Q}_n = \left\{ 2^k \ : \ k \in \{-n, -(n-1), \ldots, 0, \ldots, n-1, n\} \right\}.$$

Theorems 7.1 and 5.2 imply

Corollary 7.1. *Assume that X is bounded,*

$$|m(x) - m(z)| \leq C \cdot \|x - z\| \quad (x, z \in \mathcal{R}^d)$$

and $|Y| \leq L$ a.s. Set

$$n_l = \left\lceil \frac{n}{2} \right\rceil \quad \text{and } n_t = n - n_l.$$

*Let m_n be the kernel estimate with naive kernel and bandwidth $h \in$
$\mathcal{Q}_n$ chosen as in Theorem 7.1, where $\mathcal{Q}_n$ is defined as above. Then
$(\log n)^{(d+2)/(2d)} n^{-1/2} \leq C$ implies, for $n \geq 2$,*

$$\mathbf{E} \int |m_n(x) - m(x)|^2 \mu(dx) \leq c_1 C^{2d/(d+2)} n^{-2/(d+2)}$$

*for some constant c_1 which depends only on L, d, and the diameter of the
support of X.*

PROOF. Without loss of generality we can assume $C \leq n^{1/d}$ (otherwise,
the assertion is trivial because of boundedness of Y). Theorems 7.1 and 5.2
imply

$$\mathbf{E} \int |m_n(x) - m(x)|^2 \mu(dx)$$

$$\leq 2 \min_{h \in \mathcal{Q}_n} \mathbf{E} \int |m_{n_l}^{(h)}(x) - m(x)|^2 \mu(dx) + c \cdot \frac{1 + \log(|\mathcal{Q}_n|)}{n_t}$$

$$\leq 2 \min_{h \in \mathcal{Q}_n} \left(\hat{c} \cdot \frac{2L^2}{n_l h^d} + C^2 h^2 \right) + c \cdot \frac{1 + \log(2n + 1)}{n_t}$$

$$\leq 2 \left(\hat{c} \cdot \frac{2L^2}{n_l h_n^d} + C^2 h_n^2 \right) + c \cdot \frac{1 + \log(2n + 1)}{n_t},$$

where $h_n \in \mathcal{Q}_n$ is chosen such that

$$C^{-2/(d+2)} n^{-1/(d+2)} \leq h_n \leq 2C^{-2/(d+2)} n^{-1/(d+2)}.$$

The choices of h_n, n_l, and n_t together with $C \geq (\log n)^{(d+2)/(2d)} n^{-1/2}$ imply

$$\mathbf{E} \int |m_n(x) - m(x)|^2 \mu(dx)$$

$$\leq \tilde{c} \cdot C^{2d/(d+2)} n^{-2/(d+2)} + 4c \cdot \frac{1 + \log(2n + 1)}{n}$$

$$\leq c_1 \cdot C^{2d/(d+2)} n^{-2/(d+2)}.$$

$\square$

Similarly, one can show the following result concerning the partitioning estimate:

Corollary 7.2. *Assume that X is bounded,*

$$|m(x) - m(z)| \leq C \cdot \|x - z\| \quad (x, z \in \mathcal{R}^d)$$

and $|Y| \leq L$ a.s. Set

$$n_l = \left\lceil \frac{n}{2} \right\rceil \quad \text{and } n_t = n - n_l.$$

Let m_n be the partitioning estimate with cubic partition and grid size $h \in \mathcal{Q}_n$ chosen as in Theorem 7.1, where $\mathcal{Q}_n$ is defined as above. Then $(\log n)^{(d+2)/(2d)} n^{-1/2} \leq C$ implies, for $n \geq 2$,

$$\mathbf{E} \int |m_n(x) - m(x)|^2 \mu(dx) \leq c_2 C^{2d/(d+2)} n^{-2/(d+2)}$$

for some constant c_2 which depends only on L, d, and the diameter of the support of X.

PROOF. See Problem 7.6 $\square$

Finally we consider the k-nearest neighbor estimates. Here we can set $\mathcal{Q}_n = \{1, \ldots, n\}$, so the optimal value from Theorem 6.2 is contained in $\mathcal{Q}_n$. Immediately from Theorems 7.1 and 6.2 we can conclude

Corollary 7.3. *Assume that X is bounded,*

$$|m(x) - m(z)| \leq C \cdot \|x - z\| \quad (x, z \in \mathcal{R}^d)$$

and $|Y| \leq L$ a.s. Set

$$n_l = \left\lceil \frac{n}{2} \right\rceil \quad \text{and } n_t = n - n_l.$$

Let m_n be the k-nearest neighbor estimate with $k \in \mathcal{Q}_n = \{1, \dots, n_l\}$ chosen as in Theorem 7.1. Then $(\log n)^{(d+2)/(2d)} n^{-1/2} \leq C$ together with $d \geq 3$ implies, for $n \geq 2$,

$$\mathbf{E} \int |m_n(x) - m(x)|^2 \mu(dx) \leq c_3 C^{2d/(d+2)} n^{-2/(d+2)}$$

for some constant c_3 which depends only on L, d, and the diameter of the support of X.

Here we use for each component of X the same smoothing parameter. But the results can be extended to optimal scaling, where one uses for each component a different smoothing parameter. Here Problems 4.7 and 5.7 characterize the rate of convergence, and splitting of the data can be used to approximate the optimal scaling parameters, which depend on the underlying distribution (cf. Problem 7.7).

In Corollaries 7.1–7.3 the expected L_2 error of the estimates is bounded from above up to a constant by the corresponding minimax lower bound for (p, C)-smooth regression functions, if $p = 1$. We would like to mention two important aspects of these results: First, the definition of the estimates does not depend on C, therefore they adapt automatically to the unknown smoothness of the regression function measured by the Lipschitz constant C. Second, the bounds are valid for finite sample size. So we are able to approach the minimax lower bound not only asymptotically but even for finite sample sizes (observe that in the proof of Theorem 3.2 we have in fact shown that the lower bound is valid for finite sample size).

Approaching the minimax lower bound for fixed sample size by some constant does not imply that one can get asymptotically the minimax rate of convergence with the optimal constant in front of $n^{-2p/(2p+d)}$. But as we show in the next theorem, this goal can also be reached by splitting the sample:

Theorem 7.2. *Under the conditions of Theorem 7.1 assume that*

$$\log |\mathcal{Q}_n| \leq \tilde{c} \log n$$

and

$$\mathbf{E} \left\{ \min_{h \in \mathcal{Q}_n} \int |m_n^{(h)}(x) - m(x)|^2 \mu(dx) \right\} \leq C_{opt}(1 + o(1)) n^{-\gamma}$$

for some $0 < \gamma < 1$. Choose $\gamma < \gamma' < 1$ and set

$$n_t = \left\lceil n^{\gamma'} \right\rceil \quad and \quad n_l = n - n_t.$$

Then

$$\mathbf{E} \int |m_n(x) - m(x)|^2 \mu(dx) \leq C_{opt}(1 + o(1)) n^{-\gamma}.$$

PROOF. Theorem 7.1 implies that

$$
\mathbf{E} \int |m_n(x) - m(x)|^2 \mu(dx)
$$

$$
\leq \quad (1+\delta)\mathbf{E}\left\{ \min_{h \in \mathcal{Q}_n} \int |m_{n_l}^{(h)}(x) - m(x)|^2 \mu(dx) \right\} + c\frac{1 + \log(|\mathcal{Q}_n|)}{n_t}
$$

$$
\leq \quad (1+\delta)C_{opt}(1+o(1))n_l^{-\gamma} + c\frac{1 + \tilde{c}\log n}{n_t}
$$

$$
\leq \quad (1+\delta)C_{opt}(1+o(1))(1-o(1))^{-\gamma}n^{-\gamma} + c\frac{1 + \tilde{c}\log n}{n^{\gamma'}}
$$

$$
(\text{since } n_l = n - \lceil n^{\gamma'}\rceil \text{ and } n - n^{\gamma'} = (1 - n^{-(1-\gamma')}) \cdot n \,)
$$

$$
= \quad (1+\delta)C_{opt}(1+o(1))n^{-\gamma}.
$$

Since $\delta > 0$ is arbitrary we get that

$$
\mathbf{E} \int |m_n(x) - m(x)|^2 \mu(dx) \leq C_{opt}(1+o(1))n^{-\gamma}.
$$

$\square$

7.3   Bibliographic Notes

The bound (7.7) on the variance can be improved (see Barron (1991) or Problem 7.3).

In the proof of Theorem 7.1 we used the union bound together with Bernstein's inequality to bound the deviation between an expectation and $(1 + \delta)$ times the corresponding sample mean. By using instead Jensen's inequality, together with the bound on the exponential moment derived in the proof of Bernstein's inequality, one can improve the constants in Theorem 7.1 (cf. Hamers and Kohler (2001)).

In the context of pattern recognition and density estimation splitting the sample was investigated in Devroye (1988) and Devroye and Lugosi (2001), respectively.

In this chapter we tried to choose one estimate from a given finite collection of the estimates that is at least as good as the best of the original ones, plus a small residual. It might not always be optimal to choose one estimate from the original set of estimates. Instead, it might be useful to construct a new estimator as a function of original estimates, such as a convex combination (see, e.g., Niemirovsky (2000), and the references therein).

Problems and Exercises

PROBLEM 7.1. Prove (7.8).

HINT: Set

$$a = s^2, \ b = 2s\delta/(16L^2), \ c = 16L^2(1+\delta)s/3, \ \text{and} \ d = 2(1+\delta)^2 + \delta(1+\delta)/3.$$

Then the left-hand side of (7.8) is equal to $f(\sigma^2)$ where

$$f(u) = \frac{a + b \cdot u}{c + d \cdot u} \quad (u > 0).$$

Compute the derivative f' of f and show

$$f'(u) \neq 0 \ \text{for all} \ u > 0.$$

Use this to determine

$$\min_{u>0} f(u)$$

by considering

$$f(0) \ \text{and} \ \lim_{u \to \infty} f(u).$$

PROBLEM 7.2. Prove a weaker version of Theorem 7.1: under the conditions of Theorem 7.1,

$$\mathbf{E} \int |m_n(x) - m(x)|^2 \mu(dx)$$

$$\leq \ \mathbf{E} \int |m_{n_l}^{(\hat{h})}(x) - m(x)|^2 \mu(dx) + 8\sqrt{2}L^2 \sqrt{\frac{\log(2|\mathcal{Q}_n|)}{n_t}}.$$

HINT:

$$\int |m_n(x) - m(x)|^2 \mu(dx) - \int |m_{n_l}^{(\hat{h})}(x) - m(x)|^2 \mu(dx)$$

$$= \ \mathbf{E}\left\{ (m_{n_l}^{(H)}(X) - Y)^2 \Big| D_n \right\} - \mathbf{E}\left\{ (m_{n_l}^{(\hat{h})}(X) - Y)^2 \Big| D_n \right\}$$

$$= \ \mathbf{E}\left\{ (m_{n_l}^{(H)}(X) - Y)^2 \Big| D_n \right\} - \frac{1}{n_t} \sum_{i=n_l+1}^{n_l+n_t} (m_{n_l}^{(H)}(X_i) - Y_i)^2$$

$$+ \frac{1}{n_t} \sum_{i=n_l+1}^{n_l+n_t} (m_{n_l}^{(H)}(X_i) - Y_i)^2 - \frac{1}{n_t} \sum_{i=n_l+1}^{n_l+n_t} (m_{n_l}^{(\hat{h})}(X_i) - Y_i)^2$$

$$+ \frac{1}{n_t} \sum_{i=n_l+1}^{n_l+n_t} (m_{n_l}^{(\hat{h})}(X_i) - Y_i)^2 - \mathbf{E}\{ (m_{n_l}^{(\hat{h})}(X) - Y)^2 \big| D_n \}$$

$$\leq \ \mathbf{E}\left\{ (m_{n_l}^{(H)}(X) - Y)^2 \Big| D_n \right\} - \frac{1}{n_t} \sum_{i=n_l+1}^{n_l+n_t} (m_{n_l}^{(H)}(X_i) - Y_i)^2$$

$$+\frac{1}{n_t}\sum_{i=n_l+1}^{n_l+n_t}(m_{n_l}^{(\hat{h})}(X_i)-Y_i)^2-\mathbf{E}\left\{(m_{n_l}^{(\hat{h})}(X)-Y)^2\Big|D_n\right\}$$

$$\leq\quad 2\max_{h\in\mathcal{Q}_n}\left|\frac{1}{n_t}\sum_{i=n_l+1}^{n_l+n_t}(m_{n_l}^{(h)}(X_i)-Y_i)^2-\mathbf{E}\left\{(m_{n_l}^{(h)}(X)-Y)^2\Big|D_n\right\}\right|$$

$$=\quad 2\max_{h\in\mathcal{Q}_n}\left|\frac{1}{n_t}\sum_{i=n_l+1}^{n_l+n_t}(m_{n_l}^{(h)}(X_i)-Y_i)^2-\mathbf{E}\left\{(m_{n_l}^{(h)}(X)-Y)^2\Big|D_{n_l}\right\}\right|.$$

Use Hoeffding's inequality (cf. Lemma A.3) to conclude

$$\mathbf{P}\left\{\int|m_n(x)-m(x)|^2\mu(dx)-\int|m_{n_l}^{(\hat{h})}(x)-m(x)|^2\mu(dx)>\epsilon|D_{n_l}\right\}$$

$$\leq\quad 2|\mathcal{Q}_n|e^{-n_t\frac{\epsilon^2}{32L^4}}.$$

Compare also Problem 8.2.

PROBLEM 7.3. See Barron (1991).

(a) Show that for any random variable V with values in some interval of length B one has

$$\mathbf{Var}\{V\}\leq\frac{B^2}{4}.$$

(b) Show that the inequality (7.7) can be improved as follows: Assume $|Y|\leq L$ *a.s.* and let f be a function $f:\mathcal{R}^d\to[-L,L]$. Set

$$Z=|f(X)-Y|^2-|m(X)-Y|^2.$$

Then

$$\sigma^2=\mathbf{Var}\{Z\}\leq 8L^2\mathbf{E}\{Z\}.$$

HINT: Use

$$Z=-2(Y-m(X))\cdot(f(X)-m(X))+(f(X)-m(X))^2$$

and

$$\mathbf{E}\left\{((Y-m(X))\cdot(f(X)-m(X)))^2\right\}$$
$$=\quad\mathbf{E}\left\{(f(X)-m(X))^2\mathbf{E}\left\{(Y-m(X))^2|X\right\}\right\}$$
$$\leq\quad L^2\mathbf{E}\left\{|f(X)-m(X)|^2\right\},$$

where the last inequality follows from (a).

PROBLEM 7.4. Use Problem 7.3 to improve the constant c in Theorem 7.1.

PROBLEM 7.5. Show that if the assumptions of Theorem 7.2 are satisfied and, in addition,

$$\mathbf{E}\left\{\min_{h\in\mathcal{Q}_n}\int|m_n^{(h)}(x)-m(x)|^2\mu(dx)\right\}\geq C_{opt}(1+o(1))n^{-\gamma},$$

then

$$\lim_{n \to \infty} \frac{\mathbf{E} \int |m_n(x) - m(x)|^2 \mu(dx)}{\mathbf{E} \left\{ \min_{h \in \mathcal{Q}_n} \int |m_{n_l}^{(h)}(x) - m(x)|^2 \mu(dx) \right\}} =: 1.$$

PROBLEM 7.6. Prove Corollary 7.2.

HINT: Proceed as in the proof of Corollary 7.1, but use the bounds from Theorem 4.3 instead of those from Theorem 5.2.

PROBLEM 7.7. (a) Use splitting the data to choose the side lengths of rectangular partitioning such that the resulting estimate approaches the rate of convergence in Problem 4.7.

(b) Use splitting the data to choose the scaling for product kernel estimates such that the resulting estimate approaches the rate of convergence in Problem 5.7.

(c) Our results in Chapter 6 concerning nearest neighbor estimates used the Euclidean distance. Obviously, all the results of this chapter hold with scaling, i.e., for norms defined by

$$\|x\|^2 = \sum_{j=1}^{d} c_j |x^{(j)}|^2,$$

where $c_1, \ldots, c_d > 0$ are the scaling factors. Use splitting the data to choose the scaling factors together with k for k-NN estimates based on such norms.

8
Cross-Validation

8.1 Best Deterministic Choice of the Parameter

Let $D_n = \{(X_1, Y_1), \ldots, (X_n, Y_n)\}$ be the sample as before. Assume a finite set $\mathcal{Q}_n$ of parameters such that for every parameter $h \in \mathcal{Q}_n$ there is a regression function estimate $m_n^{(h)}(\cdot) = m_n^{(h)}(\cdot, D_n)$. Let $\bar{h}_n \in \mathcal{Q}_n$ be such that

$$\mathbf{E}\left\{\int |m_n^{(\bar{h}_n)}(x) - m(x)|^2 \mu(dx)\right\} = \min_{h \in \mathcal{Q}_n} \mathbf{E}\left\{\int |m_n^{(h)}(x) - m(x)|^2 \mu(dx)\right\},$$

where $\bar{h}_n$ is called the best deterministic choice of the parameter. Obviously, $\bar{h}_n$ is not an estimate, it depends on the unknown distribution of (X, Y), in particular on m and μ.

This best deterministic choice can be approximated by cross-validation. For every parameter $h \in \mathcal{Q}_n$ let $m_n^{(h)}$ and $m_{n,i}^{(h)}$ be the regression estimates from D_n and $D_n \setminus (X_i, Y_i)$, respectively, where

$$D_n \setminus (X_i, Y_i) = \{(X_1, Y_1), \ldots, (X_{i-1}, Y_{i-1}), (X_{i+1}, Y_{i+1}), \ldots, (X_n, Y_n)\}.$$

The cross-validation selection of h is

$$H = H_n = \arg\min_{h \in \mathcal{Q}_n} \frac{1}{n} \sum_{i=1}^{n} (m_{n,i}^{(h)}(X_i) - Y_i)^2.$$

Define the cross-validation regression estimate by

$$m_n(x) = m_n^{(H)}(x). \tag{8.1}$$

Throughout this chapter we use the notation

$$\Delta_n^{(h)} = \mathbf{E} \int |m_n^{(h)}(x) - m(x)|^2 \mu(dx).$$

In the sequel we show that H_n approximates the best deterministic choice $\bar{h} = \bar{h}_{n-1}$ for sample size $n-1$ in the sense that $\mathbf{E}\left\{\Delta_{n-1}^{(H_n)}\right\}$ approximates $\Delta_{n-1}^{(\bar{h}_{n-1})}$ with an asymptotically small correction term.

8.2 Partitioning and Kernel Estimates

Theorem 8.1 yields relations between $\mathbf{E}\left\{\Delta_{n-1}^{(H_n)}\right\}$ and $\Delta_{n-1}^{(\bar{h}_{n-1})}$.

Theorem 8.1. *Let* $|Y| \leq L < \infty$. *Choose* $m_n^{(h)}$ *of the form*

$$m_n^{(h)}(x) = \frac{\sum_{j=1}^{n} Y_j K_h(x, X_j)}{\sum_{j=1}^{n} K_h(x, X_j)}$$

where the binary valued function $K_h : \mathcal{R}^d \times \mathcal{R}^d \rightarrow \{0, 1\}$ *with* $K_h(x, x) = 1$ *fulfills the covering assumption (C) that a constant* $\rho > 0$ *depending only on* $\{K_h; h \in \cup_n \mathcal{Q}_n\}$ *exists with*

$$\int \frac{K_h(x, z)}{\int K_h(x, t)\mu(dt)} \mu(dx) \leq \rho$$

for all $z \in \mathcal{R}^d$, *all* $h \in \cup_n \mathcal{Q}_n$, *and all probability measures* μ.
(a)

$$\mathbf{E}\left\{\Delta_{n-1}^{(H_n)}\right\} \leq \Delta_{n-1}^{(\bar{h}_{n-1})} + c\sqrt{\frac{\log(|\mathcal{Q}_n|)}{n}}$$

for some constant c *depending only on* L *and* ρ.
(b) *For any* $\delta > 0$

$$\mathbf{E}\left\{\Delta_{n-1}^{(H_n)}\right\} \leq (1 + \delta)\,\Delta_{n-1}^{(\bar{h}_{n-1})} + c\frac{|\mathcal{Q}_n|}{n}\log n,$$

where c *depends only on* $\delta, L,$ *and* ρ.

The proof, which is difficult and long, will be given in Section 8.3. We recommend skipping it during the first reading.

The covering assumption (C) in Theorem 8.1 is fulfilled for kernel estimates using naive kernel and partitioning estimates (see below). Before we consider the application of Theorem 8.1 to these estimates in detail, we give some comments concerning convergence order.

Neglecting $\log n$, the correction terms in parts (a) and (b) are both of the order $n^{-1/2}$ if $|\mathcal{Q}_n| = O(n^{1/2})$. One is interested that the correction term is less than $\triangle_{n-1}^{(\overline{h}_{n-1})}$. For Lipschitz-continuous m one has

$$\triangle_{n-1}^{(\overline{h}_{n-1})} = O\left(n^{-2/(d+2)}\right)$$

in naive kernel estimation and cubic partitioning estimation according to Theorems 5.2 and 4.3, respectively, which is optimal according to Theorem 3.2. In this case, for $d \geq 3$ and $\log(|\mathcal{Q}_n|) = O(\log n)$, i.e., $|\mathcal{Q}_n| \leq n^s$ for some $s > 0$, or for $\log(|\mathcal{Q}_n|) = O(n^t)$ for some $0 < t < (d-2)/(d+2)$, part (a) yields the desired result, and for $d \geq 1$ and $\log(|\mathcal{Q}_n|) \leq c^* \log n$ with $c^* < d/(d+2)$, part (b) yields the desired result. The latter also holds if

$$\triangle_{n-1}^{(\overline{h}_{n-1})} = O(n^{-\gamma})$$

with $\gamma < 1$ near to 1, if c^* is chosen sufficiently small.

Now we give more detailed applications of Theorem 8.1 to kernel and partitioning estimates. Let $\mathcal{P}_h$ be a partition of $\mathcal{R}^d$, and denote by $m_n^{(h)}$ the partitioning estimate for this partition and sample size n. Because of the proof of Theorem 4.2 the covering assumption (C) is satisfied with $\rho = 1$:

$$\int \frac{I_{\{z \in A_n(x)\}}}{\mu(A_n(x))} \mu(dx) \;=\; \int \frac{I_{\{z \in A_n(x)\}}}{\mu(A_n(z))} \mu(dx)$$

$$=\; \frac{\mu(A_n(z))}{\mu(A_n(z))}$$

$$\leq\; 1.$$

Or, let

$$m_n^{(h)}(x) = \frac{\sum_{j=1}^n Y_j K\left(\frac{x-X_j}{h}\right)}{\sum_{j=1}^n K\left(\frac{x-X_j}{h}\right)}.$$

be the kernel estimate with bandwidth h and naive kernel K. Then according to Lemma 23.6 the covering assumption is satisfied with a ρ depending on d only.

For these estimates Theorem 8.1, together with Theorems 5.2 and 4.3, implies

Corollary 8.1. *Assume that X is bounded,*

$$|m(x) - m(z)| \leq C \cdot \|x - z\| \quad (x, z \in \mathcal{R}^d)$$

and $|Y| \leq L$ a.s.

Let m_n be the partitioning estimate with cubic partitioning and grid size $h \in \mathcal{Q}_n$ chosen as in Theorem 8.1, or let m_n be the kernel estimate with naive kernel and bandwidth $h \in \mathcal{Q}_n$ chosen as in Theorem 8.1. Let $d \geq 3$,

$$\mathcal{Q}_n = \left\{2^k \;:\; k \in \{-n, -(n-1), \ldots, 0, \ldots, n-1, n\}\right\}$$

and

$$(\log n)^{(d+2)/(4d)} n^{-(d-2)/(4d)} \le C,$$

or, let $d \ge 1$,

$$\mathcal{Q}_n = \left\{ \lceil 2^{-n^{1/4}+k} \rceil \; : \; k \in \{1, 2, \ldots, 2\lceil n^{1/4} \rceil\} \right\}$$

and

$$(\log n)^{(d+2)/(2d)} n^{-(3d-2)/(8d)} \le C.$$

Then, in each of the four cases,

$$\mathbf{E}\left\{ \Delta_{n-1}^{(H_n)} \right\} \le c_1 C^{2d/(d+2)} n^{-2/(d+2)}$$

for some constant c_1 which depends only on L, d, and the diameter of the support of X.

PROOF. See Problem 8.1 □

As in the previous chapter the results can be extended to optimal scaling and adapting to the optimal constant in front of $n^{-2/(d+2)}$. We leave the details to the reader (cf. Problems 8.4 and 8.5).

8.3   Proof of Theorem 8.1

PROOF OF (a). For $\epsilon > 0$, we show

$$\mathbf{P}\{\Delta_{n-1}^{(H_n)} - \Delta_{n-1}^{(\bar{h}_{n-1})} > \epsilon\} \le 2|\mathcal{Q}_n|e^{-n\epsilon^2/(128L^4)} + 2|\mathcal{Q}_n|e^{-n\epsilon^2/(128L^4(1+4\rho)^2)},$$

from which (a) follows (cf. Problem 8.2). Observe that, for each $h > 0$,

$$\begin{aligned}
\Delta_{n-1}^{(h)} &= \mathbf{E}\left\{ \frac{1}{n} \sum_{i=1}^{n} \left(m_{n,i}^{(h)}(X_i) - m(X_i) \right)^2 \right\} \\
&= \mathbf{E}\left\{ \frac{1}{n} \sum_{i=1}^{n} \left((m_{n,i}^{(h)}(X_i) - Y_i)^2 - (m(X_i) - Y_i)^2 \right) \right\}. \quad (8.2)
\end{aligned}$$

Therefore, because of the definition of $H = H_n$ and $\bar{h} = \bar{h}_{n-1}$,

$$\begin{aligned}
&\Delta_{n-1}^{(H)} - \Delta_{n-1}^{(\bar{h})} \\
&= \Delta_{n-1}^{(H)} - \frac{1}{n} \sum_{i=1}^{n} \left((m_{n,i}^{(H)}(X_i) - Y_i)^2 - (m(X_i) - Y_i)^2 \right) \\
&\quad + \frac{1}{n} \sum_{i=1}^{n} \left((m_{n,i}^{(H)}(X_i) - Y_i)^2 - (m(X_i) - Y_i)^2 \right)
\end{aligned}$$

$$- \frac{1}{n} \sum_{i=1}^{n} \left((m_{n,i}^{(\bar{h})}(X_i) - Y_i)^2 - (m(X_i) - Y_i)^2 \right)$$

$$+ \frac{1}{n} \sum_{i=1}^{n} \left((m_{n,i}^{(\bar{h})}(X_i) - Y_i)^2 - (m(X_i) - Y_i)^2 \right) - \Delta_{n-1}^{(\bar{h})}$$

$$\leq \quad \Delta_{n-1}^{(H)} - \frac{1}{n} \sum_{i=1}^{n} \left((m_{n,i}^{(H)}(X_i) - Y_i)^2 - (m(X_i) - Y_i)^2 \right)$$

$$+ \frac{1}{n} \sum_{i=1}^{n} \left((m_{n,i}^{(\bar{h})}(X_i) - Y_i)^2 - (m(X_i) - Y_i)^2 \right) - \Delta_{n-1}^{(\bar{h})}$$

$$\leq \quad 2 \max_{h \in \mathcal{Q}_n} \left| \Delta_{n-1}^{(h)} - \frac{1}{n} \sum_{i=1}^{n} \left((m_{n,i}^{(h)}(X_i) - Y_i)^2 - (m(X_i) - Y_i)^2 \right) \right|$$

$$= \quad 2 \max_{h \in \mathcal{Q}_n} \left| \frac{1}{n} \sum_{i=1}^{n} \left((m_{n,i}^{(h)}(X_i) - Y_i)^2 - (m(X_i) - Y_i)^2 \right. \right.$$

$$\left. \left. - \mathbf{E}\{ (m_{n,i}^{(h)}(X_i) - Y_i)^2 - (m(X_i) - Y_i)^2 \} \right) \right|.$$

Consequently,

$$\mathbf{P}\{ \Delta_{n-1}^{(H)} - \Delta_{n-1}^{(\bar{h})} > \epsilon \}$$

$$\leq \quad \mathbf{P} \left\{ \max_{h \in \mathcal{Q}_n} \left| \frac{1}{n} \sum_{i=1}^{n} \left((m_{n,i}^{(h)}(X_i) - Y_i)^2 - (m(X_i) - Y_i)^2 \right. \right. \right.$$

$$\left. \left. \left. - \mathbf{E}\{ (m_{n,i}^{(h)}(X_i) - Y_i)^2 - (m(X_i) - Y_i)^2 \} \right) \right| > \epsilon/2 \right\}$$

$$\leq \quad \sum_{h \in \mathcal{Q}_n} \mathbf{P} \left\{ \left| \frac{1}{n} \sum_{i=1}^{n} \left((m_{n,i}^{(h)}(X_i) - Y_i)^2 - (m(X_i) - Y_i)^2 \right. \right. \right.$$

$$\left. \left. \left. - \mathbf{E}\{ (m_{n,i}^{(h)}(X_i) - Y_i)^2 - (m(X_i) - Y_i)^2 \} \right) \right| > \epsilon/2 \right\}$$

$$\leq \quad \sum_{h \in \mathcal{Q}_n} \mathbf{P} \left\{ \left| \frac{1}{n} \sum_{i=1}^{n} \left((m(X_i) - Y_i)^2 - \mathbf{E}\{ (m(X_i) - Y_i)^2 \} \right) \right| > \epsilon/4 \right\}$$

$$+ \quad \sum_{h \in \mathcal{Q}_n} \mathbf{P} \left\{ \left| \frac{1}{n} \sum_{i=1}^{n} \left((m_{n,i}^{(h)}(X_i) - Y_i)^2 - \mathbf{E}\{ (m_{n,i}^{(h)}(X_i) - Y_i)^2 \} \right) \right| > \epsilon/4 \right\}.$$

By Hoeffding's inequality (Lemma A.3) the first sum on the right-hand side is upper bounded by

$$2|\mathcal{Q}_n|e^{-n\epsilon^2/(128L^4)}.$$

For the term indexed by h of the second sum we use McDiarmid's inequality (Theorem A.2). Fix $1 \leq l \leq n$. Let

$$
\begin{aligned}
D'_n &= \{(X'_1, Y'_1), \ldots, (X'_l, Y'_l), \ldots, (X'_n, Y'_n)\} \\
&= \{(X_1, Y_1), \ldots, (X'_l, Y'_l), \ldots, (X_n, Y_n)\},
\end{aligned}
$$

and define $m'_{n,j}$ as $m_{n,j}$ with D_n replaced by D'_n ($j = 1, \ldots, n$). We will show that

$$\left| \sum_{j=1}^{n} (m_{n,j}^{(h)}(X_j) - Y_j)^2 - \sum_{j=1}^{n} (m_{n,j}^{(h)\prime}(X'_j) - Y'_j)^2 \right| \leq 4L^2(1 + 4\rho). \tag{8.3}$$

Then McDiarmid's inequality yields the bound for the second sum

$$2|\mathcal{Q}_n|e^{-n\epsilon^2/(128L^4(1+4\rho)^2)}.$$

In order to prove (8.3), because of the symmetry, we can assume that $l = 1$. We obtain

$$\left| \sum_{i=1}^{n} (m_{n,i}^{(h)}(X_i) - Y_i)^2 - \sum_{i=1}^{n} (m_{n,i}^{(h)\prime}(X'_i) - Y'_i)^2 \right|$$

$$\leq \quad 4L^2 + 4L \sum_{i=2}^{n} |m_{n,i}^{(h)}(X_i) - m_{n,i}^{(h)\prime}(X_i)|. \tag{8.4}$$

In view of a bound for

$$\sum_{i=2}^{n} |m_{n,i}^{(h)}(X_i) - m_{n,i}^{(h)\prime}(X_i)|$$

we write

$$m_{n,i}^{(h)}(X_i) = \frac{\sum_{j\in\{2,\ldots,n\}\setminus\{i\}} K_h(X_i, X_j)Y_j + K_h(X_i, X_1)Y_1}{\sum_{j\in\{2,\ldots,n\}\setminus\{i\}} K_h(X_i, X_j) + K_h(X_i, X_1)}$$

and

$$m_{n,i}^{(h)\prime}(X_i) = \frac{\sum_{j\in\{2,\ldots,n\}\setminus\{i\}} K_h(X_i, X_j)Y_j + K_h(X_i, X'_1)Y'_1}{\sum_{j\in\{2,\ldots,n\}\setminus\{i\}} K_h(X_i, X_j) + K_h(X_i, X'_1)}$$

and distinguish the four cases:

$$(1) \quad K_h(X_i, X_1) = K_h(X_i, X'_1) = 0;$$
$$(2) \quad K_h(X_i, X_1) = 1, \ K_h(X_i, X'_1) = 0;$$
$$(3) \quad K_h(X_i, X_1) = 0, \ K_h(X_i, X'_1) = 1;$$
$$(4) \quad K_h(X_i, X_1) = K_h(X_i, X'_1) = 1.$$

In the first case,

$$|m_{n,i}^{(h)}(X_i) - m_{n,i}^{(h)\prime}(X_i)| = 0.$$

In the second case,

$$|m_{n,i}^{(h)}(X_i) - m_{n,i}^{(h)\prime}(X_i)| = \frac{\left| Y_1 - \dfrac{\sum_{j\in\{2,\ldots,n\}\setminus\{i\}} K_h(X_i,X_j)Y_j}{\sum_{j\in\{2,\ldots,n\}\setminus\{i\}} K_h(X_i,X_j)} \right|}{\sum_{j\in\{2,\ldots,n\}\setminus\{i\}} K_h(X_i, X_j) + 1}$$

$$\leq \frac{2L}{\sum_{j\in\{2,\ldots,n\}\setminus\{i\}} K_h(X_i, X_j) + 1}.$$

The same bound can be obtained in the third case and in the fourth case, in the fourth case because of

$$|m_{n,i}^{(h)}(X_i) - m_{n,i}^{(h)\prime}(X_i)| = \frac{|Y_1 - Y_1'|}{\sum_{j\in\{2,\ldots,n\}\setminus\{i\}} K_h(X_i, X_j) + 1}$$

$$\leq \frac{2L}{\sum_{j\in\{2,\ldots,n\}\setminus\{i\}} K_h(X_i, X_j) + 1}.$$

Using these bounds, which in each case may be multiplied by $K_h(X_i, X_1) + K_h(X_i, X_1')$, we obtain

$$\sum_{i=2}^{n} |m_{n,i}^{(h)}(X_i) - m_{n,i}^{(h)\prime}(X_i)| \leq 2L \sum_{i=2}^{n} \frac{K_h(X_i, X_1) + K_h(X_i, X_1')}{\sum_{j=2}^{n} K_h(X_i, X_j)}$$

$$= 2L \int \frac{K_h(x, X_1)}{\int K_h(x,t)\mu_{n-1}(dt)} \mu_{n-1}(dx)$$

$$+ 2L \int \frac{K_h(x, X_1')}{\int K_h(x,t)\mu_{n-1}(dt)} \mu_{n-1}(dx)$$

$$\leq 4L\rho,$$

where for the last inequality the covering assumption (C) is used for the empirical measure μ_{n-1} for the sample $\{X_2, \ldots, X_n\}$. $\square$

PROOF OF (b). By the definition of $H = H_n$ and $\overline{h} = \overline{h}_{n-1}$,

$$\triangle_{n-1}^{(H)} - (1+\delta)\,\triangle_{n-1}^{(\overline{h})}$$

$$= \triangle_{n-1}^{(H)} - \frac{1+\delta}{n} \sum_{i=1}^{n} \left((m_{n,i}^{(H)}(X_i) - Y_i)^2 - (m(X_i) - Y_i)^2 \right)$$

$$+ \frac{1+\delta}{n} \sum_{i=1}^{n} \left((m_{n,i}^{(H)}(X_i) - Y_i)^2 - (m(X_i) - Y_i)^2 \right)$$

$$- \frac{1+\delta}{n} \sum_{i=1}^{n} \left((m_{n,i}^{(\overline{h})}(X_i) - Y_i)^2 - (m(X_i) - Y_i)^2 \right)$$

$$+\frac{1+\delta}{n}\sum_{i=1}^{n}\left((m_{n,i}^{(\overline{h})}(X_i)-Y_i)^2-(m(X_i)-Y_i)^2\right)-(1+\delta)\,\triangle_{n-1}^{(\overline{h})}$$

$$\leq\ \triangle_{n-1}^{(H)}-\frac{1+\delta}{n}\sum_{i=1}^{n}\left((m_{n,i}^{(H)}(X_i)-Y_i)^2-(m(X_i)-Y_i)^2\right)$$

$$+(1+\delta)\left[\frac{1}{n}\sum_{i=1}^{n}\left((m_{n,i}^{(\overline{h})}(X_i)-Y_i)^2-(m(X_i)-Y_i)^2\right)-\triangle_{n-1}^{(\overline{h})}\right].$$

Then (8.2) yields

$$\mathbf{E}\,\triangle_{n-1}^{(H)}$$

$$\leq\ (1+\delta)\,\triangle_{n-1}^{(\overline{h})}$$

$$+\int_0^\infty\mathbf{P}\bigg\{\,\triangle_{n-1}^{(H)}-\frac{1+\delta}{n}\sum_{i=1}^{n}\Big((m_{n,i}^{(H)}(X_i)-Y_i)^2$$

$$-(m(X_i)-Y_i)^2\Big)\geq s\bigg\}ds.$$

Now by Chebyshev's inequality and

$$(m_{n,i}^{(h)}(X_i)-Y_i)^2-(m(X_i)-Y_i)^2$$

$$=(m_{n,i}^{(h)}(X_i)-m(X_i))^2+2(m_{n,i}^{(h)}(X_i)-m(X_i))\cdot(m(X_i)-Y_i)$$

we get

$$\mathbf{P}\left\{\triangle_{n-1}^{(h)}-\frac{1+\delta}{n}\sum_{i=1}^{n}\left((m_{n,i}^{(h)}(X_i)-Y_i)^2-(m(X_i)-Y_i)^2\right)\geq s\right\}$$

$$\leq\ (1+\delta)^2\frac{\mathbf{Var}\left\{\sum_{i=1}^{n}\left((m_{n,i}^{(h)}(X_i)-Y_i)^2-(m(X_i)-Y_i)^2\right)\right\}}{n^2(s+\delta\triangle_{n-1}^{(h)})^2}$$

$$\leq\ (1+\delta)^2\cdot\left(\frac{\mathbf{Var}\left\{\sum_{i=1}^{n}\left(m_{n,i}^{(h)}(X_i)-m(X_i)\right)^2\right\}}{n^2(s+\delta\triangle_{n-1}^{(h)})^2}\right.$$

$$\left.+\frac{4\mathbf{E}\left\{\sum_{i=1}^{n}\left(m_{n,i}^{(h)}(X_i)-m(X_i)\right)(m(X_i)-Y_i)\right\}^2}{n^2(s+\delta\triangle_{n-1}^{(h)})^2}\right)$$

$$\leq\ c\frac{n\,\triangle_{n-1}^{(h)}+1}{n^2(s+\delta\triangle_{n-1}^{(h)})^2}\log n,$$

with some $c > 0$, by Lemmas 8.2 and 8.3 below. Thus

$$\mathbf{E}\,\triangle_{n-1}^{(H)}$$

$$\leq\;(1+\delta)\,\triangle_{n-1}^{(\overline{h})} + \sum_{h\in\mathcal{Q}_n}\int_0^\infty \min\left\{1, c\,\frac{n\,\triangle_{n-1}^{(h)}+1}{n^2(s+\delta\triangle_{n-1}^{(h)})^2}\log n\right\}ds$$

$$\leq\;(1+\delta)\,\triangle_{n-1}^{(\overline{h})} + \sum_{h\in\mathcal{Q}_n}\int_0^\infty c\,\frac{\triangle_{n-1}^{(h)}}{n(s+\delta\triangle_{n-1}^{(h)})^2}\log n\,ds$$

$$\qquad + \sum_{h\in\mathcal{Q}_n}\int_0^\infty \min\left\{1, c\,\frac{\log n}{n^2 s^2}\right\}ds$$

$$=\;(1+\delta)\,\triangle_{n-1}^{(\overline{h})} + \frac{c}{\delta n}|\mathcal{Q}_n|\log n + \frac{2\sqrt{c}}{n}|\mathcal{Q}_n|\sqrt{\log n}.$$

This yields the assertion. $\qquad\qquad\qquad\qquad\qquad\qquad\qquad\qquad\qquad\square$

In the proof we have used Lemmas 8.2 and 8.3 below. The proof of Lemma 8.2 is based on Lemma 8.1.

Lemma 8.1. *Let* $(X_1, Y_1), \ldots, (X_n, Y_n), (\tilde{X}_n, \tilde{Y}_n)$ *be i.i.d. Then* $c > 0$ *exists with*

$$\mathbf{E}\sum_{i=1}^{n-1}\frac{|m_{n,i}^{(h)}(X_i) - m(X_i)|^2 K_h(X_i, \tilde{X}_n)}{1 + \sum_{j\in\{1,\ldots,n-1\}\setminus\{i\}} K_h(X_i, X_j)}$$

$$\leq\;c\,(\log n)\left[\mathbf{E}|m_{n,1}^{(h)}(X_1) - m(X_1)|^2 + \frac{1}{n}\right]$$

for each $h \in \mathcal{Q}_n, n \in \{2, 3, \ldots\}$.

PROOF. Set $m_{n,i} = m_{n,i}^{(h)}$ and $R_h(x) = \{t \in \mathcal{R}^d; K_h(x, t) = 1\}$. Then the left-hand side equals

$$(n-1)\int \mu(R_h(x_1))\mathbf{E}|m_{n,1}(x_1) - m(x_1)|^2 \frac{1}{1 + \sum_{j=2}^{n-1} K_h(x_1, X_j)}\mu(dx_1).$$

For $x_1 \in \mathcal{R}^d$, we note

$$|m_{n,1}(x_1) - m(x_1)|^2$$

$$=\;\left|\sum_{l=2}^n \frac{(Y_l - m(x_1))K_h(x_1, X_l)}{1 + \sum_{j\in\{2,\ldots,n\}\setminus\{l\}} K_h(x_1, X_j)} - m(x_1)I_{[K_h(x_1,X_l)=0(l=2,\ldots,n)]}\right|^2$$

$$=\;\left|\sum_{l=2}^n \frac{(Y_l - m(x_1))K_h(x_1, X_l)}{1 + \sum_{j\in\{2,\ldots,n\}\setminus\{l\}} K_h(x_1, X_j)}\right|^2 + m(x_1)^2 I_{[K_h(x_1,X_l)=0(l=2,\ldots,n)]}$$

$$=\;\sum_{l=2}^n \frac{|Y_l - m(x_1)|^2 K_h(x_1, X_l)}{[1 + \sum_{j\in\{2,\ldots,n\}\setminus\{l\}} K_h(x_1, X_j)]^2}$$

$$+ \sum_{\substack{l,l' \in \{2,\dots,n\} \\ l \neq l'}} \frac{(Y_l - m(x_1))K_h(x_1, X_l)(Y_{l'} - m(x_1))K_h(x_1, X_{l'})}{[2 + \sum_{j \in \{2,\dots,n\}\setminus\{l,l'\}} K_h(x_1, X_l)]^2}$$

$$+ m(x_1)^2 I_{[K_h(x_1,X_l)=0(l=2,\dots,n)]}. \tag{8.5}$$

We shall show the existence of a $c > 0$ with

$$(n-1)\mu(R_h(x_1))$$

$$\times \mathbf{E} \sum_{l=2}^n \frac{|Y_l - m(x_1)|^2 K_h(x_1, X_l)}{[1 + \sum_{j \in \{2,\dots,n\}\setminus\{l\}} K_h(x_1, X_j)]^2} \frac{1}{2 + \sum_{j \in \{2,\dots,n-1\}\setminus\{l\}} K_h(x_1, X_j)}$$

$$\leq c\mathbf{E} \sum_{l=2}^n \frac{|Y_l - m(x_1)|^2 K_h(x_1, X_l)}{[1 + \sum_{j \in \{2,\dots,n\}\setminus\{l\}} K_h(x_1, X_j)]^2} \tag{8.6}$$

and

$$(n-1)\mu(R_h(x_1))\mathbf{E} \sum_{l \neq l'} \frac{(Y_l - m(x_1))K_h(x_1, X_l)(Y_{l'} - m(x_1))K_h(x_1, X_{l'})}{[2 + \sum_{j \in \{2,\dots,n\}\setminus\{l,l'\}} K_h(x_1, X_l)]^2}$$

$$\times \frac{1}{3 + \sum_{j \in \{2,\dots,n\}\setminus\{l,l'\}} K_h(x_1, X_j)}$$

$$\leq c\mathbf{E} \sum_{l \neq l'} \frac{(Y_l - m(x_1))K_h(x_1, X_l)(Y_{l'} - m(x_1))K_h(x_1, X_{l'})}{[2 + \sum_{j \in \{2,\dots,n\}\setminus\{l,l'\}} K_h(x_1, X_l)]^2} \tag{8.7}$$

and

$$(n-1)\mu(R_h(x_1))m(x_1)^2\mathbf{E}I_{[K_h(x_1,X_l)=0\ (l=2,\dots,n)]} \frac{1}{1 + \sum_{j=2}^{n-1} K_h(x_1, X_j)}$$

$$\leq c\,(\log n)\left[m(x_1)^2\mathbf{E}I_{[K_h(x_1,X_l)=0\ (l=2,\dots,n)]} + \frac{1}{n} \right] \tag{8.8}$$

for all $x_1 \in \mathcal{R}^d$. These results together with (8.5) yield the assertion. Set $p = \mu(R_h(x_1))$ and let $B(n,p)$ be a binomially (n,p)–distributed random variable. In view of (8.6) and (8.7) it remains to show

$$(n-1)p\mathbf{E}\frac{1}{[1 + B(n-2,p)]^3} \leq c\mathbf{E}\frac{1}{[1 + B(n-2,p)]^2} \tag{8.9}$$

for $n \in \{3,4,\dots\}$ and, because of

$$\mathbf{E}(Y_l - m(x_1))K_h(x_1, X_l)(Y_{l'} - m(x_1))K_h(x_1, X_{l'})$$

$$= [\mathbf{E}(Y_l - m(x_1))K_h(x_1, X_l)]^2 \geq 0\ \ (l \neq l'),$$

we would like to get

$$(n-1)p\mathbf{E}\frac{1}{[2 + B(n-3,p)]^3} \leq c\mathbf{E}\frac{1}{[2 + B(n-3,p)]^2} \tag{8.10}$$

for $n \in \{4, 5, \ldots\}$, respectively. Each of these relations is easily verified via the equivalent relations

$$np \sum_{k=0}^{n} \frac{1}{(1+k)^3} \binom{n}{k} p^k (1-p)^{n-k} \le c' \sum_{k=0}^{n} \frac{1}{(1+k)^2} \binom{n}{k} p^k (1-p)^{n-k}$$

for some $c' > 0$. The latter follows from

$$\sum_{k=0}^{n} \binom{n+3}{k+3} p^{k+3}(1-p)^{n+3-(k+3)} \le \sum_{k=0}^{n} \binom{n+2}{k+2} p^{k+2}(1-p)^{n+2-(k+2)},$$

i.e.,

$$1 - p + (n+2)p \le (1-p)^2 + (n+3)p(1-p) + (n+3)(n+2)p^2/2.$$

The left-hand side of (8.8) equals

$$(n-1)m(x_1)^2 p (1-p)^{n-1}.$$

Distinguishing the cases $p \le (\log n)/n$ and $p > (\log n)/n$, we obtain the upper bounds

$$m(x_1)^2 (\log n)(1-p)^{n-1}$$

and (noticing $(1-p)^{n-1} \le e^{-(n-1)p} \le e \cdot e^{-np}$ and monotonicity of $s \cdot e^{-s}$ for $s \ge 1$)

$$L^2 enpe^{-np} \le L^2 e \frac{\log n}{n},$$

respectively. Thus (8.8) is obtained. $\qquad\square$

Lemma 8.2. *There is a constant $c^* > 0$ such that*

$$\mathbf{Var}\left(\sum_{i=1}^{n} |m_{n,i}^{(h)}(X_i) - m(X_i)|^2 \right) \le c^*(\log n)[n\mathbf{E}|m_{n,1}^{(h)}(X_1) - m(X_1)|^2 + 1]$$

for each $h \in \mathcal{Q}_n, n \in \{2, 3, \ldots\}$.

PROOF. Set $m_{n,i} = m_{n,i}^{(h)}$. For each $h \in \mathcal{Q}_n, n \in \{2, 3, \ldots\}$, and for $i \in \{1, \ldots, n\}$,

$$\mathbf{E} \sum_{i=1}^{n-1} \frac{|m_{n,i}(X_i) - m(X_i)|^2 K_h(X_i, X_n)}{1 + \sum_{j \in \{1,\ldots,n-1\}\setminus\{i\}} K_h(X_i, X_j)}$$

$$\le 2\mathbf{E} \sum_{i=1}^{n} \frac{|m_{n,i}(X_i) - m(X_i)|^2 K_h(X_i, X_n)}{1 + \sum_{j \in \{1,\ldots,n\}\setminus\{i\}} K_h(X_i, X_j)}$$

$$= \frac{2}{n}\mathbf{E} \sum_{l=1}^{n} \sum_{i=1}^{n} \frac{|m_{n,i}(X_i) - m(X_i)|^2 K_h(X_i, X_l)}{1 + \sum_{j \in \{1,\ldots,n\}\setminus\{i\}} K_h(X_i, X_j)}$$

$$\begin{aligned}
&\leq \quad \frac{2}{n}\mathbf{E}\sum_{i=1}^{n}|m_{n,i}(X_i) - m(X_i)|^2 \\
&= \quad 2\mathbf{E}|m_{n,1}(X_1) - m(X_1)|^2. \tag{8.11}
\end{aligned}$$

We use Theorem A.3 with $m = d + 1$, $Z_i = (X_i, Y_i), i = 1, \ldots, n$,

$$f(Z_1, \ldots, Z_n) = \sum_{i=1}^{n}|m_{n,i}(X_i) - m(X_i)|^2$$

and the notations there. Further we notice that $(X_1, Y_1), \ldots, (X_n, Y_n)$ are independent and identically distributed $(d + 1)$-dimensional random vectors. $\tilde{m}_{n,i}$ shall be obtained from $m_{n,i}$ via replacing (X_n, Y_n) by its copy $(\tilde{X}_n, \tilde{Y}_n)$ there, where $(X_1, Y_1), \ldots, (X_n, Y_n), (\tilde{X}_n, \tilde{Y}_n)$ are independent ($i = 1, \ldots, n - 1$). We set

$$\begin{aligned}
V_n \quad = \quad & \sum_{i=1}^{n}(m_{n,i}(X_i) - m(X_i))^2 \\
& - \left[\sum_{i=1}^{n-1}(\tilde{m}_{n,i}(X_i) - m(X_i))^2 + (m_{n,n}(\tilde{X}_n) - m(\tilde{X}_n)^2\right].
\end{aligned}$$

It suffices to show

$$\mathbf{E}V_n^2 \leq c^*(\log n)\left[\mathbf{E}\{|m_{n,1}(X_1) - m(X_1)|^2\} + \frac{1}{n}\right].$$

Let

$$\begin{aligned}
U_i \quad = \quad & m_{n,i}(X_i) - m(X_i) \\
= \quad & \sum_{l\in\{1,\ldots,n\}\setminus\{i\}} \frac{Y_l K_h(X_i, X_l)}{1 + \sum_{j\in\{1,\ldots,n\}\setminus\{i,l\}} K_h(X_i, X_j)} - m(X_i)
\end{aligned}$$

and

$$\begin{aligned}
W_i \quad = \quad & \tilde{m}_{n,i}(X_i) - m(X_i) \\
= \quad & \sum_{l\in\{1,\ldots,n-1\}\setminus\{i\}} \frac{Y_l K_h(X_i, X_l)}{1 + \sum_{j\in\{1,\ldots,n-1\}\setminus\{i,l\}} K_h(X_i, X_j) + K_h(X_i, \tilde{X}_n)} \\
& + \frac{\tilde{Y}_n K_h(X_i, \tilde{X}_n)}{1 + \sum_{j\in\{1,\ldots,n-1\}\setminus\{i\}} K_h(X_i, X_j)} - m(X_i)
\end{aligned}$$

for $i = 1, \ldots, n - 1$. Thus

$$V_n = \sum_{i=1}^{n-1}\left(U_i^2 - W_i^2\right) + |m_{n,n}(X_n) - m(X_n)|^2 - |m_{n,n}(\tilde{X}_n) - m(\tilde{X}_n)|^2.$$

Therefore

$$V_n^2 \le 3|\sum_{i=1}^{n-1}(U_i^2 - W_i^2)|^2 + 3|m_{n,n}(X_n) - m(X_n)|^4 + 3|m_{n,n}(\tilde{X}_n) - m(\tilde{X}_n)|^4.$$

$$(8.12)$$

We obtain

$$|U_i| \le 2L, \ |W_i| \le 2L,$$

$$|U_i - W_i| \le 2L \frac{K_h(X_i, X_n) + K_h(X_i, \tilde{X}_n)}{1 + \sum_{j \in \{1,\ldots,n-1\}\setminus\{i\}} K_h(X_i, X_j)}$$

for $i \in \{1, \ldots, n-1\}$, thus

$$\sum_{i=1}^{n-1} |U_i - W_i| \le 4L\rho$$

(by covering assumption (C)), then via the Cauchy–Schwarz inequality

$$\left(\sum_{i=1}^{n-1}(U_i^2 - W_i^2)\right)^2$$

$$\le \ 2\left(\sum_{i=1}^{n-1} |U_i||U_i - W_i|\right)^2 + 2\left(\sum_{i=1}^{n-1} |W_i||U_i - W_i|\right)^2$$

$$\le \ 8L\rho \sum_{i=1}^{n-1} U_i^2 |U_i - W_i| + 8L\rho \sum_{i=1}^{n-1} W_i^2 |U_i - W_i|$$

and

$$\mathbf{E}\left\{\left(\sum_{i=1}^{n-1}(U_i^2 - W_i^2)\right)^2\right\}$$

$$\le \ 16L^2\rho\mathbf{E}\left\{\sum_{i=1}^{n-1}(U_i^2 + W_i^2)\frac{K_h(X_i, X_n) + K_h(X_i, \tilde{X}_n)}{1 + \sum_{j \in \{1,\ldots,n-1\}\setminus\{i\}} K_h(X_i, X_j)}\right\}.$$

Via (8.12) and a symmetry relation with respect to (X_n, Y_n) and $(\tilde{X}_n, \tilde{Y}_n)$ we obtain

$$\mathbf{E}V_n^2$$

$$\le \ 96L^2\rho\mathbf{E}\left\{\sum_{i=1}^{n-1}|m_{n,i}(X_i) - m(X_i)|^2 \frac{K_h(X_i, X_n) + K_h(X_i, \tilde{X}_n)}{1 + \sum\limits_{j \in \{1,\ldots,n-1\}\setminus\{i\}} K_h(X_i, X_j)}\right\}$$

$$+ 24L^2\mathbf{E}\left\{(m_{n,n}(X_n) - m(X_n)^2\right\}.$$

Now the assertion follows from (8.11) and Lemma 8.1. $\square$

Lemma 8.3. *There is a constant $c^{**} > 0$ such that*

$$\mathbf{E}\left\{\left(\sum_{i=1}^{n}(m_{n,i}^{(h)}(X_i) - m(X_i))(m(X_i) - Y_i)\right)^2\right\}$$

$$\leq \quad c^{**} n \mathbf{E}\{|m_{n,1}^{(h)}(X_1) - m(X_1)|^2\}$$

for each $h \in \mathcal{Q}_n, n \in \{2,3,\ldots\}$.

PROOF. Set $m_{n,i} = m_{n,i}^{(h)}$ and

$$m_n^*(x) = \frac{\sum_{j=1}^{n} m(X_j)K_h(x, X_j)}{\sum_{j=1}^{n} K_h(x, X_j)},$$

then

$$\mathbf{E}\{|m_n(x) - m(x)|^2\}$$

$$= \mathbf{E}\left\{\left(\frac{\sum_{i=1}^{n}(Y_i - m(X_i))K_h(x, X_i)}{\sum_{i=1}^{n} K_h(x, X_i)} + m_n^*(x) - m(x)\right)^2\right\}$$

$$= \mathbf{E}\left\{\left(\sum_{i=1}^{n}\frac{(Y_i - m(X_i))K_h(x, X_i)}{1 + \sum_{l \in \{1,\ldots,n\}\setminus\{i\}} K_h(x, X_l)}\right)^2\right\} + \mathbf{E}\left\{|m_n^*(x) - m(x)|^2\right\}$$

$$= n\mathbf{E}\left\{\left(\frac{(Y_1 - m(X_1))K_h(x, X_1)}{1 + \sum_{l=2}^{n} K_h(x, X_l)}\right)^2\right\} + \mathbf{E}\left\{|m_n^*(x) - m(x)|^2\right\}. \quad (8.13)$$

We notice

$$\mathbf{E}\left\{\sum_{i=1}^{n}[(m_{n,i}(X_i) - m(X_i))(m(X_i) - Y_i)]^2\right\}$$

$$\leq \quad 4L^2\, n\mathbf{E}\{|m_{n,1}(X_1) - m(X_1)|^2\}.$$

Further, noticing

$$\mathbf{E}\{Y_l - m(X_l)|X_1,\ldots,X_n,Y_1,\ldots,Y_{l-1},Y_{l+1},\ldots,Y_n\} = 0$$

$(l = 1,\ldots,n)$, we obtain

$$\mathbf{E}\sum_{i \neq j}(m_{n,i}(X_i) - m(X_i))(m(X_i) - Y_i)$$

$$\times (m_{n,j}(X_j) - m(X_j))(m(X_j) - Y_j)$$

$$= \sum_{i \neq j}\mathbf{E}\frac{Y_j Y_i K_h(X_i, X_j)K_h(X_j, X_i)(m(X_i) - Y_i)(m(X_j) - Y_j)}{\left[1 + \sum_{l \in \{1,\ldots,n\}\setminus\{i,j\}} K_h(X_i, X_l)\right]\left[1 + \sum_{l \in \{1,\ldots,n\}\setminus\{i,j\}} K_h(X_j, X_l)\right]}$$

$$
= \quad n(n-1)\mathbf{E}\Big\{(Y_1 - m(X_1))^2(Y_2 - m(X_2))^2
$$

$$
\times \frac{K_h(X_2, X_1)K_h(X_1, X_2)}{\left[1 + \sum_{l=3}^{n} K_h(X_2, X_l)\right]\left[1 + \sum_{l=3}^{n} K_h(X_1, X_l)\right]}\Big\}
$$

$$
\leq \quad n(n-1)\mathbf{E}\left\{(Y_1 - m(X_1))^2(Y_2 - m(X_2))^2 \frac{K_h(X_2, X_1)}{\left[1 + \sum_{l=3}^{n} K_h(X_2, X_l)\right]^2}\right\}
$$

(by $ab \leq a^2/2 + b^2/2$ and symmetry)

$$
\leq \quad 4L^2 n(n-1)\mathbf{E}\frac{(Y_1 - m(X_1))^2 K_h(X_2, X_1)}{\left[1 + \sum_{l=3}^{n} K_h(X_2, X_l)\right]^2}
$$

$$
= \quad 4L^2 n(n-1)\int \mathbf{E}\frac{(Y_1 - m(X_1))^2 K_h(x, X_1)}{\left[1 + \sum_{l=2}^{n-1} K_h(x, X_l)\right]^2}\mu(dx)
$$

$$
\leq \quad 4L^2 n \int \mathbf{E}\left\{|m_{n-1}(x) - m(x)|^2\right\}\mu(dx) \qquad \text{(by (8.13))}
$$

$$
= \quad 4L^2 n \mathbf{E}\left\{|m_{n,1}(X_1) - m(X_1)|^2\right\}.
$$

□

8.4   Nearest Neighbor Estimates

Theorem 8.1 cannot be applied for a nearest neighbor estimate. Let $m_n^{(k)}$ be the k-NN estimate for sample size $n \geq 2$. Then $h = k$ can be considered as a parameter, and we choose $\mathcal{Q}_n = \{1, \ldots, n\}$. Let m_n denote the cross-validation nearest neighbor estimate, i.e., put

$$
H = H_n = \arg\min_h \frac{1}{n}\sum_{i=1}^{n}(m_{n,i}^{(h)}(X_i) - Y_i)^2
$$

and

$$
m_n = m_n^{(H)}.
$$

For the nearest neighbor estimate again we have covering (Corollary 6.1) with $\rho = \gamma_d$.

Theorem 8.2. *Assume that $|Y| \leq L$. Then, for the cross-validation nearest neighbor estimate m_n,*

$$
\mathbf{E}\{\Delta_{n-1}^{(H_n)}\} \leq \Delta_{n-1}^{(\bar{h}_{n-1})} + c\sqrt{\frac{\log n}{n}}
$$

for some constant c depending only on L and γ_d.

PROOF. See Problem 8.3. □

Theorems 8.2 and 6.2 imply

Corollary 8.2. *Assume that X is bounded,*

$$|m(x) - m(z)| \leq C \cdot \|x - z\| \quad (x, z \in \mathcal{R}^d)$$

and $|Y| \leq L$ a.s. Let m_n be the k-nearest neighbor estimate with k chosen as in Theorem 8.2. Then for $d \geq 3$ and

$$(\log n)^{(d+2)/(4d)} n^{-(d-2)/(4d)} \leq C,$$

one has

$$\mathbf{E}\{\Delta_{n-1}^{(H_n)}\} \leq c_1 C^{2d/(d+2)} n^{-2/(d+2)}$$

for some constant c_1 which depends only on L, d, and the diameter of the support of X.

PROOF. See Problem 8.6. □

8.5 Bibliographic Notes

The concept of cross-validation in statistics was introduced by Lunts and Brailovsky (1967), Allen (1974), and M. Stone (1974), for regression estimation by Clark (1975), and Wahba and Wold (1975). Further literature can be found, e.g., in Härdle (1990) and Simonoff (1996). Cross-validation for kernel and nearest neighbor estimates has been studied by Chiu (1991), Hall (1984), Härdle, Hall, and Marron (1985; 1988), Härdle and Kelly (1987), Härdle and Marron (1985; 1986), Li (1984) and Wong (1983), under certain optimality aspects. Assuming bounded Q_n, a consequence of the results of Hall (1984) and Härdle and Marron (1985; 1986) is that, for X with continuous density and for continuous m,

$$\frac{\Delta_{n-1}^{(H_n)}}{\Delta_{n-1}^{(\bar{h}_{n-1})}} \to 1 \quad a.s.$$

A corresponding result of stochastic convergence for fixed design and nearest neighbor regression is due to Li (1987). Part (b) of Theorem 8.1 was obtained by Walk (2002b). Theorem 8.2 is a slight modification of Devroye and Wagner (1979).

Problems and Exercises

PROBLEM 8.1. Prove Corollary 8.1.
 HINT: Proceed as in the proof of Corollary 7.1.

PROBLEM 8.2. Prove that, for a random variable Z,

$$\mathbf{P}\{Z > \epsilon\} \leq C e^{-cn\epsilon^2} \text{ for all } \epsilon > 0$$

$(C > 1, c > 0)$ implies that

$$\mathbf{E}Z \leq \sqrt{\frac{1 + \log C}{cn}}.$$

HINT: Without loss of generality assume $Z \geq 0$ (otherwise replace Z by $Z \cdot I_{\{Z \geq 0\}}$). Then

$$
\begin{aligned}
\mathbf{E}Z \quad &\leq \quad \sqrt{\mathbf{E}Z^2} \\
&= \quad \sqrt{\int_0^\infty \mathbf{P}\{Z^2 > \epsilon\}\, d\epsilon} \\
&\leq \quad \sqrt{\int_0^{\frac{\log C}{cn}} 1\, d\epsilon + \int_{\frac{\log C}{cn}}^\infty C e^{-cn\epsilon}\, d\epsilon} \\
&= \quad \sqrt{\frac{1 + \log C}{cn}}.
\end{aligned}
$$

PROBLEM 8.3. Prove Theorem 8.2.

HINT:

Step (1). With $h = k$, follow the line of the proof of Theorem 8.1 until (8.4):

$$
\left| \sum_{j=1}^n (m_{n,j}^{(h)}(X_j) - Y_j)^2 - \sum_{j=1}^n (m_{n,j}^{(h)\prime}(X_j') - Y_j')^2 \right|
$$

$$
\leq \quad 4L^2 + 4L \sum_{j=2}^n |m_{n,j}^{(h)}(X_j) - m_{n,j}^{(h)\prime}(X_j)|.
$$

Step (2). Apply Corollary 6.1 to show that

$$
\sum_{j=2}^n |m_{n,j}^{(k)}(X_j) - m_{n,j}^{(k)\prime}(X_j)| \leq 4L\gamma_d.
$$

According to Corollary 6.1,

$$
\sum_{j=2}^n |m_{n,j}^{(k)}(X_j) - m_{n,j}^{(k)\prime}(X_j)|
$$

$$
= \quad \sum_{j=2}^n \left| \frac{1}{k} Y_1 I_{\{X_1 \text{ among } X_1,\dots,X_{j-1},X_{j+1},\dots,X_n \text{ is one of the } k \text{ NNs of } X_j\}} \right.
$$

$$
+ \frac{1}{k} \sum_{l \in \{2,\dots,n\}-\{j\}}
$$

$$
Y_l I_{\{X_l \text{ among } X_1,\dots,X_{j-1},X_{j+1},\dots,X_n \text{ is one of the } k \text{ NNs of } X_j\}}
$$

$$-\frac{1}{k}Y_1'I_{\{x_1' \text{ among } x_1',\dots,x_{j-1},x_{j+1},\dots,x_n \text{ is one of the } k \text{ NNs of } x_j\}}$$

$$-\frac{1}{k}\sum_{l\in\{2,\dots,n\}-\{j\}}$$

$$\left. Y_l I_{\{x_l \text{ among } x_1',\dots,x_{j-1},x_{j+1},\dots,x_n \text{ is one of the } k \text{ NNs of } x_j\}}\right|$$

$$\leq \frac{1}{k}L\sum_{j=2}^{n}I_{\{x_1 \text{ among } x_1,\dots,x_{j-1},x_{j+1},\dots,x_n \text{ is one of the } k \text{ NNs of } x_j\}}$$

$$+\frac{1}{k}L\sum_{j=2}^{n}I_{\{x_1' \text{ among } x_1',\dots,x_{j-1},x_{j+1},\dots,x_n \text{ is one of the } k \text{ NNs of } x_j\}}$$

$$+\frac{1}{k}L\sum_{j=2}^{n}\sum_{l\in\{2,\dots,n\}-\{j\}}$$

$$\left| I_{\{x_l \text{ among } x_1,\dots,x_{j-1},x_{j+1},\dots,x_n \text{ is one of the } k \text{ NNs of } x_j\}} \right.$$

$$\left. -I_{\{x_l \text{ among } x_1',\dots,x_{j-1},x_{j+1},\dots,x_n \text{ is one of the } k \text{ NNs of } x_j\}}\right|$$

$$\leq L\gamma_d + L\gamma_d + 2L\gamma_d.$$

PROBLEM 8.4. (a) Use cross-validation to choose the side lengths of rectangular partitioning such that the resulting estimate approaches the rate of convergence in Problem 4.7.

(b) Use cross-validation to choose the scaling for product kernel estimates such that the resulting estimate approaches the rate of convergence in Problem 5.7.

HINT: Proceed as in Problem 7.7.

PROBLEM 8.5. Formulate and prove a version of Theorem 7.2 which uses cross-validation instead of splitting of the data.

PROBLEM 8.6. Prove Corollary 8.2.

HINT: Use Theorems 8.2 and 6.2.

PROBLEM 8.7. Show that, under the conditions of part (a) of Theorem 8.1,

$$\mathbf{E}\left\{\int |m_n^{(H_n)}(x) - m(x)|^2\mu(dx)\right\} \leq \Delta_n^{(\bar{h}_n)} + c\cdot\frac{Q_n|}{\sqrt{n}}$$

for some constant c depending only on L and ρ.

HINT: Use Theorem A.3 for the treatment of

$$\int |m_n^{(h)}(x) - m(x)|^2\mu(dx) - \Delta_n^{(h)} \quad (h \in Q_n),$$

further Problem 8.2 and Theorem 8.1 (a).

9
Uniform Laws of Large Numbers

In the least squares estimation problem we minimize the empirical L_2 risk

$$\frac{1}{n} \sum_{i=1}^{n} |f(X_i) - Y_i|^2$$

over a set of functions $\mathcal{F}_n$ depending on the sample size n. One of the main steps in proving the consistency of such estimates is to show that the empirical L_2 risk is uniformly (over $\mathcal{F}_n$) close to the L_2 risk. More precisely, we will need to show

$$\sup_{f \in \mathcal{F}_n} \left| \frac{1}{n} \sum_{i=1}^{n} |f(X_i) - Y_i|^2 - \mathbf{E}\{|f(X) - Y|^2\} \right| \to 0 \quad (n \to \infty) \quad a.s. \quad (9.1)$$

Set $Z = (X, Y)$, $Z_i = (X_i, Y_i)$ $(i = 1, \ldots, n)$, $g_f(x, y) = |f(x) - y|^2$ for $f \in \mathcal{F}_n$ and $\mathcal{G}_n = \{g_f : f \in \mathcal{F}_n\}$. Then (9.1) can be written as

$$\sup_{g \in \mathcal{G}_n} \left| \frac{1}{n} \sum_{i=1}^{n} g(Z_i) - \mathbf{E}\{g(Z)\} \right| \to 0 \quad (n \to \infty) \quad a.s.$$

Thus we are interested in bounding the distance between an average and its expectation uniformly over a set of functions. In this chapter we discuss techniques for doing this.

9.1 Basic Exponential Inequalities

For the rest of this chapter let $Z, Z_1, Z_2, \dots$ be independent and identically distributed random variables with values in $\mathcal{R}^d$, and for $n \in \mathcal{N}$ let $\mathcal{G}_n$ be a class of functions $g : \mathcal{R}^d \to \mathcal{R}$. We derive sufficient conditions for

$$\sup_{g \in \mathcal{G}_n} \left| \frac{1}{n} \sum_{i=1}^{n} g(Z_i) - \mathbf{E}\{g(Z)\} \right| \to 0 \quad (n \to \infty) \quad a.s. \tag{9.2}$$

For each fixed function g with $\mathbf{E}|g(Z)| < \infty$ the strong law of large numbers implies

$$\lim_{n \to \infty} \left| \frac{1}{n} \sum_{j=1}^{n} g(Z_j) - \mathbf{E}\{g(Z)\} \right| = 0 \quad a.s.$$

We will use Hoeffding's inequality (Lemma A.3) to extend it to sets of functions. Recall that if g is a function $g : \mathcal{R}^d \to [0, B]$, then by Hoeffding's inequality

$$\mathbf{P}\left\{ \left| \frac{1}{n} \sum_{j=1}^{n} g(Z_j) - \mathbf{E}\{g(Z)\} \right| > \epsilon \right\} \leq 2e^{-\frac{2n\epsilon^2}{B^2}},$$

which together with the union bound implies

$$\mathbf{P}\left\{ \sup_{g \in \mathcal{G}_n} \left| \frac{1}{n} \sum_{j=1}^{n} g(Z_j) - \mathbf{E}\{g(Z)\} \right| > \epsilon \right\} \leq 2|\mathcal{G}_n| e^{-\frac{2n\epsilon^2}{B^2}}. \tag{9.3}$$

Thus, for finite classes $\mathcal{G}_n$ satisfying

$$\sum_{n=1}^{\infty} |\mathcal{G}_n| e^{-\frac{2n\epsilon^2}{B^2}} < \infty \tag{9.4}$$

for all $\epsilon > 0$, (9.2) follows from (9.3) and the Borel–Cantelli lemma (see the proof of Lemma 9.1 below for details).

In our applications (9.4) is never satisfied because the cardinality of $\mathcal{G}_n$ is always infinite. But sometimes it is possible to choose finite sets $\mathcal{G}_{n,\epsilon}$ which satisfy (9.4) and

$$\left\{ \sup_{g \in \mathcal{G}_n} \left| \frac{1}{n} \sum_{j=1}^{n} g(Z_j) - \mathbf{E}\{g(Z)\} \right| > \epsilon \right\}$$

$$\subset \left\{ \sup_{g \in \mathcal{G}_{n,\epsilon}} \left| \frac{1}{n} \sum_{j=1}^{n} g(Z_j) - \mathbf{E}\{g(Z)\} \right| > \epsilon' \right\} \tag{9.5}$$

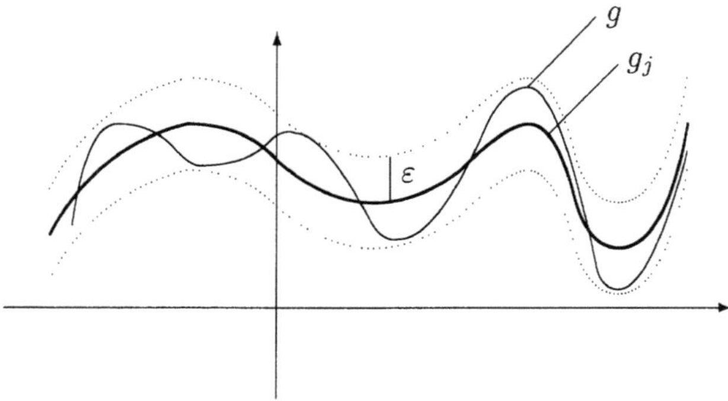

Figure 9.1. Sup norm distance between function g and member g_j of cover is less than ϵ.

for some ϵ' depending on ϵ (but not on n). Clearly, then (9.2) follows from (9.5) and (9.4) (for $\mathcal{G}_{n,\epsilon}$ instead of $\mathcal{G}_n$) with the above argument applied to $\mathcal{G}_{n,\epsilon}$.

To construct classes $\mathcal{G}_{n,\epsilon}$ which satisfy (9.5) one can use covers with respect to the supremum norm:

Definition 9.1. *Let $\epsilon > 0$ and let $\mathcal{G}$ be a set of functions $\mathcal{R}^d \to \mathcal{R}$. Every finite collection of functions $g_1, \ldots, g_N : \mathcal{R}^d \to \mathcal{R}$ with the property that for every $g \in \mathcal{G}$ there is a $j = j(g) \in \{1, \ldots, N\}$ such that*

$$\|g - g_j\|_\infty := \sup_z |g(z) - g_j(z)| < \epsilon$$

*is called an ϵ-**cover of** $\mathcal{G}$ with respect to $\| \cdot \|_\infty$.*

If $\{g_1, \ldots, g_N\}$ is an ϵ-cover of $\mathcal{G}$ with respect to $\| \cdot \|_\infty$, then $\mathcal{G}$ is a subset of the union of all $\| \cdot \|_\infty$-balls with center g_i and radius ϵ $(i = 1, \ldots, n)$. The fewer balls are needed to cover $\mathcal{G}$, the smaller $\mathcal{G}$ is in some sense.

Definition 9.2. *Let $\epsilon > 0$ and let $\mathcal{G}$ be a set of functions $\mathcal{R}^d \to \mathcal{R}$. Let $\mathcal{N}(\epsilon, \mathcal{G}, \| \cdot \|_\infty)$ be the size of the smallest ϵ-cover of $\mathcal{G}$ w.r.t. $\| \cdot \|_\infty$. Take $\mathcal{N}(\epsilon, \mathcal{G}, \| \cdot \|_\infty) = \infty$ if no finite ϵ-cover exists. Then $\mathcal{N}(\epsilon, \mathcal{G}, \| \cdot \|_\infty)$ is called an ϵ-**covering number** of $\mathcal{G}$ w.r.t. $\| \cdot \|_\infty$ and will be abbreviated to $\mathcal{N}_\infty(\epsilon, \mathcal{G})$.*

Lemma 9.1. *For $n \in \mathcal{N}$ let $\mathcal{G}_n$ be a set of functions $g : \mathcal{R}^d \to [0, B]$ and let $\epsilon > 0$. Then*

$$\mathbf{P} \left\{ \sup_{g \in \mathcal{G}_n} \left| \frac{1}{n} \sum_{j=1}^n g(Z_j) - \mathbf{E}\{g(Z)\} \right| > \epsilon \right\} \leq 2\mathcal{N}_\infty\left(\epsilon/3, \mathcal{G}_n\right) e^{-\frac{2n\epsilon^2}{9B^2}}.$$

Furthermore, if

$$\sum_{n=1}^\infty \mathcal{N}_\infty\left(\epsilon/3, \mathcal{G}_n\right) e^{-\frac{2n\epsilon^2}{9B^2}} < \infty \tag{9.6}$$

for all $\epsilon > 0$, then (9.2) holds.

Since in the above probability the supremum is taken over a possible uncountable set, there may be some measurability problems. In the book of van der Vaart and Wellner (1996) this issue is handled very elegantly using the notion of outer probability. In most of our applications it will suffice to consider countable sets of functions, therefore, here and in the sequel, we shall completely ignore this problem.

PROOF. Let $\mathcal{G}_{n,\frac{\epsilon}{3}}$ be an $\frac{\epsilon}{3}$–cover of $\mathcal{G}_n$ w.r.t. $\|\cdot\|_\infty$ of minimal cardinality. Fix $g \in \mathcal{G}_n$. Then there exists $\bar{g} \in \mathcal{G}_{n,\frac{\epsilon}{3}}$ such that $\|g - \bar{g}\|_\infty < \frac{\epsilon}{3}$. Using this one gets

$$\left| \frac{1}{n} \sum_{i=1}^n g(Z_i) - \mathbf{E}\{g(Z)\} \right|$$

$$\leq \left| \frac{1}{n} \sum_{i=1}^n g(Z_i) - \frac{1}{n} \sum_{i=1}^n \bar{g}(Z_i) \right| + \left| \frac{1}{n} \sum_{i=1}^n \bar{g}(Z_i) - \mathbf{E}\{\bar{g}(Z)\} \right|$$

$$+ |\mathbf{E}\{\bar{g}(Z)\} - \mathbf{E}\{g(Z)\}|$$

$$\leq \|g - \bar{g}\|_\infty + \left| \frac{1}{n} \sum_{i=1}^n \bar{g}(Z_i) - \mathbf{E}\{\bar{g}(Z)\} \right| + \|g - \bar{g}\|_\infty$$

$$\leq \frac{2}{3}\epsilon + \left| \frac{1}{n} \sum_{i=1}^n \bar{g}(Z_i) - \mathbf{E}\{\bar{g}(Z)\} \right|.$$

Hence,

$$\mathbf{P}\left\{ \sup_{g \in \mathcal{G}_n} \left| \frac{1}{n} \sum_{i=1}^n g(Z_i) - \mathbf{E}\{g(Z)\} \right| > \epsilon \right\}$$

$$\leq \mathbf{P}\left\{ \sup_{g \in \mathcal{G}_{n,\frac{\epsilon}{3}}} \left| \frac{1}{n} \sum_{i=1}^n g(Z_i) - \mathbf{E}\{g(Z)\} \right| > \frac{\epsilon}{3} \right\}$$

$$\overset{(9.3)}{\leq} 2\left|\mathcal{G}_{n,\frac{\epsilon}{3}}\right| e^{-\frac{2n(\frac{\epsilon}{3})^2}{B^2}} = 2\mathcal{N}_\infty\left(\frac{\epsilon}{3}, \mathcal{G}_n\right) e^{-\frac{2n\epsilon^2}{9B^2}}.$$

If (9.6) holds, then this implies

$$\sum_{n=1}^\infty \mathbf{P}\left\{ \sup_{g \in \mathcal{G}_n} \left| \frac{1}{n} \sum_{i=1}^n g(Z_i) - \mathbf{E}\{g(Z)\} \right| > \frac{1}{k} \right\} < \infty$$

for each $k \in \mathcal{N}$. Using the Borel–Cantelli lemma one concludes that, for each $k \in \mathcal{N}$,

$$\limsup_{n \to \infty} \sup_{g \in \mathcal{G}_n} \left| \frac{1}{n} \sum_{i=1}^n g(Z_i) - \mathbf{E}\{g(Z)\} \right| \leq \frac{1}{k} \quad a.s.,$$

hence also with probability one

$$\limsup_{n\to\infty} \sup_{g\in\mathcal{G}_n} \left| \frac{1}{n}\sum_{i=1}^{n} g(Z_i) - \mathbf{E}\{g(Z)\} \right| \leq \frac{1}{k} \quad \text{for all } k \in \mathcal{N},$$

which implies (9.2). $\qquad\qquad\qquad\qquad\qquad\qquad\qquad\qquad\qquad\qquad\square$

9.2　Extension to Random L_1 Norm Covers

Supremum norm covers are often too large to satisfy (9.6). But clearly, (9.6) is not necessary for the proof of (9.2). To motivate a weaker condition let us again consider (9.5). For the sake of simplicity ignore the expected value $\mathbf{E}g(Z)$ (think of it as an average $\frac{1}{n}\sum_{j=n+1}^{2n} g(Z_j)$, where $Z_1,\ldots,Z_{2n}$ are i.i.d.). So instead of a data-independent set $\tilde{\mathcal{G}}_{n,\epsilon}$ we now try to construct a set $\bar{\mathcal{G}}_{n,\epsilon}$ depending on $Z_1,\ldots,Z_{2n}$ which satisfies

$$\left\{ \sup_{g\in\mathcal{G}_n} \left| \frac{1}{n}\sum_{j=1}^{n} g(Z_j) - \frac{1}{n}\sum_{j=n+1}^{2n} g(Z_j) \right| > \epsilon \right\}$$

$$\subset \left\{ \sup_{g\in\bar{\mathcal{G}}_{n,\epsilon}} \left| \frac{1}{n}\sum_{j=1}^{n} g(Z_j) - \frac{1}{n}\sum_{j=n+1}^{2n} g(Z_j) \right| > \epsilon' \right\}$$

for some ϵ' depending on ϵ (but not on n). Then it is clear that all we need is a data-dependent cover which can approximate each $g \in \mathcal{G}_n$ with respect to

$$\|g - h\|_{2n} = \frac{1}{2n}\sum_{j=1}^{2n} |g(Z_j) - h(Z_j)|.$$

To formulate this idea introduce the following covering numbers:

Definition 9.3. *Let* $\epsilon > 0$, *let* $\mathcal{G}$ *be a set of functions* $\mathcal{R}^d \to \mathcal{R}$, $1 \leq p < \infty$, *and let* ν *be a probability measure on* $\mathcal{R}^d$. *For a function* $f : \mathcal{R}^d \to \mathcal{R}$ *set*

$$\|f\|_{L_p(\nu)} := \left\{ \int |f(z)|^p d\nu \right\}^{\frac{1}{p}}.$$

(a) *Every finite collection of functions* $g_1,\ldots,g_N : \mathcal{R}^d \to \mathcal{R}$ *with the property that for every* $g \in \mathcal{G}$ *there is a* $j = j(g) \in \{1,\ldots,N\}$ *such that*

$$\|g - g_j\|_{L_p(\nu)} < \epsilon$$

is called an ϵ-**cover** *of* $\mathcal{G}$ *with respect to* $\|\cdot\|_{L_p(\nu)}$.
(b) *Let* $\mathcal{N}(\epsilon,\mathcal{G},\|\cdot\|_{L_p(\nu)})$ *be the size of the smallest* ϵ-cover *of* $\mathcal{G}$ *w.r.t.* $\|\cdot\|_{L_p(\nu)}$. *Take* $\mathcal{N}(\epsilon,\mathcal{G},\|\cdot\|_{L_p(\nu)}) = \infty$ *if no finite* ϵ-cover *exists. Then* $\mathcal{N}(\epsilon,\mathcal{G},\|\cdot\|_{L_p(\nu)})$ *is called an* ϵ-**covering number** *of* $\mathcal{G}$ *w.r.t.* $\|\cdot\|_{L_p(\nu)}$.

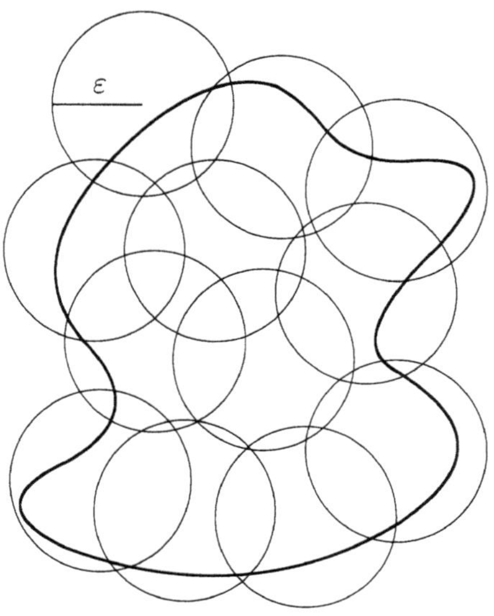

Figure 9.2. Example of ϵ-cover.

(c) *Let* $z_1^n = (z_1, \ldots, z_n)$ *be* n *fixed points in* $\mathcal{R}^d$. *Let* ν_n *be the corresponding empirical measure, i.e.,*

$$\nu_n(A) = \frac{1}{n} \sum_{i=1}^{n} I_A(z_i) \quad (A \subseteq \mathcal{R}^d).$$

Then

$$\|f\|_{L_p(\nu_n)} = \left\{ \frac{1}{n} \sum_{i=1}^{n} |f(z_i)|^p \right\}^{\frac{1}{p}}$$

and any ϵ-cover of $\mathcal{G}$ w.r.t. $\| \cdot \|_{L_p(\nu_n)}$ *will be called an L_p ϵ-**cover of** $\mathcal{G}$ **on** z_1^n *and the ϵ-covering number of $\mathcal{G}$ w.r.t.* $\| \cdot \|_{L_p(\nu_n)}$ *will be denoted by*

$$\mathcal{N}_p(\epsilon, \mathcal{G}, z_1^n).$$

In other words, $\mathcal{N}_p(\epsilon, \mathcal{G}, z_1^n)$ *is the minimal* $N \in \mathcal{N}$ *such that there exist functions* $g_1, \ldots, g_N : \mathcal{R}^d \to \mathcal{R}$ *with the property that for every* $g \in \mathcal{G}$ *there is a* $j = j(g) \in \{1, \ldots, N\}$ *such that*

$$\left\{ \frac{1}{n} \sum_{i=1}^{n} |g(z_i) - g_j(z_i)|^p \right\}^{\frac{1}{p}} < \epsilon.$$

If $Z_1^n = (Z_1, \ldots, Z_n)$ is a sequence of i.i.d. random variables, then $\mathcal{N}_1(\epsilon, \mathcal{G}, Z_1^n)$ is a random variable, whose expected value plays a central role in our problem. With these notations we can extend the first half of

Lemma 9.1 from fixed supremum norm covers to L_1 covers on a random set of points.

Theorem 9.1. *Let $\mathcal{G}$ be a set of functions $g : \mathcal{R}^d \to [0, B]$. For any n, and any $\epsilon > 0$,*

$$\mathbf{P}\left\{\sup_{g \in \mathcal{G}}\left|\frac{1}{n}\sum_{i=1}^{n}g(Z_i) - \mathbf{E}\{g(Z)\}\right| > \epsilon\right\} \le 8\mathbf{E}\left\{\mathcal{N}_1(\epsilon/8, \mathcal{G}, Z_1^n)\right\}e^{-n\epsilon^2/(128B^2)}.$$

In the proof we use symmetrization and covering arguments, which we will also later apply in many other proofs, e.g., in the proofs of Theorems 11.2, 11.4, and 11.6.

PROOF. The proof will be divided into four steps.

STEP 1. Symmetrization by a ghost sample.

Replace the expectation inside the above probability by an empirical mean based on a ghost sample $Z_1'^n = (Z_1', \ldots, Z_n')$ of i.i.d. random variables distributed as Z and independent of Z_1^n. Let g^* be a function $g \in \mathcal{G}$ such that

$$\left|\frac{1}{n}\sum_{i=1}^{n}g(Z_i) - \mathbf{E}g(Z)\right| > \epsilon,$$

if there exists any such function, and let g^* be an other arbitrary function contained in $\mathcal{G}$, if such a function doesn't exist. Note that g^* depends on Z_1^n and that $\mathbf{E}\{g^*(Z)|Z_1^n\}$ is the expectation of $g^*(Z)$ with respect to Z. Application of Chebyshev's inequality yields

$$\mathbf{P}\left\{\left|\mathbf{E}\{g^*(Z)|Z_1^n\} - \frac{1}{n}\sum_{i=1}^{n}g^*(Z_i')\right| > \frac{\epsilon}{2}\,\bigg|\, Z_1^n\right\}$$

$$\le \frac{\mathbf{Var}\{g^*(Z)|Z_1^n\}}{n(\frac{\epsilon}{2})^2} \le \frac{\frac{B^2}{4}}{n \cdot \frac{\epsilon^2}{4}} = \frac{B^2}{n\epsilon^2},$$

where we have used $0 \le g^*(Z) \le B$ which implies

$$\mathbf{Var}\{g^*(Z)|Z_1^n\} = \mathbf{Var}\left\{g^*(Z) - \frac{B}{2}\,\bigg|\, Z_1^n\right\}$$

$$\le \mathbf{E}\left\{\left|g^*(Z) - \frac{B}{2}\right|^2\,\bigg|\, Z_1^n\right\}$$

$$\le \frac{B^2}{4}.$$

Thus, for $n \ge \frac{2B^2}{\epsilon^2}$, we have

$$\mathbf{P}\left\{\left|\mathbf{E}\{g^*(Z)|Z_1^n\} - \frac{1}{n}\sum_{i=1}^{n}g^*(Z_i')\right| \le \frac{\epsilon}{2}\,\bigg|\, Z_1^n\right\} \ge \frac{1}{2}. \tag{9.7}$$

Hence

$$\mathbf{P}\left\{\sup_{g\in\mathcal{G}}\left|\frac{1}{n}\sum_{i=1}^{n}g(Z_i)-\frac{1}{n}\sum_{i=1}^{n}g(Z_i')\right|>\frac{\epsilon}{2}\right\}$$

$$\geq\mathbf{P}\left\{\left|\frac{1}{n}\sum_{i=1}^{n}g^*(Z_i)-\frac{1}{n}\sum_{i=1}^{n}g^*(Z_i')\right|>\frac{\epsilon}{2}\right\}$$

$$\geq\mathbf{P}\left\{\left|\frac{1}{n}\sum_{i=1}^{n}g^*(Z_i)-\mathbf{E}\{g^*(Z)|Z_1^n\}\right|>\epsilon,\right.$$

$$\left.\left|\frac{1}{n}\sum_{i=1}^{n}g^*(Z_i')-\mathbf{E}\{g^*(Z)|Z_1^n\}\right|\leq\frac{\epsilon}{2}\right\}.$$

The last probability can be determined by computing in a first step the corresponding probability conditioned on Z_1^n, and by averaging in a second step the result with respect to Z_1^n. Whether

$$\left|\frac{1}{n}\sum_{i=1}^{n}g^*(Z_i)-\mathbf{E}\{g^*(Z)|Z_1^n\}\right|>\epsilon$$

holds or not, depends only on Z_1^n. If it holds, then the probability above conditioned on Z_1^n is equal to

$$\mathbf{P}\left\{\left|\frac{1}{n}\sum_{i=1}^{n}g^*(Z_i')-\mathbf{E}\{g^*(Z)|Z_1^n\}\right|\leq\frac{\epsilon}{2}\bigg|Z_1^n\right\},$$

otherwise it is zero. Using this, we get

$$\mathbf{P}\left\{\left|\frac{1}{n}\sum_{i=1}^{n}g^*(Z_i)-\mathbf{E}\{g^*(Z)|Z_1^n\}\right|>\epsilon,\right.$$

$$\left.\left|\frac{1}{n}\sum_{i=1}^{n}g^*(Z_i')-\mathbf{E}\{g^*(Z)|Z_1^n\}\right|\leq\frac{\epsilon}{2}\right\}$$

$$=\mathbf{E}\left\{I_{\{|\frac{1}{n}\sum_{i=1}^{n}g^*(Z_i)-\mathbf{E}\{g^*(Z)|Z_1^n\}|>\epsilon\}}\right.$$

$$\left.\times\mathbf{P}\left\{\left|\frac{1}{n}\sum_{i=1}^{n}g^*(Z_i')-\mathbf{E}\{g^*(Z)|Z_1^n\}\right|\leq\frac{\epsilon}{2}\bigg|Z_1^n\right\}\right\}$$

$$\geq\frac{1}{2}\mathbf{P}\left\{\left|\frac{1}{n}\sum_{i=1}^{n}g^*(Z_i)-\mathbf{E}\{g^*(Z)|Z_1^n\}\right|>\epsilon\right\}$$

$$= \frac{1}{2} \mathbf{P} \left\{ \sup_{g \in \mathcal{G}} \left| \frac{1}{n} \sum_{i=1}^{n} g(Z_i) - \mathbf{E} g(Z) \right| > \epsilon \right\},$$

where the last inequality follows from (9.7). Thus we have shown that, for $n \geq \frac{2B^2}{\epsilon^2}$,

$$\mathbf{P} \left\{ \sup_{g \in \mathcal{G}} \left| \frac{1}{n} \sum_{i=1}^{n} g(Z_i) - \mathbf{E}\{g(Z)\} \right| > \epsilon \right\}$$

$$\leq 2 \mathbf{P} \left\{ \sup_{g \in \mathcal{G}} \left| \frac{1}{n} \sum_{i=1}^{n} g(Z_i) - \frac{1}{n} \sum_{i=1}^{n} g(Z_i') \right| > \frac{\epsilon}{2} \right\}.$$

STEP 2. Introduction of additional randomness by random signs.
Let $U_1, \ldots, U_n$ be independent and uniformly distributed over $\{-1, 1\}$ and independent of $Z_1, \ldots, Z_n, Z_1', \ldots, Z_n'$. Because of the independence and identical distribution of $Z_1, \ldots, Z_n, Z_1', \ldots, Z_n'$, the joint distribution of $Z_1^n, Z_1'^n$, is not affected if one randomly interchanges the corresponding components of Z_1^n and $Z_1'^n$. Hence

$$\mathbf{P} \left\{ \sup_{g \in \mathcal{G}} \left| \frac{1}{n} \sum_{i=1}^{n} \left(g(Z_i) - g(Z_i') \right) \right| > \frac{\epsilon}{2} \right\}$$

$$= \mathbf{P} \left\{ \sup_{g \in \mathcal{G}} \left| \frac{1}{n} \sum_{i=1}^{n} U_i \left(g(Z_i) - g(Z_i') \right) \right| > \frac{\epsilon}{2} \right\}$$

$$\leq \mathbf{P} \left\{ \sup_{g \in \mathcal{G}} \left| \frac{1}{n} \sum_{i=1}^{n} U_i g(Z_i) \right| > \frac{\epsilon}{4} \right\} + \mathbf{P} \left\{ \sup_{g \in \mathcal{G}} \left| \frac{1}{n} \sum_{i=1}^{n} U_i g(Z_i') \right| > \frac{\epsilon}{4} \right\}$$

$$= 2 \mathbf{P} \left\{ \sup_{g \in \mathcal{G}} \left| \frac{1}{n} \sum_{i=1}^{n} U_i g(Z_i) \right| > \frac{\epsilon}{4} \right\}.$$

STEP 3. Conditioning and introduction of a covering.
Next we condition in the last probability on Z_1^n, which is equivalent to fixing $z_1, \ldots, z_n \in \mathcal{R}^d$ and to considering

$$\mathbf{P} \left\{ \exists g \in \mathcal{G} \; : \; \left| \frac{1}{n} \sum_{i=1}^{n} U_i g(z_i) \right| > \frac{\epsilon}{4} \right\}. \tag{9.8}$$

Let $\mathcal{G}_{\frac{\epsilon}{8}}$ be an $L_1 \; \frac{\epsilon}{8}$-cover of $\mathcal{G}$ on z_1^n. Fix $g \in \mathcal{G}$. Then there exists $\bar{g} \in \mathcal{G}_{\frac{\epsilon}{8}}$ such that

$$\frac{1}{n} \sum_{i=1}^{n} |g(z_i) - \bar{g}(z_i)| < \frac{\epsilon}{8}. \tag{9.9}$$

W.l.o.g. we may assume $0 \le \bar{g}(z) \le B$ (otherwise, truncate $\bar{g}$ at 0 and B).
Then (9.9) implies

$$
\left| \frac{1}{n} \sum_{i=1}^{n} U_i g(z_i) \right| = \left| \frac{1}{n} \sum_{i=1}^{n} U_i \bar{g}(z_i) + \frac{1}{n} \sum_{i=1}^{n} U_i (g(z_i) - \bar{g}(z_i)) \right|
$$

$$
\le \left| \frac{1}{n} \sum_{i=1}^{n} U_i \bar{g}(z_i) \right| + \frac{1}{n} \sum_{i=1}^{n} |g(z_i) - \bar{g}(z_i)|
$$

$$
< \left| \frac{1}{n} \sum_{i=1}^{n} U_i \bar{g}(z_i) \right| + \frac{\epsilon}{8}.
$$

Using this we can bound the probability in (9.8) by

$$
\mathbf{P} \left\{ \exists g \in \mathcal{G}_{\frac{\epsilon}{8}} : \left| \frac{1}{n} \sum_{i=1}^{n} U_i g(z_i) \right| + \frac{\epsilon}{8} > \frac{\epsilon}{4} \right\}
$$

$$
\le |\mathcal{G}_{\frac{\epsilon}{8}}| \max_{g \in \mathcal{G}_{\frac{\epsilon}{8}}} \mathbf{P} \left\{ \left| \frac{1}{n} \sum_{i=1}^{n} U_i g(z_i) \right| > \frac{\epsilon}{8} \right\}.
$$

Choose $\mathcal{G}_{\frac{\epsilon}{8}}$ as an L_1 $\frac{\epsilon}{8}$-cover on z_1^n of minimal size. Then we have

$$
\mathbf{P} \left\{ \exists g \in \mathcal{G} : \left| \frac{1}{n} \sum_{i=1}^{n} U_i g(z_i) \right| > \frac{\epsilon}{4} \right\}
$$

$$
\le \mathcal{N}_1 \left(\frac{\epsilon}{8}, \mathcal{G}, z_1^n \right) \max_{g \in \mathcal{G}_{\frac{\epsilon}{8}}} \mathbf{P} \left\{ \left| \frac{1}{n} \sum_{i=1}^{n} U_i g(z_i) \right| > \frac{\epsilon}{8} \right\}.
$$

STEP 4. Application of Hoeffding's inequality.
In this step we bound

$$
\mathbf{P} \left\{ \left| \frac{1}{n} \sum_{i=1}^{n} U_i g(z_i) \right| > \frac{\epsilon}{8} \right\},
$$

where $z_1, \ldots, z_n \in \mathcal{R}^d$, $g : \mathcal{R}^d \to \mathcal{R}$, and $0 \le g(z) \le B$.
Since $U_1 g(z_1), \ldots, U_n g(z_n)$ are independent random variables with

$$
-B \le U_i g(z_i) \le B \quad (i = 1, \ldots, n),
$$

we have, by Hoeffding's inequality,

$$
\mathbf{P} \left\{ \left| \frac{1}{n} \sum_{i=1}^{n} U_i g(z_i) \right| > \frac{\epsilon}{8} \right\} \le 2 \exp \left(-\frac{2n \left(\frac{\epsilon}{8} \right)^2}{(2B)^2} \right) \le 2 \exp \left(-\frac{n \epsilon^2}{128 \, B^2} \right).
$$

In the case of $n \ge 2B^2/\epsilon^2$ the assertion is now implied by the four steps.
For $n < 2B^2/\epsilon^2$ the bound on the probability trivially holds, because the
right-hand side is greater than one. $\qquad \square$

9.3 Covering and Packing Numbers

From Theorem 9.1 one can derive (9.2) by an application of the Borel–Cantelli lemma, if one has suitable upper bounds for the L_1 covering numbers. These upper bounds will be derived in the sequel.

We first introduce the concept of L_p packing numbers and study the relationship between packing and covering numbers. These results will be used to obtain upper bounds on covering numbers.

Definition 9.4. *Let $\epsilon > 0$, let $\mathcal{G}$ be a set of functions $\mathcal{R}^d \to \mathcal{R}$, $1 \leq p < \infty$, and let ν be a probability measure on $\mathcal{R}^d$. Recall that, for a function $f : \mathcal{R}^d \to \mathcal{R}$,*

$$\|f\|_{L_p(\nu)} := \left\{ \int |f(z)|^p d\nu \right\}^{\frac{1}{p}}.$$

(a) *Every finite collection of functions $g_1, \ldots, g_N \in \mathcal{G}$ with*

$$\|g_j - g_k\|_{L_p(\nu)} \geq \epsilon$$

*for all $1 \leq j < k \leq N$ is called an ϵ-**packing** of $\mathcal{G}$ w.r.t. $\| \cdot \|_{L_p(\nu)}$.*
(b) *Let $\mathcal{M}(\epsilon, \mathcal{G}, \| \cdot \|_{L_p(\nu)})$ be the size of the largest ϵ-packing of $\mathcal{G}$ w.r.t. $\| \cdot \|_{L_p(\nu)}$. Take $\mathcal{M}(\epsilon, \mathcal{G}, \| \cdot \|_{L_p(\nu)}) = \infty$ if there exists an ϵ-packing of $\mathcal{G}$ w.r.t. $\| \cdot \|_{L_p(\nu)}$ of size N for every $N \in \mathcal{N}$. Then $\mathcal{M}(\epsilon, \mathcal{G}, \| \cdot \|_{L_p(\nu)})$ is called an ϵ-**packing number** of $\mathcal{G}$ w.r.t. $\| \cdot \|_{L_p(\nu)}$.*
(c) *Let $z_1^n = (z_1, \ldots, z_n)$ be n fixed points in $\mathcal{R}^d$. Let ν_n be the corresponding empirical measure, i.e.,*

$$\nu_n(A) = \frac{1}{n} \sum_{i=1}^{n} I_A(z_i) \quad (A \subseteq \mathcal{R}^d).$$

Then

$$\|f\|_{L_p(\nu_n)} = \left\{ \frac{1}{n} \sum_{i=1}^{n} |f(z_i)|^p \right\}^{\frac{1}{p}}$$

*and any ϵ-packing of $\mathcal{G}$ w.r.t. $\| \cdot \|_{L_p(\nu_n)}$ will be called an L_p ϵ-**packing of** $\mathcal{G}$ **on** z_1^n and the ϵ-packing number of $\mathcal{G}$ w.r.t. $\| \cdot \|_{L_p(\nu_n)}$ will be denoted by*

$$\mathcal{M}_p(\epsilon, \mathcal{G}, z_1^n).$$

In other words, $\mathcal{M}_p(\epsilon, \mathcal{G}, z_1^n)$ is the maximal $N \in \mathcal{N}$ such that there exist functions $g_1, \ldots, g_N \in \mathcal{G}$ with

$$\left\{ \frac{1}{n} \sum_{i=1}^{n} |g_j(z_i) - g_k(z_i)|^p \right\}^{\frac{1}{p}} \geq \epsilon$$

for all $1 \leq j < k \leq N$.

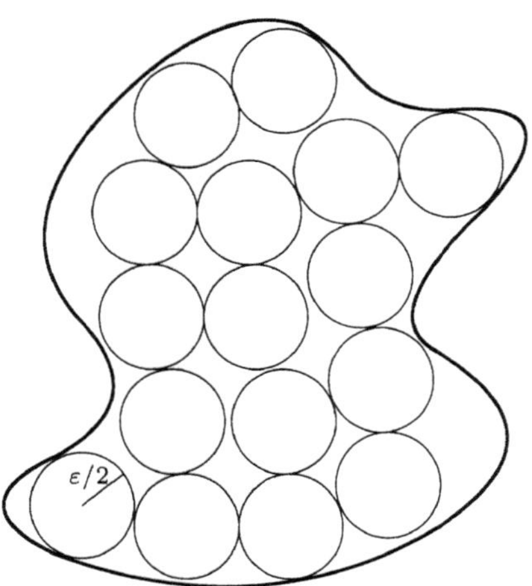

Figure 9.3. ϵ-packing.

As the next lemma shows, the L_p packing numbers are closely related to the L_p covering numbers.

Lemma 9.2. *Let $\mathcal{G}$ be a class of functions on $\mathcal{R}^d$ and let ν be a probability measure on $\mathcal{R}^d$, $p \geq 1$ and $\epsilon > 0$. Then*

$$\mathcal{M}(2\epsilon, \mathcal{G}, \|\cdot\|_{L_p(\nu)}) \leq \mathcal{N}(\epsilon, \mathcal{G}, \|\cdot\|_{L_p(\nu)}) \leq \mathcal{M}(\epsilon, \mathcal{G}, \|\cdot\|_{L_p(\nu)}).$$

In particular,

$$\mathcal{M}_p(2\epsilon, \mathcal{G}, z_1^n) \leq \mathcal{N}_p(\epsilon, \mathcal{G}, z_1^n) \leq \mathcal{M}_p(\epsilon, \mathcal{G}, z_1^n)$$

for all $z_1, \ldots, z_n \in \mathcal{R}^d$.

PROOF. Let $\{f_1, \ldots, f_l\}$ be a 2ϵ-packing of $\mathcal{G}$ w.r.t. $\|\cdot\|_{L_p(\nu)}$. Then any set

$$U_\epsilon(g) = \left\{ h : \mathcal{R}^d \to \mathcal{R} \ : \ \|h - g\|_{L_p(\nu)} < \epsilon \right\}$$

can contain at most one of the f_i's. This proves the first inequality.

In the proof of the second inequality, we assume $\mathcal{M}(\epsilon, \mathcal{G}, \|\cdot\|_{L_p(\nu)}) < \infty$ (otherwise the proof is trivial). Let $\{g_1, \ldots, g_l\}$ be an ϵ-packing of $\mathcal{G}$ w.r.t. $\|\cdot\|_{L_p(\nu)}$ of maximal cardinality $l = \mathcal{M}(\epsilon, \mathcal{G}, \|\cdot\|_{L_p(\nu)})$. Let $h \in \mathcal{G}$ be arbitrary. Then $\{h, g_1, \ldots, g_l\}$ is a subset of $\mathcal{G}$ of cardinality $l + 1$, hence cannot be an ϵ-packing of $\mathcal{G}$ with respect to $\|\cdot\|_{L_p(\nu)}$. Thus there exists $j \in \{1, \ldots, l\}$ such that

$$\|h - g_j\|_{L_p(\nu)} < \epsilon.$$

This proves that $\{g_1, \ldots, g_l\}$ is an ϵ-cover of $\mathcal{G}$ with respect to $\|\cdot\|_{L_p(\nu)}$, which implies

$$\mathcal{N}(\epsilon, \mathcal{G}, \|\cdot\|_{L_p(\nu)}) \leq l = \mathcal{M}(\epsilon, \mathcal{G}, \|\cdot\|_{L_p(\nu)}).$$

$\square$

Next we use the above lemma to bound L_2 ϵ-covering numbers on points z_1^n of balls in linear vector spaces.

Lemma 9.3. *Let $\mathcal{F}$ be a set of functions $f : \mathcal{R}^d \to \mathcal{R}$. Assume that $\mathcal{F}$ is a linear vector space of dimension D. Then one has, for arbitrary $R > 0$, $\epsilon > 0$, and $z_1, \ldots, z_n \in \mathcal{R}^d$,*

$$\mathcal{N}_2\left(\epsilon, \left\{f \in \mathcal{F} : \frac{1}{n}\sum_{i=1}^n |f(z_i)|^2 \leq R^2\right\}, z_1^n\right) \leq \left(\frac{4R + \epsilon}{\epsilon}\right)^D.$$

PROOF. For $f, g : \mathcal{R}^d \to \mathcal{R}$ set

$$<f, g>_n := \frac{1}{n}\sum_{i=1}^n f(z_i)g(z_i) \quad \text{and} \quad \|f\|_n^2 := <f, f>_n.$$

Let $\{f_1, \ldots, f_N\}$ be an arbitrary ϵ-packing of $\{f \in \mathcal{F} : \|f\|_n \leq R\}$ w.r.t. $\|\cdot\|_n$, i.e., $f_1, \ldots, f_N \in \{f \in \mathcal{F} : \|f\|_n \leq R\}$ satisfy

$$\|f_i - f_j\|_n \geq \epsilon \quad \text{for all } 1 \leq i < j \leq N.$$

Because of Lemma 9.2 it suffices to show

$$N \leq \left(\frac{4R + \epsilon}{\epsilon}\right)^D. \tag{9.10}$$

In order to show (9.10) let $B_1, \ldots, B_D$ be a basis of the linear vector space $\mathcal{F}$. Then, for any $a_1, b_1, \ldots, a_D, b_D \in \mathcal{R}$,

$$\|\sum_{j=1}^D a_j B_j - \sum_{j=1}^D b_j B_j\|_n^2$$

$$= <\sum_{j=1}^D (a_j - b_j)B_j, \sum_{j=1}^D (a_j - b_j)B_j >_n = (a - b)^T B(a - b),$$

where

$$B = (<B_j, B_k>_n)_{1 \leq j,k \leq D} \quad \text{and} \quad (a - b) = (a_1 - b_1, \ldots, a_D - b_D)^T.$$

Because of

$$a^T B a = \|\sum_{j=1}^D a_j B_j\|_n^2 \geq 0 \quad (a \in \mathcal{R}^D),$$

the symmetric matrix B is positive semidefinite, hence there exists a symmetric matrix $B^{1/2}$ such that

$$B = B^{1/2} \cdot B^{1/2}.$$

We have

$$\left\| \sum_{j=1}^{D} a_j B_j - \sum_{j=1}^{D} b_j B_j \right\|_n^2 = (a - b)^T B^{1/2} B^{1/2}(a - b) = \|B^{1/2}(a - b)\|^2,$$

where $\|\cdot\|$ is the Euclidean norm in $\mathcal{R}^D$.

Because $f_i \in \mathcal{F}$ we get

$$f_i = \sum_{j=1}^{D} a_j^{(i)} B_j$$

for some

$$a^{(i)} = (a_1^{(i)}, \ldots, a_D^{(i)})^T \in \mathcal{R}^D \quad (i = 1, \ldots, N).$$

It follows

$$\|B^{1/2} a^{(i)}\| = \|f_i\|_n \leq R$$

and

$$\|B^{1/2} a^{(i)} - B^{1/2} a^{(j)}\| = \|f_i - f_j\|_n \geq \epsilon$$

for all $1 \leq i < j \leq N$. Hence the N balls in $\mathcal{R}^d$ with centers $B^{1/2} a^{(1)}$, ..., $B^{1/2} a^{(N)}$ and radius $\epsilon/4$ are disjoint subsets of the ball with center zero and radius $R + \epsilon/4$. By comparing the volumes of the balls one gets

$$N \cdot c_D \cdot \left(\frac{\epsilon}{4}\right)^D \leq c_D \cdot \left(R + \frac{\epsilon}{4}\right)^D,$$

where c_D is the volume of a ball with radius one in $\mathcal{R}^D$. This implies the assertion. $\square$

9.4 Shatter Coefficients and VC Dimension

In this section we derive bounds for L_p covering numbers of sets of functions, which are not necessarily subsets of some finite-dimensional vector space of functions. Therefore we need the following definition:

Definition 9.5. *Let $\mathcal{A}$ be a class of subsets of $\mathcal{R}^d$ and let $n \in \mathcal{N}$.*
(a) For $z_1, \ldots, z_n \in \mathcal{R}^d$ define

$$s(\mathcal{A}, \{z_1, \ldots, z_n\}) = |\{A \cap \{z_1, \ldots, z_n\} : A \in \mathcal{A}\}|,$$

that is, $s(\mathcal{A}, \{z_1, \ldots, z_n\})$ is the number of different subsets of $\{z_1, \ldots, z_n\}$ of the form $A \cap \{z_1, \ldots, z_n\}$, $A \in \mathcal{A}$.

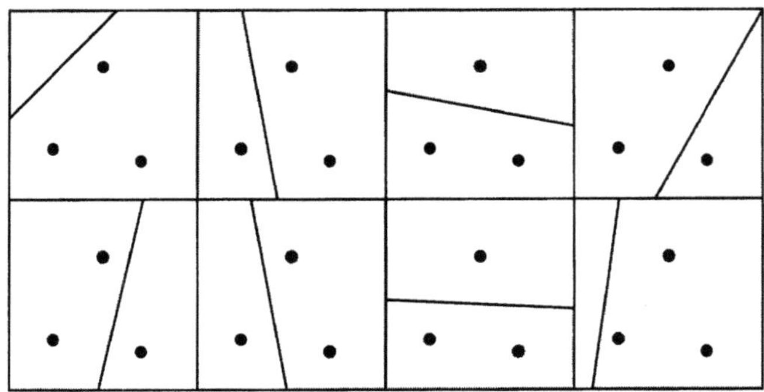

Figure 9.4. Three points can be shattered by half-spaces on the plane.

(b) *Let G be a subset of $\mathcal{R}^d$ of size n. One says that $\mathcal{A}$* **shatters** *G if $s(\mathcal{A}, G) = 2^n$, i.e., if each subset of G can be represented in the form $A \cap G$ for some $A \in \mathcal{A}$.*

(c) *The nth* **shatter coefficient** *of $\mathcal{A}$ is*

$$S(\mathcal{A}, n) = \max_{\{z_1,\ldots,z_n\} \subseteq \mathcal{R}^d} s\left(\mathcal{A}, \{z_1, \ldots, z_n\}\right).$$

That is, the shatter coefficient is the maximal number of different subsets of n points that can be picked out by sets from $\mathcal{A}$.

Clearly, $s\left(\mathcal{A}, \{z_1, \ldots, z_n\}\right) \leq 2^n$ and $S(\mathcal{A}, n) \leq 2^n$. If $S(\mathcal{A}, n) < 2^n$ then $s\left(\mathcal{A}, \{z_1, \ldots, z_n\}\right) < 2^n$ for all $z_1, \ldots, z_n \in \mathcal{R}^d$. If $s\left(\mathcal{A}, \{z_1, \ldots, z_n\}\right) < 2^n$, then $\{z_1, \ldots, z_n\}$ has a subset such that there is no set in $\mathcal{A}$ that contains exactly that subset of $\{z_1, \ldots, z_n\}$.

It is easy to see that $S(\mathcal{A}, k) < 2^k$ implies $S(\mathcal{A}, n) < 2^n$ for all $n > k$. The last time when $S(\mathcal{A}, k) = 2^k$ is important:

Definition 9.6. *Let $\mathcal{A}$ be a class of subsets of $\mathcal{R}^d$ with $\mathcal{A} \neq \emptyset$. The* **VC dimension** *(or Vapnik–Chervonenkis dimension) $V_\mathcal{A}$ of $\mathcal{A}$ is defined by*

$$V_\mathcal{A} = \sup\left\{n \in \mathcal{N} \ : \ S(\mathcal{A}, n) = 2^n\right\},$$

i.e., the VC dimension $V_\mathcal{A}$ is the largest integer n such that there exists a set of n points in $\mathcal{R}^d$ which can be shattered by $\mathcal{A}$.

Example 9.1. *The class of all intervals in $\mathcal{R}$ of the form $(-\infty, b]$ ($b \in \mathcal{R}$) fails to pick out the largest of any two distinct points, hence its VC dimension is 1. The class of all intervals in $\mathcal{R}$ of the form $(a, b]$ ($a, b \in \mathcal{R}$) shatters every two-point set but cannot pick out the largest and the smallest point of any set of three distinct points. Thus its VC dimension is 2.*

Our next two theorems, which we will use later to derive bounds on L_p packing numbers, state the suprising fact that either $S(\mathcal{A}, n) = 2^n$ for all n (in which case $V_\mathcal{A} = \infty$) or $S(\mathcal{A}, n)$ is bounded by some polynomial in n of degree $V_\mathcal{A} < \infty$.

Theorem 9.2. *Let $\mathcal{A}$ be a set of subsets of $\mathcal{R}^d$ with VC dimension $V_\mathcal{A}$. Then, for any $n \in \mathcal{N}$,*

$$S(\mathcal{A}, n) \le \sum_{i=0}^{V_\mathcal{A}} \binom{n}{i}.$$

PROOF. Let $z_1, \ldots, z_n$ be any n distinct points. Clearly, it suffices to show that

$$s(\mathcal{A}, \{z_1, \ldots, z_n\}) = |\{A \cap \{z_1, \ldots, z_n\} \ : \ A \in \mathcal{A}\}| \le \sum_{i=0}^{V_\mathcal{A}} \binom{n}{i}.$$

Denote by $F_1, \ldots, F_k$ the collection of all $k = \binom{n}{V_\mathcal{A}+1}$ subsets of $\{z_1, \ldots, z_n\}$ of size $V_\mathcal{A} + 1$. By the definition of VC dimension, $\mathcal{A}$ shatters none of the sets F_i, i.e., for each $i \in \{1, \ldots, k\}$ there exists $H_i \subseteq F_i$ such that

$$A \cap F_i \ne H_i \quad \text{for all } A \in \mathcal{A}. \tag{9.11}$$

Now $F_i \subseteq \{z_1, \ldots, z_n\}$ implies $A \cap F_i = (A \cap \{z_1, \ldots, z_n\}) \cap F_i$ and, hence, (9.11) can be rewritten as

$$(A \cap \{z_1, \ldots, z_n\}) \cap F_i \ne H_i \quad \text{for all } A \in \mathcal{A}. \tag{9.12}$$

Set

$$\mathcal{C}_0 = \{C \subseteq \{z_1, \ldots, z_n\} \ : \ C \cap F_i \ne H_i \text{ for each } i\}.$$

Then (9.12) implies

$$\{A \cap \{z_1, \ldots, z_n\} \ : \ A \in \mathcal{A}\} \subseteq \mathcal{C}_0,$$

hence it suffices to prove

$$|\mathcal{C}_0| \le \sum_{i=0}^{V_\mathcal{A}} \binom{n}{i}.$$

This is easy in one special case: If $H_i = F_i$ for every i, then

$$C \cap F_i \ne H_i \quad \Leftrightarrow \quad C \cap F_i \ne F_i \quad \Leftrightarrow \quad F_i \not\subseteq C,$$

which implies that $\mathcal{C}_0$ consists of all subsets of $\{z_1, \ldots, z_n\}$ of cardinality less than $V_\mathcal{A} + 1$, and hence

$$|\mathcal{C}_0| = \sum_{i=0}^{V_\mathcal{A}} \binom{n}{i}.$$

We will reduce the general case to the special case just treated.

For each i define

$$H_i' = (H_i \cup \{z_1\}) \cap F_i,$$

that is, augment H_i by z_1 provided z_1 is contained in F_i. Set

$$\mathcal{C}_1 = \{C \subseteq \{z_1, \ldots, z_n\} \; : \; C \cap F_i \neq H_i' \text{ for each } i\}.$$

We will show that the cardinality of $\mathcal{C}_1$ is not less than the cardinality of $\mathcal{C}_0$. (*Notice*: This is not equivalent to $\mathcal{C}_0 \subseteq \mathcal{C}_1$.)

The sets $\mathcal{C}_0$ and $\mathcal{C}_1$ can be written as disjoint unions

$$\mathcal{C}_0 = (\mathcal{C}_0 \cap \mathcal{C}_1) \cup (\mathcal{C}_0 \setminus \mathcal{C}_1), \quad \mathcal{C}_1 = (\mathcal{C}_0 \cap \mathcal{C}_1) \cup (\mathcal{C}_1 \setminus \mathcal{C}_0),$$

therefore is suffices to prove

$$|\mathcal{C}_0 \setminus \mathcal{C}_1| \leq |\mathcal{C}_1 \setminus \mathcal{C}_0|.$$

To do this, we will show that the map

$$C \mapsto C \setminus \{z_1\}$$

is one-to-one from $\mathcal{C}_0 \setminus \mathcal{C}_1$ into $\mathcal{C}_1 \setminus \mathcal{C}_0$.

Let $C \in \mathcal{C}_0 \setminus \mathcal{C}_1$. Then $C \cap F_i \neq H_i$ for each i, but $C \cap F_{i_0} = H_{i_0}'$ for some i_0. Clearly, this implies

$$H_{i_0}' = (H_{i_0} \cup \{z_1\}) \cap F_{i_0} \neq H_{i_0},$$

and hence z_1 does not belong to H_{i_0}, but belongs to F_{i_0}, H_{i_0}', and C.

Because $z_1 \in C$ for all $C \in \mathcal{C}_0 \setminus \mathcal{C}_1$, stripping of the point z_1 defines a one-to-one map.

It remains to show that $C \setminus \{z_1\} \in \mathcal{C}_1 \setminus \mathcal{C}_0$. Because of $C \cap F_{i_0} = H_{i_0}'$ and $z_1 \notin H_{i_0}$,

$$(C \setminus \{z_1\}) \cap F_{i_0} = (C \cap F_{i_0}) \setminus \{z_1\} = H_{i_0}' \setminus \{z_1\} = H_{i_0},$$

hence $C \setminus \{z_1\} \notin \mathcal{C}_0$. In addition, $C \setminus \{z_1\} \in \mathcal{C}_1$, because if $z_1 \notin F_i$, then $C \in \mathcal{C}_0$ implies

$$(C \setminus \{z_1\}) \cap F_i = C \cap F_i \neq H_i = H_i';$$

and if $z_1 \in F_i$, then z_1 is also contained in H_i', but certainly not in $C \setminus \{z_1\}$, hence

$$(C \setminus \{z_1\}) \cap F_i \neq H_i'.$$

This proves that the cardinality of $\mathcal{C}_1$ is not less than the cardinality of $\mathcal{C}_0$. By repeating this procedure $(n-1)$ times (with $z_2, z_3, \ldots$ taken instead of z_1, and starting with $\mathcal{C}_1, \mathcal{C}_2, \ldots$ instead of $\mathcal{C}_0$) one generates classes $\mathcal{C}_2$, $\mathcal{C}_3, \ldots, \mathcal{C}_n$ with

$$|\mathcal{C}_0| \leq |\mathcal{C}_1| \leq \cdots \leq |\mathcal{C}_n|.$$

The sets H_i in the definition of $\mathcal{C}_n$ satisfy $H_i = F_i$ and, hence the assertion follows from the special case already considered. $\qquad\square$

Theorem 9.3. *Let $\mathcal{A}$ be a set of subsets of $\mathcal{R}^d$ with VC dimension $V_\mathcal{A} < \infty$. Then, for all $n \in \mathcal{N}$,*

$$S(\mathcal{A}, n) \leq (n+1)^{V_\mathcal{A}},$$

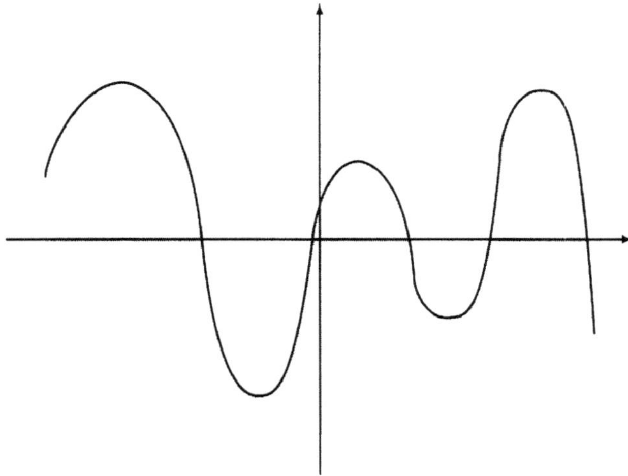

Figure 9.5. Subgraph of a function.

and, for all $n \geq V_{\mathcal{A}}$,

$$S(\mathcal{A}, n) \leq \left(\frac{e\,n}{V_{\mathcal{A}}}\right)^{V_{\mathcal{A}}}.$$

PROOF. Theorem 9.2, together with the binomial theorem, imply

$$S(\mathcal{A}, n) \leq \sum_{i=0}^{V_{\mathcal{A}}} \binom{n}{i} = \sum_{i=0}^{V_{\mathcal{A}}} \frac{n!}{(n-i)!} \cdot \frac{1}{i!} \leq \sum_{i=0}^{V_{\mathcal{A}}} n^i \cdot \binom{V_{\mathcal{A}}}{i} = (n+1)^{V_{\mathcal{A}}}.$$

If $V_{\mathcal{A}}/n \leq 1$ then, again by Theorem 9.2 and the binomial theorem,

$$
\left(\frac{V_{\mathcal{A}}}{n}\right)^{V_{\mathcal{A}}} S(\mathcal{A}, n) \;\leq\; \left(\frac{V_{\mathcal{A}}}{n}\right)^{V_{\mathcal{A}}} \sum_{i=0}^{V_{\mathcal{A}}} \binom{n}{i} \leq \sum_{i=0}^{V_{\mathcal{A}}} \left(\frac{V_{\mathcal{A}}}{n}\right)^i \binom{n}{i}
$$

$$
\leq \sum_{i=0}^{n} \left(\frac{V_{\mathcal{A}}}{n}\right)^i \binom{n}{i} = \left(1 + \frac{V_{\mathcal{A}}}{n}\right)^n \leq e^{V_{\mathcal{A}}}.
$$

$\square$

Next we use these results to derive upper bounds on L_p packing numbers. Let $\mathcal{G}$ be a class of functions on $\mathcal{R}^d$ taking their values in $[0, B]$. To bound the L_p packing number of $\mathcal{G}$ we will use the VC dimension of the set

$$\mathcal{G}^+ := \left\{ \{(z, t) \in \mathcal{R}^d \times \mathcal{R} \,;\, t \leq g(z)\} \,;\, g \in \mathcal{G} \right\}$$

of all subgraphs of functions of $\mathcal{G}$.

Theorem 9.4. *Let $\mathcal{G}$ be a class of functions $g : \mathcal{R}^d \to [0, B]$ with $V_{\mathcal{G}^+} \geq 2$, let $p \geq 1$, let ν be a probability measure on $\mathcal{R}^d$, and let $0 < \epsilon < \frac{B}{4}$. Then*

$$\mathcal{M}\left(\epsilon, \mathcal{G}, \|\cdot\|_{L_p(\nu)}\right) \leq 3 \left(\frac{2eB^p}{\epsilon^p} \log \frac{3eB^p}{\epsilon^p}\right)^{V_{\mathcal{G}^+}}.$$

PROOF. The proof is divided into four steps. In the first three steps we prove the assertion in the case $p = 1$, in the fourth step we reduce the general case to the case $p = 1$.

STEP 1. We first relate the packing number $\mathcal{M}\left(\epsilon, \mathcal{G}, \|\cdot\|_{L_1(\nu)}\right)$ to a shatter coefficient of $\mathcal{G}^+$. Set $m = \mathcal{M}\left(\epsilon, \mathcal{G}, \|\cdot\|_{L_1(\nu)}\right)$ and let $\overline{\mathcal{G}} = \{g_1, \ldots, g_m\}$ be an ϵ-packing of $\mathcal{G}$ w.r.t. $\|\cdot\|_{L_1(\nu)}$.

Let $Q_1, \ldots, Q_k \in \mathcal{R}^d$ be k independent random variables with common distribution ν. Generate k independent random variables $T_1, \ldots, T_k$ uniformly distributed on $[0, B]$, which are also independent from $Q_1, \ldots, Q_k$. Denote $R_i = (Q_i, T_i)$ $(i = 1, \ldots, k)$, $G_f = \{(z, t) : t \leq f(z)\}$ for $f : \mathcal{R}^d \to [0, B]$, and $R_1^k = \{R_1, \ldots, R_k\}$. Then

$$S(\mathcal{G}^+, k) = \max_{\{z_1,\ldots,z_k\} \in \mathcal{R}^d \times \mathcal{R}} s(\mathcal{G}^+, \{z_1, \ldots, z_k\})$$

$$\geq \mathbf{E}s(\mathcal{G}^+, \{R_1, \ldots, R_k\})$$

$$\geq \mathbf{E}s(\{G_f : f \in \overline{\mathcal{G}}\}, \{R_1, \ldots, R_k\})$$

$$\geq \mathbf{E}s(\{G_f : f \in \overline{\mathcal{G}}, \, G_f \cap R_1^k \neq G_g \cap R_1^k \text{ for all } g \in \overline{\mathcal{G}}, g \neq f\}, R_1^k)$$

$$= \mathbf{E}\left\{\sum_{f \in \overline{\mathcal{G}}} I_{\{G_f \cap R_1^k \neq G_g \cap R_1^k \text{ for all } g \in \overline{\mathcal{G}}, g \neq f\}}\right\}$$

$$= \sum_{f \in \overline{\mathcal{G}}} \mathbf{P}\{G_f \cap R_1^k \neq G_g \cap R_1^k \text{ for all } g \in \overline{\mathcal{G}}, g \neq f\}$$

$$= \sum_{f \in \overline{\mathcal{G}}} (1 - \mathbf{P}\{\exists g \in \overline{\mathcal{G}}, g \neq f, G_f \cap R_1^k = G_g \cap R_1^k\})$$

$$\geq \sum_{f \in \overline{\mathcal{G}}} (1 - m \max_{g \in \overline{\mathcal{G}}, g \neq f} \mathbf{P}\{G_f \cap R_1^k = G_g \cap R_1^k\}). \tag{9.13}$$

Fix $f, g \in \overline{\mathcal{G}}, f \neq g$. By the independence and identical distribution of $R_1, \ldots, R_k$,

$$\mathbf{P}\{G_f \cap R_1^k = G_g \cap R_1^k\}$$

$$= \mathbf{P}\{G_f \cap \{R_1\} = G_g \cap \{R_1\}, \ldots, G_f \cap \{R_k\} = G_g \cap \{R_k\}\}$$

$$= (\mathbf{P}\{G_f \cap \{R_1\} = G_g \cap \{R_1\}\})^k.$$

Now

$$\mathbf{P}\{G_f \cap \{R_1\} = G_g \cap \{R_1\}\}$$
$$= 1 - \mathbf{E}\left\{\mathbf{P}\{ G_f \cap \{R_1\} \neq G_g \cap \{R_1\}\big|Q_1 \}\right\}$$
$$= 1 - \mathbf{E}\left\{\mathbf{P}\left\{ f(Q_1) < T_1 \leq g(Q_1) \text{ or } g(Q_1) < T_1 \leq f(Q_1)\big|Q_1\right\}\right\}$$
$$= 1 - \mathbf{E}\left\{\frac{|f(Q_1) - g(Q_1)|}{B}\right\}$$
$$= 1 - \frac{1}{B}\int |f(x) - g(x)|\nu(dx)$$
$$\leq 1 - \frac{\epsilon}{B}$$

since f and g are ϵ-separated. Hence

$$\mathbf{P}\{G_f \cap R_1^k = G_g \cap R_1^k\} \leq \left(1 - \frac{\epsilon}{B}\right)^k \leq \exp\left(-\frac{\epsilon k}{B}\right)$$

which, together with (9.13), implies

$$S(\mathcal{G}^+, k) \;\geq\; \sum_{f \in \overline{\mathcal{G}}}\left(1 - m \max_{g \in \overline{\mathcal{G}}, g \neq f} \mathbf{P}\{G_f \cap R_1^k = G_g \cap R_1^k\}\right)$$
$$\geq\; m\left(1 - m\exp\left(-\frac{\epsilon k}{B}\right)\right). \tag{9.14}$$

Set $k = \left\lfloor \frac{B}{\epsilon}\log(2m)\right\rfloor$. Then

$$1 - m\exp\left(-\frac{\epsilon k}{B}\right)$$
$$\geq 1 - m\exp\left(-\frac{\epsilon}{B}\left(\frac{B}{\epsilon}\log(2m) - 1\right)\right) = 1 - \frac{1}{2}\exp\left(\frac{\epsilon}{B}\right)$$
$$\geq 1 - \frac{1}{2}\exp\left(\frac{1}{4}\right) \geq 1 - \frac{1}{2}\cdot 1.3 \geq \frac{1}{3},$$

hence

$$\mathcal{M}\left(\epsilon, \mathcal{G}, \|\cdot\|_{L_1(\nu)}\right) = m \leq 3S(\mathcal{G}^+, k), \tag{9.15}$$

where $k = \left\lfloor \frac{B}{\epsilon}\log\left(2\mathcal{M}\left(\epsilon, \mathcal{G}, \|\cdot\|_{L_1(\nu)}\right)\right)\right\rfloor$.
STEP 2. Application of Theorem 9.3.
If

$$k = \left\lfloor \frac{B}{\epsilon}\log\left(2\mathcal{M}\left(\epsilon, \mathcal{G}, \|\cdot\|_{L_1(\nu)}\right)\right)\right\rfloor \leq V_{\mathcal{G}^+}$$

then

$$\mathcal{M}\left(\epsilon, \mathcal{G}, \|\cdot\|_{L_1(\nu)}\right)$$

$$\leq \frac{1}{2}\exp\left(\frac{\epsilon(V_{\mathcal{G}^+}+1)}{B}\right) \leq \frac{e}{2}\exp(V_{\mathcal{G}^+}) \leq 3\left(\frac{2eB}{\epsilon}\log\frac{3eB}{\epsilon}\right)^{V_{\mathcal{G}^+}},$$

where we have used $0 < \epsilon \leq B/4$. Therefore it suffices to prove the assertion in the case

$$k > V_{\mathcal{G}^+}.$$

In this case we can apply Theorem 9.3 to (9.15) and conclude

$$\mathcal{M}\left(\epsilon, \mathcal{G}, \|\cdot\|_{L_1(\nu)}\right) \;\leq\; 3\left(\frac{ek}{V_{\mathcal{G}^+}}\right)^{V_{\mathcal{G}^+}}$$

$$\leq\; 3\left(\frac{eB}{\epsilon V_{\mathcal{G}^+}}\log\left(2\mathcal{M}\left(\epsilon, \mathcal{G}, \|\cdot\|_{L_1(\nu)}\right)\right)\right)^{V_{\mathcal{G}^+}}.$$

STEP 3. Let $a \in \mathcal{R}^+, b \in \mathcal{N}$, with $a \geq e$ and $b \geq 2$. We will show that

$$x \leq 3\left\{\frac{a}{b}\log(2x)\right\}^b$$

implies

$$x \leq 3(2a\log(3a))^b. \tag{9.16}$$

Setting $a = \frac{eB}{\epsilon}$ and $b = V_{\mathcal{G}^+}$ this, together with Step 2, implies the assertion in the case $p = 1$.

Note that

$$x \leq 3\left\{\frac{a}{b}\log(2x)\right\}^b$$

is equivalent to

$$(2x)^{1/b} \leq 6^{1/b}\frac{a}{b}\log(2x) = 6^{1/b}a\log((2x)^{1/b}).$$

Set $u = (2x)^{1/b}$ and $c = 6^{1/b}a$. Then $e \leq a \leq c$ and the last inequality can be rewritten

$$u \leq c\log(u). \tag{9.17}$$

We will show momentarily that this implies

$$u \leq 2c\log(c). \tag{9.18}$$

From (9.18) one easily concludes (9.16). Indeed,

$$x = \frac{1}{2}u^b \leq \frac{1}{2}(2c\log c)^b = \frac{1}{2}(2\cdot 6^{1/b}a\log(6^{1/b}a))^b \leq 3(2a\log(3a))^b,$$

where the last inequality follows from $6^{1/b} \leq 3$ for $b \geq 2$.

In conclusion we will show that (9.17) implies (9.18). Set $f_1(u) = u$ and $f_2(u) = c\log(u)$. Then it suffices to show

$$f_1(u) > f_2(u)$$

for $u > 2c \log(c)$. Because

$$f_1'(u) = 1 \geq \frac{1}{2\log(e)} \geq \frac{1}{2\log(c)} = \frac{c}{2c\log(c)} \geq \frac{c}{u} = f_2'(u)$$

for $u > 2c\log(c)$, this is equivalent to

$$f_1(2c\log(c)) > f_2(2c\log(c)).$$

This in turn is equivalent to

$$
\begin{aligned}
2c\log(c) > c\log(2c\log(c)) \quad &\Leftrightarrow \quad 2c\log(c) > c\log(2) + c\log(c) + c\log(\log(c)) \\
&\Leftrightarrow \quad c\log(c) - c\log(2) - c\log(\log(c)) > 0 \\
&\Leftrightarrow \quad \log\left(\frac{c}{2\log(c)}\right) > 0 \\
&\Leftrightarrow \quad c > 2\log(c). \quad\quad (9.19)
\end{aligned}
$$

Set $g_1(v) = v$ and $g_2(v) = 2\log(v)$. Then

$$g_1(e) = e > 2\log(e) = g_2(e)$$

and for $v \geq e$ one has

$$g_1'(v) = 1 \geq \frac{2}{v} = g_2'(v).$$

This proves

$$g_1(v) > g_2(v)$$

for $v \geq e$, which together with $c \geq e$ implies (9.19). Steps 1 to 3 imply the assertion in the case $p = 1$.

STEP 4. Let $1 < p < \infty$. Then for any $g_j, g_k \in \mathcal{G}$,

$$\|g_j - g_k\|_{L_p(\nu)}^p \leq B^{p-1}\|g_j - g_k\|_{L_1(\nu)}.$$

Therefore any ϵ-packing of $\mathcal{G}$ w.r.t. $\|\cdot\|_{L_p(\nu)}$ is also an $\frac{\epsilon^p}{B^{p-1}}$ packing of $\mathcal{G}$ w.r.t. $\|\cdot\|_{L_1(\nu)}$ which, together with the results of the first three steps, implies

$$
\begin{aligned}
\mathcal{M}\left(\epsilon, \mathcal{G}, \|\cdot\|_{L_p(\nu)}\right) &\leq \mathcal{M}\left(\frac{\epsilon^p}{B^{p-1}}, \mathcal{G}, \|\cdot\|_{L_1(\nu)}\right) \\
&\leq 3\left(\frac{2eB}{\epsilon^p/B^{p-1}}\log\frac{3eB}{\epsilon^p/B^{p-1}}\right)^{V_{\mathcal{G}+}} \\
&= 3\left(\frac{2eB^p}{\epsilon^p}\log\frac{3eB^p}{\epsilon^p}\right)^{V_{\mathcal{G}+}}.
\end{aligned}
$$

The proof is complete. $\qquad\square$

In order to derive, via Lemma 9.2 and Theorem 9.4, upper bounds on L_p packing numbers, all we need now is an upper bound on the VC dimension

$V_{\mathcal{G}+}$. We have

$$
\begin{aligned}
\mathcal{G}^+ &= \left\{ \{(z,t) \in \mathcal{R}^d \times \mathcal{R} \, : \, t \le g(z)\} \, : \, g \in \mathcal{G} \right\} \\
&\subseteq \left\{ \{(z,t) \in \mathcal{R}^d \times \mathcal{R} \, : \, \alpha \cdot t + g(z) \ge 0\} \, : \, g \in \mathcal{G}, \alpha \in \mathcal{R} \right\}.
\end{aligned}
$$

If $\mathcal{G}$ is a linear vector space of dimension K, then $\{\alpha \cdot t + g(z) \, : \, g \in \mathcal{G}, \alpha \in \mathcal{R}\}$ is a linear vector space of dimension $r = K + 1$ and by the following theorem we get $V_{\mathcal{G}+} \le r$.

Theorem 9.5. *Let $\mathcal{G}$ be an r-dimensional vector space of real functions on $\mathcal{R}^d$, and set*

$$
\mathcal{A} = \left\{ \{z \, : \, g(z) \ge 0\} \, : \, g \in \mathcal{G} \right\}.
$$

Then

$$
V_{\mathcal{A}} \le r.
$$

PROOF. It suffices to show that no set of size $r + 1$ can be shattered by sets of the form $\{z \, : \, g(z) \ge 0\}$, $g \in \mathcal{G}$.

Choose any collection $\{z_1, \ldots, z_{r+1}\}$ of distinct points from $\mathcal{R}^d$. Define the linear mapping $L : \mathcal{G} \to \mathcal{R}^{r+1}$ by

$$
L(g) = (g(z_1), \ldots, g(z_{r+1}))^T \quad (g \in \mathcal{G}).
$$

Denote the image of $\mathcal{G}$ by $L\mathcal{G}$. Clearly, $L\mathcal{G}$ is a linear subspace of the $(r+1)$-dimensional space $\mathcal{R}^{r+1}$, and the dimension of $L\mathcal{G}$ is less than or equal to the dimension of $\mathcal{G}$, i.e., it is at most r. Hence there exists a nonzero vector

$$
\gamma = (\gamma_1, \ldots, \gamma_{r+1})^T \in \mathcal{R}^{r+1},
$$

that is orthogonal to $L\mathcal{G}$, i.e., that satisfies

$$
\gamma_1 g(z_1) + \cdots + \gamma_{r+1} g(z_{r+1}) = 0 \quad \text{for all } g \in \mathcal{G}. \tag{9.20}
$$

Replacing γ by $-\gamma$ if necessary, we may assume that at least one of the γ_i's is negative. Equation (9.20) implies

$$
\sum_{i:\gamma_i \ge 0} \gamma_i g(z_i) = \sum_{i:\gamma_i < 0} (-\gamma_i) g(z_i) \quad \text{for all } g \in \mathcal{G}. \tag{9.21}
$$

Suppose there is a $g \in \mathcal{G}$ for which $\{z \, : \, g(z) \ge 0\}$ picks out precisely those points z_i with $\gamma_i \ge 0$. For this g, the left-hand side of (9.21) would be nonnegative (because $\gamma_i \ge 0$ implies $g(z_i) \ge 0$ and hence $\gamma_i g(z_i) \ge 0$), but the right-hand side would be negative (because $\gamma_i < 0$ implies $g(z_i) < 0$ and hence $-\gamma_i g(z_i) < 0$). This is a contradiction, so $\{z_1, \ldots, z_{r+1}\}$ cannot be shattered and the proof is complete. $\qquad \square$

For the sake of illustration let us compare the results which we can derive from Theorem 9.4 with Lemma 9.3. Let $\mathcal{F}$ be a linear vector space of dimension D consisting of functions $f : \mathcal{R}^d \to \mathcal{R}$ and let $\epsilon, R > 0$ and $z_1, \ldots, z_n \in \mathcal{R}^d$. Then Lemma 9.2, and Theorems 9.4 and 9.5 imply

$$\mathcal{N}_2\left(\epsilon, \{f \in \mathcal{F} \ : \ \|f\|_\infty \leq R\}, z_1^n\right) \leq 3\left(\frac{2e(2R)^2}{\epsilon^2} \cdot \log \frac{3e(2R)^2}{\epsilon^2}\right)^{D+1},$$

$$(9.22)$$

while by Lemma 9.3 we have

$$\mathcal{N}_2\left(\epsilon, \left\{f \in \mathcal{F} \ : \ \frac{1}{n}\sum_{i=1}^n |f(z_i)|^2 \leq R^2\right\}, z_1^n\right) \leq \left(\frac{4R + \epsilon}{\epsilon}\right)^D. \qquad (9.23)$$

Because of

$$\{f \in \mathcal{F} \ : \ \|f\|_\infty \leq R\} \subseteq \left\{f \in \mathcal{F} \ : \ \frac{1}{n}\sum_{i=1}^n |f(z_i)|^2 \leq R^2\right\},$$

formula (9.23) implies

$$\mathcal{N}_2\left(\epsilon, \{f \in \mathcal{F} \ : \ \|f\|_\infty \leq R\}, z_1^n\right) \leq \left(\frac{4R + \epsilon}{\epsilon}\right)^D,$$

so in this special case we get a bound similar to (9.22).

The advantage of (9.23) in comparison to (9.22) is that in some special cases (cf. Chapter 19) we will be able to apply it for bounds R on the empirical L_2 norm of the functions of similar size as ϵ, in which case the covering number will be bounded by

$$\text{const}^{D+1}.$$

On the other hand, in all of our applications the bound in (9.22) will always be much larger than the above term, because the bound on the supremum norm of the functions will always be much larger than ϵ.

The advantage of Theorem 9.4 is that it can (and will) be applied to much more general situations than Lemma 9.3.

9.5   A Uniform Law of Large Numbers

Let $Z, Z_1, Z_2, \ldots$ be independent and identically distributed random variables with values in $\mathcal{R}^d$. Let $\mathcal{G}$ be a class of functions $g : \mathcal{R}^d \to \mathcal{R}$ such that $\mathbf{E}g(Z)$ exists for each $g \in \mathcal{G}$. One says that $\mathcal{G}$ satisfies the **uniform law of large numbers (ULLN)**, if

$$\sup_{g \in \mathcal{G}} \left| \frac{1}{n}\sum_{i=1}^n g(Z_i) - \mathbf{E}g(Z) \right| \to 0 \quad (n \to \infty) \quad a.s.$$

In order to illustrate our previous results we apply them to derive the following ULLN.

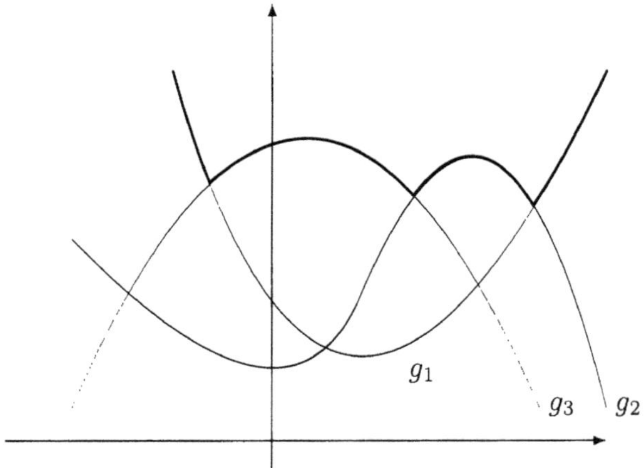

Figure 9.6. Envelope of $\{g_1, g_2, g_3\}$.

Theorem 9.6. *Let $\mathcal{G}$ be a class of functions $g : \mathcal{R}^d \to \mathcal{R}$ and let*

$$G : \mathcal{R}^d \to \mathcal{R}, \quad G(x) := \sup_{g \in \mathcal{G}} |g(x)| \quad (x \in \mathcal{R}^d)$$

be an envelope of $\mathcal{G}$. Assume $\mathbf{E}G(Z) < \infty$ and $V_{\mathcal{G}^+} < \infty$. Then

$$\sup_{g \in \mathcal{G}} \left| \frac{1}{n} \sum_{i=1}^{n} g(Z_i) - \mathbf{E}g(Z) \right| \to 0 \quad (n \to \infty) \quad a.s.$$

PROOF. For $L > 0$ set

$$\mathcal{G}_L := \left\{ g \cdot I_{\{G \leq L\}} \; : \; g \in \mathcal{G} \right\}.$$

For any $g \in \mathcal{G}$,

$$\left| \frac{1}{n} \sum_{i=1}^{n} g(Z_i) - \mathbf{E}g(Z) \right|$$

$$\leq \left| \frac{1}{n} \sum_{i=1}^{n} g(Z_i) - \frac{1}{n} \sum_{i=1}^{n} g(Z_i) \cdot I_{\{G(Z_i) \leq L\}} \right|$$

$$+ \left| \frac{1}{n} \sum_{i=1}^{n} g(Z_i) \cdot I_{\{G(Z_i) \leq L\}} - \mathbf{E}\left\{ g(Z) \cdot I_{\{G(Z) \leq L\}} \right\} \right|$$

$$+ \left| \mathbf{E}\left\{ g(Z) \cdot I_{\{G(Z) \leq L\}} \right\} - \mathbf{E}\left\{ g(Z) \right\} \right|$$

$$\leq \left| \frac{1}{n} \sum_{i=1}^{n} g(Z_i) \cdot I_{\{G(Z_i) \leq L\}} - \mathbf{E}\left\{ g(Z) \cdot I_{\{G(Z) \leq L\}} \right\} \right|$$

$$+\frac{1}{n}\sum_{i=1}^{n} G(Z_i) \cdot I_{\{G(Z_i)>L\}} + \mathbf{E}\left\{G(Z) \cdot I_{\{G(Z)>L\}}\right\},$$

which implies

$$\sup_{g \in \mathcal{G}} \left| \frac{1}{n}\sum_{i=1}^{n} g(Z_i) - \mathbf{E}g(Z) \right|$$

$$\leq \sup_{g \in \mathcal{G}_L} \left| \frac{1}{n}\sum_{i=1}^{n} g(Z_i) - \mathbf{E}g(Z) \right|$$

$$+\frac{1}{n}\sum_{i=1}^{n} G(Z_i) \cdot I_{\{G(Z_i)>L\}} + \mathbf{E}\left\{G(Z) \cdot I_{\{G(Z)>L\}}\right\}.$$

By $\mathbf{E}G(Z) < \infty$ and the strong law of large numbers we get

$$\frac{1}{n}\sum_{i=1}^{n} G(Z_i) \cdot I_{\{G(Z_i)>L\}} \to \mathbf{E}\left\{G(Z) \cdot I_{\{G(Z)>L\}}\right\} \quad (n \to \infty) \quad a.s.$$

and

$$\mathbf{E}\left\{G(Z) \cdot I_{\{G(Z)>L\}}\right\} \to 0 \quad (L \to \infty).$$

Hence, it suffices to show

$$\sup_{g \in \mathcal{G}_L} \left| \frac{1}{n}\sum_{i=1}^{n} g(Z_i) - \mathbf{E}g(Z) \right| \to 0 \quad (n \to \infty) \quad a.s. \qquad (9.24)$$

Let $\epsilon > 0$ be arbitrary. The functions in $\mathcal{G}_L$ are bounded in absolute value by L. Application of Theorem 9.1, Lemma 9.2, and Theorem 9.4 yields

$$\mathbf{P}\left\{ \sup_{g \in \mathcal{G}_L} \left| \frac{1}{n}\sum_{i=1}^{n} g(Z_i) - \mathbf{E}g(Z) \right| > \epsilon \right\}$$

$$\leq 8\mathbf{E}\left\{ \mathcal{N}_1\left(\frac{\epsilon}{8}, \mathcal{G}_L, Z_1^n\right) \right\} \cdot \exp\left(-\frac{n\epsilon^2}{128(2L)^2}\right)$$

$$\leq 8\mathbf{E}\left\{ \mathcal{M}_1\left(\frac{\epsilon}{8}, \mathcal{G}_L, Z_1^n\right) \right\} \cdot \exp\left(-\frac{n\epsilon^2}{512L^2}\right)$$

$$\leq 24\left(\frac{2e(2L)}{\epsilon/8} \cdot \log\frac{3e(2L)}{\epsilon/8} \right)^{V_{\mathcal{G}_L^+}} \cdot \exp\left(-\frac{n\epsilon^2}{512L^2}\right).$$

If $\{(x_1, y_1), \ldots, (x_k, y_k)\}$ is shattered by $\mathcal{G}_L^+$, then $|y_l| \leq G(x_l)$ $(l = 1, \ldots, k)$ and, hence, is also shattered by $\mathcal{G}^+$. Therefore $V_{\mathcal{G}_L^+} \leq V_{\mathcal{G}^+}$ which, together with the above results, yields

$$\mathbf{P}\left\{\sup_{g\in\mathcal{G}_L}\left|\frac{1}{n}\sum_{i=1}^{n}g(Z_i)-\mathbf{E}g(Z)\right|>\epsilon\right\}$$

$$\leq 24\left(\frac{2e(2L)}{\epsilon/8}\cdot\log\frac{3e(2L)}{\epsilon/8}\right)^{V_{\mathcal{G}^+}}\cdot\exp\left(-\frac{n\epsilon^2}{512L^2}\right).$$

The right-hand side is summable for each $\epsilon>0$ which, together with the Borel–Cantelli lemma, implies (9.24). $\qquad\square$

9.6 Bibliographic Notes

Theorem 9.1 is due to Pollard (1984). The symmetrization technique used in the proof of Pollard's inequality follows the ideas of Vapnik and Chervonenkis (1971), which were later extended by Dudley (1978), Pollard (1984), and Giné (1996). Theorem 9.1 is a generalization of the celebrated Vapnik-Chervonenkis inequality for uniform deviations of relative frequencies from their probabilities, see Vapnik and Chervonenkis (1971).

Various extensions of this inequality were provided in Devroye (1982a), Alexander (1984), Massart (1990), Talagrand (1994), van der Vaart and Wellner (1996), and Vapnik (1998). Various aspects of the empirical process theory and computational learning are discussed in Alexander (1984), Dudley (1984), Shorack and Wellner (1986), Pollard (1989; 1990), Ledoux and Talagrand (1991), Talagrand (1994), Giné (1996), Devroye, Györfi, and Lugosi (1996), Ledoux (1996), Gaenssler and Ross (1999), van de Geer (2000), van der Vaart and Wellner (1996), Vapnik (1998), and Bartlett and Anthony (1999). Theorem 9.2, known in the literature as the Sauer lemma, has been proved independently by Vapnik and Chervonenkis (1971), Sauer (1972), and Shelah (1972) and its extensions were studied by Szarek and Talagrand (1997), Alesker (1997), and Alon, Ben-David, and Haussler (1997).

The inequality of Theorem 9.4 is, for $p=1$, due to Haussler (1992). Theorem 9.5 was proved by Steele (1975) and Dudley (1978). There are much more general versions of uniform laws of large numbers in the literature than Theorem 9.6, see, e.g., van de Geer (2000).

Problems and Exercises

PROBLEM 9.1. (a) Let $\mathcal{A}$ be the class of all intervals in $\mathcal{R}$ of the form $(-\infty,b]$ $(b\in\mathcal{R})$. Show

$$S(\mathcal{A},n)=n+1.$$

(b) Let $\mathcal{A}$ be the class of all intervals in $\mathcal{R}$ of the form $(a,b]$ $(a,b\in\mathcal{R})$. Show

$$S(\mathcal{A},n)=\frac{n\cdot(n+1)}{2}.$$

(c) Generalize (a) and (b) to the multivariate case.

PROBLEM 9.2. (a) Show that the VC dimension of the set of all intervals in $\mathcal{R}^d$, of the form

$$(-\infty, x_1] \times \cdots \times (-\infty, x_d] \quad (x_1, \ldots, x_d \in \mathcal{R})$$

is d.

(b) Show that the VC dimension of the set of all intervals in $\mathcal{R}^d$, of the form

$$(x_1, y_1] \times \cdots \times (x_d, y_d] \quad (x_1, y_1, \ldots, x_d, y_d \in \mathcal{R}),$$

is $2 \cdot d$.

PROBLEM 9.3. (a) Determine the VC dimension of the set of all balls in $\mathcal{R}^2$.
(b) Use Lemma 9.5 to derive an upper bound for the VC dimension of the set of all balls in $\mathcal{R}^d$.

PROBLEM 9.4. Let $\mathcal{A}$ be a class of sets $A \subseteq \mathcal{R}^d$. Show for any $p \geq 1$, any $z_1, \ldots, z_n \in \mathcal{R}^d$, and any $0 < \epsilon < 1$,

$$\mathcal{N}_p\left(\epsilon, \{I_A \ : \ A \in \mathcal{A}\}, z_1^n\right) \leq s\left(\mathcal{A}, \{z_1, \ldots, z_n\}\right).$$

HINT: Use

$$\left\{\frac{1}{n} \sum_{i=1}^{n} |g_1(z_i) - g_2(z_i)|^p\right\}^{1/p} \leq \max_{i=1,\ldots,n} |g_1(z_i) - g_2(z_i)|.$$

PROBLEM 9.5. Let $Z, Z_1, Z_2, \ldots$ be i.i.d. real-valued random variables. Let F be the distribution function of Z given by

$$F(t) = \mathbf{P}\{Z \leq t\} = \mathbf{E}\{I_{(-\infty, t]}(Z)\},$$

and let F_n be the empirical distribution function of $Z_1, \ldots, Z_n$ given by

$$F_n(t) = \frac{\#\{1 \leq i \leq n \ : \ Z_i \leq t\}}{n} = \frac{1}{n} \sum_{i=1}^{n} I_{(-\infty, t]}(Z_i).$$

(a) Show, for any $0 < \epsilon < 1$,

$$\mathbf{P}\left\{\sup_{t \in \mathcal{R}} |F_n(t) - F(t)| \geq \epsilon\right\} \leq 8 \cdot (n+1) \cdot \exp\left(-\frac{n \cdot \epsilon^2}{128}\right).$$

(b) Conclude from a) the Glivenko–Cantelli theorem:

$$\sup_{t \in \mathcal{R}} |F_n(t) - F(t)| \to 0 \quad (n \to \infty) \quad a.s.$$

(c) Generalize (a) and (b) to multivariate (empirical) distribution functions.
HINT: Apply Theorem 9.1 and Problems 9.1 and 9.4.

10
Least Squares Estimates I: Consistency

In this chapter we show how one can use the techniques introduced in Chapter 9 to derive sufficient conditions for the consistency of various least squares estimates.

10.1 Why and How Least Squares?

We know from Section 1.1 that the regression function m satisfies

$$\mathbf{E}\left\{(m(X) - Y)^2\right\} = \inf_f \mathbf{E}\left\{(f(X) - Y)^2\right\},$$

where the infimum is taken over all measurable functions $f : \mathcal{R}^d \to \mathcal{R}$, thus $\mathbf{E}\left\{(m(X) - Y)^2\right\}$ can be computed by minimizing $\mathbf{E}\left\{(f(X) - Y)^2\right\}$ over all measurable functions. Clearly, this is impossible in the regression function estimation problem, because the functional to be optimized depends on the unknown distribution of (X, Y).

The idea of the least squares principle is to estimate the L_2 risk

$$\mathbf{E}\left\{(f(X) - Y)^2\right\}$$

by the empirical L_2 risk

$$\frac{1}{n} \sum_{j=1}^{n} |f(X_j) - Y_j|^2 \tag{10.1}$$

and to choose as a regression function estimate a function that minimizes this empirical L_2 risk.

If $X_1, \ldots, X_n$ are all distinct (which happens with probability 1 if X has a density), then minimizing (10.1) leads to an estimate interpolating the data $(X_1, Y_1), \ldots, (X_n, Y_n)$ and having empirical L_2 risk 0. Obviously, such an estimate will not be consistent in general.

Therefore one first chooses a "suitable" class of functions $\mathcal{F}_n$ (maybe depending on the data but, at least, depending on the sample size n) and then selects a function from this class which minimizes the empirical L_2 risk, i.e., one defines the estimate m_n by

$$m_n(\cdot) = \arg\min_{f \in \mathcal{F}_n} \frac{1}{n} \sum_{j=1}^{n} |f(X_j) - Y_j|^2, \tag{10.2}$$

which means, by definition,

$$m_n \in \mathcal{F}_n \text{ and } \frac{1}{n} \sum_{j=1}^{n} |m_n(X_j) - Y_j|^2 = \min_{f \in \mathcal{F}_n} \frac{1}{n} \sum_{j=1}^{n} |f(X_j) - Y_j|^2.$$

Here we assumed the existence of minimizing functions, though not necessarily their uniqueness. In cases where the minima do not exist, the same analysis can be carried out with functions whose error is arbitrarily close to the infimum but, for the sake of simplicity, we maintain the assumption of existence throughout the book. We will show later (see (10.4) and (10.5)) that in most of our applications the minima indeed exist.

The class of candidate functions grows as the sample size n grows. This is the "method of sieves," introduced by Grenander (1981).

The choice of $\mathcal{F}_n$ has two effects on the error of the estimate. On one hand, if $\mathcal{F}_n$ is not too "massive" (and we will later give precise conditions on $\mathcal{F}_n$ using the concepts introduced in Chapter 9), then the empirical L_2 risk will be close to the L_2 risk uniformly over $\mathcal{F}_n$. Thus the error introduced by minimizing the empirical L_2 risk instead of the L_2 risk will be small. On the other hand, because of the requirement that our estimate is contained in $\mathcal{F}_n$, it cannot be better (with respect to the L_2 error) than for the best function in $\mathcal{F}_n$. This is formulated in the following lemma:

Lemma 10.1. *Let* $\mathcal{F}_n = \mathcal{F}_n(D_n)$ *be a class of functions* $f : \mathcal{R}^d \to \mathcal{R}$ *depending on the data* $D_n = \{(X_1, Y_1), \ldots, (X_n, Y_n)\}$. *If* m_n *satisfies (10.2) then*

$$\int |m_n(x) - m(x)|^2 \mu(dx)$$

$$\leq 2 \sup_{f \in \mathcal{F}_n} \left| \frac{1}{n} \sum_{j=1}^{n} |f(X_j) - Y_j|^2 - \mathbf{E}\left\{ (f(X) - Y)^2 \right\} \right|$$

$$+ \inf_{f \in \mathcal{F}_n} \int |f(x) - m(x)|^2 \mu(dx).$$

PROOF. It follows from Section 1.1 (cf. (1.1)) that

$$\int |m_n(x) - m(x)|^2 \mu(dx) = \mathbf{E}\{|m_n(X) - Y|^2 | D_n\} - \mathbf{E}\{|m(X) - Y|^2\}$$

$$= \left(\mathbf{E}\{|m_n(X) - Y|^2 | D_n\} - \inf_{f \in \mathcal{F}_n} \mathbf{E}\{|f(X) - Y|^2\} \right)$$

$$+ \left(\inf_{f \in \mathcal{F}_n} \mathbf{E}\{|f(X) - Y|^2\} - \mathbf{E}\{|m(X) - Y|^2\} \right). \qquad (10.3)$$

By (1.1)

$$\inf_{f \in \mathcal{F}_n} \mathbf{E}\{|f(X) - Y|^2\} - \mathbf{E}\{|m(X) - Y|^2\} = \inf_{f \in \mathcal{F}_n} \int |f(x) - m(x)|^2 \mu(dx).$$

Thus all we need is an upper bound for the first term. By (10.2), one gets

$$\mathbf{E}\left\{|m_n(X) - Y|^2 \Big| D_n\right\} - \inf_{f \in \mathcal{F}_n} \mathbf{E}\left\{|f(X) - Y|^2\right\}$$

$$= \sup_{f \in \mathcal{F}_n} \left(\mathbf{E}\left\{|m_n(X) - Y|^2 \Big| D_n\right\} - \frac{1}{n} \sum_{j=1}^{n} |m_n(X_j) - Y_j|^2 \right.$$

$$+ \frac{1}{n} \sum_{j=1}^{n} |m_n(X_j) - Y_j|^2 - \frac{1}{n} \sum_{j=1}^{n} |f(X_j) - Y_j|^2$$

$$+ \frac{1}{n} \sum_{j=1}^{n} |f(X_j) - Y_j|^2 - \mathbf{E}\left\{|f(X) - Y|^2\right\} \Bigg)$$

$$\leq \sup_{f \in \mathcal{F}_n} \left(\mathbf{E}\left\{|m_n(X) - Y|^2 \Big| D_n\right\} - \frac{1}{n} \sum_{j=1}^{n} |m_n(X_j) - Y_j|^2 \right.$$

$$+ \frac{1}{n} \sum_{j=1}^{n} |f(X_j) - Y_j|^2 - \mathbf{E}\left\{|f(X) - Y|^2\right\} \Bigg)$$

$$\leq 2 \sup_{f \in \mathcal{F}_n} \left| \frac{1}{n} \sum_{j=1}^{n} |f(X_j) - Y_j|^2 - \mathbf{E}\left\{|f(X) - Y|^2\right\} \right|.$$

$$\square$$

Often the first term on the right-hand side of (10.3), i.e.,

$$\mathbf{E}\{|m_n(X) - Y|^2 | D_n\} - \inf_{f \in \mathcal{F}_n} \mathbf{E}\{|f(X) - Y|^2\},$$

is called the *estimation error*, and the second term, i.e.,

$$\inf_{f \in \mathcal{F}_n} \mathbf{E}\{|f(X) - Y|^2\} - \mathbf{E}\{|m(X) - Y|^2\} = \inf_{f \in \mathcal{F}_n} \int |f(x) - m(x)|^2 \mu(dx),$$

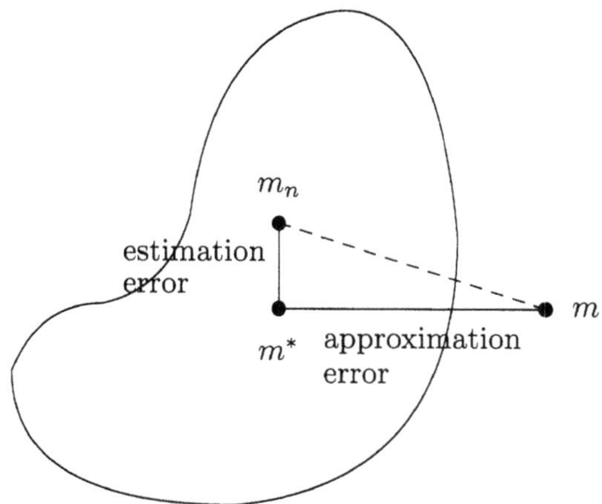

Figure 10.1. Approximation and estimation errors.

is called the *approximation error* of the estimator. The estimation error measures the distance between the L_2 risk of the estimate and the L_2 risk of the best function in $\mathcal{F}_n$. The approximation error measures how well the regression function can be approximated by functions of $\mathcal{F}_n$ in L_2 (compare Figure 10.1).

In order to get universally consistent estimates it suffices to show that both terms converge to 0 for all distributions of (X, Y) with $\mathbf{E}Y^2 < \infty$.

For the approximation error this is often quite simple: If, for example, $\mathcal{F}_n$ is nested, that is, $\mathcal{F}_n \subseteq \mathcal{F}_{n+1}$ for all n, then

$$\lim_{n \to \infty} \inf_{f \in \mathcal{F}_n} \int |f(x) - m(x)|^2 \mu(dx) = 0$$

for all measures μ and all $m \in L_2$ simply means that $\bigcup_{n=1}^{\infty} \mathcal{F}_n$ is dense in $L_2 = L_2(\mu)$ for all distributions μ. This is true, for example, if $\bigcup_{n=1}^{\infty} \mathcal{F}_n$ is dense in $C_0^{\infty}(\mathcal{R}^d)$ with respect to the sup norm $\|\cdot\|_{\infty}$ since $C_0^{\infty}(\mathcal{R}^d)$ is dense in $L_2(\mu)$ for all distributions μ and $\int |f(x) - m(x)|^2 \mu(dx) \leq \|f - m\|_{\infty}^2$ (cf. Corollary A.1).

The estimation error is more difficult. The main tools for analyzing it are exponential distribution-free inequalities for a uniform distance of the L_2 risk from empirical L_2 risk over the class $\mathcal{F}_n$, e.g., the inequalities which we have introduced in Chapter 9.

The inequalities we will use require the uniform boundedness of $|f(X) - Y|^2$ over $\mathcal{F}_n$. We will see in the next section that it suffices to show the convergence of the estimation error to 0 for bounded Y. So let us assume for the rest of this section that Y is bounded, i.e., $|Y| \leq L$ a.s. for some $L > 0$. Then, in order to ensure that $|f(X) - Y|^2$ is uniformly bounded over $\mathcal{F}_n$, one can simply choose $\mathcal{F}_n$ such that all functions in $\mathcal{F}_n$

are bounded by a constant $\beta_n > 0$, depending on the sample size n and converging to infinity (this is needed for the approximation error). A typical result which one can get in this way is the following theorem:

Theorem 10.1. *Let $\psi_1, \psi_2, \ldots : \mathcal{R}^d \to \mathcal{R}$ be bounded functions with $|\psi_j(x)| \le 1$. Assume that the set of functions*

$$\bigcup_{K=1}^{\infty} \left\{ \sum_{j=1}^{K} a_j \psi_j(x) : a_1, \ldots, a_K \in \mathcal{R} \right\}$$

is dense in $L_2(\mu)$ for any probability measure μ on $\mathcal{R}^d$. Define the regression function estimate m_n as a function minimizing the empirical L_2 risk

$$\frac{1}{n} \sum_{i=1}^{n} (f(X_i) - Y_i)^2$$

over functions $f(x) = \sum_{j=1}^{K_n} a_j \psi_j(x)$ with $\sum_{j=1}^{K_n} |a_j| \le \beta_n$. If $\mathbf{E}\{Y^2\} < \infty$, and K_n and β_n satisfy

$$K_n \to \infty, \quad \beta_n \to \infty, \quad \frac{K_n \beta_n^4 \log \beta_n}{n} \to 0 \ and \ \frac{\beta_n^4}{n^{1-\delta}} \to 0$$

for some $\delta > 0$, then $\int (m_n(x) - m(x))^2 \mu(dx) \to 0$ with probability one, i.e., the estimate is strongly universally consistent.

For the proof see Problem 10.2.

Unfortunately, the assumption that all functions in $\mathcal{F}_n$ are bounded by a constant $\beta_n > 0$, makes the computation of the estimator difficult. In most of our applications $\mathcal{F}_n$ will be defined as a set of linear combinations of some basis functions. Uniform boundedness in this case means that one has to restrict the values of the coefficients of these linear combinations, i.e., one would choose

$$\mathcal{F}_n = \left\{ \sum_{j=1}^{K_n} a_j f_{j,n} : \sum_{j=1}^{K_n} |a_j| \le \beta_n \right\}$$

for some bounded basis functions $f_{1,n}, \ldots, f_{K_n,n}$. To compute a function which minimizes the empirical L_2 risk over such a class one has to solve a quadratic minimization problem with inequality constraints for the coefficients a_j. There is no known fast algorithm which can do this.

If one doesn't require the uniform boundedness of $\mathcal{F}_n$ then the computation of a function which minimizes the empirical L_2 risk is much easier, since for arbitrary functions $f_{1,n}, \ldots, f_{K_n,n}$ and

$$\mathcal{F}_n = \left\{ \sum_{j=1}^{K_n} a_j f_{j,n} : a_j \in \mathcal{R} \ (j = 1, \ldots, K_n) \right\}$$

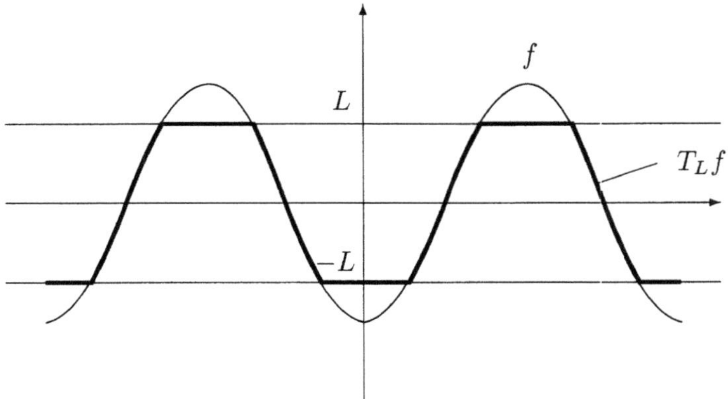

Figure 10.2. Truncation of function.

equation (10.2) is equivalent to $m_n = \sum_{j=1}^{K_n} a_j f_{j,n}$ with

$$\|\mathbf{Y} - \mathbf{Ba}\|_2^2 = \inf_{\mathbf{b} \in \mathcal{R}^{K_n}} \|\mathbf{Y} - \mathbf{Bb}\|_2^2, \tag{10.4}$$

where $\|\cdot\|_2$ is the Euclidean norm in $\mathcal{R}^{K_n}$,

$$\mathbf{Y} = (Y_1, \dots, Y_n)^T, \ \mathbf{B} = (f_{j,n}(X_i))_{i=1,\dots,n,\, j=1,\dots,K_n},$$

and

$$\mathbf{a} = (a_1, \dots, a_{K_n})^T.$$

It is well-known from numerical mathematics (cf. Stoer (1993), Chapter 4.8.1) that (10.4) is equivalent to

$$\mathbf{B}^T\mathbf{Ba} = \mathbf{B}^T\mathbf{Y} \tag{10.5}$$

and that a solution of (10.5) always exists. Thus all one has to do to compute a function which minimizes the empirical L_2 risk is to solve a system of linear equations.

Therefore, we do not require uniform boundedness of $\mathcal{F}_n$. To ensure the consistency of the estimator (cf. Problem 10.3) we truncate it after the computation, i.e., we choose $\tilde{m}_n$ such that

$$\tilde{m}_n \in \mathcal{F}_n \text{ and } \frac{1}{n}\sum_{j=1}^{n} |\tilde{m}_n(X_j) - Y_j|^2 = \min_{f \in \mathcal{F}_n} \frac{1}{n}\sum_{j=1}^{n} |f(X_j) - Y_j|^2 \tag{10.6}$$

and define the estimate m_n by truncation of $\tilde{m}_n$,

$$m_n(x) = T_{\beta_n}\tilde{m}_n(x), \tag{10.7}$$

where T_L is the truncation operator

$$T_L u = \begin{cases} u & \text{if } |u| \leq L, \\ L\,\text{sign}(u) & \text{otherwise,} \end{cases}$$

(compare Figure 10.2). The next lemma shows that this estimate behaves similarly to an estimate defined by empirical L_2 risk minimization over a class of truncated functions

$$T_{\beta_n}\mathcal{F}_n = \{T_{\beta_n}f : f \in \mathcal{F}_n\}.$$

Lemma 10.2. *Let $\mathcal{F}_n = \mathcal{F}_n(D_n)$ be a class of functions $f : \mathcal{R}^d \to \mathcal{R}$. If m_n satisfies (10.6) and (10.7) and $|Y| \leq \beta_n$ a.s., then*

$$\int |m_n(x) - m(x)|^2 \mu(dx)$$

$$\leq 2 \sup_{f \in T_{\beta_n}\mathcal{F}_n} \left| \frac{1}{n} \sum_{j=1}^{n} |f(X_j) - Y_j|^2 - \mathbf{E}\left\{(f(X) - Y)^2\right\} \right|$$

$$+ \inf_{f \in \mathcal{F}_n,\, \|f\|_\infty \leq \beta_n} \int |f(x) - m(x)|^2 \mu(dx).$$

PROOF. Using the decomposition from the proof of Lemma 10.1, with $\mathcal{F}_n$ replaced by $\{f \in \mathcal{F}_n : \|f\|_\infty \leq \beta_n\}$, one gets

$$\int |m_n(x) - m(x)|^2 \mu(dx)$$

$$= \mathbf{E}\{|m_n(X) - Y|^2|D_n\} - \inf_{f \in \mathcal{F}_n,\, \|f\|_\infty \leq \beta_n} \mathbf{E}\{|f(X) - Y|^2\}$$

$$+ \inf_{f \in \mathcal{F}_n,\, \|f\|_\infty \leq \beta_n} \int |f(x) - m(x)|^2 \mu(dx).$$

Now

$$\mathbf{E}\left\{|m_n(X) - Y|^2 \Big| D_n\right\} - \inf_{f \in \mathcal{F}_n,\, \|f\|_\infty \leq \beta_n} \mathbf{E}\left\{|f(X) - Y|^2\right\}$$

$$\leq \sup_{f \in \mathcal{F}_n,\, \|f\|_\infty \leq \beta_n} \left(\mathbf{E}\left\{|m_n(X) - Y|^2 \Big| D_n\right\} - \frac{1}{n} \sum_{j=1}^{n} |m_n(X_j) - Y_j|^2 \right.$$

$$+ \frac{1}{n} \sum_{j=1}^{n} |m_n(X_j) - Y_j|^2 - \frac{1}{n} \sum_{j=1}^{n} |\tilde{m}_n(X_j) - Y_j|^2$$

$$+ \frac{1}{n} \sum_{j=1}^{n} |\tilde{m}_n(X_j) - Y_j|^2 - \frac{1}{n} \sum_{j=1}^{n} |f(X_j) - Y_j|^2$$

$$\left. + \frac{1}{n} \sum_{j=1}^{n} |f(X_j) - Y_j|^2 - \mathbf{E}\left\{|f(X) - Y|^2\right\} \right).$$

Because of (10.6) the third term on the right-hand side is less than or equal to 0. The same is true for the second term, since if $\tilde{u}, v \in \mathcal{R}$, $|v| \leq \beta_n$, and

$u = T_{\beta_n}\tilde{u}$, then $|u - v| \leq |\tilde{u} - v|$. Therefore the assertion follows from $m_n \in T_{\beta_n}\mathcal{F}_n$ and $\{f \in \mathcal{F}_n : \|f\|_\infty \leq \beta_n\} \subseteq T_{\beta_n}\mathcal{F}_n$. $\qquad\square$

10.2 Consistency from Bounded to Unbounded Y

The aim of this section is to prove Theorem 10.2, which extends Lemma 10.2 to unbounded Y. To formulate this theorem we use the notation

$$Y_L = T_L Y$$

and

$$Y_{i,L} = T_L Y_i.$$

Theorem 10.2. *Let $\mathcal{F}_n = \mathcal{F}_n(D_n)$ be a class of functions $f : \mathcal{R}^d \to \mathcal{R}$ and assume that the estimator m_n satisfies (10.6) and (10.7).*
(a) If

$$\lim_{n\to\infty} \beta_n = \infty, \tag{10.8}$$

$$\lim_{n\to\infty} \inf_{f\in\mathcal{F}_n,\, \|f\|_\infty \leq \beta_n} \int |f(x) - m(x)|^2 \mu(dx) = 0 \quad a.s., \tag{10.9}$$

$$\lim_{n\to\infty} \sup_{f\in T_{\beta_n}\mathcal{F}_n} \left| \frac{1}{n}\sum_{j=1}^n |f(X_j) - Y_{j,L}|^2 - \mathbf{E}\left\{(f(X) - Y_L)^2\right\} \right| = 0 \tag{10.10}$$

a.s. for all $L > 0$, then

$$\lim_{n\to\infty} \int |m_n(x) - m(x)|^2 \mu(dx) = 0 \quad a.s.$$

(b) If (10.8) is fulfilled and

$$\lim_{n\to\infty} \mathbf{E}\left\{ \inf_{f\in\mathcal{F}_n,\, \|f\|_\infty \leq \beta_n} \int |f(x) - m(x)|^2 \mu(dx) \right\} = 0, \tag{10.11}$$

$$\lim_{n\to\infty} \mathbf{E}\left\{ \sup_{f\in T_{\beta_n}\mathcal{F}_n} \left| \frac{1}{n}\sum_{j=1}^n |f(X_j) - Y_{j,L}|^2 - \mathbf{E}\left\{(f(X) - Y_L)^2\right\} \right| \right\} = 0 \tag{10.12}$$

for all $L > 0$, then

$$\lim_{n\to\infty} \mathbf{E}\left\{ \int |m_n(x) - m(x)|^2 \mu(dx) \right\} = 0.$$

Observe that in the above theorem $\mathcal{F}_n$ may depend on the data, and hence

$$\inf_{f\in\mathcal{F}_n,\, \|f\|_\infty \leq \beta_n} \int |f(x) - m(x)|^2 \mu(dx)$$

is a random variable.

PROOF. (a) Because of

$$\int_{\mathcal{R}^d} |m_n(x) - m(x)|^2 \mu(dx) = \mathbf{E}\left\{ |m_n(X) - Y|^2 \big| D_n \right\} - \mathbf{E}|m(X) - Y|^2$$

it suffices to show

$$\left\{ \mathbf{E}\left\{ |m_n(X) - Y|^2 \big| D_n \right\} \right\}^{\frac{1}{2}} - \left\{ \mathbf{E}|m(X) - Y|^2 \right\}^{\frac{1}{2}} \to 0 \quad a.s. \qquad (10.13)$$

We use the decomposition

$$0 \le \left\{ \mathbf{E}\left\{ |m_n(X) - Y|^2 \big| D_n \right\} \right\}^{\frac{1}{2}} - \left\{ \mathbf{E}|m(X) - Y|^2 \right\}^{\frac{1}{2}}$$

$$= \left(\left\{ \mathbf{E}\left\{ |m_n(X) - Y|^2 \big| D_n \right\} \right\}^{\frac{1}{2}} - \inf_{f \in \mathcal{F}_n, \|f\|_\infty \le \beta_n} \left\{ \mathbf{E}|f(X) - Y|^2 \right\}^{\frac{1}{2}} \right)$$

$$+ \left(\inf_{f \in \mathcal{F}_n, \|f\|_\infty \le \beta_n} \left\{ \mathbf{E}|f(X) - Y|^2 \right\}^{\frac{1}{2}} - \left\{ \mathbf{E}|m(X) - Y|^2 \right\}^{\frac{1}{2}} \right). \qquad (10.14)$$

It follows from (10.9), by the triangle inequality, that

$$\inf_{f \in \mathcal{F}_n, \|f\|_\infty \le \beta_n} \left\{ \mathbf{E}|f(X) - Y|^2 \right\}^{\frac{1}{2}} - \left\{ \mathbf{E}|m(X) - Y|^2 \right\}^{\frac{1}{2}}$$

$$\le \inf_{f \in \mathcal{F}_n, \|f\|_\infty \le \beta_n} \left| \left\{ \mathbf{E}|f(X) - Y|^2 \right\}^{\frac{1}{2}} - \left\{ \mathbf{E}|m(X) - Y|^2 \right\}^{\frac{1}{2}} \right|$$

$$\le \inf_{f \in \mathcal{F}_n, \|f\|_\infty \le \beta_n} \left\{ \mathbf{E}|(f(X) - Y) - (m(X) - Y)|^2 \right\}^{\frac{1}{2}}$$

$$= \inf_{f \in \mathcal{F}_n, \|f\|_\infty \le \beta_n} \left\{ \int |f(x) - m(x)|^2 \mu(dx) \right\}^{\frac{1}{2}} \to 0 \quad (n \to \infty) \quad a.s.$$

Therefore for (10.13) we have to show that

$$\limsup_{n \to \infty} \left(\left\{ \mathbf{E}\left\{ |m_n(X) - Y|^2 \big| D_n \right\} \right\}^{\frac{1}{2}} \right.$$

$$\left. - \inf_{f \in \mathcal{F}_n, \|f\|_\infty \le \beta_n} \left\{ \mathbf{E}|f(X) - Y|^2 \right\}^{\frac{1}{2}} \right) \le 0 \quad a.s. \qquad (10.15)$$

To this end, let $L > 0$ be arbitrary. Because of (10.8) we can assume w.l.o.g. that $\beta_n > L$. Then

$$\left\{ \mathbf{E}\left\{ |m_n(X) - Y|^2 \big| D_n \right\} \right\}^{\frac{1}{2}} - \inf_{f \in \mathcal{F}_n, \|f\|_\infty \le \beta_n} \left\{ \mathbf{E}|f(X) - Y|^2 \right\}^{\frac{1}{2}}$$

$$= \sup_{f \in \mathcal{F}_n, \|f\|_\infty \le \beta_n} \left\{ \left\{ \mathbf{E}\left\{ |m_n(X) - Y|^2 \big| D_n \right\} \right\}^{\frac{1}{2}} - \left\{ \mathbf{E}|f(X) - Y|^2 \right\}^{\frac{1}{2}} \right\}$$

$$\leq \sup_{f \in \mathcal{F}_n, \|f\|_\infty \leq \beta_n} \left\{ \left\{ \mathbf{E}\left\{ |m_n(X) - Y|^2 \big| D_n \right\} \right\}^{\frac{1}{2}} \right.$$

$$- \left\{ \mathbf{E}\left\{ |m_n(X) - Y_L|^2 \big| D_n \right\} \right\}^{\frac{1}{2}}$$

$$+ \left\{ \mathbf{E}\left\{ |m_n(X) - Y_L|^2 \big| D_n \right\} \right\}^{\frac{1}{2}} - \left\{ \frac{1}{n} \sum_{i=1}^{n} |m_n(X_i) - Y_{i,L}|^2 \right\}^{\frac{1}{2}}$$

$$+ \left\{ \frac{1}{n} \sum_{i=1}^{n} |m_n(X_i) - Y_{i,L}|^2 \right\}^{\frac{1}{2}} - \left\{ \frac{1}{n} \sum_{i=1}^{n} |\tilde{m}_n(X_i) - Y_{i,L}|^2 \right\}^{\frac{1}{2}}$$

$$+ \left\{ \frac{1}{n} \sum_{i=1}^{n} |\tilde{m}_n(X_i) - Y_{i,L}|^2 \right\}^{\frac{1}{2}} - \left\{ \frac{1}{n} \sum_{i=1}^{n} |\tilde{m}_n(X_i) - Y_i|^2 \right\}^{\frac{1}{2}}$$

$$+ \left\{ \frac{1}{n} \sum_{i=1}^{n} |\tilde{m}_n(X_i) - Y_i|^2 \right\}^{\frac{1}{2}} - \left\{ \frac{1}{n} \sum_{i=1}^{n} |f(X_i) - Y_i|^2 \right\}^{\frac{1}{2}}$$

$$+ \left\{ \frac{1}{n} \sum_{i=1}^{n} |f(X_i) - Y_i|^2 \right\}^{\frac{1}{2}} - \left\{ \frac{1}{n} \sum_{i=1}^{n} |f(X_i) - Y_{i,L}|^2 \right\}^{\frac{1}{2}}$$

$$+ \left\{ \frac{1}{n} \sum_{i=1}^{n} |f(X_i) - Y_{i,L}|^2 \right\}^{\frac{1}{2}} - \left\{ \mathbf{E}|f(X) - Y_L|^2 \right\}^{\frac{1}{2}}$$

$$\left. + \left\{ \mathbf{E}|f(X) - Y_L|^2 \right\}^{\frac{1}{2}} - \left\{ \mathbf{E}|f(X) - Y|^2 \right\}^{\frac{1}{2}} \right\}.$$

Now we give upper bounds for the terms in each row on the right-hand side of the last inequality: The second and seventh term are bounded above by

$$\sup_{f \in T_{\beta_n} \mathcal{F}_n} \left| \left\{ \frac{1}{n} \sum_{i=1}^{n} |f(X_i) - Y_{i,L}|^2 \right\}^{\frac{1}{2}} - \left\{ \mathbf{E}|f(X) - Y_L|^2 \right\}^{\frac{1}{2}} \right|$$

(observe that $m_n \in T_{\beta_n} \mathcal{F}_n$). Because of (10.6) the fifth term is bounded above by zero. For the third term observe that, if $\tilde{u}, v \in \mathcal{R}$ with $|v| \leq \beta_n$ and $u = T_{\beta_n} \tilde{u}$, then $|u - v| \leq |\tilde{u} - v|$. Therefore the third term is also not greater than zero.

Using these upper bounds and the triangle inequality for the remaining terms one gets

$$\left\{ \mathbf{E}\left\{ |m_n(X) - Y|^2 \,\middle|\, D_n \right\} \right\}^{\frac{1}{2}} - \inf_{f \in \mathcal{F}_n, \|f\|_\infty \le \beta_n} \left\{ \mathbf{E}|f(X) - Y|^2 \right\}^{\frac{1}{2}}$$

$$\le \; 2 \cdot \left\{ \mathbf{E}|Y - Y_L|^2 \right\}^{\frac{1}{2}} + 2 \cdot \left\{ \frac{1}{n} \sum_{i=1}^n |Y_i - Y_{i,L}|^2 \right\}^{\frac{1}{2}}$$

$$+ \, 2 \cdot \sup_{f \in T_{\beta_n}\mathcal{F}_n} \left| \left\{ \frac{1}{n} \sum_{i=1}^n |f(X_i) - Y_{i,L}|^2 \right\}^{\frac{1}{2}} - \left\{ \mathbf{E}|f(X) - Y_L|^2 \right\}^{\frac{1}{2}} \right|.$$

Equation (10.10), the uniform continuity of $x \mapsto \sqrt{x}$ on $[0, \infty)$, and the strong law of large numbers imply

$$\limsup_{n \to \infty} \left(\left\{ \mathbf{E}\left\{ |m_n(X) - Y|^2 \,\middle|\, D_n \right\} \right\}^{\frac{1}{2}} - \inf_{f \in \mathcal{F}_n, \|f\|_\infty \le \beta_n} \left\{ \mathbf{E}|f(X) - Y|^2 \right\}^{\frac{1}{2}} \right)$$

$$\le 4 \cdot \left\{ \mathbf{E}|Y - Y_L|^2 \right\}^{\frac{1}{2}} \quad a.s.$$

One gets the assertion with $L \to \infty$.

(b) Because of

$$\int_{\mathcal{R}^d} |m_n(x) - m(x)|^2 \mu(dx)$$

$$= \mathbf{E}\{|m_n(X) - Y|^2 | D_n\} - \mathbf{E}\{|m(X) - Y|^2\}$$

$$= \left(\left(\mathbf{E}\{|m_n(X) - Y|^2 | D_n\}\right)^{\frac{1}{2}} - \left(\mathbf{E}\{|m(X) - Y|^2\}\right)^{\frac{1}{2}} \right)$$

$$\times \left(\left(\mathbf{E}\{|m_n(X) - Y|^2 | D_n\}\right)^{\frac{1}{2}} + \left(\mathbf{E}\{|m(X) - Y|^2\}\right)^{\frac{1}{2}} \right)$$

$$= \left(\left(\mathbf{E}\{|m_n(X) - Y|^2 | D_n\}\right)^{\frac{1}{2}} - \left(\mathbf{E}\{|m(X) - Y|^2\}\right)^{\frac{1}{2}} \right)^2$$

$$+ 2 \left(\mathbf{E}\{|m(X) - Y|^2\} \right)^{\frac{1}{2}} \left(\left(\mathbf{E}\{|m_n(X) - Y|^2 | D_n\}\right)^{\frac{1}{2}} \right.$$

$$\left. - \left(\mathbf{E}\{|m(X) - Y|^2\}\right)^{\frac{1}{2}} \right)$$

it suffices to show that

$$\mathbf{E}\left(\left(\mathbf{E}\{|m_n(X) - Y|^2 | D_n\}\right)^{\frac{1}{2}} - \left(\mathbf{E}\{|m(X) - Y|^2\}\right)^{\frac{1}{2}} \right)^2 \to 0 \quad (n \to \infty).$$

We use the same error decomposition as in (a):

$$\mathbf{E}\left(\left(\mathbf{E}\{|m_n(X) - Y|^2 | D_n\}\right)^{\frac{1}{2}} - \left(\mathbf{E}\{|m(X) - Y|^2\}\right)^{\frac{1}{2}} \right)^2$$

$$\le 2\mathbf{E}\left\{ \left(\mathbf{E}\{|m_n(X) - Y|^2 | D_n\}\right)^{\frac{1}{2}} - \inf_{f \in \mathcal{F}_n, \|f\|_\infty \le \beta_n} \left(\mathbf{E}\{|f(X) - Y|^2\}\right)^{\frac{1}{2}} \right\}^2$$

$$+ 2\mathbf{E}\left\{ \inf_{f \in \mathcal{F}_n, \|f\|_\infty \le \beta_n} \left(\mathbf{E}\{|f(X) - Y|^2\}\right)^{\frac{1}{2}} - \left(\mathbf{E}\{|m(X) - Y|^2\}\right)^{\frac{1}{2}} \right\}^2.$$

By the triangle inequality and (10.11), one gets, for the second term on the right-hand side of the last inequality,

$$2\mathbf{E}\left(\inf_{f\in\mathcal{F}_n,\|f\|_\infty\le\beta_n}\left(\mathbf{E}\{|f(X)-Y|^2\}\right)^{\frac{1}{2}}-\left(\mathbf{E}\{|m(X)-Y|^2\}\right)^{\frac{1}{2}}\right)^2$$

$$\le 2\mathbf{E}\left(\inf_{f\in\mathcal{F}_n,\|f\|_\infty\le\beta_n}\left(\mathbf{E}\{|f(X)-m(X)|^2\}\right)^{\frac{1}{2}}\right)^2$$

$$= 2\mathbf{E}\left\{\inf_{f\in\mathcal{F}_n,\|f\|_\infty\le\beta_n}\mathbf{E}\{|f(X)-m(X)|^2\}\right\}\to 0\quad(n\to\infty).$$

Thus it suffices to show that

$$\mathbf{E}\left\{\left(\mathbf{E}\{|m_n(X)-Y|^2|D_n\}\right)^{\frac{1}{2}}\right.$$

$$\left.-\inf_{f\in\mathcal{F}_n,\|f\|_\infty\le\beta_n}\left(\mathbf{E}\{|f(X)-Y|^2\}\right)^{\frac{1}{2}}\right\}^2\to 0\quad(n\to\infty).\qquad(10.16)$$

On one hand,

$$\left(\mathbf{E}\{|m_n(X)-Y|^2|D_n\}\right)^{\frac{1}{2}}-\inf_{f\in\mathcal{F}_n,\|f\|_\infty\le\beta_n}\left(\mathbf{E}\{|f(X)-Y|^2\}\right)^{\frac{1}{2}}$$

$$\ge\left(\mathbf{E}\{|m(X)-Y|^2\}\right)^{\frac{1}{2}}-\inf_{f\in\mathcal{F}_n,\|f\|_\infty\le\beta_n}\left(\mathbf{E}\{|f(X)-Y|^2\}\right)^{\frac{1}{2}}$$

$$\ge-\inf_{f\in\mathcal{F}_n,\|f\|_\infty\le\beta_n}\left(\int_{\mathcal{R}^d}|f(x)-m(x)|^2\mu(dx)\right)^{\frac{1}{2}}.$$

On the other hand, it follows from the proof of part (a) that

$$\left(\mathbf{E}\{|m_n(X)-Y|^2|D_n\}\right)^{\frac{1}{2}}-\inf_{f\in\mathcal{F}_n,\|f\|_\infty\le\beta_n}\left(\mathbf{E}\{|f(X)-Y|^2\}\right)^{\frac{1}{2}}$$

$$\le 2\left(\mathbf{E}\{|Y-Y_L|^2\}\right)^{\frac{1}{2}}+2\left(\frac{1}{n}\sum_{i=1}^n|Y_i-Y_{i,L}|^2\right)^{\frac{1}{2}}$$

$$+2\sup_{f\in T_{\beta_n}\mathcal{F}_n}\left|\left(\frac{1}{n}\sum_{i=1}^n|f(X_i)-Y_{i,L}|^2\right)^{\frac{1}{2}}-\left(\mathbf{E}\{|f(X)-Y_L|^2\}\right)^{\frac{1}{2}}\right|.$$

The inequalities $(a+b+c)^2\le 3a^2+3b^2+3c^2$ $(a,b,c\in\mathcal{R})$ and

$$(\sqrt{a}-\sqrt{b})^2\le|\sqrt{a}-\sqrt{b}|\cdot|\sqrt{a}+\sqrt{b}|=|a-b|\quad(a,b\in\mathcal{R}_+)$$

imply

$$\mathbf{E}\left\{\left(\mathbf{E}\{|m_n(X)-Y|^2|D_n\}\right)^{\frac{1}{2}}-\inf_{f\in\mathcal{F}_n,\|f\|_\infty\le\beta_n}\left(\mathbf{E}\{|f(X)-Y|^2\}\right)^{\frac{1}{2}}\right\}^2$$

$$\leq \mathbf{E}\left\{ \inf_{f\in\mathcal{F}_n, \|f\|_\infty \leq \beta_n} \int_{\mathcal{R}^d} |f(x) - m(x)|^2 \mu(dx) \right\}$$

$$+ 12\mathbf{E}\{|Y - Y_L|^2\} + 12\mathbf{E}\left\{ \frac{1}{n} \sum_{i=1}^n |Y_i - Y_{i,L}|^2 \right\}$$

$$+ 12\mathbf{E}\left\{ \sup_{f\in\mathcal{T}_{\beta_n}\mathcal{F}_n} \left| \frac{1}{n} \sum_{i=1}^n |f(X_i) - Y_{i.L}|^2 - \mathbf{E}\{|f(X) - Y_L|^2\} \right| \right\}$$

$$\rightarrow 24\mathbf{E}\{|Y - Y_L|^2\} \quad (n \rightarrow \infty),$$

where we have used (10.11), (10.12), and the strong law of large numbers. With $L \rightarrow \infty$ the assertion follows. $\qquad\square$

10.3   Linear Least Squares Series Estimates

For the sake of illustration we formulate an analogue of Theorem 10.1:

Theorem 10.3. *Let $\psi_1, \psi_2, \dots : \mathcal{R}^d \rightarrow \mathcal{R}$ be bounded functions. Assume that the set of functions*

$$\bigcup_{k=1}^\infty \left\{ \sum_{j=1}^k a_j \psi_j(x) : a_1, \dots, a_k \in \mathcal{R} \right\} \tag{10.17}$$

is dense in $L_2(\mu)$ for any probability measure μ on $\mathcal{R}^d$. Define the regression function estimate $\tilde{m}_n$ as a function minimizing the empirical L_2 risk

$$\frac{1}{n} \sum_{i=1}^n (f(X_i) - Y_i)^2$$

over $\mathcal{F}_n = \left\{ \sum_{j=1}^{K_n} a_j \psi_j(x) : a_1, \dots, a_{K_n} \in \mathcal{R} \right\}$, and put

$$m_n(x) = T_{\beta_n} \tilde{m}_n(x).$$

(a) If $\mathbf{E}\{Y^2\} < \infty$, and K_n and β_n satisfy

$$K_n \rightarrow \infty, \quad \beta_n \rightarrow \infty, \quad and \quad \frac{K_n \beta_n^4 \log \beta_n}{n} \rightarrow 0, \tag{10.18}$$

then

$$\mathbf{E} \int (m_n(x) - m(x))^2 \mu(dx) \rightarrow 0,$$

i.e., the estimate is weakly universally consistent.
(b) If $\mathbf{E}\{Y^2\} < \infty$, and K_n and β_n satisfy (10.18) and, in addition,

$$\frac{\beta_n^4}{n^{1-\delta}} \rightarrow 0 \tag{10.19}$$

for some $\delta > 0$, then

$$\int (m_n(x) - m(x))^2 \mu(dx) \to 0 \quad a.s.,$$

i.e., the estimate is strongly universally consistent.

In the proof we will use the denseness of (10.17) in $L_2(\mu)$ in order to show that the approximation error converges to zero as postulated in (10.9) and (10.11). In order to show that the estimation error converges to zero we use that the "complexity" of the space $\mathcal{F}_n$ of functions (measured by its vector space dimension K_n) is restricted by (10.18).

PROOF. Because of Theorem 10.2 it suffices to show that (10.18) implies (10.11) and (10.12), and that (10.18) and (10.19) imply (10.9) and (10.10).

PROOF OF (10.9) AND (10.11). Let $\epsilon > 0$. By assumption,

$$\bigcup_{k=1}^{\infty} \left\{ \sum_{j=1}^{k} a_j \psi_j(x) \; : \; a_1, \ldots, a_k \in \mathcal{R} \right\}$$

is dense in $L_2(\mu)$, where μ denotes the distribution of X. It follows from $\mathbf{E} Y^2 < \infty$ that $m \in L_2(\mu)$. Hence there exist $k^* \in \mathcal{N}$ and $a_1^*, \ldots, a_{k^*}^* \in \mathcal{R}$ such that

$$\int_{\mathcal{R}^d} \left| \sum_{j=1}^{k^*} a_j^* \psi_j(x) - m(x) \right|^2 \mu(dx) < \epsilon.$$

Since $\psi_1, \psi_2 \ldots$ are bounded,

$$\sup_{x \in \mathcal{R}^d} \left| \sum_{j=1}^{k^*} a_j^* \psi_j(x) \right| < \infty.$$

Using $K_n \to \infty$ $(n \to \infty)$ and $\beta_n \to \infty$ $(n \to \infty)$ one concludes that, for all $n \geq n_0(\epsilon)$,

$$\sum_{j=1}^{k^*} a_j^* \psi_j \in \left\{ f \in \mathcal{F}_n \; : \; \|f\|_\infty \leq \beta_n \right\}.$$

Hence

$$\inf_{f \in \mathcal{F}_n, \|f\|_\infty \leq \beta_n} \int_{\mathcal{R}^d} |f(x) - m(x)|^2 \mu(dx) < \epsilon$$

for $n \geq n_0(\epsilon)$. Since $\epsilon > 0$ was arbitrary, this implies (10.9) and (10.11).

PROOF OF (10.10) AND (10.12). Let $L > 0$ be arbitrary. Because of $\beta_n \to \infty$ $(n \to \infty)$ we may assume w.l.o.g. $L < \beta_n$. Set

$$Z = (X, Y), \; Z_1 = (X_1, Y_1), \ldots, \; Z_n = (X_n, Y_n),$$

and
$$\mathcal{H}_n = \left\{ h : \mathcal{R}^d \times \mathcal{R} \to \mathcal{R} \ : \ \exists f \in T_{\beta_n}\mathcal{F}_n \text{ such that } h(x,y) = |f(x) - T_L y|^2 \right\}.$$
Observe that the functions in $\mathcal{H}_n$ satisfy
$$0 \leq h(x,y) \leq 2\beta_n^2 + 2L^2 \leq 4\beta_n^2.$$
By Theorem 9.1 one has, for arbitrary $\epsilon > 0$,
$$\mathbf{P}\left\{ \sup_{f \in T_{\beta_n}\mathcal{F}_n} \left| \frac{1}{n}\sum_{i=1}^n |f(X_i) - Y_{i,L}|^2 - \mathbf{E}\{|f(X) - Y_L|^2\} \right| > \epsilon \right\}$$
$$= \mathbf{P}\left\{ \sup_{h \in \mathcal{H}_n} \left| \frac{1}{n}\sum_{i=1}^n h(Z_i) - \mathbf{E}\{h(Z)\} \right| > \epsilon \right\}$$
$$\leq 8\mathbf{E}\mathcal{N}_1\left(\frac{\epsilon}{8}, \mathcal{H}_n, Z_1^n\right) e^{-\frac{n\epsilon^2}{128(4\beta_n^2)^2}}. \tag{10.20}$$

Next we bound the covering number in (10.20). By Lemma 9.2 we can bound it by the corresponding packing number
$$\mathcal{N}_1\left(\frac{\epsilon}{8}, \mathcal{H}_n, Z_1^n\right) \leq \mathcal{M}_1\left(\frac{\epsilon}{8}, \mathcal{H}_n, Z_1^n\right).$$
Let $h_i(x,y) = |f_i(x) - T_L y|^2$ $((x,y) \in \mathcal{R}^d \times \mathcal{R})$ for some $f_i \in \mathcal{F}_n$. Then
$$\frac{1}{n}\sum_{i=1}^n |h_1(Z_i) - h_2(Z_i)|$$
$$= \frac{1}{n}\sum_{i=1}^n \left| |f_1(X_i) - T_L Y_i|^2 - |f_2(X_i) - T_L Y_i|^2 \right|$$
$$= \frac{1}{n}\sum_{i=1}^n |f_1(X_i) - f_2(X_i)| \cdot |f_1(X_i) - T_L Y_i + f_2(X_i) - T_L Y_i|$$
$$\leq 4\beta_n \frac{1}{n}\sum_{i=1}^n |f_1(X_i) - f_2(X_i)|.$$

Thus, if $\{h_1, \dots, h_l\}$ is an $\frac{\epsilon}{8}$-packing of $\mathcal{H}_n$ on Z_1^n, then $\{f_1, \dots, f_l\}$ is an $\epsilon/8(4\beta_n)$-packing of $T_{\beta_n}\mathcal{F}_n$ on X_1^n. Then
$$\mathcal{M}_1\left(\frac{\epsilon}{8}, \mathcal{H}_n, Z_1^n\right) \leq \mathcal{M}_1\left(\frac{\epsilon}{32\beta_n}, T_{\beta_n}\mathcal{F}_n, X_1^n\right). \tag{10.21}$$

By Theorem 9.4 we can bound the latter term
$$\mathcal{M}_1\left(\frac{\epsilon}{32\beta_n}, T_{\beta_n}\mathcal{F}_n, X_1^n\right)$$
$$\leq 3\left(\frac{2e(2\beta_n)}{\frac{\epsilon}{32\beta_n}} \log\left(\frac{3e(2\beta_n)}{\frac{\epsilon}{32\beta_n}} \right) \right)^{V_{T_{\beta_n}\mathcal{F}_n^+}}$$

$$= 3 \left(\frac{128 e \beta_n^2}{\epsilon} \log \left(\frac{192 e \beta_n^2}{\epsilon} \right) \right)^{V_{T_{\beta_n} \mathcal{F}_n^+}}. \tag{10.22}$$

Let $(x, y) \in \mathcal{R}^d \times \mathcal{R}$. If $y > \beta_n$, then (x, y) is contained in none of the sets $T_{\beta_n} \mathcal{F}_n^+$ and, if $y \leq -\beta_n$, then (x, y) is contained in every set of $T_{\beta_n} \mathcal{F}_n^+$. Hence, if $T_{\beta_n} \mathcal{F}_n^+$ shatters a set of points, then the y-coordinates of these points are all bounded in absolute value by β_n and $\mathcal{F}_n^+$ also shatters this set of points. This proves

$$V_{T_{\beta_n} \mathcal{F}_n^+} \leq V_{\mathcal{F}_n^+}, \tag{10.23}$$

where $V_{\mathcal{F}_n^+}$ can be bounded by Theorem 9.5. Observe that

$$\begin{aligned} \mathcal{F}_n^+ &= \{\{(x, t) \ : \ t \leq f(x)\} : f \in \mathcal{F}_n\} \\ &\subseteq \{\{(x, t) \ : \ f(x) + a_0 t \geq 0\} : f \in \mathcal{F}_n, \ a_0 \in \mathcal{R}\}. \end{aligned}$$

Now

$$\left\{ f(x) + a_0 t \ : \ f \in \mathcal{F}_n, a_0 \in \mathcal{R} \right\} = \left\{ \sum_{j=1}^{K_n} a_j \psi_j(x) + a_0 t \ : \ a_0, \dots, a_{K_n} \in \mathcal{R} \right\}$$

is a linear vector space of dimension $K_n + 1$, thus Theorem 9.5 implies

$$V_{\mathcal{F}_n^+} \leq K_n + 1. \tag{10.24}$$

Formulae (10.20)–(10.24) imply

$$\mathbf{P} \left\{ \sup_{f \in T_{\beta_n} \mathcal{F}_n} \left| \frac{1}{n} \sum_{i=1}^{n} |f(X_i) - Y_{i,L}|^2 - \mathbf{E}\{|f(X) - Y_L|^2\} \right| > \epsilon \right\}$$

$$\leq 24 \left(\frac{128 e \beta_n^2}{\epsilon} \log \left(\frac{192 e \beta_n^2}{\epsilon} \right) \right)^{K_n + 1} e^{-\frac{n \epsilon^2}{2048 \beta_n^4}}$$

$$\leq 24 \left(\frac{192 e \beta_n^2}{\epsilon} \right)^{2(K_n + 1)} e^{-\frac{n \epsilon^2}{2048 \beta_n^4}}, \tag{10.25}$$

where we have used $\log(x) \leq x - 1 \leq x$ $(x \in \mathcal{R}_+)$. Now, assume that (10.18) and (10.19) hold. Then

$$\sum_{n=1}^{\infty} \mathbf{P} \left\{ \sup_{f \in T_{\beta_n} \mathcal{F}_n} \left| \frac{1}{n} \sum_{i=1}^{n} |f(X_i) - Y_{i,L}|^2 - \mathbf{E}\{|f(X) - Y_L|^2\} \right| > \epsilon \right\}$$

$$\leq \sum_{n=1}^{\infty} 24 \cdot \exp \left(2(K_n + 1) \log \frac{192 e \beta_n^2}{\epsilon} - \frac{n \epsilon^2}{2048 \beta_n^4} \right)$$

$$= \sum_{n=1}^{\infty} 24 \cdot \exp \left(-n^{\delta} \frac{n^{1-\delta}}{\beta_n^4} \left(\frac{\epsilon^2}{2048} - \frac{2(K_n + 1) \beta_n^4 \log \frac{192 e \beta_n^2}{\epsilon}}{n} \right) \right)$$

$$< \infty,$$

where we have used that (10.18) and (10.19) imply

$$\frac{n^{1-\delta}}{\beta_n^4} \to \infty \quad (n \to \infty)$$

and

$$\frac{2(K_n + 1)\beta_n^4 \log \frac{192e\beta_n^2}{\epsilon}}{n} \to 0 \quad (n \to \infty).$$

This, together with the Borel–Cantelli lemma, proves (10.10).

Let $\tilde{Z}$ be a nonnegative random variable and let $\epsilon > 0$. Then

$$\mathbf{E}\{\tilde{Z}\} = \int_0^\infty \mathbf{P}\{\tilde{Z} > t\}\, dt \leq \epsilon + \int_\epsilon^\infty \mathbf{P}\{\tilde{Z} > t\}\, dt.$$

Using this and (10.25) one gets

$$\mathbf{E}\left\{ \sup_{f \in T_{\beta_n}\mathcal{F}_n} \left| \frac{1}{n}\sum_{i=1}^n |f(X_i) - Y_{i,L}|^2 - \mathbf{E}\left\{|f(X) - Y_L|^2\right\} \right| \right\}$$

$$\leq \epsilon + \int_\epsilon^\infty 24 \cdot \left(\frac{192e\beta_n^2}{t}\right)^{2(K_n+1)} \cdot \exp\left(-\frac{n \cdot t^2}{2048\beta_n^4}\right) dt$$

$$\leq \epsilon + 24 \cdot \left(\frac{192e\beta_n^2}{\epsilon}\right)^{2(K_n+1)} \cdot \left[-\frac{2048\beta_n^4}{n \cdot \epsilon} \cdot \exp\left(-\frac{n \cdot \epsilon \cdot t}{2048\beta_n^4}\right)\right]_{t=\epsilon}^\infty$$

$$= \epsilon + 24 \cdot \left(\frac{192e\beta_n^2}{\epsilon}\right)^{2(K_n+1)} \cdot \frac{2048\beta_n^4}{n \cdot \epsilon} \cdot \exp\left(-\frac{n \cdot \epsilon^2}{2048\beta_n^4}\right)$$

$$= \epsilon + 24 \cdot \frac{2048\beta_n^4}{n \cdot \epsilon} \cdot \exp\left(2(K_n + 1) \cdot \log\left(\frac{192e\beta_n^2}{\epsilon}\right) - \frac{n \cdot \epsilon^2}{2048\beta_n^4}\right)$$

$$\to \epsilon \quad (n \to \infty),$$

if (10.18) holds. With $\epsilon \to 0$ one gets (10.12). $\qquad\square$

10.4 Piecewise Polynomial Partitioning Estimates

Let $\mathcal{P}_n = \{A_{n,1}, A_{n,2}, \dots\}$ be a partition of $\mathcal{R}^d$ and let $\tilde{m}_n$ be the corresponding partitioning estimate, i.e.,

$$\tilde{m}_n(x) = \frac{\sum_{i=1}^n Y_i \cdot I_{\{X_i \in A_n(x)\}}}{\sum_{i=1}^n I_{\{X_i \in A_n(x)\}}},$$

where $A_n(x)$ denotes the cell $A_{n,j} \in \mathcal{P}_n$ which contains x. As we have already seen in Chapter 2, $\tilde{m}_n$ satisfies

$$\tilde{m}_n(\cdot) = \arg\min_{f \in \mathcal{F}_n} \frac{1}{n}\sum_{i=1}^n |f(X_i) - Y_i|^2,$$

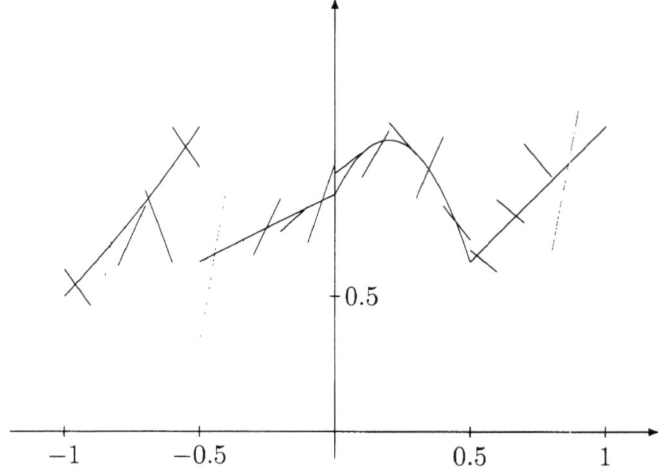

Figure 10.3. Piecewise polynomial partitioning estimate, degree $M = 1$, $h = 0.1$, L_2 error $= 0.472119$.

where $\mathcal{F}_n$ is the set of all piecewise constant functions with respect to $\mathcal{P}_n$. Hence the partitioning estimate fits (via the principle of least squares) a piecewise constant function to the data.

As we have seen in Chapter 4 the partitioning estimate does not achieve the optimal (or even a nearly optimal) rate of convergence if m is (p, C)-smooth for some $p > 1$.

A straightforward generalization of fitting a piecewise constant function to the data is to fit (via the principle of least squares) a piecewise polynomial of some fixed degree $M > 0$ to the data. We will call the resulting estimate a *piecewise polynomial partitioning estimate*. Figures 10.3–10.8 show application of this estimate to our standard data example.

In this section we show how one can use the results of this chapter to derive the consistency of such estimates. In the next chapter we will see that these estimates are able to achieve (at least up to a logarithmic factor) the optimal rate of convergence if the regression function is (p, C)-smooth, even if $p > 1$.

For simplicity we assume $X \in [0, 1]$ *a.s.*, the case of unbounded and multivariate X will be handled in Problems 10.6 and 10.7.

Let $M \in \mathcal{N}_0$ and for $n \in \mathcal{N}$ let $K_n \in \mathcal{N}$, $\beta_n \in \mathcal{R}_+$ and let $\mathcal{P}_n = \{A_{n,1}, \dots, A_{n,K_n}\}$ be a partition of $[0, 1]$ consisting of K_n cells. Let $\mathcal{G}_M$ be the set of all polynomials of degree M (or less), and set

$$\mathcal{G}_M \circ \mathcal{P}_n = \left\{ \sum_{j=1}^{K_n} p_j I_{A_{n,j}} \; : \; p_j \in \mathcal{G}_M \; (j = 1, \dots, K_n) \right\},$$

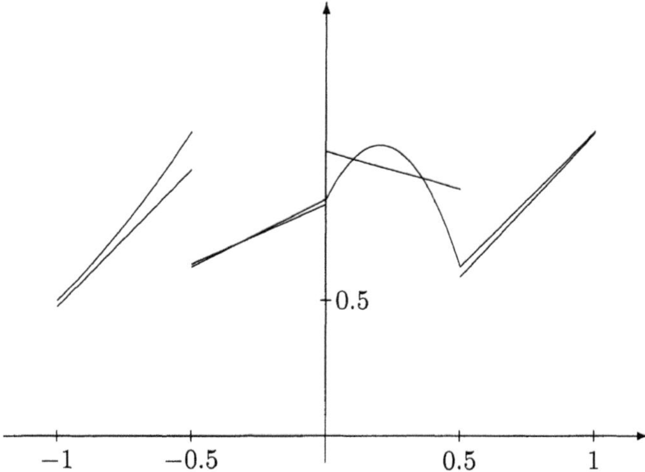

Figure 10.4. Piecewise polynomial partitioning estimate, degree $M = 1$, $h = 0.5$, L_2 error $= 0.002786$.

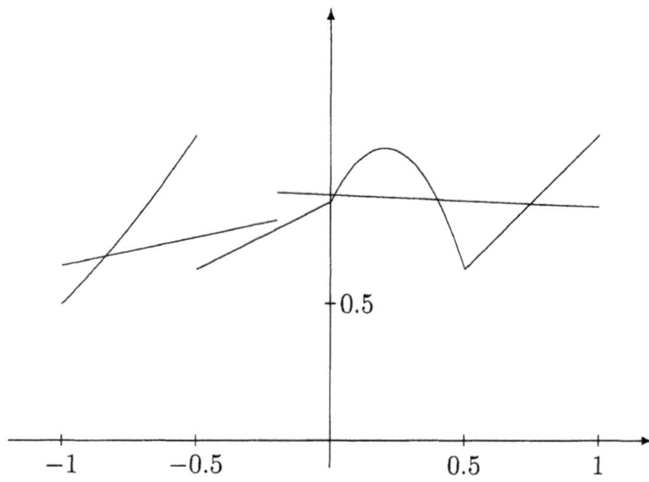

Figure 10.5. Piecewise polynomial partitioning estimate, degree $M = 1$, $h = 0.8$, L_2 error $= 0.013392$.

where $\mathcal{G}_M \circ \mathcal{P}_n$ is the set of all piecewise polynomials of degree M (or less) w.r.t. $\mathcal{P}_n$.

Define the piecewise polynomial partitioning estimate by

$$\tilde{m}_n(\cdot) = \arg \min_{f \in \mathcal{G}_M \circ \mathcal{P}_n} \frac{1}{n} \sum_{i=1}^{n} |f(X_i) - Y_i|^2 \quad \text{and} \quad m_n(\cdot) = T_{\beta_n} \tilde{m}_n(\cdot).$$

$$(10.26)$$

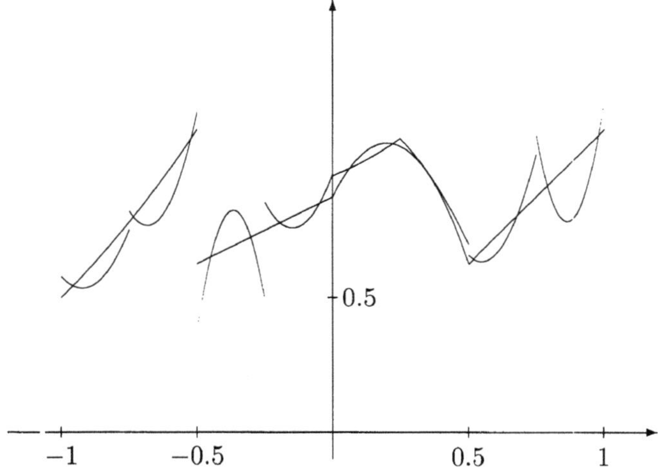

Figure 10.6. Piecewise polynomial partitioning estimate, degree $M = 2$, $h = 0.25$, L_2 error $= 0.004330$.

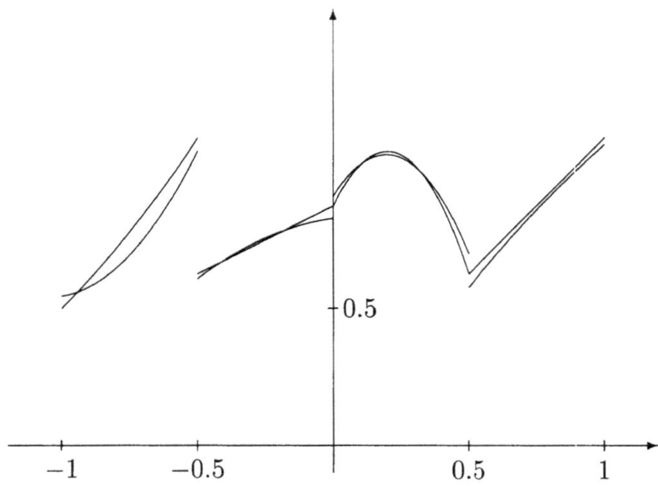

Figure 10.7. Piecewise polynomial partitioning estimate, degree $M = 2$, $h = 0.5$, L_2 error $= 0.000968$.

Theorem 10.4. *Let M, β_n, K_n, $\mathcal{P}_n$ be as above and define the estimate m_n by (10.26).*
(a) If

$$K_n \to \infty, \quad \beta_n \to \infty, \quad \text{and} \quad \frac{K_n \beta_n^4 \log \beta_n}{n} \to 0, \tag{10.27}$$

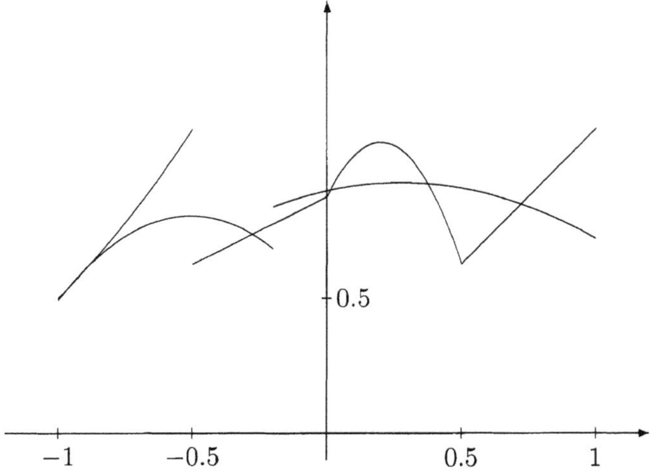

Figure 10.8. Piecewise polynomial partitioning estimate, degree $M = 2$, $h = 0.8$, L_2 error $= 0.012788$.

and

$$\max_{j=1,\ldots K_n} diam\,(A_{n,j}) \to 0 \quad (n \to \infty), \tag{10.28}$$

then

$$\mathbf{E} \int |m_n(x) - m(x)|^2 \mu(dx) \to 0 \quad (n \to \infty)$$

for all distributions of (X, Y) with $X \in [0, 1]$ a.s. and $\mathbf{E}Y^2 < \infty$.
(b) If K_n and β_n satisfy (10.27) and (10.28) and, in addition,

$$\frac{\beta_n^4}{n^{1-\delta}} \to 0 \quad (n \to \infty) \tag{10.29}$$

for some $\delta > 0$, then

$$\int |m_n(x) - m(x)|^2 \mu(dx) \to 0 \quad (n \to \infty) \quad a.s.$$

for all distributions of (X, Y) with $X \in [0, 1]$ a.s. and $\mathbf{E}Y^2 < \infty$.

For degree $M = 0$ the estimate in Theorem 10.4 is a truncated version of the partitioning estimate in Theorem 4.2. The conditions in Theorem 4.2 are weaker than in Theorem 10.4, and in Theorem 4.2 there is no truncation of the estimate required. But Theorem 10.4 is applicable to more general estimates. It follows from Problem 10.3 that in this more general context truncation of the estimate is necessary.

PROOF. Because of Theorem 10.2 it suffices to show that (10.27) and (10.28) imply

$$\inf_{f \in \mathcal{G}_M \circ \mathcal{P}_n, \|f\|_\infty \leq \beta_n} \int |f(x) - m(x)|^2 \mu(dx) \to 0 \quad (n \to \infty) \tag{10.30}$$

and

$$\mathbf{E}\left\{ \sup_{f \in T_{\beta_n} \mathcal{G}_M \circ \mathcal{P}_n} \left| \frac{1}{n} \sum_{i=1}^n |f(X_i) - Y_{i,L}|^2 - \mathbf{E}\{|f(X) - Y_L|^2\} \right| \right\} \to 0$$

$$(n \to \infty) \tag{10.31}$$

for all $L > 0$, and that (10.27), (10.28), and (10.29) imply

$$\sup_{f \in T_{\beta_n} \mathcal{G}_M \circ \mathcal{P}_n} \left| \frac{1}{n} \sum_{i=1}^n |f(X_i) - Y_{i,L}|^2 - \mathbf{E}\{|f(X) - Y_L{}^2\} \right| \to 0$$

$$(n \to \infty) \qquad a.s. \tag{10.32}$$

for all $L > 0$. The proof of (10.31) and (10.32) is left to the reader (cf. Problem 10.5). In order to show (10.30) let $\epsilon > 0$ be arbitrary. By Theorem A.1 there exists a continuous function $\bar{m}$ such that

$$\int |\bar{m}(x) - m(x)|^2 \mu(dx) \leq \frac{\epsilon}{4}.$$

Since $\bar{m}$ is uniformly continuous on the compact interval $[0,1]$, there exists a $\delta > 0$ such that $|\bar{m}(x) - \bar{m}(z)| < \sqrt{\epsilon}/2$ for all $x, z \in [0,1], |x - z| < \delta$. Choose arbitrary points $z_{n,j} \in A_{n,j}$ $(j = 1, \ldots, K_n)$ and set

$$f_n = \sum_{j=1}^{K_n} \bar{m}(z_{n,j}) \cdot I_{A_{n,j}} \in \mathcal{G}_0 \circ \mathcal{P}_n \subseteq \mathcal{G}_M \circ \mathcal{P}_n.$$

Then $z \in A_{n,j}$ and $\operatorname{diam}(A_{n,j}) < \delta$ imply

$$|f_n(z) - \bar{m}(z)| = |\bar{m}(z_{n,j}) - \bar{m}(z)| < \sqrt{\epsilon}/2.$$

Using this one gets, for n sufficiently large (i.e., for n so large that $\beta_n \geq \max_{z \in [0,1]} |\bar{m}(z)|$ and $\max_{j=1,\ldots,K_n} \operatorname{diam}(A_{n,j}) < \delta$),

$$\inf_{f \in \mathcal{G}_M \circ \mathcal{P}_n, \|f\|_\infty \leq \beta_n} \int |f(x) - m(x)|^2 \mu(dx)$$

$$\leq 2 \inf_{f \in \mathcal{G}_M \circ \mathcal{P}_n, \|f\|_\infty \leq \beta_n} \int |f(x) - \bar{m}(x)|^2 \mu(dx) + 2 \cdot \frac{\epsilon}{4}$$

$$\leq 2 \int |f_n(x) - \bar{m}(x)|^2 \mu(dx) + \frac{\epsilon}{2}$$

$$\leq 2 \sup_{x \in [0,1]} |f_n(x) - \bar{m}(x)|^2 + \frac{\epsilon}{2}$$

$$\leq 2 \cdot \frac{\epsilon}{4} + \frac{\epsilon}{2} = \epsilon.$$

With $\epsilon \to 0$ this implies (10.30). $\square$

It is possible to modify the estimate such that it is weakly and strongly universally consistent (cf. Problem 10.6). Multivariate piecewise polynomial partitioning estimates can be defined by using piecewise multivariate polynomials (cf. Problem 10.7).

10.5   Bibliographic Notes

It is well-known that solving the set of normal equations of linear least squares estimates (i.e., (10.5)) may cause serious numerical problems due to ill–conditioning. Numerical methods for solving these equations are discussed, e.g., in Daniel and Wood (1980), Farebrother (1988), and Maindonald (1984).

For least squares estimates one minimizes the empirical L_2 risk. Asymptotic properties of more general empirical risk minimization problems were studied by several authors such as Vapnik and Chervonenkis (1971), Vapnik (1982; 1998), and Haussler (1992). Minimization of the empirical L_2 risk has also become known in the statistics literature as "minimum contrast estimation," e.g., see Nemirovsky et al. (1985), van de Geer (1990), and Birgé and Massart (1993). Consistency of least squares and other minimum contrast estimates under general conditions was investigated, e.g., in Nemirovsky et al. (1983), and Nemirovsky et al. (1984).

In the context of pattern recognition many nice results concerning empirical risk minimization can be found in the book of Devroye, Györfi, and Lugosi (1996).

Theorem 10.1 is due to Lugosi and Zeger (1995). Consistency of sieves estimates has been studied, e.g., by Geman and Hwang (1982) and van de Geer and Wegkamp (1996). The latter article also contains necessary conditions for the consistency of least squares estimates.

Problems and Exercises

PROBLEM 10.1. Let $\mathcal{F}_n$ be a class of functions $f : \mathcal{R}^d \to \mathcal{R}$ and define the estimator m_n by

$$m_n(\cdot) = \arg \min_{f \in \mathcal{F}_n} \frac{1}{n} \sum_{i=1}^{n} |f(X_i) - Y_i|^2.$$

Assume $\mathbf{E}Y^2 < \infty$. Show that without truncation of the estimate the following modification of Theorem 10.2 is valid:

$$\inf_{f \in \mathcal{F}_n} \int |f(x) - m(x)|^2 \mu(dx) \to 0 \quad (n \to \infty)$$

and

$$\sup_{f \in \mathcal{F}_n} \left| \frac{1}{n} \sum_{i=1}^{n} |f(X_i) - T_L Y_i|^2 - \mathbf{E}\left\{ |f(X) - T_L Y|^2 \right\} \right| \to 0 \quad (n \to \infty) \quad a.s.$$

for all $L > 0$ imply

$$\int |m_n(x) - m(x)|^2 \mu(dx) \to 0 \quad (n \to \infty) \quad a.s.$$

PROBLEM 10.2. Prove Theorem 10.1.

HINT: Proceed as in the proof of Theorem 10.3, but apply Problem 10.1 instead of Theorem 10.2

PROBLEM 10.3. (Devroye, personal communication, 1998). Let $n \in \mathcal{N}$ be fixed. Let $\mathcal{F}_n$ be the set of all functions which are piecewise linear on a partition of $[0, 1]$ consisting of intervals. Assume that $[0, h_1]$ is one of these intervals ($h_1 > 0$). Show that the piecewise linear estimate

$$m_n(\cdot) = \arg\min_{f \in \mathcal{F}_n} \left\{ \frac{1}{n} \sum_{i=1}^{n} |f(X_i) - Y_i|^2 \right\}$$

satisfies

$$\mathbf{E} \int |m_n(x) - m(x)|^2 \mu(dx) = \infty$$

if X is uniformly distributed on $[0, 1]$, Y is $\{-1, 1\}$-valued with $\mathbf{E}Y = 0$, and X, Y are independent.

HINT:

Step (a). Let A be the event that $X_1, X_2 \in [0, h_1]$, $X_3, \ldots, X_n \in [h_1, 1]$, and $Y_1 \neq Y_2$. Then $\mathbf{P}\{A\} > 0$ and

$$\mathbf{E} \int (m_n(x) - m(x))^2 \mu(dx) = \mathbf{E} \int m_n(x)^2 dx \geq \left(\mathbf{E}\left\{ \int |m_n(x)| dx \Big| A \right\} \mathbf{P}\{A\} \right)^2.$$

Step (b). Given A, on $[0, h_1]$ the piecewise linear estimate m_n has the form

$$m_n(x) = \frac{\pm 2}{\Delta}(x - c),$$

where $\Delta = |X_1 - X_2|$ and $0 \leq c \leq h_1$. Then

$$\mathbf{E}\left\{ \int |m_n(x)| dx \Big| A \right\} \geq \mathbf{E}\left\{ \frac{2}{\Delta} \int_0^{h_1} |x - h_1/2| dx \Big| A \right\}.$$

Step (c).

$$\mathbf{E}\left\{ \frac{1}{\Delta} \Big| A \right\} = \mathbf{E}\left\{ \frac{1}{\Delta} \right\} = \infty.$$

PROBLEM 10.4. Let $\beta > 0$, let $\mathcal{G}$ be a set of functions $g : \mathcal{R}^d \to [-\beta, \beta]$, and let $\mathcal{H}$ be the set of all functions $h : \mathcal{R}^d \times \mathcal{R} \to \mathcal{R}$ defined by

$$h(x, y) = |g(x) - T_\beta y|^2 \quad ((x, y) \in \mathcal{R}^d \times \mathcal{R})$$

for some $g \in \mathcal{G}$. Show that for any $\epsilon > 0$ and any $(x,y)_1^n = ((x_1, y_1), \ldots, (x_n, y_n)) \in (\mathcal{R}^d \times [-\beta, \beta])^n$,

$$\mathcal{N}_1\left(\epsilon, \mathcal{H}, (x,y)_1^n\right) \leq \mathcal{N}_1\left(\frac{\epsilon}{4\beta}, \mathcal{G}, x_1^n\right).$$

HINT: Choose an L_1 cover of $\mathcal{G}$ on x_1^n of minimal size. Show that you can assume w.l.o.g. that the functions in this cover are bounded in absolute value by β. Use this cover as in the proof of Theorem 10.3 to construct an L_1 cover of $\mathcal{H}$ on $(x,y)_1^n$.

PROBLEM 10.5. Show that under the assumptions of Theorem 10.4, (10.31) and (10.32) hold.

HINT: Proceed as in the proof of Theorem 10.3.

PROBLEM 10.6. Modify the estimate in Theorem 10.4 in such a way that it is weakly and strongly universally consistent.

HINT: Choose $A_n \in \mathcal{R}_+$ such that A_n tends "not too quickly" to infinity. Define $\mathcal{F}_n$ as the set of all piecewise polynomials of degree M (or less) with respect to an equidistant partition of $[-A_n, A_n]$ consisting of K_n intervals, and extend these functions on the whole $\mathcal{R}$ by setting them to zero outside $[-A_n, A_n]$.

PROBLEM 10.7. Construct a multivariate version of the estimate in Problem 10.6 and show that it is strongly universally consistent.

HINT: Use functions which are equal to a multivariate polynomial of degree M (or less, in each coordinate) with respect to suitable partitions.

11

Least Squares Estimates II: Rate of Convergence

In this chapter we study the rates of convergence of least squares estimates. We separately consider linear and nonlinear estimates. The key tools in the derivation of these results are extensions of the exponential inequalities in Chapter 9, which we will also use later to define adaptive versions of the estimates.

11.1   Linear Least Squares Estimates

In this section we will study a truncated version $m_n(\cdot) = T_L \tilde{m}_n(\cdot)$ of a linear least squares estimate

$$\tilde{m}_n(\cdot) = \arg \min_{f \in \mathcal{F}_n} \frac{1}{n} \sum_{i=1}^{n} |f(X_i) - Y_i|^2, \tag{11.1}$$

where $\mathcal{F}_n$ is a linear vector space of functions $f : \mathcal{R}^d \to \mathcal{R}$, which depends on n. Examples of such estimates are linear least squares series estimates and piecewise polynomial partitioning estimates.

We are interested in the rate of convergence of

$$\|m_n - m\| = \left\{ \int |m_n(x) - m(x)|^2 \, \mu(dx) \right\}^{1/2}$$

to zero. To bound $m_n - m$ in the $L_2(\mu)$ norm, we will first bound it in the empirical norm $\|\cdot\|_n$, given by

$$\|f\|_n^2 = \int |f(x)|^2 \, \mu_n(dx) = \frac{1}{n} \sum_{i=1}^{n} |f(X_i)|^2,$$

and then show how one can bound the $L_2(\mu)$ norm by the empirical norm.

Our first result gives a bound on $\|\tilde{m}_n - m\|_n^2$. Observe that if we assume that m is bounded in absolute value by L (which we will do later), then $|m_n(x) - m(x)| \leq |\tilde{m}_n(x) - m(x)|$ for all x which implies $\|m_n - m\|_n^2 \leq \|\tilde{m}_n - m\|_n^2$.

Theorem 11.1. *Assume*

$$\sigma^2 = \sup_{x \in \mathcal{R}^d} \mathbf{Var}\{Y | X = x\} < \infty.$$

Let the estimate $\tilde{m}_n$ be defined by (11.1), where $\mathcal{F}_n$ is a linear vector space of functions $f : \mathcal{R}^d \to \mathcal{R}$ which may depend on $X_1, \ldots, X_n$. Let $K_n = K_n(X_1, \ldots, X_n)$ be the vector space dimension of $\mathcal{F}_n$. Then

$$\mathbf{E}\left\{\|\tilde{m}_n - m\|_n^2 \Big| X_1, \ldots, X_n\right\} \leq \sigma^2 \frac{K_n}{n} + \min_{f \in \mathcal{F}_n} \|f - m\|_n^2.$$

PROOF. In order to simplify the notation we will use the abbreviation

$$\mathbf{E}^*\{\cdot\} = \mathbf{E}\{\cdot | X_1, \ldots, X_n\}.$$

In the first step of the proof we show that

$$\|\mathbf{E}^*\{\tilde{m}_n\} - m\|_n^2 = \min_{f \in \mathcal{F}_n} \|f - m\|_n^2. \tag{11.2}$$

By the results of Section 10.1, (11.1) is equivalent to

$$\tilde{m}_n = \sum_{j=1}^{K_n} a_j f_{j,n},$$

where $f_{1,n}, \ldots, f_{K_n,n}$ is a basis of $\mathcal{F}_n$ and $a = (a_j)_{j=1,\ldots,K_n}$ satisfies

$$\frac{1}{n} B^T B a = \frac{1}{n} B^T Y$$

with

$$B = (f_{j,n}(X_i))_{1 \leq i \leq n, 1 \leq j \leq K_n} \quad \text{and} \quad Y = (Y_1, \ldots, Y_n)^T.$$

If we take the conditional expectation given $X_1, \ldots, X_n$, then we get

$$\mathbf{E}^*\{\tilde{m}_n\} = \sum_{j=1}^{K_n} \mathbf{E}^*\{a_j\} \cdot f_{j,n},$$

where $\mathbf{E}^*\{a\} = (\mathbf{E}^*\{a_j\})_{j=1,\dots,K_n}$ satisfies

$$\frac{1}{n}B^T B \mathbf{E}^*\{a\} = \frac{1}{n}B^T (m(X_1),\dots,m(X_n))^T.$$

There we have used

$$\mathbf{E}^*\left\{\frac{1}{n}B^T Y\right\}$$

$$= \frac{1}{n}B^T (\mathbf{E}^*\{Y_1\},\dots,\mathbf{E}^*\{Y_n\})^T$$

$$= \frac{1}{n}B^T (m(X_1),\dots,m(X_n))^T.$$

Hence, again by the results of Section 10.1, $\mathbf{E}^*\{\tilde{m}_n\}$ is the least squares estimate in $\mathcal{F}_n$ on the data $(X_1, m(X_1))$, ..., $(X_n, m(X_n))$ and therefore satisfies

$$\|\mathbf{E}^*\{\tilde{m}_n\} - m\|_n^2 = \frac{1}{n}\sum_{i=1}^{n} |\mathbf{E}^*\{\tilde{m}_n(X_i)\} - m(X_i)|^2$$

$$= \min_{f\in\mathcal{F}_n} \frac{1}{n}\sum_{i=1}^{n} |f(X_i) - m(X_i)|^2 = \min_{f\in\mathcal{F}_n} \|f - m\|_n^2.$$

Next, we observe

$$\mathbf{E}^*\{\|\tilde{m}_n - m\|_n^2\}$$

$$= \mathbf{E}^*\left\{\frac{1}{n}\sum_{i=1}^{n} |\tilde{m}_n(X_i) - m(X_i)|^2\right\}$$

$$= \mathbf{E}^*\left\{\frac{1}{n}\sum_{i=1}^{n} |\tilde{m}_n(X_i) - \mathbf{E}^*\{\tilde{m}_n(X_i)\}|^2\right\}$$

$$+ \mathbf{E}^*\left\{\frac{1}{n}\sum_{i=1}^{n} |\mathbf{E}^*\{\tilde{m}_n(X_i)\} - m(X_i)|^2\right\}$$

$$= \mathbf{E}^*\left\{\|\tilde{m}_n - \mathbf{E}^*\{\tilde{m}_n\}\|_n^2\right\} + \|\mathbf{E}^*\{\tilde{m}_n\} - m\|_n^2,$$

where the second equality follows from

$$\mathbf{E}^*\left\{\frac{1}{n}\sum_{i=1}^{n} (\tilde{m}_n(X_i) - \mathbf{E}^*\{\tilde{m}_n(X_i)\}) (\mathbf{E}^*\{\tilde{m}_n(X_i)\} - m(X_i))\right\}$$

$$= \frac{1}{n}\sum_{i=1}^{n} (\mathbf{E}^*\{\tilde{m}_n(X_i)\} - m(X_i)) \mathbf{E}^*\left\{\tilde{m}_n(X_i) - \mathbf{E}^*\{\tilde{m}_n(X_i)\}\right\}$$

$$= \frac{1}{n}\sum_{i=1}^{n} (\mathbf{E}^*\{\tilde{m}_n(X_i)\} - m(X_i)) \cdot 0$$

$$= 0.$$

Thus it remains to show

$$\mathbf{E}^* \left\{ \|\tilde{m}_n - \mathbf{E}^*\{\tilde{m}_n\}\|_n^2 \right\} \leq \sigma^2 \frac{K_n}{n}. \tag{11.3}$$

Choose a complete orthonormal system $f_1, \ldots, f_K$ in $\mathcal{F}_n$ with respect to the empirical scalar product $< \cdot, \cdot >_n$, given by

$$< f, g >_n = \frac{1}{n} \sum_{i=1}^n f(X_i) g(X_i).$$

Such a system will depend on $X_1, \ldots, X_n$, but it will always satisfy $K \leq K_n$. Then we have, on $\{X_1, \ldots, X_n\}$,

$$\text{span}\, \{f_1, \ldots, f_K\} = \mathcal{F}_n,$$

hence $\tilde{m}_n$ is also the least squares estimate of m in $\text{span}\,\{f_1, \ldots, f_K\}$. Therefore, for $x \in \{X_1, \ldots, X_n\}$,

$$\tilde{m}_n(x) = f(x)^T \frac{1}{n} B^T Y,$$

where

$$f(x) = (f_1(x), \ldots, f_K(x))^T \quad \text{and} \quad B = (f_j(X_i))_{1 \leq i \leq n, 1 \leq j \leq K}.$$

Here we have used

$$\frac{1}{n} B^T B = (< f_j, f_k >_n)_{1 \leq j,k \leq K} = (\delta_{j,k})_{1 \leq j,k \leq K}, \tag{11.4}$$

where $\delta_{j,k}$ is the Kronecker symbol, i.e., $\delta_{j,k} = 1$ for $j = k$ and $\delta_{j,k} = 0$ otherwise. Now,

$$\mathbf{E}^* \left\{ |\tilde{m}_n(x) - \mathbf{E}^*\{\tilde{m}_n(x)\}|^2 \right\}$$

$$= \mathbf{E}^* \left\{ \left| f(x)^T \frac{1}{n} B^T Y - f(x)^T \frac{1}{n} B^T (m(X_1), \ldots, m(X_n))^T \right|^2 \right\}$$

$$= \mathbf{E}^* \left\{ \left| f(x)^T \frac{1}{n} B^T (Y_1 - m(X_1), \ldots, Y_n - m(X_n))^T \right|^2 \right\}$$

$$= \mathbf{E}^* \left\{ f(x)^T \frac{1}{n} B^T \left((Y_i - m(X_i))(Y_j - m(X_j)) \right)_{1 \leq i,j \leq n} \frac{1}{n} B f(x) \right\}$$

$$= f(x)^T \frac{1}{n} B^T \left(\mathbf{E}^* \{(Y_i - m(X_i))(Y_j - m(X_j))\} \right)_{1 \leq i,j \leq n} \frac{1}{n} B f(x).$$

Since

$$\mathbf{E}^* \{(Y_i - m(X_i))(Y_j - m(X_j))\} = \delta_{i,j} \mathbf{Var}\{Y_i | X_i\}$$

we get, for any vector $c = (c_1, \ldots, c_n)^T \in \mathcal{R}^n$,

$$c^T \left(\mathbf{E}^* \left\{(Y_i - m(X_i))(Y_j - m(X_j))\right\}\right)_{1 \le i,j \le n} c$$

$$= \sum_{i=1}^{n} \mathbf{Var}\{Y_i | X_i\} c_i^2 \le \sigma^2 c^T c,$$

which, together with (11.4), implies

$$\mathbf{E}^* \left\{|\tilde{m}_n(x) - \mathbf{E}^* \tilde{m}_n(x)|^2\right\} \le \frac{\sigma^2}{n} f(x)^T f(x) = \frac{\sigma^2}{n} \sum_{j=1}^{K} |f_j(x)|^2.$$

It follows that

$$\mathbf{E}^* \left\{\|\tilde{m}_n - \mathbf{E}^*\{\tilde{m}_n\}\|_n^2\right\} = \frac{1}{n} \sum_{i=1}^{n} \mathbf{E}^* \left\{|\tilde{m}_n(X_i) - \mathbf{E}^*\{\tilde{m}_n(X_i)\}|^2\right\}$$

$$\le \frac{1}{n} \sum_{i=1}^{n} \frac{\sigma^2}{n} \sum_{j=1}^{K} |f_j(X_i)|^2$$

$$= \frac{\sigma^2}{n} \sum_{j=1}^{K} \|f_j\|_n^2$$

$$= \frac{\sigma^2}{n} K \le \frac{\sigma^2}{n} K_n.$$

$$\square$$

To bound the $L_2(\mu)$ norm $\| \cdot \|$ by the empirical norm $\| \cdot \|_n$ we will use the next theorem.

Theorem 11.2. *Let $\mathcal{F}$ be a class of functions $f : \mathcal{R}^d \to \mathcal{R}$ bounded in absolute value by B. Let $\epsilon > 0$. Then*

$$\mathbf{P}\{\exists f \in \mathcal{F} : \|f\| - 2\|f\|_n > \epsilon\} \le 3\mathbf{E}\mathcal{N}_2\left(\frac{\sqrt{2}}{24}\epsilon, \mathcal{F}, X_1^{2n}\right) \exp\left(-\frac{n\epsilon^2}{288B^2}\right).$$

PROOF. The proof will be divided in several steps.

STEP 1. Replace the $L_2(\mu)$ norm by the empirical norm defined by a ghost sample.

Let $X_1'^n = (X_{n+1}, \ldots, X_{2n})$ be a ghost sample of i.i.d. random variables distributed as X and independent of X_1^n. Define

$$\|f\|_n'^2 = \frac{1}{n} \sum_{i=n+1}^{2n} |f(X_i)|^2.$$

Let f^* be a function $f \in \mathcal{F}$ such that

$$\|f\| - 2\|f\|_n > \epsilon,$$

if there exists any such function, and let f^* be an other arbitrary function contained in $\mathcal{F}$, if such a function doesn't exist. Observe that f^* depends on X_1^n. Then

$$\mathbf{P}\left\{2\|f^*\|_n' + \frac{\epsilon}{2} > \|f^*\| \,\Big|\, X_1^n\right\}$$

$$\geq \mathbf{P}\left\{4\|f^*\|_n'^2 + \frac{\epsilon^2}{4} > \|f^*\|^2 \,\Big|\, X_1^n\right\}$$

$$= 1 - \mathbf{P}\left\{4\|f^*\|_n'^2 + \frac{\epsilon^2}{4} \leq \|f^*\|^2 \,\Big|\, X_1^n\right\}$$

$$= 1 - \mathbf{P}\left\{3\|f^*\|^2 + \frac{\epsilon^2}{4} \leq 4(\|f^*\|^2 - \|f^*\|_n'^2) \,\Big|\, X_1^n\right\}.$$

By the Chebyshev inequality

$$\mathbf{P}\left\{3\|f^*\|^2 + \frac{\epsilon^2}{4} \leq 4(\|f^*\|^2 - \|f^*\|_n'^2) \,\Big|\, X_1^n\right\}$$

$$\leq \frac{16\,\mathbf{Var}\left(\frac{1}{n}\sum_{i=n+1}^{2n} |f^*(X_i)|^2 | X_1^n\right)}{\left(3\|f^*\|^2 + \frac{\epsilon^2}{4}\right)^2}$$

$$\leq \frac{16\frac{1}{n}B^2\|f^*\|^2}{\left(3\|f^*\|^2 + \frac{\epsilon^2}{4}\right)^2}$$

$$\leq \frac{\frac{16}{3}\frac{B^2}{n}\left(3\|f^*\|^2 + \frac{\epsilon^2}{4}\right)}{\left(3\|f^*\|^2 + \frac{\epsilon^2}{4}\right)^2}$$

$$\leq \frac{16B^2}{3n} \cdot \frac{4}{\epsilon^2} = \frac{64B^2}{3\epsilon^2 n}.$$

Hence, for $n \geq \frac{64B^2}{\epsilon^2}$,

$$\mathbf{P}\left\{2\|f^*\|_n' + \frac{\epsilon}{2} > \|f^*\| \,\Big|\, X_1^n\right\} \geq \frac{2}{3}. \tag{11.5}$$

Next,

$$\mathbf{P}\left\{\exists f \in \mathcal{F} : \|f\|_n' - \|f\|_n > \frac{\epsilon}{4}\right\}$$

$$\geq \mathbf{P}\left\{2\|f^*\|_n' - 2\|f^*\|_n > \frac{\epsilon}{2}\right\}$$

$$\geq \mathbf{P}\left\{2\|f^*\|_n' + \frac{\epsilon}{2} - 2\|f^*\|_n > \epsilon, \ 2\|f^*\|_n' + \frac{\epsilon}{2} > \|f^*\|\right\}$$

$$\geq \mathbf{P}\left\{\|f^*\| - 2\|f^*\|_n > \epsilon, \ 2\|f^*\|_n' + \frac{\epsilon}{2} > \|f^*\|\right\}$$

$$= \mathbf{E}\left\{I_{\{\|f^*\|-2\|f^*\|_n>\epsilon\}} \cdot \mathbf{P}\left\{2\|f^*\|_n' + \frac{\epsilon}{2} > \|f^*\| \,\Big|\, X_1^n\right\}\right\}$$

$$\geq \frac{2}{3} \mathbf{P} \left\{ \|f^*\| - 2\|f^*\|_n > \epsilon \right\}$$

$$\text{(for } n \geq 64B^2/\epsilon^2 \text{ by } (11.5))$$

$$= \frac{2}{3} \mathbf{P} \left\{ \exists f \in \mathcal{F} : \|f\| - 2\|f\|_n > \epsilon \right\}.$$

This proves

$$\mathbf{P} \left\{ \exists f \in \mathcal{F} : \|f\| - 2\|f\|_n > \epsilon \right\} \leq \frac{3}{2} \mathbf{P} \left\{ \exists f \in \mathcal{F} : \|f\|'_n - \|f\|_n > \frac{\epsilon}{4} \right\}$$

for $n \geq 64B^2/\epsilon^2$. Observe that for $n \leq 64B^2/\epsilon^2$ the assertion is trivial, because in this case the right-hand side of the inequality in the theorem is greater than 1.

STEP 2. Introduction of additional randomness.
Let $U_1, \ldots, U_n$ be independent and uniformly distributed on $\{-1, 1\}$ and independent of $X_1, \ldots, X_{2n}$. Set

$$Z_i = \left\{ \begin{array}{ll} X_{i+n} & \text{if } U_i = 1, \\ X_i & \text{if } U_i = -1, \end{array} \right. \quad \text{and} \quad Z_{i+n} = \left\{ \begin{array}{ll} X_i & \text{if } U_i = 1, \\ X_{i+n} & \text{if } U_i = -1, \end{array} \right.$$

$(i = 1, \ldots, n)$. Because of the independence and identical distribution of $X_1, \ldots, X_{2n}$ the joint distribution of X_1^{2n} is not affected if one randomly interchanges the corresponding components of X_1^n and X_{n+1}^{2n} and is, hence, equal to the joint distribution of Z_1^{2n}. Thus

$$\mathbf{P} \left\{ \exists f \in \mathcal{F} : \|f\|'_n - \|f\|_n > \frac{\epsilon}{4} \right\}$$

$$= \mathbf{P} \left\{ \exists f \in \mathcal{F} : \left(\frac{1}{n} \sum_{i=n+1}^{2n} |f(X_i)|^2 \right)^{1/2} - \left(\frac{1}{n} \sum_{i=1}^{n} |f(X_i)|^2 \right)^{1/2} > \frac{\epsilon}{4} \right\}$$

$$= \mathbf{P} \left\{ \exists f \in \mathcal{F} : \left(\frac{1}{n} \sum_{i=n+1}^{2n} |f(Z_i)|^2 \right)^{1/2} - \left(\frac{1}{n} \sum_{i=1}^{n} |f(Z_i)|^2 \right)^{1/2} > \frac{\epsilon}{4} \right\}.$$

STEP 3. Conditioning and introduction of a covering.
Next we condition in the last probability on X_1^{2n}. Let

$$\mathcal{G} = \left\{ g_j \; : \; j = 1, \ldots, \mathcal{N}_2 \left(\frac{\sqrt{2}}{24} \epsilon, \mathcal{F}, X_1^{2n} \right) \right\}$$

be a $\frac{\sqrt{2}}{24}\epsilon$-cover of $\mathcal{F}$ w.r.t. $\|\cdot\|_{2n}$ of minimal size, where

$$\|f\|_{2n}^2 = \frac{1}{2n} \sum_{i=1}^{2n} |f(X_i)|^2.$$

W.l.o.g. we may assume $-B \le g_j(x) \le B$ for all $x \in \mathcal{R}^d$. Fix $f \in \mathcal{F}$. Then there exists a $g \in \mathcal{G}$ such that

$$\|f - g\|_{2n} \le \frac{\sqrt{2}}{24}\epsilon.$$

It follows that

$$\left\{ \frac{1}{n} \sum_{i=n+1}^{2n} |f(Z_i)|^2 \right\}^{\frac{1}{2}} - \left\{ \frac{1}{n} \sum_{i=1}^{n} |f(Z_i)|^2 \right\}^{\frac{1}{2}}$$

$$= \left\{ \frac{1}{n} \sum_{i=n+1}^{2n} |f(Z_i)|^2 \right\}^{\frac{1}{2}} - \left\{ \frac{1}{n} \sum_{i=n+1}^{2n} |g(Z_i)|^2 \right\}^{\frac{1}{2}}$$

$$+ \left\{ \frac{1}{n} \sum_{i=n+1}^{2n} |g(Z_i)|^2 \right\}^{\frac{1}{2}} - \left\{ \frac{1}{n} \sum_{i=1}^{n} |g(Z_i)|^2 \right\}^{\frac{1}{2}}$$

$$+ \left\{ \frac{1}{n} \sum_{i=1}^{n} |g(Z_i)|^2 \right\}^{\frac{1}{2}} - \left\{ \frac{1}{n} \sum_{i=1}^{n} |f(Z_i)|^2 \right\}^{\frac{1}{2}}$$

$$\le \left\{ \frac{1}{n} \sum_{i=n+1}^{2n} |g(Z_i) - f(Z_i)|^2 \right\}^{\frac{1}{2}} + \left\{ \frac{1}{n} \sum_{i=n+1}^{2n} |g(Z_i)|^2 \right\}^{\frac{1}{2}}$$

$$- \left\{ \frac{1}{n} \sum_{i=1}^{n} |g(Z_i)|^2 \right\}^{\frac{1}{2}} + \left\{ \frac{1}{n} \sum_{i=1}^{n} |g(Z_i) - f(Z_i)|^2 \right\}^{\frac{1}{2}}$$

(by triangle inequality)

$$\le \sqrt{2}\|f - g\|_{2n} + \left\{ \frac{1}{n} \sum_{i=n+1}^{2n} |g(Z_i)|^2 \right\}^{\frac{1}{2}}$$

$$- \left\{ \frac{1}{n} \sum_{i=1}^{n} |g(Z_i)|^2 \right\}^{\frac{1}{2}} + \sqrt{2}\|f - g\|_{2n}$$

(by definition of Z_1^{2n})

$$\le \frac{\epsilon}{6} + \left\{ \frac{1}{n} \sum_{i=n+1}^{2n} |g(Z_i)|^2 \right\}^{\frac{1}{2}} - \left\{ \frac{1}{n} \sum_{i=1}^{n} |g(Z_i)|^2 \right\}^{\frac{1}{2}}.$$

Using this we get

$$\mathbf{P}\left\{ \exists f \in \mathcal{F} : \left(\frac{1}{n} \sum_{i=n+1}^{2n} |f(Z_i)|^2 \right)^{1/2} - \left(\frac{1}{n} \sum_{i=1}^{n} |f(Z_i)|^2 \right)^{1/2} > \frac{\epsilon}{4} \Big| X_1^{2n} \right\}$$

$$\leq \mathbf{P}\left\{\exists g \in \mathcal{G}: \left(\frac{1}{n}\sum_{i=n+1}^{2n}|g(Z_i)|^2\right)^{1/2} - \left(\frac{1}{n}\sum_{i=1}^{n}|g(Z_i)|^2\right)^{1/2} > \frac{\epsilon}{12}\bigg|X_1^{2n}\right\}$$

$$\leq |\mathcal{G}|\cdot\max_{g\in\mathcal{G}}\mathbf{P}\left\{\left(\frac{1}{n}\sum_{i=n+1}^{2n}|g(Z_i)|^2\right)^{1/2} - \left(\frac{1}{n}\sum_{i=1}^{n}|g(Z_i)|^2\right)^{1/2} > \frac{\epsilon}{12}\bigg|X_1^{2n}\right\},$$

where

$$|\mathcal{G}| = \mathcal{N}_2\left(\frac{\sqrt{2}}{24}\epsilon, \mathcal{F}, X_1^{2n}\right).$$

STEP 4. Application of Hoeffding's inequality.
In this step we bound

$$\mathbf{P}\left\{\left(\frac{1}{n}\sum_{i=n+1}^{2n}|g(Z_i)|^2\right)^{1/2} - \left(\frac{1}{n}\sum_{i=1}^{n}|g(Z_i)|^2\right)^{1/2} > \frac{\epsilon}{12}\bigg|X_1^{2n}\right\},$$

where $g: \mathcal{R}^d \to \mathcal{R}$ satisfies $-B \leq g(x) \leq B$ for all $x \in \mathcal{R}^d$.

By definition of Z_1^{2n}, and $\sqrt{a} + \sqrt{b} \geq \sqrt{a+b}$ for all $a, b \geq 0$,

$$\left(\frac{1}{n}\sum_{i=n+1}^{2n}|g(Z_i)|^2\right)^{1/2} - \left(\frac{1}{n}\sum_{i=1}^{n}|g(Z_i)|^2\right)^{1/2}$$

$$\leq \left|\frac{\frac{1}{n}\sum_{i=n+1}^{2n}|g(Z_i)|^2 - \frac{1}{n}\sum_{i=1}^{n}|g(Z_i)|^2}{\left(\frac{1}{n}\sum_{i=n+1}^{2n}|g(Z_i)|^2\right)^{1/2} + \left(\frac{1}{n}\sum_{i=1}^{n}|g(Z_i)|^2\right)^{1/2}}\right|$$

$$\leq \frac{\left|\frac{1}{n}\sum_{i=1}^{n}U_i\left(|g(X_i)|^2 - |g(X_{i+n})|^2\right)\right|}{\left(\frac{1}{n}\sum_{i=n+1}^{2n}|g(Z_i)|^2 + \frac{1}{n}\sum_{i=1}^{n}|g(Z_i)|^2\right)^{1/2}}$$

$$= \frac{\left|\frac{1}{n}\sum_{i=1}^{n}U_i\left(|g(X_i)|^2 - |g(X_{i+n})|^2\right)\right|}{\left(\frac{1}{n}\sum_{i=1}^{2n}|g(X_i)|^2\right)^{1/2}},$$

which implies that the above probability is bounded by

$$\mathbf{P}\left\{\left|\frac{1}{n}\sum_{i=1}^{n}U_i\left(|g(X_i)|^2 - |g(X_{i+n})|^2\right)\right| > \frac{\epsilon}{12}\left(\frac{1}{n}\sum_{i=1}^{2n}|g(X_i)|^2\right)^{1/2}\bigg|X_1^{2n}\right\}.$$

By Hoeffding's inequality (cf. Lemma A.3) this in turn is bounded by

$$2\exp\left(-\frac{2n^2\frac{\epsilon^2}{144}\left(\frac{1}{n}\sum_{i=1}^{2n}|g(X_i)|^2\right)}{\sum_{i=1}^{n}4\left(|g(X_i)|^2 - |g(X_{i+n})|^2\right)^2}\right)$$

$$\leq 2 \exp\left(-\frac{n\epsilon^2 \sum_{i=1}^{2n} |g(X_i)|^2}{\sum_{i=1}^{n} 288 B^2 \left(|g(X_i)|^2 + |g(X_{i+n})|^2\right)}\right)$$

$$= 2 \exp\left(-\frac{n\epsilon^2}{288 B^2}\right),$$

where the last inequality follows from

$$\left(|g(X_i)|^2 - |g(X_{i+n})|^2\right)^2 \;\leq\; |g(X_i)|^4 + |g(X_{i+n})|^4$$
$$\leq\; B^2 \left(|g(X_i)|^2 + |g(X_{i+n})|^2\right).$$

Steps 1 to 4 imply the assertion. □

Combining Theorems 11.1 and 11.2 we get

Theorem 11.3. *Assume*

$$\sigma^2 = \sup_{x \in \mathcal{R}^d} \mathbf{Var}\{Y | X = x\} < \infty$$

and

$$\|m\|_\infty = \sup_{x \in \mathcal{R}^d} |m(x)| \leq L$$

for some $L \in \mathcal{R}_+$. Let $\mathcal{F}_n$ be a linear vector space of functions $f : \mathcal{R}^d \to \mathcal{R}$. Let K_n be the vector space dimension of $\mathcal{F}_n$. Define the estimate m_n by

$$m_n(\cdot) = T_L \tilde{m}_n(\cdot) \quad where \quad \tilde{m}_n(\cdot) = \arg\min_{f \in \mathcal{F}_n} \frac{1}{n} \sum_{i=1}^{n} |f(X_i) - Y_i|^2.$$

Then

$$\mathbf{E} \int |m_n(x) - m(x)|^2 \mu(dx)$$

$$\leq c \cdot \max\{\sigma^2, L^2\} \frac{(\log(n) + 1) \cdot K_n}{n} + 8 \inf_{f \in \mathcal{F}_n} \int |f(x) - m(x)|^2 \mu(dx),$$

for some universal constant c.

The second term on the right-hand side of the above inequality is eight times the approximation error of the estimate while the first term is an upper bound on the estimation error. Observe that in this theorem we do not assume that Y is bounded, we only assume that m is bounded and that we know this bound.

PROOF OF THEOREM 11.3. We start with the decomposition

$$\int |m_n(x) - m(x)|^2 \mu(dx)$$

$$= \left(\|m_n - m\| - 2\|m_n - m\|_n + 2\|m_n - m\|_n\right)^2$$

$$\leq \left(\max\left\{\|m_n - m\| - 2\|m_n - m\|_n, 0\right\} + 2\|m_n - m\|_n\right)^2$$

$$\leq 2\left(\max\left\{\|m_n - m\| - 2\|m_n - m\|_n, 0\right\}\right)^2 + 8\|m_n - m\|_n^2$$

$$= T_{1,n} + T_{2,n}.$$

Because of $\|m\|_\infty \leq L$ we have

$$\|m_n - m\|_n^2 \leq \|\tilde{m}_n - m\|_n^2,$$

which together with Theorem 11.1 implies

$$\begin{aligned}
\mathbf{E}\{T_{2,n}\} &\leq 8\mathbf{E}\{\mathbf{E}\{\|\tilde{m}_n - m\|_n^2 | X_1, \ldots, X_n\}\} \\
&\leq 8\sigma^2 \frac{K_n}{n} + 8\mathbf{E}\left\{\min_{f \in \mathcal{F}_n} \|f - m\|_n^2\right\} \\
&\leq 8\sigma^2 \frac{K_n}{n} + 8 \inf_{f \in \mathcal{F}_n} \int |f(x) - m(x)|^2 \mu(dx).
\end{aligned}$$

Hence it suffices to show

$$\mathbf{E}\{T_{1,n}\} \leq \tilde{c} \cdot L^2 \frac{(\log(n) + 1) \cdot K_n}{n}. \tag{11.6}$$

In order to show this, let $u > 576L^2/n$ be arbitrary. Then, by Theorem 11.2,

$$\begin{aligned}
\mathbf{P}\{T_{1,n} > u\} &= \mathbf{P}\left\{2\left(\max\left\{\|m_n - m\| - 2\|m_n - m\|_n, 0\right\}\right)^2 > u\right\} \\
&\leq \mathbf{P}\left\{\exists f \in T_L \mathcal{F}_n : \|f - m\| - 2\|f - m\|_n > \sqrt{u/2}\right\} \\
&\leq 3\mathbf{E}\mathcal{N}_2\left(\sqrt{u}/24, \mathcal{F}_n, X_1^{2n}\right) \cdot \exp\left(-\frac{n \cdot u}{576(2L)^2}\right) \\
&\leq 3\mathbf{E}\mathcal{N}_2\left(\frac{L}{\sqrt{n}}, \mathcal{F}_n, X_1^{2n}\right) \cdot \exp\left(-\frac{n \cdot u}{576(2L)^2}\right).
\end{aligned}$$

Using Lemma 9.2, Theorem 9.4, $V_{T_L \mathcal{F}_n^+} \leq V_{\mathcal{F}_n^+}$, and Theorem 9.5 we get

$$\begin{aligned}
\mathcal{N}_2\left(L/\sqrt{n}, \mathcal{F}_n, X_1^{2n}\right) &\leq 3\left(\frac{3e(2L)^2}{(L/\sqrt{n})^2}\right)^{2(K_n+1)} \\
&= 3(12en)^{2(K_n+1)}.
\end{aligned}$$

It follows, for any $u > 576L^2/n$,

$$\mathbf{P}\{T_{1,n} > u\} \leq 9 \cdot (12en)^{2(K_n+1)} \cdot \exp\left(-\frac{n \cdot u}{2304 \cdot L^2}\right).$$

We get, for any $v > 576L^2/n$,

$$\begin{aligned}
\mathbf{E}\{T_{1,n}\} &\leq v + \int_v^\infty \mathbf{P}\{T_{1,n} > t\}dt \\
&\leq v + 9 \cdot (12en)^{2(K_n+1)} \cdot \int_v^\infty \exp\left(-\frac{n \cdot t}{2304 \cdot L^2}\right) dt
\end{aligned}$$

$$= \quad v + 9 \cdot (12en)^{2(K_n+1)} \cdot \frac{2304L^2}{n} \cdot \exp\left(-\frac{n \cdot v}{2304 \cdot L^2}\right).$$

Setting

$$v = \frac{2304L^2}{n} \cdot \log\left(9(12en)^{2(K_n+1)}\right)$$

this implies (11.6), which in turn implies the assertion. $\qquad\square$

11.2   Piecewise Polynomial Partitioning Estimates

In this section we illustrate the previous results by applying them to piecewise polynomial partitioning estimates. The next lemma will be needed to bound the approximation error.

Lemma 11.1. *Let $M \in \mathcal{N}_0$, $K \in \mathcal{N}$, $C > 0$, $q \in \{0, \ldots, M\}$, $r \in (0, 1]$, and set $p = q + r$. Let $m : [0, 1] \to \mathcal{R}$ be some (p, C)-smooth function, i.e., assume that the qth derivative $m^{(q)}$ of m exists and satisfies*

$$|m^{(q)}(x) - m^{(q)}(z)| \leq C \cdot |x - z|^r \quad (x, z \in [0, 1]).$$

Then there exists a piecewise polynomial f of degree M (or less) with respect to an equidistant partition of $[0, 1]$ consisting of K intervals of length $1/K$ such that

$$\sup_{x \in [0,1]} |f(x) - m(x)| \leq \frac{1}{2^p \cdot q!} \cdot \frac{C}{K^p}.$$

PROOF. Fix $z_0 \in (0, 1)$ and let g_k be the Taylor polynomial of m of degree k around z_0 given by

$$g_k(z) = \sum_{j=0}^{k} \frac{m^{(j)}(z_0)}{j!}(z - z_0)^j,$$

where $m^{(j)}(z_0)$ is the jth derivative of m at the point z_0 ($k \in \{0, 1, \ldots, q\}$). We will show

$$|g_q(z) - m(z)| \leq \frac{1}{q!} \cdot C \cdot |z - z_0|^p \quad (z \in [0, 1]). \tag{11.7}$$

The assertion follows by choosing f on each interval as the Taylor polynomial of m around the midpoint of the interval. For $q = 0$ (11.7) follows directly from assumption m (p, C)-smooth. In order to show (11.7) in the case $q > 0$, we use the well-known integral form of the Taylor series remainder, which can be proven by induction and integration by parts,

$$m(z) - g_k(z) = \frac{1}{k!} \int_{z_0}^{z} (z - t)^k m^{(k+1)}(t) \, dt.$$

Hence, for $f = g_q$,

$$m(z) - f(z)$$

$$= m(z) - g_{q-1}(z) - \frac{m^{(q)}(z_0)}{q!}(z - z_0)^q$$

$$= \frac{1}{(q-1)!}\int_{z_0}^{z}(z-t)^{q-1}m^{(q)}(t)dt - \frac{m^{(q)}(z_0)}{(q-1)!}\int_{z_0}^{z}(z-t)^{q-1}dt$$

$$= \frac{1}{(q-1)!}\int_{z_0}^{z}(z-t)^{q-1}\cdot(m^{(q)}(t) - m^{(q)}(z_0))dt.$$

From this, and the assumption that m is (p, C)-smooth, one concludes

$$\begin{aligned}
|m(z) - f(z)| &\leq \left|\frac{1}{(q-1)!}\int_{z_0}^{z}(z-t)^{q-1}\cdot C\cdot|t-z_0|^r dt\right| \\
&\leq \frac{C\cdot|z-z_0|^r}{(q-1)!}\left|\int_{z_0}^{z}(z-t)^{q-1}dt\right| \\
&= \frac{C\cdot|z-z_0|^{q+r}}{q!}.
\end{aligned}$$

$\square$

We are now in a position to derive results on the convergence of the piecewise polynomial partitioning estimate to the regression function. First we consider the error in the empirical norm.

Corollary 11.1. *Let* $M \in \mathcal{N}_0$ *and* $K_n \in \mathcal{N}$. *Let* $\mathcal{F}_n$ *be the set of all piecewise polynomials of degree* M *(or less) w.r.t. an equidistant partition of* $[0, 1]$ *into* K_n *intervals. Let the estimate* $\tilde{m}_n$ *be defined by (11.1). Assume that the distribution of* (X, Y) *satisfies* $X \in [0, 1]$ *a.s. and*

$$\sigma^2 = \sup_{x \in \mathcal{R}^d}\mathbf{Var}\{Y|X = x\} < \infty.$$

Then

$$\mathbf{E}\left\{\|\tilde{m}_n - m\|_n^2\Big|X_1, \ldots, X_n\right\} \leq \sigma^2\frac{(M+1)\cdot K_n}{n} + \min_{f\in\mathcal{F}_n}\|f - m\|_n^2.$$

Furthermore, if m *is* (p, C)*-smooth for some* $p = q + r \leq (M+1)$, $q \in \mathcal{N}_0$, $r \in (0, 1]$, *then*

$$\mathbf{E}\left\{\|\tilde{m}_n - m\|_n^2\Big|X_1, \ldots, X_n\right\} \leq \sigma^2\frac{(M+1)\cdot K_n}{n} + \frac{1}{2^{2p}q!^2}\cdot\frac{C^2}{K_n^{2p}}$$

and for

$$K_n = \left\lceil\left(\frac{2p}{2^{2p}q!^2(M+1)}\cdot\frac{C^2 n}{\sigma^2}\right)^{1/(2p+1)}\right\rceil$$

one gets for any $C \geq \sigma/n^{1/2}$

$$\mathbf{E}\left\{\|\tilde{m}_n - m\|_n^2 \Big| X_1, \ldots, X_n\right\} \leq c_M C^{\frac{2}{2p+1}} \cdot \left(\frac{\sigma^2}{n}\right)^{\frac{2p}{2p+1}}$$

for some constant c_M depending only on M.

PROOF. $\mathcal{F}_n$ is a linear vector space of dimension $(M+1) \cdot K_n$, hence the first inequality follows from Theorem 11.1. Furthermore, Lemma 11.1, together with $X \in [0,1]$ *a.s.*, implies

$$\min_{f \in \mathcal{F}_n} \|f - m\|_n^2 \leq \min_{f \in \mathcal{F}_n} \sup_{x \in [0,1]} |f(x) - m(x)|^2 \leq \frac{1}{2^{2p}q!^2} \cdot \frac{C^2}{K_n^{2p}}.$$

From this one gets the second inequality. The definition of K_n implies the third inequality.   □

By applying Theorem 11.3 to piecewise polynomial partitioning estimates we can bound the error in the $L_2(\mu)$ norm.

Corollary 11.2. *Let $M \in \mathcal{N}_0$ and $K_n \in \mathcal{N}$. Let $\mathcal{F}_n$ be the set of all piecewise polynomials of degree M (or less) w.r.t. an equidistant partition of $[0,1]$ into K_n intervals. Define the estimate m_n by*

$$m_n(\cdot) = T_L \tilde{m}_n(\cdot) \quad where \quad \tilde{m}_n(\cdot) = \arg\min_{f \in \mathcal{F}_n} \frac{1}{n} \sum_{i=1}^{n} |f(X_i) - Y_i|^2.$$

Assume that the distribution of (X, Y) satisfies $X \in [0,1]$ a.s.,

$$\sigma^2 = \sup_{x \in \mathcal{R}^d} \mathbf{Var}\{Y|X = x\} < \infty,$$

$$\|m\|_\infty = \sup_{x \in \mathcal{R}^d} |m(x)| \leq L$$

and

$$m \ is \ (p, C) - smooth$$

for some $C > 0$, $p = q + r$, $q \in \{0, \ldots, M\}$, $r \in (0,1]$. Then

$$\mathbf{E}\left\{\int |m_n(x) - m(x)|^2 \mu(dx)\right\}$$

$$\leq c \cdot \max\{\sigma^2, L^2\} \frac{(\log(n) + 1) \cdot K_n(M + 1)}{n} + 8\frac{1}{2^{2p}q!^2} \cdot \frac{C^2}{K_n^{2p}}$$

and for

$$K_n = \left\lceil \left(\frac{C^2}{\max\{\sigma^2, L^2\}} \frac{n}{\log(n)}\right)^{1/(2p+1)} \right\rceil$$

one gets for any $C \geq \max\{\sigma, L\} \cdot (\log(n)/n)^{1/2}$

$$\mathbf{E}\left\{\int |m_n(x) - m(x)|^2 \mu(dx)\right\}$$

$$\leq c_M C^{\frac{2}{2p+1}} \cdot \left(\max\{\sigma^2, L^2\} \cdot \frac{(\log(n) + 1)}{n}\right)^{\frac{2p}{2p+1}}$$

for some constant c_M depending only on M.

PROOF. Lemma 11.1, together with $X \in [0,1]$ *a.s.*, implies

$$\inf_{f \in \mathcal{F}_n} \int |f(x) - m(x)|^2 \mu(dx) \quad \leq \quad \inf_{f \in \mathcal{F}_n} \sup_{x \in [0,1]} |f(x) - m(x)|^2$$

$$\leq \quad \frac{1}{2^{2p} q!^2} \cdot \frac{C^2}{K_n^{2p}}.$$

From this together with Theorem 11.3, one gets the first inequality. The definition of K_n implies the second inequality. $\square$

It follows from Chapter 3 that the above rate of convergence result is optimal up to the logarithmic factor $\log(n)^{2p/(2p+1)}$. For $M = 0$ the estimate in Corollary 11.2 is the partitioning estimate of Theorem 4.3. From this we know that the logarithmic factor is not necessary for $p = 1$. We will later see (cf. Chapter 19) how to get rid of the logarithmic factor for $p \neq 1$.

11.3 Nonlinear Least Squares Estimates

In this section we generalize Theorem 11.3 from linear vector spaces to general sets of functions. This will require the introduction of complicated, but extremely useful, exponential inequalities, which will be used throughout this book.

In the rest of this section we will assume $|Y| \leq L \leq \beta_n$ a.s. and the estimate m_n will be defined by

$$m_n(\cdot) = T_{\beta_n} \tilde{m}_n(\cdot) \tag{11.8}$$

and

$$\tilde{m}_n(\cdot) = \arg\min_{f \in \mathcal{F}_n} \frac{1}{n} \sum_{i=1}^{n} |f(X_i) - Y_i|^2. \tag{11.9}$$

Let us first try to apply the results which we have derived in order to show the consistency of the estimate: it follows from Lemma 10.2 that

$$\mathbf{E} \int |m_n(x) - m(x)|^2 \mu(dx)$$

is bounded by

$$2\mathbf{E} \sup_{f \in T_{\beta_n}\mathcal{F}_n} \left| \frac{1}{n}\sum_{j=1}^{n} |f(X_j) - Y_j|^2 - \mathbf{E}\left\{ (f(X) - Y)^2 \right\} \right|$$

$$+ \inf_{f \in \mathcal{F}_n,\, \|f\|_\infty \leq \beta_n} \int |f(x) - m(x)|^2 \mu(dx).$$

To bound the first term we can apply Lemma 9.1 or Theorem 9.1, where we have used Hoeffding's inequality on fixed sup-norm and random L_1 norm covers, respectively. In both cases we have bounded the probability

$$\mathbf{P}\left\{ \sup_{f \in T_{\beta_n}\mathcal{F}_n} \left| \frac{1}{n}\sum_{j=1}^{n} |f(X_j) - Y_j|^2 - \mathbf{E}\left\{ |f(X) - Y|^2 \right\} \right| > \epsilon_n \right\}$$

by some term tending to infinity as $n \to \infty$ times

$$\exp\left(-c\,\frac{n\epsilon_n^2}{\beta_n^4} \right). \tag{11.10}$$

Thus, if we want these upper bounds to converge to zero as $n \to \infty$, then ϵ_n must converge to zero not faster than $\beta_n^2\, n^{-\frac{1}{2}}$. Unfortunately, as we have seen in Chapter 3, this is far away from the optimal rate of convergence.

Therefore, to analyze the rate of convergence of the expected value of the L_2 error, we will use a different decomposition than in Section 10.1:

$$\int |m_n(x) - m(x)|^2 \mu(dx)$$

$$= \left\{ \mathbf{E}\left\{ |m_n(X) - Y|^2 \big| D_n \right\} - \mathbf{E}|m(X) - Y|^2 \right.$$

$$\left. -2\frac{1}{n}\sum_{i=1}^{n}\left\{ |m_n(X_i) - Y_i|^2 - |m(X_i) - Y_i|^2 \right\} \right\}$$

$$+2\frac{1}{n}\sum_{i=1}^{n}\left\{ |m_n(X_i) - Y_i|^2 - |m(X_i) - Y_i|^2 \right\}. \tag{11.11}$$

Let us first observe that we can obtain a nice upper bound for the expectation of the second term on the right-hand side of (11.11). For simplicity, we ignore the factor 2. By using the definition of m_n (see (11.8) and (11.9)) and $|Y| \leq \beta_n$ *a.s.* one gets

$$\mathbf{E}\left\{ \frac{1}{n}\sum_{i=1}^{n}\left\{ |m_n(X_i) - Y_i|^2 - |m(X_i) - Y_i|^2 \right\} \right\}$$

$$\leq \mathbf{E}\left\{ \frac{1}{n}\sum_{i=1}^{n}\left\{ |\tilde{m}_n(X_i) - Y_i|^2 - |m(X_i) - Y_i|^2 \right\} \right\}$$

$$= \mathbf{E}\left\{ \inf_{f\in\mathcal{F}_n} \frac{1}{n}\sum_{i=1}^{n}\{|f(X_i)-Y_i|^2-|m(X_i)-Y_i|^2\}\right\}$$

$$\leq \inf_{f\in\mathcal{F}_n} \mathbf{E}\left\{ \frac{1}{n}\sum_{i=1}^{n}\{|f(X_i)-Y_i|^2-|m(X_i)-Y_i|^2\}\right\}$$

$$= \inf_{f\in\mathcal{F}_n} \mathbf{E}|f(X)-Y|^2-\mathbf{E}|m(X)-Y|^2$$

$$= \inf_{f\in\mathcal{F}_n} \int |f(x)-m(x)|^2\mu(dx). \tag{11.12}$$

Next we derive an upper bound for the first term on the right-hand side of
(11.11). We have

$$\mathbf{P}\Big\{ \mathbf{E}\left\{|m_n(X)-Y|^2|D_n\right\}-\mathbf{E}|m(X)-Y|^2$$

$$-2\frac{1}{n}\sum_{i=1}^{n}\{|m_n(X_i)-Y_i|^2-|m(X_i)-Y_i|^2\}>\epsilon\Big\}$$

$$=\mathbf{P}\Big\{ \mathbf{E}\left\{|m_n(X)-Y|^2|D_n\right\}-\mathbf{E}|m(X)-Y|^2$$

$$-\frac{1}{n}\sum_{i=1}^{n}\{|m_n(X_i)-Y_i|^2-|m(X_i)-Y_i|^2\}$$

$$>\frac{\epsilon}{2}+\frac{1}{2}\left\{\mathbf{E}\left\{|m_n(X)-Y|^2|D_n\right\}-\mathbf{E}\,m(X)-Y|^2\right\}\Big\}$$

$$\leq\mathbf{P}\Big\{ \exists f\in T_{\beta_n}\mathcal{F}_n : \mathbf{E}|f(X)-Y|^2-\mathbf{E}|m(X)-Y|^2$$

$$-\frac{1}{n}\sum_{i=1}^{n}\{|f(X_i)-Y_i|^2-|m(X_i)-Y_i|^2\}$$

$$>\frac{\epsilon}{2}+\frac{1}{2}\left\{\mathbf{E}|f(X)-Y|^2-\mathbf{E}|m(X)-Y|^2\right\}\Big\}. \tag{11.13}$$

We know from Chapter 9 how to extend a bound on the right-hand side
of (11.13) from a fixed function to a set of functions by the use of fixed
sup-norm or random L_1 norm covers.

For simplicity let us consider for a moment the right-hand side of (11.13)
for only one fixed function $f\in T_{\beta_n}\mathcal{F}_n$. Set $Z=(X,Y)$, $Z_i=(X_i,Y_i)$
$(i=1,\ldots,n)$,

$$g(Z)=|f(X)-Y|^2-|m(X)-Y|^2$$

and

$$g(Z_i) = |f(X_i) - Y_i|^2 - |m(X_i) - Y_i|^2 \quad (i = 1, ..., n).$$

Then $g(Z), g(Z_1), ..., g(Z_n)$ are i.i.d random variables such that $|g(Z)| \leq 4\beta_n^2$ and we want to bound

$$\mathbf{P}\left\{ \mathbf{E}g(Z) - \frac{1}{n}\sum_{i=1}^{n} g(Z_i) > \frac{\epsilon}{2} + \frac{1}{2}\mathbf{E}g(Z) \right\}. \tag{11.14}$$

The main trick is that the variance of $g(Z)$ is bounded by some constant times the expectation of $g(Z)$ (compare also Problem 7.3): Indeed,

$$
\begin{aligned}
g(Z) &= (f(X) - Y + m(X) - Y)\left((f(X) - Y) - (m(X) - Y)\right) \\
&= (f(X) + m(X) - 2Y)(f(X) - m(X))
\end{aligned}
$$

and thus

$$
\begin{aligned}
\sigma^2 &= \mathbf{Var}(g(Z)) \leq \mathbf{E}g(Z)^2 \leq 16\beta_n^2\, \mathbf{E}|f(X) - m(X)|^2 \\
&= 16\beta_n^2 \left(\mathbf{E}|f(X) - Y|^2 - \mathbf{E}|m(X) - Y|^2\right) \\
&= 16\beta_n^2\, \mathbf{E}g(Z).
\end{aligned} \tag{11.15}
$$

This enables us to derive an excellent bound for (11.14) by applying the Bernstein inequality (Lemma A.2):

$$\mathbf{P}\left\{ \mathbf{E}g(Z) - \frac{1}{n}\sum_{i=1}^{n} g(Z_i) > \frac{\epsilon}{2} + \frac{1}{2}\mathbf{E}g(Z) \right\}$$

$$\leq \mathbf{P}\left\{ \mathbf{E}g(Z) - \frac{1}{n}\sum_{i=1}^{n} g(Z_i) > \frac{\epsilon}{2} + \frac{1}{2}\frac{\sigma^2}{16\beta_n^2} \right\}$$

$$\leq \exp\left(-\frac{n\{\frac{\epsilon}{2} + \frac{\sigma^2}{32\beta_n^2}\}^2}{2\sigma^2 + 2\frac{8\beta_n^2}{3}\left\{\frac{\epsilon}{2} + \frac{\sigma^2}{32\beta_n^2}\right\}} \right)$$

$$\leq \exp\left(-\frac{n\{\frac{\epsilon}{2} + \frac{\sigma^2}{32\beta_n^2}\}^2}{\left(64\beta_n^2 + \frac{16\beta_n^2}{3}\right) \cdot \left\{\frac{\epsilon}{2} + \frac{\sigma^2}{32\beta_n^2}\right\}} \right)$$

$$= \exp\left(-\frac{n\left\{\frac{\epsilon}{2} + \frac{\sigma^2}{32\beta_n^2}\right\}}{64\beta_n^2 + \frac{16}{3}\beta_n^2} \right)$$

$$\leq \exp\left(-\frac{1}{128 + \frac{32}{3}} \cdot \frac{n\epsilon}{\beta_n^2} \right).$$

The main advantage of this upper bound is that ϵ appears only in linear and not squared form as in (11.10) and therefore (for constant β_n) this upper bound converges to zero whenever

$$\epsilon n = \epsilon_n n \to \infty \quad \text{as } n \to \infty.$$

What we now need is an extension of this upper bound from the case of a fixed function to the general case of a set of functions as in (11.13). As we have mentioned before, this can be done by the use of random L_1 norm covers. The result is summarized in the next theorem.

Theorem 11.4. *Assume* $|Y| \leq B$ *a.s. and* $B \geq 1$. *Let* $\mathcal{F}$ *be a set of functions* $f : \mathcal{R}^d \to \mathcal{R}$ *and let* $|f(x)| \leq B, B \geq 1$. *Then, for each* $n \geq 1$,

$$\mathbf{P}\left\{ \exists f \in \mathcal{F} : \mathbf{E}|f(X) - Y|^2 - \mathbf{E}|m(X) - Y|^2 \right.$$

$$-\frac{1}{n}\sum_{i=1}^{n}\left\{ |f(X_i) - Y_i|^2 - |m(X_i) - Y_i|^2 \right\}$$

$$\left. \geq \epsilon \cdot (\alpha + \beta + \mathbf{E}|f(X) - Y|^2 - \mathbf{E}|m(X) - Y|^2) \right\}$$

$$\leq 14 \sup_{x_1^n} \mathcal{N}_1\left(\frac{\beta\epsilon}{20B}, \mathcal{F}, x_1^n \right) \exp\left(-\frac{\epsilon^2(1-\epsilon)\alpha n}{214(1+\epsilon)B^4} \right)$$

where $\alpha, \beta > 0$ *and* $0 < \epsilon \leq 1/2$.

We will prove this theorem in the next two sections.
Now we are ready to formulate and prove our main result.

Theorem 11.5. *Let* $n \in \mathcal{N}$ *and* $1 \leq L < \infty$. *Assume* $|Y| \leq L$ *a.s. Let the estimate* m_n *be defined by minimization of the empirical* L_2 *risk over a set of functions* $\mathcal{F}_n$ *and truncation at* $\pm L$. *Then one has*

$$\mathbf{E}\left\{ \int |m_n(x) - m(x)|^2 \mu(dx) \right\}$$

$$\leq \frac{c_1}{n} + \frac{(c_2 + c_3 \log(n))V_{\mathcal{F}_n^+}}{n} + 2 \inf_{f \in \mathcal{F}_n} \int |f(x) - m(x)|^2 \mu(dx),$$

where

$$c_1 = 24 \cdot 214 L^4 (1 + \log 42), \quad c_2 = 48 \cdot 214 L^4 \log(480eL^2),$$

and

$$c_3 = 48 \cdot 214 L^4.$$

If $\mathcal{F}_n$ is a linear vector space of dimension K_n then $V_{\mathcal{F}_n^+}$ is bounded from above by $K_n + 1$ (cf. Theorem 9.5), and we get again the bound from Theorem 11.3 with slightly different conditions. The main advantage of the above theorem compared to Theorem 11.3 is that in the above theorem

$\mathcal{F}_n$ doesn't have to be a linear vector space. But, unfortunately, we need stronger assumptions on Y: Y must be bounded while in Theorem 11.3 we only needed assumptions on the variance of Y and boundedness of the regression function.

PROOF. We use the error decomposition

$$\int |m_n(x) - m(x)|^2 \mu(dx)$$

$$= \Big\{ \mathbf{E}\{|m_n(X) - Y|^2 | D_n\} - \mathbf{E}\{|m(X) - Y|^2\}$$

$$-2 \cdot \left(\frac{1}{n} \sum_{i=1}^{n} |m_n(X_i) - Y_i|^2 - \frac{1}{n} \sum_{i=1}^{n} |m(X_i) - Y_i|^2 \right) \Big\}$$

$$+2 \cdot \left(\frac{1}{n} \sum_{i=1}^{n} |m_n(X_i) - Y_i|^2 - \frac{1}{n} \sum_{i=1}^{n} |m(X_i) - Y_i|^2 \right)$$

$$= T_{1,n} + T_{2,n}.$$

By (11.12),

$$\mathbf{E}\{T_{2,n}\} \leq 2 \inf_{f \in \mathcal{F}_n} \int_{\mathcal{R}^d} |f(x) - m(x)|^2 \mu(dx),$$

thus it suffices to show

$$\mathbf{E}\{T_{1,n}\} \leq \frac{c_1}{n} + \frac{(c_2 + c_3 \log(n)) \cdot V_{\mathcal{F}_n^+}}{n}.$$

Let $t \geq \frac{1}{n}$ be arbitrary. Then

$$\mathbf{P}\{T_{1,n} > t\}$$

$$= \mathbf{P}\Big\{ \mathbf{E}\{|m_n(X) - Y|^2 | D_n\} - \mathbf{E}\{|m(X) - Y|^2\}$$

$$-\frac{1}{n} \sum_{i=1}^{n} \{|m_n(X_i) - Y_i|^2 - |m(X_i) - Y_i|^2\}$$

$$> \frac{1}{2} \left(t + \mathbf{E}\{|m_n(X) - Y|^2 | D_n\} - \mathbf{E}\{|m(X) - Y|^2\} \right) \Big\}$$

$$\leq \mathbf{P}\Big\{ \exists f \in T_L \mathcal{F}_n : \mathbf{E}|f(X) - Y|^2 - \mathbf{E}|m(X) - Y|^2$$

$$-\frac{1}{n} \sum_{i=1}^{n} \{|f(X_i) - Y_i|^2 - |m(X_i) - Y_i|^2\}$$

$$\geq \frac{1}{2} \cdot \left(\frac{t}{2} + \frac{t}{2} + \mathbf{E}|f(X) - Y|^2 - \mathbf{E}|m(X) - Y|^2 \right) \Bigg\}$$

$$\leq 14 \sup_{x_1^n} \mathcal{N}_1 \left(\frac{t}{80L}, T_L \mathcal{F}_n, x_1^n \right) \cdot \exp \left(-\frac{n}{24 \cdot 214 L^4} t \right)$$

$$\leq 14 \sup_{x_1^n} \mathcal{N}_1 \left(\frac{1}{80L \cdot n}, T_L \mathcal{F}_n, x_1^n \right) \cdot \exp \left(-\frac{n}{24 \cdot 214 L^4} t \right),$$

where we have used Theorem 11.4 and $t \geq \frac{1}{n}$. By Lemma 9.2 and Theorem 9.4 we get, for the covering number

$$\mathcal{N}_1 \left(\frac{1}{80L \cdot n}, T_L \mathcal{F}_n, x_1^n \right) \quad \leq \quad 3 \left(\frac{2e(2L)}{\frac{1}{80L \cdot n}} \log \left(\frac{3e(2L)}{\frac{1}{80L \cdot n}} \right) \right)^{V_{T_L \mathcal{F}_n^+}}$$

$$\leq \quad 3 \left(480 e L^2 n \right)^{2 V_{T_L \mathcal{F}_n^+}}.$$

Using this and

$$V_{T_L \mathcal{F}_n^+} \leq V_{\mathcal{F}_n^+}$$

(see proof of Theorem 10.3), one gets, for arbitrary $\epsilon \geq \frac{1}{n}$,

$$\mathbf{E}\{T_{1,n}\} \quad = \int_0^\infty \mathbf{P}\{T_{1,n} > t\}\, dt \leq \epsilon + \int_\epsilon^\infty \mathbf{P}\{T_{1,n} > t\}\, dt$$

$$\leq \epsilon + \int_\epsilon^\infty 42 \left(480 e L^2 n \right)^{2 V_{\mathcal{F}_n^+}} \exp \left(-\frac{n}{24 \cdot 214 L^4} t \right)\, dt$$

$$= \epsilon + 42 \left(480 e L^2 n \right)^{2 V_{\mathcal{F}_n^+}} \frac{24 \cdot 214 L^4}{n} \exp \left(-\frac{n}{24 \cdot 214 L^4} \epsilon \right).$$

The above expression is minimized for

$$\epsilon = \frac{24 \cdot 214 L^4}{n} \log \left(42 \left(480 e L^2 n \right)^{2 V_{\mathcal{F}_n^+}} \right),$$

which yields

$$\mathbf{E}\{T_{1,n}\} \leq \frac{24 \cdot 214 L^4}{n} \cdot \left(\log(42) + 2 V_{\mathcal{F}_n^+} \log(480 e L^2 n) \right) + \frac{24 \cdot 214 L^4}{n}.$$

$$\square$$

11.4   Preliminaries to the Proof of Theorem 11.4

In the proof of Theorem 11.4, which will be given in the next section, we will need the following two auxiliary results:

Lemma 11.2. *Let $V_1, \ldots, V_n$ be i.i.d. random variables, $0 \le V_i \le B, 0 < \alpha < 1$, and $\nu > 0$. Then*

$$\mathbf{P}\left\{\frac{|\frac{1}{n}\sum_{i=1}^{n} V_i - \mathbf{E}V_1|}{\nu + \frac{1}{n}\sum_{i=1}^{n} V_i + \mathbf{E}V_1} > \alpha\right\} \le \mathbf{P}\left\{\frac{|\frac{1}{n}\sum_{i=1}^{n} V_i - \mathbf{E}V_1|}{\nu + \mathbf{E}V_1} > \alpha\right\}$$

$$< \frac{B}{4\alpha^2\nu n}.$$

PROOF. By the Chebyshev inequality we have

$$\mathbf{P}\left\{\frac{|\sum_{i=1}^{n}(V_i - \mathbf{E}V_i)|}{n\nu + n\mathbf{E}V_1} > \alpha\right\}$$

$$= \mathbf{P}\left\{|\sum_{i=1}^{n}(V_i - \mathbf{E}V_i)| > \alpha n(\nu + \mathbf{E}V_1)\right\}$$

$$\le \frac{\mathbf{E}|\sum_{i=1}^{n}(V_i - \mathbf{E}V_i)|^2}{(\alpha n(\nu + \mathbf{E}V_1))^2} = \frac{\mathbf{Var}(V_1)}{n\alpha^2(\nu + \mathbf{E}V_1)^2}. \tag{11.16}$$

Now

$$\mathbf{Var}(V_1) = \mathbf{E}\left\{(V_1 - \mathbf{E}V_1)(V_1 - \mathbf{E}V_1)\right\}$$

$$= \mathbf{E}\left\{V_1(V_1 - \mathbf{E}V_1)\right\} \le \mathbf{E}\{V_1\}(B - \mathbf{E}\{V_1\}).$$

Substituting the bound on the variance into the right-hand side of (11.16) we get

$$\mathbf{P}\left\{\frac{|\sum_{i=1}^{n}(V_i - \mathbf{E}V_i)|}{n\nu + n\mathbf{E}V_1} > \alpha\right\} \le \frac{\mathbf{E}\{V_1\}(B - \mathbf{E}\{V_1\})}{n\alpha^2(\nu + \mathbf{E}\{V_1\})^2}. \tag{11.17}$$

In order to maximize the bound in (11.17) with respect to $\mathbf{E}V_1$ consider the function $f(x) = x(B - x)/n\alpha^2(\nu + x)^2$ which attains its maximal value $\frac{B^2}{4n\alpha^2\nu(B+\nu)}$ for $x = B\nu/(B + 2\nu)$. Hence the right-hand side of (11.17) is bounded above by

$$\frac{B^2}{4\alpha^2\nu n(B + \nu)} < \frac{B}{4\alpha^2\nu n}$$

yielding the desired result. $\qquad\square$

Theorem 11.6. *Let $B \ge 1$ and let $\mathcal{G}$ be a set of functions $g : \mathcal{R}^d \to [0, B]$. Let $Z, Z_1, \ldots, Z_n$ be i.i.d. $\mathcal{R}^d$-valued random variables. Assume $\alpha > 0$, $0 < \epsilon < 1$, and $n \ge 1$. Then*

$$\mathbf{P}\left\{\sup_{g \in \mathcal{G}} \frac{\frac{1}{n}\sum_{i=1}^{n} g(Z_i) - \mathbf{E}g(Z)}{\alpha + \frac{1}{n}\sum_{i=1}^{n} g(Z_i) + \mathbf{E}g(Z)} > \epsilon\right\}$$

$$\le 4\mathbf{E}\mathcal{N}_1\left(\frac{\alpha\epsilon}{5}, \mathcal{G}, Z_1^n\right) \exp\left(-\frac{3\epsilon^2\alpha n}{40B}\right). \tag{11.18}$$

PROOF. The proof will be divided into several steps.

STEP 1. Replace the expectation inside the probability in (11.18) by an empirical mean based on a ghost sample.
Draw a "ghost" sample $Z_1'^n = (Z_1', \ldots, Z_n')$ of i.i.d. random variables distributed as Z and independent of Z_1^n. Let g^* be a function $g \in \mathcal{G}$ such that

$$\frac{1}{n} \sum_{i=1}^{n} g(Z_i) - \mathbf{E}g(Z) > \epsilon \left(\alpha + \frac{1}{n} \sum_{i=1}^{n} g(Z_i) + \mathbf{E}g(Z) \right),$$

if there exists any such function, and let g^* be an other arbitrary function contained in $\mathcal{G}$, if such a function doesn't exist. Note that g^* depends on Z_1^n.

Observe that

$$\frac{1}{n} \sum_{i=1}^{n} g(Z_i) - \mathbf{E}g(Z) > \epsilon \left(\alpha + \frac{1}{n} \sum_{i=1}^{n} g(Z_i) + \mathbf{E}g(Z) \right)$$

and

$$\frac{1}{n} \sum_{i=1}^{n} g(Z_i') - \mathbf{E}g(Z) \leq \frac{\epsilon}{4} \left(\alpha + \frac{1}{n} \sum_{i=1}^{n} g(Z_i') + \mathbf{E}g(Z) \right)$$

imply

$$\frac{1}{n} \sum_{i=1}^{n} g(Z_i) - \frac{1}{n} \sum_{i=1}^{n} g(Z_i')$$

$$> \frac{3\epsilon\alpha}{4} + \epsilon \frac{1}{n} \sum_{i=1}^{n} g(Z_i) - \frac{\epsilon}{4} \frac{1}{n} \sum_{i=1}^{n} g(Z_i') + \frac{3\epsilon}{4} \mathbf{E}g(Z),$$

which is equivalent to

$$\left(1 - \frac{5}{8}\epsilon \right) \left(\frac{1}{n} \sum_{i=1}^{n} g(Z_i) - \frac{1}{n} \sum_{i=1}^{n} g(Z_i') \right)$$

$$> \frac{3\epsilon}{8} \left(2\alpha + \frac{1}{n} \sum_{i=1}^{n} g(Z_i) + \frac{1}{n} \sum_{i=1}^{n} g(Z_i') \right) + \frac{3\epsilon}{4} \mathbf{E}g(Z).$$

Because of $0 < 1 - \frac{5}{8}\epsilon < 1$ and $\mathbf{E}g(Z) \geq 0$ the last inequality implies

$$\frac{1}{n} \sum_{i=1}^{n} g(Z_i) - \frac{1}{n} \sum_{i=1}^{n} g(Z_i') > \frac{3\epsilon}{8} \left(2\alpha + \frac{1}{n} \sum_{i=1}^{n} g(Z_i) + \frac{1}{n} \sum_{i=1}^{n} g(Z_i') \right).$$

Using this we conclude

$$\mathbf{P}\left\{\exists g \in \mathcal{G} \,:\, \frac{1}{n}\sum_{i=1}^{n} g(Z_i) - \frac{1}{n}\sum_{i=1}^{n} g(Z_i') \right.$$

$$\left. > \frac{3\epsilon}{8}\left(2\alpha + \frac{1}{n}\sum_{i=1}^{n} g(Z_i) + \frac{1}{n}\sum_{i=1}^{n} g(Z_i')\right)\right\}$$

$$\geq \mathbf{P}\left\{\frac{1}{n}\sum_{i=1}^{n} g^*(Z_i) - \frac{1}{n}\sum_{i=1}^{n} g^*(Z_i') \right.$$

$$\left. > \frac{3\epsilon}{8}\left(2\alpha + \frac{1}{n}\sum_{i=1}^{n} g^*(Z_i) + \frac{1}{n}\sum_{i=1}^{n} g^*(Z_i')\right)\right\}$$

$$\geq \mathbf{P}\left\{\frac{1}{n}\sum_{i=1}^{n} g^*(Z_i) - \mathbf{E}\{g^*(Z)|Z_1^n\} \right.$$

$$> \epsilon\left(\alpha + \frac{1}{n}\sum_{i=1}^{n} g^*(Z_i) + \mathbf{E}\{g^*(Z)|Z_1^n\}\right),$$

$$\frac{1}{n}\sum_{i=1}^{n} g^*(Z_i') - \mathbf{E}\{g^*(Z)|Z_1^n\}$$

$$\left. \leq \frac{\epsilon}{4}\left(\alpha + \frac{1}{n}\sum_{i=1}^{n} g^*(Z_i') + \mathbf{E}\{g^*(Z)|Z_1^n\}\right)\right\}$$

$$= \mathbf{E}\left\{ I_{\left\{\frac{1}{n}\sum_{i=1}^{n} g^*(Z_i) - \mathbf{E}\{g^*(Z)|Z_1^n\} > \epsilon\left(\alpha + \frac{1}{n}\sum_{i=1}^{n} g^*(Z_i) + \mathbf{E}\{g^*(Z)|Z_1^n\}\right)\right\}} \right.$$

$$\times \mathbf{P}\left\{\frac{1}{n}\sum_{i=1}^{n} g^*(Z_i') - \mathbf{E}\{g^*(Z)|Z_1^n\} \right.$$

$$\left.\left. \leq \frac{\epsilon}{4}\left(\alpha + \frac{1}{n}\sum_{i=1}^{n} g^*(Z_i') + \mathbf{E}\{g^*(Z)|Z_1^n\}\right)\,\bigg|\, Z_1^n\right\}\right\}.$$

Using Lemma 11.2 we get

$$\mathbf{P}\left\{\frac{1}{n}\sum_{i=1}^{n} g^*(Z_i') - \mathbf{E}\{g^*(Z)|Z_1^n\} \right.$$

$$\left. > \frac{\epsilon}{4}\left(\alpha + \frac{1}{n}\sum_{i=1}^{n} g^*(Z_i') + \mathbf{E}\{g^*(Z)|Z_1^n\}\right)\,\bigg|\, Z_1^n\right\}$$

$$< \frac{B}{4\left(\frac{\epsilon}{4}\right)^2 \alpha n} = \frac{4B}{\epsilon^2 \alpha n}.$$

Thus, for $n > \frac{8B}{\epsilon^2 \alpha}$, the probability inside the expectation is greater than or equal to $\frac{1}{2}$ and we can conclude

$$
\mathbf{P}\left\{ \exists g \in \mathcal{G} \; : \; \frac{1}{n}\sum_{i=1}^{n} g(Z_i) - \frac{1}{n}\sum_{i=1}^{n} g(Z_i') \right.
$$
$$
\left. > \frac{3\epsilon}{8}\left(2\alpha + \frac{1}{n}\sum_{i=1}^{n} g(Z_i) + \frac{1}{n}\sum_{i=1}^{n} g(Z_i') \right) \right\}
$$
$$
\geq \frac{1}{2}\mathbf{P}\left\{ \frac{1}{n}\sum_{i=1}^{n} g^*(Z_i) - \mathbf{E}(g^*(Z)|Z_1^n) \right.
$$
$$
\left. > \epsilon\left(\alpha + \frac{1}{n}\sum_{i=1}^{n} g^*(Z_i) + \mathbf{E}\{g^*(Z)|Z_1^n\} \right) \right\}
$$
$$
= \frac{1}{2}\mathbf{P}\left\{ \exists g \in \mathcal{G} : \frac{1}{n}\sum_{i=1}^{n} g(Z_i) - \mathbf{E}g(Z) \right.
$$
$$
\left. > \epsilon\left(\alpha + \frac{1}{n}\sum_{i=1}^{n} g(Z_i) + \mathbf{E}g(Z) \right) \right\}.
$$

This proves

$$
\mathbf{P}\left\{ \exists g \in \mathcal{G} : \frac{\frac{1}{n}\sum_{i=1}^{n} g(Z_i) - \mathbf{E}g(Z)}{\alpha + \frac{1}{n}\sum_{i=1}^{n} g(Z_i) + \mathbf{E}g(Z)} > \epsilon \right\}
$$
$$
\leq 2\mathbf{P}\left\{ \exists g \in \mathcal{G} \; : \; \frac{1}{n}\sum_{i=1}^{n} g(Z_i) - \frac{1}{n}\sum_{i=1}^{n} g(Z_i') \right.
$$
$$
\left. > \frac{3\epsilon}{8}\left(2\alpha + \frac{1}{n}\sum_{i=1}^{n} g(Z_i) + \frac{1}{n}\sum_{i=1}^{n} g(Z_i') \right) \right\}
$$

for $n > \frac{8B}{\epsilon^2 \alpha}$. Observe that for $n \leq \frac{8B}{\epsilon^2 \alpha}$ the right-hand side of (11.18) is greater than 1, and hence the assertion is trivial.

STEP 2. Introduction of additional randomness by random signs.
Let $U_1, \dots, U_n$ be independent and uniformly distributed over $\{-1, 1\}$ and independent of $Z_1, \dots, Z_n, Z_1', \dots, Z_n'$. Because of the independence and identical distribution of $Z_1, \dots, Z_n, Z_1', \dots, Z_n'$ the joint distribution of $Z_1^n, Z_1'^n$ is not affected if one randomly interchanges the corresponding components of Z_1^n and $Z_1'^n$. Clearly, this also doesn't affect $\frac{1}{n}\sum_{i=1}^{n}(g(Z_i) + g(Z_i'))$. Hence

$$\mathbf{P}\left\{\exists g \in \mathcal{G} \;:\; \frac{1}{n}\sum_{i=1}^{n}\left(g(Z_i) - g(Z_i')\right)\right.$$

$$\left. > \frac{3\epsilon}{8}\left(2\alpha + \frac{1}{n}\sum_{i=1}^{n}\left(g(Z_i) + g(Z_i')\right)\right)\right\}$$

$$= \mathbf{P}\left\{\exists g \in \mathcal{G} \;:\; \frac{1}{n}\sum_{i=1}^{n}U_i\left(g(Z_i) - g(Z_i')\right)\right.$$

$$\left. > \frac{3\epsilon}{8}\left(2\alpha + \frac{1}{n}\sum_{i=1}^{n}\left(g(Z_i) + g(Z_i')\right)\right)\right\}$$

$$\leq \mathbf{P}\left\{\exists g \in \mathcal{G} \;:\; \frac{1}{n}\sum_{i=1}^{n}U_i g(Z_i) > \frac{3\epsilon}{8}\left(\alpha + \frac{1}{n}\sum_{i=1}^{n}g(Z_i)\right)\right\}$$

$$+ \mathbf{P}\left\{\exists g \in \mathcal{G} \;:\; \frac{1}{n}\sum_{i=1}^{n}U_i g(Z_i') < -\frac{3\epsilon}{8}\left(\alpha + \frac{1}{n}\sum_{i=1}^{n}g(Z_i')\right)\right\}$$

$$= 2\mathbf{P}\left\{\exists g \in \mathcal{G} \;:\; \frac{1}{n}\sum_{i=1}^{n}U_i g(Z_i) > \frac{3\epsilon}{8}\left(\alpha + \frac{1}{n}\sum_{i=1}^{n}g(Z_i)\right)\right\},$$

where we have used the fact that $-U_i$ has the same distribution as U_i.

STEP 3. Conditioning and introduction of a covering.

Next we condition the last probability on Z_1^n, which is equivalent to fixing $z_1, \dots, z_n \in \mathcal{R}^d$ and to considering

$$\mathbf{P}\left\{\exists g \in \mathcal{G} \;:\; \frac{1}{n}\sum_{i=1}^{n}U_i g(z_i) > \frac{3\epsilon}{8}\left(\alpha + \frac{1}{n}\sum_{i=1}^{n}g(z_i)\right)\right\}. \tag{11.19}$$

Let $\delta > 0$ and let $\mathcal{G}_\delta$ be an L_1 δ-cover of $\mathcal{G}$ on z_1^n. Fix $g \in \mathcal{G}$. Then there exists a $\bar{g} \in \mathcal{G}_\delta$ such that

$$\frac{1}{n}\sum_{i=1}^{n}|g(z_i) - \bar{g}(z_i)| < \delta. \tag{11.20}$$

Without loss of generality we may assume $0 \leq \bar{g}(z) \leq B$. Formula (11.20) implies

$$\frac{1}{n}\sum_{i=1}^{n}U_i g(z_i) = \frac{1}{n}\sum_{i=1}^{n}U_i \bar{g}(z_i) + \frac{1}{n}\sum_{i=1}^{n}U_i(g(z_i) - \bar{g}(z_i))$$

$$\leq \frac{1}{n}\sum_{i=1}^{n}U_i \bar{g}(z_i) + \frac{1}{n}\sum_{i=1}^{n}|g(z_i) - \bar{g}(z_i)| < \frac{1}{n}\sum_{i=1}^{n}U_i \bar{g}(z_i) + \delta$$

and

$$\frac{1}{n}\sum_{i=1}^{n}g(z_i) \;\geq\; \frac{1}{n}\sum_{i=1}^{n}\bar{g}(z_i) - \frac{1}{n}\sum_{i=1}^{n}|g(z_i)-\bar{g}(z_i)| \geq \frac{1}{n}\sum_{i=1}^{n}\bar{g}(z_i) - \delta.$$

Using this we can bound the probability in (11.19) by

$$\mathbf{P}\left\{\exists g \in \mathcal{G}_\delta \;:\; \frac{1}{n}\sum_{i=1}^{n}U_i g(z_i) + \delta > \frac{3\epsilon}{8}\left(\alpha + \frac{1}{n}\sum_{i=1}^{n}g(z_i) - \delta\right)\right\}$$

$$\leq |\mathcal{G}_\delta| \max_{g\in\mathcal{G}_\delta}\mathbf{P}\left\{\frac{1}{n}\sum_{i=1}^{n}U_i g(z_i) > \frac{3\epsilon\alpha}{8} - \frac{3\epsilon\delta}{8} - \delta + \frac{3\epsilon}{8}\frac{1}{n}\sum_{i=1}^{n}g(z_i)\right\}.$$

Set $\delta = \frac{\epsilon\alpha}{5}$, which implies

$$\frac{3\epsilon\alpha}{8} - \frac{3\epsilon\delta}{8} - \delta \geq \frac{3\epsilon\alpha}{8} - \frac{3\epsilon\alpha}{40} - \frac{\epsilon\alpha}{5} = \frac{\epsilon\alpha}{10},$$

and choose $\mathcal{G}_\delta$ as an L_1 $\frac{\epsilon\alpha}{5}$-cover on z_1^n of minimal size. Then we have

$$\mathbf{P}\left\{\exists g \in \mathcal{G} \;:\; \frac{1}{n}\sum_{i=1}^{n}U_i g(z_i) > \frac{3\epsilon}{8}\left(\alpha + \frac{1}{n}\sum_{i=1}^{n}g(z_i)\right)\right\}$$

$$\leq \mathcal{N}_1\left(\frac{\epsilon\alpha}{5}, \mathcal{G}, z_1^n\right)\max_{g\in\mathcal{G}_{\frac{\epsilon\alpha}{5}}}\mathbf{P}\left\{\frac{1}{n}\sum_{i=1}^{n}U_i g(z_i) > \frac{\epsilon\alpha}{10} + \frac{3\epsilon}{8}\frac{1}{n}\sum_{i=1}^{n}g(z_i)\right\}.$$

STEP 4. Application of Hoeffding's inequality.
In this step we bound

$$\mathbf{P}\left\{\frac{1}{n}\sum_{i=1}^{n}U_i g(z_i) > \frac{\epsilon\alpha}{10} + \frac{3\epsilon}{8}\frac{1}{n}\sum_{i=1}^{n}g(z_i)\right\},$$

where $z_1,\ldots,z_n \in \mathcal{R}^d$, $g : \mathcal{R}^d \to \mathcal{R}$, and $0 \leq g(z) \leq B$.
$U_1 g(z_1),\ldots,U_n g(z_n)$ are independent random variables with

$$-g(z_i) \leq U_i g(z_i) \leq g(z_i) \quad (i=1,\ldots,n),$$

therefore, by Hoeffding's inequality,

$$\mathbf{P}\left\{\frac{1}{n}\sum_{i=1}^{n}U_i g(z_i) > \frac{\epsilon\alpha}{10} + \frac{3\epsilon}{8}\frac{1}{n}\sum_{i=1}^{n}g(z_i)\right\}$$

$$\leq \exp\left(-\frac{2n^2\left(\frac{\epsilon\alpha}{10} + \frac{3\epsilon}{8}\frac{1}{n}\sum_{i=1}^{n}g(z_i)\right)^2}{4\sum_{i=1}^{n}g(z_i)^2}\right)$$

$$\leq \exp\left(-\frac{n^2\left(\frac{\epsilon\alpha}{10} + \frac{3\epsilon}{8}\frac{1}{n}\sum_{i=1}^{n}g(z_i)\right)^2}{2B\sum_{i=1}^{n}g(z_i)}\right)$$

$$= \exp\left(-\frac{9\epsilon^2}{128B}\frac{\left(n\frac{4}{15}\alpha + \sum_{i=1}^n g(z_i)\right)^2}{\sum_{i=1}^n g(z_i)}\right).$$

An easy calculation shows that, for arbitrary $a > 0$, one has

$$\frac{(a+y)^2}{y} \geq \frac{(a+a)^2}{a} = 4a \quad (y \in \mathcal{R}_+).$$

This implies

$$\frac{\left(n\frac{4}{15}\alpha + \sum_{i=1}^n g(z_i)\right)^2}{\sum_{i=1}^n g(z_i)} \geq 4n\frac{4}{15}\alpha = \frac{16}{15}\alpha n$$

and, hence,

$$\mathbf{P}\left\{\frac{1}{n}\sum_{i=1}^n U_i g(z_i) > \frac{\epsilon\alpha}{10} + \frac{3\epsilon}{8}\frac{1}{n}\sum_{i=1}^n g(z_i)\right\}$$

$$\leq \exp\left(-\frac{9\epsilon^2}{128B}\frac{16}{15}\alpha n\right)$$

$$= \exp\left(-\frac{3\alpha\epsilon^2 n}{40B}\right).$$

The assertion is now implied by the four steps. $\square$

11.5   Proof of Theorem 11.4

PROOF. Let us introduce the following notation

$$Z = (X, Y),\ Z_i = (X_i, Y_i),\ i = 1,\ldots,n,$$

and

$$g_f(x, y) = |f(x) - y|^2 - |m(x) - y|^2.$$

Observe that $|f(x)| \leq B$, $|y| \leq B$, and $|m(x)| \leq B$ imply

$$-4B^2 \leq g_f(x, y) \leq 4B^2.$$

We can rewrite the probability in the theorem as follows

$$\mathbf{P}\left\{\exists f \in \mathcal{F} : \mathbf{E}g_f(Z) - \frac{1}{n}\sum_{i=1}^n g_f(Z_i) \geq \epsilon(\alpha + \beta + \mathbf{E}g_f(Z))\right\}. \qquad (11.21)$$

The proof will proceed in several steps.

STEP 1. Symmetrization by a ghost sample.
Replace the expectation on the left-hand side of the inequality in (11.21)
by the empirical mean based on the ghost sample $Z_1'^n$ of i.i.d. random

variables distributed as Z and independent of Z_1^n. Consider a function $f_n \in \mathcal{F}$ depending upon Z_1^n such that

$$\mathbf{E}\{g_{f_n}(Z)|Z_1^n\} - \frac{1}{n}\sum_{i=1}^{n} g_{f_n}(Z_i) \geq \epsilon(\alpha + \beta) + \epsilon\mathbf{E}\{g_{f_n}(Z)|Z_1^n\},$$

if such a function exists in $\mathcal{F}$, otherwise choose an arbitrary function in $\mathcal{F}$. Chebyshev's inequality, together with

$$\mathbf{Var}\{g_{f_n}(Z)|Z_1^n\} \leq 16B^2\mathbf{E}\{g_{f_n}(Z)|Z_1^n\}$$

(cf. (11.15)), imply

$$\mathbf{P}\left\{\mathbf{E}\{g_{f_n}(Z)|Z_1^n\} - \frac{1}{n}\sum_{i=1}^{n} g_{f_n}(Z_i') \right.$$
$$\left. > \frac{\epsilon}{2}(\alpha + \beta) + \frac{\epsilon}{2}\mathbf{E}\{g_{f_n}(Z)|Z_1^n\}\,\bigg|\,Z_1^n\right\}$$

$$\leq \frac{\mathbf{Var}\{g_{f_n}(Z)|Z_1^n\}}{n \cdot \left(\frac{\epsilon}{2}(\alpha + \beta) + \frac{\epsilon}{2}\mathbf{E}\{g_{f_n}(Z)|Z_1^n\}\right)^2}$$

$$\leq \frac{16B^2\mathbf{E}\{g_{f_n}(Z)|Z_1^n\}}{n \cdot \left(\frac{\epsilon}{2}(\alpha + \beta) + \frac{\epsilon}{2}\mathbf{E}\{g_{f_n}(Z)|Z_1^n\}\right)^2}$$

$$\leq \frac{16B^2}{\epsilon^2(\alpha + \beta)n},$$

where the last inequality follows from

$$f(x) = \frac{x}{(a + x)^2} \leq f(a) = \frac{1}{4a}$$

for all $x \geq 0$ and all $a > 0$. Thus, for $n > \frac{128B^2}{\epsilon^2(\alpha+\beta)}$,

$$\mathbf{P}\left\{\mathbf{E}\{g_{f_n}(Z)|Z_1^n\} - \frac{1}{n}\sum_{i=1}^{n} g_{f_n}(Z_i') \leq \frac{\epsilon}{2}(\alpha + \beta) + \frac{\epsilon}{2}\mathbf{E}\{g_{f_n}(Z)|Z_1^n\}\,\bigg|\,Z_1^n\right\}$$

$$\geq \frac{7}{8}. \tag{11.22}$$

Hence

$$\mathbf{P}\left\{\exists f \in \mathcal{F}: \frac{1}{n}\sum_{i=1}^{n} g_f(Z_i') - \frac{1}{n}\sum_{i=1}^{n} g_f(Z_i) \geq \frac{\epsilon}{2}(\alpha + \beta) + \frac{\epsilon}{2}\mathbf{E}g_f(Z)\right\}$$

$$\geq \mathbf{P}\left\{\frac{1}{n}\sum_{i=1}^{n} g_{f_n}(Z_i') - \frac{1}{n}\sum_{i=1}^{n} g_{f_n}(Z_i) \geq \frac{\epsilon}{2}(\alpha + \beta) + \frac{\epsilon}{2}\mathbf{E}\{g_{f_n}(Z)|Z_1^n\}\right\}$$

$$\geq \mathbf{P}\left\{\mathbf{E}\{g_{f_n}(Z)|Z_1^n\} - \frac{1}{n}\sum_{i=1}^{n} g_{f_n}(Z_i) \geq \epsilon(\alpha + \beta) + \epsilon\mathbf{E}\{g_{f_n}(Z)|Z_1^n\},\right.$$

$$\left.\mathbf{E}\{g_{f_n}(Z)|Z_1^n\} - \frac{1}{n}\sum_{i=1}^{n} g_f(Z_i') \leq \frac{\epsilon}{2}(\alpha + \beta) + \frac{\epsilon}{2}\mathbf{E}\{g_{f_n}(Z)|Z_1^n\}\right\}$$

$$= \mathbf{E}\left\{I_{\left\{\mathbf{E}\{g_{f_n}(Z)|Z_1^n\} - \frac{1}{n}\sum_{i=1}^{n} g_{f_n}(Z_i) \geq \epsilon(\alpha+\beta)+\epsilon\mathbf{E}\{g_{f_n}(Z)|Z_1^n\}\right\}}\right.$$

$$\left.\times \mathbf{E}\left\{I_{\left\{\mathbf{E}\{g_{f_n}(Z)|Z_1^n\} - \frac{1}{n}\sum_{i=1}^{n} g_f(Z_i') \leq \frac{\epsilon}{2}(\alpha+\beta)+\frac{\epsilon}{2}\mathbf{E}\{g_{f_n}(Z)|Z_1^n\}\right\}}\Big|Z_1^n\right\}\right\}$$

$$= \mathbf{E}\left\{I_{\{\ldots\}}\mathbf{P}\left\{\mathbf{E}\{g_{f_n}(Z)|Z_1^n\} - \frac{1}{n}\sum_{i=1}^{n} g_f(Z_i')\right.\right.$$

$$\left.\left.\leq \frac{\epsilon}{2}(\alpha + \beta) + \frac{\epsilon}{2}\mathbf{E}\{g_{f_n}(Z)|Z_1^n\}\Big|Z_1^n\right\}\right\}$$

$$\geq \frac{7}{8}\mathbf{P}\left\{\mathbf{E}\{g_{f_n}(Z)|Z_1^n\} - \frac{1}{n}\sum_{i=1}^{n} g_{f_n}(Z_i) \geq \epsilon(\alpha + \beta) + \epsilon\mathbf{E}\{g_{f_n}(Z)|Z_1^n\}\right\}$$

$$= \frac{7}{8}\mathbf{P}\left\{\exists f \in \mathcal{F} : \mathbf{E}g_f(Z) - \frac{1}{n}\sum_{i=1}^{n} g_f(Z_i) \geq \epsilon(\alpha + \beta) + \epsilon\mathbf{E}g_f(Z)\right\},$$

where the last inequality follows from (11.22). Thus we have shown that, for $n > \frac{128B^2}{\epsilon^2(\alpha+\beta)}$,

$$\mathbf{P}\left\{\exists f \in \mathcal{F} : \mathbf{E}g_f(Z) - \frac{1}{n}\sum_{i=1}^{n} g_f(Z_i) \geq \epsilon(\alpha + \beta) + \epsilon\mathbf{E}g_f(Z)\right\}$$

$$\leq \frac{8}{7}\mathbf{P}\left\{\exists f \in \mathcal{F} : \frac{1}{n}\sum_{i=1}^{n} g_f(Z_i') - \frac{1}{n}\sum_{i=1}^{n} g_f(Z_i)\right.$$

$$\left.\geq \frac{\epsilon}{2}(\alpha + \beta) + \frac{\epsilon}{2}\mathbf{E}g_f(Z)\right\}. \qquad (11.23)$$

STEP 2. Replacement of the expectation in (11.23) by an empirical mean of the ghost sample.

First we introduce additional conditions in the probability (11.23),

$$\mathbf{P}\left\{\exists f \in \mathcal{F} : \frac{1}{n}\sum_{i=1}^{n} g_f(Z_i') - \frac{1}{n}\sum_{i=1}^{n} g_f(Z_i) \geq \frac{\epsilon}{2}(\alpha + \beta) + \frac{\epsilon}{2}\mathbf{E}g_f(Z)\right\}$$

$$\leq \mathbf{P}\left\{\exists f \in \mathcal{F} : \frac{1}{n}\sum_{i=1}^{n} g_f(Z_i') - \frac{1}{n}\sum_{i=1}^{n} g_f(Z_i) \geq \frac{\epsilon}{2}(\alpha + \beta) + \frac{\epsilon}{2}\mathbf{E}g_f(Z),\right.$$

$$\left.\frac{1}{n}\sum_{i=1}^{n} g_f^2(Z_i) - \mathbf{E}g_f^2(Z) \leq \epsilon\left(\alpha + \beta + \frac{1}{n}\sum_{i=1}^{n} g_f^2(Z_i) + \mathbf{E}g_f^2(Z)\right),\right.$$

$$\frac{1}{n} \sum_{i=1}^{n} g_f^2(Z_i') - \mathbf{E} g_f^2(Z) \le \epsilon \left(\alpha + \beta + \frac{1}{n} \sum_{i=1}^{n} g_f^2(Z_i') + \mathbf{E} g_f^2(Z) \right) \Bigg\}$$

$$+ 2\mathbf{P} \left\{ \exists f \in \mathcal{F} : \frac{\frac{1}{n} \sum_{i=1}^{n} g_f^2(Z_i) - \mathbf{E} g_f^2(Z)}{\left(\alpha + \beta + \frac{1}{n} \sum_{i=1}^{n} g_f^2(Z_i) + \mathbf{E} g_f^2(Z) \right)} > \epsilon \right\}. \quad (11.24)$$

Application of Theorem 11.6 to the second probability on the right–hand side of (11.24) yields

$$\mathbf{P} \left\{ \exists f \in \mathcal{F} : \frac{\frac{1}{n} \sum_{i=1}^{n} g_f^2(Z_i) - \mathbf{E} g_f^2(Z)}{\left(\alpha + \beta + \frac{1}{n} \sum_{i=1}^{n} g_f^2(Z_i) + \mathbf{E} g_f^2(Z) \right)} > \epsilon \right\}$$

$$\le 4\mathbf{E} \mathcal{N}_1 \left(\frac{(\alpha + \beta)\epsilon}{5}, \{g_f : f \in \mathcal{F}\}, Z_1^n \right) \exp \left(-\frac{3\epsilon^2 (\alpha + \beta) n}{40(16B^4)} \right).$$

Now we consider the first probability on the right-hand side of (11.24). The second inequality inside the probability implies

$$(1 + \epsilon) \mathbf{E} g_f^2(Z) \ge (1 - \epsilon) \frac{1}{n} \sum_{i=1}^{n} g_f^2(Z_i) - \epsilon(\alpha + \beta),$$

which is equivalent to

$$\frac{1}{32B^2} \mathbf{E} g_f^2(Z) \ge \frac{1 - \epsilon}{32B^2(1 + \epsilon)} \frac{1}{n} \sum_{i=1}^{n} g_f^2(Z_i) - \epsilon \frac{(\alpha + \beta)}{32B^2(1 + \epsilon)}.$$

We can deal similarly with the third inequality. Using this and the inequality $\mathbf{E} g_f(Z) \ge \frac{1}{16B^2} \mathbf{E} g_f^2(Z) = 2\frac{1}{32B^2} \mathbf{E} g_f^2(Z)$ (see (11.15)) we can bound the first probability on the right-hand side of (11.24) by

$$\mathbf{P} \left\{ \exists f \in \mathcal{F} : \frac{1}{n} \sum_{i=1}^{n} g_f(Z_i') - \frac{1}{n} \sum_{i=1}^{n} g_f(Z_i) \right.$$

$$\ge \epsilon(\alpha + \beta)/2 + \frac{\epsilon}{2} \left(\frac{1 - \epsilon}{32B^2(1 + \epsilon)} \frac{1}{n} \sum_{i=1}^{n} g_f^2(Z_i) - \frac{\epsilon(\alpha + \beta)}{32B^2(1 + \epsilon)} \right.$$

$$\left. \left. + \frac{1 - \epsilon}{32B^2(1 + \epsilon)} \frac{1}{n} \sum_{i=1}^{n} g_f^2(Z_i') - \frac{\epsilon(\alpha + \beta)}{32B^2(1 + \epsilon)} \right) \right\}.$$

This shows

$$\mathbf{P} \left\{ \exists f \in \mathcal{F} : \frac{1}{n} \sum_{i=1}^{n} g_f(Z_i') - \frac{1}{n} \sum_{i=1}^{n} g_f(Z_i) \ge \epsilon(\alpha + \beta)/2 + \epsilon \mathbf{E} g_f(Z)/2 \right\}$$

$$\leq \mathbf{P}\left\{\exists f \in \mathcal{F} : \frac{1}{n}\sum_{i=1}^{n}(g_f(Z_i') - g_f(Z_i))\right.$$

$$\left. \geq \epsilon(\alpha+\beta)/2 - \frac{\epsilon^2(\alpha+\beta)}{32B^2(1+\epsilon)} + \frac{\epsilon(1-\epsilon)}{64B^2(1+\epsilon)}\frac{1}{n}\sum_{i=1}^{n}(g_f^2(Z_i) + g_f^2(Z_i'))\right\}$$

$$+ 8\mathbf{E}\mathcal{N}_1\left(\frac{(\alpha+\beta)\epsilon}{5}, \{g_f : f \in \mathcal{F}\}, Z_1^n\right)\exp\left(-\frac{3\epsilon^2(\alpha+\beta)n}{640B^4}\right). \qquad (11.25)$$

STEP 3. Additional randomization by random signs.

Let $U_1, \ldots, U_n$ be independent and uniformly distributed over the set $\{-1, 1\}$ and independent of $Z_1, \ldots, Z_n, Z_1', \ldots, Z_n'$. Because of the independence and identical distribution of $Z_1, \ldots, Z_n'$ the joint distribution of $Z_1^n, Z_1'^n$ is not affected by the random interchange of the corresponding components in Z_1^n and $Z_1'^n$. Therefore the first probability on the right-hand side of inequality (11.25) is equal to

$$\mathbf{P}\left\{\exists f \in \mathcal{F} : \frac{1}{n}\sum_{i=1}^{n}U_i(g_f(Z_i') - g_f(Z_i))\right.$$

$$\left. \geq \frac{\epsilon}{2}(\alpha+\beta) - \frac{\epsilon^2(\alpha+\beta)}{32B^2(1+\epsilon)} + \frac{\epsilon(1-\epsilon)}{64B^2(1+\epsilon)}\left(\frac{1}{n}\sum_{i=1}^{n}(g_f^2(Z_i) + g_f^2(Z_i'))\right)\right\}$$

and this, in turn, by the union bound, is bounded by

$$\mathbf{P}\left\{\exists f \in \mathcal{F} : \left|\frac{1}{n}\sum_{i=1}^{n}U_i g_f(Z_i')\right|\right.$$

$$\left. \geq \frac{1}{2}\left(\epsilon(\alpha+\beta)/2 - \frac{\epsilon^2(\alpha+\beta)}{32B^2(1+\epsilon)}\right) + \frac{\epsilon(1-\epsilon)}{64B^2(1+\epsilon)}\frac{1}{n}\sum_{i=1}^{n}g_f^2(Z_i')\right\}$$

$$+ \mathbf{P}\left\{\exists f \in \mathcal{F} : \left|\frac{1}{n}\sum_{i=1}^{n}U_i g_f(Z_i)\right|\right.$$

$$\left. \geq \frac{1}{2}\left(\epsilon(\alpha+\beta)/2 - \frac{\epsilon^2(\alpha+\beta)}{32B^2(1+\epsilon)}\right) + \frac{\epsilon(1-\epsilon)}{64B^2(1+\epsilon)}\frac{1}{n}\sum_{i=1}^{n}g_f^2(Z_i)\right\}$$

$$= 2\mathbf{P}\left\{\exists f \in \mathcal{F} : \left|\frac{1}{n}\sum_{i=1}^{n}U_i g_f(Z_i)\right|\right.$$

$$\left. \geq \epsilon(\alpha+\beta)/4 - \frac{\epsilon^2(\alpha+\beta)}{64B^2(1+\epsilon)} + \frac{\epsilon(1-\epsilon)}{64B^2(1+\epsilon)}\frac{1}{n}\sum_{i=1}^{n}g_f^2(Z_i)\right\}. \quad (11.26)$$

STEP 4. Conditioning and using covering.

Next we condition the probability on the right-hand side of (11.26) on Z_1^n, which is equivalent to fixing $z_1, \ldots, z_n$ and considering

$$\mathbf{P}\left\{ \exists f \in \mathcal{F} : \left| \frac{1}{n} \sum_{i=1}^{n} U_i g_f(z_i) \right| \right.$$
$$\left. \geq \epsilon(\alpha + \beta)/4 - \frac{\epsilon^2(\alpha + \beta)}{64B^2(1 + \epsilon)} + \frac{\epsilon(1 - \epsilon)}{64B^2(1 + \epsilon)} \frac{1}{n} \sum_{i=1}^{n} g_f^2(z_i) \right\}.$$

Let $\delta > 0$ and let $\mathcal{G}_\delta$ be an L_1 δ-cover of $\mathcal{G}_{\mathcal{F}} = \{g_f : f \in \mathcal{F}\}$ on $z_1, \ldots, z_n$. Fix $f \in \mathcal{F}$. Then there exists $g \in \mathcal{G}_\delta$ such that

$$\frac{1}{n} \sum_{i=1}^{n} |g(z_i) - g_f(z_i)| < \delta.$$

Without loosing generality we can assume $-4B^2 \leq g(z) \leq 4B^2$. This implies

$$\left| \frac{1}{n} \sum_{i=1}^{n} U_i g_f(z_i) \right| = \left| \frac{1}{n} \sum_{i=1}^{n} U_i g(z_i) + \frac{1}{n} \sum_{i=1}^{n} U_i (g_f(z_i) - g(z_i)) \right|$$

$$\leq \left| \frac{1}{n} \sum_{i=1}^{n} U_i g(z_i) \right| + \frac{1}{n} \sum_{i=1}^{n} |g_f(z_i) - g(z_i)|$$

$$< \left| \frac{1}{n} \sum_{i=1}^{n} U_i g(z_i) \right| + \delta$$

and

$$\frac{1}{n} \sum_{i=1}^{n} g_f^2(z_i) = \frac{1}{n} \sum_{i=1}^{n} g^2(z_i) + \frac{1}{n} \sum_{i=1}^{n} (g_f^2(z_i) - g^2(z_i))$$

$$= \frac{1}{n} \sum_{i=1}^{n} g^2(z_i) + \frac{1}{n} \sum_{i=1}^{n} (g_f(z_i) + g(z_i))(g_f(z_i) - g(z_i))$$

$$\geq \frac{1}{n} \sum_{i=1}^{n} g^2(z_i) - 8B^2 \frac{1}{n} \sum_{i=1}^{n} |g_f(z_i) - g(z_i)|$$

$$\geq \frac{1}{n} \sum_{i=1}^{n} g^2(z_i) - 8B^2\delta.$$

It follows

$$\mathbf{P}\left\{ \exists f \in \mathcal{F} : \left| \frac{1}{n} \sum_{i=1}^{n} U_i g_f(z_i) \right| \right.$$
$$\left. \geq \epsilon(\alpha + \beta)/4 - \frac{\epsilon^2(\alpha + \beta)}{64B^2(1 + \epsilon)} + \frac{\epsilon(1 - \epsilon)}{64B^2(1 + \epsilon)} \frac{1}{n} \sum_{i=1}^{n} g_f^2(z_i) \right\}$$

$$\leq \mathbf{P}\left\{\exists g \in \mathcal{G}_\delta : \left|\frac{1}{n}\sum_{i=1}^{n} U_i g(z_i)\right| + \delta\right.$$

$$\geq \epsilon(\alpha + \beta)/4 - \frac{\epsilon^2(\alpha + \beta)}{64B^2(1 + \epsilon)}$$

$$\left.+ \frac{\epsilon(1 - \epsilon)}{64B^2(1 + \epsilon)}\left(\frac{1}{n}\sum_{i=1}^{n} g^2(z_i) - 8B^2\delta\right)\right\}$$

$$\leq |\mathcal{G}_\delta| \max_{g \in \mathcal{G}_\delta} \mathbf{P}\left\{\left|\frac{1}{n}\sum_{i=1}^{n} U_i g(z_i)\right|\right.$$

$$\geq \epsilon(\alpha + \beta)/4 - \frac{\epsilon^2(\alpha + \beta)}{64B^2(1 + \epsilon)} - \delta - \delta\frac{\epsilon(1 - \epsilon)}{8(1 + \epsilon)}$$

$$\left.+ \frac{\epsilon(1 - \epsilon)}{64B^2(1 + \epsilon)}\frac{1}{n}\sum_{i=1}^{n} g^2(z_i)\right\}.$$

Next we set $\delta = \epsilon\beta/5$. This, together with $B \geq 1$ and $0 < \epsilon \leq \frac{1}{2}$, implies

$$\frac{\epsilon\beta}{4} - \frac{\epsilon^2\beta}{64B^2(1 + \epsilon)} - \delta - \delta\frac{\epsilon(1 - \epsilon)}{8(1 + \epsilon)}$$

$$= \frac{\epsilon\beta}{20} - \frac{\epsilon^2\beta}{64B^2(1 + \epsilon)} - \frac{\epsilon^2(1 - \epsilon)\beta}{40(1 + \epsilon)}$$

$$\geq 0.$$

Thus

$$\mathbf{P}\left\{\exists f \in \mathcal{F} : \left|\frac{1}{n}\sum_{i=1}^{n} U_i g_f(z_i)\right|\right.$$

$$\left.\geq \epsilon(\alpha + \beta)/4 - \frac{\epsilon^2(\alpha + \beta)}{64B^2(1 + \epsilon)} + \frac{\epsilon(1 - \epsilon)}{64B^2(1 + \epsilon)}\frac{1}{n}\sum_{i=1}^{n} g_f^2(z_i)\right\}$$

$$\leq |\mathcal{G}_{\frac{\epsilon\beta}{5}}| \max_{g \in \mathcal{G}_{\frac{\epsilon\beta}{5}}} \mathbf{P}\left\{\left|\frac{1}{n}\sum_{i=1}^{n} U_i g(z_i)\right|\right.$$

$$\left.\geq \frac{\epsilon\alpha}{4} - \frac{\epsilon^2\alpha}{64B^2(1 + \epsilon)} + \frac{\epsilon(1 - \epsilon)}{64B^2(1 + \epsilon)}\frac{1}{n}\sum_{i=1}^{n} g^2(z_i)\right\}.$$

STEP 5. Application of Bernstein's inequality.
In this step we use Bernstein's inequality to bound

$$\mathbf{P}\left\{\left|\frac{1}{n}\sum_{i=1}^{n} U_i g(z_i)\right| \geq \frac{\epsilon\alpha}{4} - \frac{\epsilon^2\alpha}{64B^2(1 + \epsilon)} + \frac{\epsilon(1 - \epsilon)}{64B^2(1 + \epsilon)}\frac{1}{n}\sum_{i=1}^{n} g^2(z_i)\right\},$$

where $z_1, \ldots, z_n \in \mathcal{R}^d \times \mathcal{R}$ are fixed and $g : \mathcal{R}^d \times \mathcal{R} \to \mathcal{R}$ satisfies $-4B^2 \le g(z) \le 4B^2$. First we relate $\frac{1}{n} \sum_{i=1}^n g^2(z_i)$ to the variance of $U_i g(z_i)$,

$$\frac{1}{n} \sum_{i=1}^n \mathbf{Var}(U_i g(z_i)) = \frac{1}{n} \sum_{i=1}^n g^2(z_i) \, \mathbf{Var}(U_i) = \frac{1}{n} \sum_{i=1}^n g^2(z_i).$$

Thus the probability above is equal to

$$\mathbf{P}\left\{ \left| \frac{1}{n} \sum_{i=1}^n V_i \right| \ge A_1 + A_2 \sigma^2 \right\},$$

where

$$V_i = U_i g(z_i), \quad \sigma^2 = \frac{1}{n} \sum_{i=1}^n \mathbf{Var}(U_i g(z_i)),$$

$$A_1 = \frac{\epsilon \alpha}{4} - \frac{\epsilon^2 \alpha}{64 B^2 (1 + \epsilon)}, \qquad A_2 = \frac{\epsilon(1 - \epsilon)}{64 B^2 (1 + \epsilon)}.$$

Observe that $V_1, \ldots, V_n$ are independent random variables satisfying $|V_i| \le |g(z_i)| \le 4B^2$ $(i = 1, \ldots, n)$, and that $A_1, A_2 \ge 0$. We have, by Bernstein's inequality,

$$\mathbf{P}\left\{ \left| \frac{1}{n} \sum_{i=1}^n V_i \right| \ge A_1 + A_2 \sigma^2 \right\}$$

$$\le 2 \exp\left(-\frac{n(A_1 + A_2 \sigma^2)^2}{2\sigma^2 + 2(A_1 + A_2 \sigma^2)\frac{8B^2}{3}} \right)$$

$$= 2 \exp\left(-\frac{nA_2^2}{\frac{16}{3} B^2 A_2} \cdot \frac{\left(\frac{A_1}{A_2} + \sigma^2 \right)^2}{\frac{A_1}{A_2} + \left(1 + \frac{3}{8B^2 A_2} \right) \sigma^2} \right)$$

$$= 2 \exp\left(-\frac{3 \cdot n \cdot A_2}{16 B^2} \cdot \frac{\left(\frac{A_1}{A_2} + \sigma^2 \right)^2}{\frac{A_1}{A_2} + \left(1 + \frac{3}{8B^2 A_2} \right) \sigma^2} \right). \qquad (11.27)$$

An easy calculation (cf. Problem 11.1) shows that, for arbitrary $a, b, u > 0$, one has

$$\frac{(a + u)^2}{a + b \cdot u} \ge \frac{\left(a + \frac{b-2}{b} a \right)^2}{a + b \frac{b-2}{b} a} = 4a \frac{b-1}{b^2}.$$

Thus setting $a = A_1/A_2$, $b = \left(1 + \frac{3}{8B^2 A_2}\right)$, $u = \sigma^2$, and using the bound above we get, for the exponent in (11.27),

$$\frac{3 \cdot n \cdot A_2}{16B^2} \cdot \frac{\left(\frac{A_1}{A_2} + \sigma^2\right)^2}{\frac{A_1}{A_2} + \left(1 + \frac{3}{8B^2 A_2}\right)\sigma^2} \geq \frac{3 \cdot n \cdot A_2}{16B^2} \cdot 4 \cdot \frac{A_1}{A_2} \frac{\frac{3}{8B^2 A_2}}{\left(1 + \frac{3}{8B^2 A_2}\right)^2}$$

$$= 18n \frac{A_1 A_2}{\left(8B^2 A_2 + 3\right)^2}.$$

Substituting the formulas for A_1 and A_2 and noticing

$$A_1 = \frac{\epsilon \alpha}{4} - \frac{\epsilon^2 \alpha}{64B^2(1+\epsilon)} \geq \frac{\epsilon \alpha}{4} - \frac{\epsilon \alpha}{64} = \frac{15\epsilon \alpha}{64}$$

we obtain

$$18n \frac{A_1 A_2}{\left(8B^2 A_2 + 3\right)^2} \geq 18n \frac{15\epsilon \alpha}{64} \cdot \frac{\epsilon(1-\epsilon)}{64B^2(1+\epsilon)} \cdot \frac{1}{\left(\frac{\epsilon(1-\epsilon)}{8(1+\epsilon)} + 3\right)^2}$$

$$\geq 18n \frac{15 \cdot \epsilon^2(1-\epsilon) \cdot \alpha}{64^2 B^2(1+\epsilon)} \cdot \frac{1}{\left(\frac{1}{32} + 3\right)^2}$$

$$= \frac{9 \cdot 15}{2 \cdot 97 \cdot 97} \cdot \frac{\epsilon^2(1-\epsilon)}{1+\epsilon} \cdot \frac{\alpha \cdot n}{B^2}$$

$$\geq \frac{\epsilon^2(1-\epsilon) \cdot \alpha \cdot n}{140B^2(1+\epsilon)}.$$

Plugging the lower bound above into (11.27) we finally obtain

$$\mathbf{P}\left\{\left|\frac{1}{n}\sum_{i=1}^{n} U_i g(z_i)\right| \geq \frac{\epsilon \alpha}{4} - \frac{\epsilon^2 \alpha}{64B^2(1+\epsilon)} + \frac{\epsilon(1-\epsilon)}{64B^2(1+\epsilon)} \frac{1}{n}\sum_{i=1}^{n} g^2(z_i)\right\}$$

$$\leq 2\exp\left(-\frac{\epsilon^2(1-\epsilon)\alpha n}{140B^2(1+\epsilon)}\right).$$

STEP 6. Bounding the covering number.

In this step we construct an L_1 $\frac{\epsilon\beta}{5}$-cover of $\{g_f : f \in \mathcal{F}\}$ on $z_1, \ldots, z_n$. Let $f_1, \ldots, f_l, l = \mathcal{N}_1(\frac{\epsilon\beta}{20B}, \mathcal{F}, x_1^n)$ be an $\frac{\epsilon\beta}{20B}$-cover of $\mathcal{F}$ on x_1^n. Without loss of generality we may assume $|f_j(x)| \leq B$ for all j. Let $f \in \mathcal{F}$ be arbitrary. Then there exists an f_j such that $\frac{1}{n}\sum_{i=1}^{n} |f(x_i) - f_j(x_i)| < \frac{\epsilon\beta}{20B}$. We have

$$\frac{1}{n}\sum_{i=1}^{n} |g_f(z_i) - g_{f_j}(z_i)|$$

$$= \frac{1}{n}\sum_{i=1}^{n} \left||f(x_i) - y_i|^2 - |m(x_i) - y_i|^2 - |f_j(x_i) - y_i|^2 + |m(x_i) - y_i|^2\right|$$

$$= \frac{1}{n} \sum_{i=1}^{n} |f(x_i) - y_i + f_j(x_i) - y_i| |f(x_i) - y_i - f_j(x_i) + y_i|$$

$$\leq 4B \frac{1}{n} \sum_{i=1}^{n} |f(x_i) - f_j(x_i)| < \frac{\epsilon \beta}{5}.$$

Thus $g_{f_1}, \ldots, g_{f_l}$ is an $\frac{\epsilon \beta}{5}$-cover of $\{g_f : f \in \mathcal{F}\}$ on z_1^n of size $\mathcal{N}_1(\frac{\epsilon \beta}{20B}, \mathcal{F}, x_1^n)$. Steps 3 through 6 imply

$$\mathbf{P} \left\{ \exists f \in \mathcal{F} : \frac{1}{n} \sum_{i=1}^{n} (g_f(Z_i') - g_f(Z_i)) \right.$$

$$\geq \frac{\epsilon}{2}(\alpha + \beta) - \frac{\epsilon^2(\alpha + \beta)}{32B^2(1 + \epsilon)}$$

$$\left. + \frac{\epsilon(1 - \epsilon)}{64B^2(1 + \epsilon)} \frac{1}{n} \sum_{i=1}^{n} (g_f^2(Z_i) + g_f^2(Z_i')) \right\}$$

$$\leq 4 \sup_{x_1^n \in (\mathcal{R}^d)^n} \mathcal{N}_1 \left(\frac{\epsilon \beta}{20B}, \mathcal{F}, x_1^n \right) \exp \left(-\frac{\epsilon^2(1 - \epsilon)\alpha n}{140B^2(1 + \epsilon)} \right).$$

STEP 7. Conclusion.
Steps 1, 2, and 6 imply, for $n > \frac{128B^2}{\epsilon^2(\alpha + \beta)}$,

$$\mathbf{P} \left\{ \exists f \in \mathcal{F} : \mathbf{E}g_f(Z) - \frac{1}{n} \sum_{i=1}^{n} g_f(Z_i) > \epsilon(\alpha + \beta + \mathbf{E}g_f(Z)) \right\}$$

$$\leq \frac{32}{7} \sup_{x_1^n} \mathcal{N}_1 \left(\frac{\epsilon \beta}{20B}, \mathcal{F}, x_1^n \right) \exp \left(-\frac{\epsilon^2(1 - \epsilon)\alpha n}{140B^2(1 + \epsilon)} \right)$$

$$+ \frac{64}{7} \sup_{x_1^n} \mathcal{N}_1 \left(\frac{\epsilon(\alpha + \beta)}{20B}, \mathcal{F}, x_1^n \right) \exp \left(-\frac{3\epsilon^2(\alpha + \beta)n}{640B^4} \right)$$

$$\leq 14 \sup_{x_1^n} \mathcal{N}_1 \left(\frac{\epsilon \beta}{20B}, \mathcal{F}, x_1^n \right) \exp \left(-\frac{\epsilon^2(1 - \epsilon)\alpha n}{214(1 + \epsilon)B^4} \right).$$

For $n \leq \frac{128B^2}{\epsilon^2(\alpha + \beta)}$ one has

$$\exp \left(-\frac{\epsilon^2(1 - \epsilon)\alpha n}{214(1 + \epsilon)B^4} \right) \geq \exp \left(-\frac{128}{214} \right) \geq \frac{1}{14},$$

and hence the assertion follows trivially. $\square$

11.6 Bibliographic Notes

Theorem 11.1 is a well-known result from fixed design regression. The bound of Theorem 11.2 is due to Pollard (1984). Results concerning the equivalence of the $L_2(\mu)$ norm and the empirical norm under regularity conditions on μ can be found in van de Geer (2000).

Theorem 11.4 has been proven by Lee, Bartlett, and Williamson (1996). The bound proven in Lemma 11.2 has been obtained by Haussler (1992) and the bound in Theorem 11.6 by Pollard (1986) and Haussler (1992).

The rate of convergence of regression estimates has been studied in many articles, see, e.g., Cox (1988), Rafajlowicz (1987), Shen and Wong (1994), Birgé and Massart (1993; 1998), and the literature cited therein.

Problems and Exercises

PROBLEM 11.1. Show that for arbitrary $a, b, u > 0$ one has

$$\frac{(a+u)^2}{a+b \cdot u} \geq \frac{\left(a + \frac{b-2}{b}a\right)^2}{a + b\frac{b-2}{b}a} = 4a\frac{b-1}{b^2}.$$

HINT: Show that the function

$$f(u) = \frac{(a+u)^2}{a+b \cdot u}$$

satisfies

$$f'(u) < 0 \text{ if } u < \frac{b-2}{b} \cdot a$$

and

$$f'(u) > 0 \text{ if } u > \frac{b-2}{b} \cdot a.$$

PROBLEM 11.2. Show that under the assumptions of Theorem 11.5 one has for arbitrary $\delta > 0$,

$$\mathbf{E}\left\{\int |m_n(x) - m(x)|^2 \mu(dx)\right\}$$

$$\leq c_{\delta,L}\frac{(1 + \log(n))V_{\mathcal{F}_n^+}}{n} + (1 + \delta) \inf_{f \in \mathcal{F}_n} \int |f(x) - m(x)|^2 \mu(dx)$$

for some constant $c_{\delta,L}$ depending only on δ and L. How does $c_{\delta,L}$ depend on δ?
HINT: Use the error decomposition

$$\int |m_n(x) - m(x)|^2 \mu(dx)$$

$$= \left\{ \mathbf{E}\{|m_n(X) - Y|^2|D_n\} - \mathbf{E}\{|m(X) - Y|^2\} \right.$$

$$\left. -(1 + \delta) \cdot \left(\frac{1}{n}\sum_{i=1}^{n} |m_n(X_i) - Y_i|^2 - \frac{1}{n}\sum_{i=1}^{n} |m(X_i) - Y_i|^2\right)\right\}$$

$$+(1+\delta)\cdot\left(\frac{1}{n}\sum_{i=1}^{n}|m_n(X_i)-Y_i|^2-\frac{1}{n}\sum_{i=1}^{n}|m(X_i)-Y_i|^2\right)$$

$$=T_{1,n}+T_{2,n}.$$

PROBLEM 11.3. Prove Theorem 11.3 using Theorems 11.1 and 11.6.

PROBLEM 11.4. Try to improve the constants in Theorem 11.4.

PROBLEM 11.5. Formulate and prove a multivariate version of Corollary 11.2.

12
Least Squares Estimates III: Complexity Regularization

12.1 Motivation

In this chapter we describe the complexity regularization principle which enables one to define least squares estimates which automatically adapt to the smoothness of the regression function.

Let us start with a motivation of the complexity regularization principle. Assume that for given data one wants to find a function $f : \mathcal{R}^d \to \mathcal{R}$ from some class $\mathcal{F}$, best describing the data by using the least squares criterion, i.e., one wants to find a function in $\mathcal{F}$ for which the empirical L_2 risk is equal to

$$\min_{f \in \mathcal{F}} \frac{1}{n} \sum_{i=1}^{n} |f(X_i) - Y_i|^2. \tag{12.1}$$

Clearly (12.1) favors large (more complex) classes $\mathcal{F}$ yielding smaller values by the least squares criterion. However, complex classes $\mathcal{F}$ lead to overfitting of the data and result in estimates which poorly fit new data (poor generalization capability because of large estimation error). In the complexity regularization approach we introduce in (12.1) an additive penalty monotonically increasing with the complexity of $\mathcal{F}$.

The motivation for the penalty used in this chapter is given by the following lemma:

Lemma 12.1. *Let $1 \leq \beta < \infty$, $\delta \in (0,1)$, and let $\mathcal{F}$ be a class of functions $f : \mathcal{R}^d \to \mathcal{R}$. Assume $|Y| \leq \beta$ a.s. Then the inequality*

$$
\int |f(x) - m(x)|^2 \mu(dx) \;\leq\; c_1 \frac{\log(\frac{42}{\delta})}{n} + (c_2 + c_3 \log(n))\frac{V_{\mathcal{F}^+}}{n}
$$

$$
+ \frac{2}{n}\sum_{i=1}^{n}\{|f(X_i) - Y_i|^2 - |m(X_i) - Y_i|^2\}
$$

$$(12.2)$$

holds simultaneously for all $f \in T_\beta \mathcal{F}$ with probability greater than or equal to $1 - \delta$, where

$$
c_1 = 5136\beta^4, \quad c_2 = 10272\beta^4 \log(480e\beta^2),
$$

and

$$
c_3 = 10272\beta^4.
$$

PROOF. Because of

$$
\int |f(x) - m(x)|^2 \mu(dx) \;=\; \mathbf{E}|f(X) - Y|^2 - \mathbf{E}|m(X) - Y|^2
$$

$$
= \; \Bigg\{ \mathbf{E}|f(X) - Y|^2 - \mathbf{E}|m(X) - Y|^2
$$

$$
- \frac{2}{n}\sum_{i=1}^{n}(|f(X_i) - Y_i|^2 - |m(X_i) - Y_i|^2)\Bigg\}
$$

$$
+ \frac{2}{n}\sum_{i=1}^{n}(|f(X_i) - Y_i|^2 - |m(X_i) - Y_i|^2)
$$

$$
=: \; T_1 + T_2,
$$

it suffices to show

$$
\mathbf{P}\left\{\exists f \in T_\beta\mathcal{F} : T_1 > c_1 \frac{\log(\frac{42}{\delta})}{n} + (c_2 + c_3 \log(n))\frac{V_{\mathcal{F}^+}}{n}\right\} \leq \delta. \qquad (12.3)
$$

Let $t \geq \frac{1}{n}$ be arbitrary. Then one has

$$
\mathbf{P}\left\{\exists f \in T_\beta\mathcal{F} : T_1 > t\right\}
$$

$$= \mathbf{P}\left\{\exists f \in T_\beta \mathcal{F} : 2\left\{\mathbf{E}|f(X) - Y|^2 - \mathbf{E}|m(X) - Y|^2\right.\right.$$

$$\left. - \frac{1}{n}\sum_{i=1}^{n}(|f(X_i) - Y_i|^2 - |m(X_i) - Y_i|^2)\right\}$$

$$\left. > t + \mathbf{E}|f(X) - Y|^2 - \mathbf{E}|m(X) - Y|^2\right\}$$

$$\le \mathbf{P}\left\{\exists f \in T_\beta \mathcal{F} : \mathbf{E}|f(X) - Y|^2 - \mathbf{E}|m(X) - Y|^2\right.$$

$$- \frac{1}{n}\sum_{i=1}^{n}(|f(X_i) - Y_i|^2 - |m(X_i) - Y_i|^2)$$

$$\left. > \frac{1}{2} \cdot (\frac{t}{2} + \frac{t}{2} + \mathbf{E}|f(X) - Y|^2 - \mathbf{E}|m(X) - Y|^2)\right\}$$

$$\le 14 \sup_{x_1^n} \mathcal{N}_1\left(\frac{\frac{1}{2}\frac{t}{2}}{20\beta}, T_\beta\mathcal{F}, x_1^n\right)\exp\left(-\frac{(\frac{1}{2})^2(1 - \frac{1}{2})\frac{t}{2}n}{214(1 + \frac{1}{2})\beta^4}\right)$$

$$\text{(by Theorem 11.4)}$$

$$\le 14 \sup_{x_1^n} \mathcal{N}_1\left(\frac{1}{80\beta n}, T_\beta\mathcal{F}, x_1^n\right)\exp\left(-\frac{t \cdot n}{8 \cdot 3 \cdot 214\beta^4}\right)$$

$$\left(\text{because of } t \ge \frac{1}{n}\right)$$

$$\le 14 \cdot 3 \left(\frac{2e(2\beta)}{\frac{1}{80\beta n}}\log\frac{3e(2\beta)}{\frac{1}{80\beta n}}\right)^{V_{T_\beta\mathcal{F}^+}}\exp\left(-\frac{t \cdot n}{5136\beta^4}\right)$$

$$\text{(by Lemma 9.2 and Theorem 9.4)}$$

$$\le 42(480e\beta^2 n)^{2V_{\mathcal{F}^+}}\exp\left(-\frac{t \cdot n}{5136\beta^4}\right),$$

where we have used the relation $V_{T_\beta\mathcal{F}^+} \le V_{\mathcal{F}^+}$.

If one defines t for a given $\delta \in (0,1)$ by

$$\delta = 42(480e\beta^2 n)^{2V_{\mathcal{F}^+}}\exp\left(-\frac{tn}{5136\beta^4}\right),$$

then it follows that

$$\frac{tn}{5136\beta^4} = \log\frac{42}{\delta} + 2V_{\mathcal{F}^+}\log(480e\beta^2 n)$$

and thus

$$
\begin{aligned}
t &= 5136\beta^4 \log\left(\frac{42}{\delta}\right)\frac{1}{n} + 2\cdot 5136\beta^4 \log(480e\beta^2 n)\frac{V_{\mathcal{F}^+}}{n} \\
&= 5136\beta^4 \log\left(\frac{42}{\delta}\right)\frac{1}{n} + \left(10272\beta^4 \log(480e\beta^2) + 10272\beta^4 \log n\right)\frac{V_{\mathcal{F}^+}}{n} \\
&= c_1\frac{\log(\frac{42}{\delta})}{n} + (c_2 + c_3 \log(n))\frac{V_{\mathcal{F}^+}}{n}.
\end{aligned}
$$

This implies (12.3) and therefore Lemma 12.1 is proved. $\qquad\square$

Now we apply (12.2) to a truncated version m_n of the least squares estimate $\tilde{m}_n$, i.e., to $m_n(\cdot) = m_n(\cdot, D_n) = T_\beta \tilde{m}_n(\cdot)$ where

$$
\tilde{m}_n(\cdot) = \tilde{m}_n(\cdot, D_n) \in \mathcal{F}
$$

minimizes the empirical L_2 error over $\mathcal{F}$:

$$
\frac{1}{n}\sum_{i=1}^{n}|\tilde{m}_n(X_i) - Y_i|^2 = \min_{f\in\mathcal{F}}\frac{1}{n}\sum_{i=1}^{n}|f(X_i) - Y_i|^2.
$$

It follows that one has, with probability greater than or equal to $1 - \delta$,

$$
\begin{aligned}
\int |m_n(x) - m(x)|^2\mu(dx) \\
\leq\; c_1\frac{\log(\frac{42}{\delta})}{n} + (c_2 + c_3 \log(n))\frac{V_{\mathcal{F}^+}}{n} \\
+ \frac{2}{n}\sum_{i=1}^{n}\{|m_n(X_i) - Y_i|^2 - |m(X_i) - Y_i|^2\}. \qquad (12.4)
\end{aligned}
$$

The idea of the complexity regularization is to choose $\mathcal{F}$ in such a way that the right-hand side of (12.4) is minimized, which is equivalent to minimizing

$$
\frac{(c_2 + c_3 \log(n))\,V_{\mathcal{F}^+}}{2}\frac{1}{n} + \frac{1}{n}\sum_{i=1}^{n}|m_n(X_i) - Y_i|^2
$$

(cf. Figure 12.1).

This will be described in detail in the next section.

12.2 Definition of the Estimate

Let $\mathcal{P}_n$ be a finite set of parameters. For $p \in \mathcal{P}_n$ let $\mathcal{F}_{n,p}$ be a set of functions $f : \mathcal{R}^d \to \mathcal{R}$ and let $pen_n(p) \in \mathcal{R}_+$ be a complexity penalty for $\mathcal{F}_{n,p}$. Let $m_{n,p}$ be a truncated least squares estimate of m in $\mathcal{F}_{n,p}$, i.e., choose

$$
\tilde{m}_{n,p}(\cdot) = \tilde{m}_{n,p}(\cdot, D_n) \in \mathcal{F}_{n,p} \qquad (12.5)
$$

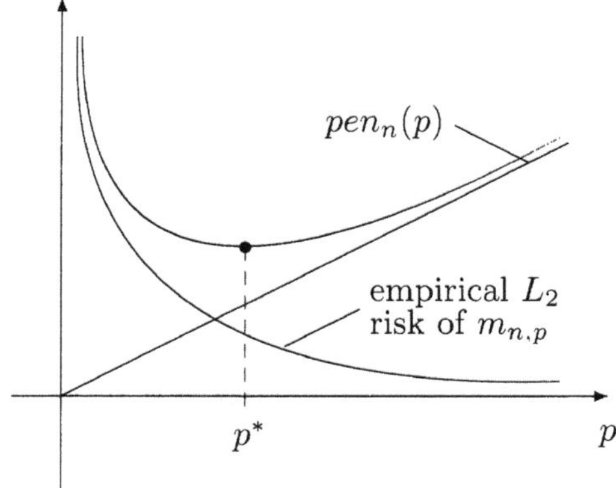

Figure 12.1. Constructing an estimate by minimizing the sum of L_2 risk and the penalty.

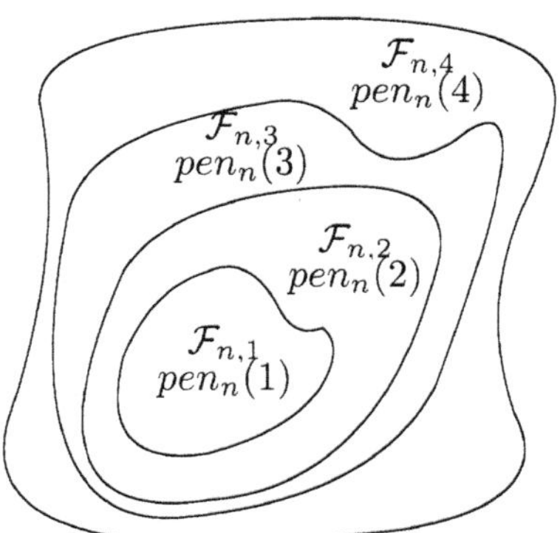

Figure 12.2. A sequence of nested classes $\mathcal{F}_{n,1} \subseteq \mathcal{F}_{,2n} \subseteq \mathcal{F}_{n,3} \subseteq \cdots$ of functions for which one will often choose an increasing sequence $pen_n(1) < pen_n(2) < pen_n(3) < \cdots$ of penalties.

which satisfies

$$\frac{1}{n} \sum_{i=1}^{n} |\tilde{m}_{n,p}(X_i) - Y_i|^2 = \min_{f \in \mathcal{F}_{n,p}} \frac{1}{n} \sum_{i=1}^{n} |f(X_i) - Y_i|^2 \qquad (12.6)$$

and set

$$m_{n,p}(\cdot) = T_{\beta_n} \tilde{m}_{n,p}(\cdot) \qquad (12.7)$$

for some $\beta_n \in \mathcal{R}_+$.

Next choose an estimate m_{n,p^*} minimizing the sum of the empirical L_2 risk of m_{n,p^*} and $pen_n(p^*)$, i.e., choose

$$p^* = p^*(D_n) \in \mathcal{P}_n \qquad (12.8)$$

such that

$$\frac{1}{n} \sum_{i=1}^{n} |m_{n,p^*}(X_i) - Y_i|^2 + pen_n(p^*)$$

$$= \min_{p \in \mathcal{P}_n} \left\{ \frac{1}{n} \sum_{i=1}^{n} |m_{n,p}(X_i) - Y_i|^2 + pen_n(p) \right\} \qquad (12.9)$$

and set

$$m_n(\cdot, D_n) = m_{n,p^*(D_n)}(\cdot, D_n). \qquad (12.10)$$

Here the penalty depends on the class of functions from which one chooses the estimate. This is in contrast to the penalized least squares estimates (cf. Chapter 20) where the penalty depends on the smoothness of the estimate.

12.3 Asymptotic Results

Our main result concerning complexity regularization is the following theorem:

Theorem 12.1. *Let* $1 \leq L < \infty$, $n \in \mathcal{N}$, *and,* $L \leq \beta_n < \infty$. *Assume* $|Y| \leq L$ *a.s. Let the estimate* m_n *be defined as above (cf. (12.5)–(12.10)) with a penalty term* $pen_n(p)$ *satisfying*

$$pen_n(p) \geq 2 \cdot 2568 \frac{\beta_n^4}{n} \left(\log(120e\beta_n^4 n) V_{\mathcal{F}_{n,p}^+} + \frac{c_p}{2} \right) \quad (p \in \mathcal{P}_n) \qquad (12.11)$$

for some $c_p \in \mathcal{R}_+$ *satisfying* $\sum_{p \in \mathcal{P}_n} e^{-c_p} \leq 1$. *Then one has*

$$\mathbf{E} \int |m_n(x) - m(x)|^2 \mu(dx)$$

$$\leq 2 \inf_{p \in \mathcal{P}_n} \left\{ pen_n(p) + \inf_{f \in \mathcal{F}_{n,p}} \int |f(x) - m(x)|^2 \mu(dx) \right\} + 5 \cdot 2568 \frac{\beta_n^4}{n}.$$

We know from Theorem 11.5 that for each fixed $p \in \mathcal{P}_n$ the estimate $m_{n,p}$ satisfies

$$\mathbf{E} \int |m_{n,p}(x) - m(x)|^2 \mu(dx)$$

$$\leq \frac{c_1}{n} + \frac{c_2 + c_3 \log(n)}{n} \cdot V_{\mathcal{F}_{n,p}^+} + 2 \inf_{f \in \mathcal{F}_{n,p}} \int |f(x) - m(x)|^2 \mu(dx).$$

According to (12.11) we can choose our penalty, in Theorem 12.1 above, up to the additional term $\beta_n^4 c_p/n$ of the form

$$\frac{c_2 + c_3 \log(n)}{n} \cdot V_{\mathcal{F}_{n,p}^+}.$$

The big advantage of the bound in Theorem 12.1 compared to Theorem 11.5 is the additional infimum over $p \in \mathcal{P}_n$, which implies that we can use the data to choose the best value for p. The additional term $\beta_n^4 c_p/n$ is the price which we have to pay for choosing p in this adaptive way.

PROOF. We use a decomposition similar to that in the proof of Lemma 12.1:

$$\int |m_n(x) - m(x)|^2 \mu(dx)$$

$$= \mathbf{E}\{|m_n(X) - Y|^2|D_n\} - \mathbf{E}|m(X) - Y|^2$$

$$= \Bigg\{ \mathbf{E}\{|m_n(X) - Y|^2|D_n\} - \mathbf{E}|m(X) - Y|^2$$

$$-\frac{2}{n}\sum_{i=1}^{n}(|m_n(X_i) - Y_i|^2 - |m(X_i) - Y_i|^2) - 2pen_n(p^*)\Bigg\}$$

$$+\frac{2}{n}\sum_{i=1}^{n}(|m_n(X_i) - Y_i|^2 - |m(X_i) - Y_i|^2) + 2pen_n(p^*)$$

$$=: T_{1,n} + T_{2,n}.$$

Because of (12.9), (12.10), $|Y_i| \le \beta_n$ a.s., and (12.6), one has

$$T_{2,n}$$

$$= 2 \inf_{p \in \mathcal{P}_n} \left\{ \frac{1}{n}\sum_{i=1}^{n} |m_{n,p}(X_i) - Y_i|^2 + pen_n(p) \right\} - \frac{2}{n}\sum_{i=1}^{n} |m(X_i) - Y_i|^2$$

$$\le 2 \inf_{p \in \mathcal{P}_n} \left\{ \frac{1}{n}\sum_{i=1}^{n} |\tilde{m}_{n,p}(X_i) - Y_i|^2 + pen_n(p) \right\} - \frac{2}{n}\sum_{i=1}^{n} |m(X_i) - Y_i|^2$$

$$= 2 \inf_{p \in \mathcal{P}_n} \left\{ \inf_{f \in \mathcal{F}_{n,p}} \left(\frac{1}{n}\sum_{i=1}^{n} |f(X_i) - Y_i|^2 \right) + pen_n(p) \right\}$$

$$-\frac{2}{n}\sum_{i=1}^{n} |m(X_i) - Y_i|^2$$

$$= 2 \inf_{p \in \mathcal{P}_n} \left\{ \inf_{f \in \mathcal{F}_{n,p}} \left(\frac{1}{n} \sum_{i=1}^{n} |f(X_i) - Y_i|^2 - \frac{1}{n} \sum_{i=1}^{n} |m(X_i) - Y_i|^2 \right) \right.$$
$$\left. + pen_n(p) \right\},$$

thus it follows

$$\mathbf{E} T_{2,n}$$
$$\leq 2 \inf_{p \in \mathcal{P}_n} \left\{ \inf_{f \in \mathcal{F}_{n,p}} \mathbf{E} \left(\frac{1}{n} \sum_{i=1}^{n} |f(X_i) - Y_i|^2 - \frac{1}{n} \sum_{i=1}^{n} |m(X_i) - Y_i|^2 \right) \right.$$
$$\left. + pen_n(p) \right\}$$
$$= 2 \inf_{p \in \mathcal{P}_n} \left\{ \inf_{f \in \mathcal{F}_{n,p}} \int |f(x) - m(x)|^2 \mu(dx) + pen_n(p) \right\}.$$

Therefore the assertion follows from

$$\mathbf{E} T_{1,n} \leq 5 \cdot 2568 \frac{\beta_n^4}{n}, \tag{12.12}$$

which we will show next.

We mimic the proof of Lemma 12.1. Let $t > 0$ be arbitrary. Then

$$\mathbf{P} \left\{ T_{1,n} > t \right\}$$
$$= \mathbf{P} \left\{ 2 \left(\mathbf{E}\{ |m_n(X) - Y|^2 | D_n \} - \mathbf{E}|m(X) - Y|^2 \right. \right.$$
$$\left. - \frac{1}{n} \sum_{i=1}^{n} (|m_n(X_i) - Y_i|^2 - |m(X_i) - Y_i|^2) \right)$$
$$\left. > t + 2pen_n(p^*) + \mathbf{E}\{ |m_n(X) - Y|^2 | D_n \} - \mathbf{E}|m(X) - Y|^2 \right\}$$
$$\leq \mathbf{P} \left\{ \exists p \in \mathcal{P}_n \exists f \in T_\beta \mathcal{F}_{n,p} : \mathbf{E}|f(X) - Y|^2 - \mathbf{E}|m(X) - Y|^2 \right.$$
$$- \frac{1}{n} \sum_{i=1}^{n} (|f(X_i) - Y_i|^2 - |m(X_i) - Y_i|^2)$$
$$\left. > \frac{1}{2} \cdot (t + 2pen_n(p) + \mathbf{E}|f(X) - Y|^2 - \mathbf{E}|m(X) - Y|^2) \right\}$$

$$\text{(because of (12.5) and (12.7))}$$

$$\leq \sum_{p \in \mathcal{P}_n} \mathbf{P}\bigg\{ \exists f \in T_\beta \mathcal{F}_{n,p} : \mathbf{E}|f(X) - Y|^2 - \mathbf{E}|m(X) - Y|^2$$

$$-\frac{1}{n} \sum_{i=1}^{n} (|f(X_i) - Y_i|^2 - |m(X_i) - Y_i|^2)$$

$$> \frac{1}{2} \cdot (t + 2pen_n(p) + \mathbf{E}|f(X) - Y|^2 - \mathbf{E}|m(X) - Y|^2) \bigg\}$$

$$\leq \sum_{p \in \mathcal{P}_n} 14 \sup_{x_1^n} \mathcal{N}_1 \left(\frac{\frac{1}{2}pen_n(p)}{20\beta_n}, T_{\beta_n} \mathcal{F}_{n,p}, x_1^n \right)$$

$$\cdot \exp \left(-\frac{(\frac{1}{2})^2 \frac{1}{2}(t + pen_n(p))n}{214(1 + \frac{1}{2})\beta_n^4} \right)$$

$$\text{(by Theorem 11.4)}$$

$$\leq \sum_{p \in \mathcal{P}_n} 14 \sup_{x_1^n} \mathcal{N}_1 \left(\frac{1}{20\beta_n n}, T_{\beta_n} \mathcal{F}_{n,p}, x_1^n \right) \exp \left(-\frac{(t + pen_n(p))n}{3 \cdot 4 \cdot 214\beta_n^4} \right)$$

$$\left(\text{since } pen_n(p) \geq \frac{2}{n} \right)$$

$$\leq \sum_{p \in \mathcal{P}_n} 14 \cdot 3 \left(\frac{2e(2\beta_n)}{\frac{1}{20\beta_n n}} \log \frac{3e(2\beta_n)}{\frac{1}{20\beta_n n}} \right)^{V_{\mathcal{F}_{n,p}^+}}$$

$$\cdot \exp \left(-\frac{pen_n(p)n}{2568\beta_n^4} \right) \exp \left(-\frac{t \cdot n}{2568\beta_n^4} \right)$$

$$\text{(by Lemma 9.2 and Theorem 9.4)}$$

$$\leq \left\{ \sum_{p \in \mathcal{P}_n} 42(120e\beta_n^2 n)^{2V_{\mathcal{F}_{n,p}^+}} \exp \left(-\frac{pen_n(p)n}{2568\beta_n^4} \right) \right\} \exp \left(-\frac{t \cdot n}{2568\beta_n^4} \right).$$

Using (12.11) one gets

$$\{\ldots\} \quad \leq \quad \sum_{p \in \mathcal{P}_n} 42(120e\beta_n^2 n)^{2V_{\mathcal{F}_{n,p}^+}} \exp \left(-2V_{\mathcal{F}_{n,p}^+} \log(120e\beta_n^2 n) - c_p \right)$$

$$= \quad \sum_{p \in \mathcal{P}_n} 42 \exp(-c_p) \leq 42,$$

thus

$$\mathbf{P}\{T_{1,n} > t\} \leq 42 \exp \left(-\frac{t \cdot n}{2568\beta_n^4} \right).$$

For arbitrary $u > 0$ it follows

$$\mathbf{E}T_{1,n} \leq \int_0^\infty \mathbf{P}\{T_{1,n} > t\}dt \leq u + \int_u^\infty \mathbf{P}\{T_{1,n} > t\}dt$$

$$\leq u + \int_u^\infty 42 \exp\left(-\frac{t \cdot n}{2568\beta_n^4}\right) dt$$

$$= u + 42 \cdot 2568\beta_n^4 \cdot \frac{1}{n} \cdot \exp\left(-\frac{u \cdot n}{2568\beta_n^4}\right).$$

Setting $u = 2568 \cdot \log(42) \cdot \frac{\beta_n^4}{n}$ one gets

$$\mathbf{E}T_{1,n} \leq 2568(1 + \log(42))\frac{\beta_n^4}{n} \leq 5 \cdot 2568\frac{\beta_n^4}{n}$$

and thus (12.12) (and also the assertion of Theorem 12.1) is proved. $\square$

REMARK. In order to choose the penalties $pen_n(p)$ such that (12.11) is satisfied, one needs an upper bound on the VC dimension of $\mathcal{F}_{n,p}^+$. Sometimes it is much easier to get bounds on covering numbers like

$$\mathcal{N}_1(\epsilon, T_{\beta_n}\mathcal{F}_{n,p}, x_1^n) \leq \mathcal{N}_1(\epsilon, T_{\beta_n}\mathcal{F}_{n,p}) \tag{12.13}$$

for all $\epsilon > 0$, $x_1^n \in (\mathcal{R}^d)^n$ rather than on VC dimension. In this case, it is possible to replace (12.11) by

$$pen_n(p) \geq 2568\frac{\beta_n^4}{n} \cdot \left(\log\left(\mathcal{N}_1\left(\frac{1}{n}, T_{\beta_n}\mathcal{F}_{n,p}\right)\right) + c_p\right) \quad (p \in \mathcal{P}_n). \tag{12.14}$$

To show that Theorem 12.1 also holds if (12.11) is replaced by (12.13) and (12.14), one observes that

$$\mathbf{P}\{T_{1,n} > t\}$$

$$\leq \sum_{p \in \mathcal{P}_n} 14 \sup_{x_1^n} \mathcal{N}_1\left(\frac{\frac{1}{2}pen_n(p)}{20\beta_n}, T_{\beta_n}\mathcal{F}_{n,p}, x_1^n\right) \exp\left(-\frac{(t + pen_n(p))n}{2568\beta_n^4}\right)$$

$$\text{(as in the proof of Theorem 12.1)}$$

$$\leq \left\{\sum_{p \in \mathcal{P}_n} 14 \cdot \mathcal{N}_1\left(\frac{1}{n}, T_{\beta_n}\mathcal{F}_{n,p}\right) \exp\left(-\frac{pen_n(p)n}{2568\beta_n^4}\right)\right\} \exp\left(-\frac{tn}{2568\beta_n^4}\right)$$

$$\left(\text{because of } pen_n(p) \geq \frac{40\beta_n}{n} \text{ and (12.13)}\right)$$

$$\leq \left\{\sum_{p \in \mathcal{P}_n} 14 \exp(-c_p)\right\} \exp\left(-\frac{tn}{2568\beta_n^4}\right) \quad \text{(because of (12.14))}$$

$$\leq 14 \exp\left(-\frac{tn}{2568\beta_n^4}\right).$$

From this one obtains the assertion as in the proof of Theorem 12.1.

12.4 Piecewise Polynomial Partitioning Estimates

Let

$$\mathcal{P}_n = \{(M, K) \in \mathcal{N}_0 \times \mathcal{N} : 0 \leq M \leq \log n, 1 \leq K \leq n\}.$$

For $(M, K) \in \mathcal{P}_n$ let $\mathcal{F}_{n,(M,K)}$ be the set of all piecewise polynomials of degree M (or less) w.r.t. an equidistant partition of $[0, 1]$ into K intervals. For the penalty we use an upper bound on the right-hand side of (12.11) given by

$$pen_n((M, K)) = \log(n)^2 \cdot \frac{K(M + 1)}{n}.$$

We get the following result which shows that, by applying complexity regularization to piecewise polynomial partitioning estimates, we get (up to a logarithmic factor) the optimal rate of convergence and can at the same time adapt to the unknown smoothness.

Corollary 12.1. *Let $1 \leq L < \infty$ and set $\beta_n = L$ ($n \in \mathcal{N}$). Let $\mathcal{P}_n$, $\mathcal{F}_{n,(M,K)}$, and $pen_n((M, K))$ be given as above and define the estimate by (12.5)–(12.10). Then, for n sufficiently large,*

$$\mathbf{E} \int |m_n(x) - m(x)|^2 \mu(dx)$$

$$\leq \min_{(M,K)\in\mathcal{P}_n} \left\{ 2\log(n)^2 \frac{K(M+1)}{n} \right.$$

$$\left. +2 \inf_{f\in\mathcal{F}_{n,(M,K)}} \int |f(x) - m(x)|^2\mu(dx) \right\} + 5 \cdot 2568\frac{L^4}{n} \quad (12.15)$$

for every distribution of (X, Y) such that $|Y| \leq L$ a.s. In particular, if $X \in [0, 1]$ a.s., $|Y| \leq L$ a.s., and m is (p, C)-smooth for some $C > 0$, $p = q + r, q \in \mathcal{N}_0, r \in (0, 1]$, then for n sufficiently large

$$\mathbf{E}\left\{\int |m_n(x) - m(x)|^2\mu(dx)\right\} \leq c \cdot C^{\frac{2}{2p+1}} \cdot \left(\frac{\log(n)^2}{n}\right)^{\frac{2p}{2p+1}} \quad (12.16)$$

for some constant c depending only on L and p.

Proof. Set

$$c_{(M,K)} = \log(|\mathcal{P}_n|) = \log\left(n(\lfloor\log(n)\rfloor + 1)\right) \quad ((M, K) \in \mathcal{P}_n).$$

Since $\mathcal{F}_{n,(M,K)}$ is a linear vector space of dimension $K(M+1)$, we get, by Theorem 9.5,

$$V_{\mathcal{F}^+_{n,(M,K)}} \leq K(M+1) + 1.$$

It follows for n sufficiently large

$$2 \cdot 2568 \frac{\beta_n^4}{n} \left(\log(120e\beta_n^4 n) \cdot V_{\mathcal{F}^+_{n,(M,K)}} + \frac{c_{(M,K)}}{2} \right)$$

$$\leq 2 \cdot 2568 \frac{L^4}{n} \left(\log(120eL^4 n) \cdot (K(M+1)+1) + \frac{c_{(M,K)}}{2} \right)$$

$$\leq \log(n)^2 \cdot \frac{K(M+1)}{n} = pen_n(M,K).$$

Hence (12.15) follows from Theorem 12.1. The proof of (12.16), which follows from (12.15) and Lemma 11.1, is left to the reader (see Problem 12.1).
$\square$

Let us compare Corollary 12.1 to Corollary 11.2. In Corollary 11.2 we have weaker assumptions on Y (Y need not to be bounded, it is only assumed that the regression function is bounded) but the estimate used there depends on the (usually unknown) smoothness of the regression function. This is in contrast to Corollary 12.1, where the estimate does not depend on the smoothness (measured by p and C) and where we derive, up to a logarithmic factor, the same rate of convergence. According to the results from Chapter 3 the derived bounds on the L_2 error are, in both corollaries, optimal up to some logarithmic factor.

12.5 Bibliographic Notes

The complexity regularization principle for the learning problem was introduced by Vapnik and Chervonenkis (1974) and Vapnik (1982) in pattern recognition as structural risk minimization. It was applied to regression estimation in Barron (1991) and was further investigated in Barron, Birgé, and Massart (1999), Krzyżak and Linder (1998), and Kohler (1998). Lugosi and Nobel (1999) investigate complexity regularization with a data-dependent penalty.

Our complexity regularization criterion is closely related to the classical C_p criterion of Mallows (1973). Results concerning this criterion and related penalized criteria can be found, e.g., in Akaike (1974), Sjibata (1976; 1981), Li (1986; 1987), Polyak and Tsybakov (1990), and Braud, Comte, and Viennet (2001).

Problems and Exercises

PROBLEM 12.1. Show that (12.15) implies (12.16).

PROBLEM 12.2. In the following we describe a modification of the estimate in
Corollary 12.1 which is weakly and strongly consistent (cf. Problem 12.4). Set

$$\mathcal{P}_n = \left\{ (M, K) \in \mathcal{N}_0 \times \mathcal{N} : 0 \leq M \leq \log(n), \log(n)^2 \leq K \leq n^{1-\delta} \right\},$$

where $0 < \delta < 1/2$. For $(M, K) \in \mathcal{P}_n$ let $\mathcal{F}_{n,(M,K)}$ be the set of all piecewise polynomials of degree M (or less) w.r.t. an equidistant partition of $[-\log(n), \log(n)]$
into K cells, and set $\beta_n = \sqrt{\log(n)}$ and $pen_n((M, K)) = \log(n)^4 \frac{K(M-1)}{n}$. Define
the estimate m_n by (12.5)–(12.10). Show that for n sufficiently large

$$\mathbf{E} \int |m_n(x) - m(x)|^2 \mu(dx)$$

$$\leq \min_{(M,K) \in \mathcal{P}_n} \left\{ 2 \log(n)^4 \frac{K(M+1)}{n} + 2 \inf_{f \in \mathcal{F}_{n.(M,K)}} \int |f(x) - m(x)|^2 \mu(dx) \right\}$$

$$+ 5 \cdot 2568 \frac{\log(n)^2}{n}$$

for every distribution of (X, Y) such that $|Y|$ is bounded *a.s.*

PROBLEM 12.3. Let m_n be defined as in Problem 12.2. Show that for every
distribution of (X, Y) which satisfies $|X|$ bounded a.s., $|Y|$ bounded a.s., and
m (p, C)-smooth one has, for n sufficiently large,

$$\mathbf{E} \left\{ \int |m_n(x) - m(x)|^2 \mu(dx) \right\} \leq c \cdot C^{\frac{2}{2p+1}} \cdot \left(\frac{\log(n)^5}{n} \right)^{\frac{2p}{2p+1}}.$$

PROBLEM 12.4. Show that the estimate defined in Problem 12.2 is weakly and
strongly universally consistent.

13
Consistency of Data-Dependent Partitioning Estimates

In this chapter we describe the so-called data-dependent partitioning estimates. We will first prove a general consistency theorem, then we will introduce several data-dependent partitioning estimates and prove their universal consistency by checking the conditions of the general consistency theorem.

13.1   A General Consistency Theorem

Initially we consider partitioning regression estimates based on data-dependent partitioning. These estimates use the data twice: First, a partition $\mathcal{P}_n = \mathcal{P}_n(D_n)$ of $\mathcal{R}^d$ is chosen according to the data, and then this partition and the data are used to define an estimate $\hat{m}_n(x)$ of $m(x)$ by averaging those Y_i for which X_i and x belong to the same cell of the partition, i.e., the estimate is defined by

$$\hat{m}_n(x) = \frac{\sum_{i=1}^{n} Y_i I_{\{X_i \in A_n(x)\}}}{\sum_{i=1}^{n} I_{\{X_i \in A_n(x)\}}}. \tag{13.1}$$

Here $A_n(x) = A_n(x, D_n)$ denotes the cell $A \in \mathcal{P}_n(D_n)$ which contains x. As usual we have used the convention $\frac{0}{0} = 0$. In (13.1) and in the following we suppress, in notation, the dependence of $A_n(x)$ on D_n.

It turns out that the estimate $\hat{m}_n$ is a least squares estimate. Indeed, define, for a given set $\mathcal{G}$ of functions $g : \mathcal{R}^d \to \mathcal{R}$ and a partition $\mathcal{P}$ of $\mathcal{R}^d$,

$$\mathcal{G} \circ \mathcal{P} = \left\{ f : \mathcal{R}^d \to \mathcal{R} \ : \ f = \sum_{A \in \mathcal{P}} g_A \cdot I_A \text{ for some } g_A \in \mathcal{G} \ (A \in \mathcal{P}) \right\}.$$

Each function in $\mathcal{G} \circ \mathcal{P}$ is obtained by applying a different function of $\mathcal{G}$ in each set of the partition $\mathcal{P}$. Let $\mathcal{G}_c$ be the set of all constant functions. Then the estimate defined by (13.1) satisfies

$$\hat{m}_n(\cdot, D_n) \in \mathcal{G}_c \circ \mathcal{P}_n \text{ and } \frac{1}{n} \sum_{i=1}^{n} |\hat{m}_n(X_i) - Y_i|^2 = \min_{f \in \mathcal{G}_c \circ \mathcal{P}_n} \frac{1}{n} \sum_{i=1}^{n} |f(X_i) - Y_i|^2 \tag{13.2}$$

(cf. Problem 2.3). Therefore we can appply the results of Chapter 10 to show the consistency of the partitioning regression estimates based on data-dependent partitioning. These results require an additional truncation of the estimate: Let $\beta_n \in \mathcal{R}_+$ with $\beta_n \to \infty$ $(n \to \infty)$ and define

$$m_n(x) = T_{\beta_n}(\hat{m}_n(x)). \tag{13.3}$$

We will need the following definition:

Definition 13.1. *Let Π be a family of partitions of $\mathcal{R}^d$. For a set $x_1^n = \{x_1, \ldots, x_n\} \subseteq \mathcal{R}^d$ let $\Delta(x_1^n, \Pi)$ be the number of distinct partitions of x_1^n induced by elements of Π, i.e., $\Delta(x_1^n, \Pi)$ is the number of different partitions $\{x_1^n \cap A \ : \ A \in \mathcal{P}\}$ of x_1^n for $\mathcal{P} \in \Pi$. The **partitioning number** $\Delta_n(\Pi)$ is defined by*

$$\Delta_n(\Pi) = \max\{\Delta(x_1^n, \Pi) \ : \ x_1, \ldots, x_n \in \mathcal{R}^d\}.$$

The partitioning number is the maximum number of different partitions of any n point set, that can be induced by members of Π.

Example 13.1. *Let Π_k be the family of all partitions of $\mathcal{R}$ into k non-empty intervals. A partition induced by an element of Π_k on a set x_1^n with $x_1 < \cdots < x_n$ is determined by natural numbers $0 \le i_1 \le \cdots \le i_{k-1} \le n$, where the $k-1$-tuple $(i_1, \ldots, i_{k-1})$ stands for the partition*

$$\{x_1, \ldots, x_{i_1}\}, \{x_{i_1+1}, \ldots, x_{i_2}\}, \ldots, \{x_{i_{k-1}+1}, \ldots, x_n\}.$$

There are $\binom{n+1+(k-1)-1}{k-1} = \binom{n+k-1}{n}$ such tuples of numbers, thus for any x_1^n one gets

$$\Delta(x_1^n, \Pi_k) = \binom{n+k-1}{n} \text{ and } \Delta_n(\Pi_k) = \binom{n+k-1}{n}. \tag{13.4}$$

Let Π be a family of finite partitions of $\mathcal{R}^d$. We will denote the maximal number of sets contained in a partition $\mathcal{P} \in \Pi$ by $M(\Pi)$, i.e., we will define

$$M(\Pi) = \max\{|\mathcal{P}| \ : \ \mathcal{P} \in \Pi\}.$$

Set

$$\Pi_n = \left\{ \mathcal{P}_n \left(\{ (x_1, y_1), \ldots, (x_n, y_n) \} \right) \; : \; (x_1, y_1), \ldots, (x_n, y_n) \in \mathcal{R}^d \times \mathcal{R} \right\}.$$
$$(13.5)$$

Π_n is a family of partitions which contains all data-dependent partitions $\mathcal{P}_n(D_n)$. The next theorems describes general conditions which imply the consistency of a data-dependent partitioning estimate. There it is first required that the set of partitions, from which the data-dependent partition is chosen, is not too "complex," i.e., that the maximal number of cells in a partition, and the logarithm of the partitioning number, are small compared to the sample size (cf. (13.7) and (13.8)). Second, it is required that the diameter of the cells of the data-dependent partition (denoted by $diam(A)$) converge in some sense to zero (cf. (13.10)).

Theorem 13.1. *Let m_n be defined by (13.1) and (13.3) and let Π_n be defined by (13.5). Assume that*

$$\beta_n \to \infty \quad (n \to \infty), \tag{13.6}$$

$$\frac{M(\Pi_n) \cdot \beta_n^4 \cdot \log(\beta_n)}{n} \to 0 \quad (n \to \infty), \tag{13.7}$$

$$\frac{\log(\Delta_n(\Pi_n)) \cdot \beta_n^4}{n} \to 0 \quad (n \to \infty), \tag{13.8}$$

$$\frac{\beta_n^4}{n^{1-\delta}} \to 0 \quad (n \to \infty) \tag{13.9}$$

for some $\delta > 0$ and

$$\inf_{S : S \subseteq \mathcal{R}^d, \mu(S) \geq 1 - \delta} \mu \left(\{ x \; : \; diam(A_n(x) \cap S) > \gamma \} \right) \to 0 \quad (n \to \infty) \quad a.s.$$
$$(13.10)$$

for all $\gamma > 0, \delta \in (0, 1)$. Then

$$\int |m_n(x) - m(x)|^2 \mu(dx) \to 0 \quad (n \to \infty) \quad a.s.$$

We will use Theorem 10.2 to prove Theorem 13.1. We need to show

$$\inf_{f \in T_{\beta_n} \mathcal{G}_c \circ \mathcal{P}_n} \int |f(x) - m(x)|^2 \mu(dx) \to 0 \quad (n \to \infty) \text{ a.s.} \tag{13.11}$$

and

$$\sup_{f \in T_{\beta_n} \mathcal{G}_c \circ \mathcal{P}_n} \left| \frac{1}{n} \sum_{i=1}^n |f(X_i) - Y_{i,L}|^2 - \mathbf{E} |f(X) - Y_L|^2 \right| \to 0 \quad (n \to \infty) \quad a.s.$$
$$(13.12)$$

for all $L > 0$. (Here we have observed that since the functions in $\mathcal{G}_c \circ \mathcal{P}_n$ are piecewise constant, $T_{\beta_n} \mathcal{G}_c \circ \mathcal{P}_n$ consists of all those functions in $\mathcal{G}_c \circ \mathcal{P}_n$ which are bounded in absolute value by β_n.)

We will use (13.10) to show (13.11). Then we will bound the left–hand side of (13.12) by

$$\sup_{f \in T_{\beta_n} \mathcal{G}_c \circ \Pi_n} \left| \frac{1}{n} \sum_{i=1}^{n} |f(X_i) - Y_{i,L}|^2 - \mathbf{E}|f(X) - Y_L|^2 \right|, \tag{13.13}$$

where

$$\mathcal{G}_c \circ \Pi_n = \bigcup_{\mathcal{P} \in \Pi_n} \mathcal{G}_c \circ \mathcal{P}$$

and apply Theorem 9.1 to (13.13). To bound the resulting covering numbers we will use the following lemma:

Lemma 13.1. *Let $1 \leq p < \infty$. Let Π be a family of partitions of $\mathcal{R}^d$ and let $\mathcal{G}$ be a class of functions $g : \mathcal{R}^d \to \mathcal{R}$. Then one has, for each $x_1, \ldots, x_n \in \mathcal{R}^d$ and for each $\epsilon > 0$,*

$$\mathcal{N}_p \left(\epsilon, \mathcal{G} \circ \Pi, x_1^n \right) \leq \Delta(x_1^n, \Pi) \left\{ \sup_{z_1, \ldots, z_m \in x_1^n, m \leq n} \mathcal{N}_p(\epsilon, \mathcal{G}, z_1^m) \right\}^{M(\Pi)} .$$

PROOF. We will use the abbreviation

$$N = \sup_{z_1, \ldots, z_m \in x_1^n, m \leq n} \mathcal{N}_p(\epsilon, \mathcal{G}, z_1^m).$$

Fix $x_1, \ldots, x_n \in \mathcal{R}^d$ and $\epsilon > 0$. Let $\mathcal{P} = \{A_j : j\} \in \Pi$ be arbitrary. Then $\mathcal{P}$ induces a partition of x_1^n consisting of sets $B_j = \{x_1, \ldots, x_n\} \cap A_j$. For each j choose an ϵ-cover of size not greater than N of $\mathcal{G}$ on B_j, i.e., choose a set $\mathcal{G}_{B_j}$ of functions $g : \mathcal{R}^d \to \mathcal{R}$ such that for each function $g \in \mathcal{G}$ there exists a function $\bar{g} \in \mathcal{G}_{B_j}$ which satisfies

$$\frac{1}{n_j} \sum_{x \in B_j} |g(x) - \bar{g}(x)|^p < \epsilon^p, \tag{13.14}$$

where $n_j = |B_j|$.

Now let $f \in \mathcal{G} \circ \Pi$ be such that $f = \sum_{A \in \bar{\mathcal{P}}} f_A I_A$ for some partition $\bar{\mathcal{P}} \in \Pi$ which induces the same partition on x_1^n as $\mathcal{P}$. Then it follows from (13.14) that for each $A \in \bar{\mathcal{P}}$ there exists some $g_{B_j} \in \mathcal{G}_{B_j}$ (with j satisfying $A \cap x_1^n = A_j \cap x_1^n = B_j$) such that

$$\frac{1}{n_j} \sum_{x \in B_j} |f_A(x) - g_{B_j}(x)|^p < \epsilon^p, \tag{13.15}$$

and for $\bar{f} = \sum_{A \in \bar{\mathcal{P}}} g_{A \cap x_1^n} I_{A \cap x_1^n}$ we get

$$\frac{1}{n} \sum_{i=1}^{n} |f(x_i) - \bar{f}(x_i)|^p \quad = \quad \frac{1}{n} \sum_{j} \sum_{x \in B_j} |f(x) - \bar{f}(x)|^p$$

$$\overset{(13.15)}{<} \quad \frac{1}{n}\sum_{j} n_j \cdot \epsilon^p = \epsilon^p.$$

Thus for each $\mathcal{P} \in \Pi$ there is an ϵ-cover of size not greater than $N^{M(\Pi)}$ for the set of all $f \in \mathcal{G} \circ \Pi$ defined for a partition which induces the same partition on x_1^n as $\mathcal{P}$. As there are at most $\Delta(x_1^n, \Pi)$ distinct partitions on x_1^n induced by members of Π, the assertion follows. $\qquad\square$

PROOF OF THEOREM 13.1. Because of Theorem 10.2 it suffices to show (13.11) and (13.12).

PROOF OF (13.11). m can be approximated arbitrarily closely in $L_2(\mu)$ by functions of $C_0^\infty(\mathcal{R}^d)$ (cf. Corollary A.1). Hence it suffices to prove (13.11) for functions $m \in C_0^\infty(\mathcal{R}^d)$. Because of (13.6) we may further assume $\|m\|_\infty \le \beta_n$.

Let $\epsilon > 0$ and $\delta \in (0,1)$. For $S \subseteq \mathcal{R}^d$ and given data D_n define $f_S \in T_{\beta_n}\mathcal{G}_c \circ \mathcal{P}_n$ by

$$f_S = \sum_{A \in \mathcal{P}(D_n)} m(z_A) \cdot I_{A\cap S}$$

for some fixed $z_A \in A$ which satisfies $z_A \in A \cap S$ if $A \cap S \ne \emptyset$ $(A \in \mathcal{P}_n)$. Choose $\gamma > 0$ such that $|m(x) - m(z)| < \epsilon$ for all $\|x - z\| < \gamma$. Then it follows that, for $z \in S$,

$$|f_S(z) - m(z)|^2 < \epsilon^2 I_{\{diam(A_n(z)\cap S)<\gamma\}} + 4\|m\|_\infty^2 I_{\{diam(A_n(z)\cap S)\ge\gamma\}}.$$

Using this one gets

$$\inf_{f\in T_{\beta_n}\mathcal{G}_c\circ\mathcal{P}_n} \int |f(x) - m(x)|^2 \mu(dx)$$

$$\le \inf_{S:\mu(S)\ge 1-\delta} \int |f_S(x) - m(x)|^2 \mu(dx)$$

$$\le \inf_{S:\mu(S)\ge 1-\delta} \left\{ \int_S |f_S(x) - m(x)|^2 \mu(dx) + 4\|m\|_\infty^2 \mu(\mathcal{R}^d \setminus S) \right\}$$

$$\le \inf_{S:\mu(S)\ge 1-\delta} \int_S |f_S(x) - m(x)|^2 \mu(dx) + 4\|m\|_\infty^2 \delta$$

$$\le \inf_{S:\mu(S)\ge 1-\delta} \int_S \left(\epsilon^2 I_{\{diam(A_n(z)\cap S)<\gamma\}} \right.$$

$$\left. +4\|m\|_\infty^2 I_{\{diam(A_n(z)\cap S)\ge\gamma\}} \right) \mu(dx)$$

$$+4\|m\|_\infty^2 \delta$$

$$\leq \epsilon^2 + 4\|m\|_\infty^2 \inf_{S:\mu(S)\geq 1-\delta} \mu\left(\{x \in \mathcal{R}^d \; : \; diam(A_n(x) \cap S) \geq \gamma\}\right)$$

$$+ 4\|m\|_\infty^2 \delta$$

$$\overset{(13.10)}{\to} \epsilon^2 + 4\|m\|_\infty^2 \delta \quad (n \to \infty) \text{ a.s.}$$

Because $\epsilon > 0$, $\delta \in (0,1)$ are arbitrary, the assertion follows.

PROOF OF (13.12). Because of $T_{\beta_n}\mathcal{G}_c \circ \mathcal{P}_n \subseteq T_{\beta_n}\mathcal{G}_c \circ \Pi_n$ it suffices to prove (13.12) with $T_{\beta_n}\mathcal{G}_c \circ \mathcal{P}_n$ replaced by $T_{\beta_n}\mathcal{G}_c \circ \Pi_n$. It follows from Theorem 9.1 and Problem 10.4 that, for $0 \leq L \leq \beta_n$,

$$\mathbf{P}\left\{\sup_{f \in T_{\beta_n}\mathcal{G}_c \circ \Pi_n} \left|\frac{1}{n}\sum_{i=1}^{n} |f(X_i) - Y_{i,L}|^2 - \mathbf{E}|f(X) - Y_L|^2\right| > \epsilon\right\}$$

$$\leq 8\mathbf{E}\mathcal{N}_1\left(\frac{\epsilon}{32\beta_n}, T_{\beta_n}\mathcal{G}_c \circ \Pi_n, X_1^n\right) \exp\left(-\frac{n\epsilon^2}{128 \cdot (4\beta_n^2)^2}\right).$$

Using Lemma 13.1 and Theorem 9.4 one gets

$$\mathcal{N}_1\left(\frac{\epsilon}{32\beta_n}, T_{\beta_n}\mathcal{G}_c \circ \Pi_n, X_1^n\right)$$

$$\leq \Delta_n(\Pi_n)\left\{\sup_{z_1,\ldots,z_m \in \{X_1,\ldots X_n\}, m \leq n} \mathcal{N}_1\left(\frac{\epsilon}{32\beta_n}, T_{\beta_n}\mathcal{G}_c, z_1^m\right)\right\}^{M(\Pi_n)}$$

$$\leq \Delta_n(\Pi_n)\left\{3\left(\frac{3e(2\beta_n)}{\frac{\epsilon}{32\beta_n}}\right)^{2V_{T_{\beta_n}\mathcal{G}_c^+}}\right\}^{M(\Pi_n)}$$

$$\leq \Delta_n(\Pi_n)\left\{\frac{333e\beta_n^2}{\epsilon}\right\}^{2M(\Pi_n)}$$

because

$$V_{T_{\beta_n}\mathcal{G}_c^+} \leq V_{\mathcal{G}_c^+} \leq 1.$$

Thus

$$\mathbf{P}\left\{\sup_{f \in T_{\beta_n}\mathcal{G}_c \circ \Pi_n} \left|\frac{1}{n}\sum_{i=1}^{n} |f(X_i) - Y_{i,L}|^2 - \mathbf{E}|f(X) - Y_L|^2\right| > \epsilon\right\}$$

$$\leq 8\Delta_n(\Pi_n)\left\{\frac{333e\beta_n^2}{\epsilon}\right\}^{2M(\Pi_n)} \exp\left(-\frac{n\epsilon^2}{2048\beta_n^4}\right)$$

$$\leq 8\exp\left(\log(\Delta_n(\Pi_n)) + 2M(\Pi_n)\log\frac{333e\beta_n^2}{\epsilon} - \frac{n\epsilon^2}{2048\beta_n^4}\right)$$

$$= 8 \exp\left(-\frac{n}{\beta_n^4}\left(\frac{\epsilon^2}{2048} - \frac{\log\left(\Delta_n(\Pi_n)\right)\beta_n^4}{n} - \frac{2M(\Pi_n)\beta_n^4 \log\frac{333 e \beta_n^2}{\epsilon}}{n}\right)\right)$$

and the assertion follows by an easy application of the Borel–Cantelli lemma. $\square$

Let $\mathcal{P}_n$ be a finite data-dependent partition and $\hat{m}_n(x)$ the corresponding partitioning estimate without truncation. Assume that there exists a positive integer k_n such that, for all $A \in \mathcal{P}_n$,

$$\mu_n(A) \geq k_n \cdot \frac{\log(n)}{n}.$$

Then

$$k_n \to \infty \quad (n \to \infty)$$

and

$$\lim_{n \to \infty} diam(A_n(X)) = 0$$

in probability imply that $\hat{m}_n$ is weakly universally consistent (Breiman et al. (1984)).

13.2 Cubic Partitions with Data-Dependent Grid Size

As a first example for a data-dependent partitioning estimate we consider a partitioning estimate based on a cubic partition with a data-dependent grid size. The partition is determined by a data-independent rectangle $[L_n, R_n)^d$, which we partition in equidistant cubic cells, i.e., we use partitions

$$\mathcal{P}_k = \left\{\mathcal{R}^d \setminus [L_n, R_n)^d\right\} \cup$$

$$\left\{[L_n + i_1 h_k, L_n + (i_1 + 1)h_k) \times \cdots \times [L_n + i_d h_k, L_n + (i_d + 1)h_k) \; : \right.$$

$$\left. i_1, \ldots, i_d \in \{0, \ldots, k-1\}\right\},$$

where $h_k = \frac{R_n - L_n}{k}$ is the grid size of the partition. This grid size is chosen in a data-dependent manner by choosing a random K which satisfies $K_{min}(n) \leq K \leq K_{max}(n)$, where $K_{min}(n), K_{max}(n) \in \mathcal{N}$ depend only on the size of the sample. We will use Theorem 13.1 to show

Theorem 13.2. *Assume*

$$L_n \to -\infty, R_n \to \infty \quad (n \to \infty), \tag{13.16}$$

$$\frac{R_n - L_n}{K_{min}(n)} \to 0 \quad (n \to \infty), \tag{13.17}$$

$$\frac{(K_{max}(n)^d + \log(n)) \cdot \beta_n^4 \log(\beta_n)}{n} \to 0 \quad (n \to \infty), \tag{13.18}$$

$$\beta_n \to \infty \quad (n \to \infty), \tag{13.19}$$

and

$$\frac{\beta_n^4}{n^{1-\delta}} \to 0 \quad (n \to \infty) \tag{13.20}$$

for some $\delta > 0$. Then any estimate m_n defined by (13.1) and (13.3) with $\mathcal{P}_n(D_n) = \mathcal{P}_K$ for some random $K = K(D_n)$ satisfying

$$K_{min}(n) \le K \le K_{max}(n) \tag{13.21}$$

is strongly universally consistent.

From this theorem we can conclude that a truncated partitioning estimate which uses a deterministic cubic partition is strongly universally consistent. We will later see (cf. Chapter 23) that suitably defined deterministic partitioning estimates are strongly universally consistent even without truncation, but the proof there will be much more involved.

One great advantage of Theorem 13.2 is that it is valid for any data-dependent choice of K. So if one restricts the range of K as in (13.21), then one can apply, e.g., splitting of the sample (cf. Chapter 7) or cross-validation (cf. Chapter 8) to choose K and gets automatically a consistent estimate. Due to the fact that K may be chosen arbitrarily from the range of (13.21) (in particular, the worst value might be chosen), the conditions there can be improved at most by some logarithmic factor. But, on the other hand, for particular rules for choosing a data-dependent K they can be relaxed, e.g., the restriction $K_{min}(n) \le K$ should not be necessary to prove consistency for a reasonable data-dependent choice of K.

PROOF. It suffices to check the conditions (13.6)–(13.10) for

$$\Pi_n = (\mathcal{P}_k)_{K_{min}(n) \le k \le K_{max}(n)}.$$

Clearly,

$$M(\Pi_n) = K_{max}(n)^d + 1. \tag{13.22}$$

To determine $\Delta_n(\Pi_n)$ fix $x_1^n \in \mathcal{R}^{d \cdot n}$. The partition, which is induced by $\mathcal{P}_k$ on x_1^n, is uniquely determined by $a_k = (a_{1,k}, \ldots, a_{n,k})$ where

$$a_{l,k} \in \{0\} \cup \{0, \ldots, K_{max}(n) - 1\}^d$$

is defined by

$$a_{l,k} = \begin{cases} 0 & \text{if } x_l \in \mathcal{R}^d \setminus [L_n, R_n)^d, \\ (i_1, \ldots, i_d) & \text{if } x_l \in [L_n + i_1 h_k, L_n + (i_1 + 1)h_k) \times \cdots \\ & \qquad\qquad \times [L_n + i_d h_k, L_n + (i_d + 1)h_k). \end{cases}$$

If $k_1 < k_2$ then $h_{k_1} > h_{k_2}$ and thus if $a_{l,k_1} = (i_1, \ldots, i_d)$ and $a_{l,k_2} = (j_1, \ldots, j_d)$ then one has $i_1 \leq j_1, \ldots, i_d \leq j_d$. Therefore if k runs from 1 up to $K_{max}(n)$ then each component of a_k can change at most $K_{max}(n)^d$ times, and it follows that $\Delta(x_1^n, \Pi_n) \leq n K_{max}(n)^d$. This implies

$$\Delta_n(\Pi_n) \leq n K_{max}(n)^d. \tag{13.23}$$

Now (13.6)–(13.9) follow easily from (13.18)–(13.23).

To show (13.10), let $\gamma > 0$ and $\delta \in (0,1)$ be arbitrary. Because of (13.16) and (13.17) one has, for sufficiently large n,

$$\mu\left([L_n, R_n)^d\right) \geq 1 - \delta \quad \text{and} \quad d \cdot \frac{R_n - L_n}{K_{min}(n)} \leq \gamma.$$

Then for sufficiently large n,

$$\inf_{S:\mu(S)\geq 1-\delta} \mu\left(\{x : diam(A_n(x) \cap S) > \gamma\}\right)$$

$$\leq \mu\left(\{x : diam(A_n(x) \cap [L_n, R_n)^d) > \gamma\}\right)$$

$$\leq \mu\left(\left\{x : d\frac{R_n - L_n}{K_{min}(n)} > \gamma\right\}\right) = \mu(\emptyset) = 0.$$

This implies (13.10), and the assertion follows from Theorem 13.1. $\qquad\square$

13.3 Statistically Equivalent Blocks

A partition is based on statistically equivalent blocks if each set of the partition contains the same number of data points.

In the sequel we only consider the case of univariate X (see Problem 13.1 for multivariate examples). For simplicity we assume that X has a density with respect to the Lebesgue measure. This implies that with probability one $X_1, \ldots, X_n$ are all distinct (this is the only point where we need the density assumption). Let $X_{(1)}, \ldots, X_{(n)}$ be the order statistics of $X_1, \ldots, X_n$, i.e., $X_{(1)} < \cdots < X_{(n)}$ and $\{X_{(1)}, \ldots, X_{(n)}\} = \{X_1, \ldots, X_n\}$ with probability one.

For univariate X, partitioning based on statistically equivalent blocks reduces to the so-called k-spacing: Let $k_n \in \mathcal{N}$ and choose a partition $\mathcal{P}_n(D_n) = \{A_1, \ldots, A_N\}$ of $\mathcal{R}$ consisting of $N = \lceil \frac{n}{k_n} \rceil$ intervals such that each interval except the rightmost interval contains exactly k_n of the X_i. So $A_1, \ldots, A_m$ are intervals such that

$$X_{((j-1)k_n+1)}, \ldots, X_{(jk_n)} \in A_j \quad (j = 1, \ldots, N-1)$$

and

$$X_{((N-1)k_n+1)}, \ldots, X_{(n)} \in A_N.$$

The exact position of the end points of the intervals is not important.

We next show strong consistency of partitioning estimates using such partitions.

Theorem 13.3. *Assume*

$$\beta_n \to \infty \quad (n \to \infty), \tag{13.24}$$

$$\frac{k_n}{n} \to 0 \quad (n \to \infty), \tag{13.25}$$

$$\frac{\beta_n^4 \log(n)}{k_n} \to 0 \quad (n \to \infty), \tag{13.26}$$

and

$$\frac{\beta_n^4}{n^{1-\delta}} \to 0 \quad (n \to \infty) \tag{13.27}$$

for some $\delta > 0$. Then any estimate m_n defined by (13.1) and (13.3) with $\mathcal{P}_n(D_n)$ defined via k_n-spacing is strongly consistent for every distribution of (X, Y) where X is an univariate random variable having a density and $\mathbf{E}Y^2 < \infty$.

PROOF. Let Π_n be the family of all partitions consisting of $\lceil \frac{n}{k_n} \rceil$ intervals. We will check the conditions of Theorem 13.1 for Π_n.

Clearly, $M(\Pi_n) = \lceil \frac{n}{k_n} \rceil$. Furthermore, by (13.4),

$$\Delta_n(\Pi_n) = \binom{n + \lceil \frac{n}{k_n} \rceil - 1}{n} \leq (n + \lceil \frac{n}{k_n} \rceil)^{\lceil \frac{n}{k_n} \rceil} \leq (2n)^{\lceil \frac{n}{k_n} \rceil}.$$

Thus (13.6)–(13.9) are implied by (13.24), (13.26), and (13.27).

To prove (13.10), fix $\gamma > 0$ and $\delta \in (0, 1)$. Choose L so large that $\mu([-L, L]) \geq 1 - \delta$. Observe that no more than $2 + \lceil \frac{2L}{\gamma} \rceil$ of the intervals $A_1, \ldots, A_N$ can satisfy $diam(A_i) > \gamma$ and $A_i \cap [-L, L] \neq \emptyset$. Hence

$$\inf_{S:\mu(S) \geq 1-\delta} \mu\left(\{x \,:\, diam(A_n(x) \cap S) > \gamma\}\right)$$

$$\leq \mu\left(\{x \,:\, diam(A_n(x) \cap [-L, L]) > \gamma\}\right)$$

$$\leq \sum_{i:diam(A_i)>\gamma, A_i \cap [-L,L] \neq \emptyset} (\mu(A_i) - \mu_n(A_i) + \mu_n(A_i))$$

$$\leq \left(\sup_{A \in \mathcal{P}, \mathcal{P} \in \Pi_n} |\mu(A) - \mu_n(A)| + \frac{k_n}{n}\right)\left(2 + \left\lceil \frac{2L}{\gamma} \right\rceil\right)$$

$$\to 0 \quad (n \to \infty) \quad a.s.,$$

where the first term in the parentheses above converges to zero by an obvious extension of the classical Glivenko–Cantelli theorem, while the second term converges to zero because of (13.25). $\qquad\square$

The concept of statistically equivalent blocks can be extended to $\mathcal{R}^d$ as follows (the so-called Gessaman rule): For fixed sample size n set $M = \lceil (n/k_n)^{1/d} \rceil$. According to the first coordinate axis, partition the data into M sets such that the first coordinates form statistically equivalent blocks. We obtain M cylindrical sets. In the same fashion, cut each of these cylindrical sets along the second axis into M statistically equivalent blocks. Continuing in the same way along the remaining coordinate axes, we obtain M^d rectangular cells, each of which (with the exception of those on the boundary) contains k_n points (see Figure 4.6). The proof of the consistency of (truncated) partitioning estimates using such a partition is left to the reader (cf. Problem 13.1).

13.4 Nearest Neighbor Clustering

A **clustering scheme** is a function $C : \mathcal{R}^d \to \mathcal{C}$, where $\mathcal{C} = \{c_1, \dots, c_k\} \subseteq \mathcal{R}^d$ is a finite set of vectors called **cluster centers**. Each clustering scheme C is associated with a partition $\mathcal{P}_C = \{A_1, \dots, A_k\}$ of $\mathcal{R}^d$ having cells $A_j = \{x : C(x) = c_j\}$. A clustering scheme C is called a **nearest neighbor clustering scheme** (NN-clustering scheme) if, for each $x \in \mathcal{R}^d$,

$$\|x - C(x)\| = \min_{c_j \in \mathcal{C}} \|x - c_j\|. \tag{13.28}$$

An example of an NN-clustering scheme in $\mathcal{R}^2$ is given in Figure 13.1.

Given only a set $\mathcal{C}$, one can use (13.28) to define a nearest neighbor clustering scheme $C : \mathcal{R}^d \to \mathcal{C}$ uniquely if one has an appropriate tie-breaking strategy. In the following we will use the tie-breaking strategy, which defines $C(x)$ such that the index j of c_j is minimal in (13.28).

We will choose the cluster centers of a clustering scheme by minimizing an empirical risk. Let $\mathbf{E}\|X\|^2 < \infty$ and define the risk of a clustering scheme C by

$$R(C) = \mathbf{E}\|X - C(X)\|^2 = \int |x - C(x)|^2 \mu(dx), \tag{13.29}$$

i.e., $R(C)$ is the expected squared distance of X to the closest cluster center of $\mathcal{C}$. Similarly, the empirical risk $R_n(C)$ of a cluster scheme is defined by

$$R_n(C) = \frac{1}{n} \sum_{i=1}^n \|X_i - C(X_i)\|^2 = \frac{1}{n} \sum_{i=1}^n \min_{j=1,\dots,k} \|X_i - c_j\|^2. \tag{13.30}$$

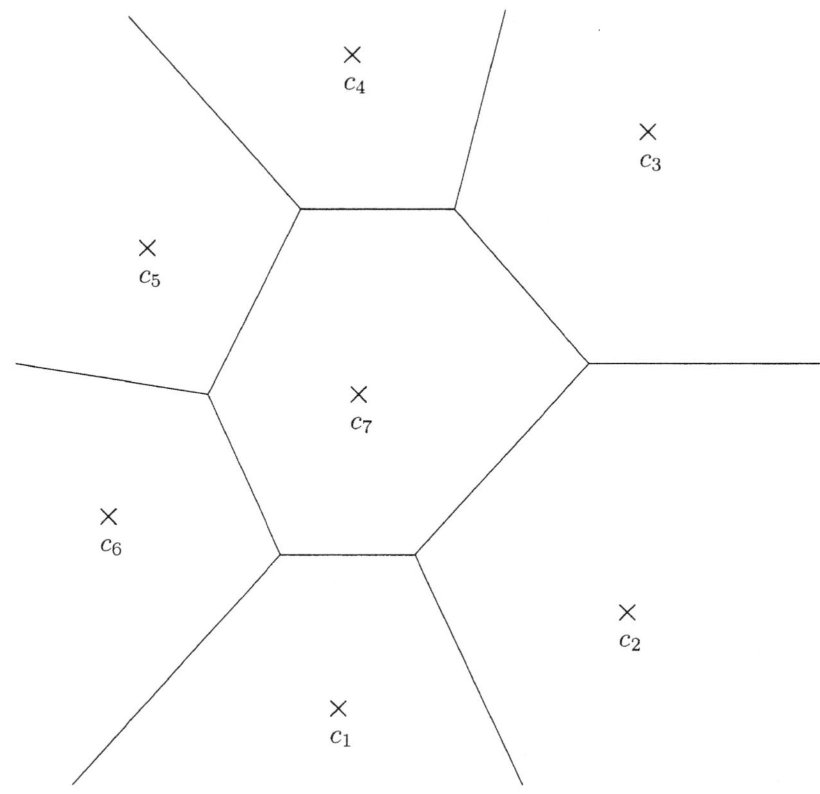

Figure 13.1. An NN-clustering scheme in $\mathcal{R}^2$.

We will consider nearest neighbor clustering schemes which minimize the empirical risk, i.e., which satisfy

$$R_n(C_n) = \min_{C\ NN-clustering\ scheme, |C(\mathcal{R}^d)| \le k_n} R_n(C). \tag{13.31}$$

Next we show that such an empirical optimal clustering scheme always exists. Fix $X_1, \ldots, X_n$, let $\mathcal{C}_n = \{c_1, \ldots, c_{k_n}\}$ and $\bar{\mathcal{C}}_n = \{\bar{c}_1, \ldots, \bar{c}_{k_n}\}$ be two sets of cluster centers and denote the corresponding clustering schemes by C and $\bar{C}$. Then

$$|R_n(C) - R_n(\bar{C})|$$

$$= \left| \frac{1}{n} \sum_{i=1}^{n} \left(\min_{j=1,\ldots,k_n} \|X_i - c_j\|^2 - \min_{j=1,\ldots,k_n} \|X_i - \bar{c}_j\|^2 \right) \right|$$

$$\leq \quad \frac{1}{n} \sum_{i=1}^{n} \max_{j=1,\dots,n} \left(\|X_i - c_j\| + \|X_i - \bar{c}_j\| \right) \cdot \|c_j - \bar{c}_j\|$$

$$\leq \quad \max_{i,j} \left(\|X_i - c_j\| + \|X_i - \bar{c}_j\| \right) \cdot \max_{j} \|c_j - \bar{c}_j\|$$

and therefore the empirical risk is (on any bounded set of cluster centers) a continuous function of the cluster centers. Because of

$$\inf_{C:\|c_j\|>L \text{ for some } j} \frac{1}{n} \sum_{i=1}^{n} \min_{j=1,\dots,k_n} \|X_i - c_j\|^2 \to \infty \quad (L \to \infty)$$

the infimal empirical risk will be obtained in a compact set and, because of the continuity, the infimal risk will really be obtained for some set of cluster centers contained in this compact set.

Using such an empirical optimal nearest neighbor clustering scheme to produce the partition of a data-dependent partitioning estimate, one gets consistent regression estimates.

Theorem 13.4. *Assume*

$$\beta_n \to \infty \quad (n \to \infty), \tag{13.32}$$

$$k_n \to \infty \quad (n \to \infty), \tag{13.33}$$

$$\frac{k_n^2 \beta_n^4 \log(n)}{n} \to 0 \quad (n \to \infty), \tag{13.34}$$

and

$$\frac{\beta_n^4}{n^{1-\delta}} \to 0 \quad (n \to \infty) \tag{13.35}$$

for some $\delta > 0$. Let m_n be defined by (13.1) and (13.3) for some data-dependent partition $\mathcal{P}_n(D_n) = \mathcal{P}_C$, where C is a k_n-nearest neighbor clustering scheme which minimizes the empirical risk. Then m_n is strongly consistent for every distribution of (X, Y) with $\mathbf{E}\|X\|^2 < \infty$ and $\mathbf{E}Y^2 < \infty$.

Remark. If one uses for the construction of the k_n-NN clustering scheme only those X_i which are contained in a data-independent rectangle $[L_n, R_n]^d$ (where $L_n \to -\infty$ and $R_n \to \infty$ not too fast for $n \to \infty$), then the resulting estimate is strongly universally consistent (i.e., then one can avoid the condition $\mathbf{E}\|X\|^2 < \infty$ in the above theorem). The details are left to the reader (cf. Problem 13.2).

We will prove Theorem 13.4 by checking the conditions of Theorem 13.1.

We will use the following lemma to show the shrinking of the cells:

Lemma 13.2. *For each $n \in \mathcal{N}$ let $C_n : \mathcal{R}^d \to \mathcal{C}_n$ minimize the empirical risk $R_n(C_n)$ over all nearest neighbor clustering schemes having k_n cluster* *centers. If $\mathbf{E}\|X\|^2 < \infty$ and $k_n \to \infty$ $(n \to \infty)$, then one has, for each $L > 0$,*

$$\max_{u \in supp(\mu) \cap [-L,L]^d} \min_{c \in \mathcal{C}_n} \|u - c\| \to 0 \quad (n \to \infty) \quad a.s. \tag{13.36}$$

PROOF.

STEP 1. We will first show

$$R_n(C_n) \to 0 \quad (n \to \infty) \quad a.s. \tag{13.37}$$

To show this, let $\{u_1, u_2, \ldots\}$ be a countable dense subset of $\mathcal{R}^d$ with $u_1 = 0$. Let $L > 0$ be arbitrary. Then

$$
\begin{aligned}
R_n(C_n) \quad &\leq \quad \frac{1}{n} \sum_{i=1}^{n} \min_{j=1,\ldots,k_n} \|X_i - u_j\|^2 \\[2mm]
&\leq \quad \frac{1}{n} \sum_{\substack{i=1,\ldots,n,\\ X_i \in [-L,L]^d}} \min_{j=1,\ldots,k_n} \|X_i - u_j\|^2 \\[1mm]
&\quad + \frac{1}{n} \sum_{\substack{i=1,\ldots,n,\\ X_i \in \mathcal{R}^d \setminus [-L,L]^d}} \min_{j=1,\ldots,k_n} \|X_i - u_j\|^2 \\[2mm]
&\leq \quad \max_{x \in [-L,L]^d} \min_{j=1,\ldots,k_n} \|x - u_j\|^2 + \frac{1}{n} \sum_{\substack{i=1,\ldots,n,\\ X_i \in \mathcal{R}^d \setminus [-L,L]^d}} \|X_i\|^2
\end{aligned}
$$

$$\text{(because of } u_1 = 0\text{)}$$

$$\to \quad 0 + \mathbf{E}\left\{\|X\|^2 I_{\{X \in \mathcal{R}^d \setminus [-L,L]^d\}}\right\} \quad (n \to \infty) \quad a.s.$$

Because of $\mathbf{E}\|X\|^2 < \infty$ one gets, for $L \to \infty$,

$$\mathbf{E}\left\{\|X\|^2 I_{\{X \in \mathcal{R}^d \setminus [-L,L]^d\}}\right\} \to 0 \quad (L \to \infty),$$

which implies (13.37).

STEP 2. Next we will show

$$\mathbf{P}\left\{\liminf_{n \to \infty} \mu_n\left(S_{u,\delta}\right) > 0 \text{ for every } u \in supp(\mu), \delta > 0\right\} = 1. \tag{13.38}$$

Let $\{v_1, v_2, \ldots\}$ be a countable dense subset of $supp(\mu)$. By definition of $supp(\mu)$ one has

$$\mu(S_{u,\delta}) > 0 \quad \text{for every } u \in supp(\mu), \delta > 0.$$

The strong law of large numbers implies, for every $i, k \in \mathcal{N}$,

$$\lim_{n \to \infty} \mu_n\left(S_{v_i, 1/k}\right) = \mu\left(S_{v_i, 1/k}\right) > 0 \quad a.s.,$$

from which it follows that, with probability one,

$$\lim_{n\to\infty} \mu_n\left(S_{v_i,1/k}\right) > 0 \text{ for every } i, k \in \mathcal{N}. \tag{13.39}$$

Fix $u \in supp(\mu)$ and $\delta > 0$. Then there exist i, k with $\frac{1}{k} < \frac{\delta}{2}$ and $u \in S_{v_i,1/k}$, which imply $S_{u,\delta} \supseteq S_{v_i,1/k}$. Hence

$$\liminf_{n\to\infty} \mu_n\left(S_{u,\delta}\right) \geq \liminf_{n\to\infty} \mu_n\left(S_{v_i,1/k}\right).$$

This together with (13.39) proves (13.38).

STEP 3. Now suppose that (13.36) doesn't hold. Then there exist $L > 0$ and $\delta > 0$ such that the event

$$\left\{\limsup_{n\to\infty} \max_{u\in supp(\mu)\cap[-L,L]^d} \min_{c\in\mathcal{C}_n} \|u - c\| > \delta\right\}$$

has probability greater than zero. On this event there exists a (random) sequence $\{n_k\}_{k\in\mathcal{N}}$ and (random) $u_{n_k} \in supp(\mu) \cap [-L, L]^d$ such that

$$\min_{c\in\mathcal{C}_{n_k}} \|u_{n_k} - c\| > \frac{\delta}{2} \quad (k \in \mathcal{N}). \tag{13.40}$$

Because $supp(\mu)\cap[-L, L]^d$ is compact one can assume w.l.o.g. (by replacing $\{n_k\}_{k\in\mathcal{N}}$ by a properly defined subsequence) $u_{n_k} \to u^*$ $(n \to \infty)$ for some (random) $u^* \in supp(\mu)$.

This implies that on this event (and therefore with probability greater than zero) one has

$$R_{n_k}(\mathcal{C}_{n_k}) \quad \geq \quad \frac{1}{n_k} \sum_{i=1,\ldots,n_k; X_i\in S_{u^*,\delta/8}} \min_{c\in\mathcal{C}_{n_k}} \|X_i - c\|^2$$

$$\geq \quad \frac{1}{n_k} \sum_{i=1,\ldots,n_k; X_i\in S_{u^*,\delta/8}} \min_{c\in\mathcal{C}_{n_k}} \left(\|u_{n_k} - c\| - \|u_{n_k} - X_i\|\right)^2$$

$$\overset{(13.40)}{\geq} \quad \frac{1}{n_k} \sum_{i=1,\ldots,n_k; X_i\in S_{u^*,\delta/8}} I_{\{u_{n_k}\in S_{u^*,\delta/8}\}} \left(\frac{\delta}{2} - \left(\frac{\delta}{8} + \frac{\delta}{8}\right)\right)^2$$

$$= \quad \frac{1}{n_k} \sum_{i=1,\ldots,n_k; X_i\in S_{u^*,\delta/8}} I_{\{u_{n_k}\in S_{u^*,\delta/8}\}} \frac{\delta^2}{16}$$

$$= \quad I_{\{u_{n_k}\in S_{u^*,\delta/8}\}} \cdot \frac{\delta^2}{16} \cdot \mu_{n_k}\left(S_{u^*,\delta/8}\right).$$

It follows that

$$\liminf_{k\to\infty} R_{n_k}(\mathcal{C}_{n_k}) \geq \frac{\delta^2}{16} \liminf_{k\to\infty} \mu_{n_k}\left(S_{u^*,\delta/8}\right) \overset{(13.38)}{>} 0 \quad a.s.,$$

which contradicts (13.37).   □

PROOF OF THEOREM 13.4. Let Π_n be the family consisting of all partitions $\mathcal{P}_C$ for some k_n-nearest neighbor clustering scheme C. Clearly,

$$M(\Pi_n) = k_n.$$

Furthermore, each set in a partition $\mathcal{P}_C$ is an intersection of at most k_n^2 hyperplanes perpendicular to one of the k_n^2 pairs of cluster centers. It follows from Theorems 9.3 and 9.5 that a hyperplane can split n points in at most $(n+1)^{d+1}$ different ways (cf. Problem 13.3). Therefore

$$\Delta_n(\Pi_n) \leq \left((n+1)^{d+1}\right)^{k_n^2} = (n+1)^{(d+1)\cdot k_n^2}.$$

Now (13.6)–(13.9) follow from (13.32)–(13.35).

To show (13.10), fix $\gamma > 0$ and $\delta \in (0,1)$. Choose L so large that $\mu([-L,L]^d) \geq 1 - \delta$. Then

$$\inf_{S:\mu(S)\geq 1-\delta} \mu\left(\{x \ : \ diam(A_n(x) \cap S) > \gamma\}\right)$$

$$\leq \mu\left(\{x \ : \ diam(A_n(x) \cap [-L,L]^d) > \gamma\}\right)$$

$$\leq \mu\left(\left\{x \ : \ 2 \max_{u \in supp(\mu) \cap [-L,L]^d} \min_{c \in \mathcal{C}_n} \|u - c\| > \gamma\right\}\right)$$

$$= 1 \cdot I_{\{2 \max_{u \in supp(\mu) \cap [-L,L]^d} \min_{c \in \mathcal{C}_n} \|u-c\| > \gamma\}} \to 0 \quad (n \to \infty) \quad a.s.$$

by Lemma 13.2. Thus Theorem 13.1 implies the assertion.   □

13.5 Bibliographic Notes

Lemmas 13.1 and 13.2 are due to Nobel (1996). Theorem 13.1 and the consistency results in Sections 13.2, 13.3, and 13.4 are extensions of results from Nobel (1996). Related results concerning classification can be found in Lugosi and Nobel (1996). Gessaman's rule, described at the end of Section 13.3, is due to Gessaman (1970).

There are several results on data-dependent partitioning estimates for nested partitions, i.e., for partitions where $\mathcal{P}_{n+1}$ is a refinement of $\mathcal{P}_n$ (maybe $\mathcal{P}_{n+1} = \mathcal{P}_n$). This refinement means that the cells of $\mathcal{P}_{n+1}$ are either cells of $\mathcal{P}_n$ or splits of some cells of $\mathcal{P}_n$. This splitting can be represented by a tree, therefore, it is often called regression trees (cf. Gordon and Olshen (1980; 1984), Breiman et al. (1984), Devroye, Györfi, and Lugosi (1996), and the references therein).

Problems and Exercises

PROBLEM 13.1. Find conditions on β_n and k_n such that the truncated data-dependent partitioning estimate, which uses a partition defined by Gessaman's rule (cf. end of Section 13.3), is strongly consistent for all distributions of (X, Y) where each component of X has a density and $\mathbf{E}Y^2 < \infty$.

PROBLEM 13.2. Show that if one uses, for the construction of the k_n-NN-clustering scheme, only those X_i, which are contained in a data-independent rectangle $[L_n, R_n]^d$ (where $L_n \to -\infty$ and $R_n \to \infty$ not too fast for $n \to \infty$), then the resulting estimate is strongly universally consistent.

PROBLEM 13.3. Show that a hyperplane can split n points in at most $(n+1)^{d+1}$ different ways.

    HINT: Use Theorems 9.3 and 9.5 to bound the shatter coefficient of the set

$$\mathcal{A} = \left\{ \{ x \in \mathcal{R}^d \; : \; a_1 x^{(1)} + \cdots + a_d x^{(d)} + a_{d+1} \geq 0 \} \; : \; a_1, \ldots, a_{d+1} \in \mathcal{R} \right\}.$$

PROBLEM 13.4. Let $M \in \mathcal{N}$ and let $\mathcal{G}_M$ be the set of all (multivariate) polynomials of degree M (or less, in each coordinate). Let $\mathcal{P}_n = \mathcal{P}_n(D_n)$ be a data-dependent partition and set

$$\mathcal{G}_M \circ \mathcal{P}_n = \left\{ f : \mathcal{R}^d \to \mathcal{R} \; : \; f = \sum_{A \in \mathcal{P}_n} g_A I_A \text{ for some } g_A \in \mathcal{G}_M \; (A \in \mathcal{P}_n) \right\}.$$

Define the estimate m_n by (13.2) and (13.3) with $\mathcal{G}_c$ replaced by $\mathcal{G}_M$. Show that (13.6)–(13.10) imply

$$\int |m_n(x) - m(x)|^2 \mu(dx) \to 0 \quad (n \to \infty) \quad a.s.$$

PROBLEM 13.5. Use Problem 13.4 to define consistent least squares estimates using piecewise polynomials with respect to data-dependent partitions, where the partitions are:
(a) cubic partitions with data-dependent grid size;
(b) statistically equivalent blocks; and
(c) defined via nearest neighbor clustering.

14

Univariate Least Squares Spline Estimates

In Chapter 10 we have introduced least squares estimates. These estimates heavily depend on the classes $\mathcal{F}_n$ of functions $f : \mathcal{R}^d \to \mathcal{R}$ over which one minimizes the empirical L_2 risk.

In Section 11.2 we have defined such classes $\mathcal{F}_n$ by choosing a partition of $\mathcal{R}^d$ and taking all piecewise polynomials with respect to that partition. The main drawback of this is that such functions are generally not smooth (e.g., not continuous) and, therefore also, the corresponding least squares estimate is generally not smooth. For the interpretability of the estimate it is often important that it is a smooth function. Moreover, using spaces of piecewise polynomials without any smoothness condition results in a high variance of the estimate in cells of the partition which contain only few of the X_i's (because on such a cell the estimate depends only on the few (X_i, Y_i)'s with X_i contained in this cell).

A remedy against this drawback of piecewise polynomials is to use sets of piecewise polynomial functions which satisfy some global smoothness condition (e.g., which are continuous). These so-called polynomial spline spaces will be investigated in this chapter.

14.1   Introduction to Univariate Splines

In the sequel we will define spaces of piecewise polynomials on an interval $[a, b)$ $(a, b \in \mathcal{R}, a < b)$ which satisfy some global smoothness conditions. To do so we choose $M \in \mathcal{N}_0$ and a partition of $[a, b)$ into intervals $[u_i, u_{i+1})$

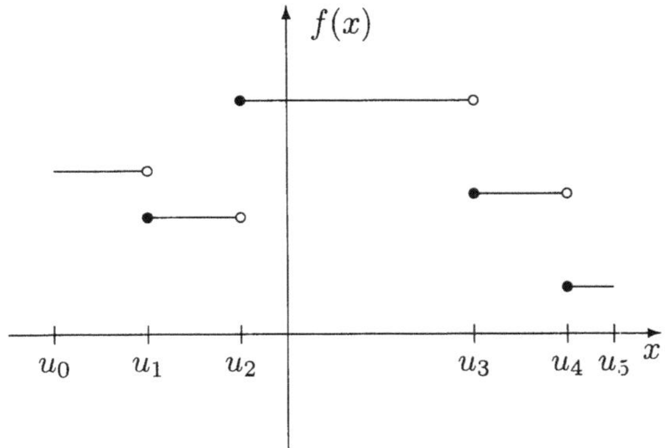

Figure 14.1. Examples for a function in $S_{u,M}([u_0, u_K))$ for $M = 0$.

$(i = 0, \ldots, K - 1)$ (where $a = u_0 < u_1 < \cdots < u_K = b)$. Then we define the corresponding spline space as the set of all functions $f : [a, b) \to \mathcal{R}$ which are $M - 1$ times continuously differentiable on $[a, b)$ and are equal to a polynomial of degree M or less on each set $[u_i, u_{i+1})$ $(i = 0, \ldots, K - 1)$.

Definition 14.1. *Let $M \in \mathcal{N}_0$ and $u_0 < u_1 < \cdots < u_K$. Set $u = \{u_j\}_{j=0,\ldots,K}$. We define the spline space $S_{u,M}([u_0, u_K))$ as*

$$S_{u,M}([u_0, u_K))$$
$$= \Big\{ f : [u_0, u_K) \to \mathcal{R} \; : \; \text{there exist polynomials } p_0, \ldots, p_{K-1} \text{ of}$$

degree M or less such that $f(x) = p_i(x)$

for $x \in [u_i, u_{i+1})$ $(i = 0, \ldots, K - 1)$

and if $M - 1 \geq 0$ then f is $M - 1$ times continuously

differentiable on $[u_0, u_K) \Big\}$.

u is called the **knot** **vector** *and M is called the* **degree** *of the* **spline** *space $S_{u,M}([u_0, u_K))$.*

Note that 0 times continuously differentiable functions are simply continuous functions.

Example 14.1. *The functions in $S_{u,0}([u_0, u_K))$ are piecewise constant (see Figure 14.1).*

Example 14.2. *The functions in $S_{u,1}([u_0, u_K))$ are piecewise linear and continuous on $[u_0, u_K)$ (see Figure 14.2).*

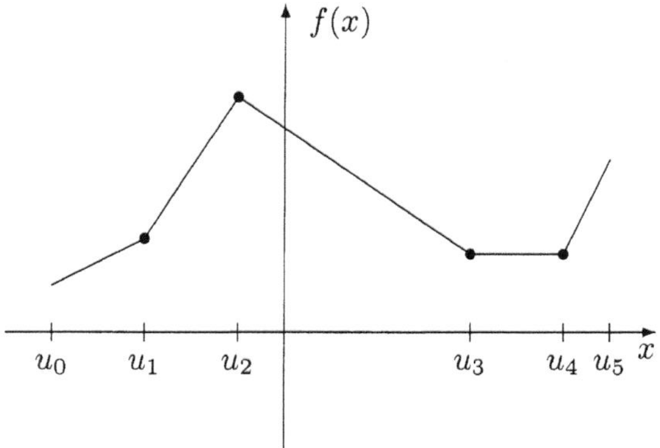

Figure 14.2. Example for a function in $S_{u,M}([u_0, u_K))$ for $M = 1$.

Clearly, $S_{u,M}([u_0, u_K))$ is a linear vector space. The next lemma presents a basis of this linear vector space. There we will use the notation

$$(x - u)_+^M = \begin{cases} (x - u)^M & \text{if } x \geq u, \\ 0 & \text{if } x < u, \end{cases}$$

and the convention $0^0 = 1$.

Lemma 14.1. *Let $M \in \mathcal{N}_0$ and $u_0 < u_1 < \cdots < u_K$. Then the set of functions*

$$\{1, x, \ldots, x^M\} \cup \{(x - u_j)_+^M \; : \; j = 1, \ldots, K - 1\} \tag{14.1}$$

is a basis of $S_{u,M}([u_0, u_K))$, i.e., for each $f \in S_{u,M}([u_0, u_K))$ there exist unique $a_0, \ldots, a_M, b_1, \ldots, b_{K-1} \in \mathcal{R}$ such that

$$f(x) = \sum_{i=0}^{M} a_i x^i + \sum_{j=1}^{K-1} b_j (x - u_j)_+^M \quad (x \in [u_0, u_K)). \tag{14.2}$$

Observe that Lemma 14.1 implies that the vector space dimension of $S_{u,M}([u_0, u_K))$ is equal to $(M + 1) + (K - 1) = M + K$.

PROOF OF LEMMA 14.1. Let us first observe that the functions in (14.1) are contained in $S_{u,M}([u_0, u_K))$. Indeed, one has

$$\frac{\partial^k}{\partial x^k}\left((x - u_j)^M\right)\Big|_{x=u_j} = M \cdot (M-1) \cdots (M-k+1) \cdot \left((x - u_j)^{M-k}\right)\Big|_{x=u_j} = 0$$

for $k = 0, \ldots, M - 1$, which implies that $(x - u_j)_+^M$ is $M - 1$ times continuously differentiable $(j = 1, \ldots, K - 1)$.

Next we will show that the functions in (14.1) are linearly independent. Let $a_0, \ldots, a_M, b_1, \ldots, b_{K-1} \in \mathcal{R}$ be arbitrary and assume

$$\sum_{i=0}^{M} a_i x^i + \sum_{j=1}^{K-1} b_j (x - u_j)_+^M = 0 \quad (x \in [u_0, u_K)). \tag{14.3}$$

For $x \in [u_0, u_1)$ one has $(x - u_j)_+^M = 0$ $(j = 1, \ldots, K-1)$, thus (14.3) implies

$$\sum_{i=0}^{M} a_i x^i = 0 \quad (x \in [u_0, u_1)). \tag{14.4}$$

Because $1, x, \ldots, x^M$ are linearly independent on each set which contains at least $M + 1$ different points it follows that $a_0 = a_1 = \cdots = a_M = 0$. This, together with (14.3), implies

$$\sum_{j=1}^{K-1} b_j (x - u_j)_+^M = 0 \quad (x \in [u_0, u_K)). \tag{14.5}$$

Setting successively $x = \frac{u_j + u_{j+1}}{2}$ $(j = 1, \ldots, K-1)$ in (14.5) one gets $b_j = 0$ for $j = 1, \ldots, K - 1$ because

$$\left(\frac{u_j + u_{j+1}}{2} - u_k \right)_+^M = 0 \quad \text{for } k > j.$$

It remains to show that for each $f \in S_{u,M}([u_0, u_K))$ there exists some $a_0, \ldots, a_M, b_1, \ldots, b_{K-1} \in \mathcal{R}$ such that (14.2) holds. Therefore we show by induction that for each $k \in \{0, \ldots, K - 1\}$ there exists some $a_0, \ldots, a_M, b_1, \ldots, b_k \in \mathcal{R}$ such that

$$f(x) = \sum_{i=0}^{M} a_i x^i + \sum_{j=1}^{k} b_j (x - u_j)_+^M \quad (x \in [u_0, u_{k+1})). \tag{14.6}$$

For $k = 0$ this clearly holds because f is a polynomial of degree M, or less, on $[u_0, u_1)$. Assume that (14.6) holds for some $k < K - 1$. Then g, defined by

$$g(x) = f(x) - \sum_{i=0}^{M} a_i x^i - \sum_{j=1}^{k} b_j (x - u_j)_+^M,$$

satisfies

$$g(x) = 0 \quad (x \in [u_0, u_{k+1})) \tag{14.7}$$

and is $M - 1$ times continuously differentiable at u_{k+1} (because of $f \in S_{u,M}([u_0, u_K)))$. Thus

$$\frac{\partial^i g(u_{k+1})}{\partial x^i} = 0 \quad \text{for } i = 0, \ldots, M - 1. \tag{14.8}$$

Furthermore, because $f \in S_{u,M}([u_0, u_K))$, g is equal to a polynomial of degree M or less on $[u_{k+1}, u_{k+2})$. Thus there exist $c_0, \ldots, c_M \in \mathcal{R}$ such that

$$g(x) = \sum_{i=0}^{M} c_i (x - u_{k+1})^i \quad (x \in [u_{k+1}, u_{k+2})).$$

Since

$$\frac{\partial^j g(u_{k+1})}{\partial x^j} = \sum_{i=j}^{M} c_i \cdot i \cdot (i-1) \cdots (i-j+1) \cdot (x - u_{k+1})^{i-j} \Big|_{x=u_{k+1}} = j! c_j$$

it follows from (14.8) that $c_0 = \cdots = c_{M-1} = 0$, thus

$$g(x) - c_M (x - u_{k+1})^M = 0 \quad (x \in [u_{k+1}, u_{k+2})).$$

This together with (14.7) implies (14.6) for $k+1$ and $b_{k+1} = c_M$. $\qquad\square$

We can use the basis from Lemma 14.1 to implement splines on a computer. To do this we represent splines as linear combinations of the basis functions and store in the computer only the coefficients of these linear combinations.

Unfortunately, the basis from Lemma 14.1 does not provide efficient spline representation on a computer. For example, because the supports of the basis functions are unbounded, evaluation of

$$f(x) = \sum_{i=0}^{M} a_i x^i + \sum_{j=1}^{K-1} b_j (x - u_j)_+^M$$

at some $x \in \mathcal{R}$ for given $a_0, \ldots, a_M, b_1, \ldots, b_{K-1}$ requires evaluation of nearly all the $(K+M)$ basis functions and thus the amount of time to do this increases with K. Therefore one prefers a basis where the support of the basis functions is as small as possible. Next we will introduce such a basis, called a B-spline basis.

The definition of B-splines depends on $2M$ additional knots $u_{-M}, \ldots,$ $u_{-1}, u_{K+1}, \ldots, u_{K+M}$ which satisfy $u_{-M} \le u_{-M+1} \le \cdots \le u_{-1} \le u_0$ and $u_K \le u_{K+1} \le \cdots \le u_{K+M}$. We will use again the notation u for the extended knot vector $u = \{u_j\}_{j=-M,\ldots,K+M}$.

Definition 14.2. *Let $M \in \mathcal{N}_0$, $u_{-M} \le \cdots \le u_{K+M}$, and set $u = \{u_j\}_{j=-M,\ldots,K+M}$. Then the B-splines $B_{j,l,u}$ of degree l and with knot vector u are recursively defined by*

$$B_{j,0,u}(x) = \begin{cases} 1 & \text{if } u_j \le x < u_{j+1}, \\ 0 & \text{if } x < u_j \text{ or } x \ge u_{j+1}, \end{cases} \tag{14.9}$$

for $j = -M, \ldots, K+M-1$, $x \in \mathcal{R}$, and

$$B_{j,l+1,u}(x) = \frac{x - u_j}{u_{j+l+1} - u_j} B_{j,l,u}(x) + \frac{u_{j+l+2} - x}{u_{j+l+2} - u_{j+1}} B_{j+1,l,u}(x) \tag{14.10}$$

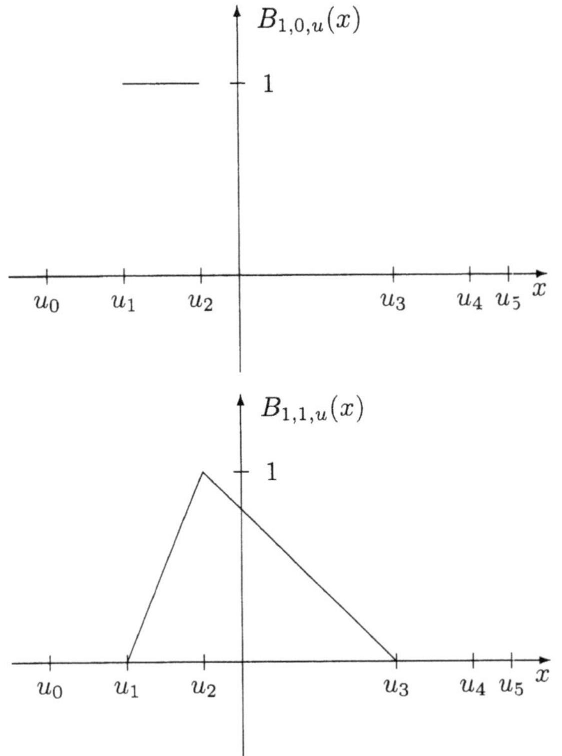

Figure 14.3. Examples for B-splines of degree 0 and 1.

for $j = -M, \ldots, K + M - l - 2$, $l = 0, \ldots, M - 1$, $x \in \mathcal{R}$.

In (14.10) $u_{j+l+1} - u_j = 0$ (or $u_{j+l+2} - u_{j+1} = 0$) implies $B_{j,l,u}(x) = 0$ (or $B_{j+1,l,u}(x) = 0$). We have used the convention $\frac{0}{0} = 0$.

Example 14.3. *For $M = 0$, $B_{j,0,u}$ is the indicator function of the set $[u_j, u_{j+1})$. Clearly, $\{B_{j,0,u} : j = 0, \ldots, K - 1\}$ is a basis of $S_{u,0}([u_0, u_K))$.*

Example 14.4. *For $M = 1$ one gets*

$$B_{j,1,u}(x) = \begin{cases} \dfrac{x - u_j}{u_{j+1} - u_j} & \text{for } u_j \le x < u_{j+1}, \\ \dfrac{u_{j+2} - x}{u_{j+2} - u_{j+1}} & \text{for } u_{j+1} \le x < u_{j+2}, \\ 0 & \text{if } x < u_j \text{ or } x \ge u_{j+2}. \end{cases}$$

The B-spline $B_{j,1,u}$ is equal to one at u_{j+1}, zero at all the other knots, and is linear between consecutive knots (the so-called hat function). For a special knot sequence this basis is illustrated in Figure 14.4. There $u_{-1} = u_0$, $K = 5$, $u_6 = u_5$, and $B_{j,1,u}$ is the function with support $[u_j, u_{j+2}]$ ($j = -1, \ldots, 4$). Observe that the support of the B-splines in Figure 14.4 is

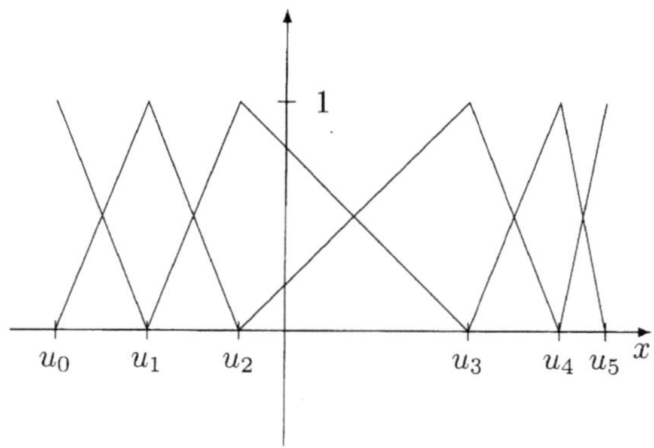

Figure 14.4. B-spline basis for degree $M = 1$.

much smaller than the support of the basis described in Lemma 14.1. It is easy to see that $\{B_{j,1,u} \; : \; j = -1, \ldots, K-1\}$ is a basis for $S_{u,1}([u_0, u_K))$.

Before we show for arbitrary M that $\{B_{j,M,u} \; : \; j = -M, \ldots, K-1\}$ is a basis for $S_{u,M}([u_0, u_K))$ we prove some useful properties of the B-splines.

Lemma 14.2. *Let $M \in \mathcal{N}_0$ and $u_{-M} \leq \cdots \leq u_0 < \cdots < u_K \leq \cdots \leq u_{K+M}$.*
(a)

$$B_{j,M,u}(x) = 0 \quad \text{for } x \notin [u_j, u_{j+M+1}) \tag{14.11}$$

for $j \in \{-M, \ldots, K-1\}$.
(b)

$$B_{j,M,u}(x) \geq 0 \tag{14.12}$$

for $x \in \mathcal{R}$ and $j \in \{-M, \ldots, K-1\}$.
(c)

$$\sum_{j=-M}^{K-1} a_j B_{j,M,u}(x)$$

$$= \sum_{j=-(M-1)}^{K-1} \left\{ a_j \frac{x - u_j}{u_{j+M} - u_j} + a_{j-1} \frac{u_{j+M} - x}{u_{j+M} - u_j} \right\} B_{j,M-1,u}(x)$$

$$\tag{14.13}$$

for $x \in [u_0, u_K)$, $a_{-M}, \ldots, a_{K-1} \in \mathcal{R}$, and $M > 0$.

PROOF. Equation (14.11) follows easily from (14.9) and (14.10) by induction on M. Then (14.12) follows from (14.11), (14.9), and (14.10). It remains

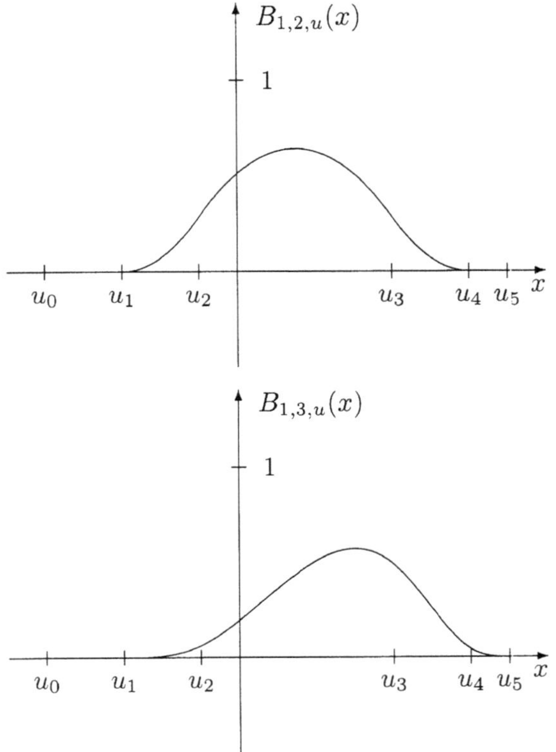

Figure 14.5. Examples for B-splines of degree 2 and 3.

to show (14.13): For $x \in [u_0, u_K)$ one has

$$\sum_{j=-M}^{K-1} a_j B_{j,M,u}(x)$$

$$\overset{(14.10)}{=} \sum_{j=-M}^{K-1} a_j \left(\frac{x - u_j}{u_{j+M} - u_j} B_{j,M-1,u}(x) \right.$$

$$\left. + \frac{u_{j+M+1} - x}{u_{j+M+1} - u_{j+1}} B_{j+1,M-1,u}(x) \right)$$

$$= \sum_{j=-(M-1)}^{K-1} \left(a_j \frac{x - u_j}{u_{j+M} - u_j} + a_{j-1} \frac{u_{j+M} - x}{u_{j+M} - u_j} \right) B_{j,M-1,u}(x)$$

$$+ a_{-M} \frac{x - u_{-M}}{u_0 - u_{-M}} B_{-M,M-1,u}(x)$$

$$+ a_{K-1} \frac{u_{K+M} - x}{u_{K+M} - u_K} B_{K,M-1,u}(x)$$

$$= \sum_{j=-(M-1)}^{K-1} \left(a_j \frac{x - u_j}{u_{j+M} - u_j} + a_{j-1} \frac{u_{j+M} - x}{u_{j+M} - u_j} \right) B_{j,M-1,u}(x)$$

where the last equality follows from (14.11). $\square$

Because of (14.11) the B-splines have bounded support. Nevertheless, the recursive definition of the B-splines seems very inconvenient for representing spline functions with the aid of these B-splines on a computer. We will show that this is not true by explaining an easy way to evaluate a linear combination of B-splines at a given point x (the so-called de Boor algorithm).

Assume that we are given the coefficients $\{a_j \ : \ j = -M, \ldots, K-1\}$ of a linear combination

$$f = \sum_{j=-M}^{K-1} a_j B_{j,M,u}$$

of B-splines and that we want to evaluate this function f at some given point $x \in [u_0, u_K)$. Then setting $a_{j,M} := a_j$ it follows from (14.13) that one has

$$f(x) \quad = \quad \sum_{j=-M}^{K-1} a_{j,M} B_{j,M,u}(x) = \sum_{j=-(M-1)}^{K-1} a_{j,M-1} B_{j,M-1,u}(x)$$

$$= \quad \cdots = \sum_{j=0}^{K-1} a_{j,0} B_{j,0,u}(x),$$

where (depending on x) the $a_{j,l}$'s are recursively defined by

$$a_{j,l-1} = a_{j,l} \frac{x - u_j}{u_{j+l} - u_j} + a_{j-1,l} \frac{u_{j+l} - x}{u_{j+l} - u_j} \tag{14.14}$$

$(j \in \{-(l-1), \ldots, K-1\}, \ l \in \{1, \ldots, M\})$.

Now let j_0 be such that $u_{j_0} \leq x < u_{j_0+1}$. Then, because of (14.9), one gets

$$f(x) = a_{j_0,0}.$$

Thus all that one has to do is to use (14.14) to compute $a_{j_0,0}$. To do this it suffices to start with $a_{j_0-M,M} = a_{j_0-M}, \ldots, a_{j_0,M} = a_{j_0}$ and to successively use (14.14) to compute $a_{j_0-l,l}, \ldots, a_{j_0,l}$ for $l = M-1, \ldots, 0$ (cf. Figure 14.6). The number of operations needed depends only on M and not on K – this is the great advantage of the B-spline basis compared with the basis of Lemma 14.1.

We will show next that $\{B_{j,M,u} \ : \ j = -M, \ldots, K-1\}$ is indeed a basis of $S_{u,M}([u_0, u_K))$. For this we will need the following lemma:

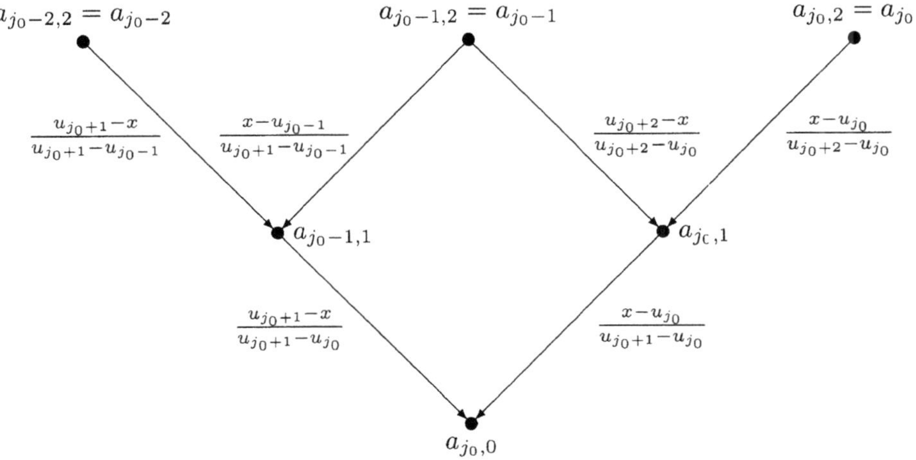

Figure 14.6. Computation of $\sum_{j=0}^{K} a_j B_{j,2,u}(x)$ for $x \in [u_{j_0}, u_{j_0+1})$.

Lemma 14.3. *Let $M \in \mathcal{N}_0$ and $u_{-M} \leq \cdots \leq u_0 < \cdots < u_K \leq \cdots \leq u_{K+M}$. For $j \in \{-M, ..., K-1\}$ and $t \in \mathcal{R}$ set*

$$\psi_{j,M}(t) = (u_{j+1} - t) \cdot \ldots \cdot (u_{j+M} - t).$$

Then

$$(x - t)^M = \sum_{j=-M}^{K-1} \psi_{j,M}(t) B_{j,M,u}(x) \tag{14.15}$$

for all $x \in [u_0, u_K)$, $t \in \mathcal{R}$.

PROOF. For $l \in \mathcal{N}$, $t \in \mathcal{R}$, set

$$\psi_{j,l}(t) = (u_{j+1} - t) \cdot \ldots \cdot (u_{j+l} - t)$$

and set $\psi_{j,0}(t) = 1$ $(t \in \mathcal{R})$. We will show

$$\sum_{j=-l}^{K-1} \psi_{j,l}(t) B_{j,l,u}(x) = (x - t) \sum_{j=-(l-1)}^{K-1} \psi_{j,l-1}(t) B_{j,l-1,u}(x) \tag{14.16}$$

for $x \in [u_0, u_K)$, $t \in \mathcal{R}$, and $l \in \{1, \ldots, M\}$. From this one obtains the assertion by

$$\sum_{j=-M}^{K-1} \psi_{j,M}(t) B_{j,M,u}(x) \overset{(14.16)}{=} (x - t) \sum_{j=-(M-1)}^{K-1} \psi_{j,M-1}(t) B_{j,M-1,u}(x)$$

$$\overset{(14.16)}{=} \quad \ldots$$

$$\overset{(14.16)}{=} \quad (x-t)^M \sum_{j=0}^{K-1} \psi_{j,0}(t) B_{j.0.u}(x)$$

$$\overset{(14.9)}{=} \quad (x-t)^M.$$

So it suffices to prove (14.16).

It follows from (14.13) that

$$\sum_{j=-l}^{K-1} \psi_{j,l}(t) B_{j,l,u}(x)$$

$$= \sum_{j=-(l-1)}^{K-1} \left(\psi_{j,l}(t)\frac{x-u_j}{u_{j+l}-u_j} + \psi_{j-1,l}(t)\frac{u_{j+l}-x}{u_{j+l}-u_j} \right) B_{j,l-1,u}(x).$$

Therefore (14.16) follows from

$$\psi_{j,l}(t)\frac{x-u_j}{u_{j+l}-u_j} + \psi_{j-1,l}(t)\frac{u_{j+l}-x}{u_{j+l}-u_j} = (x-t)\psi_{j,l-1}(t) \qquad (14.17)$$

$(x,t \in \mathcal{R}, l \in \mathcal{N}, j = -(l-1), \ldots, K-1)$, which we will show next. For fixed $t \in \mathcal{R}$ both sides in (14.17) are linear polynomials in x. Therefore it suffices to show (14.17) for $x = u_j$ and $x = u_{j+l}$. For $x = u_j$ one gets, for the left-hand side of (14.17),

$$\psi_{j,l}(t) \cdot 0 + \psi_{j-1,l}(t) \cdot 1 = (u_j-t)\cdots(u_{j+l-1}-t) = (u_j-t)\cdot\psi_{j,l-1}(t),$$

and for $x = u_{j+l}$ one gets for the left-hand side of (14.17),

$$\psi_{j,l}(t) \cdot 1 + \psi_{j-1,l}(t) \cdot 0 = (u_{j+1}-t)\cdot\ldots\cdot(u_{j+l}-t) = (u_{j+l}-t)\cdot\psi_{j,l-1}(t).$$

This implies (14.17), thus Lemma 14.3 is proved.   □

Theorem 14.1. *For $M \in \mathcal{N}_0$ and $u_{-M} \leq \cdots \leq u_0 < \cdots < u_K \leq \cdots \leq u_{K+M}$, $\{B_{j,M,u} : j = -M, \ldots, K-1\}$ restricted to $[u_0, u_K)$ is a basis of $S_{u,M}([u_0, u_K))$.*

PROOF. We will show

$$S_{u,M}([u_0, u_K)) \subseteq span\,\{B_{j,M,u} : j = -M, \ldots, K-1\}. \qquad (14.18)$$

By Lemma 14.1, $S_{u,M}([u_0, u_K))$ is a linear vector space of dimension $K+M$ and obviously $span\,\{B_{j,M,u} : j = -M, \ldots, K-1\}$ is a linear vector space of dimension less than or equal to $K+M$, therefore it follows from (14.18) that the two vector spaces are equal and that $\{B_{j,M,u} : j = -M, \ldots, K-1\}$ is a basis of them. Thus it suffices to prove (14.18). Because of Lemma 14.1, (14.18) follows from

$$p \in span\,\{B_{j,M,u} : j = -M, \ldots, K-1\} \qquad (14.19)$$

for each polynomial p of degree M or less and from

$$(x-u_k)_+^M \in span\,\{B_{j.M,u} : j = -M, \ldots, K-1\} \qquad (14.20)$$

for each $k = 1, \ldots, K - 1$. If $t_0, \ldots, t_M \in \mathcal{R}$ are pairwise distinct then each polynomial of degree M or less can be expressed as a linear combination of the polynomials $(x - t_j)^M$ $(j = 0, \ldots, M)$. Thus (14.19) follows from Lemma 14.3. We now show

$$(x - u_k)_+^M = \sum_{j=k}^{K-1} \psi_{j,M}(u_k) B_{j,M,u}(x) \quad (x \in [u_0, u_K), k \in \{1, \ldots, K-1\})$$

(14.21)

which implies (14.20).

For $x < u_k$ one has

$$(x - u_k)_+^M = 0 = \sum_{j=k}^{K-1} \psi_{j,M}(u_k) \cdot 0 \stackrel{(14.11)}{=} \sum_{j=k}^{K-1} \psi_{j,M}(u_k) B_{j,M,u}(x).$$

For $x \geq u_k$ one has

$$(x - u_k)_+^M = (x - u_k)^M \stackrel{Lemma\ 14.3}{=} \sum_{j=-M}^{K-1} \psi_{j,M}(u_k) B_{j,M,u}(x)$$

$$\stackrel{(14.11)}{=} \sum_{j=k-M}^{K-1} \psi_{j,M}(u_k) B_{j,M,u}(x) = \sum_{j=k}^{K-1} \psi_{j,M}(u_k) B_{j,M,u}(x),$$

because for $j \in \{k - M, \ldots, k - 1\}$ one has

$$\psi_{j,M}(u_k) = (u_{j+1} - u_k) \cdot \ldots \cdot (u_{j+M} - u_k) = 0.$$

$\square$

In Definition 14.2 the B-splines $B_{j,M,u}$ are defined on whole $\mathcal{R}$. Therefore also $span\{B_{j,M,u} : j = -M, \ldots, K - 1\}$ is a space of functions defined on whole $\mathcal{R}$. By Theorem 14.1 the restriction of this space of functions to $[u_0, u_K)$ is equal to $S_{u,M}([u_0, u_K))$.

To define the least squares estimates we need spaces of functions defined on whole $\mathcal{R}$, therefore we introduce

Definition 14.3. *For $M \in \mathcal{N}_0$ and $u_{-M} \leq \cdots \leq u_0 < \cdots < u_K \leq \cdots \leq u_{K+M}$ the spline space $S_{u,M}$ of functions $f : \mathcal{R} \to \mathcal{R}$ is defined by*

$$S_{u,M} = span\{B_{j,M,u} : j = -M, \ldots, K - 1\}.$$

Here M is called the **degree** *and $u = \{u_j\}_{j=-M,\ldots,K-1}$ is called the* **knot sequence** *of $S_{u,M}$.*

Remark. (a) If we restrict all the functions in $S_{u,M}$ on $[u_0, u_K)$, then we get the spline space $S_{u,M}([u_0, u_K))$.

(b) While the spline space $S_{u,M}([u_0, u_K))$ is independent of the knots $u_{-M}, \ldots, u_{-1}$ and $u_{K+1}, \ldots, u_{K+M}$, the B-spline basis and therefore also $S_{u,M}$ depends on the knots.

(c) The functions in $S_{u,M}$ are equal to a polynomial of degree less than or equal to M on each set $(-\infty, u_{-M}), [u_{-M}, u_{-M+1}), \ldots, [u_{K+M-1}, u_{K+M}),$

$[u_{K+M}, \infty)$. They are zero outside $[u_{-M}, u_{K+M})$, $M - 1$ times continuously differentiable on $[u_0, u_K)$, and in the case $u_{-M} < \cdots < u_0$ and $u_K < \cdots < u_{K+M}$ even $M - 1$ times continuously differentiable on $[u_{-M}, u_{K+M})$.

We will often need the following property of the B-splines:

Lemma 14.4. *Let* $M \in \mathcal{N}_0$ *and* $u_{-M} \leq \cdots \leq u_0 < \cdots < u_K \leq \cdots \leq u_{K+M}$. *Then one has*

$$\sum_{j=-M}^{K-1} B_{j,M,u}(x) = 1 \quad \text{for } x \in [u_0, u_K). \tag{14.22}$$

PROOF. Differentiating (14.15) M times with respect to t one gets

$$
\begin{aligned}
M! \cdot (-1)^M &= \frac{\partial^M}{\partial t^M}(x - t)^M \\
&= \sum_{j=-M}^{K-1} \frac{\partial^M}{\partial t^M}\left((u_{j+1} - t) \cdot \ldots \cdot (u_{j+M} - t)\right) B_{j,M,u}(x) \\
&= \sum_{j=-M}^{K-1} M! \cdot (-1)^M B_{j,M,u}(x),
\end{aligned}
$$

which implies the assertion. $\qquad\square$

Lemma 14.5. *Let* $0 \leq i \leq K - 1$. *Then*

$$\{B_{i-M,M,u}, \ldots, B_{i,M,u}\}$$

is a basis of $\text{span}\{1, x, \ldots, x^M\}$ *on* $[u_i, u_{i+1})$, *i.e.,*

$$\text{span}\{B_{i-M,M,u}, \ldots, B_{i,M,u}\} = \text{span}\{1, x, \ldots, x^M\} \quad \text{on } [u_i, u_{i+1}) \tag{14.23}$$

and

$$B_{i-M,M,u}, \ldots, B_{i,M,u} \text{ are linearly independent on } [u_i, u_{i+1}). \tag{14.24}$$

PROOF. Differentiating (14.15) $(M - l)$ times with respect to t one gets

$$(-1)^{M-l} M \cdot (M - 1) \cdot \ldots \cdot (l + 1)(x - t)^l = \sum_{j=-M}^{K-1} \frac{\partial^{M-l} \psi_{j,M}}{\partial t^{M-l}}(t) \cdot B_{j,M,u}(x).$$

If $x \in [u_i, u_{i+1})$, then (14.11) implies

$$B_{j,M,u}(x) = 0 \quad \text{for } j \notin \{i - M, \ldots, i\}.$$

Hence,

$$(-1)^{M-l} M \cdot (M - 1) \cdot \ldots \cdot (l + 1)x^l = \sum_{j=i-M}^{i} \frac{\partial^{M-l} \psi_{j,M}}{\partial t^{M-l}}(0) \cdot B_{j,M,u}(x),$$

which proves

$$1, x, \ldots, x^M \in span\{B_{i-M,M,u}, \ldots, B_{i,M,u}\} \quad \text{on } [u_i, u_{i+1}).$$

On the other hand, the definition of the spline spaces implies

$$B_{i-M,M,u}, \ldots, B_{i,M,u} \in span\{1, x, \ldots, x^M\} \quad \text{on } [u_i, u_{i+1}),$$

from which one concludes (14.23). Furthermore, $1, x, \ldots, x^M$ are linearly independent on $[u_i, u_{i+1})$, thus the dimension of $span\{B_{i-M,M,u}, \ldots, B_{i,M,u}\}$ is equal to the number of functions in $\{B_{i-M,M,u}, \ldots, B_{i,M,u}\}$, which implies (14.24). $\qquad\square$

By definition, the derivative of a spline function f of degree M is a spline function of degree $M - 1$ with respect to the same knot vector. Hence the derivative f' of a linear combination f of B-splines of degree M can be represented by a linear combination of B-splines of degree $M - 1$. Our next lemma shows that it is easy to compute the coefficients of f' given the coefficients of f.

Lemma 14.6. *(a) For all $j \in \{-M, \ldots, K - 1\}$ and $x \in [u_0, u_K)$,*

$$\frac{\partial}{\partial x} B_{j,M,u}(x) = \frac{M}{u_{j+M} - u_j} B_{j,M-1,u}(x) - \frac{M}{u_{j+M+1} - u_{j+1}} B_{j+1,M-1,u}(x).$$

(b) For all $x \in [u_0, u_K)$,

$$\frac{\partial}{\partial x} \sum_{j=-M}^{K-1} a_j \cdot B_{j,M,u}(x) = \sum_{j=-(M-1)}^{K-1} \frac{M}{u_{j+M} - u_j}(a_j - a_{j-1}) B_{j,M-1,u}(x).$$

PROOF. Because of $\frac{\partial}{\partial x} B_{j,M,u} \in S_{u,M-1}$ we get

$$\frac{\partial}{\partial x} B_{j,M,u}(x) = \sum_{i=-(M-1)}^{K-1} \alpha_{i,j} B_{i,M-1,u}(x) \quad (x \in [u_0, u_K))$$

for some $\alpha_{-(M-1),j}, \ldots, \alpha_{K-1,j} \in \mathcal{R}$. Let $k \leq j - 1$ or $k \geq j + M + 1$. Then

$$B_{j,M,u}(x) = 0 \quad \text{for all } x \in [u_k, u_{k+1}),$$

which implies

$$0 = \frac{\partial}{\partial x} B_{j,M,u}(x) = \sum_{i=-(M-1)}^{K-1} \alpha_{i,j} B_{i,M-1,u}(x) = \sum_{i=k-(M-1)}^{k} \alpha_{i,j} B_{i,M-1,u}(x)$$

for all $x \in [u_k, u_{k+1})$. From this and Lemma 14.5 we conclude that

$$\alpha_{k-(M-1),j} = \cdots = \alpha_{k,j} = 0$$

if $k \leq j - 1$ or $k \geq j + M + 1$, hence

$$\alpha_{-(M-1),j} = \cdots = \alpha_{j-1,j} = \alpha_{j+2,j} = \cdots = \alpha_{K-1,j} = 0$$

and therefore

$$\frac{\partial}{\partial x} B_{j,M,u}(x) = \alpha_j B_{j,M-1,u}(x) + \beta_j B_{j+1,M-1,u}(x) \quad (x \in [u_0, u_K))$$

for some $\alpha_j, \beta_j \in \mathcal{R}$. It remains to determine the explicit form of α_j and β_j.

Because of

$$0 \;=\; \frac{\partial}{\partial x} 1 = \frac{\partial}{\partial x} \sum_{j=-M}^{K-1} B_{j,M,u}(x)$$

$$=\; \sum_{j=-M}^{K-1} \left(\alpha_j B_{j,M-1,u}(x) + \beta_j B_{j+1,M-1,u}(x) \right)$$

$$=\; \sum_{j=-(M-1)}^{K-1} \left(\alpha_j + \beta_{j-1} \right) B_{j,M-1,u}(x)$$

$$\text{(because } B_{-M,M-1,u}(x) = B_{K,M-1,u}(x) = 0 \text{ for } x \in [u_0, u_K))$$

for all $x \in [u_0, u_K)$, we get

$$\beta_j = -\alpha_{j+1} \quad \text{for } j = -M, \ldots, K-2.$$

By Lemma 14.3,

$$M \cdot x^{M-1} \;=\; \frac{\partial}{\partial x} x^M = \frac{\partial}{\partial x} \sum_{j=-M}^{K-1} \psi_{j,M}(0) B_{j,M,u}(x)$$

$$=\; \sum_{j=-M}^{K-1} \psi_{j,M}(0) \left(\alpha_j B_{j,M-1,u}(x) - \alpha_{j+1} B_{j+1,M-1,u}(x) \right)$$

$$=\; \sum_{j=-(M-1)}^{K-1} \alpha_j \left(\psi_{j,M}(0) - \psi_{j-1,M}(0) \right) B_{j,M-1,u}(x)$$

$$\text{(because of } B_{-M,M-1,u}(x) = B_{K,M-1,u}(x) = 0 \text{ for } x \in [u_0, u_K))$$

for all $x \in [u_0, u_K)$. On the other hand, by applying Lemma 14.3 with degree $M-1$ instead of M, we get

$$x^{M-1} = \sum_{j=-(M-1)}^{K-1} \psi_{j,M-1}(0) B_{j,M-1,u}(x).$$

Hence,

$$M \cdot \psi_{j,M-1}(0) = \alpha_j \left(\psi_{j,M}(0) - \psi_{j-1,M}(0) \right),$$

which implies

$$\alpha_j \;=\; \frac{M \cdot \psi_{j,M-1}(0)}{\psi_{j,M}(0) - \psi_{j-1,M}(0)}$$

$$= \frac{M \cdot u_{j+1} \cdot \ldots \cdot u_{j+M-1}}{u_{j+1} \cdot \ldots \cdot u_{j+M} - u_j \cdot \ldots \cdot u_{j+M-1}}$$

$$= \frac{M}{u_{j+M} - u_j}$$

and

$$\beta_j = -\alpha_{j+1} = -\frac{M}{u_{j+M+1} - u_{j+1}}.$$

This proves (a). The assertion of (b) follows directly from (a). $\qquad\square$

14.2 Consistency

In this section we will investigate the question: How should one choose the degree and the knot sequence of a univariate spline space depending on the data in order to get universally consistent least squares estimates?

In order to give an answer to this question, we generalize Theorem 13.1, which deals with data-dependent histogram estimates, to least squares estimates using data-dependent spline spaces. One of the conditions in Theorem 13.1 was that the measure of those cells in the partition, for which the diameter does not shrink to zero, converges to zero (cf. (13.10)). The condition (14.30) below generalizes this to data-dependent spline spaces.

Theorem 14.2. *For $n \in \mathcal{N}$ let $M_{max}(n) \in \mathcal{N}$, $K_{max}(n) \in \mathcal{N}$, and $\beta_n \in \mathcal{R}_+$. Depending on D_n choose $M \in \mathcal{N}_0$, $K \in \mathcal{N}$, and $u_{-M}, \ldots, u_{K+M} \in \mathcal{R}$ such that $M \leq M_{max}(n)$, $K \leq K_{max}(n)$, and $u_{-M} \leq \cdots \leq u_0 < \cdots < u_K \leq \cdots \leq u_{K+M}$. Define the estimate m_n by*

$$\tilde{m}_n = \arg \min_{f \in \mathcal{F}_n} \frac{1}{n} \sum_{j=1}^{n} |f(X_j) - Y_j|^2 \tag{14.25}$$

and

$$m_n(x) = T_{\beta_n} \tilde{m}_n(x) \tag{14.26}$$

with $\mathcal{F}_n = S_{u,M}$.
(a) Assume that

$$\beta_n \to \infty \quad (n \to \infty), \tag{14.27}$$

$$\frac{(K_{max}(n) \cdot M_{max}(n) + M_{max}(n)^2)\beta_n^4 \log(n)}{n} \to 0 \quad (n \to \infty) \tag{14.28}$$

and

$$\frac{\beta_n^4}{n^{1-\delta}} \to 0 \quad (n \to \infty) \tag{14.29}$$

for some $\delta > 0$. If, in addition, the distribution μ of X satisfies

$$\mu\left(\left(\left\{(-\infty, u_0) \cup \bigcup_{\substack{k=1,\ldots,K, \\ u_k - u_{k-M-1} > \gamma}} [u_{k-1}, u_k) \cup [u_K, \infty)\right\} \cap [-L, L]\right)\right) \to 0 \tag{14.30}$$

$(n \to \infty)$ *a.s. for each* $L, \gamma > 0$, *or the empirical distribution* μ_n *of* $X_1, \ldots, X_n$ *satisfies*

$$\mu_n\left(\left(\left\{(-\infty, u_0) \cup \bigcup_{\substack{k=1,\ldots,K, \\ u_k - u_{k-M-1} > \gamma}} [u_{k-1}, u_k) \cup [u_K, \infty)\right\} \cap [-L, L]\right)\right) \to 0 \tag{14.31}$$

$(n \to \infty)$ *a.s. for each* $L, \gamma > 0$, *then, for* $\mathbf{E}Y^2 < \infty$,

$$\int |m_n(x) - m(x)|^2 \mu(dx) \to 0 \quad (n \to \infty) \quad a.s.$$

(b) Assume that (14.27) and (14.28) hold. If, in addition, the distribution μ of X satisfies

$$\mathbf{E}\mu\left(\left(\left\{(-\infty, u_0) \cup \bigcup_{\substack{k=1,\ldots,K, \\ u_k - u_{k-M-1} > \gamma}} [u_{k-1}, u_k) \cup [u_K, \infty)\right\} \cap [-L, L]\right)\right) \to 0 \tag{14.32}$$

$(n \to \infty)$ *for each* $L, \gamma > 0$, *or the empirical distribution* μ_n *of* $X_1, \ldots, X_n$ *satisfies*

$$\mathbf{E}\mu_n\left(\left(\left\{(-\infty, u_0) \cup \bigcup_{\substack{k=1,\ldots,K, \\ u_k - u_{k-M-1} > \gamma}} [u_{k-1}, u_k) \cup [u_K, \infty)\right\} \cap [-L, L]\right)\right) \to 0 \tag{14.33}$$

$(n \to \infty)$ *for each* $L, \gamma > 0$, *then, for* $\mathbf{E}Y^2 < \infty$,

$$\mathbf{E}\int |m_n(x) - m(x)|^2 \mu(dx) \to 0 \quad (n \to \infty).$$

Remark. (a) The left-hand sides in (14.30) and (14.31) are random variables because the degree and the knot sequence depend on the data.

(b) In the case $M = 0$ the estimate m_n is a truncated histogram estimate using a data-dependent partition. In this case the assertion follows from Theorem 13.1.

If the degree and the knot sequence are chosen such that (14.30) or (14.31) hold for every distribution μ of X, then Theorem 14.2 implies that the estimate is universally consistent. Before we prove the theorem we will give examples for such choices of the degree and the knot sequence.

In the first example we consider data-independent knots.

Example 14.5. *Let $M \in \mathcal{N}_0$. Let $L_n, R_n \in \mathcal{R}$ and $K_n \in \mathcal{N}$ $(n \in \mathcal{N})$ be such that*

$$L_n \to -\infty, \quad R_n \to \infty \quad (n \to \infty) \tag{14.34}$$

and

$$\frac{R_n - L_n}{K_n} \to 0 \quad (n \to \infty). \tag{14.35}$$

Set $K = K_n$ and

$$u_k = L_n + k \cdot \frac{R_n - L_n}{K_n} \quad (k = -M, ..., K_n + M). \tag{14.36}$$

Then (14.30) holds because, for fixed $L, \gamma > 0$,

$$\left\{ (-\infty, u_0) \cup \bigcup_{k=1,...,K,\; u_k - u_{k-M-1} > \gamma} [u_{k-1}, u_k) \cup [u_K, \infty) \right\} \cap [-L, L] = \emptyset$$

for n sufficiently large (i.e., for n so large that $u_0 = L_n < -L$, $u_K = R_n > L$, and $u_k - u_{k-M-1} = (M+1) \cdot (R_n - L_n)/K_n \leq \gamma$).

In the next example we consider data-dependent knots.

Example 14.6. *Let $M \in \mathcal{N}_0$. Let $C_n, K_n \in \mathcal{N}$, $\delta_n \geq 0$ $(n \in \mathcal{N})$ be such that*

$$\delta_n \to 0 \quad (n \to \infty) \tag{14.37}$$

and

$$\frac{C_n}{n} \to 0 \quad (n \to \infty). \tag{14.38}$$

Set $K = K_n$ and choose the knots such that there are less than C_n of the $X_1, ..., X_n$ in each of the intervals $(-\infty, u_0)$ and $[u_{K_n}, \infty)$ and such that for every $k \in \{1, ..., K_n\}$ with $u_k - u_{k-M-1} > \delta_n$ there are less than C_n of the $X_1, ..., X_n$ in $[u_{k-1}, u_k)$. Then (14.31) holds.

Indeed, let $L, \gamma > 0$. Because of (14.37) we can assume w.l.o.g. that $\delta_n < \gamma$. Then $u_k - u_{k-M-1} > \gamma$ implies $\mu_n([u_{k-1}, u_k)) \leq C_n/n$, thus

$$\mu_n\left(\left\{ (-\infty, u_0) \cup \bigcup_{\substack{k=1,...,K_n, \\ u_k - u_{k-M-1} > \gamma}} [u_{k-1}, u_k) \cup [u_{K_n}, \infty) \right\} \cap [-L, L] \right)$$

$$\leq \mu_n\left((-\infty, u_0)\right) + \sum_{u_k - u_{k-M-1} > \gamma, [u_{k-1}, u_k) \cap [-L, L] \neq \emptyset} \mu_n\left([u_{k-1}, u_k)\right)$$

$$+ \mu_n\left([u_{K_n}, \infty)\right)$$

$$\leq 2\frac{C_n}{n} + (M+1)\left(\frac{2L}{\gamma} + 2\right)\frac{C_n}{n} \to 0 \quad (n \to 0)$$

because of (14.38).

Example 14.7. *Assume* $X_1, \ldots, X_n$ *are all distinct a.s. and choose each* $\lceil \frac{n}{K_n} \rceil$*th-order statistic of* $X_1, \ldots, X_n$ *as a knot. Then each sequence* $\{m_n\}_{n \in \mathcal{N}}$ *of estimators which satisfies (14.25) and (14.26) with* $\mathcal{F}_n = S_{u,M}$ *is weakly and strongly consistent for every distribution of* (X, Y) *with* X *nonatomic and* $\mathbf{E}Y^2 < \infty$, *provided that (14.27)–(14.29) hold and*

$$K_n \to \infty \quad (n \to \infty). \tag{14.39}$$

This follows immediately from Theorem 14.2 and the previous example by setting $C_n = \lceil \frac{n}{K_n} \rceil + 1$ and $\delta_n = 0$.

PROOF OF THEOREM 14.2. **(a)** Because of Theorem 10.2 it suffices to show that, for each $L > 0$,

$$\sup_{f \in T_{\beta_n} S_{u,M}} \left| \frac{1}{n}\sum_{j=1}^n |f(X_j) - Y_{j,L}|^2 - \mathbf{E}|f(X) - Y_L|^2 \right| \to 0 \quad (n \to \infty) \quad a.s.$$

$$\tag{14.40}$$

and

$$\inf_{f \in S_{u,M}, \|f\|_\infty \leq \beta_n} \int |f(x) - m(x)|^2 \mu(dx) \to 0 \quad (n \to \infty) \quad a.s. \tag{14.41}$$

PROOF OF (14.40). Let Π_n be the family of all partitions of $\mathcal{R}$ consisting of $K_{max}(n) + 2M_{max}(n) + 2$ or less intervals and let $\mathcal{G}$ be the set of all polynomials of degree $M_{max}(n)$ or less. Then $S_{u,M} \subset \mathcal{G} \circ \Pi_n$ and, therefore, it suffices to show (14.40) with the data-dependent set $S_{u,M}$ replaced by the data-independent set $\mathcal{G} \circ \Pi_n$. $\mathcal{G}$ is a linear space of functions of dimension $M_{max}(n) + 1$, thus $V_{\mathcal{G}+} \leq M_{max}(n) + 2$ (see Theorem 9.5). By Example 13.1 the partitioning number of Π_n satisfies

$$\Delta_n(\Pi_n) \quad \leq \quad \binom{n + K_{max}(n) + 2M_{max}(n) + 1}{n}$$

$$\leq \quad (n + K_{max}(n) + 2M_{max}(n) + 1)^{K_{max}(n) + 2M_{max}(n) + 1}.$$

As in the proof of Theorem 13.1 this implies, for $0 \leq L \leq \beta_n$,

$$\mathbf{P}\left[\sup_{f \in T_{\beta_n} \mathcal{G} \circ \Pi_n} \left| \frac{1}{n}\sum_{i=1}^n |f(X_i) - Y_{i,L}|^2 - \mathbf{E}|f(X) - Y_L|^2 \right| > t \right]$$

$$\leq 8(n + K_{max}(n) + 2\, M_{max}(n) + 1)^{(K_{max}(n) + 2\, M_{max}(n) + 1)}$$

$$\times \left(\frac{333 e \beta_n^2}{t}\right)^{2(M_{max}(n)+2)(K_{max}(n)+2\,M_{max}(n)+2)} \exp\left(-\frac{n\, t^2}{2048 \beta_n^4}\right)$$

and from this, (14.28), and (14.29), one obtains the assertion by an easy application of the Borel–Cantelli lemma.

PROOF OF (14.41). $C_0^\infty(\mathcal{R})$ is dense in $L_2(\mu)$ (cf. Corollary A.1), hence it suffices to prove (14.41) for some function $m \in C_0^\infty(\mathcal{R})$. Because of (14.27) we may further assume $\|m\|_\infty \leq \beta_n$.

Define $Qm \in S_{u,M}$ by $Qm = \sum_{j=-M}^{K-1} m(u_j) \cdot B_{j,M,u}$. Then (14.12) and (14.22) imply

$$|(Qm)(x)| \leq \max_{j=-M,\ldots,K-1} |m(u_j)| \sum_{j=-M}^{K-1} B_{j,M,u}(x) \leq \|m\|_\infty \leq \beta_n,$$

thus $Qm \in \{f \in S_{u,M} : \|f\|_\infty \leq \beta_n\}$. Let $x \in [u_i, u_{i+1})$ for some $0 \leq i \leq K-1$. Then

$$|m(x) - (Qm)(x)| \overset{(14.22)}{=} \left| \sum_{j=-M}^{K-1} (m(x) - m(u_j)) B_{j,M,u}(x) \right|$$

$$\overset{(14.11)}{=} \left| \sum_{j=i-M}^{i} (m(x) - m(u_j)) B_{j,M,u}(x) \right|$$

$$\overset{(14.12)}{\leq} \max_{j=i-M,\ldots,i} |m(x) - m(u_j)| \sum_{j=i-M}^{i} B_{j,M,u}(x)$$

$$\overset{(14.22)}{\leq} \|m'\|_\infty \cdot |u_{i+1} - u_{i-M}| \leq \|m'\|_\infty h_{u,M}(x),$$

where

$$h_{u,M}(x) = \begin{cases} (u_{i+1} - u_{i-M}) & \text{if } x \in [u_i, u_{i+1}) \text{ for some } 0 \leq i \leq K-1, \\ \infty & \text{if } x \in (-\infty, u_0) \text{ or } x \in [u_k, \infty). \end{cases}$$

Using this one gets, for arbitrary $L, \gamma > 0$,

$$\inf_{f \in S_{u,M},\, \|f\|_\infty \leq \beta_n} \int |f(x) - m(x)|^2 \mu(dx)$$

$$\leq \int |(Qm)(x) - m(x)|^2 \mu(dx)$$

$$= \int_{\mathcal{R}\setminus[-L,L]} |(Qm)(x) - m(x)|^2 \mu(dx)$$

$$+ \int_{\{x\in\mathcal{R}\,:\,h_{u,M}(x)>\gamma\}\cap[-L,L]} |(Qm)(x) - m(x)|^2 \mu(dx)$$

$$+ \int_{\{x\in\mathcal{R}\,:\,h_{u,M}(x)\leq\gamma\}\cap[-L,L]} |(Qm)(x) - m(x)|^2 \mu(dx)$$

$$\leq 4\|m\|_\infty^2 \left(\mu(\mathcal{R}\setminus[-L,L]) + \mu(\{x\in\mathcal{R}\,:\,h_{u,M}(x)>\gamma\}\cap[-L,L])\right)$$
$$+\gamma^2\|m'\|_\infty^2$$

$$= 4\|m\|_\infty^2 \cdot \left(\mu(\mathcal{R}\setminus[-L,L])\right.$$

$$\left.+\mu\left(\left\{(-\infty,u_0)\cup\bigcup_{\substack{k=1,\ldots,K,\\u_k-u_{k-M-1}>\gamma}}[u_{k-1},u_k)\cup[u_K,\infty)\right\}\cap[-L,L]\right)\right)$$

$$+\gamma^2\|m'\|_\infty^2.$$

If (14.30) holds, then one gets

$$\limsup_{n\to\infty}\ \inf_{f\in S_{u,M},\|f\|_\infty\leq\beta_n}\int |f(x) - m(x)|^2 \mu(dx)$$
$$\leq 4\|m\|_\infty^2\mu(\mathcal{R}\setminus[-L,L]) + \gamma^2\|m'\|_\infty^2 \quad a.s.$$

for each $L,\gamma > 0$, and the assertion follows with $L\to\infty$ and $\gamma\to 0$. If (14.31) holds, then

$$\mu\left(\left\{(-\infty,u_0)\cup\bigcup_{\substack{k=1,\ldots,K,\\u_k-u_{k-M-1}>\gamma}}[u_{k-1},u_k)\cup[u_K,\infty)\right\}\cap[-L,L]\right)$$

$$\leq \mu_n\left(\left\{(-\infty,u_0)\cup\bigcup_{\substack{k=1,\ldots,K,\\u_k-u_{k-M-1}>\gamma}}[u_{k-1},u_k)\cup[u_K,\infty)\right\}\cap[-L,L]\right)$$

$$+ \sup_{f\in\mathcal{G}_0\circ\Pi_n}\left|\frac{1}{n}\sum_{i=1}^n f(X_i) - \mathbf{E}f(X)\right|,$$

where $\mathcal{G}_0$ consists of two functions which are constant zero and constant one, respectively. For $f\in\mathcal{G}_0\circ\Pi_n$ one has $f(x)\in\{0,1\}$ $(x\in\mathcal{R})$. This, together with (14.40), implies

$$\sup_{f\in\mathcal{G}_0\circ\Pi_n}\left|\frac{1}{n}\sum_{i=1}^n f(X_i) - \mathbf{E}f(X)\right|$$

$$= \sup_{f \in \mathcal{G}_0 \circ \Pi_n} \left| \frac{1}{n} \sum_{i=1}^{n} |f(X_i) - 0|^2 - \mathbf{E}|f(X) - 0|^2 \right| \to 0 \quad (n \to \infty) \quad a.s.,$$

thus (14.31) implies (14.30) which in turn implies the assertion.
(b) The proof of (b) is similar to the first part of the proof and is therefore omitted (cf. Problem 14.1). $\qquad\square$

14.3 Spline Approximation

In this section we will investigate how well smooth functions (e.g., continuously differentiable functions) can be approximated by spline functions. Our aim is to derive results similar to the following result for approximation with piecewise polynomials satisfying no global smoothness condition: If f is $(M+1)$ times continuously differentiable and one approximates f on an interval $[a, b)$ by partitioning the interval into subintervals $[u_j, u_{j+1})$ $(j = 0, \ldots, K-1)$ and defines q on each interval $[u_j, u_{j+1})$ as a Taylor polynomial of f of degree M around a fixed point in $[u_j, u_{j+1})$, then one has, for each $j \in \{0, \ldots, K-1\}$ and each $x \in [u_j, u_{j+1})$,

$$|f(x) - q(x)| \leq \frac{\|f^{(M+1)}\|_{\infty, [u_j, u_{j+1})}}{(M+1)!} (u_{j+1} - u_j)^{M+1}. \tag{14.42}$$

Here $f^{(M+1)}$ denotes the $(M+1)$th derivative of f, and

$$\|f^{(M+1)}\|_{\infty, [u_j, u_{j+1})} = \sup_{x \in [u_j, u_{j+1})} |f^{(M+1)}(x)|.$$

In the sequel we will use the notation $C(\mathcal{R})$ for the set of continuous functions $f : \mathcal{R} \to \mathcal{R}$. The following definitions will be useful:

Definition 14.4. *(a) For $j \in \{-M, \ldots, K-1\}$ let $Q_j : C(\mathcal{R}) \to \mathcal{R}$ be a linear mapping (i.e., $Q_j(\alpha f + \beta g) = \alpha Q_j(f) + \beta Q_j(g)$ for $\alpha, \beta \in \mathcal{R}$, $f, g \in C(\mathcal{R})$) such that $Q_j f$ depends only on the values of f in $[u_j, u_{j+M+1})$. Then the functional $Q : C(\mathcal{R}) \to S_{u,M}$ defined by*

$$Qf = \sum_{j=-M}^{K-1} (Q_j f) \cdot B_{j,M,u}$$

is called a **quasi interpolant**.
(b) A quasi interpolant Q is called **bounded** *if there exists a constant $c \in \mathcal{R}$ such that*

$$|Q_j f| \leq c \cdot \|f\|_{\infty, [u_j, u_{j+M+1})} \quad (j \in \{-M, \ldots, K-1\}, f \in C(\mathcal{R})). \tag{14.43}$$

The smallest constant c such that (14.43) holds is denoted by $\|Q\|$.
(c) A quasi interpolant Q has **order** l *if*

$$(Qp)(x) = p(x) \quad (x \in [u_0, u_K))$$

for each polynomial p of degree l or less.

Example 14.8. *Let $t_j \in [u_j, u_{j+M+1})$ for $j \in \{-M, \ldots, K-1\}$. Then*

$$Qf = \sum_{j=-M}^{K-1} f(t_j) B_{j,M,u}$$

defines a bounded quasi interpolant, which has order zero (because of (14.22)).

Example 14.9. *Let $M > 0$ and define Q by*

$$Qf = \sum_{j=-M}^{K-1} f\left(\frac{u_{j+1} + \cdots + u_{j+M}}{M}\right) B_{j,M,u}.$$

Clearly, Q is a bounded quasi interpolant. We show that Q has order 1 : Because of (14.22) it suffices to show that Q reproduces the linear polynomial $p(x) = x$. Differentiating (14.15) $M - 1$ times with respect to t and setting $t = 0$ yields

$$M! \cdot (-1)^{M-1} x = \sum_{j=-M}^{K-1} (-1)^{M-1}(M-1)! \cdot (u_{j+1} + \cdots + u_{j+M}) B_{j,M,u}(x),$$

where we have used

$$\psi_{j,M}(t) = (-1)^M t^M + (u_{j+1} + \cdots + u_{j+M})(-1)^{M-1} t^{M-1} + q(t)$$

for some polynomial q of degree less than $M - 1$. This implies $Qp = p$.

For bounded quasi interpolants of order l one can show approximation results similar to (14.42).

Theorem 14.3. *Let $M \in \mathcal{N}_0$ and $u_{-M} \leq \cdots \leq u_0 < \cdots < u_K \leq \cdots \leq u_{K+M}$. For $x \in \mathcal{R}$ set*

$$h(x) = \max_{j:u_j \leq x < u_{j+M+1}} (u_{j+M+1} - u_j)$$

(i.e., $h(x)$ is the maximal length of support of the B-splines which do not vanish at x). Let Q be a bounded quasi interpolant of order l. Then one has, for each $f \in C^{l+1}([u_{-M}, u_{K+M}))$,

$$|(Qf)(x) - f(x)| \leq \|Q\| \frac{\|f^{(l+1)}\|_{\infty,[u_{-M},u_{K+M})}}{(l+1)!} h(x)^{l+1} \quad (x \in [u_0, u_K)).$$

PROOF. Fix $x \in [u_0, u_K)$ and let p be the Taylor polynomial of f of order l about x. Then one has, for each $z \in \mathcal{R}$,

$$|f(z) - p(z)| \leq \frac{\|f^{(l+1)}\|_{\infty,[\min\{x,z\},\max\{x,z\}]}}{(l+1)!} |z - x|^{l+1}. \tag{14.44}$$

Further by definition of p one has $f(x) = p(x)$ and because Q is of order l, $(Qp)(x) = p(x)$. Using this one gets

$$|(Qf)(x) - f(x)| = |(Qf)(x) - p(x)| = |(Qf)(x) - (Qp)(x)|$$

$$= \left| \sum_{j=-M}^{K-1} Q_j f \cdot B_{j,M,u}(x) - \sum_{j=-M}^{K-1} Q_j p \cdot B_{j,M,u}(x) \right|$$

$$\leq \sum_{j=-M}^{K-1} |Q_j(f - p)| B_{j,M,u}(x) \quad \text{(because of (14.12))}$$

$$\leq \sum_{j=-M,\dots,K-1,\ u_j \leq x < u_{j+M+1}} |Q_j(f - p)| B_{j,M,u}(x)$$

$$\text{(because of (14.11))}$$

$$\leq \max_{j \in \{-M,\dots,K-1\}, u_j \leq x < u_{j+M+1}} |Q_j(f - p)|$$

$$\text{(because of (14.12) and (14.22))}$$

$$\leq \|Q\| \max_{j \in \{-M,\dots,K-1\}, u_j \leq x < u_{j+M+1}} \|f - p\|_{\infty,[u_j, u_{j+M+1})}.$$

For $z \in [u_j, u_{j+M+1})$, $u_j \leq x < u_{j+M+1}$, it follows, from (14.44), that

$$
\begin{aligned}
|f(z) - p(z)| &\leq \frac{\|f^{(l+1)}\|_{\infty,[\min\{x,z\},\max\{x,z\}]}}{(l+1)!} |z - x|^{l+1} \\[2mm]
&\leq \frac{\|f^{(l+1)}\|_{\infty,[u_{-M},u_{K+M})}}{(l+1)!} (u_{j+M+1} - u_j)^{l+1} \\[2mm]
&\leq \frac{\|f^{(l+1)}\|_{\infty,[u_{-M},u_{K+M})}}{(l+1)!} h(x)^{l+1},
\end{aligned}
$$

which implies the assertion. $\qquad\square$

This theorem yields good approximation results provided one has a quasi interpolant of high order. Clearly, for $S_{u,M}$ there cannot exist a quasi interpolant of order greater than M (because Qp is a piecewise polynomial of degree M or less which cannot be equal to a polynomial of degree greater than M on $[u_0, u_K)$). Next we show that there always exist bounded quasi interpolants $Q : C(\mathcal{R}) \to S_{u,M}$ of order M.

Theorem 14.4. *Let $M \in \mathcal{N}_0$ and $u_{-M} \leq \cdots \leq u_0 < \cdots < u_K \leq \cdots \leq u_{K+M}$. Then there exists a bounded quasi interpolant $Q : C(\mathcal{R}) \to S_{u,M}$ of order M such that $\|Q\|$ is bounded above by a constant depending only on M (and not on the knot sequence!).*

PROOF. For $j \in \{-M, \ldots, K-1\}$ let $t_{j,0}, \ldots, t_{j,M} \in [u_j, u_{j+M+1})$ and $Q_{j,0}, \ldots, Q_{j,M} \in \mathcal{R}$. Define $Q_j : C(\mathcal{R}) \to \mathcal{R}$ by

$$Q_j f = \sum_{k=0}^{M} Q_{j,k} \cdot f(t_{j,k}) \quad (f \in C(\mathcal{R})).$$

Then, clearly, $Q : C(\mathcal{R}) \to S_{u,M}$, defined by $Qf = \sum_{j=-M}^{K-1} Q_j f \cdot B_{j,M,u}$, is a quasi interpolant. We will show that it is possible to choose the $t_{j,k}$, $Q_{j,k}$ such that Q is a bounded quasi interpolant of order M with $\|Q\|$ bounded by a constant depending only on M.

First observe that Q is of order M if and only if Q reproduces for each $t \in \mathcal{R}$ the polynomial $(x-t)^M$, i.e., if

$$(x-t)^M = \sum_{j=-M}^{K-1} Q_j \left((\cdot - t)^M \right) B_{j,M,u}(x) \quad (x \in [u_0, u_K), t \in \mathcal{R}).$$

By Lemma 14.3 one has

$$(x-t)^M = \sum_{j=-M}^{K-1} \psi_{j,M}(t) B_{j,M,u}(x) \quad (x \in [u_0, u_K), t \in \mathcal{R}),$$

and thus it follows from Theorem 14.1 that Q is of order M if and only if

$$\sum_{k=0}^{M} Q_{j,k}(t_{j,k} - t)^M = (u_{j+1} - t) \cdot \ldots \cdot (u_{j+M} - t) \quad (j \in \{-M, \ldots, K-1\})$$

$$\tag{14.45}$$

for every $t \in \mathcal{R}$. The right-hand side of (14.45) is a polynomial of degree M or less (with respect to t). If the $t_{j,0}, \ldots, t_{j,M}$ are distinct, then

$$\{(t_{j,k} - t)^M \; : \; k = 0, \ldots, M\}$$

is a basis of the space of all polynomials of degree M or less and there exists uniquely determined $Q_{j,0}, \ldots, Q_{j,M} \in \mathcal{R}$ such that (14.45) holds for every $t \in \mathcal{R}$. It remains to show that the $t_{j,k}$ can be chosen such that the resulting quasi interpolant is bounded by some constant depending only on M.

Both sides of (14.45) are polynomials of degree M or less. Therefore (14.45) holds for every $t \in \mathcal{R}$ if and only if it holds for $M+1$ distinct values of t and hence is equivalent to

$$\sum_{k=0}^{M} Q_{j,k}(t_{j,k} - t_{j,l})^M = (u_{j+1} - t_{j,l}) \cdot \ldots \cdot (u_{j+M} - t_{j,l}) \tag{14.46}$$

$(j \in \{-M, \ldots, K-1\}, l \in \{0, \ldots, M\})$, provided $t_{j,0}, \ldots, t_{j,M}$ are pairwise distinct $(j \in \{-M, \ldots, K-1\})$.

Let us now choose the $t_{j,k}$ such that the resulting quasi interpolant satisfies the assertion of the theorem. For $j \in \{-M, \ldots, K-1\}$ choose

$\nu_j \in \{j, \ldots, j+M-1\}$ such that $[u_{\nu_j}, u_{\nu_j+1})$ is the largest interval $[u_i, u_{i+1})$ contained in $[u_j, u_{j+M})$ and set

$$t_{j,k} = u_{\nu_j} + \frac{k}{M}(u_{\nu_j+1} - u_{\nu_j}) \quad (k \in \{0, \ldots, M\}).$$

Then

$$t_{j,k} - t_{j,l} = (k - l) \cdot \frac{u_{\nu_j+1} - u_{\nu_j}}{M}$$

and

$$\frac{u_{j+i} - t_{j,l}}{\frac{u_{\nu_j+1} - u_{\nu_j}}{M}} = \frac{u_{j+i} - u_{\nu_j}}{\frac{u_{\nu_j+1} - u_{\nu_j}}{M}} - l.$$

Thus, in this case, (14.46) is equivalent to

$$\sum_{k=0}^{M} (k - l)^M Q_{j,k} = \left(\frac{u_{j+1} - u_{\nu_j}}{\frac{u_{\nu_j+1} - u_{\nu_j}}{M}} - l \right) \cdot \ldots \cdot \left(\frac{u_{j+M} - u_{\nu_j}}{\frac{u_{\nu_j+1} - u_{\nu_j}}{M}} - l \right) \quad (14.47)$$

($l \in \{0, \ldots, M\}$) for each $j \in \{-M, \ldots, K-1\}$.

Fix $j \in \{-M, \ldots, K-1\}$. Then (14.47) is an $(M+1) \times (M+1)$ linear equation system for the $Q_{j,0}, \ldots, Q_{j,M}$. It has a unique solution because (14.45) has a unique solution. Furthermore, its left-hand side depends only on M and by the choice of $[u_{\nu_j}, u_{\nu_j+1})$ the distance between u_{j+k} and u_{ν_j} is less than or equal to M times $u_{\nu_j+1} - u_{\nu_j}$ ($k \in \{1, \ldots, M\}$), which implies that the components of its right-hand side are bounded in absolute value by $(M^2 + M)^M$. Therefore there exists some constant c depending only on M such that the solution of (14.47) satisfies

$$|Q_{j,k}| \leq c \quad (j \in \{-M, \ldots, K-1\}, k \in \{0, \ldots, M\}).$$

Thus

$$|Q_j f| = \left| \sum_{k=0}^{M} Q_{j,k} \cdot f(t_{j,k}) \right| \leq c \cdot (M+1) \cdot \|f\|_{\infty, [u_j, u_{j+M+1})}$$

which in turn implies $\|Q\| \leq c \cdot (M+1)$. $\square$

14.4   Rate of Convergence

In this section we derive rate of convergence results for least squares estimates using spline spaces with equidistant knots. Throughout this section we will assume $supp(X)$ bounded and, for simplicity, we also assume $supp(X) \subseteq [0,1]$.

Let $M \in \mathcal{N}_0$ and let $\{K_n\}_{n \in \mathcal{N}}$ be a sequence of natural numbers. Set

$$u_k = \frac{k}{K_n} \quad (k = -M, \ldots, K_n + M).$$

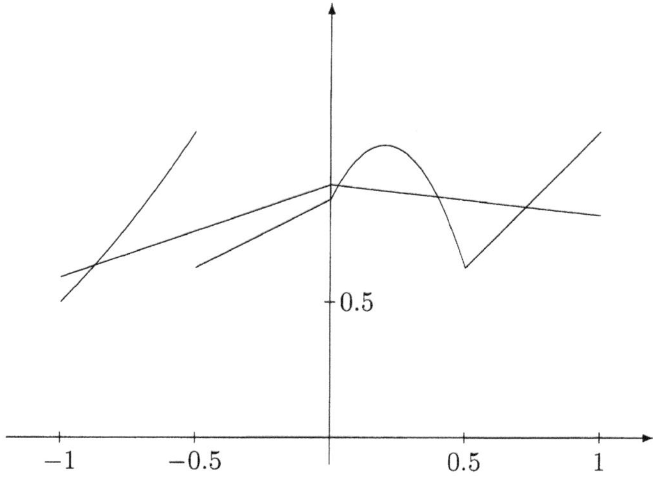

Figure 14.7. Least squares spline estimate with $K_n = 2$, $u_k = -1 + 2k/K_n$, and $M = 1$: L_2 error $= 0.013607$.

Let $S_{K_n,M}$ be the spline space with knot sequence $\{u_k\}_{k=-M}^{K_n+M}$ and degree M. Define the estimate m_n by

$$\tilde{m}_n = \arg \min_{f \in S_{K_n,M}} \frac{1}{n} \sum_{i=1}^{n} |f(X_i) - Y_i|^2$$

and set

$$m_n(x) = T_L(\tilde{m}_n(x)) \quad (x \in \mathcal{R})$$

where $L \in \mathcal{R}_+$ is some known bound on the supremum norm of the regression function.

Figures 14.7–14.9 show application of this estimate for various choices of K_n to our simulated data set of Chapter 1. A comparison of the estimate in Figure 14.9 with the piecewise polynomial partitioning estimate in Figure 10.3 shows that the least squares estimate is indeed much less oscillatory than the piecewise polynomial partitioning estimate.

Theorem 14.5. *Let the estimate m_n be defined as above. Let $p \in \{1, \ldots, M + 1\}$ and assume that*

$$\sigma^2 = \sup_{x \in \mathcal{R}} \mathbf{Var}\{Y | X = x\} < \infty,$$

$$\|m\|_\infty = \sup_{x \in \mathcal{R}} |m(x)| \leq L,$$

$$supp(X) \subseteq [0, 1],$$

and that m is p times continuously differentiable. Then there exist constants $c_2, c_3 \in \mathcal{R}_+$ which depend only on M such that

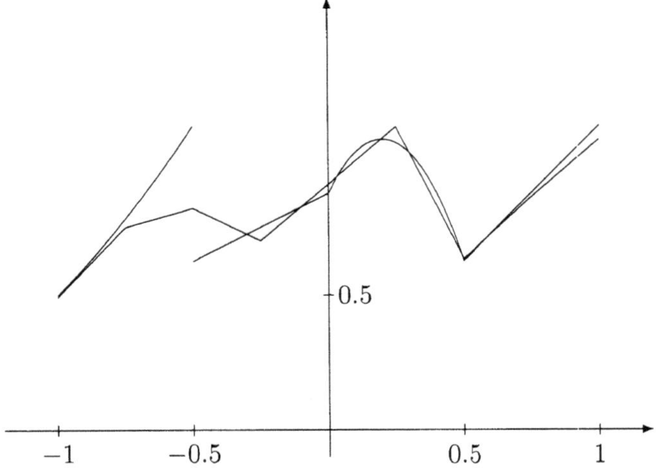

Figure 14.8. Least squares spline estimate with $K_n = 8$, $u_k = -1 + 2k/K_n$, and $M = 1$: L_2 error $= 0.004024$.

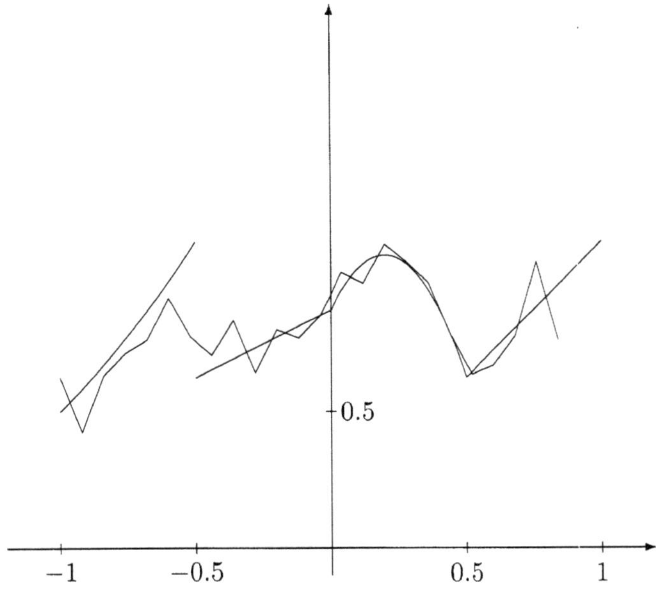

Figure 14.9. Least squares spline estimate with $K_n = 25$, $u_k = -1 + 2k/K_n$, and $M = 1$: L_2 error $= 0.013431$.

$$\mathbf{E}\{\int |m_n(x) - m(x)|^2 \mu(dx)\}$$

$$\leq c_1 \max\{\sigma^2, L^2\} \cdot \frac{(\log(n) + 1) \cdot (K_n + M + 1)}{n} + c_2 \|m^{(p)}\|_\infty^2 \left(\frac{M+1}{K_n}\right)^{2p}.$$

In particular, for $K_n = \left\lfloor \left(\frac{n}{\log(n)}\right)^{\frac{1}{2p+1}} \right\rfloor$:

$$\mathbf{E}\left\{ \int |m_n(x) - m(x)|^2 \mu(dx) \right\} \le c_3 \left(\frac{\log(n)}{n}\right)^{\frac{2p}{2p+1}}$$

for some constant $c_3 \in \mathcal{R}_+$ which depends on M, σ^2, L^2, and $\|m^{(p)}\|_\infty$.

PROOF. By Theorem 11.3 we have

$$\mathbf{E} \int |m_n(x) - m(x)|^2 \mu(dx)$$

$$\le c \cdot \max\{\sigma^2, L^2\} \frac{(\log(n) + 1) \cdot (K_n + M + 1)}{n}$$

$$+ 8 \inf_{f \in S_{K_n, M}} \int |f(x) - m(x)|^2 \mu(dx).$$

An application of Theorems 14.3 and 14.4 yields

$$8 \inf_{f \in S_{K_n, M}} \int |f(x) - m(x)|^2 \mu(dx) \;\le\; 8 \inf_{f \in S_{K_n, M}} \sup_{x \in [0,1]} |f(x) - m(x)|^2$$

$$\le \; \hat{c} \cdot \|m^{(p)}\|_\infty^2 \left(\frac{M+1}{K_n}\right)^{2p},$$

which implies the assertion. $\square$

The definition of the estimate in Theorem 14.5 depends on the smoothness p of the regression function. This can be avoided by using complexity regularization.

Set $\mathcal{P}_n = \{1, \ldots, n\}$ and for $K \in \mathcal{P}_n$ define the estimate $m_{n,K}$ by

$$\tilde{m}_{n,K} = \arg \min_{f \in S_{K,M}} \frac{1}{n} \sum_{i=1}^{n} |f(X_i) - Y_i|^2$$

and

$$m_{n,K}(x) = T_L(\tilde{m}_{n,K}(x)) \quad (x \in \mathcal{R}).$$

To penalize the complexity of $S_{K,M}$ we use the penalty

$$pen_n(K) = \frac{L^4 \log^2(n) \cdot K}{n}.$$

Set

$$K^* = \arg \min_{K \in \mathcal{P}_n} \left\{ pen_n(K) + \frac{1}{n} \sum_{i=1}^{n} |m_{n,K}(X_i) - Y_i|^2 \right\}$$

and

$$m_n(x) = m_{n,K^*}(x) \quad (x \in \mathcal{R}).$$

We have the following result:

Theorem 14.6. *Let $L > 0$, $M \in \mathcal{N}_0$, and let m_n be defined as above. Then one has, for any $p \in \{1, \ldots, M + 1\}$,*

$$\mathbf{E}\{\int |m_n(x) - m(x)|^2 \mu(dx)\} = O\left(\left(\frac{\log^2(n)}{n}\right)^{\frac{2p}{2p+1}}\right)$$

for every distribution of (X, Y) with $supp(X) \subseteq [0, 1]$, $|Y| \leq L$ a.s., and m p times continuously differentiable.

PROOF. See Problem 14.2. $\qquad\qquad\square$

Observe that the estimate in Theorem 14.6 doesn't depend on p any more.

14.5 Bibliographic Notes

Theorem 14.2 is due to Kohler (1999). The proofs of the deterministic properties of splines which we present in this and in the next chapter are based on lectures given by K. Höllig at the University of Stuttgart, and parts of these lectures have been published in Höllig (1998). Different kinds of introductions to polynomial splines and references concerning the deterministic results proven in this chapter can be found in de Boor (1978) and Schumaker (1981).

Many references concerning the application of splines in statistics can be found in the survey articles by Wegman and Wright (1983) and Agarwal (1989). Consistency of least squares splines in supremum norm was studied in Zhu (1992). In fixed design regression, Agarwal and Studden (1980) investigated the question of how to place the (nonequidistant) knots of a least squares spline estimate in order to minimize the expected error of the estimate.

Problems and Exercises

PROBLEM 14.1. Prove Theorem 14.2 (b).
   HINT: Proceed as in the proof of the first part of Theorem 14.2, but apply part (b) of Theorem 10.2.

PROBLEM 14.2. Prove Theorem 14.6.
   HINT: Apply Theorem 12.1 and use the approximation result derived in the proof of Theorem 14.5.

PROBLEM 14.3. Let $M \in \mathcal{N}_0$, $K \in \mathcal{N}$, and $u_{-M} \leq \cdots \leq u_0 < \cdots < u_K \leq \cdots \leq u_{K+M}$. For $x \in \mathcal{R}$, set

$$h(x) = \max_{j : u_j \leq x < u_{j+M+1}} (u_{j+M+1} - u_j).$$

Let $p = q + r$ for some $q \in \{0, \ldots, M\}$, $r \in (0, 1]$. Let Q be a bounded quasi interpolant of order q. Show that for every (p, C)-smooth function $f : \mathcal{R} \to \mathcal{R}$,

$$|(Qf)(x) - f(x)| \leq \|Q\| \cdot \frac{C}{q!} \cdot h(x)^p \quad (x \in [u_0, u_K)).$$

HINT: Proceed as in the proof of Theorem 14.3, but use (11.7) instead of (14.44).

PROBLEM 14.4. Use Problem 14.3 to derive a version of Theorem 14.5, where the assumption that m is p times continuously differentiable is replaced by the assumption m (p, C)-smooth.

PROBLEM 14.5. Show that the estimate in Theorem 14.6 satisfies

$$\mathbf{E} \int |m_n(x) - m(x)|^2 \mu(dx) \leq c \cdot C^{2/(2p+1)} \left(\frac{\log^2(n)}{n} \right)^{2p/(2p+1)}$$

for every $p \leq M + 1$ and every distribution of (X, Y) with $supp(X) \subseteq [0, 1]$, $|Y| \leq L$ a.s., and m (p, C)-smooth.

HINT: Apply Problem 14.3.

15

Multivariate Least Squares Spline Estimates

15.1 Introduction to Tensor Product Splines

An easy way to construct multivariate spline spaces is to use tensor products of univariate spline spaces. The tensor product of univariate functions $f_1, \ldots, f_d : \mathcal{R} \to \mathcal{R}$ is the function $f : \mathcal{R}^d \to \mathcal{R}$ defined by

$$f(x_1, \ldots, x_d) = f_1(x_1) \cdot \ldots \cdot f_d(x_d) \quad (x_1, \ldots, x_d \in \mathcal{R}).$$

We will define tensor product spline spaces as a linear span of all tensor products of univariate spline functions belonging to d fixed univariate spline spaces.

In order to avoid complicated notation we will explain this in the sequel only in the case $d = 2$. The general case can be handled in an analogous way.

Definition 15.1. *Let $M_x, M_z \in \mathcal{N}_0$, $u_{-M_x} \leq \cdots \leq u_0 < \cdots < u_K \leq \cdots \leq u_{K+M_x}$, and $v_{-M_z} \leq \cdots \leq v_0 < \cdots < v_L \leq \cdots \leq v_{L+M_z}$. Then the* **tensor product spline space** S_{u,M_x,v,M_z} *of* **degree** (M_x, M_z) *and with* **knot sequences** $u = \{u_i\}_{i=-M_x,\ldots,K+M_x}$ *and* $v = \{v_j\}_{j=-M_z,\ldots,L+M_z}$ *is defined by*

$$S_{u,M_x,v,M_z} = span\Big\{ f : \mathcal{R}^2 \to \mathcal{R} \ : \ f(x,z) = g(x)h(z)\,(x,z \in \mathcal{R})$$

$$for\ some\ g \in S_{u,M_x}, h \in S_{v,M_z} \Big\}.$$

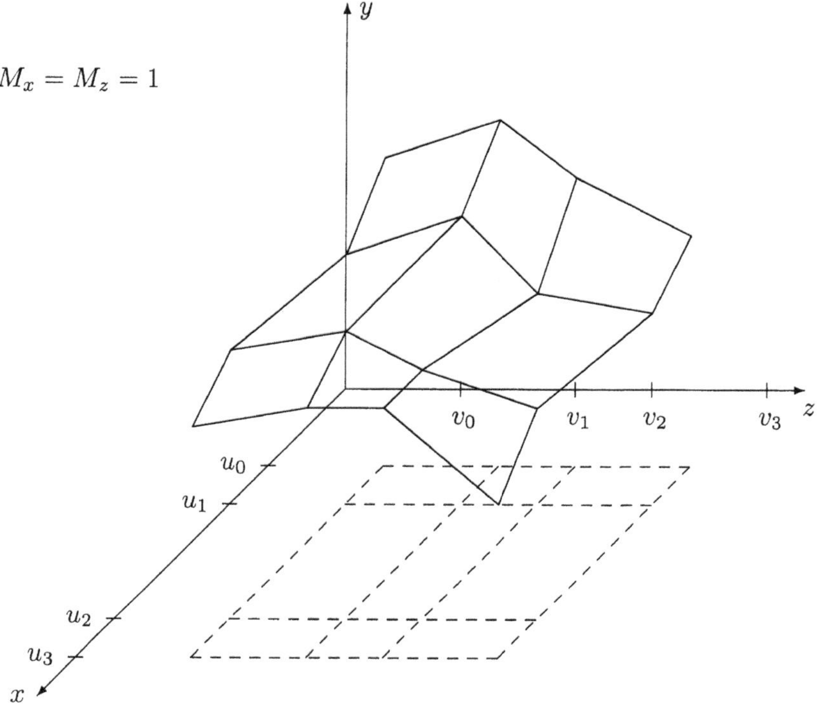

Figure 15.1. A function of $S_{u,1,v,1}$.

Example 15.1. *Let $M_x = M_z = 1$ and choose the knot sequences u and v as in Figure 15.1. The knot sequences induce a partition of $\mathcal{R}^2$. The functions in $S_{u,1,v,1}$ are continuous and are piecewise bilinear with respect to this partition.*

It follows from the definitions of S_{u,M_x} and S_{v,M_z} that on each set $[u_i, u_{i+1}) \times [v_j, v_{j+1})$ the functions in S_{u,M_x,v,M_z} are equal to a finite sum of tensor products of univariate polynomials of degree less than or equal to M_x and univariate polynomials of degree less than or equal to M_z $(-M_x \leq i \leq K + M_x - 1, -M_z \leq j \leq L + M_z - 1)$. Furthermore, they are zero outside $[u_{-M_x}, u_{K+M_x}) \times [v_{-M_z}, v_{L+M_z})$ and satisfy some global smoothness conditions depending on (M_x, M_z) and the knot sequences. For instance, if $u_{-M_x} < \cdots < u_{K+M_x}$ and $v_{-M_z} < \cdots < v_{L+M_z}$ then the partial derivatives of the functions in S_{u,M_x,v,M_z} with respect to x (z) are continuous up to order $M_x - 1$ $(M_z - 1)$.

A basis of S_{u,M_x,v,M_z} can be constructed by the use of tensor products of univariate B-splines.

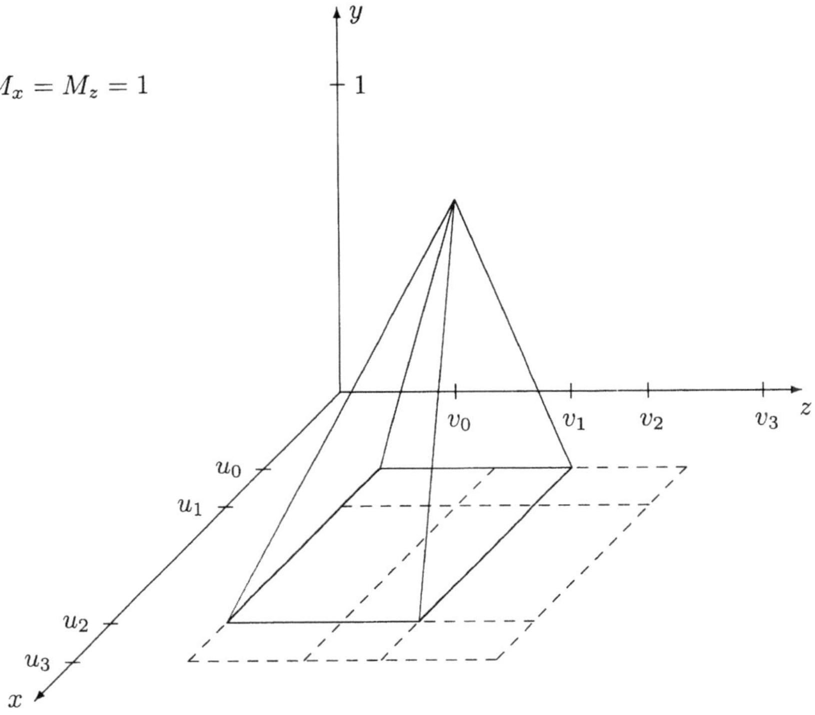

Figure 15.2. Example of a tensor product B-spline of degree $(1, 1)$.

Lemma 15.1. *For* $-M_x \leq i < K$ *and* $-M_z \leq j < L$ *define the tensor product B-spline* $B_{(i,j),M_x,u,M_z,v}$ *by*

$$B_{(i,j),M_x,u,M_z,v}(x, z) = B_{i,M_x,u}(x) \cdot B_{j,M_z,v}(z) \quad (x, z \in \mathcal{R}). \qquad (15.1)$$

Then

$$\left\{ B_{(i,j),M_x,u,M_z,v} \; : \; -M_x \leq i < K, -M_z \leq j < L \right\}$$

is a basis of S_{u,M_x,v,M_z}.

We leave the easy proof to the reader (cf. Problem 15.1).

The next lemma summarizes some useful properties of tensor product B-splines.

Lemma 15.2. *Let* $M_x, M_z \in \mathcal{N}_0$, $u_{-M_x} \leq \cdots \leq u_0 < \cdots < u_K \leq \cdots \leq u_{K+M_x}$, *and* $v_{-M_z} \leq \cdots \leq v_0 < \cdots < v_L \leq \cdots \leq v_{L+M_z}$.
(a)

$$B_{(i,j),M_x,u,M_z,v}(x, z) = 0 \quad for \; (x, z) \notin [u_i, u_{i+M_x+1}) \times [v_j, v_{j+M_z+1})$$
$$(15.2)$$

for $-M_x \le i \le K - 1$ *and* $-M_z \le j \le L - 1$.
(b)

$$B_{(i,j),M_x,u,M_z,v}(x,z) \ge 0 \quad (x, z \in \mathcal{R}) \tag{15.3}$$

for $-M_x \le i \le K - 1$ *and* $-M_z \le j \le L - 1$.
(c)

$$\sum_{i=-M_x}^{K-1} \sum_{j=-M_z}^{L-1} B_{(i,j),M_x,u,M_z,v}(x,z) = 1 \tag{15.4}$$

for $(x, z) \in [u_0, u_K) \times [v_0, v_L)$.

PROOF. The assertion follows directly from (15.1), (14.11), (14.12), and (14.22). □

If a bivariate spline function is represented as a linear combination of tensor product B-splines, then it can be evaluated as fast as in the univariate case. Indeed, assume that we are given coefficients

$$\{a_{i,j} : -M_x \le i \le K - 1, -M_z \le j \le L - 1\}$$

of a linear combination

$$f = \sum_{i=-M_x}^{K-1} \sum_{j=-M_z}^{L-1} a_{i,j} B_{(i,j),M_x,u,M_z,v}$$

of tensor product B-splines and that we want to evaluate this function at some given point $(x, z) \in [u_0, u_K) \times [v_0, v_L)$. Choose i_0, j_0 such that

$$(x, z) \in [u_{i_0}, u_{i_0+1}) \times [v_{j_0}, v_{j_0+1}).$$

Then

$$f(x,z) \overset{(15.2)}{=} \sum_{i=i_0-M_x}^{i_0} \sum_{j=j_0-M_z}^{j_0} a_{i,j} B_{(i,j),M_x,u,M_z,v}$$

$$\overset{(15.1)}{=} \sum_{i=i_0-M_x}^{i_0} \left\{ \sum_{j=j_0-M_z}^{j_0} a_{i,j} B_{j,M_z,v}(z) \right\} \cdot B_{i,M_x,u}(x).$$

Now use $M_x + 1$ times the evaluation algorithm for linear combinations of univariate B-splines described in Section 14.1 to compute

$$b_i = \sum_{j=j_0-M_z}^{j_0} a_{i,j} B_{j,M_z,v}(z) \quad (i = i_0 - M_x, \ldots, i_0).$$

Then use it once more to compute

$$f(x,z) = \sum_{i=i_0-M_x}^{i_0} b_i \cdot B_{i,M_x,u}(x).$$

As in the univariate case the number of operations needed to do this depends only on M_x and M_z (and not on K or on L).

To derive results concerning the approximation of smooth functions by tensor product splines we will again use so–called quasi interpolants. The following definition extends Definition 14.4 to the case of tensor product splines.

Definition 15.2. *(a) For $i \in \{-M_x, \ldots, K-1\}$ and $j \in \{-M_z, \ldots, L-1\}$ let $Q_{(i,j)} : C(\mathcal{R}^2) \to \mathcal{R}$ be a linear mapping such that $Q_{(i,j)}(f)$ depends only on the values of f in $[u_i, u_{i+M_x+1}) \times [v_j, v_{j+M_z+1})$. Then the functional $Q : C(\mathcal{R}^2) \to S_{u,M_x,v,M_z}$, defined by*

$$Qf = \sum_{i=-M_x}^{K-1} \sum_{j=-M_z}^{L-1} Q_{(i,j)}(f) \cdot B_{(i,j),M_x,u,M_z,v},$$

is called a **quasi interpolant**.
(b) A quasi interpolant $Q : C(\mathcal{R}^2) \to S_{u,M_x,v,M_z}$ is called **bounded** *if there exists some constant $c \in \mathcal{R}$ such that*

$$|Q_{(i,j)}(f)| \leq c \cdot \|f\|_{\infty,[u_i,u_{i+M_x+1}) \times [v_j,v_{j+M_z+1})} \tag{15.5}$$

for $i \in \{-M_x, \ldots, K-1\}$, $j \in \{-M_z, \ldots, L-1\}$, and $f \in C(\mathcal{R}^2)$. Here

$$\|f\|_{\infty,[u_i,u_{i+M_x+1}) \times [v_j,v_{j+M_z+1})} = \sup_{(x,z) \in [u_i,u_{i+M_x+1}) \times [v_j,v_{j+M_z+1})} |f(x,z)|.$$

The smallest constant c such that (15.5) holds is denoted by $\|Q\|$.
(c) A quasi interpolant Q has **order** *(l_x, l_z) if*

$$(Qp)(x,z) = p(x,z) \quad ((x,z) \in [u_0, u_K) \times [v_0, v_L))$$

for each polynomial $p : \mathcal{R}^2 \to \mathcal{R}$ of degree less than or equal to (l_x, l_z).

Here and in the following a polynomial $p : \mathcal{R}^2 \to \mathcal{R}$ of degree less than or equal to (l_x, l_z) is a finite sum of tensor products of polynomials of degree less than or equal to l_x and of polynomials of degree less than or equal to l_z.

The next theorem extends Theorem 14.3 to the case of tensor product splines.

Theorem 15.1. *Let $M_x, M_z \in \mathcal{N}_0$, $u_{-M_x} \leq \cdots \leq u_0 < \cdots < u_K \leq \cdots \leq u_{K+M_x}$, and $v_{-M_z} \leq \cdots \leq v_0 < \cdots < v_L \leq \cdots \leq v_{L+M_z}$. For $x, z \in \mathcal{R}$ set*

$$h_u(x) = \max_{i:u_i \leq x < u_{i+M_x+1}} (u_{i+M_x+1} - u_i)$$

and

$$h_v(z) = \max_{j:v_j \leq x < v_{j+M_z+1}} (v_{j+M_z+1} - v_j).$$

Let Q be a bounded quasi interpolant of order (l,l). Let $f : \mathcal{R}^2 \to \mathcal{R}$ be a function for which the partial derivatives $\frac{\partial^{l_1+l_2} f}{\partial x^{l_1} \partial z^{l_2}}$ of order $l_1 + l_2 \leq l+1$

are continuous on $[u_{-M_x}, u_{K+M_x}) \times [v_{-M_z}, v_{L+M_z})$. Then there exists a constant $c(f) \in \mathcal{R}_+$ which depends only on l and on the partial derivatives of f of order less than or equal to l such that

$$|(Qf)(x,z) - f(x,z)| \leq \|Q\| \cdot c(f) \cdot \left(h_u(x)^{l+1} + h_v(z)^{l+1}\right)$$

for all $(x,z) \in [u_0, u_K) \times [v_0, v_L)$.

PROOF. Fix $(x,z) \in [u_{i_0}, u_{i_0+1}) \times [v_{j_0}, v_{j_0+1})$. Let p be the Taylor polynomial

$$p(u,v) = \sum_{0 \leq l_1+l_2 \leq l} \frac{\partial^{l_1+l_2} f(x,z)}{\partial x^{l_1} \partial z^{l_2}} \cdot \frac{(u-x)^{l_1}(v-z)^{l_2}}{l_1! \cdot l_2!}.$$

Then the following formula holds for the remainder of the Taylor polynomial (cf. Problem 15.2)

$$\begin{aligned}
f(u,v) &- p(u,v) \\
&= \sum_{l_1+l_2=l+1} \int_0^1 (1-t)^l \frac{\partial^{l_1+l_2} f(x + t \cdot (u-x), z + t \cdot (v-z))}{\partial x^{l_1} \partial z^{l_2}} \\
&\qquad\qquad\qquad \times \frac{l}{l_1! \cdot l_2!} \cdot (u-x)^{l_1}(v-z)^{l_2}\, dt, \quad (15.6)
\end{aligned}$$

which implies

$$|f(u,v) - p(u,v)| \leq c \cdot \left(|u-x|^{l+1} + |v-z|^{l+1}\right)$$

for some constant $c \in \mathcal{R}_+$ which depends only on l and on the partial derivatives of f of order less than or equal to l. From this, $p(x,z) = f(x,z)$ and $Qp = p$, we conclude, as in the proof of Theorem 14.3,

$$\begin{aligned}
&|(Qf)(x,z) - f(x,z)| \\
=\ &|(Qf)(x,z) - p(x,z)| \\
=\ &|(Qf)(x,z) - (Qp)(x,z)| \\
\leq\ &\sum_{i=-M_x}^{K-1} \sum_{j=-M_z}^{L-1} |Q_{(i,j)}(f-p)| \cdot B_{(i,j),M_x,u,M_z,v}(x,z)
\end{aligned}$$

$$\text{(because of (15.3))}$$

$$=\ \sum_{i=i_0-M_x}^{i_0} \sum_{j=j_0-M_z}^{j_0} |Q_{(i,j)}(f-p)| \cdot B_{(i,j),M_x,u,M_z,v}(x,z)$$

$$\text{(because of (15.2))}$$

$$\leq \quad \max_{i\in\{i_0-M_x,\ldots,i_0\},\, j\in\{j_0-M_z,\ldots,j_0\}} |Q_{(i,j)}(f-p)|$$

$$\text{(because of (15.3) and (15.4))}$$

$$\leq \quad \max_{\substack{i\in\{i_0-M_x,\ldots,i_0\},\\ j\in\{j_0-M_z,\ldots,j_0\}}} \|Q\| \cdot \|f-p\|_{\infty,[u_i,u_{i+M_x+1})\times[v_j,v_{j+M_z+1})}$$

$$\leq \quad \|Q\| \cdot \|f-p\|_{\infty,[u_{i_0}-h_u(x),u_{i_0}+h_u(x))\times[v_{j_0}-h_v(z),v_{j_0}+h_v(z))}$$

$$\text{(by definition of } h_u(x) \text{ and } h_v(z))$$

$$\leq \quad \|Q\| \cdot c \cdot \left((2\,h_u(x))^{l+1} + (2\,h_v(z))^{l+1}\right).$$

This implies the assertion. $\qquad\qquad\square$.

Next we show that there always exists a bounded quasi interpolant of order (M_x, M_z).

Theorem 15.2. *Let $M_x, M_z \in \mathcal{N}_0$, $u_{-M_x} \leq \cdots \leq u_0 < \cdots < u_K \leq \cdots \leq u_{K+M_x}$, and $v_{-M_z} \leq \cdots \leq v_0 < \cdots < v_L \leq \cdots \leq v_{L+M_z}$. Then there exists a bounded quasi interpolant $Q : C(\mathcal{R}^2) \to S_{u,M_x,v,M_z}$ of order (M_x, M_z) such that $\|Q\|$ is bounded above by a constant depending only on (M_x, M_z) (and not on the knot sequences).*

PROOF. Let $t_{j,k}, Q_{j,k}$ be defined as in the proof of Theorem 14.4 for the knot sequence u and let $s_{j,k}, R_{j,k}$ be defined in the same way for the knot sequence v. Define $Q_{(i,j)} : C(\mathcal{R}^2) \to \mathcal{R}$ by

$$Q_{(i,j)}(f) = \sum_{k=0}^{M_x} \sum_{l=0}^{M_z} Q_{i,k} \cdot R_{j,l} \cdot f(t_{i,k}, s_{j,l}).$$

Then

$$Qf := \sum_{i=-M_x}^{K-1} \sum_{j=-M_z}^{L-1} Q_{(i,j)}(f) \cdot B_{(i,j),M_x,u,M_z,v}$$

defines a bounded quasi interpolant whose norm is bounded above by

$$\max_{i,j} \sum_{k=0}^{M_x} \sum_{l=0}^{M_z} |Q_{i,k}| \cdot |R_{j,l}|. \tag{15.7}$$

It follows from the proof of Theorem 14.4, that the term in (15.7) is bounded above by some constant depending only on (M_x, M_z). Because of

$$Q_{(i,j)}\left(p(\cdot) \cdot q(\cdot)\right) = \left(\sum_{k=0}^{M_x} Q_{i,k} \cdot p(t_{i,k})\right) \cdot \left(\sum_{l=0}^{M_z} R_{j,l} \cdot p(s_{j,l})\right)$$

the quasi interpolant Q has order (M_x, M_z). $\qquad\qquad\square$

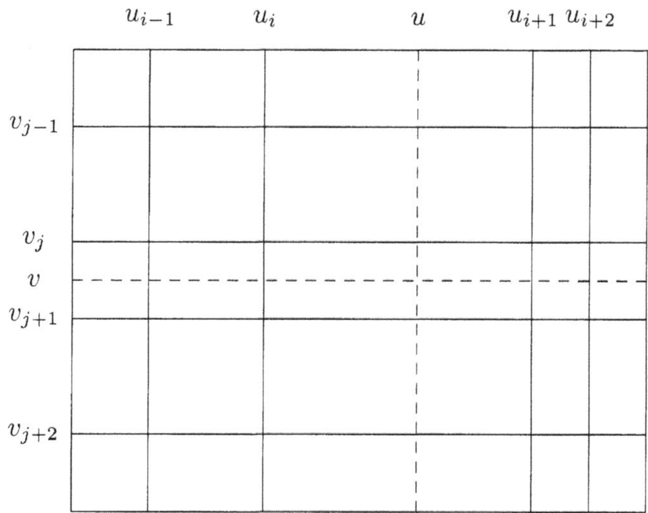

Figure 15.3. Local refinement of the partition corresponding to S_{u,M_x,v,M_z} is not possible.

15.2   Consistency

In the next two sections we will consider estimates defined by applying the principle of least squares to tensor product spline spaces.

According to the previous section one can define a space S_{u,M_x,v,M_z} of tensor product spline functions $f : \mathcal{R}^2 \to \mathcal{R}$ by choosing degrees $M_x, M_z \in \mathcal{N}_0$ and knot sequences $u = \{u_i\}_{-M_x \leq i \leq K+M_x}$ and $v = \{v_j\}_{-M_z \leq j \leq L+M_z}$. The functions in S_{u,M_x,v,M_z} are piecewise polynomials of degree less than or equal to (M_x, M_z) with respect to the partition of $\mathcal{R}^2$ consisting of rectangles $[u_i, u_{i+1}) \times [v_j, v_{j+1})$ $(-M_x \leq i < K+M_x, -M_z \leq j < L+M_z)$.

It follows from Theorem 15.1 that one should choose the size of a rectangle $[u_i, u_{i+1}) \times [v_j, v_{j+1})$ small in order to obtain good approximations of smooth functions by functions of S_{u,M_x,v,M_z} on that rectangle. The sizes of the rectangles $[u_i, u_{i+1}) \times [v_j, v_{j+1})$ are determined by the knot sequences u and v. Unfortunately, one cannot control them locally, i.e., one cannot choose them small in one subset of $\mathcal{R}^2$ without influencing them in the rest of $\mathcal{R}^2$. To see this, assume that you want to make the size of one fixed rectangle $[u_{i_0}, u_{i_0+1}) \times [v_{j_0}, v_{j_0+1})$ smaller. You can do this by introducing a new knot between u_{i_0} and u_{i_0+1} (or by introducing a new knot between v_{j_0} and v_{j_0+1}). But this also splits all the rectangles $[u_{i_0}, u_{i_0+1}) \times [v_j, v_{j+1})$ $(-M_z \leq j < L + M_z)$ (or all rectangles $[u_i, u_{i+1}) \times [v_{j_0}, v_{j_0+1})$ $(-M_x \leq i < K + M_x)$), see Figure 15.3.

Of course it is possible to control the size of the rectangles $[u_i, u_{i+1}) \times [v_j, v_{j+1})$ globally, i.e., to choose them all together too small or too large.

For this it is sufficient to use equidistant knot sequences and, therefore, we will consider, in the next two sections, only tensor product spline spaces with equidistant knot sequences.

In the following we will define a space of tensor product spline functions $f : \mathcal{R}^d \to \mathcal{R}$. This will depend on parameters $L_n, R_n \in \mathcal{R}$, $K \in \mathcal{N}$, and $M \in \mathcal{N}_0$.

Let $B^1_{i,M,K}$ be the univariate B-spline with degree M, knot sequence $\{L_n + i \cdot \frac{R_n - L_n}{K}\}_{i \in \mathcal{Z}}$, and support

$$[L_n + i \cdot \frac{R_n - L_n}{K}, L_n + (i + M + 1) \cdot \frac{R_n - L_n}{K}].$$

For $i_1, \ldots, i_d \in \mathcal{Z}$ define the tensor product B-spline

$$B^d_{(i_1,\ldots,i_d),M,K} : \mathcal{R}^d \to \mathcal{R}$$

by

$$B^d_{(i_1,\ldots,i_d),M,K}(x_1,\ldots,x_d) = B^1_{i_1,M,K}(x_1) \cdot \ldots \cdot B^1_{i_d,M,K}(x_d)$$

$(x_1,\ldots,x_d \in \mathcal{R})$. Then the tensor product spline space $S_{M,K}([L_n, R_n]^d)$ is defined as

$$S_{M,K}([L_n, R_n]^d)$$
$$= \ \ span\left\{ B^d_{(i_1,\ldots,i_d),M,K} \ : \ supp(B^d_{(i_1,\ldots,i_d),M,K}) \cap [L_n, R_n]^d \neq \emptyset \right\}.$$

The next theorem gives conditions on the parameters of $S_{M,K}([L_n, R_n]^d)$ which imply that the resulting least squares estimates are strongly universally consistent. There only M and K depend on the data.

Theorem 15.3. *For* $n \in \mathcal{N}$ *let* $L_n, R_n \in \mathcal{R}$, $M_{max}(n) \in \mathcal{N}_0$, $K_{min}(n), K_{max}(n) \in \mathcal{N}$, *and* $\beta_n \in \mathcal{R}_+$, *such that* $L_n \leq R_n$, $K_{min}(n) \leq K_{max}(n)$,

$$\beta_n \to \infty \quad (n \to \infty) \tag{15.8}$$

$$L_n \to -\infty \quad (n \to \infty), \ R_n \to \infty \quad (n \to \infty), \tag{15.9}$$

$$(M_{max}(n) + 1) \cdot \frac{R_n - L_n}{K_{min}(n)} \to 0 \quad (n \to \infty), \tag{15.10}$$

$$\frac{(M_{max}(n) + 1)^d (K_{max}(n) + M_{max}(n))^d \beta_n^4 \log(n)}{n} \to 0 \quad (n \to \infty), \tag{15.11}$$

and

$$\frac{\beta_n^4}{n^{1-\delta}} \to 0 \quad (n \to \infty) \tag{15.12}$$

for some $\delta > 0$.

Depending on D_n choose $M \in \mathcal{N}_0$ and $K \in \mathcal{N}$ such that $M \le M_{max}(n)$ and $K_{min}(n) \le K \le K_{max}(n)$, and define the estimate m_n by

$$\tilde{m}_n(\cdot) = \arg \min_{f \in S_{M,K}([L_n,R_n]^d)} \frac{1}{n} \sum_{i=1}^{n} |f(X_i) - Y_i|^2 \tag{15.13}$$

and

$$m_n = T_{\beta_n} \tilde{m}_n. \tag{15.14}$$

Then m_n is strongly universally consistent.

PROOF. Because of Theorem 10.2 it suffices to show

$$\inf_{f \in S_{M,K}([L_n,R_n]^d),\|f\|_\infty \le \beta_n} \int |f(x) - m(x)|^2 \mu(dx) \to 0 \quad (n \to \infty) \quad a.s. \tag{15.15}$$

and

$$\sup_{f \in T_{\beta_n} S_{M,K}([L_n,R_n]^d)} \left| \frac{1}{n} \sum_{i=1}^{n} |f(X_i) - Y_{i,L}|^2 - \mathbf{E}|f(X) - Y_L|^2 \right| \to 0 \tag{15.16}$$

$(n \to \infty)$ *a.s.* for every $L > 0$.

PROOF OF (15.15). $C_0^\infty(\mathcal{R}^d)$ is dense in $L_2(\mu)$ (cf. Corollary A.1), hence it suffices to prove (15.15) for functions $m \in C_0^\infty(\mathcal{R}^d)$. Because of (15.8) one may further assume that $\|m\|_\infty \le \beta_n$.

Set $I = \{(i_1, \ldots, i_d) \in \mathcal{Z}^d : supp(B_{(i_1,\ldots,i_d),M,K}^d) \cap [L_n, R_n]^d \ne \emptyset\}$. For $i \in I$ choose $u_i \in supp(B_{i,M,K}^d)$ and set

$$Qm = \sum_{i \in I} m(u_i) \cdot B_{i,M,K}^d.$$

Then $\|Qm\|_\infty \le \|m\|_\infty$, which implies

$$\inf_{f \in S_{M,K}([L_n,R_n]^d),\|f\|_\infty \le \beta_n} \int |f(x) - m(x)|^2 \mu(dx)$$

$$\le \int |(Qm)(x) - m(x)|^2 \mu(dx)$$

$$\le 4\|m\|_\infty^2 \mu\left(\mathcal{R}^d \setminus [L_n, R_n]^d\right) + \int_{[L_n,R_n]^d} |(Qm)(x) - m(x)|^2 \mu(dx)$$

$$\le 4\|m\|_\infty^2 \mu\left(\mathcal{R}^d \setminus [L_n, R_n]^d\right) + \sup_{x \in [L_n,R_n]^d} |(Qm)(x) - m(x)|^2.$$

Formula (15.9) implies

$$4\|m\|_\infty^2 \mu\left(\mathcal{R}^d \setminus [L_n, R_n]^d\right) \to 0 \quad (n \to \infty).$$

Furthermore, for $x \in [L_n, R_n]^d$, one gets

$$|(Qm)(x) - m(x)|$$

$$\leq \sum_{i \in I} |m(u_i) - m(x)| \cdot B^d_{i,M,K}(x)$$

$$= \sum_{i \in I, x \in supp(B^d_{i,M,K})} |m(u_i) - m(x)| \cdot B^d_{i,M,K}(x)$$

$$\leq \sup_{\substack{u,v \in \mathcal{R}^d, \\ \|u-v\|_\infty \leq (M+1) \cdot \frac{R_n - L_n}{K}}} |m(u) - m(v)| \cdot \sum_{i \in I, x \in supp(B^d_{i,M,K})} B^d_{i,M,K}(x)$$

$$\leq \sup_{\substack{u,v \in \mathcal{R}^d, \\ \|u-v\|_\infty \leq (M_{max}(n)+1) \cdot \frac{R_n - L_n}{K_{min}(n)}}} |m(u) - m(v)|,$$

from which we conclude

$$\sup_{x \in [L_n, R_n]^d} |(Qm)(x) - m(x)|^2$$

$$\leq \sup_{\substack{u,v \in \mathcal{R}^d, \\ \|u-v\|_\infty \leq (M_{max}(n)+1) \cdot \frac{R_n - L_n}{K_{min}(n)}}} |m(u) - m(v)| \to 0 \quad (n \to \infty)$$

because of (15.10) and $m \in C_0^\infty(\mathcal{R}^d)$.

PROOF OF (15.16). Let $\mathcal{G}_n$ be the set of all polynomials $p : \mathcal{R}^d \to \mathcal{R}$ of degree $M_{max}(n)$ or less in each coordinate. Each function in $S_{M,K}([L_n, R_n]^d)$ is, on each rectangle

$$\left[L_n + i_1 \frac{R_n - L_n}{K}, L_n + (i_1 + 1)\frac{R_n - L_n}{K} \right) \times \cdots$$

$$\times \left[L_n + i_d \frac{R_n - L_n}{K}, L_n + (i_d + 1)\frac{R_n - L_n}{K} \right)$$

$(i_1, \ldots, i_d \in \{-M_{max}(n), \ldots, K + M_{max}(n) - 1\})$, equal to a polynomial contained in $\mathcal{G}_n$. Denote the family of all partitions consisting of such rectangles (and one additional set) for some $K \leq K_{max}(n)$ by Π_n. Then $S_{M,K}([L_n, R_n]^d) \subseteq \mathcal{G}_n \circ \Pi_n$ and it suffices to prove (15.16) with the data-dependent set $S_{M,K}([L_n, R_n]^d)$ replaced by the data-independent set $\mathcal{G}_n \circ \Pi_n$.

$\mathcal{G}_n$ is a linear vector space of functions of dimension $(M_{max}(n)+1)^d$, thus $V_{\mathcal{G}_n^+} \leq (M_{max}(n)+1)^d + 1$. Furthermore, Π_n is a family of cubic partitions consisting of $(K_{max}(n) + 2 M_{max}(n))^d + 1$ or less sets. As in the proof of (13.23) one gets

$$\Delta_n(\Pi_n) \leq n \cdot (K_{max}(n) + 2 M_{max}(n))^d.$$

From this, (15.11), and (15.12), one obtains the assertion as in the proof of Theorem 13.1.

$$\square$$

15.3   Rate of Convergence

In this section we derive rate of convergence results for least squares estimates using tensor product spline spaces with equidistant knots. Throughout this section we will assume $supp(X)$ bounded, for notational simplicity even $supp(X) \subseteq [0,1]^d$.

Let $M \in \mathcal{N}_0$ and let $\{K_n\}_{n \in \mathcal{N}}$ be a sequence of natural numbers. In the previous section we have introduced spline spaces $S_{M,K_n}([0,1]^d)$. With this notation define the estimate m_n by

$$\tilde{m}_n = \arg \min_{f \in S_{M,K_n}([0,1]^d)} \frac{1}{n} \sum_{i=1}^{n} |f(X_i) - Y_i|^2$$

and set

$$m_n(x) = T_L(\tilde{m}_n(x)) \quad (x \in \mathcal{R}^d).$$

Using Theorems 11.3 and 15.1 one can show

Theorem 15.4. *Let the estimate m_n be defined as above. Let $p \in \{1, \dots, M+1\}$ and assume that*

$$\sigma^2 = \sup_{x \in [0,1]^d} \mathbf{Var}\{Y|X = x\} < \infty,$$

$$\|m\|_\infty = \sup_{x \in [0,1]^d} |m(x)| \le L,$$

$$supp(X) \subseteq [0,1]^d,$$

and that m is p times continuously differentiable. Then there exist constants $c_1, c_2 \in \mathcal{R}_+$ which depend only on σ^2, L, d, M, and the supremum norms of the partial derivatives of m of order less than or equal to p such that

$$\mathbf{E} \int |m_n(x) - m(x)|^2 \mu(dx)$$

$$\le c_1 \frac{(\log(n) + 1) \cdot (K_n + M + 1)^d}{n} + c_2 \left(\frac{M+1}{K_n}\right)^{2p}.$$

The proof is left to the reader (cf. Problem 15.3).

Corollary 15.1. *Let $p \in \mathcal{N}$, set $M = p - 1$,*

$$K_n = \left\lfloor \left(\frac{n}{\log(n)}\right)^{\frac{1}{2p+d}} \right\rfloor$$

and define the estimate m_n as above. Then one has

$$\mathbf{E}\{\int |m_n(x) - m(x)|^2 \mu(dx)\} = O\left(\left(\frac{\log(n)}{n}\right)^{\frac{2p}{2p+d}}\right)$$

for every distribution of (X, Y) with

$$\sigma^2 = \sup_{x \in [0,1]^d} \mathbf{Var}\{Y|X = x\} < \infty,$$

$$\|m\|_\infty = \sup_{x \in [0,1]^d} |m(x)| \leq L,$$

$$supp(X) \subseteq [0,1]^d,$$

and m p times continuously differentiable.

The definition of the estimate in Corollary 15.1 depends on the smoothness p of the regression function. This can be avoided by using complexity regularization.

Set $\mathcal{P}_n = \{1, \ldots, n\}$ and for $K \in \mathcal{P}_n$ define the estimate $m_{n,K}$ by

$$\tilde{m}_{n,K} = \arg\min_{f \in S_{M,K}([0,1]^d)} \frac{1}{n} \sum_{i=1}^n |f(X_i) - Y_i|^2$$

and

$$m_{n,K}(x) = T_L(\tilde{m}_{n,K}(x)) \quad (x \in \mathcal{R}^d).$$

To penalize the complexity of $S_{K,M}$ we use the penalty term

$$pen_n(K) = \frac{\log^2(n) \cdot K^d}{n}.$$

Set

$$K^* = \arg\min_{K \in \mathcal{P}_n} \left\{ pen_n(K) + \frac{1}{n} \sum_{i=1}^n |m_{n,K}(X_i) - Y_i|^2 \right\}$$

and

$$m_n(x) = m_{n,K^*}(x) \quad (x \in \mathcal{R}^d).$$

Theorem 12.1 implies the following result:

Theorem 15.5. *Let $L \in \mathcal{R}_+$, $M \in \mathcal{N}_0$, and let m_n be defined as above. Then one has, for any $p \in \{1, \ldots, M+1\}$,*

$$\mathbf{E}\{\int |m_n(x) - m(x)|^2 \mu(dx)\} = O\left(\left(\frac{\log^2(n)}{n}\right)^{\frac{2p}{2p+d}}\right)$$

for every distribution of (X, Y) with $supp(X) \subseteq [0,1]^d$, $|Y| \leq L$ a.s., and m p times continuously differentiable.

Observe that the estimate in Theorem 15.5 doesn't depend on p any more.

15.4   Bibliographic Notes

For references concerning the deterministic properties of splines presented in this chapter we refer to Section 14.5. Estimates which are based on an adaptive choice of multivariate spline spaces, constructed by stepwise insertion and deletion of knots, are investigated in Friedman (1991) and Stone et al. (1997).

Problems and Exercises

PROBLEM 15.1. Prove Lemma 15.1.

PROBLEM 15.2. Prove (15.6).
   HINT: Apply the identity

$$g(1) = \sum_{j=0}^{l} \frac{g^{(j)}(0)}{j!} \cdot (1-0)^j + \frac{1}{l!} \int_0^1 (1-t)^l g^{(l+1)}(t)\, dt$$

to

$$g(t) = f(x + t \cdot (u - x), z + t \cdot (v - z)).$$

PROBLEM 15.3. Prove Theorem 15.4.
   HINT: Proceed as in the proof of Theorem 14.5.

PROBLEM 15.4. Prove Theorem 15.5.
   HINT: Use Theorem 12.1.

16
Neural Networks Estimates

16.1   Neural Networks

Artificial neural networks (ANN) have been motivated by the desire to model the human brain by a computer. ANN consist of computational units called neurons. They receive a number of input signals and produce an output signal. Originally proposed neuron models produced binary output signals resembling the response of biological neurons which are firing or produce an output when their input activation exceeds some threshold. Neurons in ANN are also interconnected as are neurons in the brain. But, here, the similarities between ANN and biological neural networks end. The number of neurons in the brain and the degree of their connectivity is much higher than in ANN. Furthermore, the learning process in the brain is entirely different from the learning of parameters in ANN. The artificial neuron is a very simplified model of a real neuron which consists of dendrites (connections) and an axon (central cell). The electrophysiological processes taking place in a real neuron can be modeled to some degree by dynamical models which are not incorporated into the static artificial neuron. In spite of these shortcomings, ANN have become very useful in designing learning algorithms for artificial intelligence.

A simple model of a neuron was introduced in the seminal paper of McCulloch and Pitts (1943) who modeled the neuron by a binary thresholding device. The *McCulloch–Pitts* model shown in Figure 16.1 consists of a linear weighted combination of the inputs followed by the threshold function. A McCulloch–Pitts neuron has been applied in pattern recognition by Rosen-

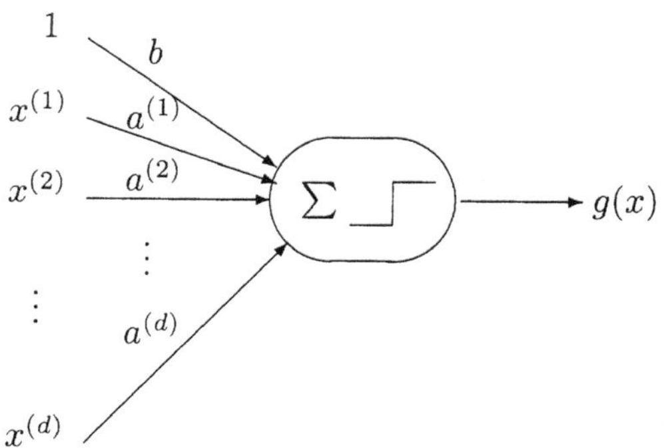

Figure 16.1. A McCulloch–Pitts neuron.

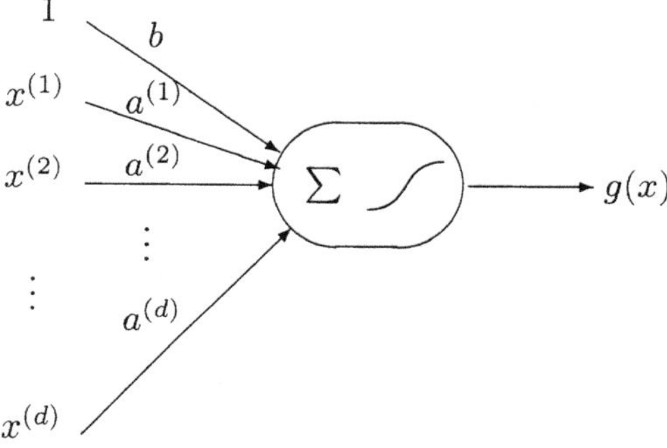

Figure 16.2. An artificial neuron.

blatt (1958) who also proposed a *perceptron* learning algorithm and proved
its convergence, see Rosenblatt (1962).

By replacing the threshold function in the McCulloch-Pitts neuron
model by a sigmoid function we obtain an artificial neuron illustrated in
Figure 16.2. It is formally defined as follows.

An *artificial neuron* is a real-valued function on $\mathcal{R}^d$ given by

$$g(x) = \sigma(a^T x + b),$$

where $x \in \mathcal{R}^d$ is an input vector, $a^T = (a^{(1)}, \ldots, a^{(d)}) \in \mathcal{R}^d, b \in \mathcal{R}$, are the
weights, and $\sigma(x) : \mathcal{R} \to [0, 1]$ is referred to as a sigmoid function.

Perceptrons can solve linear pattern recognition problems but fail to
solve trivial nonlinear pattern recognition problems such as discriminating

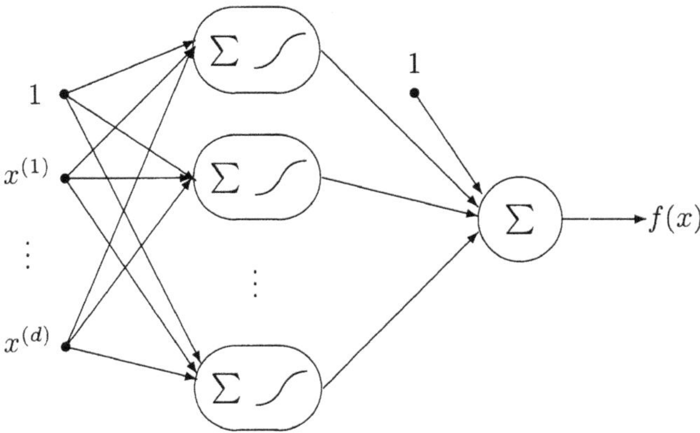

Figure 16.3. Neural network with one hidden layer.

two classes represented by the vertices of a square, where points from the same class are the end points of a diagonal (exclusive-or problem – see Minsky and Pappert (1969)). The limitations of perceptrons were obviated by adding additional layers of neurons called hidden layers.

A *feedforward neural network* with a hidden layer of k neurons is a real-valued function on $\mathcal{R}^d$ of the form

$$f(x) = \sum_{i=1}^{k} c_i \sigma(a_i^T x + b_i) + c_0,$$

where $\sigma : \mathcal{R} \to [0,1]$ is called a *sigmoid* function and $a_1, \ldots, a_k \in \mathcal{R}^d, b_1, \ldots, b_k, c_0, \ldots, c_k \in \mathcal{R}$ are the parameters that specify the network.

Definition 16.1. *A sigmoid function σ is called a squashing function if it is nondecreasing,* $\lim_{x \to -\infty} \sigma(x) = 0$ *and* $\lim_{x \to \infty} \sigma(x) = 1$.

Since squashing functions are nondecreasing they have at most a countable number of discontinuities and are therefore measurable. Some examples of squashing functions, illustrated in Figure 16.3, are given below:

- threshold squasher

$$\sigma(x) = I_{\{x \in [0,\infty)\}};$$

- ramp squasher

$$\sigma(x) = x I_{\{x \in [0,1]\}} + I_{\{x \in (1,\infty)\}};$$

- cosine squasher

$$\sigma(x) = (1 + \cos(x + 3\pi/2))\frac{1}{2} I_{\{x \in [-\pi/2, \pi/2]\}} + I_{\{x \in (\pi/2, \infty)\}};$$

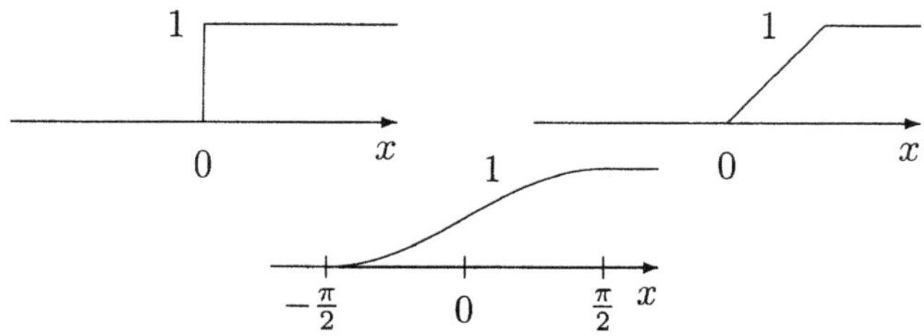

Figure 16.4. Threshold, ramp, and cosine squashers.

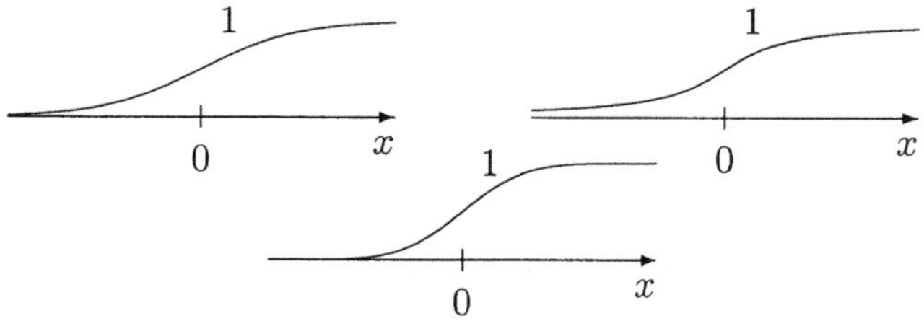

Figure 16.5. Logistic, arctan and Gaussian squashers.

- logistic squasher

$$\sigma(x) = \frac{1}{1 + \exp(-x)};$$

- arctan squasher

$$\sigma(x) = \frac{1}{2} + \frac{1}{\pi} \arctan(x);$$

- Gaussian squasher

$$\sigma(x) = \frac{1}{\sqrt{2\pi}} \int_{-\infty}^{x} \exp(-y^2/2)\,dy.$$

16.2 Consistency

We now consider neural network regression function estimators. Given the training sequence $D_n = \{(X_1, Y_1), \ldots, (X_n, Y_n)\}$ of n i.i.d. copies of (X, Y) the parameters of the network are chosen to minimize the empirical L_2 risk

$$\frac{1}{n} \sum_{j=1}^{n} |f(X_j) - Y_j|^2. \tag{16.1}$$

However, in order to obtain consistency, we again, as in the previous chapters, restrict the range of some of the parameters. With a constraint on the

c_i's, we minimize (16.1) for the class of neural networks

$$\mathcal{F}_n = \left\{ \sum_{i=1}^{k_n} c_i \sigma(a_i^T x + b_i) + c_0 : k_n \in \mathcal{N},\ a_i \in \mathcal{R}^d,\ b_i \in \mathcal{R},\ \sum_{i=0}^{k_n} |c_i| \leq \beta_n \right\}$$

(16.2)

and obtain $m_n \in \mathcal{F}_n$ satisfying

$$\frac{1}{n} \sum_{j=1}^n |m_n(X_j) - Y_j|^2 = \min_{f \in \mathcal{F}_n} \frac{1}{n} \sum_{j=1}^n |f(X_j) - Y_j|^2.$$

(16.3)

We assume that the minimum in (16.3) exists although it may not be unique. If the minimum does not exist we can carry out our analysis with functions whose empirical L_2 risk is arbitrarily close to the infimum.

Regarding the computational aspect there is no practical algorithm for optimizing (16.3). A practical steepest descent algorithm called *backpropagation* iteratively converges to a local minimum of the empirical L_2 risk. However, there is no guarantee that the global minimum will be reached. The backpropagation algorithm has been introduced by Rumelhart and McClelland (1986).

The next consistency theorem states that with certain restrictions imposed on β_n and k_n, the empirical L_2 risk minimization provides universally consistent neural network estimates.

Theorem 16.1. *Let $\mathcal{F}_n$ be the class of neural networks with the squashing function σ defined in (16.2) and let m_n be the network that minimizes the empirical L_2 risk in $\mathcal{F}_n$. If k_n and β_n satisfy*

$$k_n \to \infty, \quad \beta_n \to \infty \quad and \quad \frac{k_n \beta_n^4 \log(k_n \beta_n^2)}{n} \to 0,$$

then $\mathbf{E} \int (m_n(x)) - m(x))^2 \mu(dx) \to 0 \quad (n \to \infty)$ for all distributions of (X, Y) with $\mathbf{E}|Y|^2 < \infty$, that is, m_n is weakly universally consistent. If, in addition, there exists $\delta > 0$ such that $\beta_n^4 / n^{1-\delta} \to 0$, then $\int (m_n(x)) - m(x))^2 \mu(dx) \to 0$ almost surely, that is, m_n is strongly universally consistent.

In order to handle the approximation error, we need a denseness theorem for feedforward neural networks. We will start with an approximation result in sup norm because of its importance in the neural network field. In the following approximation lemma we will take the approach similar to Hornik, Stinchcombe, and White (1989) and make use of the Stone–Weierstrass theorem.

Lemma 16.1. *Let σ be a squashing function and let K be a compact subset of $\mathcal{R}^d$. Then, for every continuous function $f : \mathcal{R}^d \to \mathcal{R}$ and every $\epsilon > 0$,* *there exists a neural network*

$$h(x) = \sum_{i=1}^{k} c_i \sigma(a_i^T x + b_i) + c_0 \quad (a_i \in \mathcal{R}^d, b_i, c_i \in \mathcal{R}),$$

such that

$$\sup_{x \in K} |f(x) - h(x)| < \epsilon.$$

PROOF. The proof will be divided into several steps. First we use the Stone–Weierstrass theorem to show that networks of the form

$$\sum_{i=1}^{k} c_i \cos(a_i^T x + b_i)$$

are uniformly dense on compacts in the space of continuous functions. Then we show that cosine functions can be approximated in sup norm by networks with the cosine squasher. In the final step of the proof we show that networks with the cosine squasher can be approximated in sup norm by networks with an arbitrary squashing function σ.

STEP 1. Consider a class of cosine networks

$$\mathcal{F} = \left\{ \sum_{i=1}^{k} c_i \cos(a_i^T x + b_i), k \in \mathcal{N}, a_i \in \mathcal{R}^d, \ b_i, c_i \in \mathcal{R} \quad (i = 1, \ldots, k) \right\}.$$

Observe that $\mathcal{F}$ is an algebra since it is closed under addition, multiplication, and scalar multiplication. Closedness under multiplication follows by repeated application of the trigonometric identity

$$\cos a \cos b = \frac{1}{2}(\cos(a+b) + \cos(a-b)).$$

We say that the family $\mathcal{F}$ *separates* points on K if for every $x, y \in K, x \neq y$ there is a function $f \in \mathcal{F}$ such that $f(x) \neq f(y)$. The family $\mathcal{F}$ *vanishes at no point* of K if there exists $f \in \mathcal{F}$ such that for every $x \in K$, $f(x) \neq 0$. We will now apply the Stone–Weierstrass theorem, see Rudin (1964). It states that every algebra defined in the space $C_K(\mathcal{R})$ of real continuous functions on a compact set K, that separates points on K and vanishes at no point of K, is dense on $C_K(\mathcal{R})$ in sup norm. In other words, the uniform closure of $\mathcal{F}$ consists of all real continuous functions on K. Clearly $\mathcal{F}$ is an algebra on K. $\mathcal{F}$ separates points on K and vanishes at no point of K. To see this pick $x, y \in K, x \neq y$. For a suitable $a \in \mathcal{R}^d$ and $b = 0$ we have $\cos(a^T x + 0) \neq \cos(a^T y + 0)$. Observe that a should be chosen such that $a^T x \in (-\pi, \pi)$, $a^T y \in (-\pi, \pi)$, and $a^T x \neq a^T y$. Next pick b such that $\cos(b) \neq 0$ and set a to zero. Clearly for all $x \in K \cos(a^T x + b) = \cos(b) \neq 0$. This implies that $\mathcal{F}_n$ is uniformly dense in the space of real continuous functions on K.

STEP 2. Let $c(x)$ be the cosine squasher of Section 16.1. We now show that we can reconstruct $\cos(u)$ on any compact interval $[-M, M]$ by

$$f(u) = \sum_{i=1}^{l} \gamma_i c(\alpha_i u + \beta_i) \quad (l \in \mathcal{N}, \alpha_i, \beta_i, \gamma_i \in \mathcal{R}).$$

Varying the parameters α_i, β_i, and γ_i we can shift $f(u)$ along the $x-$ and $y-$ axes, thus it suffices to show that we can reconstruct $2(\cos(u) + 1)$ on any compact interval by functions $f(u)$. It is easy to see that

$$(\cos(u) + 1)I_{\{u \in [-\pi, \pi]\}} = 2(c(u + \pi/2) - c(u - \pi/2)) \tag{16.4}$$

(see Figure 16.6).

By adding a finite number of shifted functions (16.4) we can reconstruct the cosine on any compact interval $[-M, M]$. If the squashing function is the cosine squasher then the conclusion of the lemma follows immediately from Steps 1 and 2.

STEP 3. Next we show that, for every $\epsilon > 0$ and arbitrary squashing function σ, there is a neural network $h_\epsilon(u) = \sum_{i=1}^{k} c_i \sigma(a_i u + b_i) + c_0$ such that

$$\sup_{u \in \mathcal{R}} |h_\epsilon(u) - c(u)| < \epsilon. \tag{16.5}$$

Pick an arbitrary ϵ $(0 < \epsilon < 1)$ and k such that $1/k < \epsilon/4$. Since σ is a squashing function there is an $M > 0$ such that

$$\sigma(-M) < \epsilon/(2k) \tag{16.6}$$

and

$$\sigma(M) > 1 - \epsilon/(2k). \tag{16.7}$$

Next by the continuity and monotonicity of c we can find constants $r_1, \ldots, r_k$ such that

$$c(r_i) = i/k, \quad i = 1, \ldots, k - 1,$$

and

$$c(r_k) = 1 - 1/2k \tag{16.8}$$

(see Figure 16.7).

For $1 \le i \le k-1$ pick a_i, b_i such that $a_i r_i + b_i = -M$ and $a_i r_{i+1} + b_i = M$, i.e., $a_i = \frac{2M}{r_{i+1} - r_i}$ and $b_i = \frac{M(r_i + r_{i+1})}{r_i - r_{i+1}}$. Thus $a_i u + b_i$ is the line through the points $(r_i, -M)$ and (r_{i+1}, M) and clearly $a_i > 0$ $(i = 1, \ldots, k - 1)$. We will now verify that the network

$$h_\epsilon(u) = \frac{1}{k} \sum_{j=1}^{k-1} \sigma(a_j u + b_j)$$

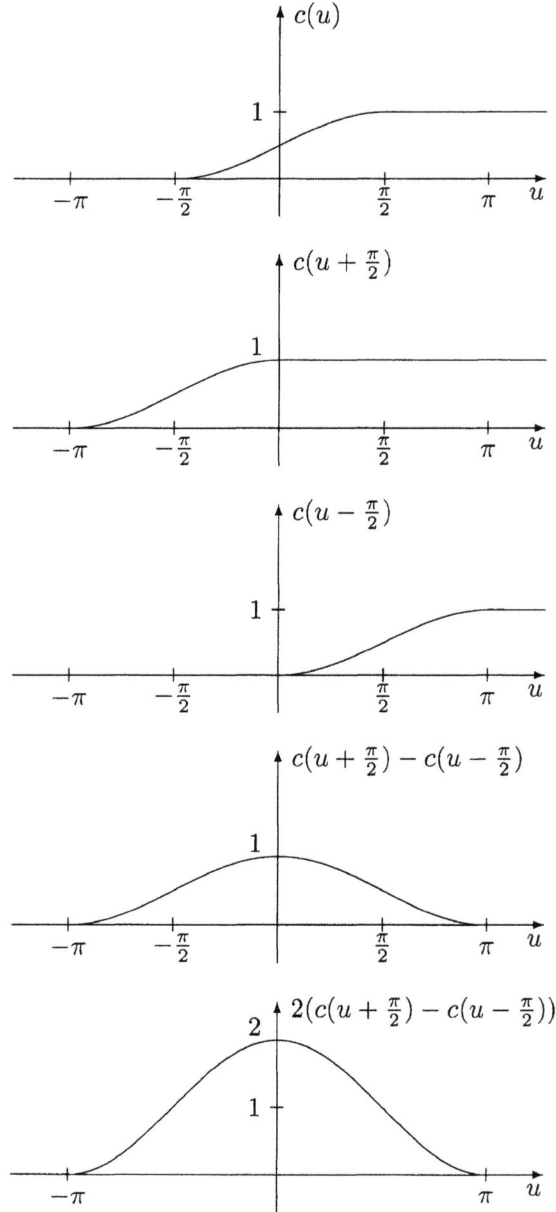

Figure 16.6. Reconstruction of cosine by cosine squashers.

approximates $c(u)$ in sup norm, that is,

$$|c(u) - h_\epsilon(u)| < \epsilon$$

on each of the subintervals $(-\infty, r_1], (r_1, r_2], \ldots, (r_k, \infty)$.

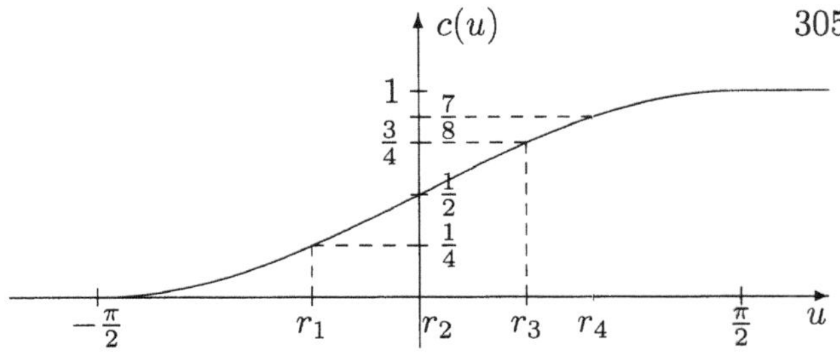

Figure 16.7. Construction of the sequence $r_1 \ldots r_k$.

Let $i \in \{1, \ldots, k-1\}$ and $u \in (r_i, r_{i+1}]$. Then $\frac{i}{k} \le c(u) \le \frac{i+1}{k}$. For $j \in \{1, \ldots, i-1\}$ we have

$$\sigma(a_j u + b_j) \ge \sigma(a_j r_{j+1} + b_j)$$
$$= \sigma(M) \ge 1 - \epsilon/(2k)$$

and for $j \in \{i+1, \ldots, k-1\}$ we have

$$\sigma(a_j u + b_j) \le \sigma(a_j r_j + b_j) = \sigma(-M) < \epsilon/(2k).$$

Split the sum in $h_\epsilon(u)$ into three components:

$$\frac{1}{k} \sum_{j=1}^{i-1} \sigma(a_j u + b_j), \quad \frac{1}{k}\sigma(a_i u + b_i), \quad \text{and} \quad \frac{1}{k} \sum_{j=i+1}^{k-1} \sigma(a_j u + b_j).$$

Applying the inequalities above and bounding each component yields

$$|c(u) - h_\epsilon(u)|$$

$$\le \left| c(u) - \frac{1}{k} \sum_{j=1}^{i-1} \sigma(a_j u + b_j) \right| + \frac{1}{k}\sigma(a_i u + b_i) + \frac{1}{k} \sum_{j=i+1}^{k-1} \sigma(a_j u + b_j)$$

$$\le \left| c(u) - \frac{i-1}{k} \right| + \left| \frac{i-1}{k} - \frac{1}{k} \sum_{j=1}^{i-1} \sigma(a_j u + b_j) \right|$$

$$+ \frac{1}{k}\sigma(a_i u + b_i) + \frac{1}{k} \sum_{j=i+1}^{k-1} \sigma(a_j u + b_j)$$

$$\le \left(\frac{i+1}{k} - \frac{i-1}{k} \right) + \left(\frac{i-1}{k} - \frac{1}{k}(i-1)(1 - \frac{\epsilon}{2k}) \right)$$

$$+ \frac{1}{k} + \frac{1}{k}(k-1-i)\frac{\epsilon}{2k}$$

$$= \frac{2}{k} + \frac{1}{k}(i-1)\frac{\epsilon}{2k} + \frac{1}{k} + \frac{1}{k}(k-1-i)\frac{\epsilon}{2k}$$

$$\le \frac{3}{k} + \frac{\epsilon}{2k}$$

$$\leq \frac{3}{4}\epsilon + \epsilon^2/8$$
$$< \frac{7}{8}\epsilon < \epsilon.$$

For $u \in (-\infty, r_1]$, we have, by $c(r_1) = 1/k$ and (16.6),

$$|c(u) - h_\epsilon(u)|$$
$$\leq \max\left\{\frac{1}{k}, \frac{k-1}{k}\frac{\epsilon}{2k}\right\}$$
$$< \max\{\epsilon/4, \epsilon/8\} < \epsilon.$$

Similarly, for $u \in (r_k, \infty)$, (16.8) and (16.7) yield

$$|c(u) - h_\epsilon(u)| \quad = |1 - c(u) - (1 - h_\epsilon(u))|$$
$$\leq \max\{|1 - c(u)|, |1 - h_\epsilon(u)|\}$$
$$= \max\left\{1 - \left(1 - \frac{1}{2k}\right), 1 - \frac{k-1}{k}\left(1 - \frac{\epsilon}{2k}\right)\right\}$$
$$= \max\left\{\frac{1}{2k}, \frac{1}{k} + \frac{\epsilon}{2k}\left(1 - \frac{1}{k}\right)\right\}$$
$$\leq \max\left\{\frac{1}{2k}, \frac{1}{k} + \frac{\epsilon}{2k}\right\}$$
$$< \max\left\{\frac{\epsilon}{8}, \frac{\epsilon}{4} + \frac{\epsilon^2}{8}\right\}$$
$$< \epsilon.$$

STEP 4. Steps 2 and 3 imply that, for every $\epsilon > 0$, every $M > 0$, and arbitrary squashing function σ, there exists a neural network $\overline{C}_{M,\epsilon}(u) = \sum_{i=1}^k \overline{c}_i \sigma(\overline{a}_i u + \overline{b}_i) + \overline{c}_0$ such that

$$\sup_{u \in [-M,M]} |\overline{C}_{M,\epsilon}(u) - \cos(u)| < \epsilon. \tag{16.9}$$

STEP 5. Let $g(x) = \sum_{i=1}^k \tilde{c}_i \cos(\tilde{a}_i^T x + \tilde{b}_i)$ be any cosine network. We show that, for arbitrary squashing function σ, arbitrary compact set $K \subset \mathcal{R}^d$, and for any $\epsilon > 0$, there exists a network $s(x) = \sum_{i=1}^k c_i \sigma(a_i^T x + b_i)$ such that

$$\sup_{x \in K} |s(x) - g(x)| < \epsilon.$$

Since K is compact and functions $a_i^T x + b_i$ $(i = 1, \ldots, k)$ are continuous there is a finite $M > 0$ such that $\sup_{x \in K} |a_i^T x + b_i| \leq M$ $(i = 1, \ldots, k)$. Let $c = \sum_{i=1}^k |\tilde{c}_i|$.

Using the results of Step 4 we have

$$\sup_{x \in K} \left| \sum_{i=1}^{k} \tilde{c}_i \cos(\tilde{a}_i^T x + \tilde{b}_i) - \sum_{i=1}^{k} \tilde{c}_i \overline{C}_{M,\epsilon/c}(\tilde{a}_i^T x + \tilde{b}_i) \right|$$

$$\leq \sum_{i=1}^{k} |\tilde{c}_i| \sup_{u \in [-M,M]} |\cos(u) - \overline{C}_{M,\epsilon/c}(u)| < \frac{c\epsilon}{c} = \epsilon.$$

Lemma 16.1 follows from the inequality above, Step 1, and the triangle inequality. $\square$

The next lemma gives a denseness result in $L_2(\mu)$ for any probability measure μ.

Lemma 16.2. *Let σ be a squashing function. Then for every probability measure μ on $\mathcal{R}^d$, every measurable function $f : \mathcal{R}^d \to \mathcal{R}$ with $\int |f(x)|^2 \mu(dx) < \infty$, and every $\epsilon > 0$, there exists a neural network*

$$h(x) = \sum_{i=1}^{k} c_i \sigma(a_i^T x + b_i) + c_0 \quad (k \in \mathcal{N}, a_i \in \mathcal{R}^d, b_i, c_i \in \mathcal{R})$$

such that

$$\int |f(x) - h(x)|^2 \mu(dx) < \epsilon.$$

PROOF. Let

$$\mathcal{F}_k = \left\{ \sum_{i=1}^{k} c_i \sigma(a_i^T x + b_i) + c_0 : k \in \mathcal{N}, a_i \in \mathcal{R}^d, \ b_i, c_i \in \mathcal{R} \right\}.$$

We need to show that $\mathcal{F} = \bigcup_{k \in \mathcal{N}} \mathcal{F}_k$ is dense in $L_2(\mu)$. Suppose, to the contrary, that this is not the case. Then there exists $g \in L_2(\mu), g \neq 0$, i.e., $\int_{\mathcal{R}^d} |g(\nu)|^2 \mu(d\nu) > 0$ such that g is orthogonal to any $f \in \mathcal{F}$, i.e.,

$$\int_{\mathcal{R}^d} f(x)g(x)\mu(dx) = 0$$

for all $f \in \mathcal{F}$. In particular,

$$\int_{\mathcal{R}^d} \sigma(a^T x + b)g(x)\mu(dx) = 0 \qquad (16.10)$$

for all $a \in \mathcal{R}^d, b \in \mathcal{R}$. In the remainder of the proof we will show that (16.10), together with the assumption that σ is a squashing function, implies $g = 0$, which contradicts $g \neq 0$. Consider the Fourier transform $\hat{g}$ of g defined by

$$\hat{g}(u) = \int_{\mathcal{R}^d} \exp(iu^T v)g(v)\mu(dv)$$

$$= \int_{\mathcal{R}^d} \cos(u^T v)g(v)\mu(dv) + i \int_{\mathcal{R}^d} \sin(u^T v)g(v)\mu(dv).$$

By the uniqueness of the Fourier transform it suffices to show that

$$\hat{g} = 0$$

(see Hewitt and Ross (1970, Theorem 31.31)). We start by showing that

$$\int_{\mathcal{R}^d} \cos(u^T v) g(v)\mu(dv) = 0 \quad (u \in \mathcal{R}^d), \tag{16.11}$$

using the uniform approximation of cosine by the univariate neural networks derived in the proof of Lemma 16.1 and the fact that g is orthogonal to these univariate neural networks (see (16.10)).

Let $0 < \epsilon < 1$ be arbitrary. Choose $N \in \mathcal{N}$ odd such that

$$\int_{\{w \in \mathcal{R}^d : |u^T w| > N\pi\}} |g(v)|\mu(dv) < \epsilon/6. \tag{16.12}$$

Let c be the cosine squasher. It follows from Step 2 in the proof of Lemma 16.1 that we can find $K \in \mathcal{N}, a_i, b_i, c_i \in \mathcal{R}$, such that

$$\cos(z) - \sum_{i=1}^{K} c_i c(a_i z + b_i) = 0$$

for $z \in [-N\pi, N\pi]$ and

$$\sum_{i=1}^{K} c_i c(a_i z + b_i) = -1$$

for $z \in \mathcal{R} \setminus [-N\pi, N\pi]$.

According to Step 3 in the proof of Lemma 16.1 there exists a neural network

$$C(z) = \sum_{j=1}^{L} \gamma_j \sigma(\alpha_j z + \beta_j)$$

such that

$$\sup_{z \in \mathcal{R}} |c(z) - C(z)|$$

$$= \sup_{z \in \mathcal{R}} |c(z) - \sum_{j=1}^{L} \gamma_j \sigma(\alpha_j z + \beta_j)|$$

$$< \frac{\epsilon}{2 \sum_{i=1}^{K} |c_i|} \int |g(v)|\mu(dv).$$

Then, for $z \in [-N\pi, N\pi]$,

$$\left| \cos(z) - \sum_{i=1}^{K} c_i C(a_i z + b_i) \right|$$

$$= \left| \sum_{i=1}^{K} c_i c(a_i z + b_i) - \sum_{i=1}^{K} c_i C(a_i z + b_i) \right|$$

$$\leq \sum_{i=1}^{K} |c_i| \frac{\epsilon}{2 \sum_{j=1}^{K} |c_j| \int |g(v)| dv}$$

$$= \frac{\epsilon}{2 \int |g(v)| \mu(dv)} \tag{16.13}$$

and, for $z \notin [-N\pi, N\pi]$,

$$\left| \cos(z) - \sum_{i=1}^{K} c_i C(a_i z + b_i) \right|$$

$$\leq 2 + \left| \sum_{i=1}^{K} c_i c(a_i z + b_i) - \sum_{i=1}^{K} c_i C(a_i z + b_i) \right|$$

$$\leq 2 + \frac{\epsilon}{2} \leq 3. \tag{16.14}$$

Thus we have

$$\left| \int_{\mathcal{R}^d} \cos(u^T v) g(v) \mu(dv) \right|$$

$$\leq \left| \int_{\mathcal{R}^d} (\cos(u^T v) - \sum_{i=1}^{K} c_i C(a_i u^T v + b_i)) g(v) \mu(dv) \right|$$

$$+ \left| \int_{\mathcal{R}^d} (\sum_{i=1}^{K} c_i C(a_i u^T v + b_i)) g(v) \mu(dv) \right|$$

$$= T_1 + T_2.$$

It follows from (16.10) that

$$T_2 = \left| \int_{\mathcal{R}^d} \left(\sum_{i=1}^{K} c_i C(a_i u^T v + b_i) \right) g(v) \mu(dv) \right|$$

$$= \left| \sum_{i=1}^{K} c_i \sum_{j=1}^{L} \gamma_j \int_{\mathcal{R}^d} \sigma(\alpha_j (a_i u^T v + b_i) + \beta_j) g(v) \mu(dv) \right|$$

$$= 0.$$

Furthermore, by (16.13), (16.14), and (16.12),

$$T_1 \quad \leq \left| \int_{\{w \in \mathcal{R}^d : |u^T w| \leq N\pi\}} \left(\cos(u^T v) - \sum_{i=1}^{K} c_i C(a_i u^T v + b_i)\right) g(v) \mu(dv) \right|$$

$$+ \left| \int_{\{w \in \mathcal{R}^d : |u^T w| > N\pi\}} \left(\cos(u^T v) - \sum_{i=1}^{K} c_i C(a_i u^T v + b_i)\right) g(v) \mu(dv) \right|$$

$$\leq \frac{\epsilon}{2 \int_{\mathcal{R}^d} |g(v)| \mu(dv)} \int_{\{w \in \mathcal{R}^d : |u^T w| \leq N\pi\}} |g(v)| \mu(dv)$$

$$+ 3 \int_{\{w \in \mathcal{R}^d : |u^T w| > N\pi\}} |g(v)| \mu(dv)$$

$$\leq \frac{\epsilon}{2} + \frac{\epsilon}{2} = \epsilon$$

and (16.11) follows from $\epsilon \to 0$.

Similarly, we can show

$$\int_{\mathcal{R}^d} \sin(u^T v) g(v) \mu(dv) = 0 \quad (u \in \mathcal{R}^d)$$

and the proof is complete. $\square$

In the proof of Theorem 16.1 we need several results concerning properties of covering numbers. The first lemma describes the VC dimension of graphs of compositions of functions.

Lemma 16.3. *Let $\mathcal{F}$ be a family of real functions on $\mathcal{R}^m$, and let $g : \mathcal{R} \to \mathcal{R}$ be a fixed nondecreasing function. Define the class $\mathcal{G} = \{g \circ f : f \in \mathcal{F}\}$. Then*

$$V_{\mathcal{G}^+} \leq V_{\mathcal{F}^+}.$$

PROOF. Let $(s_1, t_1), \ldots, (s_n, t_n)$ be such that they are shattered by $\mathcal{G}^+$. Then there exist functions $f_1, \ldots, f_{2^n} \in \mathcal{F}$ such that the binary vector

$$\left(I_{\{g(f_j(s_1)) \geq t_1\}}, \ldots, I_{\{g(f_j(s_n)) \geq t_n\}} \right)$$

takes on all 2^n values if $j = 1, \ldots, 2^n$. For all $1 \leq i \leq n$ define the numbers

$$u_i = \min_{1 \leq j \leq 2^n} \{f_j(s_i) : g(f_j(s_i)) \geq t_i\}$$

and

$$l_i = \max_{1 \leq j \leq 2^n} \{f_j(s_i) : g(f_j(s_i)) < t_i\}.$$

By the monotonicity of g, $u_i > l_i$, which implies $l_i < (u_i + l_i)/2 < u_i$. Furthermore,

$$g(f_j(s_i)) \geq t_i \implies f_j(s_i) \geq u_i \implies f_j(s_i) > \frac{u_i + l_i}{2}$$

and, likewise,

$$g(f_j(s_i)) < t_i \implies f_j(s_i) \le l_i \implies f_j(s_i) < \frac{u_i - l_i}{2}.$$

Thus the binary vector

$$\left(I_{\{f_j(s_1) \ge \frac{u_1 + l_1}{2}\}}, \dots, I_{\{f_j(s_n) \ge \frac{u_n + l_n}{2}\}} \right)$$

takes on the same values as

$$\left(I_{\{g(f_j(s_1)) \ge t_1\}}, \dots, I_{\{g(f_j(s_n)) \ge t_n\}} \right)$$

for every $j \le 2^n$. Therefore, the pairs

$$\left(s_1, \frac{u_1 + l_1}{2} \right), \dots, \left(s_n, \frac{u_n + l_n}{2} \right)$$

are shattered by $\mathcal{F}^+$, which proves the lemma. $\quad\square$

The next two lemmas are about the covering numbers of classes of functions whose members are the sums or products of functions from other classes.

Lemma 16.4. *Let $\mathcal{F}$ and $\mathcal{G}$ be two families of real functions on $\mathcal{R}^m$. If $\mathcal{F} \oplus \mathcal{G}$ denotes the set of functions $\{f + g : f \in \mathcal{F}, g \in \mathcal{G}\}$, then for any $z_1^n \in \mathcal{R}^{n \cdot m}$ and $\epsilon, \delta > 0$, we have*

$$\mathcal{N}_1(\epsilon + \delta, \mathcal{F} \oplus \mathcal{G}, z_1^n) \le \mathcal{N}_1(\epsilon, \mathcal{F}, z_1^n) \mathcal{N}_1(\delta, \mathcal{G}, z_1^n).$$

PROOF. Let $\{f_1, \dots, f_K\}$ and $\{g_1, \dots, g_L\}$ be an ϵ-cover and a δ-cover of $\mathcal{F}$ and $\mathcal{G}$, respectively, on z_1^n of minimal size. Then, for every $f \in \mathcal{F}$ and $g \in \mathcal{G}$, there exist $k \in \{1, \dots, K\}$ and $l \in \{1, \dots, L\}$ such that

$$\frac{1}{n} \sum_{i=1}^{n} |f(z_i) - f_k(z_i)| < \epsilon,$$

and

$$\frac{1}{n} \sum_{i=1}^{n} |g(z_i) - g_l(z_i)| < \delta.$$

By the triangle inequality

$$\frac{1}{n} \sum_{i=1}^{n} |f(z_i) + g(z_i) - (f_k(z_i) + g_l(z_i))|$$

$$\le \frac{1}{n} \sum_{i=1}^{n} |f(z_i) - f_k(z_i)| + \frac{1}{n} \sum_{i=1}^{n} |g(z_i) - g_l(z_i)| \le \epsilon + \delta$$

which proves that $\{f_k + g_l : 1 \le k \le K, 1 \le l \le L\}$ is an $(\epsilon + \delta)$-cover of $\mathcal{F} \oplus \mathcal{G}$ on z_1^n. $\quad\square$

Lemma 16.5. *Let $\mathcal{F}$ and $\mathcal{G}$ be two families of real functions on $\mathcal{R}^m$ such that $|f(x)| \leq M_1$ and $|g(x)| \leq M_2$ for all $x \in \mathcal{R}^m$, $f \in \mathcal{F}$, $g \in \mathcal{G}$. If $\mathcal{F} \odot \mathcal{G}$ denotes the set of functions $\{f \cdot g : f \in \mathcal{F}, g \in \mathcal{G}\}$ then, for any $z_1^n \in \mathcal{R}^{n \cdot m}$ and $\epsilon, \delta > 0$ we have*

$$\mathcal{N}_1(\epsilon + \delta, \mathcal{F} \odot \mathcal{G}, z_1^n) \leq \mathcal{N}_1(\epsilon/M_2, \mathcal{F}, z_1^n)\mathcal{N}_1(\delta/M_1, \mathcal{G}, z_1^n).$$

PROOF. Let $\{f_1, \ldots, f_K\}$ and $\{g_1, \ldots, g_L\}$ be an ϵ/M_2-cover and a δ/M_1-cover of $\mathcal{F}$ and $\mathcal{G}$, respectively, on z_1^n of minimal size. Then, for every $f \in \mathcal{F}$ and $g \in \mathcal{G}$, there exist $k \in \{1, \ldots, K\}$ and $l \in \{1, \ldots, L\}$ such that $|f_k(z)| \leq M_1$, $|g_l(z)| \leq M_2$, and

$$\frac{1}{n} \sum_{i=1}^n |f(z_i) - f_k(z_i)| < \epsilon/M_2,$$

and

$$\frac{1}{n} \sum_{i=1}^n |g(z_i) - g_l(z_i)| < \delta/M_1.$$

We have, by the triangle inequality,

$$\frac{1}{n} \sum_{i=1}^n |f(z_i)g(z_i) - f_k(z_i)g_l(z_i)|$$

$$= \frac{1}{n} \sum_{i=1}^n |f(z_i)(g_l(z_i) + g(z_i) - g_l(z_i)) - f_k(z_i)g_l(z_i)|$$

$$\leq \frac{1}{n} \sum_{i=1}^n |g_l(z_i)(f(z_i) - f_k(z_i))| + \frac{1}{n} \sum_{i=1}^n |f(z_i)(g(z_i) - g_l(z_i))|$$

$$\leq M_2 \frac{1}{n} \sum_{i=1}^n |f(z_i) - f_k(z_i)| + M_1 \frac{1}{n} \sum_{i=1}^n |g(z_i) - g_l(z_i)| \leq \epsilon + \delta$$

which implies that $\{f_k g_l : 1 \leq k \leq K, 1 \leq l \leq L\}$ is an $(\epsilon + \delta)$-cover of $\mathcal{F} \odot \mathcal{G}$ on z_1^n. $\square$

PROOF OF THEOREM 16.1 . We can mimic the proof of Theorem 10.3. It is only the bound on the covering numbers that requires additional work. The argument in Chapter 10 implies that the approximation error, $\inf_{f \in \mathcal{F}_n} \int |f(x) - m(x)|^2 \mu(dx)$, converges to zero as $k_n, \beta_n \to \infty$, if the union of the $\mathcal{F}_n$'s is dense in $L_2(\mu)$ for every μ (Lemma 16.2).

To handle the estimation error, we use Theorem 10.2, which implies that we can assume $|Y| \leq L$ almost surely, for some L, and then we have to show that

$$\sup_{f \in \mathcal{F}_n} \left| \mathbf{E}|f(X) - Y|^2 - \frac{1}{n} \sum_{j=1}^n |f(X_j) - Y_j|^2 \right| \to 0 \quad a.s..$$

We proceed as in the proof of Theorem 10.3. Define

$$Z = (X, Y), \ Z_1 = (X_1, Y_1), \ldots, \ Z_n = (X_n, Y_n)$$

and

$$\mathcal{H}_n = \left\{ h : \mathcal{R}^d \times \mathcal{R} \to \mathcal{R} \ : \ \exists f \in \mathcal{F}_n \text{ such that } h(x, y) = |f(x) - y|^2 \right\}.$$

We may assume $\beta_n \geq L$ so that functions in $\mathcal{H}_n$ satisfy

$$0 \leq h(x, y) \leq 2\beta_n^2 + 2L^2 \leq 4\beta_n^2.$$

Using the bound of Theorem 9.1, we have, for arbitrary $\epsilon > 0$,

$$\mathbf{P}\left\{ \sup_{f \in \mathcal{F}_n} \left| \frac{1}{n} \sum_{i=1}^{n} |f(X_i) - Y_i|^2 - \mathbf{E}\{|f(X) - Y|^2\} \right| > \epsilon \right\}$$

$$= \mathbf{P}\left\{ \sup_{h \in \mathcal{H}_n} \left| \frac{1}{n} \sum_{i=1}^{n} h(Z_i) - \mathbf{E}\{h(Z)\} \right| > \epsilon \right\}$$

$$\leq 8\mathbf{E}\mathcal{N}_1\left(\frac{\epsilon}{8}, \mathcal{H}_n, Z_1^n \right) e^{-\frac{n\epsilon^2}{128(4\beta_n^2)^2}}. \tag{16.15}$$

Next we bound the covering number in (16.15). Let $h_i(x, y) = |f_i(x) - y|^2$ $((x, y) \in \mathcal{R}^d \times \mathcal{R})$ for some $f_i \in \mathcal{F}_n$. Mimicking the derivation in the proof of Theorem 10.3 we get

$$\frac{1}{n} \sum_{i=1}^{n} |h_1(Z_i) - h_2(Z_i)|$$

$$= \frac{1}{n} \sum_{i=1}^{n} \left| |f_1(X_i) - Y_i|^2 - |f_2(X_i) - Y_i|^2 \right|$$

$$= \frac{1}{n} \sum_{i=1}^{n} |f_1(X_i) - f_2(X_i)| \cdot |f_1(X_i) - Y_i + f_2(X_i) - Y_i|$$

$$\leq 4\beta_n \frac{1}{n} \sum_{i=1}^{n} |f_1(X_i) - f_2(X_i)|.$$

Thus

$$\mathcal{N}_1\left(\frac{\epsilon}{8}, \mathcal{H}_n, Z_1^n \right) \leq \mathcal{N}_1\left(\frac{\epsilon}{32\beta_n}, \mathcal{F}_n, X_1^n \right). \tag{16.16}$$

Define the following classes of functions:

$$\mathcal{G}_1 = \{ a^T x + b : a \in \mathcal{R}^d, b \in \mathcal{R} \},$$

$$\mathcal{G}_2 = \{ \sigma(a^T x + b) : a \in \mathcal{R}^d, b \in \mathcal{R} \},$$

$$\mathcal{G}_3 = \{ c\sigma(a^T x + b) : a \in \mathcal{R}^d, b \in \mathcal{R}, c \in [-\beta_n, \beta_n] \},$$

where $\mathcal{G}_1$ is a linear vector space of dimension $d+1$, thus Theorem 9.5 implies

$$V_{\mathcal{G}_1^+} \leq d+2$$

(also see the proof of Theorem 9.4). Since σ is a nondecreasing function, Lemma 16.3 implies that

$$V_{\mathcal{G}_2^+} \leq d+2.$$

Thus, by Theorem 9.4,

$$\begin{aligned}
&\mathcal{N}_1\left(\epsilon, \mathcal{G}_2, X_1^n\right) \\
&\leq 3\left(\frac{2e}{\epsilon}\log\frac{3e}{\epsilon}\right)^{d+2} \\
&\leq 3\left(\frac{3e}{\epsilon}\right)^{2d+4}.
\end{aligned}$$
(16.17)

By Lemma 16.5,

$$\begin{aligned}
&\mathcal{N}_1\left(\epsilon, \mathcal{G}_3, X_1^n\right) \\
&\leq \mathcal{N}_1\left(\epsilon/2, \{c : |c| \leq \beta_n\}, X_1^n\right)\mathcal{N}_1\left(\frac{\epsilon}{2\beta_n}, \mathcal{G}_2, X_1^n\right) \\
&\leq \frac{4\beta_n}{\epsilon}3\left(\frac{3e}{\frac{\epsilon}{2\beta_n}}\right)^{2d+4} \\
&\leq \left(\frac{12e\beta_n}{\epsilon}\right)^{2d+5}.
\end{aligned}$$
(16.18)

Upon applying Lemma 16.4 we obtain the bound on the covering number of $\mathcal{F}_n$,

$$\begin{aligned}
&\mathcal{N}_1\left(\epsilon, \mathcal{F}_n, X_1^n\right) \\
&\leq \mathcal{N}_1\left(\frac{\epsilon}{k_n+1}, \{c_0 : |c_0| \leq \beta_n\}, X_1^n\right)\cdot\left(\mathcal{N}_1\left(\frac{\epsilon}{k_n+1}, \mathcal{G}_2, X_1^n\right)\right)^{k_n} \\
&\leq \frac{2\beta_n(k_n+1)}{\epsilon}\left(\frac{12e\beta_n}{\left(\frac{\epsilon}{k_n+1}\right)}\right)^{(2d+5)k_n} \\
&\leq \left(\frac{12e\beta_n(k_n+1)}{\epsilon}\right)^{(2d+5)k_n+1}.
\end{aligned}$$
(16.19)

Using the bound (16.16) together with (16.19) on the right-hand side of inequality (16.15) we obtain

$$\mathbf{P}\left\{ \sup_{f \in T_{\beta_n} \mathcal{F}_n} \left| \frac{1}{n} \sum_{i=1}^{n} |f(X_i) - Y_i|^2 - \mathbf{E}|f(X) - Y|^2 \right| > \epsilon \right\}$$

$$\leq 8 \left(\frac{384 e \beta_n^2 (k_n + 1)}{\epsilon} \right)^{(2d+5)k_n + 1} e^{-n\epsilon^2/128 \cdot 2^4 \beta_n^4}.$$

As in the proof of Theorem 10.3 this implies the final result. $\qquad\square$

16.3 Rate of Convergence

Consider the class of neural networks with k neurons and with bounded output weights

$$\mathcal{F}_{n,k} = \left\{ \sum_{i=1}^{k} c_i \sigma(a_i^T x + b_i) + c_0 : k \in \mathcal{N}, a_i \in \mathcal{R}^d, \ b_i, c_i \in \mathcal{R}, \ \sum_{i=0}^{k} |c_i| \leq \beta_n \right\}. \tag{16.20}$$

Let $\mathcal{F} = \bigcup_{n,k} \mathcal{F}_{n,k}$ and let $\overline{\mathcal{F}}$ be the closure of $\mathcal{F}$ in $L^2(\mu)$.

In this section we will assume that $|Y| \leq L \leq \beta_n$ a.s. and we will examine how fast $\mathbf{E} \int |m_n(x) - m(x)|^2 \mu(dx)$ converges to zero.

We use complexity regularization to control the size of the neural network estimate. Let

$$f_{n.k} = \arg\min_{f \in \mathcal{F}_{n.k}} \frac{1}{n} \sum_{i=1}^{n} |f(X_i) - Y_i|^2,$$

that is, $f_{n,k}$ minimizes the empirical risk for n training samples over $\mathcal{F}_{n,k}$. We assume the existence of such a minimizing function for each k and n. The penalized empirical L_2 risk is defined for each $f \in \mathcal{F}_{n,k}$ as

$$\frac{1}{n} \sum_{i=1}^{n} |f(X_i) - Y_i|^2 + pen_n(k).$$

Following the ideas presented in Chapter 12 we use an upper-bound $\mathcal{N}_1(1/n, \mathcal{F}_{n,k})$ such that

$$\mathcal{N}_1(1/n, \mathcal{F}_{n,k}, x_1^n) \leq \mathcal{N}_1(1/n, \mathcal{F}_{n,k})$$

for all $n \in \mathcal{N}, x_1^n \in (\mathcal{R}^d)^n$ and we choose $pen_n(k)$ as

$$pen_n(k) \geq 2568 \frac{\beta_n^4}{n} \cdot (\log \mathcal{N}_1(1/n, \mathcal{F}_{n,k}) + t_k) \tag{16.21}$$

for some $t_k \in \mathcal{R}_+$ with $\sum_k e^{-t_k} \leq 1$. For instance, t_k can be chosen as $t_k = 2\log(k) + t_0, t_0 \geq \sum_{k=1}^{\infty} k^{-2}$. Our estimate m_n is then defined as the

$f_{n,k}$ minimizing the penalized empirical risk over all classes, i.e.,

$$m_n = f_{n,k^*(n)} \tag{16.22}$$

$$k^*(n) = \arg\min_{k \geq 1} \left(\frac{1}{n} \sum_{i=1}^{n} |f_{n,k}(X_i) - Y_i|^2 + pen_n(k) \right).$$

The next theorem is an immediate consequence of Theorem 12.1 and (12.14).

Theorem 16.2. *Let $1 \leq L < \infty, n \in \mathcal{N}$, and let $L \leq \beta_n < \infty$. Assume $|Y| \leq L$ a. s., and let the neural network estimate m_n with squashing function σ be defined by minimizing the penalized empirical risk as in (16.22) with the penalty satisfying condition (16.21) for some $t_k \in \mathcal{R}_+$ such that $\sum_k e^{-t_k} \leq 1$. Then we have*

$$\mathbf{E} \int |m_n(x) - m(x)|^2 \mu(dx)$$

$$\leq 2 \min_k \left\{ pen_n(k) + \inf_{f \in \mathcal{F}_{n,k}} \int |f(x) - m(x)|^2 \mu(dx) \right\}$$

$$+ 5 \cdot 2568 \frac{\beta_n^4}{n}. \tag{16.23}$$

Before presenting further results we need to review the basic properties of the Fourier transform, see, e.g., Rudin (1966). The Fourier transform $\hat{F}$ of a function $f \in L_1(\mathcal{R}^d)$ is given by

$$\hat{F}(\omega) = \frac{1}{(2\pi)^{d/2}} \int_{\mathcal{R}^d} e^{-i\omega^T x} f(x) dx \quad (\omega \in \mathcal{R}^d).$$

If $\hat{F} \in L_1(\mathcal{R}^d)$ then the inverse formula

$$f(x) = \int_{\mathcal{R}^d} e^{i\omega^T x} \hat{F}(\omega) d\omega$$

or, equivalently,

$$f(x) = f(0) + \int_{\mathcal{R}^d} (e^{i\omega^T x} - 1)\hat{F}(\omega) d\omega \tag{16.24}$$

holds almost everywhere with respect to the Lebesgue measure. The Fourier transform $\hat{F}$ of a function f is a complex function which can be written

$$\begin{aligned}
\hat{F}(\omega) \quad &= |\hat{F}(\omega)| \cos(\theta(\omega)) + i \cdot |\hat{F}(\omega)| \sin(\theta(\omega)) \\
&= |\hat{F}(\omega)| e^{i\theta(\omega)} \\
&= \mathrm{Re}(\hat{F}(\omega)) + i \cdot \mathrm{Im}(\hat{F}(\omega)),
\end{aligned}$$

where

$$F(\omega) := |\hat{F}(\omega)| = \sqrt{\mathrm{Re}^2(\hat{F}(\omega)) + \mathrm{Im}^2(\hat{F}(\omega))}$$

is the magnitude of $\hat{F}(\omega)$ and

$$\theta(\omega) = \arctan \frac{\mathrm{Im}(\hat{F}(\omega))}{\mathrm{Re}(\hat{F}(\omega))}$$

is the phase angle. Observe that by the Taylor expansion

$$e^{i\omega^T x} = 1 + e^{i\theta} \cdot i\omega^T x,$$

where θ is between 0 and $\omega^T x$. We then obtain

$$|e^{i\omega^T x} - 1| \leq |\omega^T x| \leq \|\omega\| \cdot \|x\|.$$

Consider the class of functions $\mathcal{F}_C$ for which (16.24) holds on $\mathcal{R}^d$ and, in addition,

$$\int_{\mathcal{R}^d} \|\omega\| F(\omega) d\omega \leq C, 0 < C < \infty. \tag{16.25}$$

A class of functions satisfying (16.25) is a subclass of functions with Fourier transform having a finite first absolute moment, i.e., $\int_{\mathcal{R}^d} \|\omega\| F(\omega) d\omega < \infty$ (these functions are continuously differentiable on $\mathcal{R}^d$). The next theorem provides the rate of convergence for the neural network estimate with a general squashing function.

Theorem 16.3. *Let $|Y| \leq L$ a.s. with and let $1 \leq L < \infty$, and let $m \in \mathcal{F}_C$. The neural network estimate m_n is defined by minimizing the penalized empirical risk in (16.22) with squashing function σ and the penalty satisfying condition (16.21) for some $t_k \in \mathcal{R}_+$ such that $\sum_k e^{-t_k} \leq 1$. Then, if $3rC + L \leq \beta_n$,*

$$\mathbf{E} \int_{S_r} |m_n(x) - m(x)|^2 \mu(dx)$$

$$\leq 2 \min_k \left\{ 2568 \frac{\beta_n^4}{n} \left((2d+6)k \log(18e\beta_n n) + t_k \right) \right.$$

$$\left. + \inf_{f \in \mathcal{F}_{n,k}} \int_{S_r} |f(x) - m(x)|^2 \mu(dx) \right\} + 5 \cdot 2568 \frac{\beta_n^4}{n}, \tag{16.26}$$

where S_r is a ball with radius r centered at 0. In particular, upon choosing $t_k = 2\log k + t_0, t_0 \geq \sum_{k=1}^{\infty} k^{-2}$, and $\beta_n \to \infty$, we get

$$\mathbf{E} \int_{S_r} |m_n(x) - m(x)|^2 \mu(dx) = O\left(\beta_n^2 \left(\frac{\log(\beta_n n)}{n} \right)^{1/2} \right). \tag{16.27}$$

If $\beta_n < \mathrm{const} < \infty$ then

$$pen(k) = 2568 \frac{\beta_n^4}{n} \cdot \left((2d+6)k \log(18e\beta_n n) + t_k \right)$$

$$= O\left(\frac{k \log(n)}{n} \right)$$

and the rate in (16.27) becomes

$$\mathbf{E} \int_{S_r} |m_n(x) - m(x)|^2 \mu(dx) = O\left(\sqrt{\frac{\log(n)}{n}}\right).$$

In the proof of Theorem 16.3 we will need a refinement of Lemmas 16.4 and 16.5. It is provided in the lemma below.

Lemma 16.6. *Let $B > 0, \mathcal{G}_1, \ldots, \mathcal{G}_k$ be classes of real functions $f : \mathcal{R}^d \to [-B, B]$, and define $\mathcal{F}$ as*

$$\mathcal{F} = \left\{\sum_{i=1}^{k} w_i f_i : (w_1, \ldots, w_k) \in \mathcal{R}^k, \sum_{i=1}^{k} |w_i| \le b, f_i \in \mathcal{G}_i, i = 1, \ldots, k\right\}.$$

Then we have, for any $z_1^n \in \mathcal{R}^{d \cdot n}$ and any $\eta, \delta > 0$,

$$\mathcal{N}_1(\eta + \delta, \mathcal{F}, z_1^n) \le \left(\frac{Be(b + 2\delta/B)}{\delta}\right)^k \prod_{i=1}^{k} \mathcal{N}_1(\eta/(b + 2\delta), \mathcal{G}_i, z_1^n).$$

PROOF. Let

$$S_b = \{w \in \mathcal{R}^k : \sum_{i=1}^{k} |w_i| \le b\}$$

and assume that $S_{b,\delta}$ is a finite subset of $\mathcal{R}^k$ with the covering property

$$\max_{w \in S_b} \min_{x \in S_{b,\delta}} \|w - x\|_1 \le \delta,$$

where $\|y\|_1 = |y^{(1)}| + \ldots + |y^{(k)}|$ denotes the l_1 norm of any $y = (y^{(1)}, \ldots, y^{(k)})^T \in \mathcal{R}^k$. Also, let the $\mathcal{G}_i(\eta)$ be an L_1 η-cover of $\mathcal{G}_i$ on z_1^n of minimal size, that is, each $\mathcal{G}_i(\eta)$ has cardinality $\mathcal{N}_1(\eta, \mathcal{G}_i, z_1^n)$ and

$$\min_{g \in \mathcal{G}_i(\eta)} \|f - g\|_n \le \eta$$

for all $f \in \mathcal{G}_i$, where $\|f\|_n = \{\frac{1}{n} \sum_{i=1}^{n} |f(z_i)|^2\}^{1/2}$. Let $f \in \mathcal{F}$ be given by $f = \sum_{i=1}^{k} w_i f_i$, and choose $x \in S_{b,\delta}$ and $\widehat{f_i} \in \mathcal{G}_i(\eta)$ with $\|w - x\|_1 \le \delta$ and $\|f_i - \widehat{f_i}\|_n \le \eta$, $i = 1, \ldots, k$. Since $\|f_i\|_n \le B$ for all i, we have

$$\|f - \sum_{i=1}^{k} x_i \widehat{f_i}\|_n \le \|\sum_{i=1}^{k} w_i f_i - \sum_{i=1}^{k} x_i f_i\|_n + \|\sum_{i=1}^{k} x_i f_i - \sum_{i=1}^{k} x_i \widehat{f_i}\|_n$$

$$\le \sum_{i=1}^{k} |w_i - x_i| \cdot \|f_i\| + \sum_{i=1}^{k} |x_i| \cdot \|f_i - \widehat{f_i}\|_n$$

$$\le \delta B + \eta b_\delta,$$

where $b_\delta = \max_{x \in S_{b,\delta}} \|x\|_1$. It follows that a set of functions of cardinality

$$|S_{b,\delta}| \cdot \prod_{i=1}^{k} \mathcal{N}_1(\eta, \mathcal{G}_i, z_1^n)$$

is a $(\delta B + \eta b_\delta)$-cover of $\mathcal{F}$. Thus we only need to bound the cardinality of $S_{b,\delta}$.

The obvious choice for $S_{b,\delta}$ is a rectangular grid spaced at width $2\delta/k$. For a given collection of grid points we define partition of space into Voronoi regions such that each region consists of all points closer to a given grid point than to any other grid point. Define $S_{b,\delta}$ as the points on the grid whose l_1 Voronoi regions intersect S_b. These Voronoi regions (and the associated grid points) are certainly contained in $S_{b+2\delta}$. To get a bound on the number of points in $S_{b,\delta}$ we can divide the volume of simplex $S_{b+2\delta}$ by the volume of a cube with side length $2\delta/k$. The volume of $S_{b+2\delta}$ is the volume of a ball with radius $r = b + 2\delta$ in $L_1(R^k)$ norm. It can be directly calculated by

$$2^k \int_0^r \int_0^{r-x_1} \cdots \int_0^{r-x_1-\cdots-x_{k-1}} dx_1 dx_2 \ldots dx_k = (2(b+2\delta))^k/k!.$$

Thus the cardinality of $S_{b+2\delta}$ is bounded above by

$$\frac{(2(b+2\delta))^k}{k!} \left(\frac{2\delta}{k}\right)^{-k} = \frac{1}{k!}\left(\frac{k(b+2\delta)}{\delta}\right)^k \leq \left(\frac{e(b+2\delta)}{\delta}\right)^k,$$

where in the last inequality we used Stirling's formula

$$k! = \sqrt{2\pi k}\, k^k e^{-k} e^{\theta(k)}, |\theta(k)| \leq 1/12k.$$

Since $b_\delta \leq b + 2\delta$, we have

$$\mathcal{N}_1(\delta B + \eta(b+2\delta), \mathcal{F}, z_1^n) \leq \left(\frac{e(b+2\delta)}{\delta}\right)^k \prod_{i=1}^{k} \mathcal{N}_1(\eta, \mathcal{G}_i, z_1^n),$$

or

$$\mathcal{N}_1(\delta + \eta, \mathcal{F}, z_1^n) = \mathcal{N}_1\left(\frac{\delta}{B}B + \frac{\eta}{b+2\delta}(b+2\delta), \mathcal{F}, z_1^n\right)$$

$$\leq \left(\frac{e(b+2\delta/B)}{\delta/B}\right)^k \prod_{i=1}^{k} \mathcal{N}_1\left(\frac{\eta}{b+2\delta}, \mathcal{G}_i, z_1^n\right)$$

$$= \left(\frac{Be(b+2\delta/B)}{\delta}\right)^k \prod_{i=1}^{k} \mathcal{N}_1(\eta/(b+2\delta), \mathcal{G}_i, z_1^n).$$

$\square$

PROOF OF THEOREM 16.3 . Following the ideas given in the proof of Theorem 12.1 we only need to find a bound on the covering number of $\mathcal{F}_{n,k}$. In order to get the postulated rate we need a stronger bound on the covering number than in (16.19). Using (16.17), Lemmas 16.4 and 16.6, and setting $\epsilon = \delta$, we get for $\beta_n \geq 2\epsilon$

$$\mathcal{N}_1(3\epsilon, \mathcal{F}_{n,k}, X_1^n)$$

$$\leq \mathcal{N}_1(\epsilon, \{c_0 : |c_0| \leq \beta_n\}, X_1^n) \cdot \left(\frac{e(\beta_n + 2\epsilon)}{\epsilon}\right)^k \prod_{i=1}^k \mathcal{N}_1(\epsilon/(\beta_n + 2\epsilon), \mathcal{G}_2, X_1^n)$$

$$\leq \frac{2\beta_n}{\epsilon} \cdot \left(\frac{e(\beta_n + 2\epsilon)}{\epsilon}\right)^{k+1} \prod_{i=1}^k \mathcal{N}_1(\epsilon/(\beta_n + 2\epsilon), \mathcal{G}_2, X_1^n)$$

$$\leq \left(\frac{2e\beta_n}{\epsilon}\right)^k \left(3\left(\frac{3e(\beta_n + 2\epsilon)}{\epsilon}\right)^{(2d+4)}\right)^k$$

$$\leq \left(\frac{6e\beta_n}{\epsilon}\right)^{(2d+6)k}.$$

Here, again, $\mathcal{G}_2 = \{\sigma(a^T x + b) : a \in \mathcal{R}^d, b \in \mathcal{R}\}$. Hence, for n sufficiently large,

$$2568\frac{\beta_n^4}{n} \cdot (\log \mathcal{N}_1(1/n, \mathcal{F}_{n,k}) + t_k)$$

$$\leq 2568\frac{\beta_n^4}{n} \cdot \left(\log\left(\frac{6e\beta_n}{\left(\frac{1}{3n}\right)}\right)^{(2d+6)k} + t_k\right)$$

$$= 2568\frac{\beta_n^4}{n} \cdot ((2d+6)k\log(18e\beta_n n) + t_k)$$

$$= pen_n(k).$$

Now (16.26) follows from Theorem 16.2. Note that upon choosing $t_k = 2\log(k) + t_0, t_0 \geq \sum_{k=1}^{\infty} k^{-2}$ we get

$$pen(k) = O\left(\frac{\beta_n^4 k \log(\beta_n n)}{n}\right)$$

and (16.27) follows from (16.26) and Lemma 16.8 below, which implies

$$\inf_{f \in \mathcal{F}_k} \int_{S_r} |f(x) - m(x)|^2 \mu(dx) = O(1/k).$$

$\square$

The next lemma plays a key role in deriving the rate of approximation with neural networks. It describes the rate of approximation of convex combinations in the $L_2(\mu)$. Let $\|f\|_S^2$ denote $\int_S f^2(x)\mu(dx)$. In our presentation we will follow Barron's (1993) approach.

Lemma 16.7. *Let*

$$\Phi = \{\phi_a : a \in \mathcal{A}, \|\phi_a\|_{S_r}^2 \leq B^2\} \subset L_2(\mu),$$

where S_r is the ball with radius r centered at 0 and $\mathcal{A} \subset \mathcal{R}^m$ is a set of parameters. Let h be any real-valued measurable function on $\mathcal{A}$ with $\int_{\mathcal{A}} |h(a)|da \in (0, \infty)$ such that a representation

$$f(x) = \int_{\mathcal{A}} \phi_a(x)h(a)da + c \tag{16.28}$$

($c \in \mathcal{R}$) is valid for all $|x| < r$.

Hence, for every $k \in \mathcal{N}$, there exists

$$f_k(x) = \sum_{j=1}^{k} c_j \phi_{a_j}(x) + c$$

such that

$$\|f - f_k\|_{S_r}^2 < \frac{B^2}{k} \left(\int_{\mathcal{A}} |h(a)|da \right)^2 \tag{16.29}$$

and the coefficients c_j can be chosen such that $\sum_{j=1}^{k} |c_j| \leq \int_{\mathcal{A}} |h(a)|da$, where $\int |h(a)|da$ is the total variation of the signed measure $h(a)da$.

PROOF. Assuming $|x| < r$, we have by, (16.28),

$$f(x) = \int_{\mathcal{A}} \phi_a(x)h(a)da + c$$
$$= \int_{\mathcal{A}} sgn(h(a)) \cdot \phi_a(x)|h(a)|da + c,$$

where $sgn(x)$ denotes the sign of x. Let Q be the probability measure on $\mathcal{A}$ defined by the density $|h(\cdot)|/ \int_{\mathcal{A}} |h(a)|da$ with respect to the Lebesgue measure. Let $D = \int_{\mathcal{A}} |h(a)|da$. Thus

$$f(x) = D \cdot \mathbf{E}_Q\{sgn(h(A))\phi_A(x)\} + c,$$

where A is a random variable with distribution Q. Let $A_1, \ldots, A_k$ be i.i.d. random variables with distribution Q independent of A, and let

$$f_k(x) = \frac{D}{k} \sum_{j=1}^{k} sgn(h(A_j))\phi_{A_j}(x) + c.$$

By (16.28) and the Fubini theorem

$$\mathbf{E}_Q \|f - f_k\|_{S_r}^2$$
$$= \mathbf{E}_Q \int_{S_r} (f(x) - f_k(x))^2 \mu(dx)$$

$$= \int_{S_r} \mathbf{E}_Q (f(x) - f_k(x))^2 \mu(dx)$$

$$= \int_{S_r} \mathbf{E}_Q \left(f(x) - \left(\frac{D}{k} \sum_{j=1}^{k} sgn(h(A_j)) \phi_{A_j}(x) + c \right) \right)^2 \mu(dx)$$

$$= \int_{S_r} \mathbf{E}_Q \left(D \cdot \mathbf{E}_Q \{ sgn(h(A)) \phi_A(x) \} + c \right.$$

$$\left. - \left(\frac{D}{k} \sum_{j=1}^{k} sgn(h(A_j)) \phi_{A_j}(x) + c \right) \right)^2 \mu(dx)$$

$$= D^2 \int_{S_r} \mathbf{Var}_Q \left(\frac{1}{k} \sum_{j=1}^{k} sgn(h(A_j)) \phi_{A_j}(x) \right) \mu(dx)$$

$$= \frac{D^2}{k} \int_{S_r} \mathbf{Var}_Q \left(sgn(h(A)) \phi_A(x) \right) \mu(dx).$$

Since

$$\mathbf{E}_Q(sgn(h(A)) \phi_A(x)) = \frac{1}{D}(f(x) - c)$$

without loss of generality we may alter the constant c in f and f_k such that $\mathbf{E}_Q(sgn(h(A)) \phi_A(x)) > 0$ on a set of μ-measure greater than zero. On this set $\mathbf{Var}_Q \left(sgn(h(A)) \phi_A(x) \right) < \mathbf{E}_Q(\phi_A^2(x))$ with strict inequality.

Hence, by the Fubini theorem,

$$\mathbf{E}_Q \| f - f_k \|_{S_r}^2 \ < \ \frac{D^2}{k} \int_{S_r} \mathbf{E}_Q(\phi_A^2(x)) \mu(dx)$$

$$= \ \frac{D^2}{k} \mathbf{E}_Q \left(\int_{S_r} \phi_A^2(x) \mu(dx) \right)$$

$$\leq \ \frac{B^2 D^2}{k}.$$

Since

$$\mathbf{E}_Q \| f - f_k \|_{S_r}^2 < \frac{B^2 D^2}{k}$$

we can find $a_1, \ldots, a_k$ such that

$$\| f - f_k \|_{S_r}^2 < \frac{B^2 D^2}{k} = \frac{B^2}{k} \left(\int_A |h(a)| da \right)^2.$$

The coefficients in f_k are $\frac{D}{k} sgn(h(a_j))$, hence their absolute values sum to at most D. $\qquad \square$

The next lemma gives the rate of approximation for neural nets with squashing functions.

Lemma 16.8. *Let σ be a squashing function. Then, for every probability measure μ on $\mathcal{R}^d$, every measurable $f \in \mathcal{F}_C$ and every $k \geq 1$, there exists a neural network f_k in*

$$\mathcal{F}_k = \left\{ \sum_{i=1}^{k} c_i \sigma(a_i^T x + b_i) + c_0; \quad k \in \mathcal{N},\ a_i \in \mathcal{R}^d,\ b_i, c_i \in \mathcal{R} \right\} \qquad (16.30)$$

such that

$$\int_{S_r} (f(x) - f_k(x))^2 \mu(dx) \leq \frac{(2rC)^2}{k}. \qquad (16.31)$$

The coefficients of the linear combination in (16.30) may be chosen so that $\sum_{i=0}^{k} |c_i| \leq 3rC + f(0)$.

PROOF. Consider the class of multiples of sigmoids and linear functions

$$G_\sigma = \{c\sigma(a^T x + b) : |c| \leq 2rC, a \in \mathcal{R}^d, b \in \mathcal{R}\}.$$

For $\mathcal{F} \subset L_2(\mu, S_r)$, let $\overline{\mathcal{F}}$ denote the closure of the convex hull of $\mathcal{F}$ in $L_2(\mu, S_r)$, where $L_2(\mu, S_r)$ denotes the space of functions such that $\int_{S_r} f^2(x)\mu(dx) < \infty$.

The key idea of the proof is to show that for every $f \in \mathcal{F}_C$, $f(x) - f(0) \in \overline{G}_\sigma$. The rate postulated in (16.31) will then follow from Lemma 16.7. Let $\Omega = \{\omega \in \mathcal{R}^d : \omega \neq 0\}$. From the properties of the Fourier transform and the fact that f is real valued, it follows that

$$f(x) - f(0) \qquad (16.32)$$

$$= \text{Re} \int_\Omega (e^{i\omega \cdot x} - 1)\hat{F}(\omega)d\omega$$

$$= \text{Re} \int_\Omega (e^{i\omega \cdot x} - 1)e^{i\theta(\omega)}F(\omega)d\omega$$

$$= \int_\Omega (\cos(\omega \cdot x + \theta(\omega)) - \cos(\theta(\omega)))F(\omega)d\omega$$

$$= \int_\Omega \frac{(\cos(\omega \cdot x + \theta(\omega)) - \cos(\theta(\omega)))}{\|\omega\|} \|\omega\| F(\omega)d\omega$$

$$= \int_\Omega g(x, \omega)\|\omega\| F(\omega)d\omega \qquad (16.33)$$

for every $x \in S_r$, where

$$g(x, \omega) = \frac{(\cos(\omega \cdot x + \theta(\omega)) - \cos(\theta(\omega)))}{\|\omega\|}.$$

Let G_{step} be the class of step functions $\phi(t, a)$ defined as

$$G_{step} = \{\phi_{(t,a)}(x) = I^*(a \cdot x - \|a\|t) : a \in \mathcal{R}^d, t \in \mathcal{R}\},$$

where

$$I^*(x) = \begin{cases} 0 & , \quad x < 0, \\ \sigma(0) & , \quad x = 0, \\ 1 & , \quad x > 0. \end{cases}$$

By the identity $\int \sin(ax+b)dx = -1/a\cos(ax+b) + const$ we get

$$g(x,\omega) = \frac{(\cos(\|\omega\|\frac{\omega\cdot x}{\|\omega\|} + \theta(\omega)) - \cos(\theta(\omega)))}{\|\omega\|}$$

$$= \int_{\frac{\omega\cdot x}{\|\omega\|}}^{0} \sin(\|\omega\|t + \theta(\omega))dt$$

$$= I. \tag{16.34}$$

Consider two cases.

CASE 1. $\frac{\omega\cdot x}{\|\omega\|} \geq 0$.

$$I = -\int_{0}^{\frac{\omega\cdot x}{\|\omega\|}} \sin(\|\omega\|t + \theta(\omega))dt$$

$$= -\int_{0}^{r} I^*\left(\frac{\omega\cdot x}{\|\omega\|} - t\right) \sin(\|\omega\|t + \theta(\omega))dt$$

since $|\frac{\omega\cdot x}{\|\omega\|}| \leq |x| \leq r$. On the other hand,

$$\int_{-r}^{0} I^*\left(t - \frac{\omega\cdot x}{\|\omega\|}\right) \sin(\|\omega\|t + \theta(\omega))dt = \int_{-r}^{0} 0\ dt = 0.$$

CASE 2. $\frac{\omega\cdot x}{\|\omega\|} < 0$. Then

$$I = \int_{\frac{\omega\cdot x}{\|\omega\|}}^{0} \sin(\|\omega\|t + \theta(\omega))dt$$

$$= \int_{-r}^{0} I^*(t - \frac{\omega\cdot x}{\|\omega\|}) \sin(\|\omega\|t + \theta(\omega))dt.$$

On the other hand,

$$\int_{0}^{r} I^*\left(\frac{\omega\cdot x}{\|\omega\|} - t\right) \sin(\|\omega\|t + \theta(\omega))dt = \int_{0}^{r} 0\ dt = 0.$$

Hence we obtain for the right-hand side of (16.34),

$$I = -\int_{0}^{r} I^*\left(\frac{\omega\cdot x}{\|\omega\|} - t\right) \sin(\|\omega\|t + \theta(\omega))dt$$

$$+ \int_{-r}^{0} I^*\left(t - \frac{\omega\cdot x}{\|\omega\|}\right) \sin(\|\omega\|t + \theta(\omega))dt$$

$$= -\int_{0}^{r} I^*\left(\frac{\omega\cdot x}{\|\omega\|} - t\right) \sin(\|\omega\|t + \theta(\omega))dt$$

$$+ \int_{-r}^{0} \left(1 - I^* \left(\frac{\omega \cdot x}{\|\omega\|} - t\right)\right) \sin(\|\omega\|t + \theta(\omega))dt$$

$$= \int_{-r}^{0} \sin(\|\omega\|t + \theta(\omega))dt - \int_{-r}^{r} I^* \left(\frac{\omega \cdot x}{\|\omega\|} - t\right) \sin(\|\omega\|t + \theta(\omega))dt$$

$$= s(\omega) - \int_{-r}^{r} I^*(\frac{\omega \cdot x}{\|\omega\|} - t) \sin(\|\omega\|t + \theta(\omega))dt$$

$$= s(\omega) - \int_{-r}^{r} \phi_{(t,\omega)}(x) \sin(\|\omega\|t + \theta(\omega))dt,$$

where $s(\omega) = \int_{-r}^{0} \sin(\|\omega\|t + \theta(\omega))dt$. Substituting the above into (16.33) we get

$$f(x) - f(0)$$
$$= \int_{\mathcal{R}^d} s(\omega)\|\omega\|F(\omega)d\omega$$
$$- \int_{[-r,r] \times \mathcal{R}^d} \phi_{(t,\omega)}(x) \sin(\|\omega\|t + \theta(\omega))\|\omega\|F(\omega)d(t,\omega)$$
$$= c - \int_{[-r,r] \times \mathcal{R}^d} \phi_{(t,\omega)}(x)\nu(t,\omega)d(t,\omega),$$

where $c = \int s(\omega)\|\omega\|F(\omega)d\omega$ and

$$\nu(t,\omega)d(t,\omega) = \sin(\|\omega\|t + \theta(\omega))\|\omega\|F(\omega)d(t,\omega)$$

is a signed measure with total variation bounded from above

$$\int_{[-r,r] \times \mathcal{R}^d} |\sin(\|\omega\|t + \theta(\omega))\|\omega\|F(\omega)|d(t,\omega)$$

$$\leq 2r \int_{\mathcal{R}^d} \|\omega\|F(\omega)d\omega = 2rC.$$

Thus by Lemma 16.7 there exists a linear combination of functions from G_{step} plus a constant $f_k(x) = \sum_{j=1}^{k} c_j \phi_{(t_j,a_j)}(x) + c_0$ which approximates $f(x)$ in $L_2(\mu, S_r)$, i.e.,

$$\int_{S_r} \|f(x) - f_k(x)\|^2 \mu(dx) < \frac{4r^2C^2}{k}. \tag{16.35}$$

In order to complete the proof we approximate a step function with the sigmoid. Note that

$$\phi_{(t,\omega)}(x) = I^* \left(\frac{\omega \cdot x}{\|\omega\|} - t\right)$$

$$= \lim_{L \to \infty} \sigma \left(L \left(\frac{\omega \cdot x}{\|\omega\|} - t\right)\right) \tag{16.36}$$

for every $\omega \in \mathcal{R}^d, x \in \mathcal{R}^d, t \in \mathcal{R}$ including the case $\frac{\omega \cdot x}{\|\omega\|} - t = 0$ when the limit in (16.36) is equal to $\sigma(0)$. This means $\overline{G}_\sigma \subset \overline{G}_{step}$. By the Lebesgue dominated convergence theorem, (16.36) holds in $L_2(\mu, S_r)$ sense, and using (16.35) together with the triangle inequality we obtain

$$\limsup_{L \to \infty} \left\| f(x) - \sum_{j=1}^{k} c_j \sigma \left(L \left(\frac{a_j \cdot x}{\|a_j\|} - t \right) - c_0 \right) \right\|_{S_r}$$

$$\leq \left\| f(x) - \left(\sum_{j=1}^{k} c_j \phi_{(t_j, a_j)}(x) - c_0 \right) \right\|_{S_r}$$

$$+ \sum_{j=1}^{k} |c_j| \limsup_{L \to \infty} \left\| \phi_{(t_j, a_j)}(x) - \sigma \left(L \left(\frac{a_j \cdot x}{\|a_j\|} - t \right) \right) \right\|_{S_r}$$

$$\leq \frac{2rC}{\sqrt{k}}.$$

As to the size of the coefficients, Lemma 16.7 states

$$\sum_{j=1}^{k} |c_j| \leq \int_{[-r,r] \times \mathcal{R}^d} |\nu(t, \omega)| d(t, \omega) \leq 2rC,$$

and

$$\begin{aligned} |c_0| &= |c + f(0)| \\ &\leq \int |s(\omega)| \|\omega\| F(\omega) d\omega + |f(0)| \\ &\leq rC + |f(0)|, \end{aligned}$$

hence

$$\sum_{j=0}^{k} |c_j| \leq 3rC + |f(0)|.$$

This completes the proof. $\qquad\qquad\qquad\qquad\qquad\qquad\qquad\qquad\square$

16.4   Bibliographic Notes

An early account of feedforward neural networks is provided in Nilsson (1965). More recent monographs include Hertz, Krogh, and Palmer (1991), Ripley (1996), Devroye, Györfi, and Lugosi (1996) and Bartlett and Anthony (1999). A large number of papers have been devoted to the theoretical analysis of neural network regression estimates. For distributions, where both X and Y are of bounded support, White (1990) proved L_2 consistency in probability for certain estimators. Unlike in Section 16.2 the range of the a_i's and b_i's in White (1990; 1991) had to be restricted.

Almost sure consistency for the same class of distributions can be obtained by using Haussler's (1992) results. Mielniczuk and Tyrcha (1993) obtained L_2 consistency for arbitrary sigmoids. Universal consistency of network classifiers with threshold sigmoid functions was shown by Faragó and Lugosi (1993). Barron (1991; 1994) applied the complexity regularization principle to regression estimation by neural networks. The consistency of the neural network regression estimate presented in Theorem 16.1 has been investigated by Lugosi and Zeger (1995).

Cybenko (1989), Hornik, Stinchcombe, and White (1989), and Funahashi (1989) proved independently, that, on compact sets, feedforward neural networks with one hidden layer are dense with respect to the supremum norm in the set of continuous functions. In other words, every continuous function on $\mathcal{R}^d$ can be approximated arbitrarily closely uniformly over any compact set by functions realized by neural networks. For a survey of such denseness results we refer the reader to Barron (1989) and Hornik (1993).

Cybenko (1989) proved the approximation result in sup norm for networks with continuous squashing functions through application of the Hahn–Banach theorem and the Riesz representation theorem. In their uniform approximation proof Hornik, Stinchcombe, and White (1989) made use of the Stone–Weierstrass theorem (see Rudin (1964)). Funahashi (1989) proved the same result using properties of the Fourier transform. The L_2 approximation result is due to Hornik (1991). Its simplified version is presented in Lemma 16.2.

The rate of approximation of convex combinations in $L_2(\mu)$, given in Lemma 16.7, and the L_2 approximation rate for neural networks with squashing functions, given in Lemma 16.8, follow Barron (1993). The rates of approximation have also been studied by Mhaskar (1996) and Maiorov and Meir (2000).

The rates of L_2 convergence for sigmoidal neural networks have been studied by Barron (1994) and McCaffrey and Gallant (1994). Barron used complexity regularization on the discrete set of parameters and imposed a Lipschitz condition on the sigmoid to obtain the rate in the class of functions covered by the approximation Lemma 16.8. Barron's results have been extended by McCaffrey and Gallant (1994) by not discretizing the parameter space. They considered networks with cosine squasher and applied them to functions in Sobolev spaces, i.e., the spaces of functions for which $f^{(r-1)}$ is absolutely continuous, $r = 1, 2, \ldots$, and $f^{(r)} \in L_p(\mu), p \geq 1$.

The bound on the VC dimension of the graphs of compositions of functions (Lemma 16.3) has been investigated by Nolan and Pollard (1987) and Dudley (1987). The properties of covering numbers described in Lemmas 16.4 and 16.5 can be found in Noland and Pollard (1987), Pollard (1990) and in Devroye, Györfi, and Lugosi (1996).

Problems and Exercises

PROBLEM 16.1. (Hornik (1989)). Prove that Lemma 16.1 is valid for neural networks with continuous nonconstant sigmoids.

HINT. Use the Stone–Weierstrass theorem.

PROBLEM 16.2. (Hornik (1993)). Consider $\sigma(a^T x + b)$, $a \in \mathcal{A}, b \in \mathcal{B}$. Extend Lemma 16.1 so that it holds for any Riemann integrable and non-polynomial sigmoid on some $\mathcal{A}$ containing a neighborhood of the origin, and on some non-degenerate compact interval $\mathcal{B}$.

PROBLEM 16.3. (Hornik (1991)). Prove that Lemma 16.2 remains true if σ is bounded and non-constant.

PROBLEM 16.4. Define $C_f = \int_{\mathcal{R}^d} \|\omega\| F(\omega) d\omega$ (see (16.25)). Let $f(x) = g(\|x\|)$, i. e., f is a radial function. Show that $C_f = V_d \int_0^\infty r^d |\hat{F}(r)| dr$, where V_d is the volume of $d-1$-dimensional unit sphere in $\mathcal{R}^d$. Prove that if $f(x) = \exp(-\|x\|^2/2)$, i. e., f is the Gaussian function then $C_f \leq d^{1/2}$.

17
Radial Basis Function Networks

17.1 Radial Basis Function Networks

The definition of a radial basis function network (RBF network) with one hidden layer and at most k nodes for a fixed function $K : \mathcal{R}_+ \to \mathcal{R}$, called a kernel, is given by the equation

$$f(x) = \sum_{i=1}^{k} w_i K\left(\|x - c_i\|_{A_i}\right) + w_0, \tag{17.1}$$

where

$$\|x - c_i\|_{A_i}^2 = [x - c_i]^T A_i [x - c_i],$$

$w_0, w_1, \ldots, w_k \in [-b, b]$, $c_1, \ldots, c_k \in \mathcal{R}^d$, and $A_1, \ldots, A_k$ are $(d \times d)$-dimensional positive semidefinite matrices (i. e., $a^T A_i a \geq 0$ for all $a \in \mathcal{R}^d$) and $b > 0$ (we allow $b = \infty$). The weights w_i, c_i, A_i are parameters of the RBF network and $K\left(\|x - c_i\|_{A_i}\right)$ is the radial basis function (**Figure 17.1**).

There are two different types of kernels used in applications: increasing kernels, i.e., kernels such that $K(x) \to \infty$ as $x \to \infty$ and decreasing kernels, i.e., kernels such that $K(x) \to 0$ as $x \to \infty$. The increasing kernels play an important role in approximation theory (see Section 17.4 for details). All our theoretical results discussed in this chapter concern only decreasing kernels. It is an open problem whether similar results remain true for increasing kernels.

Common choices for decreasing kernels are:

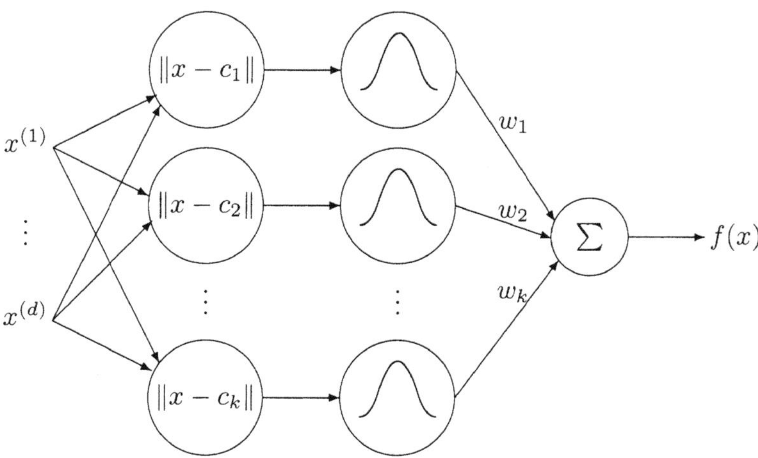

Figure 17.1. Radial basis network with one hidden layer.

- $K(x) = I_{\{x \in [0,1]\}}$ (window);

- $K(x) = \max\{(1 - x^2), 0\}$ (truncated parabolic);

- $K(x) = e^{-x^2}$ (Gaussian); and

- $K(x) = e^{-x}$ (exponential).

For a given fixed kernel K, there are three sets of parameters of RBF networks:

(i) the weight vectors of the output layer of an RBF network w_i ($i = 1, \cdots, k$),

(ii) the center vectors c_i ($i = 1, \cdots, k$), and

(iii) $d \times d$ positive semidefinite matrices A_i ($i = 1, \cdots, k$) determining the size of the receptive field of the basis functions $K\left(\|x - c_i\|_{A_i}\right)$.

The last two sets constitute the weights of the hidden layer of an RBF network. The most common choice for $K(x)$ is the Gaussian kernel, $K(x) = e^{-x^2}$, which leads to

$$K\left(\|x - c_i\|_{A_i}\right) = e^{-[x-c_i]^T A_i [x-c_i]}.$$

For a specific $K(x)$, e.g., Gaussian, the size, shape, and orientation of the receptive field of a node are determined by the matrix A_i. When $A_i = \frac{1}{\sigma_i^2} I$, the shape is a hyperspherical ball with radius σ_i. When $A_i = diag[\sigma_{i,1}^{-2}, \ldots, \sigma_{i,d}^{-2}]$, the shape of the receptive field is an elliptical ball with each axis coinciding with a coordinate axis; the lengths of the axes of the ellipsoid are determined by $\sigma_{i,1}, \ldots, \sigma_{i,d}$. When A_i is nondiagonal but

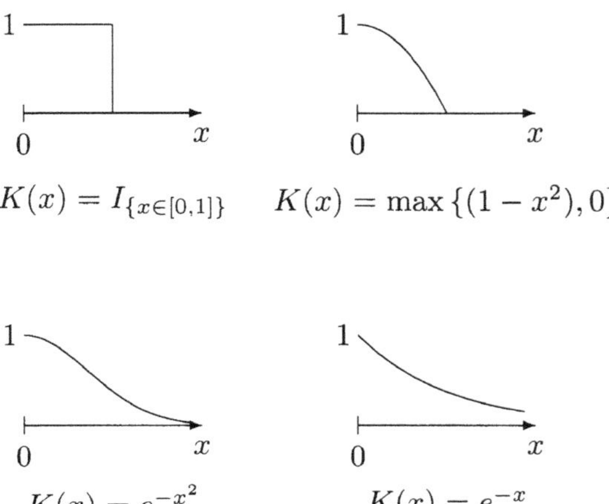

Figure 17.2. Window, truncated parabolic, Gaussian, and exponential kernels.

symmetric, we have $A_i = R_i^T D_i R_i$ where D_i is a diagonal matrix which determines the shape and size of the receptive field and R_i is a rotation matrix which determines the orientation of the receptive field.

There is a close relationship between RBF networks and smoothing splines (see Chapter 20). Consider the multivariate penalized least squares problem in which we minimize

$$\frac{1}{n}\sum_{i=1}^{n}|g(X_i) - Y_i|^2 + \lambda_n \cdot J_k^2(g) \tag{17.2}$$

over the Sobolev space $W^k(\mathcal{R}^d)$ consisting of functions g whose weak derivatives of order k are contained in $L_2(\mathcal{R}^d)$. The complexity of g is penalized by $\lambda_n \cdot J_k^2(g)$, where

$$J_k^2(g) = \int_{\mathcal{R}^d} \exp(||s||^2/\beta)|\tilde{g}(s)|^2 ds$$

and $\tilde{g}$ denotes the Fourier transform of g. The minimization of (17.2) leads to

$$K(||x||) = \exp(-\beta||x||^2). \tag{17.3}$$

Note that (17.3) is a Gaussian kernel which satisfies the conditions of Theorems 17.1 and 17.2.

Another interesting property of an RBF networks which distinguishes them from neural networks with squashing functions is that the center vectors can be selected as cluster centers of the input data. In practical applications, the number of clusters is usually much smaller than the number of data points resulting in RBF networks of smaller complexity than multilayer neural networks.

The problem of determining the specific values of parameters from the training sequence $D_n = \{(X_1, Y_1), \ldots, (X_n, Y_n)\}$ consisting of n i.i.d. copies of (X, Y) is called *learning* or *training*. The most common parameter learning strategies are:

- cluster input vectors $X_i (i = 1, \ldots, n)$ and set center vectors c_i $(i = 1, \ldots, k)$ to cluster centers. Remaining parameters are determined by minimizing the empirical L_2 risk on D_n. If the elements of the covariance matrices A_i $(i = 1, \ldots, k)$ are chosen arbitrarily then finding the output weights w_i $(i = 1, \ldots, k)$ by the least squares method is an easy linear problem, see Section 10.1;

- choose from D_n a random k-element subset

$$D'_n = \{(X'_1, Y'_1), \ldots, (X'_k, Y'_k)\}$$

  of samples and assign $X'_i \to c_i, Y'_i \to w_i$ $(i = 1, \ldots, k)$. The elements of the covariance matrices $A_i, i = 1, \ldots, k$ are chosen arbitrarily; and

- choose all the parameters of the network by minimizing the empirical L_2 risk.

Computationally, the last strategy is the most costly. In practice, parameters of RBF networks are learned by the steepest descent backpropagation algorithm. This algorithm is not guaranteed to find a global minimum of the empirical L_2 risk.

In this chapter we will focus our attention on RBF networks trained by minimizing the empirical L_2 error subject to size restrictions on the output weights. We will study the asymptotic properties of such networks. The empirical risk minimization is computationally very complex and there are no efficient algorithms which can find parameters of multidimensional RBF networks minimizing the empirical L_2 risk.

17.2   Consistency

Given the training set D_n, our estimate of the regression function $m(x) = \mathbf{E}\{Y|X = x\}$ is an RBF network m_n which minimizes the empirical L_2 risk

$$\frac{1}{n} \sum_{j=1}^{n} |f(X_j) - Y_j|^2.$$

More specifically, for each n we fix Θ_n as the set of parameters defined by

$$\Theta_n = \left\{ \theta = (w_0, \dots, w_{k_n}, c_1, \dots, c_{k_n}, A_1, \dots, A_{k_n}) : \sum_{i=0}^{k_n} |w_i| \le b_n \right\},$$

and we choose our regression estimator m_n from the class

$$\begin{aligned}
\mathcal{F}_n \quad &= \{ f_\theta : \theta \in \Theta_n \} \\
&= \left\{ \sum_{i=1}^{k_n} w_i K \left(\|x - c_i\|_{A_i} \right) + w_0 : \sum_{i=0}^{k_n} |w_i| \le b_n \right\}
\end{aligned} \qquad (17.4)$$

with m_n satisfying

$$\frac{1}{n} \sum_{j=1}^{n} |m_n(X_j) - Y_j|^2 = \min_{f \in \mathcal{F}_n} \frac{1}{n} \sum_{j=1}^{n} |f(X_j) - Y_j|^2. \qquad (17.5)$$

We assume the existence of a minimum in (17.5). If the minimum does not exist we can work with functions whose empirical L_2 risk is close to the infimum.

Thus, m_n, the optimal estimator is sought among RBF networks consisting of at most k_n neurons satisfying the weight constraint $\sum_{i=0}^{k_n} |w_i| \le b_n$. The number of allowable nodes k_n will be a function of the training set size n, to be specified later. If we assume that $|K(u)| \le K^*$ for all $u \ge 0$, and let $k^* = \max\{K^*, 1\}$, then these constraints and the Cauchy–Schwarz inequality imply that, for any $\theta \in \Theta_n$ and $x \in \mathcal{R}^d$,

$$\begin{aligned}
|f_\theta(x)|^2 \quad &= \left| \sum_{i=1}^{k_n} w_i K \left(\|x - c_i\|_{A_i} \right) + w_0 \right|^2 \\
&\le \left(\sum_{i=0}^{k_n} |w_i| \right) \left(\sum_{i=1}^{k_n} |w_i| K^2 \left(\|x - c_i\|_{A_i} \right) + w_0 \right) \\
&\le b_n^2 k^{*2}.
\end{aligned} \qquad (17.6)$$

Our analysis of RBF networks will be confined to networks with the *regular radial* kernels defined next.

Definition 17.1. *Kernel $K : [0, \infty) \to \mathcal{R}$ is a regular radial kernel if it is nonnegative, monotonically decreasing, left continuous, $\int_{\mathcal{R}^d} K(\|x\|)dx \ne 0$, and $\int_{\mathcal{R}^d} K(\|x\|)dx < \infty$, where $\| \cdot \|$ is the Euclidean norm on.*

Note that a regular radial kernel is bounded, i.e., $K(x) \le k^*$ $(x \in [0, \infty))$ for some finite constant k^*. All the kernels in Figure 17.2 are regular radial kernels. The next result describes the consistency properties of m_n.

Theorem 17.1. *Let $|Y| \le L < \infty$ a.s. Consider a family $\mathcal{F}_n$ of RBF networks defined by (17.4), with $k_n \ge 1$, and let K be a regular radial* *kernel. If*

$$k_n, b_n \to \infty$$

and

$$k_n b_n^4 \log(k_n b_n^2)/n \to 0$$

as $n \to \infty$, then the RBF network m_n minimizing the empirical L_2 risk over $\mathcal{F}_n = \{f_\theta : \theta \in \Theta_n\}$ is weakly universally consistent. If, in addition,

$$\frac{b_n^4}{n^{1-\delta}} \to 0 \quad (n \to \infty)$$

for some $\delta > 0$, then m_n is strongly universally consistent.

PROOF. Following the proofs of Theorems 10.2 and 10.3, for weak consistency, it suffices to show

$$\inf_{f \in \mathcal{F}_n} \int |f(x) - m(x)|^2 \mu(dx) \to 0 \quad (n \to \infty) \tag{17.7}$$

and

$$\mathbf{E}\left\{ \sup_{f \in \mathcal{F}_n} \left| \frac{1}{n} \sum_{i=1}^{n} |f(X_i) - Y_i|^2 - \mathbf{E}\{|f(X) - Y|^2\} \right| \right\} \to 0 \quad (n \to \infty)$$

for bounded Y, and for strong consistency it suffices to show (17.7) and

$$\sup_{f \in \mathcal{F}_n} \left| \frac{1}{n} \sum_{i=1}^{n} |f(X_i) - Y_i|^2 - \mathbf{E}\{|f(X) - Y|^2\} \right| \to 0 \quad (n \to \infty) \quad a.s.$$

for bounded Y. Here we do not need to consider truncated functions because the functions in $\mathcal{F}_n$ are bounded. The last three limit relations follow from Lemmas 17.1 and 17.2 below. $\square$

We first consider the approximation error for a subset of the family of RBF networks in (17.1) by constraining A_i to be diagonal with the equal elements. Letting $A_i = h_i^{-2}I$, (17.1) becomes

$$f_\theta(x) = \sum_{i=1}^{k} w_i K\left(\left\| \frac{x - c_i}{h_i} \right\|^2 \right) + w_0, \tag{17.8}$$

where $\theta = (w_0, \ldots, w_k, c_1, \ldots, c_k, h_1, \ldots, h_k)$ is the vector of parameters, $w_0, \ldots, w_k \in \mathcal{R}$, $h_1, \ldots, h_k \in \mathcal{R}$, and $c_1, \ldots, c_k \in \mathcal{R}^d$.

We will show in Lemma 17.1 that $\bigcup_{k=1}^{\infty} \mathcal{F}_k$ is dense in $L_2(\mu)$ for any probability measure μ on $\mathcal{R}^d$ and for RBF networks with regular radial kernels. Consequently, the approximation error $\inf_{f \in \mathcal{F}_n} \int |f(x) - m(x)|^2 \mu(dx)$ converges to zero when $k_n \to \infty$.

Lemma 17.1. *Assume that K is a regular radial kernel. Let μ be an arbitrary probability measure on $\mathcal{R}^d$. Then the RBF networks given by (17.8)* are dense in $L_2(\mu)$. In particular, if $m \in L_2(\mu)$, then, for any $\epsilon > 0$, there exist parameters $\theta = (w_0, \ldots, w_k, c_1, \ldots, c_k, h_1, \ldots, h_k)$ such that

$$\int_{\mathcal{R}^d} |f_\theta(x) - m(x)|^2 \mu(dx) < \epsilon. \tag{17.9}$$

PROOF. Let $\|f\|$ denote the $L_2(\mu)$ norm of any $f \in L_2(\mu)$. By Theorem A.1 we have that, for any $m \in L_2(\mu)$ and any $\epsilon > 0$, there exists a continuous g supported on a compact set Q such that

$$\|m - g\|_{L_2} < \frac{\epsilon}{2}. \tag{17.10}$$

Set $\hat{K}(x) = K(\|x\|^2)/\int K(\|x\|^2)dx$, $\hat{K}_h(x) = \frac{1}{h^d}\hat{K}(\|x\|/h)$, and define

$$\sigma_h(x) = \int_{\mathcal{R}^d} g(y)\hat{K}_h(x - y)dy.$$

First we show that $\lim_{h\to 0} \sigma_h(x) = g(x)$ for all $x \in \mathcal{R}^d$. Since g is uniformly continuous for each $\delta > 0$ we can find a $\gamma > 0$, such that

$$|g(x - y) - g(x)| < \delta$$

whenever $|y| < \gamma$. We have

$$
\begin{aligned}
|\sigma_h(x) - g(x)| \quad &= |\int_{\mathcal{R}^d} (g(y) - g(x))\hat{K}_h(x - y)dy| \\
&= |\int_{\mathcal{R}^d} (g(x - y) - g(x))\hat{K}_h(y)dy| \\
&\leq \delta \int_{|y|<\gamma} \hat{K}_h(y)dy + \int_{|y|\geq\gamma} |g(x - y) - g(x)|\hat{K}_h(y)dy \\
&\leq \delta \int_{\mathcal{R}^d} \hat{K}(y)dy + 2 \sup_{x\in\mathcal{R}^d} |g(x)| \int_{|y|\geq\gamma/h} \hat{K}(y)dy
\end{aligned}
$$

and the last term goes to zero as $h \to 0$ since $\int K(\|x\|)dx < \infty$. We also have $|\sigma_h(x)| \leq B\int |\hat{K}(x)|dx < \infty$ for all $x \in \mathcal{R}^d$, where $B = \sup_{x\in\mathbb{R}^d} |g(x)|$. By the dominated convergence theorem, and since $\delta > 0$ is arbitrary, $\lim_{h\to 0} \|g - \sigma_h\|_{L_2(\mu)} = 0$. Consequently, we can choose $h > 0$ such that

$$\|g - \sigma_h\|_{L_2} < \frac{\epsilon}{4}. \tag{17.11}$$

In the rest of the proof, using a probabilistic argument similar to the one by Barron (1993) for Lemma 17.4, we will demonstrate that there exists an f_θ which approximates σ_h within $\epsilon/4$ in $L_2(\mu)$ norm.

First assume that $g(x) \geq 0$ and is not identically zero for all x, and define the probability density function φ by

$$\varphi(x) = g(x)/\int g(y)dy.$$

Then $\sigma_h(x) = \mathbf{E}\{\tilde{K}(x, Z)\}$, where $\tilde{K}(x, y) = \hat{K}_h(x - y) \int g(y)dy$ and Z has density φ.

Let $Z_1, Z_2, \ldots$ be an i.i.d. sequence of random variables with each Z_i having density φ. By the strong law of large numbers, for all x,

$$\lim_{k \to \infty} \frac{1}{k} \sum_{i=1}^{k} \tilde{K}(x, Z_i) = \sigma_h(x) \quad a.s.,$$

therefore, by the dominated convergence theorem,

$$\int \left(\frac{1}{k} \sum_{i=1}^{k} \tilde{K}(x, Z_i) - \sigma_h(x) \right)^2 \mu(dx) \to 0 \quad a.s.$$

as $k \to \infty$. Thus there exists a sequence $(z_1, z_2, \ldots)$ for which

$$\int \left(\frac{1}{k} \sum_{i=1}^{k} \tilde{K}(x, z_i) - \sigma_h(x) \right)^2 \mu(dx) < \frac{\epsilon}{4}$$

for k large enough.

Since $\frac{1}{k} \sum_{i=1}^{k} \tilde{K}(x, z_i)$ is an RBF network in the form of (17.8), this implies the existence of an f_θ such that

$$\|\sigma_h - f_\theta\|_{L_2} < \frac{\epsilon}{4}. \tag{17.12}$$

To generalize (17.12) for arbitrary g we use the decomposition $g(x) = g^+(x) - g^-(x)$, where g^+ and g^- denote the positive and negative parts of g, respectively. Then

$$\sigma_h(x) = \sigma_h^{(1)}(x) - \sigma_h^{(2)}(x) = \int_{\mathcal{R}^d} g^+(y)\hat{K}(x - z)dz - \int_{\mathcal{R}^d} g^-(z)\hat{K}(x - z)dz.$$

We can approximate $\sigma_h^{(1)}$ and $\sigma_h^{(2)}$ separately as in (17.12) above by $f_{\theta^{(1)}}$ and $f_{\theta^{(2)}}$, respectively. Then for $f_\theta = f_{\theta^{(1)}} - f_{\theta^{(2)}}$ we get

$$\|\sigma_h - f_\theta\|_{L_2} < \frac{\epsilon}{2}. \tag{17.13}$$

We conclude from (17.10), (17.11), and (17.13) that

$$\|m - f_\theta\|_{L_2(\mu)} < \epsilon$$

which proves (17.9). Note that the above proof also establishes the first statement of the theorem, namely that $\{f_\theta : \theta \in \Theta\}$ is dense in $L_2(\mu)$. $\square$

Next we consider

$$\sup_{f \in \mathcal{F}_n} \left| \frac{1}{n} \sum_{i=1}^{n} |f(X_i) - Y_i|^2 - \mathbf{E}\{|f(X) - Y|^2\} \right|.$$

We have the following result:

Lemma 17.2. *Assume $|Y| \leq L < \infty$ a.s. Consider a family of RBF networks defined by (17.4), with $k = k_n \geq 1$. Assume that K is a regular radial kernel. If*

$$k_n, b_n \to \infty$$

and

$$k_n b_n^4 \log(k_n b_n^2)/n \to 0$$

as $n \to \infty$, then

$$\mathbf{E}\left\{ \sup_{f \in \mathcal{F}_n} \left| \frac{1}{n} \sum_{i=1}^{n} |f(X_i) - Y_i|^2 - \mathbf{E}\{|f(X) - Y|^2\} \right| \right\} \to 0 \quad (n \to \infty)$$

for all distributions of (X, Y) with Y bounded. If, in addition,

$$\frac{b_n^4}{n^{1-\delta}} \to 0 \quad (n \to \infty)$$

for some $\delta > 0$, then

$$\sup_{f \in \mathcal{F}_n} \left| \frac{1}{n} \sum_{i=1}^{n} |f(X_i) - Y_i|^2 - \mathbf{E}\{|f(X) - Y|^2\} \right| \to 0 \quad (n \to \infty) \quad a.s.$$

PROOF. Let K be bounded by k^*. Without loss of generality we may assume $L^2 \leq b_n^2 k^{*2}$ and $b_n \geq 1$. If $|y| \leq L$, then by (17.6) the functions $h(x, y) = (f(x) - y)^2$, $f \in \mathcal{F}_n$, are bounded above by

$$h(x, y) \leq 4\max\{|f(x)|^2, |y|^2\} \leq 4\max\{b_n^2 k^{*2}, L^2\} \leq 4b_n^2 k^{*2}. \tag{17.14}$$

Define the family of functions

$$\mathcal{H}_n = \{h : \mathcal{R}^{d+1} \to \mathcal{R} \;:\; h(x, y) = (f(x) - T_L y)^2$$
$$((x, y) \in \mathcal{R}^{d+1}) \text{ for some } f \in \mathcal{F}_n\}, \tag{17.15}$$

where T_L is the usual truncation operator. Thus each member of $\mathcal{H}_n$ maps $\mathcal{R}^{d+1}$ into $\mathcal{R}$. Hence

$$\sup_{f \in \mathcal{F}_n} \left| \frac{1}{n} \sum_{j=1}^{n} |f(X_j) - Y_j|^2 - \mathbf{E}|f(X) - Y|^2 \right|$$

$$= \sup_{h \in \mathcal{H}_n} \left| \frac{1}{n} \sum_{i=1}^{n} h(X_i, Y_i) - \mathbf{E}h(X, Y) \right|,$$

and for all $h \in \mathcal{H}_n$ we have $|h(x, y)| \leq 4b_n^2 k^{*2}$ for all $(x, y) \in \mathcal{R}^d \times \mathcal{R}$. Using Pollard's inequality, see Theorem 9.1, we obtain

$$\mathbf{P}\left\{ \sup_{f \in \mathcal{F}_n} \left| \frac{1}{n} \sum_{j=1}^{n} |f(X_j) - Y_j|^2 - \mathbf{E}|f(X) - Y|^2 \right| > \epsilon \right\}$$

$$= \mathbf{P} \left\{ \sup_{h \in \mathcal{H}_n} \left| \frac{1}{n} \sum_{i=1}^{n} h(X_i, Y_i) - \mathbf{E} h(X, Y) \right| > \epsilon \right\}$$

$$\leq 8 \mathbf{E} \left\{ \mathcal{N}_1(\epsilon/8, \mathcal{H}_n, Z_1^n) \right\} e^{-n\epsilon^2/128(4k^{*2}b_n^2)^2}. \tag{17.16}$$

In the remainder of the proof we obtain an upper bound on the L_1 covering number $\mathcal{N}_1(\epsilon/8, \mathcal{H}_n, z_1^n)$, which will be independent of z_1^n.

Let f_1 and f_2 be two real functions on $\mathcal{R}^d$ satisfying $|f_i(x)|^2 \leq b_n^2 k^{*2}$ ($i = 1, 2$) for all $x \in \mathcal{R}^d$. Then for $h_1(x, y) = (f_1(x) - y)^2$ and $h_2(x, y) = (f_2(x) - y)^2$, and any $z_1^n = ((x_1, y_1), \ldots, (x_n, y_n))$ with $|y_i| \leq L$ ($i = 1, \ldots, n$) we have

$$\frac{1}{n} \sum_{i=1}^{n} |h_1(x_i, y_i) - h_2(x_i, y_i)| = \frac{1}{n} \sum_{i=1}^{n} |(f_1(x_i) - y_i)^2 - (f_2(x_i) - y_i)^2|$$

$$= \frac{1}{n} \sum_{i=1}^{n} |f_1(x_i) - f_2(x_i)| \cdot |f_1(x_i) + f_2(x_i) - 2y_i|$$

$$\leq 4k^* b_n \frac{1}{n} \sum_{i=1}^{n} |f_1(x_i) - f_2(x_i)|. \tag{17.17}$$

Since $|f(x)|^2 \leq b_n^2 k^{*2}$ ($x \in \mathcal{R}^d$) for all $f \in \mathcal{F}_n$, functions in the cover $f_1, \ldots, f_l$, $l = \mathcal{N}_1(\epsilon, \mathcal{F}_n, x_1^n)$, (Definition 9.3) can be chosen so that they also satisfy $|f_i(x)|^2 \leq b_n^2 k^{*2}$ ($x \in \mathcal{R}^d$) ($i = 1, \ldots, l$). Combining this with (17.17) we conclude that, for n large enough,

$$\mathcal{N}_1 \left(\frac{\epsilon}{8}, \mathcal{H}_n, z_1^n \right) \leq \mathcal{N}_1 \left(\frac{\epsilon}{32k^* b_n}, \mathcal{F}_n, x_1^n \right).$$

We can use Lemmas 16.4 and 16.5 to relate the covering numbers of the class of functions in $\mathcal{F}_n$ to covering numbers of the class

$$\mathcal{G} = \{ K(\|x - c\|_A) : c \in \mathcal{R}^d \}.$$

We have

$$\mathcal{N}_1 \left(\frac{\epsilon}{32k^* b_n}, \mathcal{F}_n, x_1^n \right)$$

$$\leq \prod_{i=1}^{k_n} \mathcal{N}_1 \left(\frac{\epsilon}{32k^* b_n (k_n + 1)}, \{ w \cdot g \: : \: g \in \mathcal{G}, |w| \leq b_n \}, x_1^n \right)$$

$$\times \mathcal{N}_1 \left(\frac{\epsilon}{32k^* b_n (k_n + 1)}, \{ w \: : \: |w| \leq b_n \}, x_1^n \right)$$

$$\leq \prod_{i=1}^{k_n} \left(\mathcal{N}_1 \left(\frac{\epsilon}{64k^* b_n^2 (k_n + 1)}, \mathcal{G}, x_1^n \right) \right.$$

$$\times \mathcal{N}_1 \left(\frac{\epsilon}{64k^{*2}b_n(k_n+1)}, \{w : |w| \le b_n\}, x_1^n \right) \Bigg)$$

$$\times \mathcal{N}_1 \left(\frac{\epsilon}{32k^*b_n(k_n+1)}, \{w : |w| \le b_n\}, x_1^n \right)$$

$$\le \prod_{i=1}^{k_n} \left(\frac{2b_n}{\frac{\epsilon}{64k^{*2}b_n(k_n+1)}} \mathcal{N}_1 \left(\frac{\epsilon}{64k^*b_n^2(k_n+1)}, \mathcal{G}, x_1^n \right) \right)$$

$$\cdot \frac{2b_n}{\frac{\epsilon}{32k^*b_n(k_n+1)}}$$

$$= \left(\frac{128k^{*2}b_n^2(k_n+1)}{\epsilon} \right)^{k+1} \left(\mathcal{N}_1 \left(\frac{\epsilon}{64k^*b_n^2(k_n+1)}, \mathcal{G}, x_1^n \right) \right)^k. \qquad (17.18)$$

To bound $\mathcal{N}_1(\epsilon/(64k^*b_n^2(k_n+1)), \mathcal{G}, x_1^n)$ we will use Theorem 9.4 relating covering numbers of $\mathcal{G}$ to the VC dimension of graph sets of functions in $\mathcal{G}$ (see Figure 17.3).

Since K is left continuous and monotone decreasing we have

$$K\left(\sqrt{[x-c]^T A[x-c]} \right) \ge t \quad \text{if and only if} \quad [x-c]^T A[x-c] \le \varphi^2(t),$$

where $\varphi(t) = \max\{y : K(y) \ge t\}$. Equivalently, (x, t) must satisfy

$$x^T Ax - x^T(Ac + A^T c) + c^T Ac - \varphi^2(t) \le 0.$$

Consider now the set of real functions on $\mathcal{R}^{d+1}$ defined for any $(x, s) \in \mathcal{R}^d \times \mathcal{R}$ by

$$g_{A,\alpha,\beta,\gamma}(x, s) = x^T Ax + x^T \alpha + \gamma + \beta s,$$

where A ranges over all $(d \times d)$-matrices, and $\alpha \in \mathcal{R}^d$, $\beta, \gamma \in \mathcal{R}$ are arbitrary. The collection $\{g_{A,\alpha,\beta,\gamma}\}$ is a $(d^2 + d + 2)$-dimensional vector space of functions. Thus the class of sets of the form $\{(x, s) : g_{A,\alpha,\beta,\gamma}(x, s) \le 0\}$ has VC dimension at most $d^2 + d + 2$ by Theorem 9.5. Clearly, if for a given collection of points $\{(x_i, t_i)\}$ a set $\{(x, t) : g(x) \ge t\}$, $g \in \mathcal{G}$ picks out the points $(x_{i_1}, t_{i_1}), \ldots, (x_{i_l}, t_{i_l})$, then there exist A, α, β, γ such that $\{(x, s) : g_{A,\alpha,\beta,\gamma}(x, s) \ge 0\}$ picks out only the points $(x_{i_1}, \varphi^2(t_{i_1})), \ldots, (x_{i_l}, \varphi^2(t_{i_l}))$. This shows that $V_{\mathcal{G}^+} \le d^2 + d + 2$. Theorem 9.4 implies

$$\mathcal{N}_1(\epsilon/(64k^*b_n^2(k_n+1)), \mathcal{G}, x_1^n)$$

$$\le 3 \left(\frac{6ek^*}{\epsilon/(64k^*b_n^2(k_n+1))} \right)^{2(d^2+d+2)}$$

$$\le 3 \left(\frac{384ek^{*2}b_n^2(k_n+1)}{\epsilon} \right)^{2(d^2+d+2)},$$

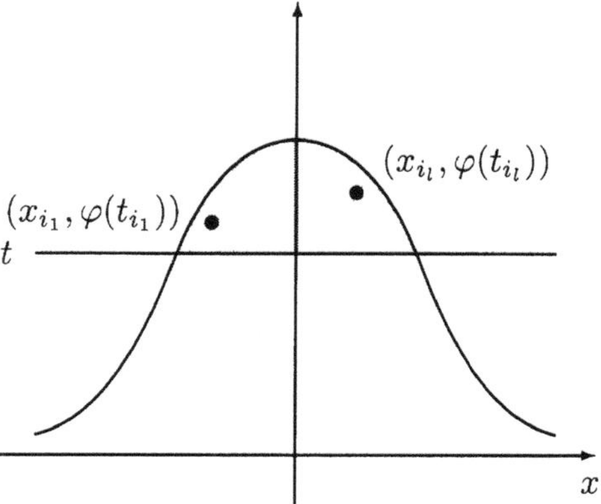

Figure 17.3. Graph set of $\mathcal{G}$.

from which, upon substitution into (17.18), we obtain for n large enough

$$\mathcal{N}_1\left(\frac{\epsilon}{32k^*b_n}, \mathcal{F}_n, x_1^n\right)$$

$$\leq 3^k\left(\frac{384ek^{*2}b_n^2(k_n+1)}{\epsilon}\right)^{(2d^2+2d+3)k+1}$$

$$\leq \left(\frac{C_1b_n^2k_n}{\epsilon}\right)^{C_2k_n}.$$

Collecting the results above we finally obtain, with appropriate constants C_1, C_2, and C_3, depending on k^*, d, and k^*, respectively, the following bound implied by inequality (17.16):

$$\mathbf{P}\left\{\sup_{f\in\mathcal{F}_n}\left|\frac{1}{n}\sum_{j=1}^n|f(X_j)-Y_j|^2 - \mathbf{E}|f(X)-Y|^2\right| > \epsilon\right\}$$

$$\leq 8\left(\frac{C_1b_n^2k_n}{\epsilon}\right)^{C_2k_n}e^{-n\epsilon^2/C_3(b_n^2)^2}$$

$$= 8\exp\left(-\frac{n}{(b_n)^4}[\epsilon^2/C_3 - \frac{C_2k_nb_n^4}{n}\log\frac{C_1b_n^2k_n}{\epsilon}]\right).$$

As in the proof of Theorem 10.3 this inequality implies Lemma 17.2.   □

17.3   Rate of Convergence

Consider the RBF networks given by (17.1) with weights satisfying the constraint $\sum_{i=0}^k |w_i| \leq \beta_n$ for a fixed $b > 0$. Thus the kth candidate class

$\mathcal{F}_k$ for the function estimation task is defined as the class of networks with k nodes

$$\mathcal{F}_{n,k} = \left\{ \sum_{i=1}^{k} w_i K\left(\|x - c_i\|_{A_i}\right) + w_0 : \sum_{i=0}^{k} |w_i| \le \beta_n \right\}. \tag{17.19}$$

In the sequel we develop bounds on the expected L_2 error of complexity regularized RBF networks.

Clearly, $|f(x)| \le k^* \beta_n$ for all $x \in \mathcal{R}^d$ and $f. \in \mathcal{F}_{n,k}$. It will be assumed that for each k we are given a finite, almost sure uniform upper bound on the random covering numbers $\mathcal{N}_1(\epsilon, \mathcal{F}_{n,k}, X_1^n)$, where $X_1^n = (X_1, \dots, X_n)$. Denoting this upper bound by $\mathcal{N}_1(\epsilon, \mathcal{F}_{n,k})$, we have

$$\mathcal{N}_1(\epsilon, \mathcal{F}_{n,k}, X_1^n) \le \mathcal{N}_1(\epsilon, \mathcal{F}_{n,k}) \quad \text{a.s.}$$

Note that we have suppressed the possible dependence of this bound on the distribution of X. Finally assume that $|Y| \le L < \infty$ a.s.

We define the complexity penalty of the kth class for n training samples as any nonnegative number $pen_n(k)$ satisfying

$$pen_n(k) \ge 2568 \frac{\beta_n^4}{n} \cdot \left(\log \mathcal{N}_1(1/n, \mathcal{F}_{n,k}) + t_k\right), \tag{17.20}$$

where the nonnegative constants $t_k \in \mathcal{R}_+$ satisfy Kraft's inequality $\sum_{k=1}^{\infty} e^{-t_k} \le 1$. As in Chapter 16 the coefficients t_k may be chosen as $t_k = 2\log k + t_0$ with $t_0 \ge \log\left(\sum_{k \ge 1} k^{-2}\right)$. The penalty is defined in a similar manner as in inequality (12.14).

We can now define our estimate. Let

$$m_{n,k} = \arg\min_{f \in \mathcal{F}_{n,k}} \frac{1}{n} \sum_{i=1}^{n} |f(X_i) - Y_i|^2,$$

that is, $m_{n,k}$ minimizes the empirical L_2 risk for n training samples over $\mathcal{F}_{n,k}$. (We assume the existence of such a minimizing function for each k and n.) The penalized empirical L_2 risk is defined for each $f \in \mathcal{F}_{n,k}$ as

$$\frac{1}{n} \sum_{i=1}^{n} |f(X_i) - Y_i|^2 + pen_n(k).$$

Our estimate m_n is then defined as the $m_{n,k}$ minimizing the penalized empirical risk over all classes

$$m_n = m_{n,k^*}, \tag{17.21}$$

where

$$k^* = \arg\min_{k \ge 1} \left(\frac{1}{n} \sum_{i=1}^{n} |m_{n,k}(X_i) - Y_i|^2 + pen_n(k) \right).$$

It is easy to see that the class $\mathcal{F} = \bigcup_k \mathcal{F}_k$ of RBF networks given by (17.19) is convex. Let $\overline{\mathcal{F}}$ be the closure of $\mathcal{F}$ in $L_2(\mu)$. We have the following theorem for the estimate (17.21).

Theorem 17.2. *Let $1 \leq L < \infty, n \in \mathcal{N}$, and let $L \leq \beta_n < \infty$. Suppose, furthermore, that $|Y| \leq L < \infty$ a.s. Let K be a regular radial kernel. Assume that the penalty satisfies (17.20) for some t_k such that $\sum_{k=1}^{\infty} e^{-t_k} \leq 1$. Then the RBF regression estimate chosen by the complexity regularization satisfies for n sufficiently large*

$$
\mathbf{E} \int |m_n(x) - m(x)|^2 \mu(dx)
$$

$$
\leq 2 \min_{k \geq 1} \left(2568 \cdot \frac{\beta_n^4}{n} \left(((2d^2 + 2d + 6)k + 5) \log 12ek^* \beta_n n + t_k \right) \right.
$$

$$
\left. + \inf_{f \in \mathcal{F}_k} \mathbf{E} \int |f(x) - m(x)|^2 \mu(dx) \right) + 5 \cdot 2568 \frac{\beta_n^4}{n}. \tag{17.22}
$$

If $m \in \overline{\mathcal{F}}$, then

$$
\mathbf{E} \int |m_n(x) - m(x)|^2 \mu(dx) = O\left(\beta_n^2 \left(\frac{\log(\beta_n n)}{n} \right)^{1/2} \right). \tag{17.23}
$$

Note that if $\beta_n < const < \infty$ then

$$
pen(k) = O\left(\frac{k \log n}{n} \right)
$$

and the rate in (17.23) becomes

$$
\mathbf{E} \int |m_n(x) - m(x)|^2 \mu(dx) = O\left(\sqrt{\frac{\log n}{n}} \right).
$$

In the proof of Theorem 17.2 we will use Lemma 17.3 which will provide a stronger bound on the covering numbers of RBF networks than (17.18). Lemma 17.4 will provide the approximation error rate.

Lemma 17.3. *Assume that K is a regular radial kernel. Then, for $0 < \epsilon < k^*/4$,*

$$
\mathcal{N}_1(\epsilon, \mathcal{F}_k, X_1^n)
$$

$$
\leq 3^k \frac{4\beta_n(\beta_n + \epsilon)}{\epsilon} \left(\frac{2ek^*(\beta_n + \epsilon/k^*)}{\epsilon} \right)^{k+1} \left(\frac{6ek^*(\beta_n + \epsilon)}{\epsilon} \right)^{2k(d^2+d+2)}.
$$

If, in addition, $\beta_n \geq \epsilon$, then

$$
\mathcal{N}_1(\epsilon, \mathcal{F}_k, X_1^n)
$$

$$
\leq \frac{3^k 8 \beta_n^2}{\epsilon} \left(\frac{12ek^* \beta_n}{\epsilon} \right)^{(2d^2+2d+5)k+1}.
$$

PROOF. Let $\mathcal{G}$ be the collection of functions $[x - c]^T A[x - c]$ parametrized by nonnegative definite matrices A and $c \in \mathcal{R}^d$. Also let

$$\mathcal{K} = \{K(\sqrt{g}(\cdot)) : g \in \mathcal{G}\} = \{\overline{K}(g(\cdot)) : g \in \mathcal{G}\},$$

where $\overline{K}(x) = K(\sqrt{x}), x \in \mathcal{R}$, is a monotone decreasing function since it is a composition of the monotone increasing function $\sqrt{x}$ and a monotone decreasing function $K(x)$. Since $\mathcal{G}$ spans a $(d^2 + d + 1)$-dimensional vector space, then by Theorem 9.5 the collection of sets

$$\mathcal{G}^+ = \{\{(x,t) : g(x) - t \geq 0\} : g \in \mathcal{G}\}$$

has VC dimension $V_{\mathcal{G}+} \leq d^2 + d + 2$. Since $\overline{K}$ is monotone decreasing, it follows from Lemma 16.3 that $V_{\mathcal{K}+} \leq d^2 + d + 2$, where the families of sets $\mathcal{K}^+$ are defined just as $\mathcal{G}^+$ with $\mathcal{K}$ in place of $\mathcal{G}$. Since $0 \leq K(x) \leq k^*$, $x \in \mathcal{R}$, Theorem 9.4 implies, for all $f \in \mathcal{K}$ and $\epsilon < k^*/4$,

$$\mathcal{N}_1(\epsilon, \mathcal{K}, X_1^n) \leq 3 \left(\frac{2ek^*}{\epsilon} \log \left(\frac{3ek^*}{\epsilon} \right) \right)^{V_{\mathcal{K}+}} \leq 3 \left(\frac{3ek^*}{\epsilon} \right)^{2V_{\mathcal{K}+}}.$$

It follows that

$$\mathcal{N}_1(\epsilon, \mathcal{K}, X_1^n) \leq 3 \left(\frac{3ek^*}{\epsilon} \right)^{2(d^2+d+2)}$$

for $\epsilon < k^*/4$. Since $\mathcal{F}_k$ is defined as

$$\mathcal{F}_k = \{\sum_{i=1}^{k} w_i f_i + w_0 : \sum_{i=0}^{k} |w_i| \leq \beta_n, f_i \in \mathcal{K}\},$$

we obtain, from Lemma 16.6 with $\eta = \delta = \epsilon/2$ and $B = k^*$,

$$\mathcal{N}_1(\epsilon, \mathcal{F}_k, X_1^n)$$

$$\leq \left(\frac{2ek^*(\beta_n + \epsilon/k^*)}{\epsilon} \right)^{k+1} (\mathcal{N}_1(\epsilon/2(\beta_n + \epsilon), \mathcal{K}, X_1^n))^k$$

$$\times \mathcal{N}_1(\epsilon/2(\beta_n + \epsilon), \{w_0 : |w_0| \leq \beta_n\}, X_1^n)$$

$$\leq \left(\frac{2ek^*(\beta_n + \epsilon/k^*)}{\epsilon} \right)^{k+1} \left(3 \left(\frac{6ek^*(\beta_n + \epsilon)}{\epsilon} \right)^{2(d^2+d+2)} \right)^k \cdot \frac{4\beta_n(\beta_n + \epsilon)}{\epsilon}$$

$$\leq 3^k \frac{4\beta_n(\beta_n + \epsilon)}{\epsilon} \left(\frac{2ek^*(\beta_n + \epsilon/k^*)}{\epsilon} \right)^{k+1} \left(\frac{6ek^*(\beta_n + \epsilon)}{\epsilon} \right)^{2k(d^2+d+2)}$$

$$\leq \frac{3^k 8\beta_n^2}{\epsilon} \left(\frac{12ek^*\beta_n}{\epsilon} \right)^{(2d^2+2d+5)k+1}$$

$$\square$$

The next lemma, by Barron (1993) needed in the proof of Theorem 17.2 describes the rate of approximation of convex combinations in the Hilbert space.

Lemma 17.4. *Denote by $\overline{\mathcal{F}}$ the closure of the convex hull of the set $\mathcal{F} = \bigcup_k \mathcal{F}_k$ in $L_2(\mu)$ with norm $||\cdot|| = ||\cdot||_{L_2(\mu)}$. Assume $||f|| \leq b$ for each $f \in \mathcal{F}$, and let $f^* \in \overline{\mathcal{F}}$. Then for every $k \geq 1$ and every $c > 2(b^2 - ||f^*||^2)$, there is an f_k in the convex hull of k points in $\mathcal{F}$ such that*

$$||f_k - f^*||^2_{L^2(\mu)} \leq \frac{c}{k}.$$

PROOF. Pick $k \geq 1$ and $\delta > 0$. Choose a function $\overline{f}$ in the convex hull of $\mathcal{F}$ such that

$$||\overline{f} - f^*|| \leq \delta/k.$$

Thus $\overline{f} = \sum_{i=1}^m \alpha_i \overline{f}_i$, with $\alpha_i \geq 0, \sum_{i=1}^m \alpha_i = 1, \overline{f}_i \in \mathcal{F}$ for m sufficiently large. Let X be randomly drawn from the set $\{1, \ldots, m\}$ with

$$\mathbf{P}\{g_X = \overline{f}_i\} = \mathbf{P}\{X = i\} = \alpha_i \quad (i = 1, \ldots, m)$$

and let $X_1, \ldots, X_k$ be independently drawn from the same distribution as X. Set $f_k = \frac{1}{k}\sum_{i=1}^k g_{X_i}$. Then

$$\mathbf{E}f_k = \frac{1}{k}\sum_{i=1}^k \mathbf{E}g_{X_i} = \frac{1}{k}\sum_{i=1}^k \mathbf{E}\sum_{j=1}^m \overline{f}_j I_{[X_i=j]} = \frac{1}{k}\sum_{i=1}^k \sum_{j=1}^m \alpha_j \overline{f}_j = \frac{1}{k}\sum_{i=1}^k \overline{f} = \overline{f}.$$

Next

$$\mathbf{E}||f_k - \overline{f}||^2 = \mathbf{E} < f_k - \overline{f}, f_k - \overline{f} >$$

$$= \mathbf{E} < \frac{1}{k}\sum_{i=1}^k g_{X_i} - \overline{f}, \frac{1}{k}\sum_{i=1}^k g_{X_i} - \overline{f} >$$

$$= \mathbf{E} < \frac{1}{k}\sum_{i=1}^k g_{X_i} - \mathbf{E}\frac{1}{k}\sum_{i=1}^k g_{X_i}, \frac{1}{k}\sum_{i=1}^k g_{X_i} - \mathbf{E}\frac{1}{k}\sum_{i=1}^k g_{X_i} >$$

$$= \frac{1}{k^2}\mathbf{E}\sum_{i,j=1}^k < g_{X_i} - \mathbf{E}g_{X_i}, g_{X_j} - \mathbf{E}g_{X_j} >$$

$$= \frac{1}{k^2}\sum_{i,j=1}^k \sum_{l,p=1}^m \mathbf{E} < \overline{f}_l(I_{[X_i=l]} - \alpha_l), \overline{f}_p(I_{[X_j=p]} - \alpha_p) >$$

$$= \frac{1}{k^2}\sum_{i=1}^k \sum_{l,p=1}^m \mathbf{E} < \overline{f}_l(I_{[X_i=l]} - \alpha_l), \overline{f}_p(I_{[X_i=p]} - \alpha_p) >$$

$$= \frac{1}{k^2}\sum_{i=1}^k \mathbf{E}||g_{X_i} - \mathbf{E}g_{X_i}||^2$$

$$= \frac{1}{k}\mathbf{E}||g_{X_1} - \overline{f}||^2.$$

Since the boundedness of g implies boundedness of g, we obtain

$$\frac{1}{k}\mathbf{E}\|g_{X_1} - \overline{f}\|^2$$

$$= \frac{1}{k}\mathbf{E} < g_{X_1} - \overline{f}, g_{X_1} - \overline{f} >$$

$$= \frac{1}{k}\left(\mathbf{E}\|g_{X_1}\|^2 + \|\overline{f}\|^2 - 2\mathbf{E} < g_{X_1}, \overline{f} >\right)$$

$$= \frac{1}{k}\left(\mathbf{E}\|g_{X_1}\|^2 + \|\overline{f}\|^2 - 2\sum_{i=1}^{m}\mathbf{E}\left(< \overline{f}_i, \overline{f} > \cdot I_{\{X_1=i\}}\right)\right)$$

$$= \frac{1}{k}\left(\mathbf{E}\|g_{X_1}\|^2 + \|\overline{f}\|^2 - 2\sum_{i=1}^{m}\alpha_i < \overline{f}_i, \overline{f} >\right)$$

$$= \frac{1}{k}\left(\mathbf{E}\|g_{X_1}\|^2 - \|\overline{f}\|^2\right)$$

$$\leq \frac{1}{k}\left(b^2 - \|\overline{f}\|^2\right).$$

We have thus bounded $\mathbf{E}\|f_k - \overline{f}\|^2$ by $\frac{1}{k}\left(b^2 - \|\overline{f}\|^2\right)$ which implies that there exist $g_1, \ldots, g_k \in \mathcal{F}$ and f_k in the convex hull of $g_1, \ldots, g_k$ such that

$$\|f_k - \overline{f}\|^2 \leq \frac{1}{k}\left(b^2 - \|\overline{f}\|^2\right).$$

Thus by the triangle inequality

$$\|f_k - f^*\|^2 \leq 2\|f_k - \overline{f}\|^2 + 2\|\overline{f} - f^*\|^2$$

$$\leq \frac{2(b^2 - \|\overline{f}\|^2)}{k} + \frac{2\delta}{k}.$$

The conclusion of the lemma follows by choosing δ sufficiently small. $\square$

PROOF OF THEOREM 17.2. Using the bound for $\mathcal{N}_1(\epsilon, \mathcal{F}_k, X_1^n)$ from Lemma 17.3 we obtain from (17.20)

$$2568\frac{\beta_n^4}{n}\left(\log \mathcal{N}_1(1/n, \mathcal{F}_{n,k}) + t_k\right)$$

$$\leq 2568\frac{\beta_n^4}{n}\left(\log \frac{3^k 8\beta_n^2}{\epsilon}\left(\frac{12ek^*\beta_n}{\epsilon}\right)^{(2d^2+2d+5)k+1} + t_k\right)$$

$$\leq 2568\frac{\beta_n^4}{n}\left(((2d^2 + 2d + 6)k + 5)\log 12ek^*\beta_n n + t_k\right)$$

$$= pen_n(k). \tag{17.24}$$

The penalty bound in (12.14) implies

$$\mathbf{E}\int |m_n(x) - m(x)|^2 \mu(dx)$$

$$\leq 2\min_{k\geq 1}\left(pen_n(k) + \inf_{f\in\mathcal{F}_k}\mathbf{E}\int |f(x) - m(x)|^2 \mu(dx)\right) + 5\cdot 2568\frac{\beta_n^4}{n},$$

from which and from (17.24) for n sufficiently large we obtain (17.22). Note that upon choosing $t_k = 2\log(k) + t_0, t_0 \geq \sum_{k=1}^{\infty} k^{-2}$ we get

$$pen(k) = O\left(\frac{\beta_n^4 k \log(\beta_n n)}{n}\right).$$

In order to obtain the second statement of Theorem 17.2 we note that the class $\cup_k\mathcal{F}_k$ is convex if the $\mathcal{F}_k$ are the collections of RBF networks defined in (17.1). We can thus apply, to the right-hand side of (17.22), Lemma 17.4 which states that there exists an RBF network $f_k \in \mathcal{F}_k$ such that

$$\|f_k - m\|_{L_2}^2 \leq \frac{c_1}{k}$$

for some $c_1 > 2((k^*b)^2 - \|m\|^2)$. Substituting this bound into (17.22) we obtain

$$\mathbf{E}\int |m_n(x) - m(x)|^2 \mu(dx)$$

$$\leq 2\min_{k\geq 1}\left(\frac{c_1}{k} + c\frac{\beta_n^4 k \log(\beta_n n)}{n}\right) + 5\cdot 2568\frac{\beta_n^4}{n}$$

$$= O\left(\beta_n^2 \left(\frac{\log(\beta_n n)}{n}\right)^{1/2}\right),$$

and (17.23) follows. $\square$

The above convergence rate results hold in the case when the regression function is a member of the $L_2(\mu)$ closure of $\mathcal{F} = \bigcup \mathcal{F}_k$, where

$$\mathcal{F}_k = \left\{\sum_{i=1}^{k} w_i K\left(\|x - c_i\|_{A_i}\right) + w_0 : \sum_{i=0}^{k} |w_i| \leq b\right\}. \qquad (17.25)$$

In other words, m should be such that for all $\epsilon > 0$ there exists a k and a member f of $\mathcal{F}_k$ with $\|f - m\|_{L_2} < \epsilon$. The precise characterization of $\overline{\mathcal{F}}$ remains largely unsolved. However, based on the work of Girosi and Anzellotti (1992) we can describe a large class of functions that is *contained* in $\overline{\mathcal{F}}$.

Let $H(x, t)$ be a bounded, real-valued, and measurable function of two variables $x \in \mathcal{R}^d$ and $t \in \mathcal{R}^n$. Suppose that ν is a signed measure on $\mathcal{R}^n$ with finite total variation $\|\nu\|$, where $\|\nu\| = \nu^+(\mathcal{R}^d) + \nu^-(\mathcal{R}^d)$ and ν^+ and ν^- are positive and negative parts of ν, respectively (see, e.g., Rudin (1966)). If $g(x)$ is defined as

$$g(x) = \int_{\mathcal{R}^n} H(x, t)\nu(dt),$$

then $g \in L_2(\mu)$ for any probability measure μ on $\mathcal{R}^d$. One can reasonably expect that g can be approximated well by functions $f(x)$ of the form

$$f(x) = \sum_{i=1}^{k} w_i H(x, t_i),$$

where $t_1, \ldots, t_k \in \mathcal{R}^n$ and $\sum_{i=1}^{k} |w_i| \leq \|\nu\|$. The case $n = d$ and $H(x, t) = G(x - t)$ has been investigated by Girosi and Anzellotti (1993), where a detailed description of function spaces arising from the different choices of the basis function G is given. Girosi (1994) extends this approach to approximation by convex combinations of translates and dilates of a Gaussian function. In general, one can prove the following:

Lemma 17.5. *Let*

$$g(x) = \int_{\mathcal{R}^n} H(x, t)\nu(dt), \qquad (17.26)$$

where $H(x, t)$ and ν are as above. Define, for each $k \geq 1$, the class of functions

$$\mathcal{G}_k = \left\{ f(x) = \sum_{i=1}^{k} w_i H(x, t_i) : \sum_{i=0}^{k} |w_i| \leq \|\nu\| \right\}.$$

Then, for any probability measure μ on $\mathcal{R}^d$ and for any $1 \leq p < \infty$, the function g can be approximated in $L_2(\mu)$ arbitrarily closely by members of $\mathcal{G} = \bigcup \mathcal{G}_k$, i.e.,

$$\inf_{f \in \mathcal{G}_k} \|f - g\|_{L_2(\mu)} \to 0 \quad as \quad k \to \infty.$$

In other words, $g \in \overline{\mathcal{G}}$.

The proof of the lemma is similar to the proof of Lemma 17.1 and is left as an exercise (see Problem 17.1).

It is worth mentioning that in general, the closure of $\bigcup_k \mathcal{G}_k$ is richer than the class of functions having representation as in (17.26).

To apply the lemma for RBF networks (17.1), let $n = d^2 + d$, $t = (A, c)$, and $H(x, t) = K(\|x - c_i\|_A)$. Note that $\overline{\mathcal{F}}$ contains all the functions g with the integral representation

$$g(x) = \int_{\mathcal{R}^{d^2 + d}} K(\|x - c_i\|_A)\, \nu(dc, dA), \qquad (17.27)$$

for which $\|\nu\| \leq b$, where b is the constraint on the weights as in (17.25). The approximation result of Lemma 17.4 for functions of the form (17.27) in the special case of X having bounded support and absolutely continuous ν is a direct consequence of Lemma 17.4. In this case the rates of approximation in both lemmas are the same. One important example of a class of functions g obtainable in this manner has been given by Girosi (1994). He used the

Gaussian basis function

$$H(x,t) = H(x,c,\sigma) = \exp\left(-\frac{\|x-c\|^2}{\sigma}\right),$$

where $c \in \mathcal{R}^d$, $\sigma > 0$, and $t = (c,\sigma)$. The results by Stein (1970) imply that members of the Bessel potential space of order $2m > d$ have an integral representation in the form of (17.26) with this $H(x,t)$, and that they can be approximated by functions of the form

$$f(x) = \sum_{i=1}^{k} w_i \exp\left(-\frac{\|x-c_i\|^2}{\sigma_i}\right) \tag{17.28}$$

in sup norm and, thus, in $L_2(\mu)$. The space of functions thus obtained includes the Sobolev space $H^{2m,1}$ of functions whose weak derivatives up to order $2m$ are in $L_1(\mathcal{R}^d)$. Note that the RBF networks considered in Theorem 17.2 contain (17.28) as a special case.

17.4   Increasing Kernels and Approximation

Increasing kernels, i.e., kernels such that $K(x) \to \infty$ as $x \to \infty$, play important role in approximation. Common choices for increasing kernels are:

- $K(x) = x$ (linear);

- $K(x) = x^3$ (cubic)

- $K(x) = \sqrt{x^2 + c^2}$ (multiquadric);

- $K(x) = x^{2n+1}$ (thin-plate spline), $n \geq 1$; and

- $K(x) = x^{2n} \log x$ (thin-plate spline), $n \geq 1$.

There is a close relationship between RBF networks and smoothing splines (see Chapter 20). Consider the multivariate penalized least squares problem in which we minimize

$$\frac{1}{n} \sum_{i=1}^{n} |g(X_i) - Y_i|^2 + \lambda_n \cdot J_k^2(g) \tag{17.29}$$

over the Sobolev space $W^k(\mathcal{R}^d)$ consisting of functions g whose weak derivatives of order k are contained in $L_2(\mathcal{R}^d)$. The complexity of g is penalized by $\lambda_n \cdot J_k^2(g)$, where

$$J_k^2(g) = \sum_{\alpha_1,\dots,\alpha_d \in \mathcal{N}_0,\, \alpha_1+\cdots+\alpha_d=k} \frac{k!}{\alpha_1! \cdot \ldots \cdot \alpha_d!} \int_{\mathcal{R}^d} \left|\frac{\partial^k g(x)}{\partial x_1^{\alpha_1} \ldots \partial x_d^{\alpha_d}}\right|^2 dx.$$

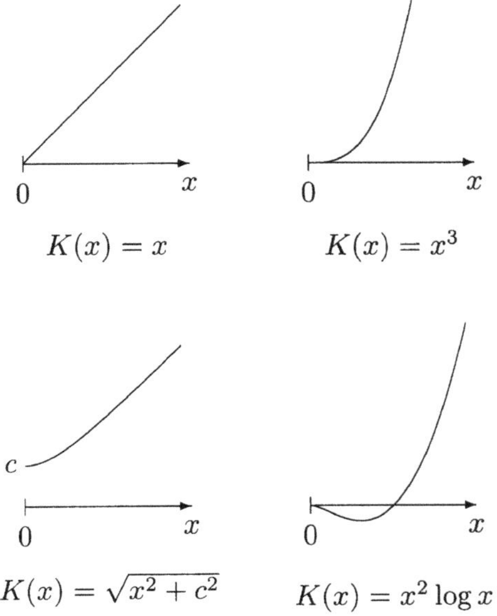

Figure 17.4. Linear, cubic, multiquadric, thin-plate spline kernels.

Using techniques from functional analysis one can show that a function which minimizes (17.29) over $W^k(\mathcal{R}^d)$ always exists. In addition, one can calculate such a function as follows.

Let $l = \binom{d+k-1}{d}$ and let $\phi_1, \ldots, \phi_l$ be all monomials $x_1^{\alpha_1} \cdot \ldots \cdot x_d^{\alpha_d}$ of total degree $\alpha_1 + \ldots + \alpha_d$ less than k. Depending on k and d define

$$K(z) = \begin{cases} \|z\|^{2k-d} \cdot \log(\|z\|) & \text{if} \quad d \text{ is even,} \\ \|z\|^{2k-d} & \text{if} \quad d \text{ is odd.} \end{cases}$$

Let $z_1, \ldots, z_N$ be the distinct values of $X_1, \ldots, X_n$. Then there exists a function of the form

$$g^*(x) = \sum_{i=1}^{N} \mu_i K(x - z_i) + \sum_{j=1}^{l} \nu_j \phi_j(x) \tag{17.30}$$

which minimizes (17.29) over $W^k(\mathcal{R}^d)$ (see Chapter 20 for further details). The kernel K in (17.30) is the thin-plate spline kernel, which is an increasing kernel.

Radial functions (primarily increasing ones) are also encountered in interpolation problems where one looks for radial function interpolants of the form

$$f(x) = \sum_{i=1}^{n} c_i K(\|x - X_i\|) + p_m(x)$$

with polynomial $p_m(x)$ on $\mathcal{R}^d$ of degree less than m interpolating data (X_i, Y_i) $(i = 1, \ldots, n)$. Typical radial functions used in interpolation are multiquadrics, shifted surface splines, and thin-plate splines.

17.5   Bibliographic Notes

Radial Basis Function (RBF) Networks have been introduced by Broomhead and Lowe (1988) and Moody and Darken (1989). Powell (1987) and Dyn (1989) described applications of RBF networks in interpolation and approximation. The universal approximation ability of RBF networks was studied by Park and Sandberg (1991; 1993) and Krzyżak, Linder, and Lugosi (1996). Lemma 17.1 which is due to Krzyżak, Linder, and Lugosi (1996) generalizes the approximation results of Park and Sandberg (1991; 1993) who showed that if $K(\|x\|) \in L_1(\lambda) \cap L_p(\lambda)$ and $\int K(\|x\|) \neq 0$, then the class of RBF networks defined in (17.8) is dense in $L_p(\lambda)$ for $p \geq 1$. Approximation Lemma 17.4 is due to Barron (1993).

Poggio and Girosi (1990), Chen, Cowan, and Grant (1991), Krzyżak et al. (1996; 1998) and Xu, Krzyżak, and Yuille (1994) investigated the issues of learning and estimation. RBF networks with centers learned by clustering were studied by Xu, Krzyżak, and Oja (1993), with randomly sampled centers by Xu, Krzyżak, and Yuille (1994). RBF networks with parameters learned by minimizing the empirical L_2 risk were investigated by Krzyżak et al. (1996; 1998).

The steepest descent backpropagation weight training algorithm has been proposed by Rumelhart, Hinton and Williams (1986) and applied to RBF training by Chen, Cowan, and Grant (1991). Convergence rates of RBF approximation schemes have been shown to be comparable with those for neural networks by Girosi and Anzellotti (1992; 1993). Niyogi and Girosi (1996) studied the tradeoff between approximation and estimation errors and provided an extensive review of the problem. L_p error rates with $1 \leq p < \infty$ were established by Krzyżak and Linder (1998) who showed Theorem 17.2.

RBF networks with optimal MISE radial functions were investigated by Krzyżak (2001) and normalized RBF networks in Krzyżak and Niemann (2001), and Krzyżak and Schäfer (2002). Radial functions were used in interpolation by Powell (1987; 1992), Dyn (1989), and Light (1992).

Problems and Exercises

PROBLEM 17.1. Prove Lemma 17.5.
  HINT: Use the probabilistic argument from the proof of Lemma 17.1.

PROBLEM 17.2. Prove the following generalization of Lemma 17.1.

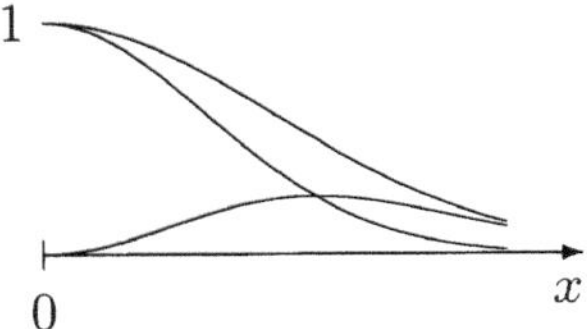

Figure 17.5. Decomposition of the kernel of bounded variation into the difference of monotonically decreasing kernels.

Suppose $K : \mathcal{R} \to \mathcal{R}$ is bounded and

$$K(\|x\|) \in L_1(\lambda) \cap L_p(\lambda)$$

for some $p \in [1, \infty)$ and assume that $\int K(\|x\|)\, dx \neq 0$. Let μ be an arbitrary probability measure on $\mathcal{R}^d$ and let $q \in (0, \infty)$. Then the RBF networks in the form (17.8) are dense in both $L_q(\mu)$ and $L_p(\lambda)$. In particular, if $m \in L_q(\mu) \cap L_p(\lambda)$, then for any ϵ there exists a $\theta = (w_0, \ldots, w_k, b_1, \ldots, b_k, c_1, \ldots, c_k)$ such that

$$\int_{\mathcal{R}^d} |f_\theta(x) - m(x)|^q \mu(dx) < \epsilon \quad \text{and} \quad \int_{\mathcal{R}^d} |f_\theta(x) - m(x)|^p dx < \epsilon.$$

PROBLEM 17.3. Show Theorem 17.2 for the bounded kernel K of bounded variation.

PROBLEM 17.4. Prove the following generalization of Lemma 17.3.

Assume that $|K(x)| \leq 1$ for all $x \in \mathcal{R}$, and suppose that K has total variation $V < \infty$. Then

$$\mathcal{N}_1(\epsilon, \mathcal{F}_k, X_1^n) \leq 3^k \frac{4b(b+\epsilon)}{\epsilon} \left(\frac{2e(b+\epsilon)}{\epsilon} \right)^{k+1} \left(\frac{6eV(b+\epsilon)}{\epsilon} \right)^{4k(d^2+d+2)}$$

HINT: Since K is of bounded variation it can be written as the difference of two monotone decreasing functions: $K = K_1 - K_2$ (see Figure 17.5).

Let $\mathcal{G}$ be the collection of functions $[x-c]^T A[x-c]$ parametrized by $c \in \mathcal{R}^d$ and the nonnegative definite matrix A. Also, let $\widehat{\mathcal{F}}_i = \{K_i(g(\cdot)) : g \in \mathcal{G}\}$ $(i = 1, 2)$ and let $\mathcal{F} = \{K(g(\cdot)) : g \in \mathcal{G}\}$. By Lemma 16.4 for the covering number of sums of families of functions, we have

$$N_1(\epsilon, \mathcal{F}, X_1^n) \leq \mathcal{N}_1(\epsilon/2, \widehat{\mathcal{F}}_1, X_1^n)\mathcal{N}_1(\epsilon/2, \widehat{\mathcal{F}}_2, X_1^n)$$

because $\mathcal{F} \subset \{f_1 - f_2 : f_1 \in \widehat{\mathcal{F}}_1, f_2 \in \widehat{\mathcal{F}}_2\}$. Since $\mathcal{G}$ spans a (d^2+d+1)-dimensional vector space, by Lemma 9.5 the collection of sets

$$\mathcal{G}^+ = \{\{(x,t) : g(x) - t \geq 0\} : g \in \mathcal{G}\}$$

has VC dimension $V_{\mathcal{G}^+} \leq d^2 + d + 2$. Since K_i is monotone, it follows from Lemma 16.3 that $V_{\widehat{\mathcal{F}}_i^+} \leq d^2 + d + 2$, where the families of sets $\widehat{\mathcal{F}}_i^+$ are defined just as $\mathcal{G}^+$ with $\widehat{\mathcal{F}}_i$ in place of $\mathcal{G}$. Let V_1 and V_2 be the total variations of K_1 and K_2, respectively. Then $V = V_1 + V_2$ and $0 \leq K_i(x) + \alpha_i \leq V_i$, $x \in \mathcal{R}$ $(i = 1, 2)$ for suitably chosen constants α_1 and α_2. Lemma 9.4 implies $0 \leq f(x) \leq B$ for all

$f \in \mathcal{F}$ and x, thus

$$\mathcal{N}_1(\epsilon, \mathcal{F}, X_1^n) \leq 3 \left(\frac{2eB}{\epsilon} \log \left(\frac{3eB}{\epsilon} \right) \right)^{V_{\mathcal{F}^+}} \leq 3 \left(\frac{3eB}{\epsilon} \right)^{2V_{\mathcal{F}^+}}.$$

Show that this implies that

$$\mathcal{N}_1(\epsilon, \mathcal{F}, X_1^n) \leq 9 \left(\frac{3eV}{\epsilon} \right)^{4(d^2 + d + 2)}.$$

Mimic the rest of the proof of Lemma 17.3 to obtain the conclusion.

18

Orthogonal Series Estimates

Orthogonal series estimates use the estimates of coefficients of a series expansion to reconstruct the regression function. In this chapter we focus on nonlinear orthogonal series estimates, where one applies a nonlinear transformation (thresholding) to the estimated coefficients. The most popular orthogonal series estimates use wavelets. We start our discussion of orthogonal series estimates by describing the motivation for using these wavelet estimates.

18.1   Wavelet Estimates

We introduce orthogonal series estimates in the context of regression estimation with fixed, equidistant design, which is the field where they have been applied most successfully. Here one gets data $(x_1, \tilde{Y}_1)$, ..., $(x_n, \tilde{Y}_n)$ according to the model

$$\tilde{Y}_i = m(x_i) + \epsilon_i, \tag{18.1}$$

where x_1, ..., x_n are fixed (nonrandom) equidistant points in $[0, 1]$, ϵ_1, ..., ϵ_n are i.i.d. random variables with $\mathbf{E}\epsilon_1 = 0$ and $\mathbf{E}\epsilon_1^2 < \infty$, and m is a function $m : [0, 1] \to \mathcal{R}$ (cf. Section 1.9).

Recall that λ denotes the Lebesgue measure on $[0, 1]$. Assume $m \in L_2(\lambda)$ and let $\{f_j\}_{j \in \mathcal{N}}$ be an orthonormal basis in $L_2(\lambda)$, i.e.,

$$< f_j, f_k >_\lambda = \int_0^1 f_j(x) f_k(x) \, dx = \delta_{j,k} \quad (j, k \in \mathcal{N}),$$

and each function in $L_2(\lambda)$ can be approximated arbitrarily well by linear combinations of the $\{f_j\}_{j\in\mathcal{N}}$. Then m can be represented by its Fourier series with respect to $\{f_j\}_{j\in\mathcal{N}}$:

$$m = \sum_{j=1}^{\infty} c_j f_j \quad \text{where } c_j =< m, f_j >_\lambda= \int_0^1 m(x)f_j(x)\,dx. \qquad (18.2)$$

In orthogonal series estimation we use estimates of the coefficients of the series expansion (18.2) in order to reconstruct the regression function.

In the model (18.1), where the $x_1, \ldots, x_n$ are equidistant in $[0,1]$, the coefficients c_j can be estimated by

$$\hat{c}_j = \frac{1}{n}\sum_{i=1}^{n}\tilde{Y}_i f_j(x_i) \quad (j \in \mathcal{N}). \qquad (18.3)$$

If (18.1) holds, then

$$\hat{c}_j = \frac{1}{n}\sum_{i=1}^{n}m(x_i)f_j(x_i) + \frac{1}{n}\sum_{i=1}^{n}\epsilon_i f_j(x_i),$$

where, hopefully,

$$\frac{1}{n}\sum_{i=1}^{n}m(x_i)f_j(x_i) \approx \int_0^1 m(x)f_j(x)\,dx = c_j$$

and

$$\frac{1}{n}\sum_{i=1}^{n}\epsilon_i f_j(x_i) \approx \mathbf{E}\left\{\frac{1}{n}\sum_{i=1}^{n}\epsilon_i f_j(x_i)\right\} = 0.$$

The traditional way of using these estimated coefficients to construct an estimate m_n of m is to truncate the series expansion at some index $\tilde{K}$ and to plug in the estimated coefficients:

$$m_n = \sum_{j=1}^{\tilde{K}} \hat{c}_j f_j. \qquad (18.4)$$

Here one tries to choose $\tilde{K}$ such that the set of functions $\{f_1,\ldots,f_{\tilde{K}}\}$ is the "best" among all subsets $\{f_1\}, \{f_1,f_2\}, \{f_1,f_2,f_3,\ldots\}$ of $\{f_j\}_{j\in\mathcal{N}}$ in view of the error of the estimate (18.4). This implicitly assumes that the most important information about m is contained in the first $\tilde{K}$ coefficients of the series expansion (18.2).

A way of overcoming this assumption was proposed by Donoho and Johnstone (1994). It consists of thresholding the estimated coefficients, e.g., to use all those coefficients whose absolute value is greater than some threshold

δ_n (so-called hard thresholding). This leads to estimates of the form

$$m_n = \sum_{j=1}^{K} \eta_{\delta_n}(\hat{c}_j) f_j, \tag{18.5}$$

where K is usually much larger than $\tilde{K}$ in (18.4), $\delta_n > 0$ is a threshold, and

$$\eta_{\delta_n}(c) = \begin{cases} c & \text{if } |c| > \delta_n, \\ 0 & \text{if } |c| \le \delta_n. \end{cases} \tag{18.6}$$

As we will see in Section 18.3 this basically tries to find the "best" of all subsets of $\{f_1, \ldots, f_K\}$ in view of the estimate (18.5).

The most popular choice for the orthogonal system $\{f_j\}_{j \in \mathcal{N}}$ are the so-called wavelet systems, where the f_j are constructed by translation of a so-called father wavelet and by translation and dilatation of a so-called mother wavelet. For these wavelet systems the series expansion (18.2) of many functions contains only a few nonzero coefficients. This together with choosing a subset of the orthonormal system by hard thresholding leads to estimates which achieve a nearly optimal minimax rate of convergence for a variety of function spaces (e.g., Hölder, Besov, etc.) (for references see Section 18.7). In particular, these estimates are able to adapt to local irregularities (e.g., jump discontinuities) of the regression function, a property which classical linear smoothers like kernel estimators with fixed bandwidth do not have.

Motivated by the success of these estimates for fixed design regression, similar estimates were also applied for random design regression, where one has i.i.d. data (X_1, Y_1), $\ldots$, (X_n, Y_n). One difficulty to overcome here is to find a reasonable way to estimate the coefficients c_j. If X is uniformly distributed on $[0, 1]$, then one can use the same estimate as for fixed, equidistant $x_1, \ldots, x_n$:

$$\hat{c}_j = \frac{1}{n} \sum_{i=1}^{n} Y_i f_j(X_i),$$

because, in this case,

$$\mathbf{E}\{\hat{c}_j\} = \mathbf{E}\left\{\mathbf{E}\{Y_1 f_j(X_1)|X_1\}\right\} = \mathbf{E}\{m(X_1) f_j(X_1)\} = c_j.$$

Clearly, this is not a reasonable estimate if X is not uniformly distributed on $[0, 1]$. In this case, it was suggested in the literature to use the data (X_1, Y_1), $\ldots$, (X_n, Y_n) to construct new, equidistant data $(x_1, \tilde{Y}_1)$, $\ldots$, $(x_n, \tilde{Y}_n)$, where $x_1, \ldots, x_n$ are equidistant in $[0, 1]$ and $\tilde{Y}_i$ is an estimate of $m(x_i)$, and then to apply (18.3) to these new data (for references, see Section 18.7). Results concerning the rate of convergence of these estimates have only been derived under the assumption that X has a density with respect to the Lebesgue–Borel measure, which is bounded away from infinity on $[0, 1]$. If this assumption is true, then the L_2 error can be bounded by some constant times

$$\int_{[0,1]} |m_n(x) - m(x)|^2 \, dx,$$

and the last term can be expressed as the sum of squares of the coefficients of the series expansion of $m_n - m$ with respect to the orthonormal system in $L_2(\lambda)$. Hence, if one estimates the coefficients of the series expansion of m in a proper way, then this automatically leads to estimates with small L_2 error. This is no longer true if μ is not "close" to the uniform distribution. Then it is not clear whether nearly correct estimation of the coefficients leads to a small L_2 error

$$\int |m_n(x) - m(x)|^2 \mu(dx),$$

because in the above term one integrates with respect to μ and not with respect to λ.

18.2 Empirical Orthogonal Series Estimates

If μ is not "close" to the uniform distribution, then a natural approach is to estimate an orthonormal expansion of m in $L_2(\mu)$. Clearly, this is not possible, because μ (i.e., the distribution of X) is unknown in an application.

What we do in the sequel is to use an orthonormal series expansion of m in $L_2(\mu_n)$ rather than in $L_2(\lambda)$, where μ_n is the empirical measure of the $X_1, \ldots, X_n$, i.e.,

$$\mu_n(A) = \frac{1}{n} \sum_{i=1}^{n} I_A(X_i) \quad (A \subseteq \mathcal{R}).$$

We will call the resulting estimates *empirical orthogonal series estimates*.

For $f, g : [0,1] \to \mathcal{R}$ define

$$< f, g >_n = \frac{1}{n} \sum_{i=1}^{n} f(X_i) g(X_i) \text{ and } \|f\|_n^2 = < f, f >_n.$$

In Section 18.4 we will describe a way to construct an orthonormal system $\{f_j\}_{j=1,\ldots,K}$ in $L_2(\mu_n)$, i.e., functions $f_1, \ldots, f_K : [0,1] \to \mathcal{R}$ which satisfy

$$< f_j, f_k >_n = \delta_{j,k} \quad (j, k = 1, \ldots, K).$$

Given such an orthonormal system, the best approximation with respect to $\| \cdot \|_n$ of m by functions in $span\{f_1, \ldots, f_K\}$ is given by

$$\sum_{j=1}^{K} c_j f_j \text{ where } c_j = < m, f_j >_n = \frac{1}{n} \sum_{i=1}^{n} m(X_i) f_j(X_i). \tag{18.7}$$

We will estimate the coefficients in (18.7) by

$$\hat{c}_j = \frac{1}{n} \sum_{i=1}^{n} Y_i f_j(X_i), \qquad (18.8)$$

and use hard thresholding to construct the estimate

$$\tilde{m}_n = \sum_{j=1}^{K} \eta_{\delta_n}(\hat{c}_j) f_j \qquad (18.9)$$

of m, where $\delta_n > 0$ is the threshold and η_{δ_n} is defined by (18.6). Finally, we truncate the estimate at some data-independent height β_n, i.e., we set

$$m_n(x) = (T_{\beta_n}\tilde{m}_n)(x) = \begin{cases} \beta_n & \text{if} \quad \tilde{m}_n(x) > \beta_n, \\ \tilde{m}_n(x) & \text{if} \quad -\beta_n \leq \tilde{m}_n(x) \leq \beta_n, \\ -\beta_n & \text{if} \quad \tilde{m}_n(x) < -\beta_n, \end{cases} \qquad (18.10)$$

where $\beta_n > 0$ and $\beta_n \to \infty$ $(n \to \infty)$.

Figures 18.1–18.3 show applications of this estimate to our standard data example with different thresholds.

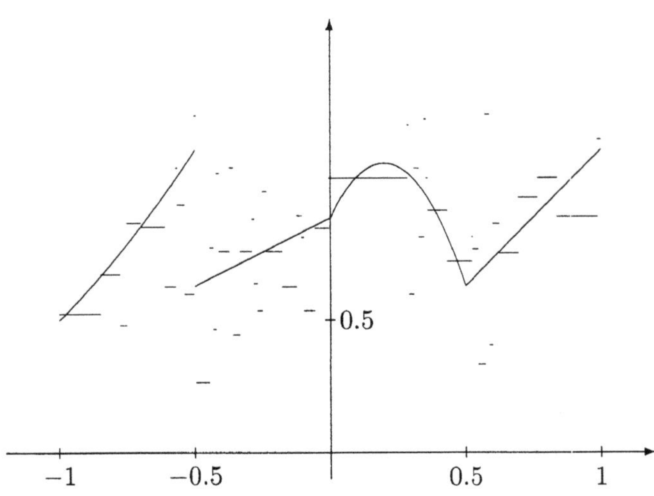

Figure 18.1. L_2 error $= 0.021336$, $\delta = 0.02$.

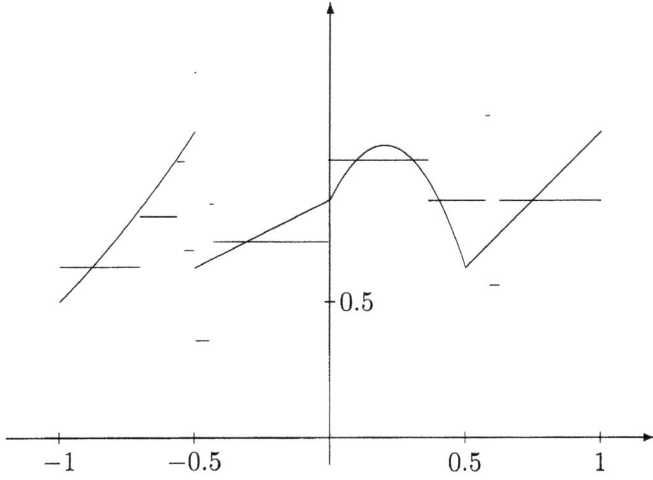

Figure 18.2. L_2 error $= 0.012590$, $\delta = 0.04$.

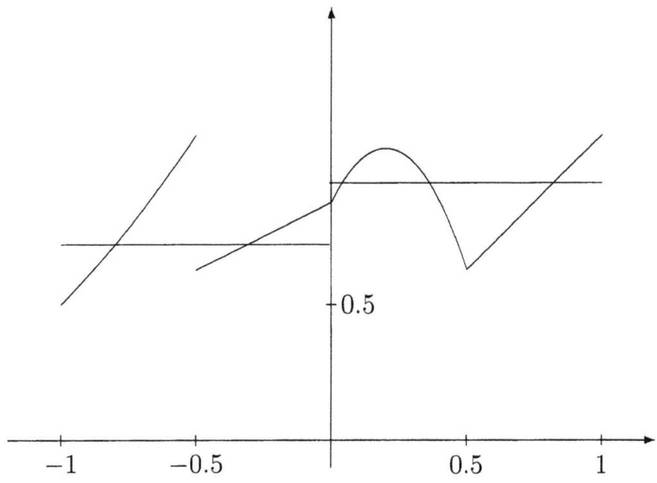

Figure 18.3. L_2 error $= 0.014011$, $\delta = 0.06$.

18.3   Connection with Least Squares Estimates

Let $\{f_j\}_{j=1,\ldots,K}$ be a family of functions $f_j : \mathcal{R} \to \mathcal{R}$. For $J \subseteq \{1,\ldots,K\}$ define $\mathcal{F}_{n,J}$ as the linear span of those f_j's with $j \in J$, i.e.,

$$\mathcal{F}_{n,J} = \left\{ \sum_{j \in J} a_j f_j \; : \; a_j \in \mathcal{R} \; (j \in J) \right\}. \tag{18.11}$$

Recall that the least squares estimate $\tilde{m}_{n,J}$ of m in $\mathcal{F}_{n,J}$ is defined by

$$\tilde{m}_{n,J} \in \mathcal{F}_{n,J} \quad \text{and} \quad \frac{1}{n}\sum_{i=1}^{n}|\tilde{m}_{n,J}(X_i) - Y_i|^2 = \min_{f \in \mathcal{F}_{n,J}} \frac{1}{n}\sum_{i=1}^{n}|f(X_i) - Y_i|^2.$$

$$(18.12)$$

Using (18.11) this can be rewritten as

$$\tilde{m}_{n,J} = \sum_{j \in J} a_j^* f_j$$

for some $a^* = \{a_j^*\}_{j \in J} \in \mathcal{R}^{|J|}$ which satisfies

$$\frac{1}{n}\|\mathbf{B}a^* - \mathbf{Y}\|_2^2 = \min_{a \in \mathcal{R}^{|J|}} \frac{1}{n}\|\mathbf{B}a - \mathbf{Y}\|_2^2, \qquad (18.13)$$

where

$$\mathbf{B} = (f_j(X_i))_{1 \leq i \leq n, j \in J} \quad \text{and} \quad \mathbf{Y} = (Y_1, \ldots, Y_n)^T.$$

As we have mentioned in Chapter 10, (18.13) is equivalent to

$$\frac{1}{n}\mathbf{B}^T\mathbf{B}a^* = \frac{1}{n}\mathbf{B}^T\mathbf{Y}, \qquad (18.14)$$

which is the so-called normal equation of the least squares problem.

Now consider the special case that $\{f_j\}_{j=1,\ldots,K}$ is orthonormal in $L_2(\mu_n)$. Then

$$\frac{1}{n}\mathbf{B}^T\mathbf{B} = (< f_j, f_k >_n)_{j,k \in J} = (\delta_{j,k})_{j,k \in J},$$

and therefore the solution of (18.14) is given by

$$a_j^* = \frac{1}{n}\sum_{i=1}^{n} f_j(X_i)Y_i \quad (j \in J). \qquad (18.15)$$

Define a_j^* by (18.15) for all $j \in \{1, \ldots, K\}$, and set

$$\hat{J} = \left\{ j \in \{1, \ldots, K\} \; : \; |a_j^*| > \delta_n \right\}. \qquad (18.16)$$

Then the orthogonal series estimate $\tilde{m}_n$ defined in Section 18.2 satisfies

$$\tilde{m}_n = \tilde{m}_{n,\hat{J}},$$

and so does the least squares estimate of m in $\mathcal{F}_{n,\hat{J}}$ as well. So hard thresholding can be considered as a way of choosing one of the 2^K least squares estimates $\tilde{m}_{n,J}$ ($J \subseteq \{1, \ldots, K\}$).

In Chapter 12 we have introduced complexity regularization as another method to use the data to select one estimate of a family of least squares estimates. There one minimizes the sum of the empirical L_2 risk and a penalty term, where the penalty term is approximately equal to the number

of parameters divided by n. To apply it in this context, we define the penalty
term by

$$pen_n(J) = c_n \frac{|J|}{n} \quad (J \subseteq \{1, \ldots, K\}),$$

where $c_n > 0$ is defined below, and choose

$$J^* \subseteq \{1, \ldots, K\}$$

such that

$$\frac{1}{n} \sum_{i=1}^{n} |\tilde{m}_{n,J^*}(X_i) - Y_i|^2 + pen_n(J^*)$$

$$= \min_{J \subseteq \{1,\ldots,K\}} \left\{ \frac{1}{n} \sum_{i=1}^{n} |\tilde{m}_{n,J}(X_i) - Y_i|^2 + pen_n(J) \right\}. \quad (18.17)$$

For properly defined c_n, $\hat{J}$ defined by (18.16) minimizes (18.17). To see
this, observe that, for $f = \sum_{j=1}^{K} b_j f_j$,

$$\frac{1}{n} \sum_{i=1}^{n} \left(f(X_i) - \tilde{m}_{n,\{1,\ldots,K\}}(X_i) \right) \cdot \left(\tilde{m}_{n,\{1,\ldots,K\}}(X_i) - Y_i \right)$$

$$= \frac{1}{n} \left(b^T - a^{*T} \right) \mathbf{B}^T \left(\mathbf{B} a^* - \mathbf{Y} \right)$$

$$= \frac{1}{n} \left(b^T - a^{*T} \right) \cdot \left(\mathbf{B}^T \mathbf{B} a^* - \mathbf{B}^T Y \right)$$

$$= \frac{1}{n} \left(b^T - a^{*T} \right) \cdot 0 \quad \text{(because of (18.14))}$$

$$= 0,$$

which, together with the orthonormality of $\{f_j\}_{j=1,\ldots,K}$, implies

$$\frac{1}{n} \sum_{i=1}^{n} |f(X_i) - Y_i|^2$$

$$= \frac{1}{n} \sum_{i=1}^{n} |\tilde{m}_{n,\{1,\ldots,K\}}(X_i) - Y_i|^2 + \frac{1}{n} \sum_{i=1}^{n} |f(X_i) - \tilde{m}_{n,\{1,\ldots,K\}}(X_i)|^2$$

$$= \frac{1}{n} \sum_{i=1}^{n} |\tilde{m}_{n,\{1,\ldots,K\}}(X_i) - Y_i|^2 + \sum_{j=1}^{K} |b_j - a_j^*|^2.$$

Hence

$$\frac{1}{n} \sum_{i=1}^{n} |\tilde{m}_{n,J}(X_i) - Y_i|^2 + pen_n(J)$$

$$= \frac{1}{n} \sum_{i=1}^{n} |\tilde{m}_{n,\{1,\dots,K\}}(X_i) - Y_i|^2 + \sum_{j \notin J} |a_j^*|^2 + c_n \frac{|J|}{n}$$

$$= \frac{1}{n} \sum_{i=1}^{n} |\tilde{m}_{n,\{1,\dots,K\}}(X_i) - Y_i|^2 + \sum_{j=1}^{K} \left\{ |a_j^*|^2 I_{\{j \notin J\}} + \frac{c_n}{n} I_{\{j \in J\}} \right\}$$

$$\geq \frac{1}{n} \sum_{i=1}^{n} |\tilde{m}_{n,\{1,\dots,K\}}(X_i) - Y_i|^2 + \sum_{j=1}^{K} \min \left\{ |a_j^*|^2, \frac{c_n}{n} \right\}.$$

Setting $c_n = n \delta_n^2$ one gets

$$\min \left\{ |a_j^*|^2, \frac{c_n}{n} \right\} = |a_j^*|^2 I_{\{|a_j^*|^2 \leq \delta_n^2\}} + \frac{c_n}{n} I_{\{|a_j^*|^2 > \delta_n^2\}}$$

$$= |a_j^*|^2 I_{\{j \notin \hat{J}\}} + \frac{c_n}{n} I_{\{j \in \hat{J}\}} \quad (j \in \{1, \dots, K\}).$$

Collecting the above results we get

$$\frac{1}{n} \sum_{i=1}^{n} |\tilde{m}_{n,J}(X_i) - Y_i|^2 + pen_n(J)$$

$$\geq \frac{1}{n} \sum_{i=1}^{n} |\tilde{m}_{n,\{1,\dots,K\}}(X_i) - Y_i|^2 + \sum_{j=1}^{K} \left\{ |a_j^*|^2 I_{\{j \notin \hat{J}\}} + \frac{c_n}{n} I_{\{j \in \hat{J}\}} \right\}$$

$$= \frac{1}{n} \sum_{i=1}^{n} |\tilde{m}_{n,\hat{J}}(X_i) - Y_i|^2 + pen_n(\hat{J}).$$

This proves that (18.17) is minimized by $\hat{J}$.

It follows that (18.12) and (18.17) is an alternative way to define the estimate of Section 18.2. Observe that it is difficult to compute the estimate using (18.17), because there one has to minimize the penalized empirical L_2 error over 2^K, i.e., exponential many function spaces. On the other hand, it is easy to compute the estimate if one uses the definition of Section 18.2. The estimate can be computed in $O(n \log n)$ time (cf. Problem 18.1).

However, the definition given in this section will be useful in proving the asymptotic properties of the estimate.

18.4 Empirical Orthogonalization of Piecewise Polynomials

Next we define an orthonormal system in $L_2(\mu_n)$ by orthonormalizing piecewise polynomials. Fix $X_1, \dots, X_n$, and denote their different values by $x_1, \dots, x_{n'}$, i.e.,

$$n' \leq n, \quad \{x_1, \dots, x_{n'}\} = \{X_1, \cdots, X_n\}, \quad \text{and} \quad x_1 < \cdots < x_{n'}.$$

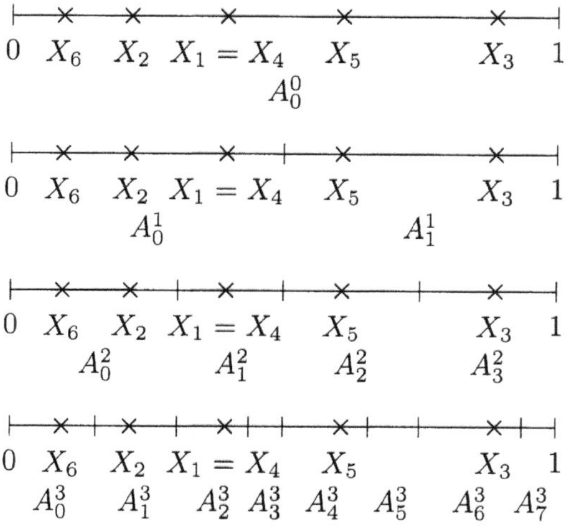

Figure 18.4. Example for construction of $\mathcal{P}_l$ ($l \in \{0, 1, 2, 3\}$).

For nonatomic μ we will have $n' = n$ with probability one.

We start by defining partitions $\mathcal{P}^l$ of $[0, 1]$ ($l \in \mathcal{N}_0$). Each $\mathcal{P}^l$ consists of 2^l intervals A_0^l, ..., $A_{2^l-1}^l$. Depending on x_1, ..., $x_{n'}$ they are recursively defined as follows: set $A_0^0 = [0, 1]$ and $\mathcal{P}^0 = \{[0, 1]\}$. Given $\mathcal{P}^l = \{A_0^l, \ldots, A_{2^l-1}^l\}$, define $\mathcal{P}^{l+1} = \{A_0^{l+1}, \ldots, A_{2^{l+1}-1}^{l+1}\}$ by subdividing each interval A_j^l into two intervals A_{2j}^{l+1}, A_{2j+1}^{l+1}, such that each of these two intervals contains nearly the same number of the x_1, ..., $x_{n'}$, i.e.,

$$A_j^l = A_{2j}^{l+1} \cup A_{2j+1}^{l+1}, \quad A_{2j}^{l+1} \cap A_{2j+1}^{l+1} = \emptyset,$$

and

$$\left| |\{i : x_i \in A_{2j}^{l+1}\}| - |\{i : x_i \in A_{2j+1}^{l+1}\}| \right| \leq 1.$$

This is always possible because the $x_1, \ldots, x_{n'}$ are pairwise distinct.

Using these nested partitions $\mathcal{P}^0$, $\mathcal{P}^1$, ..., we define the nested spaces of piecewise polynomials V_0^M, V_1^M, ..., where $M \in \mathcal{N}_0$ denotes the degree of the polynomials. Let V_l^M be the set of all piecewise polynomials of degree not greater than M with respect to $\mathcal{P}^l$, i.e.,

$$V_l^M = \left\{ f(x) = \sum_{j=0}^{2^l-1} \sum_{k=0}^{M} a_{j,k} x^k \cdot I_{A_j^l}(x) : a_{j,k} \in \mathcal{R} \right\}.$$

Clearly, $V_0^M \subseteq V_1^M \subseteq V_2^M \subseteq \cdots$. We will construct an orthonormal basis of

$$V_{\lceil \log_2(n) \rceil}^M$$

in $L_2(\mu_n)$.

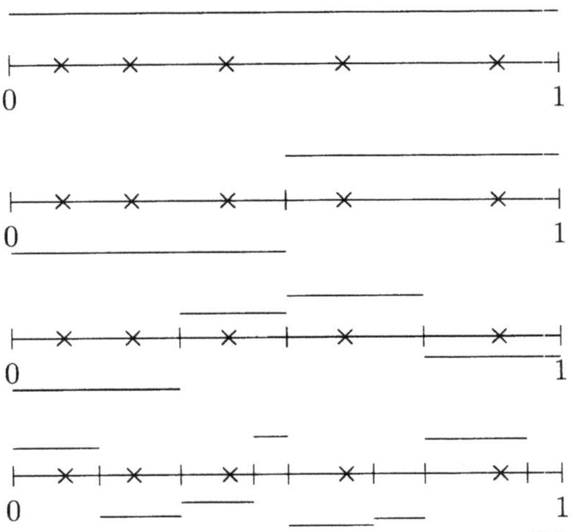

Figure 18.5. Example of functions from V_l^0, $l \in \{0, 1, 2, 3\}$.

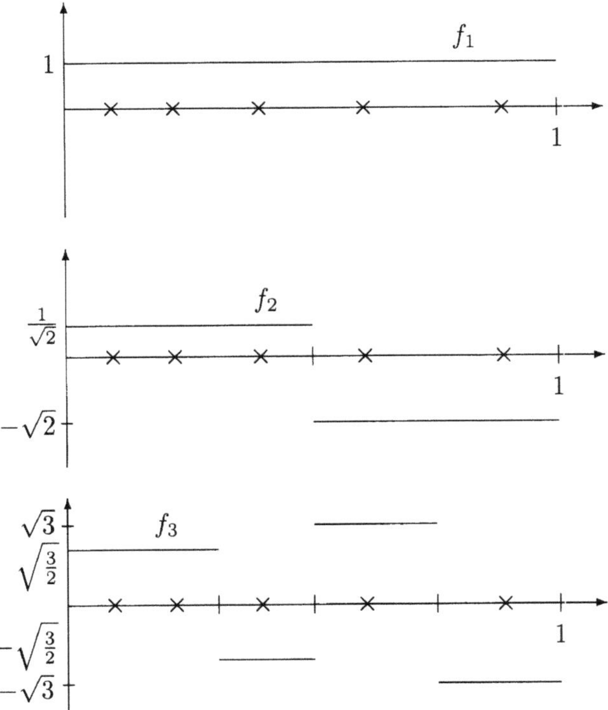

Figure 18.6. Construction of piecewise constant orthonormal functions.

To do this, we first decompose V_{l+1}^M into an orthogonal sum of spaces $U_{l+1,0}^M, \ldots, U_{l+1,2^l-1}^M$, i.e, we construct orthogonal spaces $U_{l+1,0}^M, \ldots,$

$U_{l+1,2^l-1}^M$ with the property that the set of all functions of the form $\sum_{j=0}^{2^l-1} f_j$ with $f_j \in U_{l+1,j}^M$ is equal to V_{l+1}^M. Observe that each $f \in V_{l+1}^M$ can be written as a sum

$$f = \sum_{j=0}^{2^l-1} f_j \quad \text{where } f_j = f \cdot I_{A_j^l} \in V_{l+1}^M.$$

Clearly, the supports of the $f_0, \ldots, f_{2^l-1}$ are all disjoint, which implies that the $f_0, \ldots, f_{2^l-1}$ are orthogonal with respect to $<,>_n$. Hence,

$$V_{l+1}^M = \bigoplus_{j=0}^{2^l-1} U_{l+1,j}^M \quad \text{with } U_{l+1,j}^M = \{f \cdot I_{A_j^l} \;:\; f \in V_{l+1}^M\}$$

is an orthogonal decomposition of V_{l+1}^M.

Let $\mathcal{B}_{l+1,j}^M$ be an orthonormal basis of the orthogonal complement of

$$V_l^M \cap U_{l+1,j}^M = \left\{ f \cdot I_{A_j^l} \;:\; f \in V_l^M \right\}$$

$$= \left\{ \sum_{k=0}^M a_k x^k I_{A_j^l} \;:\; a_0, \ldots, a_M \in \mathcal{R} \right\}$$

in

$$U_{l+1,j}^M = \left\{ f \cdot I_{A_j^l} \;:\; f \in V_{l+1}^M \right\},$$

i.e., of the set of all functions in $U_{l+1,j}^M$ which are orthogonal to $V_l^M \cap U_{l+1,j}^M$. Such an orthonormal basis can be computed easily: Assume g is an element of the orthogonal complement of $V_l^M \cap U_{l+1,j}^M$ in $U_{l+1,j}^M$. Then $g \in U_{l+1,j}^M$, which implies

$$g(x) = \sum_{k=0}^M a_k x^k \cdot I_{A_{2j}^{l+1}}(x) + \sum_{k=0}^M b_k x^k \cdot I_{A_{2j+1}^{l+1}}(x) \quad (x \in [0,1]) \qquad (18.18)$$

for some $a_0, \ldots, a_M, b_0, \ldots, b_M \in \mathcal{R}$. Furthermore, g is orthogonal to $V_l^M \cap U_{l+1,j}^M$, which is equivalent to assuming that g is orthogonal (with respect to $<,>_n$) to

$$1 \cdot I_{A_j^l}, \quad x \cdot I_{A_j^l}, \ldots, \quad x^M \cdot I_{A_j^l}.$$

This leads to a homogeneous linear equation system for the coefficients $a_0, \ldots, b_M$ of g. Hence all the functions of the orthogonal complement of $V_l^M \cap U_{l+1,j}^M$ in $U_{l+1,j}^M$ can be computed by solving a linear equation system, and an orthonormal basis of this orthogonal complement can be computed by orthonormalizing the solutions of this linear equation system with respect to the scalar product induced by $<,>_n$.

Set

$$\mathcal{B}_{l+1}^M = \mathcal{B}_{l+1,0}^M \cup \cdots \cup \mathcal{B}_{l+1,2^l-1}^M.$$

Then it is easy to see that $\mathcal{B}^M_{l+1}$ is an orthonormal basis of the orthogonal complement of V^M_l in V^M_{l+1} (cf. Problem 18.2). Choose an arbitrary orthonormal basis $\mathcal{B}^M_0$ of V^M_0. Then

$$\mathcal{B} = \mathcal{B}^M_0 \cup \cdots \cup \mathcal{B}^M_{\lceil \log_2(n) \rceil}$$

is an orthonormal basis of $V^M_{\lceil \log_2(n) \rceil}$ in $L_2(\mu_n)$. This is the orthonormal system we use for the estimate defined in Section 18.2, i.e., we set

$$\{f_j\}_{j=1,\ldots,K} = \mathcal{B}.$$

Let $\mathcal{P}$ be an arbitrary partition of $[0, 1]$ consisting of intervals. The main property of the orthonormal system $\{f_j\}_{j=1,\ldots,K}$ defined above is that any piecewise polynomial of degree M (or less) with respect to $\mathcal{P}$ can be represented in $L_2(\mu_n)$ by a linear combination of only slightly more than $|\mathcal{P}|$ of the f_j's. More precisely,

Lemma 18.1. *Let $\{f_1, \ldots, f_K\}$ be the family of functions constructed above.*
(a) Each f_j is a piecewise polynomial of degree M (or less) with respect to a partition consisting of four or less intervals.
(b) Let $\mathcal{P}$ be a finite partition of $[0, 1]$ consisting of intervals, and let f be a piecewise polynomial of degree M (or less) with respect to this partition $\mathcal{P}$. Then there exist coefficients $a_0, \ldots, a_K \in \mathcal{R}$ such that

$$f(X_i) = \sum_{j=1}^{K} a_j f_j(X_i) \quad (i = 1, \ldots, n)$$

and

$$
\begin{aligned}
|\{j \,:\, a_j \neq 0\}| &\leq (M+1)(\lceil \log_2(n) \rceil + 1) \cdot |\mathcal{P}| \\
&\leq 2(M+1)(\log(n) + 1) \cdot |\mathcal{P}|.
\end{aligned}
$$

PROOF. (a) For $f_i \in \mathcal{B}^M_0$ the assertion is trivial. If $f_i \notin \mathcal{B}^M_0$, then $f_i \in \mathcal{B}^M_{l+1,j}$ for some $0 \leq l < \log_2(n)$, $j \leq 2^l$, and the assertion follows from (18.18).
(b) The second inequality follows from

$$\lceil \log_2(n) \rceil + 1 \leq \frac{\log(n)}{\log(2)} + 2 \leq 2(\log(n) + 1),$$

hence it suffices to show the first inequality.

By construction each interval of the partition $\mathcal{P}^{\lceil \log_2(n) \rceil}$ contains at most one of the $X_1, \ldots, X_n$, which implies that there exists $\bar{f} \in V^M_{\lceil \log_2(n) \rceil}$ such that

$$f(X_i) = \bar{f}(X_i) \quad (i = 1, \ldots, n).$$

Choose $a_1, \ldots, a_K \in \mathcal{R}$ such that

$$\bar{f} = \sum_{j=1}^{K} a_j f_j.$$

Then

$$f(X_i) = \sum_{j=1}^{K} a_j f_j(X_i) \quad (i = 1, \ldots, n).$$

Since $\{f_1, \ldots, f_K\}$ are orthonormal w.r.t. $< \cdot, \cdot >_n$ we get

$$< f, f_k >_n = \sum_{j=1}^{K} a_j < f_j, f_k >_n = a_k.$$

Hence it suffices to show

$$\left| \left\{ \tilde{f} \in \mathcal{B}_{l+1}^M \; : \; < f, \tilde{f} >_n \neq 0 \right\} \right| \leq (M+1) \cdot |\mathcal{P}| \quad (0 \leq l < \log_2(n)).$$

There are at most $|\mathcal{P}| - 1$ indices j such that f is not equal to a polynomial of degree M (or less) on A_j^l. Since each $\tilde{f} \in \mathcal{B}_{l+1,j}^M$ is orthogonal to every function which is equal to a polynomial on A_j^l we get

$$\left| \left\{ \tilde{f} \in \mathcal{B}_{l+1}^M \; : \; < f, \tilde{f} >_n \neq 0 \right\} \right| \; \leq \; \sum_{\substack{j \, : \, f \text{ is not equal to a polynomial} \\ \text{of degree } M \text{ (or less) on } A_j^l}} |\mathcal{B}_{l+1,j}^M|$$

$$\leq \; (|\mathcal{P}| - 1) \cdot (M+1),$$

which implies the assertion. $\qquad\qquad\qquad\qquad\qquad\qquad\qquad\qquad\square$

18.5 Consistency

In this section we study the consistency of our orthogonal series estimate. For simplicity we only consider the case $X \in [0,1]$ *a.s.* It is straightforward to modify the definition of the estimate such that the resulting estimate is weakly and strongly universally consistent for univariate X (cf. Problem 18.5).

In order to be able to show strong consistency of the estimate, we need the following slight modification of its definition. Let $\alpha \in (0, \frac{1}{2})$. Depending on the data let the functions f_j and the estimated coefficients $\hat{c}_j$ be defined as in Sections 18.2 and 18.4. Denote by $(\hat{c}_{(1)}, f_{(1)}), \ldots, (\hat{c}_{(K)}, f_{(K)})$ a permutation of $(\hat{c}_1, f_1), \ldots, (\hat{c}_K, f_K)$ with the property

$$|\hat{c}_{(1)}| \geq |\hat{c}_{(2)}| \geq \cdots \geq |\hat{c}_{(K)}|. \tag{18.19}$$

Define the estimate $\tilde{m}_n$ by

$$\tilde{m}_n = \sum_{j=1}^{\min\{K,\lfloor n^{1-\alpha}\rfloor\}} \eta_{\delta_n}(\hat{c}_{(j)}) f_{(j)}. \tag{18.20}$$

This ensures that $\tilde{m}_n$ is a linear combination of no more than $n^{1-\alpha}$ of the f_j's. As in Section 18.3 one can show that (18.20) implies

$$\tilde{m}_n = \tilde{m}_{n,J^*} \text{ for some } J^* \subseteq \{1,\ldots,K\}, \tag{18.21}$$

where J^* satisfies $|J^*| \le n^{1-\alpha}$ and

$$\frac{1}{n}\sum_{i=1}^{n}|\tilde{m}_{n,J^*}(X_i) - Y_i|^2 + pen_n(J^*)$$

$$= \min_{\substack{J\subseteq\{1,\ldots,K\},\\ |J|\le n^{1-\alpha}}} \left\{\frac{1}{n}\sum_{i=1}^{n}|\tilde{m}_{n,J}(X_i) - Y_i|^2 + pen_n(J)\right\}. \tag{18.22}$$

With this modification of the estimate we are able to show:

Theorem 18.1. *Let $M \in \mathcal{N}_0$ be fixed. Let the estimate m_n be defined by (18.8), (18.19), (18.20), and (18.10) with $\beta_n = \log(n)$ and $\delta_n \le \frac{1}{(\log(n)+1)^2}$. Then*

$$\int |m_n(x) - m(x)|^2 \mu(dx) \to 0 \quad (n \to \infty) \quad a.s.$$

and

$$\mathbf{E}\int |m_n(x) - m(x)|^2 \mu(dx) \to 0 \quad (n \to \infty)$$

for every distribution of (X,Y) with $X \in [0,1]$ a.s. and $\mathbf{E}Y^2 < \infty$.

PROOF. We will only show the almost sure convergence of the L_2 error to zero, the weak consistency can be derived in a similar way using arguments from the proof of Theorem 10.2 (cf. Problem 18.4).

Let $L > 0$. Set $Y_L = T_L Y$, $Y_{1,L} = T_L Y_1$, $\ldots$, $Y_{n,L} = T_L Y_n$. Let $\mathcal{F}_n$ be the set of all piecewise polynomials of degree M (or less) with respect to a partition of $[0,1]$ consisting of $4n^{1-\alpha}$ or less intervals, let $\mathcal{G}_M$ be the set of all polynomials of degree M (or less), let $\mathcal{P}_n$ be an equidistant partition of $[0,1]$ into $\lceil \log(n) \rceil$ intervals, and denote by $\mathcal{G}_M \circ \mathcal{P}_n$ the set of all piecewise polynomials of degree M (or less) w.r.t. $\mathcal{P}_n$.

In the first step of the proof we show that the assertion follows from

$$\inf_{f\in\mathcal{G}_M\circ\mathcal{P}_n, \|f\|_\infty \le \log(n)} \int |f(x) - m(x)|^2 \mu(dx) \to 0 \quad (n \to \infty) \tag{18.23}$$

and

$$\sup_{f \in T_{\log(n)} \mathcal{F}_n} \left| \frac{1}{n} \sum_{i=1}^{n} |f(X_i) - Y_{i,L}|^2 - \mathbf{E}\{|f(X) - Y_L|^2\} \right| \to 0 \quad (n \to \infty)$$

(18.24)

a.s. for every $L > 0$. In the second and third steps, we will show (18.23) and (18.24), respectively.

So, assume temporarily that (18.23) and (18.24) hold. Because of

$$\int |m_n(x) - m(x)|^2 \mu(dx) = \mathbf{E}\left\{ |m_n(X) - Y|^2 \big| D_n \right\} - \mathbf{E}|m(X) - Y|^2$$

it suffices to show

$$\left\{ \mathbf{E}\left\{ |m_n(X) - Y|^2 \big| D_n \right\} \right\}^{\frac{1}{2}} - \left\{ \mathbf{E}|m(X) - Y|^2 \right\}^{\frac{1}{2}} \to 0 \quad a.s. \qquad (18.25)$$

We use the decomposition

$$\begin{aligned}
0 \; &\leq \; \left\{ \mathbf{E}\left\{ |m_n(X) - Y|^2 \big| D_n \right\} \right\}^{\frac{1}{2}} - \left\{ \mathbf{E}|m(X) - Y|^2 \right\}^{\frac{1}{2}} \\[2mm]
&= \; \left(\left\{ \mathbf{E}\left\{ |m_n(X) - Y|^2 \big| D_n \right\} \right\}^{\frac{1}{2}} \right. \\[4mm]
&\qquad\qquad \left. - \inf_{\substack{f \in \mathcal{G}_M \circ \mathcal{P}_n, \\ \|f\|_\infty \leq \log(n)}} \left\{ \mathbf{E}|f(X) - Y|^2 \right\}^{\frac{1}{2}} \right) \\[4mm]
&\quad + \left(\inf_{\substack{f \in \mathcal{G}_M \circ \mathcal{P}_n, \\ \|f\|_\infty \leq \log(n)}} \left\{ \mathbf{E}|f(X) - Y|^2 \right\}^{\frac{1}{2}} - \left\{ \mathbf{E}|m(X) - Y|^2 \right\}^{\frac{1}{2}} \right).
\end{aligned}$$

(18.26)

It follows from (18.23) by the triangle inequality that the second term of (18.26) converges to zero. Therefore for (18.25) it suffices to show

$$\limsup_{n \to \infty} \left(\left\{ \mathbf{E}\left\{ |m_n(X) - Y|^2 \big| D_n \right\} \right\}^{\frac{1}{2}} - \inf_{\substack{f \in \mathcal{G}_M \circ \mathcal{P}_n, \\ \|f\|_\infty \leq \log(n)}} \left\{ \mathbf{E}|f(X) - Y|^2 \right\}^{\frac{1}{2}} \right) \leq 0$$

(18.27)

a.s. To this end, let $L > 0$ be arbitrary. We can assume w.l.o.g. that $\log(n) > L$. Then

$$\begin{aligned}
&\left\{ \mathbf{E}\left\{ |m_n(X) - Y|^2 \big| D_n \right\} \right\}^{\frac{1}{2}} - \inf_{f \in \mathcal{G}_M \circ \mathcal{P}_n, \|f\|_\infty \leq \log(n)} \left\{ \mathbf{E}|f(X) - Y|^2 \right\}^{\frac{1}{2}} \\[3mm]
&= \sup_{f \in \mathcal{G}_M \circ \mathcal{P}_n, \|f\|_\infty \leq \log(n)} \left\{ \mathbf{E}\left\{ |m_n(X) - Y|^2 \big| D_n \right\} \right\}^{\frac{1}{2}} - \left\{ \mathbf{E}|f(X) - Y|^2 \right\}^{\frac{1}{2}} \\[3mm]
&\leq \sup_{f \in \mathcal{G}_M \circ \mathcal{P}_n, \|f\|_\infty \leq \log(n)} \left\{ \left\{ \mathbf{E}\left\{ |m_n(X) - Y|^2 \big| D_n \right\} \right\}^{\frac{1}{2}} \right.
\end{aligned}$$

$$- \left\{ \mathbf{E} \left\{ |m_n(X) - Y_L|^2 \,\middle|\, D_n \right\} \right\}^{\frac{1}{2}}$$

$$+ \left\{ \mathbf{E} \left\{ |m_n(X) - Y_L|^2 \,\middle|\, D_n \right\} \right\}^{\frac{1}{2}} - \left\{ \frac{1}{n} \sum_{i=1}^{n} |m_n(X_i) - Y_{i,L}|^2 \right\}^{\frac{1}{2}}$$

$$+ \left\{ \frac{1}{n} \sum_{i=1}^{n} |m_n(X_i) - Y_{i,L}|^2 \right\}^{\frac{1}{2}} - \left\{ \frac{1}{n} \sum_{i=1}^{n} |\tilde{m}_n(X_i) - Y_{i,L}|^2 \right\}^{\frac{1}{2}}$$

$$+ \left\{ \frac{1}{n} \sum_{i=1}^{n} |\tilde{m}_n(X_i) - Y_{i,L}|^2 \right\}^{\frac{1}{2}} - \left\{ \frac{1}{n} \sum_{i=1}^{n} |\tilde{m}_n(X_i) - Y_i|^2 \right\}^{\frac{1}{2}}$$

$$+ \left\{ \frac{1}{n} \sum_{i=1}^{n} |\tilde{m}_n(X_i) - Y_i|^2 \right\}^{\frac{1}{2}} - \left\{ \frac{1}{n} \sum_{i=1}^{n} |f(X_i) - Y_i|^2 \right\}^{\frac{1}{2}}$$

$$+ \left\{ \frac{1}{n} \sum_{i=1}^{n} |f(X_i) - Y_i|^2 \right\}^{\frac{1}{2}} - \left\{ \frac{1}{n} \sum_{i=1}^{n} |f(X_i) - Y_{i,L}|^2 \right\}^{\frac{1}{2}}$$

$$+ \left\{ \frac{1}{n} \sum_{i=1}^{n} |f(X_i) - Y_{i,L}|^2 \right\}^{\frac{1}{2}} - \left\{ \mathbf{E}|f(X) - Y_L|^2 \right\}^{\frac{1}{2}}$$

$$+ \left\{ \mathbf{E}|f(X) - Y_L|^2 \right\}^{\frac{1}{2}} - \left\{ \mathbf{E}|f(X) - Y|^2 \right\}^{\frac{1}{2}} \Bigg\}.$$

Now we give upper bounds for the terms in each row of the right-hand side of the last inequality.

The second and seventh terms are bounded above by

$$\sup_{f \in T_{\log(n)} \mathcal{F}_n} \left| \left\{ \frac{1}{n} \sum_{i=1}^{n} |f(X_i) - Y_{i,L}|^2 \right\}^{\frac{1}{2}} - \left\{ \mathbf{E}|f(X) - Y_L|^2 \right\}^{\frac{1}{2}} \right|$$

(observe $m_n = T_{\log(n)} \tilde{m}_n$ and $\tilde{m}_n \in \mathcal{F}_{n,J^*} \subseteq \mathcal{F}_n$). For the third term observe that if $x, y \in \mathcal{R}$ with $|y| \leq \log(n)$ and $z = T_{\log(n)} x$, then $|z - y| \leq |x - y|$. Therefore the third term is not greater than zero.

Next we bound the fifth term. Fix $f \in \mathcal{G}_M \circ \mathcal{P}_n$. By definition of $\mathcal{P}_n$ and Lemma 18.1 there exists $\bar{J} \subseteq \{1, \ldots, n\}$ and $\bar{f} \in \mathcal{F}_{n,\bar{J}}$ such that

$$f(X_i) = \bar{f}(X_i) \quad (i = 1, \ldots, n) \quad \text{and} \quad |\bar{J}| \leq 2(M+1)(\log(n) + 1)^2.$$

This, together with (18.22), implies

$$\frac{1}{n} \sum_{i=1}^{n} |\tilde{m}_n(X_i) - Y_i|^2 - \frac{1}{n} \sum_{i=1}^{n} |f(X_i) - Y_i|^2$$

$$= \frac{1}{n} \sum_{i=1}^{n} |\tilde{m}_n(X_i) - Y_i|^2 - \frac{1}{n} \sum_{i=1}^{n} |\bar{f}(X_i) - Y_i|^2$$

$$\leq pen_n(\bar{J})$$

$$\leq n\delta_n^2 \frac{2(M+1)(\log(n)+1)^2}{n}.$$

Using these upper bounds and the triangle inequality for the remaining terms one gets

$$\left\{ \mathbf{E}\left\{ |m_n(X) - Y|^2 \,\middle|\, D_n \right\} \right\}^{\frac{1}{2}} - \inf_{f \in \mathcal{G}_M \circ \mathcal{P}_n, \|f\|_\infty \leq \log(n)} \left\{ \mathbf{E}|f(X) - Y|^2 \right\}^{\frac{1}{2}}$$

$$\leq \ 2 \cdot \left\{ \mathbf{E}|Y - Y_L|^2 \right\}^{\frac{1}{2}} + 2 \cdot \left\{ \frac{1}{n} \sum_{i=1}^{n} |Y_i - Y_{i,L}|^2 \right\}^{\frac{1}{2}}$$

$$+ 2(M+1)\delta_n^2 (\log(n)+1)^2$$

$$+ 2 \cdot \sup_{f \in T_{\log(n)}\mathcal{F}_n} \left| \left\{ \frac{1}{n} \sum_{i=1}^{n} |f(X_i) - Y_{i,L}|^2 \right\}^{\frac{1}{2}} - \left\{ \mathbf{E}|f(X) - Y_L|^2 \right\}^{\frac{1}{2}} \right|.$$

Because of (18.24), $\delta_n \leq \frac{1}{(\log(n)+1)^2}$, and the strong law of large numbers

$$\limsup_{n \to \infty} \left(\left\{ \mathbf{E}\left\{ |m_n(X) - Y|^2 \,\middle|\, D_n \right\} \right\}^{\frac{1}{2}} - \inf_{\substack{f \in \mathcal{G}_M \circ \mathcal{P}_n, \\ \|f\|_\infty \leq \log(n)}} \left\{ \mathbf{E}|f(X) - Y|^2 \right\}^{\frac{1}{2}} \right)$$

$$\leq 4 \cdot \left\{ \mathbf{E}|Y - Y_L|^2 \right\}^{\frac{1}{2}} \quad a.s.$$

With $L \to \infty$ one gets the assertion.

In the second step we prove (18.23). Since m can be approximated arbitrarily closely in $L_2(\mu)$ by continuously differentiable functions, we may assume w.l.o.g. that m is continuously differentiable. For each $A \in \mathcal{P}_n$ choose some $x_A \in A$ and set $f^* = \sum_{A \in \mathcal{P}_n} m(x_A)I_A$. Then $f^* \in \mathcal{G}_M \circ \mathcal{P}_n$, and for n sufficiently large (i.e., for n such that $\|m\|_\infty \leq \log(n)$) we get

$$\inf_{f \in \mathcal{G}_M \circ \mathcal{P}_n, \|f\|_\infty \leq \log(n)} \int |f(x) - m(x)|^2 \mu(dx)$$

$$\leq \sup_{x \in [0,1]} |f^*(x) - m(x)|^2$$

$$\leq \frac{c}{\log^2(n)} \to 0 \quad (n \to \infty),$$

where c is some constant depending on the first derivative of m.

In the third step we prove (18.24). Let $L > 0$. W.l.o.g. we may assume $L \leq \log(n)$. Set

$$\mathcal{H}_n := \left\{ h : \mathcal{R} \times \mathcal{R} \to \bar{\mathcal{R}} \ : \ h(x,y) = |f(x) - T_L y|^2 \quad ((x,y) \in \mathcal{R}^d \times \mathcal{R}) \right.$$

$$\left. \text{for some } f \in T_{\log(n)}\mathcal{F}_n \right\}.$$

For $h \in \mathcal{H}_n$ one has $0 \leq h(x,y) \leq 4\log(n)^2$ $((x,y) \in \mathcal{R} \times \mathcal{R})$. Using the notion of covering numbers and Theorem 9.1, one concludes

$$\mathbf{P}\left\{ \sup_{f \in T_{\log(n)}\mathcal{F}_n} \left| \frac{1}{n}\sum_{i=1}^{n} |f(X_i) - Y_{i,L}|^2 - \mathbf{E}|f(X) - Y_L|^2 \right| > t \right\}$$

$$= \mathbf{P}\left\{ \sup_{h \in \mathcal{H}_n} \left| \frac{1}{n}\sum_{i=1}^{n} h(X_i, Y_i) - \mathbf{E}h(X,Y) \right| > t \right\}$$

$$\leq 8\,\mathbf{E}\left\{ \mathcal{N}_1\left(\frac{t}{8}, \mathcal{H}_n, (X,Y)_1^n \right) \right\} \exp\left(-\frac{nt^2}{2048\log(n)^4} \right). \qquad (18.28)$$

Next we bound the covering number in (18.28). Observe first that if

$$h_j(x,y) = |f_j(x) - T_{\log(n)}y|^2 \quad ((x,y) \in \mathcal{R} \times \mathcal{R})$$

for some functions f_j bounded in absolute value by $\log(n)$ $(j = 1,2)$, then

$$\frac{1}{n}\sum_{i=1}^{n} |h_1(X_i, Y_i) - h_2(X_i, Y_i)|$$

$$= \frac{1}{n}\sum_{i=1}^{n} |f_1(X_i) - T_{\log(n)}Y_i + f_2(X_i) - T_{\log(n)}Y_i| \cdot |f_1(X_i) - f_2(X_i)|$$

$$\leq 4\log(n)\frac{1}{n}\sum_{i=1}^{n} |f_1(X_i) - f_2(X_i)|.$$

Thus

$$\mathcal{N}_1\left(\frac{t}{8}, \mathcal{H}_n, (X,Y)_1^n \right) \leq \mathcal{N}_1\left(\frac{t}{32\log(n)}, T_{\log(n)}\mathcal{F}_n, X_1^n \right). \qquad (18.29)$$

Using the notion of VC dimension and partitioning numbers, Theorem 9.4, and Lemma 13.1, one gets

$$\mathcal{N}_1\left(\frac{t}{32\log(n)}, T_{\log(n)}\mathcal{F}_n, X_1^n \right)$$

$$\leq \Delta_n(\mathcal{P})\left\{ \sup_{z_1,\ldots,z_l \in X_1^n, l \leq n} \mathcal{N}_1\left(\frac{t}{32\log(n)}, T_{\log(n)}\mathcal{G}_M, z_1^l \right) \right\}^{4n^{1-\alpha}}$$

$$\leq \Delta_n(\mathcal{P})\left\{ 3\frac{6e\log(n)}{\frac{t}{32\log(n)}} \right\}^{2V_{T_{\log(n)}\mathcal{G}_M^+}4n^{1-\alpha}}$$

$$= \Delta_n(\mathcal{P})\left\{ \frac{576e\log^2(n)}{t} \right\}^{2V_{T_{\log(n)}\mathcal{G}_M^+}4n^{1-\alpha}}, \qquad (18.30)$$

where $\mathcal{P}$ is the set of all partitions of $[0,1]$ consisting of $4n^{1-\alpha}$ or less intervals. Example 13.1 in Chapter 13 implies

$$
\begin{aligned}
\Delta_n(\mathcal{P}) &\leq \left(n + 4n^{1-\alpha}\right)^{4n^{1-\alpha}} \\
&\leq (5n)^{4n^{1-\alpha}}.
\end{aligned}
\tag{18.31}
$$

Furthermore, one easily concludes, from the definition of the VC dimension, that

$$
V_{T_{\log(n)}\mathcal{G}_M^+} \leq V_{\mathcal{G}_M^+},
$$

which, together with Theorem 9.5, implies

$$
V_{T_{\log(n)}\mathcal{G}_M^+} \leq M + 2.
\tag{18.32}
$$

It follows from (18.28)–(18.32),

$$
\mathbf{P}\left\{ \sup_{f \in T_{\log(n)}\mathcal{F}_n} \left| \frac{1}{n} \sum_{i=1}^{n} |f(X_i) - Y_{i,L}|^2 - \mathbf{E}|f(X) - Y_L|^2 \right| > t \right\}
$$

$$
\leq 8(5n)^{4n^{1-\alpha}} \left(\frac{576e \log^2(n)}{t} \right)^{8(M+2)n^{1-\alpha}} \exp\left(-\frac{nt^2}{2048 \log^4(n)} \right),
$$

from which one gets the assertion by an application of the Borel–Cantelli lemma. $\qquad\square$

18.6　Rate of Convergence

In this section we study the rate of convergence of our orthogonal series estimate. For simplicity we assume that Y is bounded, i.e., $|Y| \leq L$ *a.s.* for some $L \in \mathcal{R}_+$, and that we know this bound L. Instead of truncating the estimate at $\log(n)$ we truncate it at L, i.e., we set $\beta_n = L$ $(n \in \mathcal{N})$.

In order to illustrate the next theorem let us compare our estimate with an ideal (but not practical) estimate defined by fitting a piecewise polynomial to the data least squares, where the partition is chosen optimally for the (unknown) underlying distribution.

Denote by $\mathcal{P}_k$ the family of all partitions of $[0,1]$ consisting of k intervals, and let $\mathcal{G}_M$ be the set of all polynomials of degree M (or less). For a partition $\mathcal{P} \in \mathcal{P}_k$ let

$$
\mathcal{G}_M \circ \mathcal{P} = \left\{ f = \sum_{A \in \mathcal{P}} g_A I_A \ : \ g_A \in \mathcal{G}_M \ (A \in \mathcal{P}) \right\}
$$

be the set of all piecewise polynomials of degree M (or less) with respect to $\mathcal{P}$, and set

$$
\mathcal{G}_M \circ \mathcal{P}_k = \bigcup_{\mathcal{P} \in \mathcal{P}_k} \mathcal{G}_M \circ \mathcal{P}.
$$

For $k \in \mathcal{N}$ and $\mathcal{P} \in \mathcal{P}_k$ consider the estimation of m by a piecewise polynomial contained in $\mathcal{G}_M \circ \mathcal{P}$. Clearly, the estimate cannot approximate m better than the "best" piecewise polynomial in $\mathcal{G}_M \circ \mathcal{P}$, hence its L_2 error is not smaller than

$$\inf_{f \in \mathcal{G}_M \circ \mathcal{P}} \int |f(x) - m(x)|^2 \mu(dx).$$

Furthermore, least squares fitting of the piecewise polynomial to the data requires estimating $(M + 1) \cdot k$ parameters, which induces an error of at least

$$\frac{(M + 1) \cdot k}{n}.$$

Thus, if one chooses $k \in \mathcal{N}$ and $\mathcal{P} \in \mathcal{P}_k$ optimally for the distribution of (X, Y), then the L_2 error of the estimate will be at least

$$\min_{k \in \mathcal{N}} \left\{ \frac{(M + 1) \cdot k}{n} + \inf_{f \in \mathcal{G}_M \circ \mathcal{P}_k} \int |f(x) - m(x)|^2 \mu(dx) \right\}.$$

The next theorem states that for bounded Y our estimate achieves this error bound up to a logarithmic factor.

Theorem 18.2. *Let $L \in \mathcal{R}_+$, $n \in \mathcal{N}$, and let the estimate m_n be defined by (18.8), (18.19), (18.20), and (18.10) with $\beta_n = L$ and*

$$\delta_n = \sqrt{c \frac{\log^2(n)}{n}}$$

where $c > 0$ is some arbitrary constant. Then there exists a constant $\bar{c}$, which depends only on L and c, such that

$$\mathbf{E} \int |m_n(x) - m(x)|^2 \mu(dx)$$

$$\leq \min_{1 \leq k \leq \frac{n^{1-\alpha}}{2(M+1)(\log(n)+1)}} \left\{ 4c(\log(n) + 1)^3 \frac{(M + 1)k}{n} \right.$$

$$\left. + 2 \inf_{f \in \mathcal{G}_M \circ \mathcal{P}_k} \int |f(x) - m(x)|^2 \mu(dx) \right\}$$

$$+ \bar{c} \frac{\log(n)}{n}$$

for every distribution of (X, Y) with $|Y| \leq L$ a.s.

Before proving the theorem we give two applications. First we consider the case when m is a piecewise polynomial. Then the result above implies that the estimate achieves the parametric rate n^{-1} up to a logarithmic factor.

Corollary 18.1. *Let the estimate be defined as in Theorem 18.2. Then*

$$\mathbf{E} \int |m_n(x) - m(x)|^2 \mu(dx) \in O\left(\frac{\log^3(n)}{n}\right)$$

for every distribution of (X, Y) with $X \in [0, 1]$ a.s., $|Y| \leq L$ a.s., and m equal to a piecewise polynomial of degree M (or less) with respect to a finite partition consisting of intervals.

As a second application we consider the estimation of a piecewise smooth function. Let $C > 0$ and $p = q + r$ for some $q \in \mathcal{N}_0$ and $0 < r \leq 1$. Recall that a function $m : [0, 1] \to \mathcal{R}$ is called (p, C)-smooth if its qth derivative $m^{(q)}$ exists and satisfies

$$|m^{(q)}(x) - m^{(q)}(z)| \leq C|x - z|^r \tag{18.33}$$

for all $x, z \in [0, 1]$. It is called piecewise (p, C)-smooth with respect to a partition $\mathcal{P} = \{I_j\}_j$ of $[0, 1]$ consisting of intervals, if its qth derivative exists on each interval I_j and satisfies (18.33) for all $x, z \in I_j$ and all j.

Corollary 18.2. *Let $n \in \mathcal{N}$. Let the estimate be defined as in Theorem 18.2, and let $0 < p \leq M + 1$ be arbitrary. Then there exists a constant $\bar{c}$, which depends only on L and c, such that for n sufficiently large*

$$\mathbf{E} \int |m_n(x) - m(x)|^2 \mu(dx) \leq \bar{c} \cdot C^{\frac{2}{2p+1}} \left(\frac{\log^3(n)}{n}\right)^{\frac{2p}{2p+1}}$$

for every distribution of (X, Y) with $X \in [0, 1]$ a.s., $|Y| \leq L$ a.s., and m piecewise (p, C)-smooth with respect to a finite partition $\mathcal{P}$, which consists of intervals.

PROOF. Let $\mathcal{P}_n$ be a partition which is a refinement of $\mathcal{P}$ and which consists of

$$|\mathcal{P}_n| = |\mathcal{P}| + \left\lceil \left(\frac{C^2 n}{\log^3(n)}\right)^{\frac{1}{2p+1}} \right\rceil$$

intervals of size not exceeding $(\frac{\log^3(n)}{C^2 n})^{\frac{1}{2p+1}}$. By approximating m on each interval of $\mathcal{P}_n$ by a Taylor polynomial of degree M it follows from Lemma 11.1 that

$$\inf_{f \in \mathcal{G}_M \circ \mathcal{P}_{|\mathcal{P}_n|}} \int |f(x) - m(x)|^2 \mu(dx) \leq \inf_{f \in \mathcal{G}_M \circ \mathcal{P}_n} \sup_{x \in [0,1]} |f(x) - m(x)|^2$$

$$\leq C^2 \left(\frac{\log^3(n)}{C^2 n}\right)^{\frac{2p}{2p+1}}.$$

This together with Theorem 18.2 implies the assertion. $\square$

PROOF OF THEOREM 18.2. We will use the following error decomposition:

$$\int |m_n(x) - m(x)|^2 \mu(dx)$$

$$= \left\{ \mathbf{E}\left\{|m_n(X) - Y|^2 | D_n\right\} - \mathbf{E}|m(X) - Y|^2 \right.$$

$$-2\left(\frac{1}{n}\sum_{i=1}^{n}|m_n(X_i) - Y_i|^2 - \frac{1}{n}\sum_{i=1}^{n}|m(X_i) - Y_i|^2 + pen_n(J^*)\right) \right\}$$

$$+2\left(\frac{1}{n}\sum_{i=1}^{n}|m_n(X_i) - Y_i|^2 - \frac{1}{n}\sum_{i=1}^{n}|m(X_i) - Y_i|^2 + pen_n(J^*)\right)$$

$$=: T_{1,n} + T_{2,n}. \tag{18.34}$$

In the first step of the proof we show

$$\mathbf{E}T_{2,n} \leq 2 \min_{1 \leq k \leq \frac{n^{1-\alpha}}{2(M+1)(\log(n)+1)}} \left\{ 2c(\log(n)+1)^3 \frac{(M+1)k}{n} \right.$$

$$\left. + \inf_{f \in \mathcal{G}_M \circ \mathcal{P}_k} \int |f(x) - m(x)|^2 \mu(dx) \right\}.$$

$$\tag{18.35}$$

To this end, we conclude from (18.10), $|Y| \leq L$ *a.s.* and (18.22)

$$T_{2,n}$$

$$\leq 2\left(\frac{1}{n}\sum_{i=1}^{n}|\tilde{m}_n(X_i) - Y_i|^2 + pen_n(J^*) - \frac{1}{n}\sum_{i=1}^{n}|m(X_i) - Y_i|^2\right)$$

$$= 2 \min_{\substack{J \subseteq \{1,\dots,K\}, \\ |J| \leq n^{1-\alpha}}} \left\{ \frac{1}{n}\sum_{i=1}^{n}|\tilde{m}_{n,J}(X_i) - Y_i|^2 + pen_n(J) \right.$$

$$\left. -\frac{1}{n}\sum_{i=1}^{n}|m(X_i) - Y_i|^2 \right\}$$

$$= 2 \min_{\substack{J \subseteq \{1,\dots,K\}, \\ |J| \leq n^{1-\alpha}}} \left\{ \min_{f \in \mathcal{F}_{n,J}} \frac{1}{n}\sum_{i=1}^{n}|f(X_i) - Y_i|^2 + pen_n(J) \right.$$

$$\left. -\frac{1}{n}\sum_{i=1}^{n}|m(X_i) - Y_i|^2 \right\}$$

$$= 2 \min_{1 \le k \le n^{1-\alpha}} \left\{ \min_{\substack{J \subseteq \{1,\dots,K\},\, |J|=k,\\ f \in \mathcal{F}_{n,J}}} \frac{1}{n} \sum_{i=1}^{n} |f(X_i) - Y_i|^2 + \delta_n^2 k \right.$$

$$\left. - \frac{1}{n} \sum_{i=1}^{n} |m(X_i) - Y_i|^2 \right\}.$$

By Lemma 18.1,

$$\min_{f \in \mathcal{G}_M \circ \mathcal{P}_k} \frac{1}{n} \sum_{i=1}^{n} |f(X_i) - Y_i|^2 \ge \min_{\substack{J \subseteq \{1,\dots,K\},\\ |J| \le 2(M+1)(\log(n)+1)k,\, f \in \mathcal{F}_{n,J}}} \frac{1}{n} \sum_{i=1}^{n} |f(X_i) - Y_i|^2,$$

hence

$$\mathbf{E} T_{2,n}$$

$$\le \mathbf{E} \left\{ 2 \min_{1 \le k \le \frac{n^{1-\alpha}}{2(M+1)(\log(n)+1)}} \left\{ \min_{f \in \mathcal{G}_M \circ \mathcal{P}_k} \frac{1}{n} \sum_{i=1}^{n} |f(X_i) - Y_i|^2 \right.\right.$$

$$\left.\left. + 2\delta_n^2 (M+1)(\log(n)+1)k - \frac{1}{n} \sum_{i=1}^{n} |m(X_i) - Y_i|^2 \right\} \right\}$$

$$\le 2 \min_{1 \le k \le \frac{n^{1-\alpha}}{2(M+1)(\log(n)+1)}} \left\{ 2\delta_n^2 (M+1)(\log(n)+1)k \right.$$

$$\left. + \inf_{f \in \mathcal{G}_M \circ \mathcal{P}_k} \int |f(x) - m(x)|^2 \mu(dx) \right\}.$$

With $\delta_n^2 = c \log^2(n)/n$ this implies the assertion of the first step.

In the second step we show

$$\mathbf{E} T_{2,n} \le \bar{c} \frac{\log(n)}{n}. \tag{18.36}$$

Let $t > 0$ be arbitrary. By definition of m_n, $m_n \in T_L \mathcal{F}_{n,J^*} \subseteq T_L \mathcal{G}_M \circ \mathcal{P}_{4|J^*|}$. Using this and $pen_n(J) = \delta_n^2 |J|$ one gets

$$\mathbf{P}\{T_{1,n} > t\}$$

$$= \mathbf{P} \left\{ \mathbf{E}\{|m_n(X) - Y|^2 | D_n\} - \mathbf{E}|m(X) - Y|^2 \right.$$

$$- \frac{1}{n} \sum_{i=1}^{n} \{|m_n(X_i) - Y_i|^2 - |m(X_i) - Y_i|^2\}$$

$$\left. > \frac{1}{2}\left(t + 2\,pen_n(J^*) + \mathbf{E}\{|m_n(X) - Y|^2 | D_n\} - \mathbf{E}|m(X) - Y|^2\right) \right\}$$

$$\leq \mathbf{P}\Big\{ \exists 1 \leq k \leq 4n^{1-\alpha}, \ \exists f \in T_L\mathcal{G}_M \circ \mathcal{P}_k :$$

$$\mathbf{E}|f(X) - Y|^2 - \mathbf{E}|m(X) - Y|^2$$

$$-\frac{1}{n}\sum_{i=1}^{n}\{|f(X_i) - Y_i|^2 - |m(X_i) - Y_i|^2\}$$

$$> \frac{1}{2}\left(t + 2\delta_n^2 k + \mathbf{E}|f(X) - Y|^2 - \mathbf{E}|m(X) - Y|^2\right)\Big\}$$

$$\leq \sum_{1\leq k\leq 4n^{1-\alpha}} \mathbf{P}\left\{\exists f \in T_L\mathcal{G}_M \circ \mathcal{P}_k : \ldots\right\}.$$

To bound the above probability we use the notion of covering numbers and Theorem 11.4. This implies

$$\mathbf{P}\{T_{1,n} > t\}$$

$$\leq \sum_{1\leq k\leq 4n^{1-\alpha}} 14 \sup_{x_1^n} \mathcal{N}_1\left(\frac{\frac{1}{2}\delta_n^2 k}{20L}, T_L\mathcal{G}_M \circ \mathcal{P}_k, x_1^n\right)$$

$$\times \exp\left(-\frac{\frac{1}{8}(t + \delta_n^2 k)n}{214(1 + 1/2)L^4}\right)$$

$$= \left\{\sum_{1\leq k\leq 4n^{1-\alpha}} 14 \sup_{x_1^n} \mathcal{N}_1\left(\ldots\right)\exp\left(-\frac{\delta_n^2 kn}{2568L^4}\right)\right\}\exp\left(-\frac{tn}{2568L^4}\right).$$

Proceeding as in the third step of the proof of Theorem 18.1 one gets, for the covering number,

$$\sup_{x_1^n}\mathcal{N}_1\left(\frac{\frac{1}{2}\delta_n^2 k}{20L}, T_L\mathcal{G}_M \circ \mathcal{P}_k, x_1^n\right) \leq (5n)^k\left\{3\frac{6eL}{\frac{\delta_n^2 k}{40L}}\right\}^{2(M+2)k}.$$

This, together with $\delta_n = \sqrt{c\frac{\log^2(n)}{n}}$, implies

$$14\sup_{x_1^n}\mathcal{N}_1\left(\frac{\frac{1}{2}\delta_n^2 k}{20L}, T_L\mathcal{G}_M \circ \mathcal{P}_k, x_1^n\right)\exp\left(-\frac{\delta_n^2 kn}{2568L^4}\right) \leq \tilde{c}$$

for some constant $\tilde{c} > 0$ depending only on L and c. This proves

$$\mathbf{P}\{T_{1,n} > t\} \leq 4\tilde{c}\cdot n\cdot\exp\left(-\frac{n}{2568L^4}t\right).$$

For arbitrary $u > 0$ it follows that

$$\mathbf{E}T_{1,n} \leq u + \int_u^\infty \mathbf{P}\{T_{1,n} > t\}\,dt = u + 4\tilde{c}\cdot 2568L^4\exp\left(-\frac{n}{2568L^4}u\right).$$

By setting

$$u = \frac{2568L^4}{n} \log(\tilde{c} \cdot n)$$

one gets (18.36) which, in turn (together with (18.34) and (18.35)), implies the assertion of Theorem 18.2.   □

18.7   Bibliographic Notes

For general introductions to wavelets see, e.g., Chui (1992), Daubechies (1992), or Meyer (1993). A description of various statistical applications of wavelets can be found in Härdle et al. (1998).

In the context of fixed design regression it was shown by Donoho and Johnstone (1994), Donoho et al. (1995), and Donoho and Johnstone (1998) that orthogonal series estimates using thresholding and wavelets achieve a nearly optimal minimax rate of convergence for a variety of function spaces (e.g., Hölder, Besov, etc.).

Motivated by the success of these estimates, several different ways of applying them to random design regression were proposed. In most of them one uses the given data to construct new, equidistant data and then applies the wavelet estimates for fixed design to these new data. Construction of the new, equidistant data can be done e.g., by binning (see Antoniadis, Grégoire, and Vial (1997)) or by using interpolation methods (see Hall and Turlach (1997), Neumann and Spokoiny (1995), and Kovac and Silverman (2000)). Under regularity conditions on the distribution of X it was shown, in Hall and Turlach (1997) and Neumann and Spokoiny (1995), that these estimates are able to adapt to a variety of inhomogeneous smoothness assumptions and achieve, up to a logarithmic factor, the corresponding optimal minimax rate of convergence.

To avoid regularity conditions on the design we used, in this chapter, the approach described in Kohler (2002a; 2000a). In particular, Theorems 18.1 and 18.2 are due to Kohler (2002a; 2000a). There we analyzed orthogonal series estimates by considering them as least squares estimates using complexity regularization. In the context of fixed design regression, Engel (1994) and Donoho (1997) defined and analyzed data-dependent partitioning estimates (i.e., special least squares estimates) by considering them as orthogonal series estimates.

Problems and Exercises

PROBLEM 18.1. (a) Show that the orthogonal system defined in Section 18.4 can be computed in $O(n \cdot \log(n))$ time.

(b) Show that the estimate defined in Sections 18.2 and 18.4 can be computed in $O(n \cdot \log(n))$ time.

PROBLEM 18.2. Let V_l^M, $U_{l+1,j}^M$, and $\mathbf{B}_{l+1,j}^M$ be defined as in Section 18.4. Show that

$$\mathbf{B}_{l+1}^M = \mathbf{B}_{l+1,0}^M \cup \cdots \cup \mathbf{B}_{l+1,2^l-1}^M$$

is the basis of the orthogonal complement of V_l^M in V_{l+1}^M.

HINT: Show that:

(1) the functions in $\mathbf{B}_{l+1}^M$ are contained in V_{l+1}^M;

(2) the functions in $\mathbf{B}_{l+1}^M$ are orthogonal to each function in V_l^M;

(3) the functions in $\mathbf{B}_{l+1}^M$ are orthonormal; and

(4) if $f \in V_{l+1}^M$ and f is orthogonal to each function in V_l^M, then $f \cdot I_{A_j^l}$ can be represented as a linear combination of the functions from $\mathbf{B}_{l+1,j}^M$ (and hence f can be represented as a linear combination of the functions from $\mathbf{B}_{l+1}^M$).

PROBLEM 18.3. Show that (18.20) implies (18.21), (18.22), and $|J^*| \leq n^{1-\alpha}$.

PROBLEM 18.4. Prove the second part of Theorem 18.1, i.e., show that the expected L_2 error converges to zero for every distribution of (X, Y) with $X \in [0, 1]$ a.s. and $\mathbf{E}Y^2 < \infty$.

HINT: Show that the assertion follows from (18.23) and

$$\mathbf{E}\left\{ \sup_{f \in T_{\log(n)}\mathcal{F}_n} \left| \frac{1}{n}\sum_{i=1}^{n} |f(X_i) - Y_{i,L}|^2 - \mathbf{E}\{|f(X) - Y_L|^2\} \right| \right\} \to 0 \quad (n \to \infty).$$

PROBLEM 18.5. Modify the definition of the orthogonal series estimate in such a way that the resulting estimate is weakly and strongly universally consistent for univariate X.

HINT: Proceed similarly as in Problem 10.6.

19

Advanced Techniques from Empirical Process Theory

In Chapter 9 we used techniques from empirical process theory to bound differences between expectations and averages uniformly over some function space. We used the resulting inequalities (Theorems 9.1, 11.4, and 11.6) to analyze least squares estimates. Unfortunately, the rate of convergence results we have obtained thus far are optimal only up to a logarithmic factor.

In the first three sections of this chapter we use some advanced techniques from empirical process theory to derive sharper inequalities. These results are rather technical, but will be extremely useful. The main result is Theorem 19.3 which we will use in Section 19.4 to derive the optimal rate of convergence for linear least squares estimates, for example, for suitably defined piecewise polynomial partitioning estimates. Furthermore, we will use Theorem 19.3 in Chapter 21 to analyze the rate of convergence of penalized least squares estimates.

19.1 Chaining

In Chapter 9 we used finite covers of the underlying function spaces of some fixed size. In this section we apply instead the so-called chaining technique, which introduces a sequence of covers of increasing cardinality. There one has to control a sum of covering numbers instead of one fixed covering number, and this sum will be bounded above by an integral of covering numbers (cf. (19.2)).

Theorem 19.1. *Let $L \in \mathcal{R}_+$ and let $\epsilon_1, \ldots, \epsilon_n$ be independent random variables with expectation zero and values in $[-L, L]$. Let $z_1, \ldots, z_n \in \mathcal{R}^d$, let $R > 0$, and let $\mathcal{F}$ be a class of functions $f : \mathcal{R}^d \to \mathcal{R}$ with the property*

$$\|f\|_n^2 := \frac{1}{n} \sum_{i=1}^{n} |f(z_i)|^2 \leq R^2 \quad (f \in \mathcal{F}). \tag{19.1}$$

Then

$$\sqrt{n}\delta \geq 48\sqrt{2}L \int_{\frac{\delta}{8L}}^{\frac{R}{2}} (\log \mathcal{N}_2(u, \mathcal{F}, z_1^n))^{1/2} \, du \tag{19.2}$$

and

$$\sqrt{n}\delta \geq 36R \cdot L$$

imply

$$\mathbf{P}\left\{ \sup_{f \in \mathcal{F}} \left| \frac{1}{n} \sum_{i=1}^{n} f(z_i) \cdot \epsilon_i \right| > \delta \right\} \leq 5 \exp\left(-\frac{n\delta^2}{2304 L^2 R^2} \right).$$

For $|\mathcal{F}| = 1$ the inequality above follows from Hoeffding's inequality (see the proof of Theorem 19.1 for details). For finite $\mathcal{F}$, Hoeffding's inequality (cf. Lemma A.3), (19.1), and the union bound imply

$$\mathbf{P}\left\{ \sup_{f \in \mathcal{F}} \left| \frac{1}{n} \sum_{i=1}^{n} f(z_i) \cdot \epsilon_i \right| > \delta \right\} \leq |\mathcal{F}| \cdot 2 \exp\left(-\frac{n\delta^2}{4L^2 R^2} \right),$$

from which one can conclude that for

$$\sqrt{n}\delta \geq 2\sqrt{2}L \cdot R \cdot (\log |\mathcal{F}|)^{1/2} \tag{19.3}$$

one has

$$\mathbf{P}\left\{ \sup_{f \in \mathcal{F}} \left| \frac{1}{n} \sum_{i=1}^{n} f(z_i) \cdot \epsilon_i \right| > \delta \right\} \leq 2 \exp\left(\log |\mathcal{F}| - \frac{n\delta^2}{8L^2 R^2} - \frac{n\delta^2}{8L^2 R^2} \right)$$

$$\leq 2 \exp\left(-\frac{n\delta^2}{8L^2 R^2} \right).$$

This is the way the above theorem is formulated. There (19.3) is replaced by (19.2), which will allow us later to derive the optimal rate of convergence for linear least squares estimates.

In the proof we will introduce finite covers to replace the supremum over the possible infinite set $\mathcal{F}$ by a maximum over some finite set. Compared with the inequalities described in Chapter 9 the main new idea is to introduce a finite sequence of finer and finer covers of $\mathcal{F}$ (the so-called chaining technique) instead of one fixed cover. This will allow us to represent any $f \in \mathcal{F}$ by

$$(f - f^S) + (f^S - f^{S-1}) + \cdots + (f^1 - f^0) + f^0,$$

where $f^0, f^1, \ldots, f^S$ are elements of the covers which approximate f better and better. f^S will be such a good approximation that the first term can be neglected. The other terms will lead to a sum of probabilities involving maxima over the finite covers which we will bound as above using the union bound and Hoeffding's inequality.

PROOF OF THEOREM 19.1. For $R \leq \delta/(2L)$ we get, by the Cauchy–Schwarz inequality,

$$\sup_{f \in \mathcal{F}} \left| \frac{1}{n} \sum_{i=1}^{n} f(z_i) \cdot \epsilon_i \right| \leq \sup_{f \in \mathcal{F}} \|f\|_n \cdot \sqrt{\frac{1}{n} \sum_{i=1}^{n} \epsilon_i^2} \leq R \cdot 2L \leq \delta,$$

so we can assume w.l.o.g. $R > \delta/(2L)$.

For $s \in \mathcal{N}_0$ let $\{f_1^s, \ldots, f_{N_s}^s\}$ be a $\| \cdot \|_n$-cover of $\mathcal{F}$ of radius $\frac{R}{2^s}$ of size

$$N_s = \mathcal{N}_2\left(\frac{R}{2^s}, \mathcal{F}, z_1^n\right).$$

Because of (19.1) we can assume w.l.o.g. $f_1^0 = 0$ and $N_0 = 1$. For $f \in \mathcal{F}$ choose

$$f^s \in \{f_1^s, \ldots, f_{N_s}^s\}$$

such that

$$\|f - f^s\|_n \leq \frac{R}{2^s}.$$

Set

$$S = \min\left\{s \geq 1 : \frac{R}{2^s} \leq \frac{\delta}{2L}\right\}.$$

Because of

$$f = f - f^0 = f - f^S + \sum_{s=1}^{S}(f^s - f^{s-1})$$

we get, by definition of S and the Cauchy–Schwarz inequality,

$$\left| \frac{1}{n} \sum_{i=1}^{n} f(z_i) \cdot \epsilon_i \right|$$

$$= \left| \frac{1}{n} \sum_{i=1}^{n} (f(z_i) - f^S(z_i)) \cdot \epsilon_i + \sum_{s=1}^{S} \frac{1}{n} \sum_{i=1}^{n} (f^s(z_i) - f^{s-1}(z_i)) \cdot \epsilon_i \right|$$

$$\leq \left| \frac{1}{n} \sum_{i=1}^{n} (f(z_i) - f^S(z_i)) \cdot \epsilon_i \right| + \sum_{s=1}^{S} \left| \frac{1}{n} \sum_{i=1}^{n} (f^s(z_i) - f^{s-1}(z_i)) \cdot \epsilon_i \right|$$

$$\leq \|f - f^S\|_n \cdot \sqrt{\frac{1}{n} \sum_{i=1}^{n} \epsilon_i^2} + \sum_{s=1}^{S} \left| \frac{1}{n} \sum_{i=1}^{n} (f^s(z_i) - f^{s-1}(z_i)) \cdot \epsilon_i \right|$$

$$\leq \frac{R}{2^S} \cdot L + \sum_{s=1}^{S} \left| \frac{1}{n} \sum_{i=1}^{n} (f^s(z_i) - f^{s-1}(z_i)) \cdot \epsilon_i \right|$$

$$\leq \frac{\delta}{2} + \sum_{s=1}^{S} \left| \frac{1}{n} \sum_{i=1}^{n} (f^s(z_i) - f^{s-1}(z_i)) \cdot \epsilon_i \right|.$$

Therefore, for any $\eta_1, \dots, \eta_S \geq 0$, $\eta_1 + \dots + \eta_S \leq 1$,

$$\mathbf{P} \left\{ \sup_{f \in \mathcal{F}} \left| \frac{1}{n} \sum_{i=1}^{n} f(z_i) \cdot \epsilon_i \right| > \delta \right\}$$

$$\leq \mathbf{P} \left\{ \exists f \in \mathcal{F} : \frac{\delta}{2} + \sum_{s=1}^{S} \left| \frac{1}{n} \sum_{i=1}^{n} (f^s(z_i) - f^{s-1}(z_i)) \cdot \epsilon_i \right| > \frac{\delta}{2} + \sum_{s=1}^{S} \eta_s \cdot \frac{\delta}{2} \right\}$$

$$\leq \sum_{s=1}^{S} \mathbf{P} \left\{ \exists f \in \mathcal{F} : \left| \frac{1}{n} \sum_{i=1}^{n} (f^s(z_i) - f^{s-1}(z_i)) \cdot \epsilon_i \right| > \eta_s \cdot \frac{\delta}{2} \right\}$$

$$\leq \sum_{s=1}^{S} N_s \cdot N_{s-1} \cdot \max_{f \in \mathcal{F}} \mathbf{P} \left\{ \left| \frac{1}{n} \sum_{i=1}^{n} (f^s(z_i) - f^{s-1}(z_i)) \cdot \epsilon_i \right| > \eta_s \cdot \frac{\delta}{2} \right\}.$$

Fix $s \in \{1, \dots, S\}$ and $f \in \mathcal{F}$. The random variables

$$(f^s(z_1) - f^{s-1}(z_1)) \cdot \epsilon_1, \dots, (f^s(z_n) - f^{s-1}(z_n)) \cdot \epsilon_n$$

are independent, have zero mean, and take values in

$$\left[-L \cdot |f^s(z_i) - f^{s-1}(z_i)|, L \cdot |f^s(z_i) - f^{s-1}(z_i)| \right] \quad (i = 1, \dots, n).$$

Therefore

$$\frac{1}{n} \sum_{i=1}^{n} \left(2L \cdot |f^s(z_i) - f^{s-1}(z_i)| \right)^2 \;=\; 4L^2 \|f^s - f^{s-1}\|_n^2$$

$$\leq\; 4L^2 \left(\|f^s - f\|_n + \|f - f^{s-1}\|_n \right)^2$$

$$\leq\; 4L^2 \left(\frac{R}{2^s} + \frac{R}{2^{s-1}} \right)^2 = 36 \frac{R^2 L^2}{2^{2s}},$$

together with Hoeffding's inequality (cf. Lemma A.3), implies

$$\mathbf{P} \left[\left| \frac{1}{n} \sum_{i=1}^{n} (f^s(z_i) - f^{s-1}(z_i)) \cdot \epsilon_i \right| > \eta_s \cdot \frac{\delta}{2} \right] \leq 2 \exp \left(-\frac{2n(\frac{\eta_s \delta}{2})^2}{36 \frac{R^2 L^2}{2^{2s}}} \right).$$

It follows that

$$\mathbf{P} \left[\sup_{f \in \mathcal{F}} \left| \frac{1}{n} \sum_{i=1}^{n} f(z_i) \cdot \epsilon_i \right| > \delta \right] \;\leq\; \sum_{s=1}^{S} 2N_s^2 \exp \left(-\frac{n\delta^2 \eta_s^2 2^{2s}}{72 R^2 L^2} \right)$$

$$=\; \sum_{s=1}^{S} 2 \exp \left(2 \log N_s - \frac{n\delta^2 \eta_s^2 2^{2s}}{72 R^2 L^2} \right).$$

In order to get rid of the covering number N_s, we choose η_s such that

$$2 \log N_s \leq \frac{1}{2} \cdot \frac{n \delta^2 \eta_s^2 2^{2s}}{72 R^2 L^2}, \tag{19.4}$$

which is equivalent to

$$\eta_s \geq \bar{\eta}_s := \frac{12 \cdot \sqrt{2} \cdot R \cdot L}{2^s \delta \sqrt{n}} \cdot \{\log N_s\}^{1/2}.$$

More precisely, we set

$$\eta_s := \max \left\{ \frac{2^{-s} \sqrt{s}}{4}, \bar{\eta}_s \right\}.$$

Because of

$$\sum_{s=1}^{S} \frac{2^{-s} \sqrt{s}}{4} \leq \frac{1}{8} \sum_{s=1}^{\infty} s \cdot (\frac{1}{2})^{s-1} = \frac{1}{8} \frac{1}{(1 - \frac{1}{2})^2} = \frac{1}{2}$$

and

$$\begin{aligned}
\sum_{s=1}^{S} \bar{\eta}_s &= \sum_{s=1}^{S} \frac{24\sqrt{2}L}{\delta\sqrt{n}} \cdot \frac{R}{2^{s+1}} \left\{ \log \mathcal{N}_2 \left(\frac{R}{2^s}, \mathcal{F}, z_1^n \right) \right\}^{1/2} \\
&\leq \sum_{s=1}^{S} \frac{24\sqrt{2}L}{\delta\sqrt{n}} \cdot \int_{R/2^{s+1}}^{R/2^s} \{\log \mathcal{N}_2(u, \mathcal{F}, z_1^n)\}^{1/2} du \\
&= \frac{24\sqrt{2}L}{\delta\sqrt{n}} \cdot \int_{R/2^{S+1}}^{R/2} \{\log \mathcal{N}_2(u, \mathcal{F}, z_1^n)\}^{1/2} du \\
&\leq \frac{1}{2},
\end{aligned}$$

where the last inequality follows from (19.2) and

$$\frac{1}{4} \cdot \frac{R}{2^{S-1}} \geq \frac{1}{4} \cdot \frac{\delta}{2L} = \frac{\delta}{(8L)},$$

we get

$$\sum_{s=1}^{S} \eta_s \leq \sum_{s=1}^{S} \frac{2^{-s} \sqrt{s}}{4} + \sum_{s=1}^{S} \bar{\eta}_s \leq 1.$$

Furthermore, $\eta_s \geq \bar{\eta}_s$ implies (19.4), from which we can conclude

$$\mathbf{P} \left[\sup_{f \in \mathcal{F}} \left| \frac{1}{n} \sum_{i=1}^{n} f(z_i) \cdot \epsilon_i \right| > \delta \right]$$

$$\leq \sum_{s=1}^{S} 2 \exp \left(-\frac{n \delta^2 \eta_s^2 2^{2s}}{144 L^2 R^2} \right)$$

$$\leq \sum_{s=1}^{S} 2 \exp\left(-\frac{n\delta^2}{16 \cdot 144 L^2 R^2} \cdot s\right)$$

$$\leq \frac{2}{1 - \exp\left(-\frac{n\delta^2}{16 \cdot 144 L^2 R^2}\right)} \cdot \exp\left(-\frac{n\delta^2}{16 \cdot 144 L^2 R^2}\right).$$

Now,

$$\frac{n\delta^2}{16 \cdot 144 L^2 R^2} \geq \frac{36^2 L^2 R^2}{16 \cdot 144 L^2 R^2} = \frac{9}{16}$$

yields

$$\frac{2}{1 - \exp\left(-\frac{n\delta^2}{16 \cdot 144 L^2 R^2}\right)} \leq \frac{2}{1 - \exp\left(-\frac{9}{16}\right)} \leq 5$$

which in turn implies the assertion. $\qquad\square$

19.2   Extension of Theorem 11.6

Theorem 11.6 implies that for

$$n\alpha\epsilon^2 \geq \frac{80B}{3} \log \mathbf{E}\mathcal{N}_1\left(\frac{\alpha\epsilon}{5}, \mathcal{F}, Z_1^n\right) \tag{19.5}$$

one has

$$\mathbf{P}\left\{\sup_{f\in\mathcal{F}} \frac{\frac{1}{n}\sum_{i=1}^{n} f(Z_i) - \mathbf{E}\{f(Z)\}}{\alpha + \mathbf{E}\{f(Z)\} + \frac{1}{n}\sum_{i=1}^{n} f(Z_i)} > \epsilon\right\} \leq 4 \exp\left(-\frac{3n\alpha\epsilon^2}{80B}\right).$$

This is the way in which the next result is formulated. There (19.5) is replaced by a condition on the integral of the logarithm of the covering number.

Theorem 19.2. *Let Z, Z_1, ..., Z_n be independent and identically distributed random variables with values in $\mathcal{R}^d$. Let $K \geq 1$ and let $\mathcal{F}$ be a class of functions $f : \mathcal{R}^d \to [0, K]$. Let $0 < \epsilon < 1$ and $\alpha > 0$. Assume that*

$$\sqrt{n}\epsilon\sqrt{\alpha} \geq 576\sqrt{K} \tag{19.6}$$

and that, for all $z_1, \ldots, z_n \in \mathcal{R}^d$ and all $\delta \geq \alpha K/2$,

$$\frac{\sqrt{n}\epsilon\delta}{192\sqrt{2}K} \geq \int_{\frac{\epsilon\delta}{32K}}^{\sqrt{\delta}} \left(\log \mathcal{N}_2\left(u, \left\{f \in \mathcal{F} : \frac{1}{n}\sum_{i=1}^{n} f(z_i) \leq \frac{4\delta}{K}\right\}, z_1^n\right)\right)^{1/2} du. \tag{19.7}$$

Then

$$\mathbf{P}\left\{\sup_{f\in\mathcal{F}} \frac{\left|\mathbf{E}\{f(Z)\} - \frac{1}{n}\sum_{i=1}^{n} f(Z_i)\right|}{\alpha + \mathbf{E}\{f(Z)\} + \frac{1}{n}\sum_{i=1}^{n} f(Z_i)} > \epsilon\right\} \leq 15 \exp\left(-\frac{n\alpha\epsilon^2}{128 \cdot 2304K}\right). \tag{19.8}$$

PROOF. The proof will be divided into four steps.

STEP 1. Replace the expectation inside the probability in (19.8) by an empirical mean.

Draw a "ghost" sample $Z_1'^n = (Z_1', \ldots, Z_n')$ of i.i.d. random variables distributed as Z_1 and independent of Z_1^n. Let $f^* = f^*(Z_1^n)$ be a function $f \in \mathcal{F}$ such that

$$\left| \frac{1}{n} \sum_{i=1}^{n} f(Z_i) - \mathbf{E}\{f(Z)\} \right| > \epsilon \left(\alpha + \frac{1}{n} \sum_{i=1}^{n} f(Z_i) + \mathbf{E}\{f(Z)\} \right),$$

if there exists any such function, and let f^* be an other arbitrary function contained in $\mathcal{F}$, if such a function doesn't exist.

Observe that

$$\left| \frac{1}{n} \sum_{i=1}^{n} f(Z_i) - \mathbf{E}\{f(Z)\} \right| > \epsilon \left(\alpha + \frac{1}{n} \sum_{i=1}^{n} f(Z_i) + \mathbf{E}\{f(Z)\} \right)$$

and

$$\left| \frac{1}{n} \sum_{i=1}^{n} f(Z_i') - \mathbf{E}\{f(Z)\} \right| < \frac{\epsilon}{2} \left(\alpha + \frac{1}{n} \sum_{i=1}^{n} f(Z_i') + \mathbf{E}\{f(Z)\} \right)$$

imply

$$\left| \frac{1}{n} \sum_{i=1}^{n} f(Z_i) - \frac{1}{n} \sum_{i=1}^{n} f(Z_i') \right| > \frac{\epsilon \alpha}{2} + \epsilon \frac{1}{n} \sum_{i=1}^{n} f(Z_i) - \frac{\epsilon}{2} \frac{1}{n} \sum_{i=1}^{n} f(Z_i') + \frac{\epsilon}{2} \mathbf{E}\{f(Z)\},$$

which is equivalent to

$$\left| \frac{1}{n} \sum_{i=1}^{n} f(Z_i) - \frac{1}{n} \sum_{i=1}^{n} f(Z_i') \right| - \frac{3}{4} \epsilon \left(\frac{1}{n} \sum_{i=1}^{n} f(Z_i) - \frac{1}{n} \sum_{i=1}^{n} f(Z_i') \right)$$

$$> \frac{\epsilon \alpha}{2} + \frac{\epsilon}{4} \frac{1}{n} \sum_{i=1}^{n} f(Z_i) + \frac{\epsilon}{4} \frac{1}{n} \sum_{i=1}^{n} f(Z_i') + \frac{\epsilon}{2} \mathbf{E}\{f(Z)\}.$$

Because of $0 < 1 + \frac{3}{4}\epsilon < 2$ and $\mathbf{E}\{f(Z)\} \geq 0$ this in turn implies

$$\left| \frac{1}{n} \sum_{i=1}^{n} f(Z_i) - \frac{1}{n} \sum_{i=1}^{n} f(Z_i') \right| > \frac{\epsilon}{8} \left(2\alpha + \frac{1}{n} \sum_{i=1}^{n} f(Z_i) + \frac{1}{n} \sum_{i=1}^{n} f(Z_i') \right).$$

Using this one gets

$$\mathbf{P}\left\{ \exists f \in \mathcal{F} : \left| \frac{1}{n} \sum_{i=1}^{n} f(Z_i) - \frac{1}{n} \sum_{i=1}^{n} f(Z_i') \right| \right.$$

$$\left. > \frac{\epsilon}{8} \left(2\alpha + \frac{1}{n} \sum_{i=1}^{n} f(Z_i) + \frac{1}{n} \sum_{i=1}^{n} f(Z_i') \right) \right\}$$

$$\geq \mathbf{P}\left\{\left|\frac{1}{n}\sum_{i=1}^{n}f^*(Z_i)-\frac{1}{n}\sum_{i=1}^{n}f^*(Z_i')\right|\right.$$
$$\left. > \frac{\epsilon}{8}\left(2\alpha+\frac{1}{n}\sum_{i=1}^{n}f^*(Z_i)+\frac{1}{n}\sum_{i=1}^{n}f^*(Z_i')\right)\right\}$$

$$\geq \mathbf{P}\left\{\left|\frac{1}{n}\sum_{i=1}^{n}f^*(Z_i)-\mathbf{E}\{f^*(Z)|Z_1^n\}\right|\right.$$
$$\left. > \epsilon\left(\alpha+\frac{1}{n}\sum_{i=1}^{n}f^*(Z_i)+\mathbf{E}\{f^*(Z)|Z_1^n\}\right),\right.$$
$$\left|\frac{1}{n}\sum_{i=1}^{n}f^*(Z_i')-\mathbf{E}\{f^*(Z)|Z_1^n\}\right|$$
$$\left. < \frac{\epsilon}{2}\left(\alpha+\frac{1}{n}\sum_{i=1}^{n}f^*(Z_i')+\mathbf{E}\{f^*(Z)|Z_1^n\}\right)\right\}$$

$$\geq \mathbf{E}\left\{I_{\left\{\left|\frac{1}{n}\sum_{i=1}^{n}f^*(Z_i)-\mathbf{E}\{f^*(Z)|Z_1^n\}\right|>\epsilon\left(\alpha+\frac{1}{n}\sum_{i=1}^{n}f^*(Z_i)+\mathbf{E}\{f^*(Z)|Z_1^n\}\right)\right\}}\right.$$
$$\times\mathbf{P}\left\{\left|\frac{1}{n}\sum_{i=1}^{n}f^*(Z_i')-\mathbf{E}\{f^*(Z)|Z_1^n\}\right|\right.$$
$$\left.\left. < \frac{\epsilon}{2}\left(\alpha+\frac{1}{n}\sum_{i=1}^{n}f^*(Z_i')+\mathbf{E}\{f^*(Z)|Z_1^n\}\right)\,\middle|\,Z_1^n\right\}\right\}.$$

By Lemma 11.2 and $n > \frac{100K}{\epsilon^2\alpha}$ (which follows from (19.6)) we get that the probability inside the expectation is bounded from below by

$$1-\frac{K}{4(\frac{\epsilon}{2})^2\alpha n}=1-\frac{K}{\epsilon^2\alpha n}\geq 0.99,$$

and we can conclude

$$\mathbf{P}\left\{\exists f\in\mathcal{F}:\left|\frac{1}{n}\sum_{i=1}^{n}f(Z_i)-\frac{1}{n}\sum_{i=1}^{n}f(Z_i')\right|\right.$$
$$\left. > \frac{\epsilon}{8}\left(2\alpha+\frac{1}{n}\sum_{i=1}^{n}f(Z_i)+\frac{1}{n}\sum_{i=1}^{n}f(Z_i')\right)\right\}$$

$$\geq 0.99\mathbf{P}\left\{\left|\frac{1}{n}\sum_{i=1}^{n}f^{*}(Z_i) - \mathbf{E}\{f^{*}(Z)|Z_1^n\}\right|\right.$$

$$\left. > \epsilon\left(\alpha + \frac{1}{n}\sum_{i=1}^{n}f^{*}(Z_i) + \mathbf{E}\{f^{*}(Z)|Z_1^n\}\right)\right\}$$

$$= 0.99\mathbf{P}\left\{\exists f \in \mathcal{F} : \left|\frac{1}{n}\sum_{i=1}^{n}f(Z_i) - \mathbf{E}\{f(Z)\}\right|\right.$$

$$\left. > \epsilon\left(\alpha + \frac{1}{n}\sum_{i=1}^{n}f(Z_i) + \mathbf{E}\{f(Z)\}\right)\right\}.$$

This proves

$$\mathbf{P}\left\{\sup_{f\in\mathcal{F}}\frac{\left|\frac{1}{n}\sum_{i=1}^{n}f(Z_i) - \mathbf{E}\{f(Z)\}\right|}{\alpha + \frac{1}{n}\sum_{i=1}^{n}f(Z_i) + \mathbf{E}\{f(Z)\}} > \epsilon\right\}$$

$$\leq \frac{100}{99}\mathbf{P}\left\{\exists f \in \mathcal{F} : \left|\frac{1}{n}\sum_{i=1}^{n}f(Z_i) - \frac{1}{n}\sum_{i=1}^{n}f(Z_i')\right|\right.$$

$$\left. > \frac{\epsilon}{8}\left(2\alpha + \frac{1}{n}\sum_{i=1}^{n}f(Z_i) + \frac{1}{n}\sum_{i=1}^{n}f(Z_i')\right)\right\}.$$

STEP 2. Introduction of additional randomness by random signs.
Let $U_1, \ldots, U_n$ be independent and uniformly distributed over $\{-1,1\}$ and independent of $Z_1,\ldots,Z_n,\ Z_1',\ldots,Z_n'$. Because of the independence and identical distribution of $Z_1,\ldots,Z_n,\ Z_1',\ldots,Z_n'$, the joint distribution of $Z_1^n, Z_1'^n$ doesn't change if one randomly interchanges the corresponding components of Z_1^n and $Z_1'^n$. Hence

$$\mathbf{P}\left\{\exists f \in \mathcal{F} : \left|\frac{1}{n}\sum_{i=1}^{n}f(Z_i) - \frac{1}{n}\sum_{i=1}^{n}f(Z_i')\right|\right.$$

$$\left. > \frac{\epsilon}{8}\left(2\alpha + \frac{1}{n}\sum_{i=1}^{n}f(Z_i) + \frac{1}{n}\sum_{i=1}^{n}f(Z_i')\right)\right\}$$

$$= \mathbf{P}\left\{\exists f \in \mathcal{F} : \left|\frac{1}{n}\sum_{i=1}^{n}U_i\cdot(f(Z_i) - f(Z_i'))\right|\right.$$

$$\left. > \frac{\epsilon}{8}\left(2\alpha + \frac{1}{n}\sum_{i=1}^{n}f(Z_i) + \frac{1}{n}\sum_{i=1}^{n}f(Z_i')\right)\right\}$$

$$\leq 2\mathbf{P}\left\{\exists f \in \mathcal{F} : \left|\frac{1}{n}\sum_{i=1}^{n}U_if(Z_i)\right| > \frac{\epsilon}{8}\left(\alpha + \frac{1}{n}\sum_{i=1}^{n}f(Z_i)\right)\right\}.$$

STEP 3. Peeling.

In this step we use the so-called peeling technique motivated in van de Geer (2000), Chapter 5.

$$\mathbf{P}\left\{\exists f \in \mathcal{F}: \left|\frac{1}{n}\sum_{i=1}^{n}U_{i}f(Z_{i})\right| > \frac{\epsilon}{8}\left(\alpha + \frac{1}{n}\sum_{i=1}^{n}f(Z_{i})\right)\right\}$$

$$\leq \sum_{k=1}^{\infty}\mathbf{P}\left\{\exists f \in \mathcal{F}: I_{\{k\neq 1\}}2^{k-1}\alpha \leq \frac{1}{n}\sum_{i=1}^{n}f(Z_{i}) < 2^{k}\alpha,\right.$$

$$\left.\left|\frac{1}{n}\sum_{i=1}^{n}U_{i}f(Z_{i})\right| > \frac{\epsilon}{8}\left(\alpha + \frac{1}{n}\sum_{i=1}^{n}f(Z_{i})\right)\right\}$$

$$\leq \sum_{k=1}^{\infty}\mathbf{P}\left\{\exists f \in \mathcal{F}: \frac{1}{n}\sum_{i=1}^{n}f(Z_{i}) \leq 2^{k}\alpha, \left|\frac{1}{n}\sum_{i=1}^{n}U_{i}f(Z_{i})\right| > \frac{\epsilon}{8}\alpha 2^{k-1}\right\}.$$

STEP 4. Application of Theorem 19.1.

Next we condition inside the above probabilities on $Z_{1}, \ldots, Z_{n}$, which is equivalent to considering, for $z_{1}, \ldots, z_{n} \in \mathcal{R}^{d}$ and $k \in \mathcal{N}$,

$$\mathbf{P}\left\{\exists f \in \mathcal{F}: \frac{1}{n}\sum_{i=1}^{n}f(z_{i}) \leq 2^{k}\alpha, \left|\frac{1}{n}\sum_{i=1}^{n}U_{i}f(z_{i})\right| > \frac{\epsilon}{8}\alpha 2^{k-1}\right\}. \quad (19.9)$$

By the assumptions of Theorem 19.2 (use $\delta = \alpha 2^{k-2}K$ in (19.7)) we have

$$\frac{\sqrt{n}\frac{\epsilon}{8}\alpha 2^{k-1}}{48\sqrt{2}}$$

$$\geq \int_{\frac{\epsilon\alpha 2^{k-2}}{32}}^{\sqrt{\alpha 2^{k}K}/2}\left(\log\mathcal{N}_{2}\left(u, \left\{f \in \mathcal{F}: \frac{1}{n}\sum_{i=1}^{n}f(z_{i}) \leq \alpha 2^{k}\right\}, z_{1}^{n}\right)\right)^{1/2}du$$

and

$$\sqrt{n}\frac{\epsilon}{8}\alpha 2^{k-1} \geq 36\sqrt{\alpha 2^{k}K}.$$

Furthermore, $\frac{1}{n}\sum_{i=1}^{n}f(z_{i}) \leq \alpha 2^{k}$ implies

$$\frac{1}{n}\sum_{i=1}^{n}f(z_{i})^{2} \leq \alpha 2^{k}K.$$

Hence Theorem 19.1, with $R^{2} = \alpha 2^{k}K, L = 1, \delta = \frac{\epsilon}{8}\alpha 2^{k-1}$, implies that (19.9) is bounded from above by

$$5\exp\left(-\frac{n\left(\frac{\epsilon}{16}\alpha 2^{k}\right)^{2}}{2304(\alpha 2^{k}K)}\right) = 5\exp\left(-\frac{n\epsilon^{2}\alpha}{16^{2}\cdot 2304K}2^{k}\right).$$

This proves

$$\sum_{k=1}^{\infty} \mathbf{P}\left\{\exists f \in \mathcal{F} : \frac{1}{n}\sum_{i=1}^{n} f(Z_i) \le 2^k \alpha, \left|\frac{1}{n}\sum_{i=1}^{n} U_i f(Z_i)\right| > \frac{\epsilon}{8}\alpha 2^{k-1}\right\}$$

$$\le \sum_{k=1}^{\infty} 5 \exp\left(-\frac{n\epsilon^2 \alpha}{16^2 \cdot 2304 K} 2k\right)$$

$$= 5 \frac{1}{1 - \exp\left(-\frac{n\epsilon^2 \alpha}{128 \cdot 2304 K}\right)} \exp\left(-\frac{n\epsilon^2 \alpha}{128 \cdot 2304 K}\right)$$

$$\le \frac{5}{1 - \exp\left(-\frac{9}{8}\right)} \exp\left(-\frac{n\epsilon^2 \alpha}{128 \cdot 2304 K}\right).$$

Steps 1 to 4 imply the assertion. $\square$

19.3   Extension of Theorem 11.4

In this section we show the following modification of Theorem 11.4. This theorem will enable us to obtain optimal rates of convergence for the piecewise polynomial partitioning estimates in Section 19.4.

Theorem 19.3. *Let Z, Z_1, $\ldots$, Z_n be independent and identically distributed random variables with values in $\mathcal{R}^d$. Let $K_1, K_2 \ge 1$ and let $\mathcal{F}$ be a class of functions $f : \mathcal{R}^d \to \mathcal{R}$ with the properties*

$$|f(z)| \le K_1 \quad (z \in \mathcal{R}^d) \quad \text{and} \quad \mathbf{E}\{f(Z)^2\} \le K_2 \mathbf{E}\{f(Z)\}.$$

Let $0 < \epsilon < 1$ and $\alpha > 0$. Assume that

$$\sqrt{n}\epsilon\sqrt{1-\epsilon}\sqrt{\alpha} \ge 288 \max\{2K_1, \sqrt{2K_2}\} \tag{19.10}$$

and that, for all $z_1, \ldots, z_n \in \mathcal{R}^d$ and all $\delta \ge \frac{\alpha}{8}$,

$$\frac{\sqrt{n}\epsilon(1-\epsilon)\delta}{96\sqrt{2}\max\{K_1, 2K_2\}}$$

$$\ge \int_{\frac{\epsilon \cdot (1-\epsilon) \cdot \delta}{16 \max\{K_1, 2K_2\}}}^{\sqrt{\delta}}$$

$$\times \left(\log \mathcal{N}_2\left(u, \left\{f \in \mathcal{F} : \frac{1}{n}\sum_{i=1}^{n} f(z_i)^2 \le 16\delta\right\}, z_1^n\right)\right)^{1/2} du. \tag{19.11}$$

Then

$$\mathbf{P}\left\{\sup_{f\in\mathcal{F}}\frac{|\mathbf{E}\{f(Z)\}-\frac{1}{n}\sum_{i=1}^{n}f(Z_i)|}{\alpha+\mathbf{E}\{f(Z)\}}>\epsilon\right\}$$

$$\leq 60\exp\left(-\frac{n\,\alpha\,\epsilon^2(1-\epsilon)}{128\cdot 2304\max\{K_1^2,K_2\}}\right). \tag{19.12}$$

PROOF. The proof is divided into six steps.

STEP 1. Replace the expectation in the nominator of (19.12) by an empirical mean.

Draw a "ghost" sample $Z_1'^n = (Z_1',\ldots,Z_n')$ of i.i.d. random variables distributed as Z_1 and independent of Z_1^n. Let $f^* = f^*(Z_1^n)$ be a function $f\in\mathcal{F}$ such that

$$\left|\frac{1}{n}\sum_{i=1}^{n}f(Z_i)-\mathbf{E}\{f(Z)\}\right|>\epsilon(\alpha+\mathbf{E}\{f(Z)\}),$$

if there exists any such function, and let f^* be an other arbitrary function contained in $\mathcal{F}$, if such a function doesn't exist.

We have

$$\mathbf{P}\left\{\exists f\in\mathcal{F}:\left|\frac{1}{n}\sum_{i=1}^{n}f(Z_i)-\frac{1}{n}\sum_{i=1}^{n}f(Z_i')\right|>\frac{\epsilon}{2}\alpha+\frac{\epsilon}{2}\mathbf{E}\{f(Z)\}\right\}$$

$$\geq\mathbf{P}\left\{\left|\frac{1}{n}\sum_{i=1}^{n}f^*(Z_i)-\frac{1}{n}\sum_{i=1}^{n}f^*(Z_i')\right|>\frac{\epsilon}{2}\alpha+\frac{\epsilon}{2}\mathbf{E}\{f^*(Z)|Z_1^n\}\right\}$$

$$\geq\mathbf{P}\left\{\left|\frac{1}{n}\sum_{i=1}^{n}f^*(Z_i)-\mathbf{E}\{f^*(Z)|Z_1^n\}\right|>\epsilon\alpha+\epsilon\mathbf{E}\{f^*(Z)|Z_1^n\},\right.$$

$$\left.\left|\frac{1}{n}\sum_{i=1}^{n}f^*(Z_i')-\mathbf{E}\{f^*(Z)|Z_1^n\}\right|\leq\frac{\epsilon}{2}\alpha+\frac{\epsilon}{2}\mathbf{E}\{f^*(Z)|Z_1^n\}\right\}$$

$$=\mathbf{E}\left\{I_{\{|\frac{1}{n}\sum_{i=1}^{n}f^*(Z_i)-\mathbf{E}\{f^*(Z)|Z_1^n\}|>\epsilon\alpha+\epsilon\mathbf{E}\{f^*(Z)|Z_1^n\}\}}\right.$$

$$\times\mathbf{P}\left\{\left|\frac{1}{n}\sum_{i=1}^{n}f^*(Z_i')-\mathbf{E}\{f^*(Z)|Z_1^n\}\right|\right.$$

$$\left.\left.\leq\frac{\epsilon}{2}\alpha+\frac{\epsilon}{2}\mathbf{E}\{f^*(Z)|Z_1^n\}\bigg|Z_1^n\right\}\right\}.$$

Chebyshev's inequality, together with

$$0\leq\mathbf{E}\left\{f^*(Z)^2|Z_1^n\right\}\leq K_2\mathbf{E}\left\{f^*(Z)|Z_1^n\right\}$$

and $n \geq \frac{20 K_2}{\epsilon^2 \alpha}$ (which follows from (19.10)), implies

$$\mathbf{P}\left\{\left|\frac{1}{n}\sum_{i=1}^{n} f^*(Z_i') - \mathbf{E}\{f^*(Z)|Z_1^n\}\right| > \frac{\epsilon}{2}\alpha + \frac{\epsilon}{2}\mathbf{E}\{f^*(Z)|Z_1^n\}\,\Big|\,Z_1^n\right\}$$

$$\leq \frac{K_2 \mathbf{E}\{f^*(Z)|Z_1^n\}}{n\left(\frac{\epsilon}{2}\alpha + \frac{\epsilon}{2}\mathbf{E}\{f^*(Z)|Z_1^n\}\right)^2} \leq \frac{K_2 \mathbf{E}\{f^*(Z)|Z_1^n\}}{n \cdot 2 \cdot \frac{\epsilon}{2}\alpha \cdot \frac{\epsilon}{2}\mathbf{E}\{f^*(Z)|Z_1^n\}}$$

$$\leq \frac{2K_2}{n\epsilon^2\alpha} \leq \frac{1}{10}.$$

Thus

$$\mathbf{P}\left\{\exists f \in \mathcal{F}: \left|\frac{1}{n}\sum_{i=1}^{n} f(Z_i) - \frac{1}{n}\sum_{i=1}^{n} f(Z_i')\right| > \frac{\epsilon}{2}\alpha + \frac{\epsilon}{2}\mathbf{E}\{f(Z)\}\right\}$$

$$\geq \frac{9}{10}\mathbf{P}\left\{\left|\frac{1}{n}\sum_{i=1}^{n} f^*(Z_i) - \mathbf{E}\{f^*(Z)|Z_1^n\}\right| > \epsilon\alpha + \epsilon\mathbf{E}\{f^*(Z)|Z_1^n\}\right\}$$

$$= \frac{9}{10}\mathbf{P}\left\{\exists f \in \mathcal{F}: \left|\frac{1}{n}\sum_{i=1}^{n} f(Z_i) - \mathbf{E}\{f(Z)\}\right| > \epsilon\alpha + \epsilon\mathbf{E}\{f(Z)\}\right\},$$

which implies

$$\mathbf{P}\left\{\sup_{f\in\mathcal{F}} \frac{\left|\frac{1}{n}\sum_{i=1}^{n} f(Z_i) - \mathbf{E}\{f(Z)\}\right|}{\alpha + \mathbf{E}\{f(Z)\}} > \epsilon\right\}$$

$$\leq \frac{10}{9}\mathbf{P}\left\{\exists f \in \mathcal{F}: \left|\frac{1}{n}\sum_{i=1}^{n} f(Z_i) - \frac{1}{n}\sum_{i=1}^{n} f(Z_i')\right| > \frac{\epsilon}{2}\alpha + \frac{\epsilon}{2}\mathbf{E}\{f(Z)\}\right\}.$$

STEP 2. Replace the expectation in the above probability by an empirical mean.

Using $\mathbf{E}\{f(Z)\} \geq \frac{1}{K_2}\mathbf{E}\{f(Z)^2\}$ we get

$$\mathbf{P}\left\{\exists f \in \mathcal{F}: \left|\frac{1}{n}\sum_{i=1}^{n} f(Z_i) - \frac{1}{n}\sum_{i=1}^{n} f(Z_i')\right| > \frac{\epsilon}{2}\alpha + \frac{\epsilon}{2}\mathbf{E}\{f(Z)\}\right\}$$

$$\leq \mathbf{P}\left\{\exists f \in \mathcal{F}: \left|\frac{1}{n}\sum_{i=1}^{n} f(Z_i) - \frac{1}{n}\sum_{i=1}^{n} f(Z_i')\right| > \frac{\epsilon}{2}\alpha + \frac{\epsilon}{2}\frac{1}{K_2}\mathbf{E}\{f(Z)^2\}\right\}$$

$$\leq \mathbf{P}\left\{\exists f \in \mathcal{F}: \left|\frac{1}{n}\sum_{i=1}^{n} f(Z_i) - \frac{1}{n}\sum_{i=1}^{n} f(Z_i')\right| > \frac{\epsilon}{2}\alpha + \frac{\epsilon}{2}\frac{1}{K_2}\mathbf{E}\{f(Z)^2\},\right.$$

$$\left.\frac{1}{n}\sum_{i=1}^{n} f(Z_i)^2 - \mathbf{E}\{f(Z)^2\} \leq \epsilon\left(\alpha + \frac{1}{n}\sum_{i=1}^{n} f(Z_i)^2 + \mathbf{E}\{f(Z)^2\}\right)\right\},$$

$$\frac{1}{n}\sum_{i=1}^{n} f(Z_i')^2 - \mathbf{E}\{f(Z)^2\} \le \epsilon \left(\alpha + \frac{1}{n}\sum_{i=1}^{n} f(Z_i')^2 + \mathbf{E}\{f(Z)^2\}\right)\Bigg\}$$

$$+2\mathbf{P}\left\{\sup_{f\in\mathcal{F}} \frac{\left|\frac{1}{n}\sum_{i=1}^{n} f(Z_i)^2 - \mathbf{E}\{f(Z)^2\}\right|}{\alpha + \frac{1}{n}\sum_{i=1}^{n} f(Z_i)^2 + \mathbf{E}\{f(Z)^2\}} > \epsilon\right\}.$$

The second inequality in the first probability on the right-hand side of the above inequality is equivalent to

$$\mathbf{E}\left\{f(Z)^2\right\} \ge \frac{1}{1+\epsilon}\left(-\epsilon\alpha + (1-\epsilon)\frac{1}{n}\sum_{i=1}^{n} f(Z_i)^2\right),$$

which implies

$$\frac{1}{2}\frac{\epsilon}{2}\frac{1}{K_2}\mathbf{E}\{f(Z)^2\} \ge -\frac{\epsilon^2}{4(1+\epsilon)K_2}\alpha + \frac{1-\epsilon}{4(1+\epsilon)K_2}\epsilon\frac{1}{n}\sum_{i=1}^{n} f(Z_i)^2.$$

This together with a similar argument applied to the third inequality and

$$\frac{\epsilon\alpha}{2} - \frac{\epsilon^2\alpha}{2(1+\epsilon)K_2} \ge \frac{\epsilon\alpha}{4}$$

(which follows from $\epsilon/(K_2(1+\epsilon)) \le 1/2$ for $0 < \epsilon < 1$ and $K_2 \ge 1$) yields

$$\mathbf{P}\left\{\exists f \in \mathcal{F} : \left|\frac{1}{n}\sum_{i=1}^{n} f(Z_i) - \frac{1}{n}\sum_{i=1}^{n} f(Z_i')\right| > \frac{\epsilon}{2}\alpha + \frac{\epsilon}{2}\mathbf{E}\{f(Z)\}\right\}$$

$$\le \mathbf{P}\left\{\exists f \in \mathcal{F} : \left|\frac{1}{n}\sum_{i=1}^{n} f(Z_i) - \frac{1}{n}\sum_{i=1}^{n} f(Z_i')\right|\right.$$

$$\left. > \frac{\epsilon\alpha}{4} + \frac{(1-\epsilon)\epsilon}{4(1+\epsilon)K_2}\left(\frac{1}{n}\sum_{i=1}^{n} f(Z_i)^2 + \frac{1}{n}\sum_{i=1}^{n} f(Z_i')^2\right)\right\}$$

$$+2\mathbf{P}\left\{\sup_{f\in\mathcal{F}} \frac{\left|\frac{1}{n}\sum_{i=1}^{n} f(Z_i)^2 - \mathbf{E}\{f(Z)^2\}\right|}{\alpha + \frac{1}{n}\sum_{i=1}^{n} f(Z_i)^2 + \mathbf{E}\{f(Z)^2\}} > \epsilon\right\}. \tag{19.13}$$

STEP 3. Application of Theorem 19.2.
Next we apply Theorem 19.2 to the second probability on the right-hand side of the above inequality. $\{f^2 : f \in \mathcal{F}\}$ is a class of functions with values in $[0, K_1^2]$. In order to be able to apply Theorem 19.2 to it, we need

$$\sqrt{n}\epsilon\sqrt{\alpha} \ge 576K_1 \tag{19.14}$$

and that, for all $\bar{\delta} \ge \alpha K_1^2/2$ and all $z_1, \ldots, z_n \in \mathcal{R}^d$,

$$\frac{\sqrt{n}\epsilon\bar{\delta}}{192\sqrt{2}K_1^2}$$

$$\geq \int_{\frac{\epsilon\bar{\delta}}{32K_1^2}}^{\sqrt{\bar{\delta}}} \left(\log \mathcal{N}_2 \left(u, \left\{ f^2 \, : \, f \in \mathcal{F}, \frac{1}{n}\sum_{i=1}^n f(z_i)^2 \leq \frac{4\bar{\delta}}{K_1^2} \right\}, z_1^n \right) \right)^{\frac{1}{2}} du.$$

$$(19.15)$$

Inequality (19.14) is implied by (19.10). Furthermore, for arbitrary functions $f_1, f_2 : \mathcal{R}^d \to [-K_1, K_1]$, one has

$$\frac{1}{n}\sum_{i=1}^n \left| f_1(z_i)^2 - f_2(z_i)^2 \right|^2 = \frac{1}{n}\sum_{i=1}^n |f_1(z_i) + f_2(z_i)|^2 \cdot |f_1(z_i) - f_2(z_i)|^2$$

$$\leq (2K_1)^2 \frac{1}{n}\sum_{i=1}^n |f_1(z_i) - f_2(z_i)|^2 ,$$

which implies

$$\mathcal{N}_2 \left(u, \left\{ f^2 \, : \, f \in \mathcal{F}, \frac{1}{n}\sum_{i=1}^n f(z_i)^2 \leq \frac{4\bar{\delta}}{K_1^2} \right\}, z_1^n \right)$$

$$\leq \mathcal{N}_2 \left(\frac{u}{2K_1}, \left\{ f \in \mathcal{F} \, : \, \frac{1}{n}\sum_{i=1}^n f(z_i)^2 \leq \frac{4\bar{\delta}}{K_1^2} \right\}, z_1^n \right).$$

Hence (19.15) follows from (19.11) (set $\delta = \bar{\delta}/(4K_1^2)$ in (19.11)).

Thus the assumptions of Theorem 19.2 are satisfied and we can conclude

$$\mathbf{P} \left\{ \sup_{f \in \mathcal{F}} \frac{\left| \frac{1}{n}\sum_{i=1}^n f(Z_i)^2 - \mathbf{E}\{f(Z)^2\} \right|}{\alpha + \frac{1}{n}\sum_{i=1}^n f(Z_i)^2 + \mathbf{E}\{f(Z)^2\}} > \epsilon \right\}$$

$$\leq 15 \exp \left(-\frac{n\epsilon^2\alpha}{128 \cdot 2304 K_1^2} \right).$$

STEP 4. Introduction of additional randomness by random signs.
Let $U_1, \ldots, U_n$ be independent and uniformly distributed over $\{-1, 1\}$ and independent of $Z_1, \ldots, Z_n, Z_1', \ldots, Z_n'$. Because of the independence and identical distribution of $Z_1, \ldots, Z_n, Z_1', \ldots, Z_n'$, the joint distribution of $Z_1^n, Z_1'^n$ doesn't change if one randomly interchanges the corresponding components of Z_1^n and $Z_1'^n$. Hence

$$\mathbf{P} \left\{ \exists f \in \mathcal{F} : \left| \frac{1}{n}\sum_{i=1}^n f(Z_i) - \frac{1}{n}\sum_{i=1}^n f(Z_i') \right| \right.$$

$$\left. > \frac{\epsilon\alpha}{4} + \frac{(1-\epsilon)\epsilon}{4(1+\epsilon)K_2} \left(\frac{1}{n}\sum_{i=1}^n f(Z_i)^2 + \frac{1}{n}\sum_{i=1}^n f(Z_i')^2 \right) \right\}$$

$$= \mathbf{P}\left\{\exists f \in \mathcal{F}: \left|\frac{1}{n}\sum_{i=1}^{n} U_i\left(f(Z_i) - f(Z_i')\right)\right|\right.$$

$$\left. > \frac{\epsilon\alpha}{4} + \frac{(1-\epsilon)\epsilon}{4(1+\epsilon)K_2}\left(\frac{1}{n}\sum_{i=1}^{n} f(Z_i)^2 + \frac{1}{n}\sum_{i=1}^{n} f(Z_i')^2\right)\right\}$$

$$\leq 2\mathbf{P}\left\{\exists f \in \mathcal{F}: \left|\frac{1}{n}\sum_{i=1}^{n} U_i f(Z_i)\right| > \frac{\epsilon\alpha}{8} + \frac{(1-\epsilon)\epsilon}{4(1+\epsilon)K_2}\frac{1}{n}\sum_{i=1}^{n} f(Z_i)^2\right\}.$$

STEP 5. Peeling.
We have

$$\mathbf{P}\left\{\exists f \in \mathcal{F}: \left|\frac{1}{n}\sum_{i=1}^{n} U_i f(Z_i)\right| > \frac{\epsilon\alpha}{8} + \frac{(1-\epsilon)\epsilon}{4(1+\epsilon)K_2}\frac{1}{n}\sum_{i=1}^{n} f(Z_i)^2\right\}$$

$$\leq \sum_{k=1}^{\infty}\mathbf{P}\left\{\exists f \in \mathcal{F}:\right.$$

$$I_{\{k\neq 1\}}2^{k-1}\frac{K_2(1+\epsilon)\alpha}{2(1-\epsilon)} \leq \frac{1}{n}\sum_{i=1}^{n} f(Z_i)^2 \leq 2^k\frac{K_2(1+\epsilon)\alpha}{2(1-\epsilon)},$$

$$\left.\left|\frac{1}{n}\sum_{i=1}^{n} U_i f(Z_i)\right| > \frac{\epsilon\alpha}{8} + \frac{(1-\epsilon)\epsilon}{4(1+\epsilon)K_2}\frac{1}{n}\sum_{i=1}^{n} f(Z_i)^2\right\}$$

$$\leq \sum_{k=1}^{\infty}\mathbf{P}\left\{\exists f \in \mathcal{F}:\right.$$

$$\left.\frac{1}{n}\sum_{i=1}^{n} f(Z_i)^2 \leq 2^k\frac{K_2(1+\epsilon)\alpha}{2(1-\epsilon)}, \left|\frac{1}{n}\sum_{i=1}^{n} U_i f(Z_i)\right| > \frac{\epsilon\alpha}{8}2^{k-1}\right\}.$$

STEP 6. Application of Theorem 19.1.
Next we condition inside the above probabilities on $Z_1, \ldots, Z_n$, which is equivalent to considering, for $z_1, \ldots, z_n \in \mathcal{R}^d$ and $k \in \mathcal{N}$,

$$\mathbf{P}\left\{\exists f \in \mathcal{F}: \frac{1}{n}\sum_{i=1}^{n} f(z_i)^2 \leq 2^k\frac{K_2(1+\epsilon)\alpha}{2(1-\epsilon)}, \left|\frac{1}{n}\sum_{i=1}^{n} U_i f(z_i)\right| > \frac{\epsilon\alpha}{8}2^{k-1}\right\}.$$

$$(19.16)$$

The assumptions of Theorem 19.3 imply (use $\delta = 2^k\frac{K_2(1+\epsilon)\alpha}{8(1-\epsilon)}$ in (19.11))

$$
\frac{\sqrt{n}\,\frac{\epsilon\alpha}{8}2^{k-1}}{48\sqrt{2}}
$$

$$
\geq \int_{\frac{\epsilon\alpha 2^{k}}{8\cdot 16}}^{\sqrt{\frac{2^{k}K_2(1+\epsilon)\alpha}{2(1-\epsilon)}}/2}
$$

$$
\times \left(\log\mathcal{N}_2\left(u, \left\{ f\in\mathcal{F} \;:\; \frac{1}{n}\sum_{i=1}^{n} f(z_i)^2 \leq \frac{2^{k}K_2(1+\epsilon)\alpha}{2(1-\epsilon)} \right\}, z_1^n \right) \right)^{\frac{1}{2}} du
$$

and

$$
\sqrt{n}\,\frac{\epsilon\alpha}{8}2^{k-1} \geq 36\sqrt{\frac{2^{k}K_2(1+\epsilon)\alpha}{2(1-\epsilon)}}.
$$

Hence we can conclude by Theorem 19.1 that (19.16) is bounded from above by

$$
5\exp\left(-\frac{n\left(\frac{\epsilon\alpha}{8}2^{k-1}\right)^2}{2304\frac{2^{k}K_2(1+\epsilon)\alpha}{2(1-\epsilon)}} \right) = 5\exp\left(-\frac{n\epsilon^2(1-\epsilon)\alpha}{64\cdot 2304(1+\epsilon)K_2}2^{k-1} \right).
$$

It follows that

$$
\sum_{k=1}^{\infty}\mathbf{P}\left\{ \exists f\in\mathcal{F} : \frac{1}{n}\sum_{i=1}^{n} f(Z_i)^2 \leq 2^{k}\frac{K_2(1+\epsilon)\alpha}{2(1-\epsilon)}, \right.
$$

$$
\left. \left|\frac{1}{n}\sum_{i=1}^{n} U_i f(Z_i)\right| > \frac{\epsilon\alpha}{8}2^{k-1} \right\}
$$

$$
\leq \sum_{k=1}^{\infty} 5\exp\left(-\frac{n\epsilon^2(1-\epsilon)\alpha}{64\cdot 2304(1+\epsilon)K_2}2^{k-1} \right)
$$

$$
\leq \sum_{k=1}^{\infty} 5\exp\left(-\frac{n\epsilon^2(1-\epsilon)\alpha}{64\cdot 2304(1+\epsilon)K_2}\cdot k \right)
$$

$$
= \frac{5}{1-\exp\left(-\frac{n\epsilon^2(1-\epsilon)\alpha}{64\cdot 2304(1+\epsilon)K_2}\right)}\exp\left(-\frac{n\epsilon^2(1-\epsilon)\alpha}{64\cdot 2304(1+\epsilon)K_2} \right)
$$

$$
\leq \frac{5}{1-\exp\left(-\frac{9}{16}\right)}\exp\left(-\frac{n\epsilon^2(1-\epsilon)\alpha}{64\cdot 2304(1+\epsilon)K_2} \right).
$$

Steps 1 to 6 imply the assertion. $\qquad\square$

19.4 Piecewise Polynomial Partitioning Estimates

In this section we use Theorem 19.3 to show that suitably defined piecewise polynomial partitioning estimates achieve for bounded Y the optimal minimax rate of convergence for estimating (p, C)–smooth regression functions. In the case of Y bounded this improves the convergence rates of Section 11.2 by a logarithmic factor.

For simplicity we assume $X \in [0, 1]$ *a.s.* and $|Y| \leq L$ *a.s.* for some $L \in \mathcal{R}_+$. We will show in Problem 19.1 how to derive similar results for multivariate X.

Recall that the piecewise polynomial partitioning estimate is defined by minimizing the empirical L_2 risk over the set $\mathcal{F}_{K,M}$ of all piecewise polynomials of degree M (or less) with respect to an equidistant partition of $[0, 1]$ into K intervals. In Section 11.2 we truncated the estimate in order to ensure that it is bounded in absolute value. Here we impose instead a bound on the supremum norm of the functions which we consider during minimization of the empirical L_2 risk. More precisely, set

$$m_{n,(K,M)}(\cdot) = \arg \min_{f \in \mathcal{F}_{K,M}(L+1)} \frac{1}{n} \sum_{i=1}^{n} |f(X_i) - Y_i|^2,$$

where

$$\mathcal{F}_{K,M}(L+1) = \left\{ f \in \mathcal{F}_{K,M} \ : \ \sup_{x \in [0,1]} |f(x)| \leq L+1 \right\}$$

is the set of all piecewise polynomials of $\mathcal{F}_{K,M}$ bounded in absolute value by $L+1$. Observe that by our assumptions the regression function is bounded in absolute value by L and that it is reasonable to fit a function to the data for which a similar bound (which we have chosen to be $L+1$) is valid. In contrast to the truncation used in Section 11.2 this ensures that the estimate is contained in a linear vector space which will allow us to apply the bound on covering numbers of balls in linear vector spaces in Lemma 9.3 in order to evaluate the conditions of Theorem 19.3.

It is not clear whether there exists any practical algorithm to compute the above estimate. The problem is the bound on the supremum norm of the functions in $\mathcal{F}_{K,M}(L+1)$ which makes it difficult to represent the functions by a linear combination of some fixed basis functions. In Problem 19.4 we will show how to change the definition of the estimate in order to get an estimate which can be computed much easier.

The next theorem gives a bound on the expected L_2 error of $m_{n,(K,M)}$:

Theorem 19.4. *Let $M \in \mathcal{N}_0$, $K \in \mathcal{N}$, and $L > 0$ and let the estimate $m_{n,(K,M)}$ be defined as above. Then*

$$\mathbf{E} \int |m_{n,(K,M)}(x) - m(x)|^2 \mu(dx)$$

$$\leq c_1 \cdot \frac{(M+1)K}{n} + 2 \inf_{f \in \mathcal{F}_{K,M}(L+1)} \int |f(x) - m(x)|^2 \mu(dx)$$

for every distribution of (X,Y) with $|Y| \leq L$ a.s. Here c_1 is a constant which depends only on L.

Before we prove this theorem we study its consequences.

Corollary 19.1. *Let $C, L > 0$ and $p = k + \beta$ with $k \in \mathcal{N}_0$ and $\beta \in (0, 1]$. Set $M = k$ and $K_n = \lceil C^{2/(2p+1)} n^{1/(2p+1)} \rceil$. Then there exists a constant c_2 which depends only on p and L such that, for all $n \geq \max\left\{ C^{1/p}, C^{-2} \right\}$,*

$$\mathbf{E} \int |m_{n,(K_n,M)}(x) - m(x)|^2 \mu(dx) \leq c_2 C^{\frac{2}{2p+1}} n^{-\frac{2p}{2p+1}}.$$

for every distribution of (X,Y) with $X \in [0,1]$ a.s., $|Y| \leq L$ a.s., and m (p, C)-smooth.

PROOF. By Lemma 11.1 there exists a piecewiese polynomial $g \in \mathcal{F}_{K_n,M}$ such that

$$\sup_{x \in [0,1]} |g(x) - m(x)| \leq \frac{1}{2^p k!} \cdot \frac{C}{K_n^p}.$$

In particular,

$$\sup_{x \in [0,1]} |g(x)| \leq \sup_{x \in [0,1]} |m(x)| + \frac{1}{2^p k!} \cdot \frac{C}{K_n^p} \leq L + (C \cdot n^{-p})^{1/(2p+1)} \leq L + 1,$$

hence $g \in \mathcal{F}_{K_n,M}(L+1)$. Here the third inequality follows from the condition $n \geq C^{1/p}$. Finally,

$$\inf_{f \in \mathcal{F}_{K_n,M}(L+1)} \int |f(x) - m(x)|^2 \mu(dx) \leq \sup_{x \in [0,1]} |g(x) - m(x)|^2$$

$$\leq \frac{1}{(2^p k!)^2} \cdot C^2 \frac{1}{K_n^{2p}}. \qquad (19.17)$$

This together with Theorem 19.4 implies the assertion. $\square$

In the proof of Theorem 19.4 we will apply the following lemma:

Lemma 19.1. *Let (X,Y), (X_1,Y_1), ..., (X_n,Y_n) be independent and identically distributed $(\mathcal{R} \times \mathcal{R})$-valued random variables. Assume $|Y| \leq L$ a.s. for some $L \geq 1$. Let $K \in \mathcal{N}$ and $M \in \mathcal{N}_0$. Then one has, for any*

$$\alpha \geq c_3 \frac{(M+1)\cdot K}{n},$$

$$\mathbf{P}\bigg\{\exists f \in \mathcal{F}_{K,M}(L+1) : \mathbf{E}\{|f(X)-Y|^2 - |m(X)-Y|^2\}$$

$$-\frac{1}{n}\sum_{i=1}^{n}\{|f(X_i)-Y_i|^2 - |m(X_i)-Y_i|^2\}$$

$$> \frac{1}{2}\bigg(\alpha + \mathbf{E}\{|f(X)-Y|^2 - |m(X)-Y|^2\}\bigg)\bigg\}$$

$$\leq 60\exp\left(-\frac{n\,\alpha}{128\cdot 2304\cdot 800\cdot L^4}\right). \tag{19.18}$$

Here c_3 is a constant which depends only on L.

PROOF. Set $Z = (X,Y)$ and $Z_i = (X_i, Y_i)$ $(i = 1,\ldots,n)$. For $f : \mathcal{R}^d \to \mathcal{R}$ define $g_f : \mathcal{R}^d \times \mathcal{R} \to \mathcal{R}$ by

$$g_f(x,y) = (|f(x)-y|^2 - |m(x)-y|^2)\cdot I_{\{y\in[-L,L]\}}.$$

Set

$$\mathcal{G} = \{g_f : f \in \mathcal{F}_{K,M}(L+1)\}.$$

Then the right-hand side of (19.18) can be written as

$$\mathbf{P}\left\{\exists g \in \mathcal{G} : \frac{\mathbf{E}\{g(Z)\} - \frac{1}{n}\sum_{i=1}^{n}g(Z_i)}{\alpha + \mathbf{E}\{g(Z)\}} > \frac{1}{2}\right\}.$$

By definition, the functions in $\mathcal{F}_{K,M}(L+1)$ are bounded in absolute value by $L+1$, hence

$$|g(z)| \leq 2(L+1)^2 + 2L^2 \leq 10L^2 \quad (z \in \mathcal{R}^d \times \mathcal{R})$$

for all $g \in \mathcal{G}$. Furthermore,

$$\mathbf{E}\{g_f(Z)^2\} = \mathbf{E}\left\{\big(|f(X)-Y|^2 - |m(X)-Y|^2\big)^2\right\}$$

$$= \mathbf{E}\left\{|(f(X)+m(X)-2Y)\cdot(f(X)-m(X))|^2\right\}$$

$$\leq (L+1+3L)^2\mathbf{E}\left\{|f(X)-m(X)|^2\right\}$$

$$= 25L^2\mathbf{E}\{g_f(Z)\} \quad (f \in \mathcal{F}_{K,M}(L+1)).$$

The functions in $\mathcal{F}_{K,M}(L+1)$ are piecewise polynomials of degree M (or less) with respect to a partition of $[0,1]$ consisting of K intervals. Hence, $f \in \mathcal{F}_{K,M}(L+1)$ implies that f^2 is a piecewise polynomial of degree $2M$ (or less) with respect to the same partition. This together with

$$g_f(x,y) = \big(f^2(x) - 2f(x)y + y^2 - |m(x)-y|^2\big)\cdot I_{\{y\in[-L,L]\}}$$

yields that $\mathcal{G}$ is a subset of a linear vector space of dimension

$$D = K \cdot ((2M + 1) + (M + 1)) + 1,$$

thus, by Lemma 9.3,

$$\log \mathcal{N}_2 \left(u, \left\{ g \in \mathcal{G} \ : \ \frac{1}{n} \sum_{i=1}^n g(z_i)^2 \leq 16\delta \right\}, z_1^n \right) \leq D \cdot \log \frac{16\sqrt{\delta} + u}{u}.$$

Because of

$$\begin{aligned}
\int_0^{\sqrt{\delta}} \left(D \cdot \log \frac{16\sqrt{\delta} + u}{u} \right)^{\frac{1}{2}} du &= \sqrt{D}\sqrt{\delta} \int_1^\infty \frac{\sqrt{\log(1 + 16v)}}{v^2} dv \\
&\leq \sqrt{D}\sqrt{\delta} \int_1^\infty \frac{\sqrt{16v}}{v^2} dv \\
&\leq 8\sqrt{D}\sqrt{\delta}
\end{aligned}$$

the assumptions of Theorem 19.3 are satisfied whenever

$$\sqrt{n}\frac{1}{4}\delta \geq 96 \cdot \sqrt{2} \cdot 50L^2 \cdot 8\sqrt{D}\sqrt{\delta}$$

for all $\delta \geq \alpha/8$, which is implied by

$$\alpha \geq c_3 \frac{(M + 1) \cdot K}{n}.$$

Thus we can conclude from Theorem 19.3

$$\mathbf{P} \left\{ \exists g \in \mathcal{G} : \frac{\mathbf{E}\{g(Z)\} - \frac{1}{n} \sum_{i=1}^n g(Z_i)}{\alpha + \mathbf{E}\{g(Z)\}} > \frac{1}{2} \right\}$$

$$\leq 60 \exp \left(-\frac{n\,\alpha\,\frac{1}{4}(1 - \frac{1}{2})}{128 \cdot 2304 \cdot 100L^4} \right).$$

$\square$

PROOF OF THEOREM 19.4. We use the error decomposition

$$\int |m_{n,(K,M)}(x) - m(x)|^2 \mu(dx)$$

$$= \mathbf{E} \left\{ |m_{n,(K,M)}(X) - Y|^2 | D_n \right\} - \mathbf{E}\{|m(X) - Y|^2\}$$

$$= T_{1,n} + T_{2,n},$$

where

$$T_{1,n} = 2\frac{1}{n} \sum_{i=1}^n \{|m_{n,(K,M)}(X_i) - Y_i|^2 - |m(X_i) - Y_i|^2\}$$

and

$$T_{2,n} = \mathbf{E} \left\{ |m_{n,(K,M)}(X) - Y|^2 | D_n \right\} - \mathbf{E}\{|m(X) - Y|^2\} - T_{1,n}.$$

By definition of the estimate,

$$T_{1,n} = 2 \min_{f \in \mathcal{F}_{K,M}(L+1)} \frac{1}{n} \sum_{i=1}^{n} \{|f(X_i) - Y_i|^2 - |m(X_i) - Y_i|^2\}$$

which implies

$$\mathbf{E}T_{1,n} \leq 2 \inf_{f \in \mathcal{F}_{K,M}(L+1)} \int |f(x) - m(x)|^2 \mu(dx). \tag{19.19}$$

Furthermore, Lemma 19.1 implies that, for $t \geq c_3 \frac{(M+1) \cdot K}{n}$,

$$\mathbf{P}\{T_{2,n} > t\}$$

$$\leq \mathbf{P}\Bigg\{ \exists f \in \mathcal{F}_{K,M}(L+1) : 2\mathbf{E}\{|f(X) - Y|^2 - |m(X) - Y|^2\}$$

$$- \frac{2}{n} \sum_{i=1}^{n} \{|f(X_i) - Y_i|^2 - |m(X_i) - Y_i|^2\}$$

$$> t + \mathbf{E}\{|f(X) - Y|^2 - |m(X) - Y|^2\} \Bigg\}$$

$$\leq 60 \exp\left(-\frac{nt}{c_4}\right).$$

Hence,

$$\mathbf{E}\{T_{2,n}\}$$

$$\leq \int_0^\infty \mathbf{P}\{T_{2,n} > t\}\, dt \leq c_3 \frac{(M+1) \cdot K}{n} + \int_{c_3 \frac{(M+1)K}{n}}^{\infty} \mathbf{P}\{T_{2,n} > t\}\, dt$$

$$\leq c_3 \frac{(M+1) \cdot K}{n} + \frac{60 c_4}{n} \exp\left(-\frac{c_3}{c_4}(M+1) \cdot K\right).$$

This together with (19.19) implies the assertion. $\square$

The estimate in Corollary 19.1 depends on the smoothness (p, C) of the regression function, which is usually unknown in an application. As in Chapter 12 this can be avoided by using the method of complexity regularization.

Let $M_0 \in \mathcal{N}_0$ and set

$$\mathcal{P}_n = \{(K, M) \in \mathcal{N} \times \mathcal{N}_0 : 1 \leq K \leq n \text{ and } 0 \leq M \leq M_0\}.$$

For $(K, M) \in \mathcal{P}_n$ define $m_{n,(K,M)}$ as above. Depending on the data $(X_1, Y_1), \ldots, (X_n, Y_n)$ choose

$$(K^*, M^*) \in \mathcal{P}_n$$

such that

$$\frac{1}{n} \sum_{i=1}^{n} |m_{n,(K^*,M^*)}(X_i) - Y_i|^2 + pen_n(K^*, M^*)$$

$$= \min_{(K,M)\in\mathcal{P}_n} \left\{ \frac{1}{n} \sum_{i=1}^{n} |m_{n,(K,M)}(X_i) - Y_i|^2 + pen_n(K, M) \right\},$$

where

$$pen_n(K, M) = c_3 \cdot \frac{(M+1)\cdot K}{n}$$

is a penalty term penalizing the complexity of $\mathcal{F}_{K,M}(L+1)$ and c_3 is the constant from Lemma 19.1 above. Set

$$m_n(x, (X_1, Y_1), \ldots, (X_n, Y_n)) = m_{n,(K^*,M^*)}(x, (X_1, Y_1), \ldots, (X_n, Y_n)).$$

The next theorem provides the bound on the expected L_2 error of the estimate. The bound is obtained by means of complexity regularization.

Theorem 19.5. *Let $M_0 \in \mathcal{N}$, $L > 0$, and let the estimate be defined as above. Then*

$$\mathbf{E} \int |m_n(x) - m(x)|^2 \mu(dx)$$

$$\leq \min_{(K,M)\in\mathcal{P}_n} \left\{ 2c_3 \cdot \frac{(M+1)\cdot K}{n} \right.$$

$$\left. +2 \inf_{f\in\mathcal{F}_{K,M}(L+1)} \int |f(x) - m(x)|^2 \mu(dx) \right\} + \frac{c_5}{n}$$

for every distribution of (X, Y) with $|Y| \leq L$ a.s. Here c_5 is a constant which depends only on L and M_0.

Before we prove Theorem 19.5 we study its consequences. Our first corollary shows that if the regression function is contained in the set $\mathcal{F}_{K,M}(L+1)$ of bounded piecewise polynomials, then the expected L_2 error of the estimate converges to zero with the parametric rate $1/n$.

Corollary 19.2. *Let $L > 0$, $M \leq M_0$, and $K \in \mathcal{N}$. Then*

$$\mathbf{E} \int |m_n(x) - m(x)|^2 \mu(dx) = O\left(\frac{1}{n}\right)$$

for every distribution of (X, Y) with $|Y| \leq L$ a.s. and $m \in \mathcal{F}_{K,M}(L+1)$.

PROOF. The assertion follows directly from Theorem 19.5. $\qquad\square$

Next we study estimation of (p, C)-smooth regression functions.

Corollary 19.3. *Let $L > 0$ be arbitrary. Then, for any $p = k + \beta$ with $k \in \mathcal{N}_0, k \leq M_0, \beta \in (0, 1]$, and any $C > 0$,*

$$\mathbf{E} \int |m_n(x) - m(x)|^2 \mu(dx) = O\left(C^{\frac{2}{2p+1}} n^{-\frac{2p}{2p+1}}\right)$$

for every distribution of (X, Y) with $X \in [0, 1]$ a.s., $|Y| \leq L$ a.s., and m (p, C)-smooth.

PROOF. The assertion follows directly from Theorem 19.5 and (19.17). $\square$

According to Chapter 3 the rate in Corollary 19.3 is the optimal minimax rate of convergence for estimation of (p, C)-smooth regression functions.

PROOF OF THEOREM 19.5. We start with the error decomposition

$$\int |m_n(x) - m(x)|^2 \mu(dx)$$
$$= \mathbf{E}\left\{|m_{n,(K^*,M^*)}(X) - Y|^2 | D_n\right\} - \mathbf{E}\{|m(X) - Y|^2\}$$
$$= T_{1,n} + T_{2,n},$$

where

$$T_{1,n} = 2\frac{1}{n}\sum_{i=1}^{n}\{|m_{n,(K^*,M^*)}(X_i) - Y_i|^2 - |m(X_i) - Y_i|^2\} + 2pen_n(K^*, M^*)$$

and

$$T_{2,n} = \mathbf{E}\left\{|m_{n,(K^*,M^*)}(X) - Y|^2 | D_n\right\} - \mathbf{E}\{|m(X) - Y|^2\} - T_{1,n}.$$

By definition of the estimate,

$$T_{1,n} = 2\min_{(K,M)\in\mathcal{P}_n}\left\{\min_{f\in\mathcal{F}_{K,M}(L+1)}\frac{1}{n}\sum_{i=1}^{n}|f(X_i) - Y_i|^2 + pen_n(K, M)\right\}$$
$$- \frac{2}{n}\sum_{i=1}^{n}|m(X_i) - Y_i|^2,$$

which implies

$$\mathbf{E}T_{1,n}$$
$$\leq 2\min_{(K,M)\in\mathcal{P}_n}\left\{pen_n(K, M) + \inf_{f\in\mathcal{F}_{K,M}(L+1)}\int |f(x) - m(x)|^2\mu(dx)\right\}.$$
$$(19.20)$$

Hence it suffices to show

$$\mathbf{E}T_{2,n} \leq \frac{c_5}{n}. \qquad (19.21)$$

To this end, let $t > 0$ be arbitrary. Then Lemma 19.1 implies

$$\mathbf{P}\{T_{2,n} > t\}$$

$$\leq \sum_{K=1}^{n} \sum_{M=0}^{M_0} \mathbf{P}\Bigg\{ \exists f \in \mathcal{F}_{K,M}(L+1) : 2\mathbf{E}\{|f(X) - Y|^2 - |m(X) - Y|^2\}$$

$$-\frac{2}{n}\sum_{i=1}^{n}\{|f(X_i) - Y_i|^2 - |m(X_i) - Y_i|^2\}$$

$$> t + 2 \cdot pen_n(M, K) + \mathbf{E}\{|f(X) - Y|^2 - |m(X) - Y|^2\}\Bigg\}$$

$$\leq \sum_{K=1}^{n} \sum_{M=0}^{M_0} 60 \exp\left(-\frac{n\,(t + 2 \cdot pen_n(K, M))}{c_6}\right)$$

$$= \exp\left(-\frac{n\,t}{c_6}\right) \cdot \sum_{K=1}^{n} \sum_{M=0}^{M_0} 60 \exp\left(-\frac{2c_3}{c_6} \cdot (M+1)K\right)$$

$$\leq (M_0 + 1) \cdot c_7 \cdot \exp\left(-\frac{n\,t}{c_6}\right).$$

Hence

$$\mathbf{E}T_{2,n} \leq \int_0^{\infty} \mathbf{P}\{T_{2,n} > t\}\, dt \leq \frac{c_6 c_7 \,(M_0 + 1)}{n}.$$

$\square$

19.5 Bibliographic Notes

Theorem 19.1 is due to van de Geer (1990). Applications of empirical process theory in statistics are described in van der Vaart and Wellner (1996) and van de Geer (2000), which are also excellent sources for learning many useful techniques from empirical process theory.

Theorems 19.2, 19.3, 19.4, and 19.5 are due to Kohler (2000a; 2000b).

The rate of convergence results in this chapter require the boundedness of (X, Y). For the empirical L_2 error

$$\|m_n - m\|_2^2 = \frac{1}{n}\sum_{i=1}^{n}|m_n(X_i) - m(X_i)|^2$$

one can derive the rate of convergence results for bounded X and unbounded Y if one assumes the existence of an exponential moment of Y (such results can be found, e.g., in van de Geer (2000)). One can use Theorem 19.2 to bound the L_2 error by some constant times the empirical L_2 error, which leads to rate of convergence results for bounded X and unbounded Y. It is an open problem whether one can also show similar results if X is unbounded, e.g., if X is normally distributed (cf. Question 1 in Stone (1982)).

Problems and Exercises

PROBLEM 19.1. Let $L > 0$ be arbitrary. Let $\mathcal{F}^{(d)}_{K,M}$ be the set of all multivariate piecewise polynomials of degree M (or less, in each coordinate) with respect to an equidistant partition of $[0,1]^d$ into K^d cubes, set

$$\mathcal{F}^{(d)}_{K,M}(L+1) = \left\{ f \in \mathcal{F}^{(d)}_{K,M} \ : \ \sup_{x \in [0,1]^d} |f(x)| \le L+1 \right\}$$

and define the estimate $m_{n,(K,M)}$ by

$$m_{n,(K,M)} = \arg \min_{f \in \mathcal{F}^{(d)}_{K,M}(L+1)} \frac{1}{n} \sum_{i=1}^{n} |f(X_i) - Y_i|^2.$$

Show that there exists a constant c depending only on L and d such that

$$\mathbf{E} \int |m_{n,(K,M)}(x) - m(x)|^2 \mu(dx)$$

$$\le c \cdot \frac{(M+1)^d K^d}{n} + 2 \inf_{f \in \mathcal{F}^{(d)}_{K,M}(L+1)} \int |f(x) - m(x)|^2 \mu(dx)$$

for all distributions of (X,Y) with $X \in [0,1]^d$ a.s. and $|Y| \le L$ a.s.

HINT: Proceed as in the proof of Theorem 19.4.

PROBLEM 19.2. Let $p = k + \beta$ for some $k \in \mathcal{N}_0$ and $\beta \in (0,1]$. Show that for a suitable choice of the parameters $K = K_n$ and M the estimate in Problem 19.1 satisfies

$$\mathbf{E} \int |m_{n,(K_n,M)}(x) - m(x)|^2 \mu(dx) \le c' C^{\frac{2d}{2p+d}} n^{-\frac{2p}{2p+d}}$$

for every distribution of (X,Y) with $X \in [0,1]^d$ a.s., $|Y| \le L$ a.s., and m (p,C)-smooth.

PROBLEM 19.3. Use complexity regularization to choose the parameters of the estimate in Problem 19.2.

PROBLEM 19.4. Let $\{B_{j,K_n,M} \ : \ j = -M, \ldots, K_n - 1\}$ be the B-spline basis of the univariate spline space $S_{K_n,M}$ introduced in Section 14.4. Let $c > 0$ be a constant specified below and set

$$\bar{S}_{K_n,M} := \left\{ \sum_{j=-M}^{K_n-1} a_j B_{j,K_n,M} \ : \ \sum_{j=-M}^{K_n-1} |a_j| \le c \cdot (L+1) \right\}.$$

Show that if one chooses c in a suitable way then the least squares estimate

$$m_{n,(K_n,M)}(\cdot) = \arg \min_{f \in \bar{S}_{K_n,M}} \frac{1}{n} \sum_{i=1}^{n} |f(X_i) - Y_i|^2$$

satisfies the bounds in Theorem 19.4 (with $\mathcal{F}_{K,M}(L+1)$ replaced by $\bar{S}_{K_n,M}$) and in Corollary 19.1.

HINT: According to de Boor (1978) there exists a constant $c > 0$ such that, for all $a_j \in \mathcal{R}$,

$$\sum_{j=-M}^{K_n-1} |a_j| \leq c \cdot \sup_{x \in [0,1]} \left| \sum_{j=-M}^{K_n-1} a_j B_{j,K_n,M}(x) \right|.$$

PROBLEM 19.5. Use complexity regularization to choose the parameters of the estimate in Problem 19.4.

PROBLEM 19.6. Assume $(X,Y) \in \mathcal{R} \times [-L, L]$ a.s. Let $\mathcal{F}$ be a set of functions $f : \mathcal{R} \to \mathcal{R}$ and assume that $\mathcal{F}$ is a subset of a linear vector space of dimension K. Let m_n be the least squares estimate

$$m_n(\cdot) = \arg \min_{f \in \mathcal{F}} \frac{1}{n} \sum_{i=1}^{n} |f(X_i) - Y_i|^2$$

and set

$$m_n^*(\cdot) = \arg \min_{f \in \mathcal{F}} \frac{1}{n} \sum_{i=1}^{n} |f(X_i) - m(X_i)|^2.$$

Show that, for all $\delta > 0$,

$$\mathbf{P}\left\{ \frac{1}{n} \sum_{i=1}^{n} |m_n(X_i) - m(X_i)|^2 > 2\delta + 18 \min_{f \in \mathcal{F}} \frac{1}{n} \sum_{i=1}^{n} |f(X_i) - m(X_i)|^2 \Big| X_1^n \right\}$$

$$\leq \mathbf{P}\left\{ \delta < \frac{1}{n} \sum_{i=1}^{n} |m_n(X_i) - m(X_i)|^2 \right.$$

$$\left. \leq \frac{1}{n} \sum_{i=1}^{n} (m_n(X_i) - m_n^*(X_i)) \cdot (Y_i - m(X_i)) \Big| X_1^n \right\}.$$

Use the peeling technique and Theorem 19.1 to show that, for $\delta \geq c \cdot \frac{K}{n}$, the last probability is bounded by

$$c' \exp(-n\delta/c')$$

and use this result to derive a rate of convergence result for

$$\frac{1}{n} \sum_{i=1}^{n} |m_n(X_i) - m(X_i)|^2.$$

PROBLEM 19.7. Apply the chaining and the peeling technique in the proof of Theorem 11.2 to derive a version of Theorem 11.2 where one uses integrals of covering numbers of balls in the function space as in Theorem 19.3. Use this result together with Problem 19.6 to give a second proof for Theorem 19.5.

20
Penalized Least Squares Estimates I: Consistency

In the definition of least squares estimates one introduces sets of functions depending on the sample size over which one minimizes the empirical L_2 risk. Restricting the minimization of the empirical L_2 risk to these sets of functions prevents the estimates from adapting too well to the given data. If an estimate adapts too well to the given data, then it is not suitable for predicting new, independent data. Penalized least squares estimates use a different strategy to avoid this problem: instead of restricting the class of functions they add a penalty term to the empirical L_2 risk which penalizes the roughness of a function and minimize the sum of the empirical L_2 risk and this penalty term basically over all functions. The most popular examples are smoothing spline estimates, where the penalty term is chosen proportional to an integral over a squared derivative of the function. In contrast to the complexity regularization method introduced in Chapter 12, the penalty here depends directly on the function considered and not only on a set of functions in which this function is defined.

In this chapter we study penalized least squares estimates. In Section 20.1 we explain how the univariate estimate is defined and how it can be computed efficiently. The proofs there are based on some optimality properties of spline interpolants, which are the topic of Section 20.2. Results about the consistency of univariate penalized least squares estimates are contained in Section 20.3. In Sections 20.4 and 20.5 we show how one can extend the previous results to the multivariate case.

20.1 Univariate Penalized Least Squares Estimates

In this section we will discuss the univariate penalized least squares estimates. In order to simplify the notation we will assume $X \in (0,1)$ *a.s.* In Problem 20.6 we will describe a modification of the estimate which leads to universally consistent estimates.

Fix the data D_n and let $k \in \mathcal{N}$ and $\lambda_n > 0$. The univariate penalized least squares estimate (or smoothing spline estimate) is defined as a function g which minimizes

$$\frac{1}{n} \sum_{i=1}^{n} |g(X_i) - Y_i|^2 + \lambda_n \int_0^1 |g^{(k)}(x)|^2 \, dx, \qquad (20.1)$$

where

$$\frac{1}{n} \sum_{i=1}^{n} |g(X_i) - Y_i|^2$$

is the empirical L_2 risk of g, which measures how well the function g is adapted to the training data, while

$$\lambda_n \int_0^1 |g^{(k)}(x)|^2 \, dx \qquad (20.2)$$

penalizes the roughness of g. As mentioned in Problem 20.1, the function which minimizes (20.1) satisfies

$$g^{(k)}(x) = 0 \text{ for } x < \min\{X_1, \ldots, X_n\} \text{ or } x > \max\{X_1, \ldots, X_n\},$$

therefore it doesn't matter if one replaces the penalty term (20.2) by $\lambda_n \int_{-\infty}^{\infty} |g^{(k)}(x)|^2 \, dx$.

If one minimizes (20.1) with respect to all (measurable) function, where the kth derivative is square integrable, it turns out that, for $k > 1$, the minimum is achieved by some function which is k times continuously differentiable. This is not true for $k = 1$. But in order to simplify the notation we will always denote the set of functions which contains the minima of (20.1) by $C^k(\mathcal{R})$. For $k > 1$, this is the set of all k times continuously differentiable functions $f : \mathcal{R} \to \mathcal{R}$, but, for $k = 1$, the exact definition of this set is more complicated.

It is not obvious that an estimate defined by minimizing (20.1) with respect to all k times differentiable functions is of any practical use. But as we will see in this section, it is indeed possible to compute such a function efficiently. More precisely, we will show in the sequel that there exists a spline function f of degree $2k - 1$ which satisfies

$$\frac{1}{n} \sum_{i=1}^{n} |f(X_i) - Y_i|^2 + \lambda_n \int_0^1 |f^{(k)}(x)|^2 \, dx$$

$$= \min_{g \in C^k(\mathcal{R})} \left\{ \frac{1}{n} \sum_{i=1}^{n} |g(X_i) - Y_i|^2 + \lambda_n \int_0^1 |g^{(k)}(x)|^2 \, dx \right\},$$

and which, furthermore, can be computed efficiently. Here $C^k(\mathcal{R})$ denotes the set of all k times differentiable functions $g : \mathcal{R} \to \mathcal{R}$.

Set $M = 2k - 1$ and choose a knot vector $\{u_j\}_{j=-M}^{K+M}$ such that

$$\{u_1, \ldots, u_{K-1}\} = \{X_1, \ldots, X_n\}$$

and

$$u_{-M} < \cdots < u_0 < 0 < u_1 < \cdots < 1 < u_K < \cdots < u_{K+M}.$$

If $X_1, \ldots, X_n$ are all distinct, then $K - 1 = n$ and $u_1, \ldots, u_{K-1}$ is a permutation of $X_1, \ldots, X_n$.

The following optimality property of spline interpolants will be proven in Section 20.2:

Lemma 20.1. *Let* $N \in \mathcal{N}$, $0 < z_1 < \cdots < z_N < 1$, *and* $k \leq N$. *Let* $g : \mathcal{R} \to \mathcal{R}$ *be an arbitrary* k *times differentiable function. Set* $M = 2k - 1$ *and* $K = N + 1$. *Define the knot vector* $u = \{u_j\}_{j=-M}^{K+M}$ *of the spline space* $S_{u,M}$ *by setting*

$$u_j = z_j \quad (j = 1, \ldots, N)$$

and by choosing

$$u_{-M} < u_{-M+1} < \cdots < u_0 < 0, \ 1 < u_K < u_{K+1} < \cdots < u_{K+M}$$

arbitrary.

Then there exists a spline function $f \in S_{u,M}$ *such that*

$$f(z_i) = g(z_i) \quad (i = 1, \ldots, N) \tag{20.3}$$

and

$$\int_0^1 |f^{(k)}(z)|^2 dz \leq \int_0^1 |g^{(k)}(z)|^2 dz. \tag{20.4}$$

According to Lemma 20.1 for each $g \in C^k(\mathcal{R})$ there exists $\bar{g} \in S_{u,M}$ such that

$$g(X_i) = \bar{g}(X_i) \quad (i = 1, \ldots, n)$$

and

$$\int_0^1 |\bar{g}^{(k)}(x)|^2 \, dx \leq \int_0^1 |g^{(k)}(x)|^2 \, dx.$$

Because of the first condition,

$$\frac{1}{n} \sum_{i=1}^{n} |\bar{g}(X_i) - Y_i|^2 = \frac{1}{n} \sum_{i=1}^{n} |g(X_i) - Y_i|^2,$$

and thus

$$\frac{1}{n}\sum_{i=1}^{n}|\bar{g}(X_i)-Y_i|^2+\lambda_n\int_0^1|\bar{g}^{(k)}(x)|^2\,dx$$

$$\leq\frac{1}{n}\sum_{i=1}^{n}|g(X_i)-Y_i|^2+\lambda_n\int_0^1|g^{(k)}(x)|^2\,dx.$$

This proves

$$\min_{g\in C^k(\mathcal{R})}\left\{\frac{1}{n}\sum_{i=1}^{n}|g(X_i)-Y_i|^2+\lambda_n\int_0^1|g^{(k)}(x)|^2\,dx\right\}$$

$$=\min_{g\in S_{u,M}}\left\{\frac{1}{n}\sum_{i=1}^{n}|g(X_i)-Y_i|^2+\lambda_n\int_0^1|g^{(k)}(x)|^2\,dx\right\},$$

therefore, it suffices to minimize the penalized empirical L_2 risk only over the finite-dimensional spline space $S_{u,M}$.

Let $g\in S_{u,M}$ be arbitrary. Then g can be written as

$$g=\sum_{j=-M}^{K-1}a_j\cdot B_{j,M,u},$$

and by Lemma 14.6, $g^{(k)}$ can be written as

$$g^{(k)}=\sum_{j=-(k-1)}^{K-1}b_j\cdot B_{j,k-1,u},$$

where one can compute the $b_j's$ from the $a_j's$ by repeatedly taking differences and multiplying with some constants depending on the knot sequence u. Hence $b=\{b_j\}$ is a linear transformation of $a=\{a_j\}$ and we have, for some $(K+k-1)\times(K+2k-1)$ matrix D,

$$b=Da.$$

It follows that

$$\int_0^1|g^{(k)}(x)|^2\,dx\;=\;\sum_{i,j=-(k-1)}^{K-1}b_ib_j\int_0^1 B_{i,k-1,u}(x)\cdot B_{j,k-1,u}(x)\,dx$$

$$=\;b^TCb$$

$$=\;a^TD^TCDa,$$

where

$$C=\left(\int_0^1 B_{i,k-1,u}(x)\cdot B_{j,k-1,u}(x)\,dx\right)_{i,j=-(k-1),\dots,K-1}.$$

In addition,

$$\frac{1}{n}\sum_{i=1}^{n}|g(X_i) - Y_i|^2 = \frac{1}{n}\|Ba - Y\|_2^2 = \frac{1}{n}\left(a^T B^T - Y^T\right)(Ba - Y),$$

where

$$B = (B_{j,2k-1,u}(X_i))_{i=1,\ldots,n;\,j=-(2k-1),\ldots,K-1} \quad \text{and} \quad Y = (Y_1,\ldots,Y_n)^T.$$

Hence,

$$\frac{1}{n}\sum_{i=1}^{n}|g(X_i) - Y_i|^2 + \lambda_n \int_0^1 |g^{(k)}(x)|^2\,dx$$

$$= \frac{1}{n}\left(a^T B^T - Y^T\right)(Ba - Y) + \lambda_n a^T D^T C D a \qquad (20.5)$$

and it suffices to show that there always exists an $a^* \in \mathcal{R}^{K+2k-1}$ which minimizes the right-hand side of the last equation.

The right-hand side of (20.5) is equal to

$$\frac{1}{n}a^T B^T Ba - 2\frac{1}{n}Y^T Ba + \frac{1}{n}Y^T Y + \lambda_n a^T D^T C D a$$

$$= a^T\left(\frac{1}{n}B^T B + \lambda_n D^T C D\right)a - 2\frac{1}{n}Y^T Ba + \frac{1}{n}Y^T Y.$$

Next we show that the matrix

$$A = \frac{1}{n}B^T B + \lambda_n D^T C D$$

is positive definite. Indeed,

$$a^T A a = \frac{1}{n}\|Ba\|_2^2 + \lambda_n (Da)^T C D a \geq 0$$

because

$$b^T C b = \int_0^1 \left|\sum_{j=-(k-1)}^{K-1} b_j B_{j,k-1,u}(x)\right|^2\,dx \geq 0.$$

Furthermore, $a^T A a = 0$ implies $Ba = 0$ and $a^T D^T C D a = 0$ from which one concludes that $g = \sum_{j=-(2k-1)}^{K-1} a_j B_{j,2k-1,u}$ satisfies

$$g(X_i) = 0 \quad (i = 1,\ldots,n)$$

and

$$g^{(k)}(x) = 0 \quad \text{for all } x \in [0,1].$$

The last condition implies that g is a polynomial of degree $k - 1$ (or less). If we assume $|\{X_1,\ldots,X_n\}| \geq k$, then we get $g = 0$ and hence $a = 0$.

In particular, A is regular, i.e., A^{-1} exists. Using this we get

$$a^T A a - 2\frac{1}{n} Y^T B a + \frac{1}{n} Y^T Y$$

$$= \left(a - A^{-1}\frac{1}{n} B^T Y \right)^T A \left(a - A^{-1}\frac{1}{n} B^T Y \right) + \frac{1}{n} Y^T Y$$

$$- \frac{1}{n^2} Y^T B A^{-1} B^T Y.$$

The last two terms do not depend on a, and because A is positive definite the first term is minimal for

$$a = A^{-1}\frac{1}{n} B^T Y.$$

This proves that the right-hand side of (20.5) is minimized by the unique solution of the linear system of equations

$$\left(\frac{1}{n} B^T B + \lambda_n D^T C D \right) a = \frac{1}{n} B^T Y. \tag{20.6}$$

We summarize our result in the next theorem.

Theorem 20.1. *For any fixed data* $(X_1, Y_1), \ldots, (X_n, Y_n)$ *such that* $X_1, \ldots, X_n \in (0, 1)$, $|\{X_1, \ldots, X_n\}| \geq k$, *and* $Y_1, \ldots, Y_n \in \mathcal{R}$, *the spline function*

$$g = \sum_{j=-(2k-1)}^{K-1} a_j B_{j,2k-1,u}$$

with knot vector $u = \{u_j\}$, *which satisfies*

$$u_{-M} < \cdots < u_0 < 0, \ \{u_1, \ldots, u_{K-1}\} = \{X_1, \ldots, X_n\}$$

and

$$1 < u_K < \cdots < u_{K+M},$$

and coefficient vector a *which is the solution of (20.6), minimizes*

$$\frac{1}{n} \sum_{i=1}^{n} |f(X_i) - Y_i|^2 + \lambda_n \int_0^1 |f^{(k)}(x)|^2 \, dx$$

with respect to all $f \in C^k(\mathcal{R})$. *The solution of (20.6) is unique.*

In order to compute the estimate one has to solve the system of linear equations (20.6). In particular, one has to compute the matrices B, C, and D. The calculation of

$$B = (B_{j,2k-1,u}(X_i))_{i=1,\ldots,n; j=-(2k-1),\ldots,K-1}$$

requires the evaluation of the B-splines $B_{j,2k-1,u}$ at the points X_i, which can be done via the recursive evaluation algorithm described in Chapter 14.

The matrix $D = D_k^{(M,K)}$ transforms the coefficients of a linear combination of B-splines into the coefficients of its kth derivative. By Lemma 14.6 one gets that, for $k = 1$, the matrix $D_1^{(M,K)}$ is given by

$$\begin{pmatrix} \dfrac{-M}{u_1 - u_{-M+1}} & \dfrac{M}{u_1 - u_{-M+1}} & 0 & \cdots & 0 \\[2mm] 0 & \dfrac{-M}{u_2 - u_{-M+2}} & \dfrac{M}{u_2 - u_{-M+2}} & \cdots & 0 \\[2mm] 0 & 0 & \dfrac{M}{u_3 - u_{-M+3}} & \cdots & 0 \\[2mm] & & & \cdots & \\[2mm] 0 & 0 & 0 & \cdots & 0 \\[2mm] 0 & 0 & 0 & \cdots & \dfrac{M}{u_{K-1+M} - u_{K-1}} \end{pmatrix}.$$

For general k one has

$$D_k^{(M,K)} = D_1^{(M-(k-1),K)} \cdot \ldots \cdot D_1^{(M-1,K)} \cdot D_1^{(M,K)}.$$

The matrix

$$C = \left(\int_0^1 B_{i,k-1,u}(x) B_{j,k-1,u}(x)\, dx \right)_{i,j=-(k-1),\ldots,K-1}$$

consists of inner products of B-splines. Now

$$\int_0^1 B_{i,k-1,u}(x) B_{j,k-1,u}(x)\, dx = \int_0^{u_1} B_{i,k-1,u}(x) \cdot B_{j,k-1,u}(x)\, dx$$

$$+ \sum_{l=1}^{K-2} \int_{u_l}^{u_{l+1}} B_{i,k-1,u}(x) B_{j,k-1,u}(x)\, dx$$

$$+ \int_{u_{K-1}}^1 B_{i,k-1,u}(x) \cdot B_{j,k-1,u}(x)\, dx$$

is a sum of K integrals of polynomials of degree $2(k-1)$. Using a linear transformation it suffices to compute K integrals of polynomials of degree $2(k-1)$ over $[-1,1]$, which can be done by Gauss quadrature. Recall that by the Gaussian quadrature there exist points

$$-1 < z_1 < z_2 < \cdots < z_k < 1$$

(symmetric about zero) and positive weights $w_1, \ldots, w_k$ such that

$$\int_{-1}^1 p(x)\, dx = \sum_{j=1}^k w_j p(z_j)$$

for every polynomial of degree $2k - 1$ (or less). Table 20.1 lists points $z_1, \ldots, z_k$ and weights $w_1, \ldots, w_k$ for various values of k.

In Theorem 20.1 we ignored the case $|\{X_1, \ldots, X_n\}| < k$. This case is rather trivial, because if $|\{X_1, \ldots, X_n\}| \le k$, then one can find a polynomial p of degree $k - 1$ (or less) which interpolates at each X_i the average of those

Table 20.1. Parameters for Gauss quadrature.

k	w_i	z_i
1	$w_1 = 2$	$z_1 = 0$
2	$w_1 = w_2 = 1$	$z_2 = -z_1 = 0.57773502...$
3	$w_1 = w_3 = 5/9$	$z_3 = -z_1 = 0.7745966692...$
	$w_2 = 8/9$	$z_3 = -z_2 = 0$

Y_j for which $X_j = X_i$. It is easy to see that such a polynomial minimizes

$$\frac{1}{n} \sum_{i=1}^{n} |g(X_i) - Y_i|^2$$

with respect to all functions $g : \mathcal{R} \to \mathcal{R}$ (cf. Problem 2.1). In addition, $p^{(k)}(x) = 0$ for all $x \in \mathcal{R}$, which implies that this polynomial also minimizes

$$\frac{1}{n} \sum_{i=1}^{n} |g(X_i) - Y_i|^2 + \lambda_n \int_0^1 |g^{(k)}(x)|^2 \, dx.$$

The polynomial is not uniquely determined if $|\{X_1, \ldots, X_n\}| < k$, hence Theorem 20.3 does not hold in this case.

20.2 Proof of Lemma 20.1

Let $t_{-M} \leq t_{-M+1} \leq \cdots \leq t_{K-1}$. Denote the number of occurrences of t_j in the sequence $t_{j+1}, \ldots, t_{K-1}$ by $\#t_j$, i.e., set

$$\#t_j = |\{i > j : t_i = t_j\}| \, .$$

In order to prove Lemma 20.1 we need results concerning the following interpolation problem: Given $\{t_j\}_{j=-M}^{K-1}$, $f_{-M}, \ldots, f_{K-1} \in \mathcal{R}$, and a spline space $S_{u,M}$, find an unique spline function $f \in S_{u,M}$ such that

$$\frac{\partial^{\#t_j} f(t_j)}{\partial x^{\#t_j}} = f_j \quad (j = -M, \ldots, K - 1). \tag{20.7}$$

If $t_{-M} < \cdots < t_{K-1}$, then $\#t_j = 0$ for all j and (20.7) is equivalent to

$$f(t_j) = f_j \quad (j = -M, \ldots, K - 1),$$

i.e., we are looking for functions which interpolate the points

$$(t_j, f_j) \quad (j = -M, \ldots, K - 1).$$

By using t_j with $\#t_j > 0$ one can also specify values of derivatives of f at t_j.

   If we represent f by a linear combination of B-splines, then (20.7) is equivalent to a system of linear equations for the coefficients of this linear combination. The vector space dimension of $S_{u,M}$ is equal to $K+M$, hence there are as many equations as free variables. Our next theorem describes under which condition this system of linear equations has a unique solution.

Theorem 20.2. (SCHOENBERG–WHITNEY THEOREM). *Let* $M \in \mathcal{N}_0$, $K \in \mathcal{N}$, $u_{-M} < \cdots < u_{K+M}$, *and* $t_{-M} \le \cdots \le t_{K-1}$. *Assume* $\#t_j \le M$ *and, if* $t_j = u_i$ *for some* i, $\#t_j \le M - 1$ $(j = -M, \ldots, K-1)$.

Then the interpolation problem (20.7) has an unique solution for any $f_{-M}, \ldots, f_{K-1} \in \mathcal{R}$ *if and only if*

$$B_{j,M,u}(t_j) > 0 \quad \text{for all } j = -M, \ldots, K-1.$$

For $M > 0$ the condition $B_{j,M,u}(t_j) > 0$ is equivalent to $u_j < t_j < u_{j+M+1}$ (cf. Lemma 14.2).

PROOF. The proof is left to the reader (cf. Problems 20.2 and 20.3). $\square$

Next we show the existence of a spline interpolant. In Lemma 20.3 we will use condition (20.9) to prove that this spline interpolant minimizes $\int_a^b |f^{(k)}(x)|^2 dx$.

Lemma 20.2. *Let* $n \in \mathcal{N}$, $a < x_1 < \cdots < x_n < b$, *and* $k \le n$. *Set* $M = 2k - 1$ *and* $K = n + 1$. *Define the knot vector* $u = \{u_j\}_{j=-M}^{K+M}$ *of the spline space* $S_{u,M}$ *by setting*

$$u_j = x_j \quad (j = 1, \ldots, n)$$

and by choosing arbitrary

$$u_{-M} < u_{-M+1} < \cdots < u_0 < a, \ b < u_K < u_{K+1} < \cdots < u_{K+M}.$$

Let $f_j \in \mathcal{R}$ $(j = 1, \ldots, n)$.

Then there exists an unique spline function $f \in S_{u,M}$ *such that*

$$f(x_i) = f_i \quad (i = 1, \ldots, n) \tag{20.8}$$

and

$$f^{(l)}(a) = f^{(l)}(b) = 0 \quad (l = k, k+1, \ldots, 2k - 1 = M). \tag{20.9}$$

PROOF. Represent any function $f \in S_{u,M}$ by its B-spline coefficients. By Lemma 14.6, (20.8) and (20.9) are equivalent to a nonhomogeneous linear equation system for the B-spline coefficients. The dimension of the spline space $S_{u,M}$ is equal to $K + M = n + 1 + (2k - 1) = n + 2k$, hence the number of rows of this equation system is equal to the number of columns. A unique solution of such a nonhomogeneous linear equation system exists if and only if the corresponding homogeneous equation system doesn't have a nontrivial solution. Therefore it suffices to show: If $f \in S_{u,M}$ satisfies (20.8) and (20.9) with $f_i = 0$ $(i = 1, \ldots, n)$, then $f = 0$.

Let $f \in S_{u,M}$ be such that (20.8) and (20.9) hold with $f_i = 0$ $(i = 1, \ldots, n)$. By the theorem of Rolle,

$$f(x_i) = 0 = f(x_{i+1})$$

implies that there exists $t_i \in (x_i, x_{i+1})$ such that

$$f'(t_i) = 0.$$

We can show by induction that for any $l \in \{0, \ldots, k\}$ there exists $t_j^{(l)} \in (x_j, x_{j+l})$ such that $f^{(l)} \in S_{u,M-l}$ satisfies

$$f^{(l)}(t_j^{(l)}) = 0 \quad (j = 1, \ldots, n - l). \tag{20.10}$$

Hence, $f^{(k)} \in S_{u,M-k} = S_{u,k-1}$ satisfies

$$f^{(k)}(t_j^{(k)}) = 0 \quad (j = 1, \ldots, n - k)$$

and

$$f^{(l)}(a) = f^{(l)}(b) = 0 \quad (l = k, \ldots, 2k - 1).$$

Here $t_j^{(k)} \in (x_j, x_{j+k}) = (u_j, u_{j+k})$ and thus the assumptions of the Schoenberg–Whitney theorem are fulfilled. The Schoenberg–Whitney theorem implies

$$f^{(k)} = 0.$$

From this and (20.10) one concludes successively

$$f^{(l)} = 0 \quad (l = k - 1, \ldots, 0).$$

$$\square$$

Our next lemma implies that the spline interpolant of Lemma 20.2 minimizes $\int_a^b |g^{(k)}(x)|^2 dx$ with respect to all k times differentiable functions. This proves Lemma 20.1.

Lemma 20.3. *Let g be an arbitrary k times differentiable function such that*

$$g(x_i) = f_i \quad (i = 1, \ldots, n).$$

Let f be the spline function of Lemma 20.2 satisfying

$$f(x_i) = f_i \quad (i = 1, \ldots, n)$$

and

$$f^{(l)}(a) = f^{(l)}(b) = 0 \quad (l = k, k + 1, \ldots, 2k - 1 = M).$$

Then

$$\int_a^b |f^{(k)}(x)|^2 \, dx \le \int_a^b |g^{(k)}(x)|^2 \, dx.$$

PROOF.

$$\int_a^b |g^{(k)}(x)|^2 \, dx = \int_a^b |f^{(k)}(x)|^2 \, dx + \int_a^b |g^{(k)}(x) - f^{(k)}(x)|^2 \, dx$$

$$+2\int_a^b f^{(k)}(x)(g^{(k)}(x) - f^{(k)}(x))\,dx.$$

Integration by parts yields

$$\int_a^b f^{(k)}(x)(g^{(k)}(x) - f^{(k)}(x))\,dx$$

$$= \left[f^{(k)}(x)(g^{(k-1)}(x) - f^{(k-1)}(x))\right]_{x=a}^{b}$$

$$-\int_a^b f^{(k+1)}(x)(g^{(k-1)}(x) - f^{(k-1)}(x))\,dx$$

$$= -\int_a^b f^{(k+1)}(x)(g^{(k-1)}(x) - f^{(k-1)}(x))\,dx$$

$$\text{(because } f^{(k)}(a) = f^{(k)}(b) = 0)$$

$$= \ldots$$

$$= (-1)^{k-1}\int_a^b f^{(2k-1)}(x)(g'(x) - f'(x))\,dx$$

$$= (-1)^{k-1}\left(f^{(2k-1)}(a)\int_a^{x_1} (g'(x) - f'(x))\,dx\right.$$

$$+ \sum_{k=1}^{n-1} f^{(2k-1)}(x_k)\int_{x_k}^{x_{k+1}} (g'(x) - f'(x))\,dx$$

$$\left.+ f^{(2k-1)}(b)\int_{x_n}^b (g'(x) - f'(x))\,dx\right)$$

$$\text{(because of } f^{(2k-1)} \text{ is piecewise constant)}$$

$$= (-1)^{k-1}\left(0 \cdot \int_a^{x_1} (g'(x) - f'(x))\,dx + \sum_{k=1}^{n-1} f^{(2k-1)}(x_k) \cdot 0\right.$$

$$\left.+ 0 \cdot \int_{x_n}^b (g'(x) - f'(x))\,dx\right)$$

$$\text{(observe } f(x_k) = g(x_k) \quad (k = 1, \ldots, n))$$

$$= 0.$$

Hence,

$$\int_a^b |g^{(k)}(x)|^2\,dx \;=\; \int_a^b |f^{(k)}(x)|^2\,dx + \int_a^b |g^{(k)}(x) - f^{(k)}(x)|^2\,dx$$

$$\geq \int_a^b |f^{(k)}(x)|^2\,dx.$$

$\square$

20.3 Consistency

In order to simplify the notation we will assume in the sequel that $X \in (0,1)$ a.s. Problem 20.6 shows that after a minor modification of the estimate this assumption is no longer necessary and the resulting estimate is universally consistent.

Let $k \in \mathcal{N}$ and let the estimate $\tilde{m}_n$ be defined via

$$\tilde{m}_n(\cdot) = \arg\min_{f \in C^k(\mathcal{R})} \left\{ \frac{1}{n} \sum_{i=1}^{n} |f(X_i) - Y_i|^2 + \lambda_n J_k^2(f) \right\}, \qquad (20.11)$$

where $\lambda_n > 0$,

$$J_k^2(f) = \int_0^1 |f^{(k)}(x)|^2 \, dx$$

and $C^k(\mathcal{R})$ is the set of all k times differentiable functions $f : \mathcal{R} \to \mathcal{R}$. In order to show consistency of the estimate $\tilde{m}_n$, we will use the following error decomposition:

$$\int |\tilde{m}_n(x) - m(x)|^2 \mu(dx)$$

$$= \mathbf{E}\{|\tilde{m}_n(X) - Y|^2 | D_n\} - \mathbf{E}\{|m(X) - Y|^2\}$$

$$= \mathbf{E}\{|\tilde{m}_n(X) - Y|^2 | D_n\} - \frac{1}{n} \sum_{i=1}^{n} |\tilde{m}_n(X_i) - Y_i|^2$$

$$+ \frac{1}{n} \sum_{i=1}^{n} |\tilde{m}_n(X_i) - Y_i|^2 - \frac{1}{n} \sum_{i=1}^{n} |m(X_i) - Y_i|^2$$

$$+ \frac{1}{n} \sum_{i=1}^{n} |m(X_i) - Y_i|^2 - \mathbf{E}\{|m(X) - Y|^2\}$$

$$=: T_{1,n} + T_{2,n} + T_{3,n}.$$

By the strong law of large numbers,

$$T_{3,n} = \frac{1}{n} \sum_{i=1}^{n} |m(X_i) - Y_i|^2 - \mathbf{E}\{|m(X) - Y|^2\} \to 0 \quad (n \to \infty) \quad a.s.$$

Furthermore, if $m \in C^k(\mathcal{R})$, then the definition of $\tilde{m}_n$ implies

$$T_{2,n} \;\leq\; \frac{1}{n} \sum_{i=1}^{n} |\tilde{m}_n(X_i) - Y_i|^2 + \lambda_n J_k^2(\tilde{m}_n) - \frac{1}{n} \sum_{i=1}^{n} |m(X_i) - Y_i|^2$$

$$\leq\; \frac{1}{n} \sum_{i=1}^{n} |m(X_i) - Y_i|^2 + \lambda_n J_k^2(m) - \frac{1}{n} \sum_{i=1}^{n} |m(X_i) - Y_i|^2$$

$$= \lambda_n J_k^2(m).$$

Hence, if $m \in C^k(\mathcal{R})$ and if we choose λ_n such that

$$\lambda_n \to 0 \quad (n \to \infty),$$

then

$$\limsup_{n \to \infty} T_{2,n} \le 0.$$

Here the assumption $m \in C^k(\mathcal{R})$ can be avoided by approximating m by a smooth function (which we will do in the proof of Theorem 20.3 below).

Thus, in order to obtain strong consistency of the estimate (20.11), we basically have to show

$$T_{1,n} = \mathbf{E}\{|\tilde{m}_n(X) - Y|^2 | D_n\} - \frac{1}{n} \sum_{i=1}^n |\tilde{m}_n(X_i) - Y_i|^2 \to 0 \quad (n \to \infty) \quad a.s.$$
$$(20.12)$$

By definition of the estimate,

$$\frac{1}{n} \sum_{i=1}^n |\tilde{m}_n(X_i) - Y_i|^2 + \lambda_n J_k^2(\tilde{m}_n)$$

$$\le \frac{1}{n} \sum_{i=1}^n |0 - Y_i|^2 + \lambda_n \cdot 0$$

$$= \frac{1}{n} \sum_{i=1}^n |Y_i|^2 \to \mathbf{E}\{Y^2\} \quad (n \to \infty) \quad a.s.,$$

which implies that, with probability one,

$$J_k^2(\tilde{m}_n) \le \frac{2}{\lambda_n} \mathbf{E}\{Y^2\} \tag{20.13}$$

for n sufficiently large. Set

$$\mathcal{F}_n := \left\{ f \in C^k(\mathcal{R}) \ : \ J_k^2(f) \le \frac{2}{\lambda_n} \mathbf{E}\{Y^2\} \right\}.$$

Then (with probability one) $\tilde{m}_n \in \mathcal{F}_n$ for n sufficiently large and therefore (20.12) follows from

$$\sup_{f \in \mathcal{F}_n} \left| \mathbf{E}\{|f(X) - Y|^2\} - \frac{1}{n} \sum_{i=1}^n |f(X_i) - Y_i|^2 \right| \to 0 \quad (n \to \infty) \quad a.s.$$
$$(20.14)$$

To show (20.14) we will use the results of Chapter 9. Recall that these results require that the random variables $|f(X) - Y|^2$ ($f \in \mathcal{F}_n$) be bounded uniformly by some constant, which may depend on n. To ensure this, we will truncate our estimate, i.e., we will set

$$m_n(x) = T_{\beta_n} \tilde{m}_n(x) = \begin{cases} \beta_n & \text{if} \quad \tilde{m}_n(x) > \beta_n, \\ \tilde{m}_n(x) & \text{if} \quad -\beta_n \leq \tilde{m}_n(x) \leq \beta_n, \\ -\beta_n & \text{if} \quad \tilde{m}_n(x) < -\beta_n, \end{cases} \qquad (20.15)$$

where $\beta_n > 0$, $\beta_n \to \infty$ $(n \to \infty)$. All that we need then is an upper bound on the covering number of

$$\left\{ T_{\beta_n} f \; : \; f \in C^k(\mathcal{R}) \text{ and } J_k^2(f) \leq \frac{2}{\lambda_n} \mathbf{E}\{Y^2\} \right\}.$$

This bound will be given in the next lemma.

Lemma 20.4. *Let $L, c > 0$ and set*

$$\mathcal{F} = \left\{ T_L f \; : \; f \in C^k(\mathcal{R}) \text{ and } J_k^2(f) \leq c \right\}.$$

Then, for any $1 \leq p < \infty$, $0 < \delta < 4L$, and $x_1, \ldots, x_n \in [0,1]$,

$$\mathcal{N}_p\left(\delta, \mathcal{F}, x_1^n\right) \leq \left(\frac{9 \cdot e \cdot 4^p L^p n}{\delta^p} \right)^{8(k+2) \cdot \left((\sqrt{c}/\delta)^{\frac{1}{k}} + 1 \right)}.$$

PROOF. In the first step of the proof we approximate the functions of $\mathcal{F}$ in the supremum norm by piecewise polynomials. In the second step we bound the covering number of these piecewise polynomials. Fix $g = T_L f \in \mathcal{F}$ where $f \in C^k(\mathcal{R})$, $J_k^2(f) \leq c$. Choose $\tilde{K}$ and

$$0 = u_0 < u_1 < \cdots < u_{\tilde{K}} = 1$$

such that

$$\int_{u_i}^{u_{i+1}} |f^{(k)}(x)|^2 \, dx = c \cdot \left(\frac{\delta}{2\sqrt{c}} \right)^{\frac{1}{k}} \qquad (i = 0, 1, \ldots, \tilde{K} - 2),$$

and

$$\int_{u_{\tilde{K}-1}}^{u_{\tilde{K}}} |f^{(k)}(x)|^2 \, dx \leq c \cdot \left(\frac{\delta}{2\sqrt{c}} \right)^{\frac{1}{k}}.$$

Then

$$c \geq \int_0^1 |f^{(k)}(x)|^2 \, dx = \sum_{i=0}^{\tilde{K}-1} \int_{u_i}^{u_{i+1}} |f^{(k)}(x)|^2 \, dx \geq (\tilde{K} - 1) \cdot c \cdot \left(\frac{\delta}{2\sqrt{c}} \right)^{\frac{1}{k}},$$

which implies

$$\tilde{K} \leq \left(\frac{2\sqrt{c}}{\delta} \right)^{\frac{1}{k}} + 1 \leq 2 \left(\frac{\sqrt{c}}{\delta} \right)^{\frac{1}{k}} + 1.$$

By a refinement of the partition $\{[u_0, u_1), \ldots, [u_{\tilde{K}-1}, u_{\tilde{K}}]\}$ one can construct points

$$0 = v_0 < v_1 < \cdots < v_K = 1$$

such that

$$\int_{v_i}^{v_{i+1}} |f^{(k)}(x)|^2 \, dx \le c \cdot \left(\frac{\delta}{2\sqrt{c}}\right)^{\frac{1}{k}} \qquad (i = 0, 1, \ldots, K-1),$$

$$|v_{i+1} - v_i| \le \left(\frac{\delta}{2\sqrt{c}}\right)^{\frac{1}{k}} \qquad (i = 0, 1, \ldots, K-1)$$

and

$$K \le 4 \left(\frac{\sqrt{c}}{\delta}\right)^{\frac{1}{k}} + 2.$$

Let p_i be the Taylor polynomial of degree $k-1$ of f about v_i. For $v_i \le x \le v_{i+1}$ one gets, by the Cauchy–Schwarz inequality,

$$
\begin{aligned}
|f(x) - p_i(x)|^2 &= \left|\frac{1}{(k-1)!} \int_{v_i}^{x} (t - v_i)^{k-1} f^{(k)}(t) \, dt\right|^2 \\
&\le \frac{1}{(k-1)!^2} \int_{v_i}^{x} (t - v_i)^{2k-2} \, dt \cdot \int_{v_i}^{x} |f^{(k)}(t)|^2 \, dt \\
&\le \frac{1}{(k-1)!^2} \frac{(v_{i+1} - v_i)^{2k-1}}{2k-1} \cdot \int_{v_i}^{v_{i+1}} |f^{(k)}(t)|^2 \, dt \\
&\le \frac{1}{(k-1)!^2 (2k-1)} \left(\frac{\delta}{2\sqrt{c}}\right)^{\frac{2k-1}{k}} \cdot c \cdot \left(\frac{\delta}{2\sqrt{c}}\right)^{\frac{1}{k}} \\
&\le \frac{\delta^2}{4}.
\end{aligned}
$$

Let $\mathcal{G}_{k-1}$ be the set of all polynomials of degree less than or equal to $k-1$ and let Π be the family of all partitions of $[0,1]$ into $K \le 4\left(\sqrt{c}/\delta\right)^{1/k} + 2$ intervals. For $\pi \in \Pi$ let $\mathcal{G}_{k-1} \circ \pi$ be the set of all piecewise polynomials of degree less than or equal to $k-1$ with respect to π and let $\mathcal{G}_{k-1} \circ \Pi$ be the union of all sets $\mathcal{G}_{k-1} \circ \pi$ $(\pi \in \Pi)$. We have shown that for each $g \in \mathcal{F}$ there exists $p \in \mathcal{G}_{k-1} \circ \Pi$ such that

$$\sup_{x \in [0,1]} |g(x) - T_L p(x)| \le \frac{\delta}{2}.$$

This implies

$$\mathcal{N}_p\left(\delta, \mathcal{F}, x_1^n\right) \le \mathcal{N}_p\left(\frac{\delta}{2}, T_L \mathcal{G}_{k-1} \circ \Pi, x_1^n\right). \tag{20.16}$$

To bound the last covering number, we use Lemma 13.1, (13.4) (together with the observation that the partition number no longer increases as soon as the number of intervals is greater than the number of points), and the usual bounds for covering numbers of linear vector spaces (Lemma 9.2 and

Theorems 9.4 and 9.5). This yields

$$\mathcal{N}_p\left(\frac{\delta}{2}, T_L\mathcal{G}_{k-1}\circ\Pi, x_1^n\right)$$

$$\leq (2n)^{4(\sqrt{c}/\delta)^{\frac{1}{k}}+2}\left\{\sup_{z_1,\ldots,z_l\in x_1^n, l\leq n}\mathcal{N}_p\left(\frac{\delta}{2}, T_L\mathcal{G}_{k-1}, z_1^l\right)\right\}^{4(\sqrt{c}/\delta)^{\frac{1}{k}}+2}$$

$$\leq (2n)^{4(\sqrt{c}/\delta)^{\frac{1}{k}}+2}\cdot\left(3\cdot\left(\frac{3e(2L)^p}{(\delta/2)^p}\right)^{2(k+1)}\right)^{4(\sqrt{c}/\delta)^{\frac{1}{k}}+2}.$$

The assertion follows from this and (20.16).   □

Using this lemma and the results of Chapter 9 we next show an auxiliary result needed to prove the consistency of m_n.

Lemma 20.5. *Let $k\in\mathcal{N}$ and for $n\in\mathcal{N}$ choose $\beta_n,\lambda_n>0$, such that*

$$\beta_n\to\infty\quad(n\to\infty),\tag{20.17}$$

$$\frac{\beta_n^4}{n^{1-\delta}}\to 0\quad(n\to\infty)\tag{20.18}$$

for some $0<\delta<1$,

$$\lambda_n\to 0\quad(n\to\infty)\tag{20.19}$$

and

$$\lambda_n\frac{1}{\beta_n^2}\left(\frac{n}{\beta_n^4\log(n)}\right)^{2k}\to\infty\quad(n\to\infty).\tag{20.20}$$

Let the estimate m_n be defined by (20.11) and (20.15). Then

$$\mathbf{E}\{|m_n(X)-Y|^2|D_n\}-\frac{1}{n}\sum_{i=1}^{n}|m_n(X_i)-Y_i|^2\to 0\quad(n\to\infty)\quad a.s.$$

for every distribution of (X,Y) with $X\in(0,1)$ a.s. and $|Y|$ bounded a.s.

PROOF. Without loss of generality we assume $|Y|\leq L\leq\beta_n$ a.s. By definition of the estimate and the strong law of large numbers

$$\frac{1}{n}\sum_{i=1}^{n}|\tilde{m}_n(X_i)-Y_i|^2+\lambda_n J_k^2(\tilde{m}_n)\leq\frac{1}{n}\sum_{i=1}^{n}|0-Y_i|^2+\lambda_n\cdot 0\to\mathbf{E}Y^2\quad(n\to\infty)$$

a.s., which implies that with probability one we have, for n sufficiently large,

$$m_n\in\mathcal{F}_n=\left\{T_{\beta_n}f\ :\ f\in C^k(\mathcal{R})\text{ and }J_k^2(f)\leq\frac{2\mathbf{E}Y^2}{\lambda_n}\right\}.$$

Hence it suffices to show

$$\sup_{g \in \mathcal{G}_n} \left| \mathbf{E}\{g(X,Y)\} - \frac{1}{n}\sum_{i=1}^{n} g(X_i, Y_i) \right| \to 0 \quad (n \to \infty) \quad a.s., \qquad (20.21)$$

where $\mathcal{G}_n = \left\{ g : \mathcal{R}^d \times \mathcal{R} \to \mathcal{R} \; : \; g(x,y) = |f(x) - T_L y|^2 \text{ for some } f \in \mathcal{F}_n \right\}$.
If $g_j(x,y) = |f_j(x) - T_L y|^2$ $((x,y) \in \mathcal{R}^d \times \mathcal{R})$ for some function f_j bounded in absolute value by β_n $(j = 1, 2)$, then

$$\frac{1}{n}\sum_{i=1}^{n} |g_1(X_i, Y_i) - g_2(X_i, Y_i)| \le 4\beta_n \frac{1}{n}\sum_{i=1}^{n} |f_1(X_i) - f_2(X_i)|,$$

which implies

$$\mathcal{N}_1\left(\frac{\epsilon}{8}, \mathcal{G}_n, (X,Y)_1^n\right) \le \mathcal{N}_1\left(\frac{\epsilon}{32\beta_n}, \mathcal{F}_n, X_1^n\right).$$

Using this, Theorem 9.1, and Lemma 20.4, one gets, for $t > 0$ arbitrary,

$$\mathbf{P}\left\{ \sup_{g \in \mathcal{G}_n} \left| \mathbf{E}\{g(X,Y)\} - \frac{1}{n}\sum_{i=1}^{n} g(X_i, Y_i) \right| > t \right\}$$

$$\le 8 \left(\frac{c_1 \beta_n n}{\frac{\epsilon}{32\beta_n}} \right)^{c_2\left(\frac{\sqrt{2 \cdot \mathbf{E}Y^2}32\beta_n}{\sqrt{\lambda_n}\epsilon}\right)^{\frac{1}{k}} + c_3} \exp\left(-\frac{n\epsilon^2}{128(4\beta_n^2)^2} \right)$$

$$= 8\exp\left(\left(c_2 \left(\frac{\sqrt{2 \cdot \mathbf{E}Y^2}32}{\epsilon} \frac{\beta_n}{\sqrt{\lambda_n}} \right)^{\frac{1}{k}} + c_3 \right) \cdot \log\left(\frac{32 c_1 \beta_n^2 n}{\epsilon} \right) \right.$$

$$\left. - \frac{n\epsilon^2}{2048\beta_n^4} \right)$$

$$\le 8\exp\left(-\frac{1}{2} \cdot \frac{n\epsilon^2}{2048\beta_n^4} \right)$$

for n sufficiently large, where we have used that (20.18) and (20.20) imply

$$\frac{(\beta_n/\sqrt{\lambda_n})^{\frac{1}{k}} \log(\beta_n^2 n)}{n/\beta_n^4} = \frac{\beta_n^4 \log(\beta_n^2 n)}{n} \cdot \left(\frac{\beta_n}{\sqrt{\lambda_n}} \right)^{\frac{1}{k}}$$

$$= \left(\left(\frac{\beta_n^2}{\lambda_n} \right) \cdot \left(\frac{\beta_n^4 \log(\beta_n^2 n)}{n} \right)^{2k} \right)^{\frac{1}{2k}} \to 0 \quad (n \to \infty).$$

From this and (20.18) one gets the assertion by an application of the Borel–Cantelli lemma. $\qquad \square$

We are now ready to formulate and prove our main result, which shows that m_n is consistent for all distributions of (X,Y) satisfying $X \in (0,1)$ a.s. and $\mathbf{E}Y^2 < \infty$.

Theorem 20.3. *Let $k \in \mathcal{N}$ and for $n \in \mathcal{N}$ choose $\beta_n, \lambda_n > 0$, such that (20.17)–(20.20) hold. Let the estimate m_n be defined by (20.11) and (20.15). Then*

$$\int |m_n(x) - m(x)|^2 \mu(dx) \to 0 \quad (n \to \infty) \quad a.s.$$

for every distribution of (X, Y) with $X \in (0, 1)$ a.s. and $\mathbf{E}Y^2 < \infty$.

PROOF. Let $L, \epsilon > 0$ be arbitrary, set $Y_L = T_L Y$ and $Y_{i,L} = T_L Y_i$ $(i = 1, \dots, n)$. Because of Corollary A.1 we can choose $g_\epsilon \in C^k(\mathcal{R})$ such that

$$\int |m(x) - g_\epsilon(x)|^2 \mu(dx) < \epsilon \quad \text{and} \quad J_k^2(g_\epsilon) < \infty.$$

We use the following error decomposition:

$$\int |m_n(x) - m(x)|^2 \mu(dx) = \mathbf{E}\{|m_n(X) - Y|^2 | D_n\} - \mathbf{E}\{|m(X) - Y|^2\}$$

$$= \mathbf{E}\{|m_n(X) - Y|^2 | D_n\} - (1 + \epsilon)\mathbf{E}\{|m_n(X) - Y_L|^2 | D_n\}$$

$$+ (1 + \epsilon)\left(\mathbf{E}\{|m_n(X) - Y_L|^2 | D_n\} - \frac{1}{n}\sum_{i=1}^{n} |m_n(X_i) - Y_{i,L}|^2\right)$$

$$+ (1 + \epsilon)\left(\frac{1}{n}\sum_{i=1}^{n} |m_n(X_i) - Y_{i,L}|^2 - \frac{1}{n}\sum_{i=1}^{n} |\tilde{m}_n(X_i) - Y_{i,L}|^2\right)$$

$$+ (1 + \epsilon)\frac{1}{n}\sum_{i=1}^{n} |\tilde{m}_n(X_i) - Y_{i,L}|^2 - (1 + \epsilon)^2 \frac{1}{n}\sum_{i=1}^{n} |\tilde{m}_n(X_i) - Y_i|^2$$

$$+ (1 + \epsilon)^2 \left(\frac{1}{n}\sum_{i=1}^{n} |\tilde{m}_n(X_i) - Y_i|^2 - \frac{1}{n}\sum_{i=1}^{n} |g_\epsilon(X_i) - Y_i|^2\right)$$

$$+ (1 + \epsilon)^2 \left(\frac{1}{n}\sum_{i=1}^{n} |g_\epsilon(X_i) - Y_i|^2 - \mathbf{E}\{|g_\epsilon(X) - Y|^2\}\right)$$

$$+ (1 + \epsilon)^2 \left(\mathbf{E}\{|g_\epsilon(X) - Y|^2\} - \mathbf{E}\{|m(X) - Y|^2\}\right)$$

$$+ ((1 + \epsilon)^2 - 1)\mathbf{E}\{|m(X) - Y|^2\}$$

$$= \sum_{j=1}^{8} T_{j,n}.$$

Because of $(a + b)^2 \leq (1 + \epsilon)a^2 + (1 + \frac{1}{\epsilon})b^2$ $(a, b > 0)$ and the strong law of large numbers we get

$$T_{1,n} = \mathbf{E}\{|(m_n(X) - Y_L) + (Y_L - Y)|^2 | D_n\}$$

$$-(1+\epsilon)\mathbf{E}\{|m_n(X) - Y_L|^2 | D_n\}$$

$$\leq \quad (1+\frac{1}{\epsilon})\mathbf{E}\{|Y - Y_L|^2\}$$

and

$$T_{4,n} \quad \leq \quad (1+\epsilon)(1+\frac{1}{\epsilon})\frac{1}{n}\sum_{i=1}^{n}|Y_i - Y_{i,L}|^2$$

$$\rightarrow \quad (1+\epsilon)(1+\frac{1}{\epsilon})\mathbf{E}\{|Y - Y_L|^2\} \quad (n \rightarrow \infty) \quad a.s.$$

By Lemma 20.5,

$$T_{2,n} \rightarrow 0 \quad (n \rightarrow \infty) \quad a.s.$$

Furthermore, if $x, y \in \mathcal{R}$ with $|y| \leq \beta_n$ and $z = T_{\beta_n}x$, then $|z - y| \leq |x - y|$, which implies

$$T_{3,n} \leq 0 \quad \text{for } n \text{ sufficiently large.}$$

It follows from the definition of the estimate and (20.19) that

$$T_{5,n} \leq (1+\epsilon)^2(\lambda_n J_k^2(g_\epsilon) - \lambda_n J_k^2(\tilde{m}_n)) \leq (1+\epsilon)^2\lambda_n J_k^2(g_\epsilon) \rightarrow 0$$

$(n \rightarrow \infty)$. By the strong law of large numbers,

$$T_{6,n} \rightarrow 0 \quad (n \rightarrow \infty) \quad a.s.$$

Finally,

$$T_{7,n} = (1+\epsilon)^2 \int |g_\epsilon(x) - m(x)|^2 \mu(dx) \leq (1+\epsilon)^2\epsilon.$$

Using this, one concludes

$$\limsup_{n \rightarrow \infty} \int |m_n(x) - m(x)|^2 \mu(dx)$$

$$\leq (2+\epsilon)(1+\frac{1}{\epsilon})\mathbf{E}\{|Y - Y_L|^2\} + (1+\epsilon)^2\epsilon + (2\epsilon + \epsilon^2)\mathbf{E}\{|m(X) - Y|^2\}$$

$a.s.$ With $L \rightarrow \infty$ and $\epsilon \rightarrow 0$, the result follows. $\square$

20.4 Multivariate Penalized Least Squares Estimates

In this section we briefly comment on how to extend the results of Section 20.1 to multivariate estimates. The exact definition of the estimate, the proof of its existence, and the derivation of a computation algorithm requires several techniques from functional analysis which are beyond the scope of this book. Therefore we will just summarize the results without proofs.

To define multivariate penalized least squares estimates we use the empirical L_2 risk

$$\frac{1}{n} \sum_{i=1}^{n} |g(X_i) - Y_i|^2$$

which measures how well the function $g : \mathcal{R}^d \to \mathcal{R}$ is adapted to the training data. The roughness of g is penalized by $\lambda_n \cdot J_k^2(g)$, where

$$J_k^2(g) \;=\; \sum_{i_1,\dots,i_k \in \{1,\dots,d\}} \int_{\mathcal{R}^d} \left| \frac{\partial^k g(x)}{\partial x_{i_1} \dots \partial x_{i_k}} \right|^2 dx$$

$$= \sum_{\alpha_1,\dots,\alpha_d \in \mathcal{N}_0,\, \alpha_1 + \cdots + \alpha_d = k} \frac{k!}{\alpha_1! \cdot \ldots \cdot \alpha_d!} \int_{\mathcal{R}^d} \left| \frac{\partial^k g(x)}{\partial x_1^{\alpha_1} \dots \partial x_d^{\alpha_d}} \right|^2 dx.$$

Hence the multivariate penalized least squares estimate minimizes

$$\frac{1}{n} \sum_{i=1}^{n} |g(X_i) - Y_i|^2 + \lambda_n \cdot J_k^2(g). \tag{20.22}$$

It is not obvious over which function space one should minimize (20.22). In order to ensure that the penalty term $J_k^2(g)$ exists one needs to assume that the partial derivatives

$$\frac{\partial^k g}{\partial x_{i_1} \dots \partial x_{i_k}}$$

exist. The proof of the existence of a function which minimizes (20.22) is based on the fact that the set of functions considered here is a Hilbert space. Therefore one doesn't require that the derivatives exist in the classical sense and uses so-called weak derivatives instead. Without going into detail we mention that one mimizes (20.22) over the Sobolev space $W^k(\mathcal{R}^d)$ consisting of all functions whose weak derivatives of order k are contained in $L^2(\mathcal{R}^d)$.

In general, the point evaluation (and thus the empirical L_2 risk) of functions from $W^k(\mathcal{R}^d)$ is not well-defined because the values of these functions are determined only outside a set of Lebesgue measure zero. In the sequel we will always assume that $2k > d$. Under this condition one can show that the functions in $W^k(\mathcal{R}^d)$ are continuous and point evaluation is well-defined.

Using techniques from functional analysis one can show that a function which minimizes (20.22) over $W^k(\mathcal{R}^d)$ always exists. In addition, one can calculate such a function by the following algorithm:

Let $l = \binom{d+k-1}{d}$ and let $\phi_1, \dots, \phi_l$ be all monomials $x_1^{\alpha_1} \cdot \ldots \cdot x_d^{\alpha_d}$ of total degree $\alpha_1 + \cdots + \alpha_d$ less than k. Depending on k and d define

$$K(z) = \begin{cases} \Theta_{k,d} \|z\|^{2k-d} \cdot \log(\|z\|) & \text{if } d \text{ is even,} \\ \Theta_{k,d} \|z\|^{2k-d} & \text{if } d \text{ is odd,} \end{cases}$$

where

$$\Theta_{k,d} = \begin{cases} \dfrac{(-1)^{k+d/2+1}}{2^{2k-1}\pi^{d/2}(k-1)!\cdot(k-d/2)!} & \text{if } d \text{ is even,} \\[2mm] \dfrac{\Gamma(d/2-k)}{2^{2k}\pi^{d/2}(k-1)!} & \text{if } d \text{ is odd.} \end{cases}$$

Let $z_1,\dots,z_N$ be the distinct values of $X_1,\dots,X_n$, and let n_i be the number of occurrences of z_i in $X_1,\dots,X_n$. Then there exists a function of the form

$$g^*(x) = \sum_{i=1}^{N} \mu_i K(x-z_i) + \sum_{j=1}^{l} \nu_j \phi_j(x) \tag{20.23}$$

which minimizes (20.22) over $W^k(\mathcal{R}^d)$. Here $\mu_1,\dots,\mu_N,\nu_1,\dots,\nu_l \in \mathcal{R}$ are solutions of the linear system of equations

$$\lambda_n \mu_i + \frac{n_i}{n}\sum_{j=1}^{N} K(z_i-z_j) + \frac{n_i}{n}\sum_{j=1}^{l}\nu_j\phi_j(z_i) \;=\; \frac{1}{n}\sum_{j:X_j=z_i} Y_j$$
$$(i=1,\dots,N)$$

$$\sum_{j=1}^{N}\mu_j\phi_m(z_j) \;=\; 0 \quad (m=1,\dots,l).$$

The solution of the above system of linear equations is unique if there is no polynomial $p \neq 0$ of total degree less than k that vanishes at all points $X_1,\dots,X_n$.

If there exists a polynomial $p \neq 0$ of total degree less than k which vanishes at all points $X_1,\dots,X_n$, then one can add this polynomial to any function g without changing the value of (20.22). Therefore in this case the minimization of (20.22) does not lead to a unique function. However, one can show in this case that any solution of the above linear system of equations defined via (20.23) yields a function which minimizes (20.22) over $W^k(\mathcal{R}^d)$.

20.5 Consistency

In the sequel we show the consistency of the multivariate penalized least squares estimates defined by

$$\tilde{m}_n(\cdot) = \arg\min_{f\in W^k(\mathcal{R}^d)} \left(\frac{1}{n}\sum_{i=1}^{n} |f(X_i)-Y_i|^2 + \lambda_n J_k^2(f) \right) \tag{20.24}$$

and

$$m_n(x) = T_{\log(n)}\tilde{m}_n(x). \tag{20.25}$$

Here

$$J_k^2(g) = \sum_{i_1,\dots,i_k \in \{1,\dots,d\}} \int_{\mathcal{R}^d} \left| \frac{\partial^k g(x)}{\partial x_{i_1} \dots \partial x_{i_k}} \right|^2 dx$$

is the penalty term for the roughness of the function $f : \mathcal{R}^d \to \mathcal{R}$, $\lambda_n > 0$ is the smoothing parameter of the estimate, and $W^k(\mathcal{R}^d)$ is the Sobolev space consisting of all functions whose weak derivatives of order k are contained in $L^2(\mathcal{R}^d)$ (cf. Section 20.4).

The following covering result, which is an extension of Lemma 20.4 to $\mathcal{R}^d$, plays a key role in proving the consistency of the multivariate penalized least squares estimate.

Lemma 20.6. *Let $L, c > 0$ and set*

$$\mathcal{F} = \left\{ T_L f \ : \ f \in W^k(\mathcal{R}^d) \text{ and } J_k^2(f) \le c \right\}.$$

Then for any $0 < \delta < L$, $1 \le p < \infty$, and $x_1, \dots, x_n \in [0,1]^d$,

$$\mathcal{N}_p(\delta, \mathcal{F}, x_1^n) \le \left(c_1 \frac{L^p n}{\delta^p} \right)^{c_2 \left(\frac{\sqrt{c}}{\delta} \right)^{\frac{d}{k}} + c_3},$$

where c_1, c_2, $c_3 \in \mathcal{R}_+$ are constants which only depend on k and d.

The proof, which is similar to the proof of Lemma 20.4, is left to the reader (cf. Problem 20.9).

From this covering result we easily get the consistency of multivariate penalized least squares estimates:

Theorem 20.4. *Let $k \in \mathcal{N}$ with $2k > d$. For $n \in \mathcal{N}$ choose $\lambda_n > 0$ such that*

$$\lambda_n \to 0 \quad (n \to \infty) \tag{20.26}$$

and

$$\frac{n \cdot \lambda_n^{\frac{d}{2k}}}{\log(n)^7} \to \infty \quad (n \to \infty). \tag{20.27}$$

Let the estimate m_n be defined by (20.24) and (20.25). Then

$$\int |m_n(x) - m(x)|^2 \mu(dx) \to 0 \quad (n \to \infty) \quad a.s.$$

for every distribution of (X, Y) with $\|X\|_2$ bounded a.s. and $\mathbf{E}Y^2 < \infty$.

The proof is left to the reader (cf. Problem 20.10)

20.6 Bibliographic Notes

Various applications of penalized modeling in statistics can be found, e.g., in Wahba (1990), Green and Silverman (1994), Eubank (1999), and Eggermont and LaRiccia (2001).

The principle of penalized modeling, in particular smoothing splines, goes back to Whittaker (1923), Schoenberg (1964), and Reinsch (1967); see Wahba (1990) or Eubank (1999) for additional references.

The results in Sections 20.1 and 20.2 are standard results in the theory of deterministic splines, although the existence and computation of smoothing splines is often shown for $k = 2$ only. References concerning the determinist theory of splines can be found in Section 14.5. For results concerning the existence and computation of multivariate penalized least squares estimates, see Duchon (1976), Cox (1984), or Wahba (1990).

The proof of the existence of penalized least squares estimates (i.e., the proof of the existence of a function which minimizes the penalized empirical L_2 risk), is often based on tools from functional analysis, in particular, on the theory of so-called reproducing kernel Hilbert spaces. This also leads to a different way of analysis of penalized least squares estimates, for details, see Wahba (1990).

Lemma 20.4 and Theorem 20.3 are based on van de Geer (1987) (in particular on Lemma 3.3.1 there). The generalization of these results to the multivariate case (i.e., Lemma 20.6 and Theorem 20.4) is due to Kohler and Krzyżak (2001).

Mammen and van de Geer (1997) considered penalized least squares estimates defined by using a penalty on the total variation of the function.

Problems and Exercises

PROBLEM 20.1. Show that if one replaces

$$\int_0^1 |f^{(k)}(x)|^2 \, dx$$

by

$$\int_L^R |f^{(k)}(x)|^2 \, dx$$

in the definition of the penalized least squares estimate for some

$$-\infty \leq L \leq \min\{X_1, \dots, X_n\} \leq \max\{X_1, \dots, X_n\} \leq R \leq \infty,$$

then the values of the estimate on $[\min\{X_1, \dots, X_n\}, \max\{X_1, \dots, X_n\}]$ do not change.

HINT: Apply Lemma 20.2 with

$$a = \min\{X_1, \dots, X_n\} - \epsilon \text{ and } b = \max\{X_1, \dots, X_n\} + \epsilon$$

and show that a function $f \in C^k(\mathcal{R})$, which minimizes

$$\frac{1}{n} \sum_{i=1}^{n} |f(X_i) - Y_i|^2 + \lambda \int_L^R |f^{(k)}(x)|^2 \, dx,$$

satisfies $f^{(k)}(x) = 0$ for $L < x < \min\{X_1, \ldots, X_n\} - \epsilon$ and for $\max\{X_1, \ldots, X_n\} + \epsilon < x < R$.

PROBLEM 20.2. Consider the interpolation problem of Theorem 20.2. Show that

$$B_{j,M,u}(t_j) = 0 \text{ for some } j \in \{-M, \ldots, K-1\}$$

implies that the solution of the interpolation problem is not unique.

HINT: First prove the assertion for $M = 0$. Then assume $M > 0$ and consider the cases $t_j \leq u_j$ and $t_j \geq u_{j+M+1}$. Show that $t_j \leq u_j$ implies $B_{k,M,u}(t_i) = 0$ for $i \leq j$ and $k \geq j$. Conclude that in this case the matrix $(B_{k,M,u}(t_i))_{i,k}$ is not regular. Argue similarly in the case $t_j \geq u_{j+M+1}$.

PROBLEM 20.3. Consider the interpolation problem of Theorem 20.2. Show that

$$B_{j,M,u}(t_j) > 0 \text{ for all } j \in \{-M, \ldots, K-1\}$$

implies that the solution of the interpolation problem is unique.

HINT: Show that the matrix $(B_{k,M,u}(t_i))_{i,k}$ is regular. Prove this first for $M = 0$. Then proceed by induction.

PROBLEM 20.4. Show that the univariate penalized least squares estimates is, without truncation, in general not consistent, even if X and Y are bounded.

HINT: Consider random variables X with $\mathbf{P}\{X = 0\} = \frac{1}{2}$ and $\mathbf{P}\{X \leq x\} = \frac{1+x}{2}$ for $x \in [0,1]$ and Y independent of X with $\mathbf{P}\{Y = -1\} = \mathbf{P}\{Y = +1\} = \frac{1}{2}$. Hence $m(x) = 0$ for all x. Now draw an i.i.d. sample $(X_1, Y_1), \ldots, (X_n, Y_n)$ from the distribution of (X, Y). Show that if the event

$$A := \{X_1 = \cdots = X_{n-1} = 0; Y_1, \ldots, Y_{n-1} = -1; X_n \neq 0; Y_n = 1\}$$

occurs, then the smoothing spline m_n obtained with penalty J_k^2 for $k \geq 2$ is the straight line through $(0, -1)$ and $(X_n, 1)$, $m_n(x) = -1 + \frac{2x}{X_n}$. Use this to conclude that the L_2 error satisfies

$$\mathbf{E} \int |m_n(x) - m(x)|^2 \mathbf{P}_X(dx) \geq \mathbf{E}\left(I_A \cdot \frac{1}{2} \int_0^1 \left| -1 + \frac{2x}{X_n} \right|^2 dx \right)$$

$$= \frac{p}{2} \cdot \mathbf{E}\left(I_{\{X_n \neq 0\}} \cdot \frac{X_n}{6} \left(\left(-1 + \frac{2}{X_n} \right)^3 + 1 \right) \right)$$

$$= \frac{p}{4} \int_0^1 \frac{u}{6} \left(\left(-1 + \frac{2}{u} \right)^3 + 1 \right) du$$

$$= \infty,$$

where $p = \mathbf{P}\{X_1 = \cdots = X_{n-1} = 0; Y_1 = \cdots = Y_{n-1} = -1; Y_n = 1\} > 0$.

PROBLEM 20.5. Let $A \geq 1$ and $x_1, \ldots, x_n \in [-A, A]$. Show that, for any $c > 0$, $L > 0$, and $0 < \delta < L$,

$$\mathcal{N}_1\left(\delta, \left\{ T_L f \; : \; f \in C^k(\mathcal{R}) \text{ and } \int_{-A}^A |f^{(k)}(x)|^2 dx \leq c \right\}, x_1^n \right)$$

$$\leq \left(c_1 \frac{L}{\delta}\right)^{(2A+2)\left(c_2\left(\frac{\sqrt{c}}{\delta}\right)^{1/k}+c_3\right)}$$

for some constants $c_1, c_2, c_3 \in \mathcal{R}$ which only depend on k.

HINT: Apply Lemma 20.4 on intervals $[i, i+1]$ with $[i, i+1] \cap [-A, A] \neq \emptyset$.

PROBLEM 20.6. Define the estimate m_n by

$$\tilde{m}_n(\cdot) = \arg \min_{f \in C^k(\mathcal{R})} \left(\frac{1}{n} \sum_{i=1}^{n} |f(X_i) - Y_i|^2 I_{\{X_i \in [-\log(n), \log(n)]\}} + \lambda_n \int_{-\infty}^{\infty} |f^{(k)}(x)|^2 \, dx \right)$$

and

$$m_n(x) = T_{\log(n)} \tilde{m}_n(x) \cdot I_{\{x \in [-\log(n), \log(n)]\}}.$$

Show that m_n is strongly universally consistent provided

$$\lambda_n \to 0 \quad (n \to \infty) \quad \text{and} \quad n\lambda_n \to \infty \quad (n \to \infty).$$

HINT: Use the error decomposition

$$\int |m_n(x) - m(x)|^2 \mu(dx) = \int_{\mathcal{R} \setminus [-\log(n), \log(n)]} |m_n(x) - m(x)|^2 \mu(dx)$$
$$+ \mathbf{E}\{|m_n(X) - Y|^2 \cdot I_{\{X \in [-\log(n), \log(n)]\}} \mid D_n\}$$
$$- \mathbf{E}\{|m(X) - Y|^2 \cdot I_{\{X \in [-\log(n), \log(n)]\}}\}.$$

PROBLEM 20.7. Show that Theorem 20.3 can also be proved via the truncation argument which we have used in Theorem 10.2.

PROBLEM 20.8. Prove that the result of Theorem 20.3 still holds if the smoothing parameter λ_n of the penalized least squares estimate depends on the data and (20.19) and (20.20) hold with probability one.

PROBLEM 20.9. Prove Lemma 20.6.

HINT:

Step 1: Partition $[0, 1]^d$ into d-dimensional rectangles $A_1, \ldots, A_K$ with the following properties:

(i) $\int_{A_i} \sum_{\alpha_1 + \cdots + \alpha_d = k} \frac{k!}{\alpha_1! \cdots \alpha_d!} \left| \frac{\partial^k f}{\partial x_1^{\alpha_1} \ldots \partial x_d^{\alpha_d}}(x) \right|^2 dx \leq c(\delta/\sqrt{c})^{d/k}$,
$(i = 1, \ldots, K)$;

(ii) $\sup_{x, z \in A_i} \|x - z\|_\infty \leq (\delta/\sqrt{c})^{1/k}$ $(i = 1, \ldots, K)$; and

(iii) $K \leq ((\sqrt{c}/\delta)^{1/k} + 1)^d + (\sqrt{c}/\delta)^{d/k}$.

To do this, start by dividing $[0, 1]^d$ into $\tilde{K} \leq ((\sqrt{c}/\delta)^{1/k} + 1)^d$ equi-volume cubes $B_1, \ldots, B_{\tilde{K}}$ of side length $(\delta/\sqrt{c})^{1/k}$. Then partition each cube B_i into d-dimensional rectangles $B_{i,1}, \ldots, B_{i,l_i}$ such that, for $j = 1, \ldots, l_i - 1$,

$$\int_{B_{i,j}} \sum_{\alpha_1 + \cdots + \alpha_d = k} \frac{k!}{\alpha_1! \cdot \ldots \cdot \alpha_d!} \left| \frac{\partial^k f}{\partial x_1^{\alpha_1} \ldots \partial x_d^{\alpha_d}}(x) \right|^2 dx = c(\delta/\sqrt{c})^{d/k},$$

and, for $j = l_i$,

$$\int_{B_{i,j}} \sum_{\alpha_1 + \cdots + \alpha_d = k} \frac{k!}{\alpha_1! \cdot \ldots \cdot \alpha_d!} \left| \frac{\partial^k f}{\partial x_1^{\alpha_1} \ldots \partial x_d^{\alpha_d}}(x) \right|^2 dx \leq c(\delta/\sqrt{c})^{d/k}.$$

Step 2: Approximate f on each rectangle A_i by a polynomial of total degree $k - 1$. Fix $1 \leq i \leq K$. Use the Sobolev integral identity, see Oden and Reddy (1976), Theorem 3.6, which implies that there exists a polynomial p_i of total degree not exceeding $k-1$ and an infinitely differentiable bounded function $Q_\alpha(x,y)$ such that, for all $x \in A_i$,

$$|f(x) - p_i(x)|$$
$$= \int_{A_i} \frac{1}{\|x - z\|_2^{d-k}} \sum_{\alpha_1 + \cdots + \alpha_d = k} Q_\alpha(x, z) \left(\frac{\partial^k f}{\partial x_1^{\alpha_1} \ldots \partial x_d^{\alpha_d}} \right)(z)\, dz.$$

Use this to conclude

$$\mathcal{N}_p((\sqrt{c_0}d^{(k-d)} + 1)\delta, \mathcal{F}, x_1^n) \leq \mathcal{N}_p(\delta, T_L\mathcal{G}, x_1^n),$$

where $T_L\mathcal{G} = \{T_L g : g \in \mathcal{G}\}$ and $\mathcal{G}$ is the set of all piecewise polynomials of total degree less than or equal to $k - 1$ with respect to a rectangular partition of $[0, 1]^d$ consisting of at most $K \leq (2^d + 1)(\sqrt{c}/\delta)^{d/k} + 2^d$ rectangles.

Step 3: Use the results of Chapters 9 and 13 to bound $\mathcal{N}_p(\delta, T_L\mathcal{G}, x_1^n)$.

PROBLEM 20.10. Prove Theorem 20.4.

HINT: Proceed as in the proof of Theorem 20.3, but use Lemma 20.6 instead of Lemma 20.4.

PROBLEM 20.11. The assumption $\|X\|_2$ bounded *a.s.* in Theorem 20.4 may be dropped if we slightly modify the estimate. Define $\tilde{m}_n$ by

$$\tilde{m}_n(\cdot) = \arg \min_{f \in W^k(\mathcal{R}^d)} \left(\frac{1}{n} \sum_{i=1}^n |f(X_i) - Y_i|^2 \cdot I_{[-\log(n),\log(n)]^d}(X_i) + \lambda_n J_k^2(f) \right)$$

and set $m_n(x) = T_{\log(n)} \tilde{m}_n(x) \cdot I_{[-\log(n),\log(n)]^d}(x)$. Show that m_n is strongly consistent for all distributions of (X, Y) with $\mathbf{E}Y^2 < \infty$, provided $2k > d$ and suitable modifications of (20.26)–(20.27) hold.

HINT: Use

$$\int |m_n(x) - m(x)|^2 \mu(dx) = \int_{\mathcal{R}^d \setminus [-\log(n),\log(n)]^d} |m_n(x) - m(x)|^2 \mu(dx)$$
$$+ \mathbf{E}\{|m_n(X) - Y|^2 \cdot I_{[-\log(n),\log(n)]^d}(X)|D_n\}$$
$$- \mathbf{E}\{|m(X) - Y|^2 \cdot I_{[-\log(n),\log(n)]^d}(X)\}.$$

21
Penalized Least Squares Estimates II: Rate of Convergence

In this chapter we study the rate of convergence of penalized least squares estimates. In Section 21.1 the smoothing parameter is chosen depending on the smoothness of the regression function. In Section 21.2 we use the complexity regularization principle to define penalized least squares estimates which automatically adapt to the smoothness of the regression function.

21.1 Rate of Convergence

Our main result in this section is

Theorem 21.1. *Let* $1 \leq L < \infty$, $n \in \mathcal{N}$, $\lambda_n > 0$, *and* $k \in \mathcal{N}$. *Define the estimate* m_n *by*

$$\tilde{m}_n(\cdot) = \arg\min_{f \in C^k(\mathcal{R})} \left\{ \frac{1}{n} \sum_{i=1}^{n} |f(X_i) - Y_i|^2 + \lambda_n J_k^2(f) \right\} \tag{21.1}$$

and

$$m_n(x) = T_L \tilde{m}_n(x) \quad (x \in \mathcal{R}), \tag{21.2}$$

where

$$J_k^2(f) = \int_0^1 |f^{(k)}(x)|^2 dx.$$

Then there exist constants $c_1, c_2 \in \mathcal{R}$ which depend only on k and L such that

$$\mathbf{E}\left\{\int |m_n(x) - m(x)|^2 \mu(dx)\right\}$$

$$\leq 2\lambda_n J_k^2(m) + c_1 \cdot \frac{\log(n)}{n \cdot \lambda_n^{1/(2k)}} + c_2 \cdot \frac{\log(n)}{n} \qquad (21.3)$$

for every distribution of (X, Y) with $X \in [0,1]$ a.s., $|Y| \leq L$ a.s., and $m \in C^k(\mathcal{R})$. In particular, for any constant $c_3 > 0$ and for

$$\lambda_n = c_3 \left(\frac{\log(n)}{n \cdot J_k^2(m)}\right)^{2k/(2k+1)}$$

there exists a constant c_4 such that $J_k^2(m) \geq \log(n)/n$ implies

$$\mathbf{E}\left\{\int |m_n(x) - m(x)|^2 \mu(dx)\right\}$$

$$\leq c_4 J_k^2(m)^{1/(2k+1)} \cdot \left(\frac{\log(n)}{n}\right)^{2k/(2k+1)}. \qquad (21.4)$$

If m is (p, C)-smooth with $p = k$, then the $(k-1)$th derivative of m is Lipschitz continuous with Lipschitz constant C. If, as in the proof of Theorem 3.2, m is in addition k times differentiable, then the kth derivative of m is bounded by C which implies

$$J_k^2(m)^{1/(2k+1)} \cdot \left(\frac{\log(n)}{n}\right)^{2k/(2k+1)} \leq C^{2/(2k+1)} \cdot \left(\frac{\log(n)}{n}\right)^{2k/(2k+1)}.$$

From this we see that the rate of convergence in Theorem 21.1 is optimal up to a logarithmic factor (cf. Theorem 3.2).

The advantage of the bound (21.4), compared with our previous bounds for (p, C)-smooth regression functions, is that in some sense $J_k^2(m) \leq C^2$ is a much weaker condition than m (k, C)-smooth, because in the latter the Lipschitz function C is independent of x, so the function satisfies the same smoothness condition on whole $[0, 1]$. But if $J_k^2(m) \leq C^2$ then the kth derivative is allowed to vary in such a way that the squared average is bounded by C.

PROOF. Inequality (21.4) is an easy consequence of (21.3), therefore we prove only (21.3). In order to simplify the notation in the proof we will abbreviate various constants which depend only on k and L by $c_5, c_6, \ldots$.

We use the error decomposition

$$\int |m_n(x) - m(x)|^2 \mu(dx) = \mathbf{E}\left\{|m_n(X) - Y|^2 | D_n\right\} - \mathbf{E}\left\{|m(X) - Y|^2\right\}$$

$$= T_{1,n} + T_{2,n},$$

where

$$T_{1,n} = 2 \left(\frac{1}{n} \sum_{i=1}^{n} \left\{ |m_n(X_i) - Y_i|^2 - |m(X_i) - Y_i|^2 \right\} + \lambda_n J_k^2(\tilde{m}_n) \right)$$

and

$$T_{2,n} = \mathbf{E}\left\{ |m_n(X) - Y|^2 | D_n \right\} - \mathbf{E}\left\{ |m(X) - Y|^2 \right\} - T_{1,n}.$$

Let us first observe that it is easy to bound $T_{1,n}$: Assuming $|Y_i| \leq L$ $(i = 1, \ldots, n)$, (21.2), and $m \in C^k(\mathcal{R})$ we get

$$\begin{aligned}
T_{1,n} &\leq& 2 \left(\frac{1}{n} \sum_{i=1}^{n} |\tilde{m}_n(X_i) - Y_i|^2 + \lambda_n J_k^2(\tilde{m}_n) - \frac{1}{n} \sum_{i=1}^{n} |m(X_i) - Y_i|^2 \right) \\
&\leq& 2 \left(\frac{1}{n} \sum_{i=1}^{n} |m(X_i) - Y_i|^2 + \lambda_n J_k^2(m) - \frac{1}{n} \sum_{i=1}^{n} |m(X_i) - Y_i|^2 \right) \\
&=& 2\lambda_n J_k^2(m).
\end{aligned}$$

Hence it suffices to show

$$\mathbf{E}\{T_{2,n}\} \leq c_1 \cdot \frac{\log(n)}{n \cdot \lambda_n^{1/(2k)}} + c_2 \cdot \frac{\log(n)}{n}. \tag{21.5}$$

In order to do this, we fix $t > 0$ and analyze

$$\mathbf{P}\{T_{2,n} > t\}$$

$$= \mathbf{P}\Bigg\{ 2\mathbf{E}\left\{ |m_n(X) - Y|^2 - |m(X) - Y|^2 | D_n \right\}$$

$$- 2\frac{1}{n} \sum_{i=1}^{n} \left\{ |m_n(X_i) - Y_i|^2 - |m(X_i) - Y_i|^2 \right\}$$

$$> t + 2\lambda_n J_k^2(\tilde{m}_n) + \mathbf{E}\left\{ |m_n(X) - Y|^2 - |m(X) - Y|^2 | D_n \right\} \Bigg\}.$$

The above probability depends on the random function m_n, which makes the analysis difficult. But we know from Chapter 10 how to get rid of this kind of randomness: we can bound the probability above by a probability where the deterministic functions are taken from some deterministic set in which m_n is contained. A simple way to do this would be to assume

$$m_n \in \left\{ T_L g \; : \; g \in C^k(\mathcal{R}), J_k^2(g) \leq \frac{2L^2}{\lambda_n} \right\} \quad a.s.$$

for n sufficiently large (cf. (20.13)), which implies that $\mathbf{P}\{T_{2,n} > t\}$ is bounded from above by

$$\mathbf{P}\bigg\{ \exists f = T_L g : g \in C^k(\mathcal{R}), J_k^2(g) \le \frac{2L^2}{\lambda_n} :$$

$$\mathbf{E}\left\{ |f(X) - Y|^2 - |m(X) - Y|^2 \right\}$$

$$- \frac{1}{n} \sum_{i=1}^{n} \left\{ |f(X_i) - Y_i|^2 - |m(X_i) - Y_i|^2 \right\}$$

$$> \frac{1}{2} \cdot (t + 2\lambda_n J_k^2(g) + \mathbf{E}\left\{ |f(X) - Y|^2 - |m(X) - Y|^2) \right\} \bigg\}.$$

Unfortunately, this bound is not sharp enough to get the right rate of convergence.

To get a better rate we use the peeling technique (cf. Chapter 19): $\tilde{m}_n$ is contained in $C^k(\mathcal{R})$, which implies that, for some $l \in \mathcal{N}_0$, we have

$$2^l t \cdot I_{\{l \neq 0\}} \le 2\lambda_n J_k^2(\tilde{m}_n) < 2^{l+1} t.$$

From this, together with the union bound, we conclude

$$\mathbf{P}\{T_{2,n} > t\}$$

$$\le \sum_{l=0}^{\infty} \mathbf{P}\bigg\{ \exists f = T_L g : g \in C^k(\mathcal{R}), 2^l t \cdot I_{\{l \neq 0\}} \le 2\lambda_n J_k^2(g) < 2^{l+1} t :$$

$$\mathbf{E}\left\{ |f(X) - Y|^2 - |m(X) - Y|^2 \right\}$$

$$- \frac{1}{n} \sum_{i=1}^{n} \left\{ |f(X_i) - Y_i|^2 - |m(X_i) - Y_i|^2 \right\}$$

$$> \frac{1}{2} \cdot (t + 2\lambda_n J_k^2(g) + \mathbf{E}\left\{ |f(X) - Y|^2 - |m(X) - Y|^2 \right\}) \bigg\}$$

$$\le \sum_{l=0}^{\infty} \mathbf{P}\bigg\{ \exists f = T_L g : g \in C^k(\mathcal{R}), J_k^2(g) < \frac{2^l t}{\lambda_n} :$$

$$\mathbf{E}\left\{ |f(X) - Y|^2 - |m(X) - Y|^2 \right\}$$

$$- \frac{1}{n} \sum_{i=1}^{n} \left\{ |f(X_i) - Y_i|^2 - |m(X_i) - Y_i|^2 \right\}$$

$$> \frac{1}{2} \cdot (2^l t + \mathbf{E}\left\{ |f(X) - Y|^2 - |m(X) - Y|^2 \right\}) \bigg\}.$$

Fix $l \in \mathcal{N}_0$. We will show momentarily that we can find constants c_5, c_6, c_7 such that, for $t \ge c_5 \frac{\log(n)}{n \lambda_n^{1/(2k)}} + c_6 \frac{\log(n)}{n}$,

$$\mathbf{P}\left\{\exists f = T_L g : g \in C^k(\mathcal{R}), J_k^2(g) < \frac{2^l t}{\lambda_n} : \right.$$

$$\mathbf{E}\left\{|f(X) - Y|^2 - |m(X) - Y|^2\right\}$$

$$-\frac{1}{n}\sum_{i=1}^{n}\left\{|f(X_i) - Y_i|^2 - |m(X_i) - Y_i|^2\right\}$$

$$\left. > \frac{1}{2} \cdot \left(2^l t + \mathbf{E}\left\{|f(X) - Y|^2 - |m(X) - Y|^2\right\}\right)\right\}$$

$$\leq 60 \cdot \exp\left(-c_7 n \cdot t \cdot 2^l\right). \tag{21.6}$$

This inequality implies the assertion, because we can conclude from (21.6) that, for $t \geq c_5 \frac{\log(n)}{n\lambda_n^{1/(2k)}} + c_6 \frac{\log(n)}{n}$,

$$\mathbf{P}\{T_{2,n} > t\} \leq \sum_{l=0}^{\infty} 60 \cdot \exp\left(-c_7 n \cdot t \cdot 2^l\right) \leq c_8 \exp\left(-c_7 n \cdot t\right),$$

which implies

$$\mathbf{E}\{T_{2,n}\}$$

$$\leq \int_0^{c_5 \frac{\log(n)}{n\lambda_n^{1/(2k)}} + c_6 \frac{\log(n)}{n}} 1\, dt + \int_{c_5 \frac{\log(n)}{n\lambda_n^{1/(2k)}} + c_6 \frac{\log(n)}{n}}^{\infty} c_8 \exp\left(-c_7 n \cdot t\right)\, dt$$

$$\leq c_1 \frac{\log(n)}{n\lambda_n^{1/(2k)}} + c_2 \frac{\log(n)}{n}.$$

So it remains to prove (21.6).

Inequality (21.6) follows directly from Theorem 19.3 provided we can show that the assumptions in Theorem 19.3 are satisfied. Set

$$\mathcal{F} = \left\{ f : \mathcal{R} \times \mathcal{R} \to \mathcal{R} \; : \; f(x, y) = |T_L g(x) - T_L y|^2 - |m(x) - T_L y|^2 \right.$$

$$\left. ((x, y) \in [0, 1] \times \mathcal{R}) \text{ for some } g \in C^k(\mathcal{R}), \; J_k^2(g) \leq \frac{2^l t}{\lambda_n} \right\}$$

and

$$Z = (X, Y), \; Z_i = (X_i, Y_i) \quad (i = 1, \ldots, n).$$

Then the left-hand side of (21.6) can be rewritten as

$$\mathbf{P}\left\{\sup_{f \in \mathcal{F}} \frac{\mathbf{E}f(Z) - \frac{1}{n}\sum_{i=1}^{n} f(Z_i)}{2^l t + \mathbf{E}f(Z)} > \frac{1}{2}\right\}.$$

Hence it suffices to show that for the set $\mathcal{F}$ of functions, $\alpha = 2^l t$, $\epsilon = \frac{1}{2}$, and suitable values of K_1 and K_2 the assumptions of Theorem 19.3 are satisfied.

We first determine K_1 and K_2. For $f \in \mathcal{F}$ we have

$$|f(z)| \leq 4L^2 \quad (z \in [0,1] \times \mathcal{R})$$

and

$$
\begin{aligned}
\mathbf{E}|f(Z)|^2 &= \mathbf{E}\{|(T_L g)(X) - Y|^2 - |m(X) - Y|^2|^2\} \\
&= \mathbf{E}\{|((T_L g)(X) - Y) - (m(X) - Y)|^2 \\
&\qquad\qquad \times |((T_L g)(X) - Y) + (m(X) - Y)|^2\} \\
&\leq 16L^2 \mathbf{E}|(T_L g)(X) - m(X)|^2 \\
&= 16L^2 \mathbf{E} f(Z).
\end{aligned}
$$

So we can choose $K_1 = 4L^2$ and $K_2 = 16L^2$.

Condition (19.10) follows from $t \geq c_5 \frac{\log(n)}{n \lambda_n^{1/(2k)}} + c_6 \frac{\log(n)}{n}$ so it remains to show that (19.11) holds.

In order to bound the covering number we observe

$$
\frac{1}{n} \sum_{i=1}^{n} \left| \left(|T_L g_1(x_i) - T_L y_i|^2 - |m(x_i) - T_L y_i|^2 \right) \right.
$$

$$
\left. - \left(|T_L g_2(x_i) - T_L y_i|^2 - |m(x_i) - T_L y_i|^2 \right) \right|^2
$$

$$
= \frac{1}{n} \sum_{i=1}^{n} \left| |T_L g_1(x_i) - T_L y_i|^2 - |T_L g_2(x_i) - T_L y_i|^2 \right|^2
$$

$$
= \frac{1}{n} \sum_{i=1}^{n} |T_L g_1(x_i) - T_L g_2(x_i)|^2 \cdot |T_L g_1(x_i) + T_L g_2(x_i) - 2 T_L y_i|^2
$$

$$
\leq 16L^2 \frac{1}{n} \sum_{i=1}^{n} |T_L g_1(x_i) - T_L g_2(x_i)|^2,
$$

which implies

$$
\mathcal{N}_2(u, \mathcal{F}, z_1^n) \leq \mathcal{N}_2 \left(\frac{u}{4L}, \left\{ T_L g \ : \ g \in C^k(\mathcal{R}), J_k^2(g) \leq \frac{2^l t}{\lambda_n} \right\}, x_1^n \right).
$$

This together with Lemma 20.4 implies, for any $u \geq \frac{1}{n}$,

$$
\log \mathcal{N}_2 \left(u, \left\{ f \in \mathcal{F} \ : \ \frac{1}{n} \sum_{i=1}^{n} f(z_i)^2 \leq 16\delta \right\}, z_1^n \right)
$$

$$
\leq \log \mathcal{N}_2(u, \mathcal{F}, z_1^n)
$$

$$
\leq \log \mathcal{N}_2 \left(\frac{u}{4L}, \left\{ T_L g \ : \ g \in C^k(\mathcal{R}), J_k^2(g) \leq \frac{2^l t}{\lambda_n} \right\}, x_1^n \right)
$$

$$\leq 8(k+2)\left(\left(\frac{\sqrt{2^l t/\lambda_n}}{u/(4L)}\right)^{1/k}+1\right)\log\left(\frac{144eL^2 n}{u^2/(16L^2)}\right)$$

$$\leq c_9 \log(n)\cdot\left(\left(\frac{2^l t}{\lambda_n}\right)^{1/2k}u^{-1/k}+1\right),$$

hence, for $\delta \geq \alpha/8 \geq 2048L^2/n$,

$$\int_{\delta/2048L^2}^{\sqrt{\delta}}\left(\log\mathcal{N}_2\left(u,\left\{f\in\mathcal{F}:\frac{1}{n}\sum_{i=1}^n f(z_i)^2\leq 16\delta\right\},z_1^n\right)\right)^{\frac{1}{2}}du$$

$$\leq\sqrt{c_9\log(n)}\cdot\int_0^{\sqrt{\delta}}\left(\left(\frac{2^l t}{\lambda_n}\right)^{1/(4k)}u^{-1/(2k)}+1\right)du$$

$$=c_{10}\sqrt{\log(n)}\left(\left(\frac{2^l t}{\lambda_n}\right)^{1/(4k)}\delta^{1/2-1/(4k)}+\delta^{1/2}\right).$$

Hence (19.11) is implied by

$$\sqrt{n}\delta\geq c_{11}\sqrt{\log(n)}\left(\left(\frac{2^l t}{\lambda_n}\right)^{1/(4k)}\delta^{1/2-1/(4k)}+\delta^{1/2}\right)$$

for all $\delta\geq\alpha/8=2^{l-3}t$.

Since

$$\frac{\sqrt{n}2^{l-3}t}{c_{11}\sqrt{\log(n)}\left(\frac{2^l t}{\lambda_n}\right)^{1/(4k)}(2^{l-3}t)^{1/2-1/(4k)}}$$

$$=\frac{1}{c_{11}8^{1/(4k)}}\cdot\frac{\sqrt{n}\lambda_n^{1/(4k)}}{\sqrt{\log(n)}}\cdot(2^{l-3}t)^{1/2}$$

$$\geq\frac{1}{2}$$

provided

$$t\geq c_{12}\frac{\log(n)}{n\cdot\lambda_n^{1/(2k)}}$$

and since

$$\frac{\sqrt{n}2^{l-3}t}{c_{11}\sqrt{\log(n)}(2^{l-3}t)^{1/2}}=\frac{1}{c_{11}}\cdot\frac{\sqrt{n}}{\sqrt{\log(n)}}\cdot(2^{l-3}t)^{1/2}\geq\frac{1}{2}$$

provided

$$t\geq c_{13}\frac{\log(n)}{n},$$

this in turn is implied by the assumption $t \geq c_5 \frac{\log(n)}{n\lambda_n^{1/(2k)}} + c_6 \frac{\log(n)}{n}$. $\square$

21.2 Application of Complexity Regularization

In this section we will use complexity regularization to automatically adapt to smoothness of the estimated regression function. In the sequel we will assume that (X, Y) takes with probability one only values in some bounded subset of $\mathcal{R} \times \mathcal{R}$. Without loss of generality this bounded subset is $[0, 1] \times [-L, L]$, i.e.,

$$(X, Y) \in [0, 1] \times [-L, L] \quad a.s.$$

for some $L \in \mathcal{R}_+$.

Let $k \in \mathcal{N}$ and $\lambda \in \mathcal{R}_+$. First define the smoothing spline estimate $\tilde{m}_{n,(k,\lambda)}$ by

$$\tilde{m}_{n,(k,\lambda)}(\cdot) = \arg \min_{f \in C^k([0,1])} \left(\frac{1}{n} \sum_{i=1}^n |f(X_i) - Y_i|^2 + \lambda J_k^2(f) \right). \quad (21.7)$$

The estimate $\tilde{m}_{n,(k,\lambda)}$ depends on the parameters $k \in \mathcal{N}$ and $\lambda \in \mathcal{R}_+$. We next describe how one can use the data D_n to choose these parameters by complexity regularization (see Chapter 12): in Lemma 21.1 below we derive an upper bound on L_2 error of (a truncated version of) the estimate $\tilde{m}_{n,(k,\lambda)}$. We then choose (k^*, λ^*) by minimizing this upper bound.

Lemma 21.1. *Let* $1 \leq L < \infty$, $\lambda \in \mathcal{R}_+$, *and* $\eta \in [0, 1]$. *Then for* n *sufficiently large one has, with probability greater than or equal to* $1 - \eta$,

$$\int |T_L \tilde{m}_{n,(k,\lambda)}(x) - m(x)|^2 \mu(dx)$$

$$\leq L^4 \frac{\log(n)}{n} + 2 \frac{L^5 (\log(n))^2}{n \cdot \lambda^{1/(2k)}} + 2 \left\{ \frac{1}{n} \sum_{i=1}^n |T_L \tilde{m}_{n,(k,\lambda)}(X_i) - Y_i|^2 \right.$$

$$\left. + \lambda J_k^2(\tilde{m}_{n,(k,\lambda)}) - \frac{1}{n} \sum_{i=1}^n |m(X_i) - Y_i|^2 \right\}$$

for every distribution of (X, Y) *with* $(X, Y) \in [0, 1] \times [-L, L]$ *almost surely.*

The proof is similar to the proof of Theorem 21.2 below and is therefore omitted (see Problem 21.1).

The basic idea in this chapter is to choose the parameters of the estimate by minimizing the upper bound given in the above lemma. This will now be described in detail.

Set

$$\mathcal{K} := \left\{ 1, \ldots, \lfloor (\log(n))^{1/(2)} \rfloor \right\}$$

and

$$\Lambda_n := \left\{ \frac{\log(n)}{2^n}, \frac{\log(n)}{2^{n-1}}, \ldots, \frac{\log(n)}{1} \right\}.$$

For $(k, \lambda) \in \mathcal{K} \times \Lambda_n$ define $m_{n,(k,\lambda)}$ by

$$m_{n,(k,\lambda)}(x) = T_L \tilde{m}_{n,(k,\lambda)}(x) \quad (x \in \mathcal{R}).$$

Depending on the data D_n we choose from the family of estimates

$$\left\{ m_{n,(k,\lambda)} \; : \; (k, \lambda) \in \mathcal{K} \times \Lambda_n \right\}$$

the estimate that minimizes the upper bound in Lemma 21.1. More precisely, we choose

$$(k^*, \lambda^*) = (k^*(D_n), \lambda^*(D_n)) \in \mathcal{K} \times \Lambda_n$$

such that

$$\frac{1}{n} \sum_{i=1}^{n} |m_{n,(k^*,\lambda^*)}(X_i) - Y_i|^2 + \lambda^* J_{k^*}^2(\tilde{m}_{n,(k^*,\lambda^*)}) + pen_n(k^*, \lambda^*)$$

$$= \min_{(k,\lambda) \in \mathcal{K} \times \Lambda_n} \left\{ \frac{1}{n} \sum_{i=1}^{n} |m_{n,(k,\lambda)}(X_i) - Y_i|^2 + \lambda J_k^2(\tilde{m}_{n,(k,\lambda)}) + pen_n(k, \lambda) \right\},$$

where

$$pen_n(k, \lambda) = \frac{L^5 (\log(n))^2}{n \cdot \lambda^{1/(2k)}} \quad ((k, \lambda) \in \mathcal{K} \times \Lambda_n),$$

and define our adaptive smoothing spline estimate by

$$m_n(x) = m_n(x, D_n) = m_{n,(k^*(D_n),\lambda^*(D_n))}(x, D_n). \tag{21.8}$$

An upper bound on the L_2 error of the estimate is given in the next theorem.

Theorem 21.2. *Let m_n be the estimate defined by (21.8).*
(a)

$$\mathbf{E} \int |m_n(x) - m(x)|^2 \mu(dx) = O \left(\frac{(\log(n))^2}{n} \right)$$

for any $p \in \mathcal{N}$ and any distribution of (X, Y) with $(X, Y) \in [0, 1] \times [-L, L]$ almost surely, $m \in C^p([0, 1])$ and $J_p^2(m) = 0$.
(b)

$$\mathbf{E} \int |m_n(x) - m(x)|^2 \mu(dx) = O \left((J_p^2(m))^{\frac{1}{2p+1}} (\log n)^2 n^{-\frac{2p}{2p+1}} \right)$$

for any $p \in \mathcal{N}$ and any distribution of (X, Y) with $(X, Y) \in [0, 1] \times [-L, L]$ almost surely, $m \in C^p([0, 1])$ and $0 < J_p^2(m) < \infty$.

If we compare Theorem 21.2 (b) with Theorem 21.1 we see that the estimate above achieves up to a logarithmic factor the same rate of convergence as the estimate in Theorem 21.1, although it doesn't use parameters which depend on the smoothness of the regression function (measured by k and $J_k^2(m)$). In this sense it is able to adapt automatically to the smoothness of m.

PROOF. Without loss of generality we assume $p \in \mathcal{K}$. We start with the error decomposition

$$\int |m_n(x) - m(x)|^2 \mu(dx) = T_{1,n} + T_{2,n},$$

where

$$
\begin{aligned}
T_{1,n} \quad = \quad & \mathbf{E}\left[|m_n(X) - Y|^2 | D_n\right] - \mathbf{E}(|m(X) - Y|^2) \\
& -2\left\{\frac{1}{n}\sum_{i=1}^{n} |m_n(X_i) - Y_i|^2 + \lambda^* J_{k^*}^2(\tilde{m}_{n,(k^*,\lambda^*)})\right. \\
& \qquad\qquad \left. -\frac{1}{n}\sum_{i=1}^{n} |m(X_i) - Y_i|^2 + pen_n(k^*, \lambda^*)\right\}
\end{aligned}
$$

and

$$
\begin{aligned}
T_{2,n} \quad = \quad & 2\left\{\frac{1}{n}\sum_{i=1}^{n} |m_n(X_i) - Y_i|^2 + \lambda^* J_{k^*}^2(\tilde{m}_{n,(k^*,\lambda^*)})\right. \\
& \qquad\qquad \left. -\frac{1}{n}\sum_{i=1}^{n} |m(X_i) - Y_i|^2 + pen_n(k^*, \lambda^*)\right\}.
\end{aligned}
$$

STEP 1. We show

$$T_{2,n} \leq 2 \inf_{\lambda \in \Lambda_n} \left\{\lambda J_p^2(m) + pen_n(p, \lambda)\right\}. \tag{21.9}$$

By definition of m_n, the Lipschitz property of T_L, $|Y_i| \leq L$ almost surely, which implies $Y_i = T_L Y_i$ ($i = 1, \ldots, n$) almost surely, and $m \in C^p([0,1])$ we have

$$
\begin{aligned}
T_{2,n} \quad \leq \quad & 2\inf_{\lambda \in \Lambda_n}\left\{\frac{1}{n}\sum_{i=1}^{n} |T_L\tilde{m}_{n,(p,\lambda)}(X_i) - Y_i|^2 + \lambda J_p^2(\tilde{m}_{n,(p,\lambda)})\right. \\
& \qquad\qquad \left. -\frac{1}{n}\sum_{i=1}^{n} |m(X_i) - Y_i|^2 + pen_n(p, \lambda)\right\} \\
\leq \quad & 2\inf_{\lambda \in \Lambda_n}\left\{\frac{1}{n}\sum_{i=1}^{n} |\tilde{m}_{n,(p,\lambda)}(X_i) - Y_i|^2 + \lambda J_p^2(\tilde{m}_{n,(p,\lambda)})\right. \\
& \qquad\qquad \left. -\frac{1}{n}\sum_{i=1}^{n} |m(X_i) - Y_i|^2 + pen_n(p, \lambda)\right\}
\end{aligned}
$$

$$\leq\; 2\inf_{\lambda\in\Lambda_n}\left\{\frac{1}{n}\sum_{i=1}^{n}|m(X_i)-Y_i|^2+\lambda J_p^2(m)\right.$$

$$\left.-\frac{1}{n}\sum_{i=1}^{n}|m(X_i)-Y_i|^2+pen_n(p,\lambda)\right\}$$

$$=\; 2\inf_{\lambda\in\Lambda_n}\left\{\lambda J_p^2(m)+pen_n(p,\lambda)\right\}.$$

STEP 2. Let $t>0$ be arbitrary. We will now show

$$\mathbf{P}\{T_{1,n}>t\}$$

$$\leq\sum_{(k,\lambda)\in\mathcal{K}\times\Lambda_n}\sum_{l=1}^{\infty}\mathbf{P}\left\{\exists f=T_L g,\, g\in C^k([0,1]),\, J_k^2(g)\leq\frac{2^l pen_n(k,\lambda)}{\lambda}:\right.$$

$$\mathbf{E}|f(X)-Y|^2-\mathbf{E}|m(X)-Y|^2$$

$$-\frac{1}{n}\sum_{i=1}^{n}\{|f(X_i)-Y_i|^2-|m(X_i)-Y_i|^2\}$$

$$\left.>\frac{1}{2}\cdot(t+2^l pen_n(k,\lambda)+\mathbf{E}|f(X)-Y|^2-\mathbf{E}|m(X)-Y|^2)\right\}.$$

This follows from

$$\mathbf{P}\{T_{1,n}>t\}$$

$$\leq\mathbf{P}\left\{\mathbf{E}\left[|m_{n,(k^*,\lambda^*)}(X)-Y|^2\big|D_n\right]-\mathbf{E}\left[|m(X)-Y|^2\right]\right.$$

$$-\frac{1}{n}\sum_{i=1}^{n}\{|m_{n,(k^*,\lambda^*)}(X_i)-Y_i|^2-|m(X_i)-Y_i|^2\}$$

$$>\frac{1}{2}\left(t+2\lambda^* J_{k^*}^2(\tilde{m}_{n,(k^*,\lambda^*)})+2pen_n(k^*,\lambda^*)\right.$$

$$\left.\left.+\mathbf{E}\left[|m_{n,(k^*,\lambda^*)}(X)-Y|^2\big|D_n\right]-\mathbf{E}\left[|m(X)-Y|^2\right]\right)\right\}$$

$$\leq \sum_{(k,\lambda)\in\mathcal{K}\times\Lambda_n} \mathbf{P}\Bigg\{ \exists f = T_L g, g \in C^k([0,1]) :$$

$$\mathbf{E}|f(X)-Y|^2 - \mathbf{E}|m(X)-Y|^2$$

$$-\frac{1}{n}\sum_{i=1}^{n}\{|f(X_i)-Y_i|^2 - |m(X_i)-Y_i|^2\}$$

$$> \frac{1}{2}\cdot(t + 2\lambda J_k^2(g) + 2pen_n(k,\lambda)$$

$$+\mathbf{E}|f(X)-Y|^2 - \mathbf{E}|m(X)-Y|^2)\Bigg\}$$

$$\leq \sum_{(k,\lambda)\in\mathcal{K}\times\Lambda_n}\sum_{l=1}^{\infty}\mathbf{P}\Bigg\{ \exists f = T_L g, g \in C^k([0,1]),$$

$$2^l pen_n(k,\lambda) \leq 2\lambda J_k^2(g) + 2pen_n(k,\lambda) < 2^{l+1} pen_n(k,\lambda) :$$

$$\mathbf{E}|f(X)-Y|^2 - \mathbf{E}|m(X)-Y|^2$$

$$-\frac{1}{n}\sum_{i=1}^{n}\{|f(X_i)-Y_i|^2 - |m(X_i)-Y_i|^2\}$$

$$> \frac{1}{2}\cdot(t + 2\lambda J_k^2(g) + 2pen_n(k,\lambda)$$

$$+\mathbf{E}|f(X)-Y|^2 - \mathbf{E}|m(X)-Y|^2)\Bigg\}$$

$$\leq \sum_{(k,\lambda)\in\mathcal{K}\times\Lambda_n}\sum_{l=1}^{\infty}\mathbf{P}\Bigg\{ \exists f = T_L g, g \in C^k([0,1]), J_k^2(g) \leq 2^l\frac{pen_n(k,\lambda)}{\lambda} :$$

$$\mathbf{E}|f(X)-Y|^2 - \mathbf{E}|m(X)-Y|^2$$

$$-\frac{1}{n}\sum_{i=1}^{n}\{|f(X_i)-Y_i|^2 - |m(X_i)-Y_i|^2\}$$

$$> \frac{1}{2}\cdot(t + 2^l pen_n(k,\lambda) + \mathbf{E}|f(X)-Y|^2 - \mathbf{E}|m(X)-Y|^2)\Bigg\}.$$

STEP 3. Fix $(k, \lambda) \in \mathcal{K} \times \Lambda_n$ and $l \in \mathcal{N}$. As in the proof of Theorem 21.1 (cf. (21.6)) one can show that, for n sufficiently large,

$$\mathbf{P}\Bigg\{\exists f = T_L g, g \in C^k([0, 1]), J_k^2(g) \leq \frac{2^l pen_n(k, \lambda)}{\lambda} :$$

$$\mathbf{E}|f(X) - Y|^2 - \mathbf{E}|m(X) - Y|^2$$

$$-\frac{1}{n}\sum_{i=1}^n \{|f(X_i) - Y_i|^2 - |m(X_i) - Y_i|^2\}$$

$$> \frac{1}{2} \cdot (t + 2^l pen_n(k, \lambda) + \mathbf{E}|f(X) - Y|^2 - \mathbf{E}|m(X) - Y|^2)\Bigg\}$$

$$\leq c_3 \exp\left(-c_4 \frac{n \cdot (t + 2^l pen_n(k, \lambda))}{L^4}\right) \tag{21.10}$$

(cf. Problem 21.2).

STEP 4. Next we demonstrate, for n sufficiently large,

$$\mathbf{E}T_{1,n} \leq \frac{c_5}{n}.$$

Using the results of Steps 2 and 3 we get, for n sufficiently large,

$$
\begin{aligned}
\mathbf{E}T_{1,n} \quad &\leq \quad \int_0^\infty \mathbf{P}\{T_{1,n} > t\}dt \\[2mm]
&\leq \quad \sum_{(k,\lambda)\in\mathcal{K}\times\Lambda_n} \sum_{l=1}^\infty \int_0^\infty c_3 \exp\left(-\frac{c_4 n \cdot (t + 2^l pen_n(k, \lambda))}{L^4}\right) dt \\[2mm]
&= \quad \sum_{(k,\lambda)\in\mathcal{K}\times\Lambda_n} \sum_{l=1}^\infty \exp\left(-\frac{c_4 n 2^l pen_n(k, \lambda)}{L^4}\right) \cdot \frac{c_3 \cdot L^4}{c_4 n} \\[2mm]
&\leq \quad \sum_{(k,\lambda)\in\mathcal{K}\times\Lambda_n} \sum_{l=1}^\infty \exp\left(-c_4 2^l L \cdot (\log(n))^{2-1/(2k)}\right) \cdot \frac{c_3 L^4}{c_4 n} \\[2mm]
&\leq \quad c_6 n \cdot (\log(n))^{1/(2)} \exp\left(-2\log(n)\right) \cdot \frac{c_3 L^4}{c_4 n} \\[2mm]
&\leq \quad \frac{c_5}{n}.
\end{aligned}
$$

STEP 5. We now conclude the proof. By the results of Steps 1 and 4 we get, for n sufficiently large,

$$\mathbf{E} \int |m_n(x) - m(x)|^2 \mu(dx)$$

$$\leq 2 \inf_{\lambda \in \Lambda_n} \left\{ \lambda \cdot J_p^2(m) + \frac{L^5(\log(n))^2}{n \cdot \lambda^{1/(2p)}} \right\} + \frac{c_5}{n}.$$

Clearly, this implies the assertion of part (a). Concerning (b), assume $0 < J_p^2(m) < \infty$ and set

$$\lambda^* = \left(\frac{L^5(\log(n))^2}{n \cdot J_p^2(m)} \right)^{2p/(2p+1)}.$$

Then for n sufficiently large there exists $\bar\lambda \in \Lambda_n$ such that

$$\lambda^* \leq \bar\lambda \leq 2\lambda^*.$$

It follows that

$$\mathbf{E} \int |m_n(x) - m(x)|^2 \mu(dx)$$

$$\leq 2 \left\{ \bar\lambda \cdot J_p^2(m) + \frac{L^5(\log(n))^2}{n \cdot \bar\lambda^{1/(2p)}} \right\} + \frac{c_5}{n}$$

$$\leq 2 \left\{ 2\lambda^* \cdot J_p^2(m) + \frac{L^5(\log(n))^2}{n \cdot (\lambda^*)^{1/(2p)}} \right\} + \frac{c_5}{n}$$

$$\leq 6 \cdot (J_p^2(m))^{1/(2p+1)} \left(\frac{L^5(\log(n))^2}{n} \right)^{2p/(2p+1)} + \frac{c_5}{n}$$

$$= O\left((J_p^2(m))^{1/(2p+1)} (\log(n))^2 n^{-2p/(2p+1)} \right).$$

$$\square$$

21.3   Bibliographic notes

Theorem 21.2 is due to Kohler, Krzyżak, and Schäfer (2002). In the context of fixed design regression the rate of convergence of (univariate) smoothing spline estimates was investigated in Rice and Rosenblatt (1983), Shen (1998), Speckman (1985), and Wahba (1975). Cox (1984) studied the rate of convergence of multivariate penalized least squares estimates. Application of complexity regularization to smoothing spline estimates for fixed design regression was conisidered in van de Geer (2001).

Problems and Exercises

PROBLEM 21.1. Prove Lemma 21.1.

HINT: Start with the error decomposition

$$\int |m_{n,(k,\lambda)}(x) - m(x)|^2 \mu(dx) = T_{1,n} + T_{2,n},$$

where

$$
\begin{aligned}
T_{1,n} \;=\; & \mathbf{E}\left[|m_{n,(k,\lambda)}(X) - Y|^2 | D_n\right] - \mathbf{E}(|m(X) - Y|^2) \\[2mm]
& -2\left\{ \frac{1}{n} \sum_{i=1}^{n} |m_{n,(k,\lambda)}(X_i) - Y_i|^2 + \lambda J_k^2(\tilde{m}_{n,(k,\lambda)}) \right.\\[2mm]
& \qquad\qquad \left. -\frac{1}{n} \sum_{i=1}^{n} |m(X_i) - Y_i|^2 + pen_n(k,\lambda) \right\}
\end{aligned}
$$

and

$$
\begin{aligned}
T_{2,n} \;=\; & 2\left\{ \frac{1}{n} \sum_{i=1}^{n} |m_{n,(k,\lambda)}(X_i) - Y_i|^2 + \lambda J_k^2(\tilde{m}_{n,(k,\lambda)}) \right.\\[2mm]
& \qquad\qquad \left. -\frac{1}{n} \sum_{i=1}^{n} |m(X_i) - Y_i|^2 + pen_n(k,\lambda) \right\}.
\end{aligned}
$$

As in the proof of Theorem 21.2 show, for n sufficiently large and any $t > 0$,

$$\mathbf{P}\{T_{1,n} > t\} \leq \sum_{l=1}^{\infty} c_3 \exp\left(-c_4 \frac{n(t + 2^l pen_n(k,\lambda))}{L^4}\right) \leq c_3 \exp\left(-c_4 \frac{n \cdot t}{L^4}\right).$$

Conclude

$$\mathbf{P}\left\{ T_{1,n} > L^4 \frac{\log(n)}{n} \right\} \leq c_3 \cdot \exp(-c_4 \log(n)) \leq \eta$$

for n sufficiently large, which implies the assertion.

PROBLEM 21.2. Prove (21.10).

HINT: Show that, for $(k,\lambda) \in \mathcal{K} \times \Lambda_n$,

$$pen_n(k,\lambda) \geq c_1 \frac{\log n}{n \cdot \lambda^{1/(2k)}} + c_2 \frac{\log n}{n}$$

and apply (21.6).

PROBLEM 21.3. Formulate and prove a multivariate version of Theorem 21.1.

PROBLEM 21.4. Formulate and prove a multivariate version of Lemma 21.1. Use it to define adaptive penalized least squares estimates for multivariate data and formulate and prove a multivariate version of Theorem 21.2.

22
Dimension Reduction Techniques

We know from Chapter 2 that the estimation of a regression function is especially difficult if the dimension of X is large. One consequence of this is that the optimal minimax rate of convergence $n^{-2k/(2k+d)}$ for the estimation of a k times differentiable regression function converges to zero rather slowly if the dimension d of X is large compared to k. The only possibility of circumventing this so-called curse of dimensionality is to impose additional assumptions on the regression functions. Such assumptions will be discussed in this chapter.

In the classical linear model one assumes

$$Y = (X, \beta) + \epsilon,$$

where $\beta \in \mathcal{R}^d$, $\mathbf{E}\epsilon = 0$, and X, ϵ are independent. Here

$$m(x) = (x, \beta) = \sum_{j=1}^{d} \beta_j x^{(j)}$$

is a linear function of the components of x. This rather restrictive parametric assumption can be generalized in various ways.

For **additive models**, one assumes that $m(x)$ is a sum of univariate functions $m_j : \mathcal{R} \to \mathcal{R}$ applied to the components of x, i.e.,

$$m(x) = \sum_{j=1}^{d} m_j(x^{(j)}).$$

In **projection pursuit** one generalizes this further by assuming that $m(x)$ is a sum of univariate functions m_j applied to projections of x onto various directions $\beta_j \in \mathcal{R}^d$:

$$m(x) = \sum_{j=1}^{K} m_j\left((x, \beta_j)\right).$$

For **single index models** one assumes $K = 1$, i.e., one assumes that the regression function is given by

$$m(x) = F((x, \beta)),$$

where $F : \mathcal{R} \to \mathcal{R}$ and $\beta \in \mathcal{R}^d$. In the literature, additive and single index models are called semiparametric models.

In the next three sections we discuss estimates which use the above assumptions to simplify the regression estimation problem.

22.1 Additive Models

In this section we assume that the regression function is an additive function of its components, i.e.,

$$m(x) = \sum_{j=1}^{d} m_j(x^{(j)}).$$

This assumption can be used to simplify the problem of regression estimation by fitting only functions to the data which have the same additive structure. We have seen already two principles which can be used to fit a function to the data: least squares and penalized least squares. In the sequel we will use the least squares principle to construct an estimate of an additive regression function.

Assume that $X \in [0,1]^d$ a.s. Let $M \in \mathcal{N}_0$ and $K_n \in \mathcal{N}$. Let $\mathcal{F}_n^{(1)}$ be the set of all piecewise polynomials of degree M (or less) w.r.t. an equidistant partition of $[0,1]$ into K_n intervals, and put

$$\mathcal{F}_n = \left\{ f : \mathcal{R}^d \to \mathcal{R} \ : \ f(x) = \sum_{j=1}^{d} f_j(x^{(j)}) \text{ for some } f_j \in \mathcal{F}_n^{(1)} \right\}.$$

Let

$$\hat{m}_n = \arg\min_{f \in \mathcal{F}_n} \frac{1}{n} \sum_{i=1}^{n} (Y_i - f(X_i))^2 .$$

Then define the estimate by

$$m_n(x) = T_L \hat{m}_n(x).$$

With a slight modification of the proof of Corollary 11.2 we can get

Theorem 22.1. *Let $C > 0$, $p = q + r$, $q \in \{0, \ldots, M\}$, $r \in (0, 1]$. Assume that the distribution of (X, Y) satisfies $X \in [0, 1]^d$ a.s.,*

$$\sigma^2 = \sup_{x \in [0,1]^d} \mathbf{Var}\{Y | X = x\} < \infty,$$

$$\|m\|_\infty = \sup_{x \in [0,1]^d} |m(x)| \leq L,$$

and

$$m(x) = m_1(x^{(1)}) + \cdots + m_d(x^{(d)})$$

for some (p, C)-smooth functions $m_j : [0, 1] \to \mathcal{R}$. Then

$$\mathbf{E}\left\{ \int |m_n(x) - m(x)|^2 \mu(dx) \right\}$$

$$\leq c \cdot \max\{\sigma^2, L^2\} \frac{(\log(n) + 1) \cdot d \cdot K_n \cdot (M + 1)}{n} + 8 \frac{d^2}{2^{2p} q!^2} \cdot \frac{C^2}{K_n^{2p}}$$

and for

$$K_n = \left\lceil \left(\frac{C^2}{\max\{\sigma^2, L^2\}} \frac{n}{\log(n)} \right)^{1/(2p+1)} \right\rceil$$

one gets, for n sufficiently large,

$$\mathbf{E}\left\{ \int |m_n(x) - m(x)|^2 \mu(dx) \right\}$$

$$\leq c_{M,d} C^{\frac{2}{2p+1}} \cdot \left(\max\{\sigma^2, L^2\} \cdot \frac{(\log(n) + 1)}{n} \right)^{\frac{2p}{2p+1}}$$

for some constant $c_{M,d}$ depending only on M and d.

Notice that the above rate of convergence does not depend on the dimension d.

PROOF. Lemma 11.1, together with $X \in [0, 1]^d$ a.s., implies

$$\inf_{f \in \mathcal{F}_n} \int |f(x) - m(x)|^2 \mu(dx)$$

$$= \inf_{f_j \in \mathcal{F}_n^{(1)}} \int |\sum_{j=1}^d f_j(x^{(j)}) - \sum_{j=1}^d m_j(x^{(j)})|^2 \mu(dx)$$

$$\leq d \cdot \inf_{f_j \in \mathcal{F}_n^{(1)}} \int \sum_{j=1}^d |f_j(x^{(j)}) - m_j(x^{(j)})|^2 \mu(dx)$$

$$\leq d \cdot \inf_{f_j \in \mathcal{F}_n(1)} \sum_{j=1}^{d} \sup_{x^{(j)} \in [0,1]} |f_j(x^{(j)}) - m_j(x^{(j)})|^2$$

$$\leq \frac{d^2}{2^{2p} q!^2} \cdot \frac{C^2}{K_n^{2p}}.$$

From this together with Theorem 11.3 one gets the first inequality. The definition of K_n implies the second inequality. $\qquad \square$

If we compare Theorem 22.1 with Corollary 11.2, we see that the assumption, that the regression function is additive, enables us to derive in the multivariate regression problem the same rate of convergence as in the univariate regression problem.

A straightforward generalization of the above result is to fit, instead of a sum of univariate functions applied to one of the components of X, a sum of functions of $d^* < d$ of the components of X to the data. In this case one can show that if the regression function itself is a sum of functions of $d^* < d$ of the components of X, and if these functions are all (p, C)-smooth, then the L_2 error converges to zero with the rate $n^{-2p/(2p+d^*)}$ (instead of $n^{-2p/(2p+d)}$) (cf. Problem 22.1). Here the rate of convergence is again independent of the dimension d of X.

22.2 Projection Pursuit

In projection pursuit one assumes that $m(x)$ is a sum of univariate functions, where each of these univariate functions is applied to the projection of x onto some vector $\beta_j \in \mathcal{R}^d$:

$$m(x) = \sum_{j=1}^{K} m_j \left((x, \beta_j) \right). \tag{22.1}$$

This is a generalization of additive models in such a way that the components of X are replaced by the projections (X, β_j). As we know from Lemma 16.2, any regression function can be approximated arbitrarily closely by functions of the form (22.1), hence the assumption (22.1) is much less restrictive than the additive model. But, on the other hand, the fitting of a function of the form (22.1) to the data is much more complicated than the fitting of an additive function.

We again use the principle of least squares to construct an estimate of the form (22.1). Assume $X \in [0, 1]^d$ *a.s.* Let $M \in \mathcal{N}_0$, $K_n \in \mathcal{N}$, and let $\mathcal{F}_n$ be the set of all piecewise polynomials of degree M (or less) w.r.t. an equidistant partition of $[-1, 1]$ into K_n intervals. In order to simplify the computation of the covering number we use, in the definition of the estimate, only those functions from $\mathcal{F}_n$, which are bounded in absolute value by some constant $B > 0$ and which are Lipschitz-continuous with some constant $A_n > 0$ (i.e., which satisfy $|f(x) - f(z)| \le A_n |x - z|$ for all $x, z \in [-1, 1]$). Let $\mathcal{F}_n(B, A_n)$ be the subset of $\mathcal{F}_n$ consisting of all these functions.

We define our estimate by

$$m_n(x) = \sum_{j=1}^{K} g_j^* \left((x, b_j^*) \right),$$

where

$$(g_1^*, b_1^*, \ldots, g_K^*, b_K^*)$$

$$= \arg \min_{g_1, b_1, \ldots, g_K, b_K : g_j \in \mathcal{F}_n(B, A_n), \|b_j\| \le 1/\sqrt{d}} \frac{1}{n} \sum_{i=1}^{n} \left| Y_i - \sum_{j=1}^{K} g_j \left((X_i, b_j) \right) \right|^2.$$

The assumption $X \in [0, 1]^d$ a.s., together with $\|b_j\| \le 1/\sqrt{d}$, implies $(X_i, b_j) \in [-1, 1]$, so $g_j((X_i, b_j))$ is defined.

Theorem 22.2. *Let $L > 0$, $C > 0$, $p = q + r$, $q \in \{0, \ldots, M\}$, $r \in (0, 1]$. Assume $X \in [0, 1]^d$ a.s., $|Y| \le L$ a.s., and*

$$m(x) = \sum_{j=1}^{K} m_j \left((x, \beta_j) \right) \tag{22.2}$$

for some (p, C)-smooth functions $m_j : [-1, 1] \to \mathcal{R}$ and some $\beta_j \in \mathcal{R}^d$, $\|\beta_j\| \le 1/\sqrt{d}$. Choose A_n such that

$$A_n \to \infty \quad (n \to \infty) \quad and \quad \frac{A_n}{\log(n)} \to 0 \quad (n \to \infty),$$

set $B = L + 1$, $M = q$, $K_n = \lceil (C^2 n / \log(n))^{1/(2p+1)} \rceil$, and define the estimate as above. Then there exists a constant $c > 0$ depending only on $L, M, d,$ and K such that, for n sufficiently large,

$$\mathbf{E} \left\{ \int |m_n(x) - m(x)|^2 \mu(dx) \right\} \le c \cdot C^{\frac{2}{2p+1}} \cdot \left(\frac{\log(n)}{n} \right)^{\frac{2p}{2p+1}}.$$

In the proof we will apply the techniques introduced in Chapter 11 for the analysis of nonlinear least squares estimates. These results require that Y is bounded rather than that the conditional variance of Y given X is bounded as in Theorem 22.1. As in Theorem 22.1 we get, in Theorem 22.2 for the multivariate regression estimation problem up to a logarithmic factor, the same rate of convergence as for the univariate regression estimation problem. But this time the assumption on the structure of the regression function is much less restrictive than in the additive model (cf. Lemma 16.2).

PROOF. In order to prove Theorem 22.2 we apply our standard techniques for the analysis of least squares estimates introduced in Chapter 11. The only two new things we need are bounds on the covering numbers and on the approximation error.

We use the error decomposition

$$\int |m_n(x) - m(x)|^2 \mu(dx) = T_{1,n} + T_{2,n},$$

where

$$T_{2,n} = 2 \cdot \frac{1}{n} \sum_{i=1}^{n} \left(|m_n(X_i) - Y_i|^2 - |m(X_i) - Y_i|^2 \right)$$

and

$$T_{1,n} = \mathbf{E}\left\{ |m_n(X) - Y|^2 - |m(X) - Y|^2 \big| D_n \right\} - T_{2,n}.$$

By (11.12),

$$\mathbf{E}\{T_{2,n}\}$$

$$\leq 2 \inf_{g_j, b_j : g_j \in \mathcal{F}_n(L+1, A_n), \|b_j\| \leq 1/\sqrt{d}} \int \left| \sum_{j=1}^{K} g_j((x, b_j)) - m(x) \right|^2 \mu(dx)$$

and by the assumptions on m we get

$$\left| \sum_{j=1}^{K} g_j((x, b_j)) - m(x) \right|^2 = \left| \sum_{j=1}^{K} (g_j((x, b_j)) - m_j((x, \beta_j))) \right|^2$$

$$\leq K \cdot \sum_{j=1}^{K} |g_j((x, b_j)) - m_j((x, \beta_j))|^2.$$

Set $b_j = \beta_j$ and choose g_j according to Lemma 11.1 (observe that for n sufficiently large we have $g_j \in \mathcal{F}_n(L+1, A_n)$). Then

$$|g_j((x, b_j)) - m_j((x, \beta_j))| \leq \sup_{u \in [-1,1]} |g(u) - m_j(u)|$$

$$\leq \frac{1}{2^p q!} \cdot \frac{C}{(K_n/2)^p}$$

which implies that, for n sufficiently large, we have

$$\mathbf{E}\{T_{2,n}\} \leq 2 \cdot K^2 \cdot \frac{1}{q!^2} \cdot \frac{C^2}{K_n^{2p}}. \tag{22.3}$$

Next we bound $\mathbf{E}\{T_{1,n}\}$. Let $\mathcal{G}_n$ be the set of all functions

$$g(x) = \sum_{j=1}^{K} g_j((x, b_j)) \quad (g_j \in \mathcal{F}_n(L+1, A_n), \|b_j\| \leq 1/\sqrt{d}).$$

As in the proof of Theorem 11.5 one gets, for arbitrary $t \geq 1/n$,

$$\mathbf{P}\left\{T_{1,n} > t\right\}$$

$$\leq 14 \sup_{x_1^n} \mathcal{N}_1\left(\frac{1}{80K \cdot (L+1) \cdot n}, \mathcal{G}_n, x_1^n\right)$$

$$\times \exp\left(-\frac{n}{24 \cdot 214 \cdot ((L+1)K)^4} \cdot t\right).$$

Next we bound the covering number. Let $\mathcal{H}_n$ be the set of all functions

$$h(x) = f((x,b)) \quad (f \in \mathcal{F}_n(L+1, A_n), \|b\| \leq 1/\sqrt{d}).$$

By Lemma 16.4 we get

$$\mathcal{N}_1\left(\frac{1}{80K \cdot (L+1) \cdot n}, \mathcal{G}_n, x_1^n\right) \leq \left(\mathcal{N}_1\left(\frac{1}{80K^2 \cdot (L+1) \cdot n}, \mathcal{H}_n, x_1^n\right)\right)^K.$$

Set $\delta = 1/(80K^2(L+1)n)$. Choose $b_1, \ldots, b_N \in \mathcal{R}^d$ such that for each $b \in \mathcal{R}^d$, $\|b\| \leq 1/\sqrt{d}$, there exists $j \in \{1, \ldots, N\}$ with

$$\|b - b_j\| \leq \frac{\delta}{2A_n\sqrt{d}}$$

and such that

$$N \leq \left(\frac{2A_n d}{\delta}\right)^d.$$

Then the Lipschitz-continuity of $f \in \mathcal{F}_n(L+1, A_n)$ implies, for $x \in [0,1]^d$,

$$|f((x,b)) - f((x,b_j))| \leq A_n|(x, b-b_j)| \leq A_n\sqrt{d} \cdot \|b-b_j\| \leq \frac{\delta}{2}.$$

This proves

$$\mathcal{N}_1(\delta, \mathcal{H}_n, x_1^n) \leq \sum_{j=1}^{N} \mathcal{N}_1\left(\frac{\delta}{2}, \{f((x,b_j)) \ : \ f \in \mathcal{F}_n(L+1, A_n)\}, x_1^n\right).$$

Since

$$\{f((x,b_j)) \ : \ f \in \mathcal{F}_n(L+1, A_n)\}$$

is a subspace of a linear vector space of dimension $K_n(M+1)$ we get, by Theorems 9.4 and 9.5,

$$\begin{aligned}
\mathcal{N}_1(\delta, \mathcal{H}_n, x_1^n) &\leq \sum_{j=1}^{N} 3\left(\frac{3e(2(L+1))}{\delta/2}\right)^{2(K_n(M+1)+1)} \\
&\leq \left(\frac{2A_n d}{\delta}\right)^d 3\left(\frac{3e(2(L+1))}{\delta/2}\right)^{2K_n(M+1)+2} \\
&\leq 3\left(\frac{12e(A_n + L + 1)d}{\delta}\right)^{2K_n(M+1)+d+2}.
\end{aligned}$$

Summarizing the above results we have

$$\mathcal{N}_1\left(\frac{1}{80K\cdot L\cdot n},\mathcal{G}_n,x_1^n\right)$$

$$\leq 3^K\left(\frac{12e(A_n+L+1)d}{1/(80K^2(L+1)n)}\right)^{(2K_n(M+1)+d+2)\cdot K}$$

$$\leq 3^K\left(960eK^2(L+1)(A_n+L+1)d\cdot n\right)^{(2K_n(M+1)+d+2)\cdot K}.$$

As in the proof of Theorem 11.5 one concludes from this

$$\mathbf{E}\{T_{1,n}\}\leq c\cdot\frac{K_n\log(n)}{n}. \tag{22.4}$$

The assertion follows from (22.3), (22.4), and the definition of K_n. $\qquad\square$

As long as the m_j are not linear functions, the functions of the form (22.1) are not linear in β_j. Therefore computation of the estimate of Theorem 22.2 above requires solving a nonlinear least squares problem, which is not possible in practice. What one can do instead is to use a stepwise approach to construct a similar estimate.

Assume that

$$Y = m(X) + \epsilon,$$

where $m : \mathcal{R}^d \to \mathcal{R}$, $\mathbf{E}\epsilon = 0$, and X,ϵ are independent. Assume, furthermore, that m has the form (22.1). Then

$$Y - \sum_{j=2}^{K} m_j((X,\beta_j)) = m((X,\beta_1)) + \epsilon$$

and m_1 is the regression function to the random vector $(\tilde{X},\tilde{Y})$, where

$$\tilde{X} = (X,\beta_1), \quad \tilde{Y} = Y - \sum_{j=2}^{K} m_j((X,\beta_j)).$$

Hence if we know all β_j and m_j except m_1, then we can compute an estimate of m_1 by applying an arbitrary univariate regression estimate to the data $(\tilde{X}_1,\tilde{Y}_1)$, ..., $(\tilde{X}_n,\tilde{Y}_n)$. In addition, we can do this for various values of β_1, use, e.g., cross-validation to estimate the L_2 risk of the corresponding estimates and choose β_1 such that the estimated L_2 risk of the corresponding estimate is as small as possible. In this way we can compute one of the (m_j,β_j) as soon as we know all of the other (m_k,β_k) $(k\neq j)$. By doing this in a stepwise manner one gets an algorithm for fitting a function of the form (22.1) to the data. For details, see, e.g., Hastie et al. (2001).

22.3 Single Index Models

In this section we study the so-called single index model

$$m(x) = F((x, \beta)),   \tag{22.5}$$

where $\beta \in \mathcal{R}^d$ and the function $F : \mathcal{R} \to \mathcal{R}$ can be arbitrary. We consider it as a special case of projection pursuit with $K = 1$.

The obvious disadvantage of considering only $K = 1$ is that not any function can be approximated arbitrarily closely by functions of the form (22.5). But, on the other hand, setting $K = 1$ simplifies the iterative algorithm described above. Furthermore, an estimate of the form (22.5) can easily be interpreted: The estimates changes only in direction β_j, and the way it changes in this direction is described by the univariate function F. In contrast, for projection pursuit, the estimate is a sum of functions which changes in various directions β_j, and although each of these functions can be plotted to visualize the way they look, it is hard to imagine how the sum of all these functions behaves.

In the sequel we consider again least squares estimates as in Section 22.2. Assume $X \in [0, 1]^d$ a.s., let $\mathcal{F}_n(L + 1, A_n)$ be defined as in Section 22.2, and set

$$m_n(x) = g^*((x, b^*)),$$

where

$$(g^*, b^*) = \arg \min_{g \in \mathcal{F}_n(L+1, A_n), \|b\| \le 1/\sqrt{d}} \frac{1}{n} \sum_{i=1}^{n} |Y_i - g((X_i, b))|^2.$$

Setting $K = 1$ in Theorem 22.2 we get

Corollary 22.1. *Let $L > 0$, $C > 0$, $p = q + r$, $q \in \{0, \dots, M\}$, $r \in (0, 1]$. Assume $X \in [0, 1]^d$ a.s., $|Y| \le L$ a.s., and*

$$m(x) = F((x, \beta))   \tag{22.6}$$

for some (p, C)-smooth function $F : [-1, 1] \to \mathcal{R}$ and some $\beta \in \mathcal{R}^d$, $\|\beta\| \le 1/\sqrt{d}$. Choose A_n such that

$$A_n \to \infty \quad (n \to \infty) \quad and \quad \frac{A_n}{\log(n)} \to 0 \quad (n \to \infty),$$

set $M = q$, $K_n = \lceil (C^2 n / \log(n))^{1/(2p+1)} \rceil$, and define the estimate as above. Then there exists a constant $c > 0$ depending only on $L, M,$ and d such that, for n sufficiently large,

$$\mathbf{E}\left\{ \int |m_n(x) - m(x)|^2 \mu(dx) \right\} \le c \cdot C^{\frac{2}{2p+1}} \cdot \left(\frac{\log(n)}{n} \right)^{\frac{2p}{2p+1}}.$$

22.4 Bibliographic Notes

The additive model and its generalizations have been investigated by Andrews and Whang (1990), Breiman (1993), Breiman and Friedman (1985), Burman (1990), Chen (1991), Hastie and Tibshirani (1990), Bickel et al. (1993), Huang (1998), Kohler (1998), Linton (1997), Linton and Härdle (1996), Linton and Nielsen (1995), Newey (1994), Stone (1985; 1994), and Wahba et al. (1995).

Projection pursuit was proposed by Friedman and Tukey (1974) and specialized to regression estimation by Friedman and Stuetzle (1981).

In the literature there are conditions for the unique identification of β in the single index model, see, e.g., Horowitz (1998), Ichimura (1993), Manski (1988), Powel (1994), Powel, Stock, and Stoker (1989), and Stoker (1991). Consistency and the rate of convergence were proved under conditions on the underlying distributions, like: M has some derivatives, X has a smooth density, etc. (Amemiya (1985), Davidson and MacKinnon (1993), Gallant (1987), Ichimura (1993), and Robinson (1987; 1988)).

For the single index model a simple estimate is based on the observation that, for differentiable F,

$$\text{grad } m(x) = \text{grad } F((x, \beta)) = \beta F'((x, \beta)),$$

therefore,

$$\mathbf{E}\{\text{grad } m(X)\} = \beta \mathbf{E}\{F'((X, \beta))\},$$

so if the expected gradient can be estimated with a good rate then we get a multiple of β. This is the principle of the average derivative estimate (Härdle and Stoker (1989)).

Concerning some general results on dimension reduction we refer to Hall (1988), Hristache et al. (2001), Nicoleris and Yatracos (1997), Samarov (1993) and Zhang (1991).

Problems and Exercises

PROBLEM 22.1. Assume that the regression function is a sum of the (p, C)-smooth functions of $d^* < d$ of the components of X. Use the principle of least squares to construct an estimate that fits a sum of the multivariate piecewise polynomials of $d^* < d$ of the components of X to the data. Show that the L_2 error of this estimate converges to zero with the rate $(\log(n)/n)^{-2p/(2p+d^*)}$ if the parameters of this estimate are chosen in a suitable way (cf. Kohler (1998)).

PROBLEM 22.2. Use the complexity regularization principle to define an adaptive version of the estimate in Theorem 22.1.

PROBLEM 22.3. Modify the definition of the estimate in Theorem 22.2 in such a way that the resulting estimate is weakly and strongly universally consistent.

PROBLEM 22.4. Use the complexity regularization principle to define an adaptive version of the estimate in Theorem 22.2.

23

Strong Consistency of Local Averaging Estimates

23.1 Partitioning Estimates

For a statistician the individual development of an estimation sequence is also of interest, therefore, in this chapter we discuss the strong consistency of local averaging estimates. Consider first the partitioning estimate in the case of bounded Y, the strong universal consistency of a modified partitioning estimate, and finally the strong universal consistency of the original partitioning estimate. The notations of Chapter 4 will be used.

Theorem 23.1. *Under the conditions (4.1) and (4.2) the partitioning estimate is strongly consistent if $|Y| \leq L$ with probability one for some $L < \infty$.*

For the proof we shall use the following special version of the Banach-Steinhaus theorem for integral operators in $L_1(\mu)$.

Theorem 23.2. *Let $K_n(x, z)$ be functions on $\mathcal{R}^d \times \mathcal{R}^d$ satisfying the following conditions:*
   (i)   There is a constant $c > 0$ such that, for all n,

$$\int |K_n(x, z)| \mu(dx) \leq c$$

   for μ-almost all z.
   (ii) There is a constant $D \geq 1$ such that

$$\int |K_n(x, z)| \mu(dz) \leq D$$

for all x and n.

(iii) *For all $a > 0$,*

$$\lim_{n\to\infty} \int \int |K_n(x,z)| I_{\{\|x-z\|>a\}} \mu(dz)\mu(dx) = 0.$$

(iv)

$$\lim_{n\to\infty} ess\ sup_x \left| \int K_n(x,z)\mu(dz) - 1 \right| = 0.$$

Then, for all $m \in L_1(\mu)$,

$$\lim_{n\to\infty} \int \left| m(x) - \int K_n(x,z)m(z)\mu(dz) \right| \mu(dx) = 0.$$

PROOF. The set of the continuous functions of compact support is dense in $L_1(\mu)$ by Theorem A.1, so choose a continuous function of compact support (thus uniformly continuous and bounded) $\tilde{m}$ such that

$$\int |m(x) - \tilde{m}(x)|\mu(dx) < \epsilon.$$

Then

$$\int \left| m(x) - \int K_n(x,z)m(z)\mu(dz) \right| \mu(dx)$$

$$\leq \int |m(x) - \tilde{m}(x)|\mu(dx)$$

$$+ \int \left| \tilde{m}(x)(1 - \int K_n(x,z)\mu(dz)) \right| \mu(dx)$$

$$+ \int \left| \int K_n(x,z)(\tilde{m}(x) - \tilde{m}(z))\mu(dz) \right| \mu(dx)$$

$$+ \int \left| \int K_n(x,z)(\tilde{m}(z) - m(z))\mu(dz) \right| \mu(dx)$$

$$= I_1 + I_2 + I_3 + I_4.$$

By the choice of $\tilde{m}$,

$$I_1 < \epsilon.$$

By condition (iv),

$$I_2 \leq \sup_u \left| 1 - \int K_n(u,z)\mu(dz) \right| \int |\tilde{m}(x)|\mu(dx) \to 0,$$

where $\tilde{m}$ is uniformly continuous, therefore it is possible to choose $\delta > 0$ such that $\|x - z\| < \delta$ implies $|\tilde{m}(x) - \tilde{m}(z)| < \epsilon$. Let $S_{x,\delta}$ be the sphere centered at x with radius δ and denote its complement by $S^c_{x,\delta}$. Then

$$
\begin{aligned}
I_3 \;\le\;& \int\!\!\int_{S_{x,\delta}} |K_n(x,z)||\tilde{m}(x)-\tilde{m}(z)|\mu(dz)\mu(dx) \\
&+ \int\!\!\int_{S^c_{x,\delta}} |K_n(x,z)||\tilde{m}(x)-\tilde{m}(z)|\mu(dz)\mu(dx) \\
\le\;& \epsilon \int\!\!\int_{S_{x,\delta}} |K_n(x,z)|\mu(dz)\mu(dx) \\
&+ 2\sup_x |\tilde{m}(x)| \int\!\!\int_{S^c_{x,\delta}} |K_n(x,z)|\mu(dz)\mu(dx),
\end{aligned}
$$

therefore, by (ii) and (iii),

$$
\limsup_{n\to\infty} I_3 \le \epsilon D.
$$

For the last term apply (i),

$$
\begin{aligned}
I_4 \;\le\;& \int\!\!\int |K_n(x,z)|\mu(dx)|\tilde{m}(z)-m(z)|\mu(dz) \\
\le\;& \sup_{u,n} \int |K_n(x,u)|\mu(dx) \int |\tilde{m}(z)-m(z)|\mu(dz) \\
\le\;& c\epsilon.
\end{aligned}
$$

$\square$

We need two additional lemmas. Set

$$
m_n^*(x) = \frac{\sum_{i=1}^n Y_i I_{\{X_i \in A_n(x)\}}}{n\mu(A_n(x))}.
$$

Lemma 23.1. *Under the conditions (4.1) and (4.2),*

$$
\mathbf{E}\left\{ \int |m(x)-m_n^*(x)|\mu(dx) \right\} \to 0.
$$

PROOF. By the triangle inequality

$$
\mathbf{E}\left\{ \int |m(x)-m_n^*(x)|\mu(dx) \right\}
$$
$$
\le \int |m(x)-\mathbf{E}m_n^*(x)|\mu(dx) + \mathbf{E}\left\{ \int |m_n^*(x)-\mathbf{E}m_n^*(x)|\mu(dx) \right\}.
$$

The first term on the right-hand side is called the bias, and the second term is called the variation of m_n^*. Introduce the notation

$$
K_n(x,z) = \sum_{j=1}^{\infty} I_{\{x \in A_{n,j},\, z \in A_{n,j}\}} = I_{\{z \in A_n(x)\}} = I_{\{x \in A_n(z)\}}
$$

and

$$K_n^*(x, z) = \frac{K_n(x, z)}{\int K_n(x, u)\mu(du)} = \frac{I_{\{z \in A_n(x)\}}}{\mu(A_n(x))}.$$

Then

$$\mathbf{E}m_n^*(x) = \int K_n^*(x, z)m(z)\mu(dz).$$

It is easy to see that conditions (ii) and (iv) of Theorem 23.2 are fulfilled for the bias. A simple argument shows that

$$\begin{aligned}
\int K_n^*(x, z)\mu(dx) &= \int \frac{I_{\{z \in A_n(x)\}}}{\mu(A_n(x))}\mu(dx) \\
&= \int \frac{I_{\{x \in A_n(z)\}}}{\mu(A_n(z))}\mu(dx) \\
&= 1
\end{aligned} \tag{23.1}$$

for μ-almost all z, therefore (i) holds. To verify (iii) let $a > 0$ and S a sphere centered at the origin. Then

$$\begin{aligned}
&\int\int_{\{z:\|x-z\|>a\}} K_n^*(x, z)\mu(dz)\mu(dx) \\
&= \sum_j \int_{A_{n,j}} \int_{\{z:\|x-z\|>a\}} K_n^*(x, z)\mu(dz)\mu(dx) \\
&= \sum_j \int_{A_{n,j}} \int_{\{z:\|x-z\|>a\}} \frac{I_{\{z \in A_{n,j}\}}}{\mu(A_{n,j})}\mu(dz)\mu(dx) \\
&= \sum_{j:A_{n,j}\cap S \neq \emptyset} \int_{A_{n,j}} \frac{\mu(\{z : \|x - z\| > a\} \cap A_{n,j})}{\mu(A_{n,j})}\mu(dx) \\
&\quad + \sum_{j:A_{n,j}\cap S = \emptyset} \int_{A_{n,j}} \frac{\mu(\{z : \|x - z\| > a\} \cap A_{n,j})}{\mu(A_{n,j})}\mu(dx).
\end{aligned}$$

By (4.1) the first term on the right-hand side is zero for sufficiently large n since the maximal diameter of $A_{n,j}$ becomes smaller than a. The second term is less than $\sum_{j:A_{n,j}\cap S = \emptyset} \mu(A_{n,j}) \leq \mu(S^c)$. So the bias tends to 0. Next consider the variation term. Let l_n be the number of cells of the partition

$\mathcal{P}_n$ that intersect S. Then

$$\mathbf{E}\left\{\int |m_n^*(x) - \mathbf{E}m_n^*(x)|\mu(dx)\right\}$$

$$= \sum_j \mathbf{E}\left\{\int_{A_{n,j}} \left|\frac{\sum_{i=1}^n Y_i I_{\{X_i \in A_{n,j}\}}}{n\mu(A_{n,j})} - \mathbf{E}\frac{\sum_{i=1}^n Y_i I_{\{X_i \in A_{n,j}\}}}{n\mu(A_{n,j})}\right| \mu(dx)\right\}$$

$$= \frac{1}{n} \sum_{j:A_{n,j}\cap S\neq\emptyset} \mathbf{E}\left\{\left|\sum_{i=1}^n Y_i I_{\{X_i \in A_{n,j}\}} - \mathbf{E}\sum_{i=1}^n Y_i I_{\{X_i \in A_{n,j}\}}\right|\right\}$$

$$+ \frac{1}{n} \sum_{j:A_{n,j}\cap S=\emptyset} \mathbf{E}\left\{\left|\sum_{i=1}^n Y_i I_{\{X_i \in A_{n,j}\}} - \mathbf{E}\sum_{i=1}^n Y_i I_{\{X_i \in A_{n,j}\}}\right|\right\}$$

and, therefore,

$$\mathbf{E}\left\{\int |m_n^*(x) - \mathbf{E}m_n^*(x)|\mu(dx)\right\}$$

$$\leq \frac{1}{n} \sum_{j:A_{n,j}\cap S\neq\emptyset} \sqrt{\mathbf{E}\left\{\left(\sum_{i=1}^n Y_i I_{\{X_i \in A_{n,j}\}} - \mathbf{E}\sum_{i=1}^n Y_i I_{\{X_i \in A_{n,j}\}}\right)^2\right\}}$$

$$+ \frac{1}{n} \sum_{j:A_{n,j}\cap S=\emptyset} 2Ln\mu(A_{n,j})$$

$$\leq \frac{1}{n} \sum_{j:A_{n,j}\cap S\neq\emptyset} \sqrt{nL^2\mu(A_{n,j})} + 2L\mu(S^c)$$

$$\leq \sum_{j:A_{n,j}\cap S\neq\emptyset} L\sqrt{\frac{\mu(A_{n,j})}{n}} + 2L\mu(S^c)$$

$$\leq Ll_n\sqrt{\frac{\frac{1}{l_n}\sum_{j:A_{n,j}\cap S\neq\emptyset}\mu(A_{n,j})}{n}} + 2L\mu(S^c)$$

$$\text{(by Jensen's inequality)}$$

$$\leq L\sqrt{\frac{l_n}{n}} + 2L\mu(S^c) \to 2L\mu(S^c),$$

by the use of (4.2). $\qquad\square$

Lemma 23.2. *Let (4.1) and (4.2) hold. Then, for each $\epsilon > 0$,*

$$\mathbf{P}\left\{\int |m(x) - m_n^*(x)|\mu(dx) > \epsilon\right\} \leq e^{-n\epsilon^2/(32L^2)}$$

if n is large enough.

PROOF. We begin with the decomposition

$$|m(x) - m_n^*(x)|$$
$$= \mathbf{E}|m(x) - m_n^*(x)| + \Big(|m(x) - m_n^*(x)| - \mathbf{E}|m(x) - m_n^*(x)|\Big).$$

$$(23.2)$$

The first term on the right-hand side of (23.2) converges in L_1 by Lemma 23.1. We use Theorem A.2 to obtain an exponential bound for the second term on the right-hand side of (23.2). Fix the training data $(x_1, y_1), \ldots, (x_n, y_n) \in \mathcal{R}^d \times [-L, L]$, and replace (x_i, y_i) by $(\widehat{x}_i, \widehat{y}_i)$ thus changing the value of $m_n^*(x)$ to $m_{n,i}^*(x)$. Then $m_n^*(x) - m_{n,i}^*(x)$ differs from zero only on $A_n(x_i)$ and $A_n(\widehat{x}_i)$, and thus

$$\left| \int |m(x) - m_n^*(x)|\mu(dx) - \int |m(x) - m_{n,i}^*(x)|\mu(dx) \right|$$
$$\leq \int |m_n^*(x) - m_{n,i}^*(x)|\mu(dx)$$
$$\leq \left(\frac{2L}{n\mu(A_n(x_i))}\mu(A_n(x_i)) + \frac{2L}{n\mu(A_n(\widehat{x}_i))}\mu(A_n(\widehat{x}_i)) \right)$$
$$\leq \frac{4L}{n}.$$

By Theorem A.2, we have that, for sufficiently large n,

$$\mathbf{P}\left\{ \int |m(x) - m_n^*(x)|\mu(dx) > \epsilon \right\}$$
$$\leq \mathbf{P}\left\{ \int |m(x) - m_n^*(x)|\mu(dx) - \mathbf{E} \int |m(x) - m_n^*(x)|\mu(dx) > \frac{\epsilon}{2} \right\}$$
$$\leq e^{-n\epsilon^2/(32L^2)}.$$

$\square$

PROOF OF THEOREM 23.1. Because of

$$|m_n(x) - m(x)|^2 \leq 2L|m_n(x) - m(x)|$$

it suffices to show that

$$\lim_{n \to \infty} \int |m_n(x) - m(x)|\mu(dx) = 0$$

with probability one. Introduce the notation

$$B_n = \left\{ x : \sum_{i=1}^{n} K_n(x, X_i) > 0 \right\},$$

where $K_n(x, z) = I_{\{z \in A_n(x)\}}$, i.e., B_n is the set of x's whose cell is nonempty. Write

$$\int |m_n(x) - m(x)| \mu(dx)$$

$$\leq \int |m_n(x) - m_n^*(x)| \mu(dx) + \int |m_n^*(x) - m(x)| \mu(dx).$$

By Lemma 23.2 and the Borel-Cantelli lemma

$$\int |m_n^*(x) - m(x)| \mu(dx) \to 0$$

with probability one. On the other hand, if $x \in B_n$, then

$$|m_n^*(x) - m_n(x)|$$

$$= \left| \frac{\sum_{i=1}^n K_n(x, X_i) Y_i}{n \int K_n(x, z) \mu(dz)} - \frac{\sum_{i=1}^n K_n(x, X_i) Y_i}{\sum_{i=1}^n K_n(x, X_i)} \right|$$

$$\leq L \sum_{i=1}^n K_n(x, X_i) \left| \frac{1}{n \int K_n(x, z) \mu(dz)} - \frac{1}{\sum_{i=1}^n K_n(x, X_i)} \right|$$

$$= L \left| \frac{\sum_{i=1}^n K_n(x, X_i)}{n \int K_n(x, z) \mu(dz)} - 1 \right|$$

$$= L |M_n^*(x) - 1|,$$

where $M_n^*(x)$ is the special form of $m_n^*(x)$ for $Y \equiv 1$. If $x \in B_n^c$, then

$$|m_n^*(x) - m_n(x)| = 0 \leq L |M_n^*(x) - 1|.$$

Therefore, by Lemma 23.2,

$$\int |m_n(x) - m_n^*(x)| \mu(dx) \leq L \int |M_n^*(x) - 1| \mu(dx) \to 0$$

with probability one, and the proof is complete. $\square$

It is not known if under (4.1) and (4.2) the standard partitioning estimate is strongly universally consistent. We will prove the universal consistency under some additional mild conditions. In Theorem 23.3 the partitioning estimate is modified such that the estimate is 0 if there are few points in the actual cell, while in Theorem 23.4 we don't change the partition too frequently.

Consider the following modification of the standard partitioning estimate:

$$m_n'(x) = \begin{cases} \dfrac{\sum_{i=1}^n Y_i I_{\{X_i \in A_n(x)\}}}{\sum_{i=1}^n I_{\{X_i \in A_n(x)\}}} & \text{if } \sum_{i=1}^n I_{\{X_i \in A_n(x)\}} > \log n, \\ 0 & \text{otherwise.} \end{cases}$$

Theorem 23.3. *Assume (4.1). If for each sphere S centered at the origin*

$$\lim_{n \to \infty} \frac{|\{j : A_{n,j} \cap S \neq \emptyset\}| \log n}{n} = 0, \qquad (23.3)$$

then m_n' is strongly universally consistent.

The following lemma is sometimes useful to extend consistencies from bounded Y's to unbounded Y's.

Lemma 23.3. *Let m_n be a local averaging regression function estimate with subprobability weights $\{W_{n,i}(x)\}$ that is strongly consistent for all distributions of (X, Y) such that Y is bounded with probability one. Assume that there is a constant c such that for all Y with $\mathbf{E}\{Y^2\} < \infty$,*

$$\limsup_{n \to \infty} \sum_{i=1}^{n} Y_i^2 \int W_{n,i}(x) \mu(dx) \leq c \mathbf{E}\{Y^2\} \quad \textit{with probability one.}$$

Then m_n is strongly universally consistent.

PROOF. Fix $\epsilon > 0$. Choose $L > 0$ such that

$$\mathbf{E}\{|Y_L - Y|^2\} < \epsilon,$$

where

$$Y_L = \begin{cases} L & \text{if } Y > L, \\ Y & \text{if } -L \leq Y \leq L, \\ -L & \text{if } Y < -L. \end{cases}$$

For $j \in \{1, \ldots, n\}$ set

$$Y_{j,L} = \begin{cases} L & \text{if } Y_j > L, \\ Y_j & \text{if } -L \leq Y_j \leq L, \\ -L & \text{if } Y_j < -L. \end{cases}$$

Let m_L and $m_{n,L}$ be the functions m and m_n when Y and $\{Y_j\}$ are replaced by Y_L and $\{Y_{j,L}\}$. Then

$$\int (m_n(x) - m(x))^2 \mu(dx)$$

$$\leq 3 \left(\int (m_n(x) - m_{n,L}(x))^2 \mu(dx) + \int (m_{n,L}(x) - m_L(x))^2 \mu(dx) \right.$$

$$\left. + \int (m_L(x) - m(x))^2 \mu(dx) \right).$$

Because of the conditions, for all L,

$$\int (m_{n,L}(x) - m_L(x))^2 \mu(dx) \to 0 \quad \text{with probability one.}$$

By Jensen's inequality,

$$
\int (m_L(x) - m(x))^2 \mu(dx)) \;=\; \mathbf{E}\left\{ (\mathbf{E}\{Y_L|X\} - \mathbf{E}\{Y|X\})^2 \right\}
$$
$$
\leq\; \mathbf{E}\left\{ (Y_L - Y)^2 \right\}
$$
$$
<\; \epsilon
$$

by the choice of L. We may apply another version of Jensen's inequality, together with the fact that the weights are subprobability weights, to bound the first term:

$$
\limsup_{n\to\infty} \int (m_n(x) - m_{n,L}(x))^2 \mu(dx)
$$
$$
=\; \limsup_{n\to\infty} \int \left(\sum_{i=1}^{n} W_{n,i}(x)(Y_i - Y_{i,L}) \right)^2 \mu(dx)
$$
$$
\leq\; \limsup_{n\to\infty} \int \sum_{i=1}^{n} W_{n,i}(x)(Y_i - Y_{i,L})^2 \mu(dx)
$$
$$
=\; \limsup_{n\to\infty} \sum_{i=1}^{n} \int W_{n,i}(x)\mu(dx)(Y_i - Y_{i,L})^2
$$
$$
\leq\; c\mathbf{E}\left\{ (Y - Y_L)^2 \right\} \quad \text{(by the condition of the theorem)}
$$
$$
<\; c\epsilon \quad \text{with probability one,}
$$

by the choice of L. $\qquad\square$

PROOF OF THEOREM 23.3. For fixed S and $c = e^2$, let

$$
\begin{aligned}
I_n &= \{j : A_{n,j} \cap S \neq 0\}, \\
J_n &= \{j : \mu_n(A_{n,j}) > \log n/n\}, \\
L_n &= \{j : \mu(A_{n,j}) > c\log n/n\}, \\
F_n &= \bigcup_{j \in I_n \cap J_n^c} A_{n,j}.
\end{aligned}
$$

First we show that, for $|Y| \leq L$,

$$
\int |m_n'(x) - m(x)|\mu(dx) \to 0 \ \text{ a.s.} \tag{23.4}
$$

By Theorem 23.1,

$$
\int |m_n(x) - m(x)|\mu(dx) \to 0 \ \text{ a.s.,}
$$

so we need that

$$
\int |m_n'(x) - m_n(x)|\mu(dx) \to 0 \ \text{ a.s.}
$$

Because of

$$\int |m'_n(x) - m_n(x)|\mu(dx) \le 2L(\mu(S^c) + \mu(F_n))$$

it suffices to show that

$$\mu(F_n) \to 0 \quad \text{a.s.}$$

One has

$$\mu(F_n) \le \sum_{j \in I_n \cap L_n^c} \mu(A_{n,j}) + \sum_{j \in I_n \cap L_n \cap J_n^c} \mu(A_{n,j})$$

$$\le c\frac{\log n}{n}|I_n| + \sum_{j \in I_n \cap L_n} \mu(A_{n,j})I_{\{c\mu_n(A_{n,j}) < \mu(A_{n,j})\}}.$$

The first term on the right-hand side tends to zero by condition (23.3). For the second term, Lemma A.1 with $\epsilon = \mu(A)/c$ implies

$$\mathbf{P}\{c\mu_n(A) < \mu(A)\} = \mathbf{P}\left\{n\mu_n(A) < \frac{n\mu(A)}{c}\right\}$$

$$\le e^{-n(\mu(A) - \frac{\mu(A)}{c} + \frac{\mu(A)}{c}\log\frac{1}{c})}$$

$$= e^{-n\mu(A)(1 - \frac{1}{c} - \frac{\log c}{c})}.$$

Thus, for any $0 < \epsilon < 1$,

$$\mathbf{P}\left\{\sum_{j \in I_n \cap L_n} \mu(A_{n,j})I_{\{c\mu_n(A_{n,j}) < \mu(A_{n,j})\}} > \epsilon\right\}$$

$$\le \mathbf{P}\left\{\sum_{j \in I_n \cap L_n} \mu(A_{n,j})I_{\{c\mu_n(A_{n,j}) < \mu(A_{n,j})\}} > \epsilon \sum_{j \in I_n \cap L_n} \mu(A_{n,j})\right\}$$

$$\le \sum_{j \in I_n \cap L_n} \mathbf{P}\left\{I_{\{c\mu_n(A_{n,j}) < \mu(A_{n,j})\}} > \epsilon\right\}$$

$$= \sum_{j \in I_n \cap L_n} \mathbf{P}\{c\mu_n(A_{n,j}) < \mu(A_{n,j})\}$$

$$\le \sum_{j \in I_n \cap L_n} e^{-n\mu(A_{n,j})(1 - \frac{1}{c} - \frac{\log c}{c})}$$

$$\le \sum_{j \in I_n} e^{-(\log n) \cdot [c(1 - \frac{1}{c} - \frac{\log c}{c})]}$$

$$= |I_n|n^{-c(1 - \frac{1}{c} - \frac{\log c}{c})} = |I_n|n^{-(e^2 - 3)},$$

therefore by condition (23.3) we have, for sufficiently large n,

$$\mathbf{P}\left\{ \sum_{j \in I_n \cap L_n} \mu(A_{n,j}) I_{\{c\mu_n(A_{n,j}) < \mu(A_{n,j})\}} > \epsilon \right\} < n^{-e^2 + 4},$$

which is summable. This proves (23.4).

Now apply Lemma 23.3, the second condition of which is satisfied if

$$\limsup_n \left\{ \max_i n \int W_{n,i}(x)\mu(dx) \right\} \le c \quad \text{a.s.}$$

Let c be as above,

$$B_n = \bigcup_{j \in J_n \cap L_n} A_{n,j}$$

and

$$D_n = \bigcup_{j \in J_n} A_{n,j}.$$

If $J_n \ne \emptyset$, then

$$\max_i n \int W_{n,i}(x)\mu(dx) = \max_{i : A_n(X_i) \subset D_n} \frac{\mu(A_n(X_i))}{\mu_n(A_n(X_i))},$$

and if $J_n = \emptyset$, then $m'_n = 0$ and

$$\max_i n \int W_{n,i}(x)\mu(dx) = 0,$$

therefore

$$\mathbf{P}\left\{\max_i n \int W_{n,i}(x)\mu(dx) > c\right\}$$

$$= \mathbf{P}\left\{J_n \neq \emptyset, \max_{i:A_n(X_i) \subset D_n} \frac{\mu(A_n(X_i))}{\mu_n(A_n(X_i))} > c\right\}$$

$$\leq n \cdot \mathbf{P}\left\{A_n(X_1) \subset D_n, \frac{\mu(A_n(X_1))}{\mu_n(A_n(X_1))} > c\right\}$$

$$= n\mathbf{P}\left\{A_n(X_1) \subset B_n, \frac{\mu(A_n(X_1))}{\mu_n(A_n(X_1))} > c\right\}$$

$$\leq n \sum_{j \in L_n} \mathbf{P}\left\{X_1 \in A_{n,j}, \frac{\mu(A_{n,j})}{\mu_n(A_{n,j})} > c\right\}$$

$$= n \sum_{j \in L_n} \mathbf{P}\left\{X_1 \in A_{n,j}, \frac{n\mu(A_{n,j})}{1 + \sum_{i=2}^n I_{\{X_i \in A_{n,j}\}}} > c\right\}$$

$$= n \sum_{j \in L_n} \mu(A_{n,j})\mathbf{P}\left\{\frac{n\mu(A_{n,j})}{1 + \sum_{i=2}^n I_{\{X_i \in A_{n,j}\}}} > c\right\}$$

$$\leq n \sum_{j \in L_n} \mu(A_{n,j})\mathbf{P}\left\{\frac{\mu(A_{n,j})}{\mu_n(A_{n,j})} > c\right\}$$

$$\leq n \sum_{j \in L_n} \mu(A_{n,j})e^{-n\mu(A_{n,j})(1-\frac{1}{c}-\frac{\log c}{c})}$$

$$\leq n \sum_{j} \mu(A_{n,j})e^{-c\cdot(\log n)\cdot(1-\frac{1}{c}-\frac{\log c}{c})}$$

$$\leq n^{-e^2+4},$$

which is again summable. $\square$

A sequence $\mathcal{P}_n$ of partitions of $\mathcal{R}^d$ by Borel sets is called nested if $A_{n+1}(x) \subseteq A_n(x)$ for all $n \in \mathcal{N}$, $x \in \mathcal{R}^d$.

Theorem 23.4. *Let m_n be a sequence of partitioning estimates with partition sequence $\mathcal{P}_n$ satisfying (4.1) and (4.2). Let indices $n_1, n_2, \ldots$ satisfy $n_{k+1} \geq D \cdot n_k$ for some fixed $D > 1$. Assume that either $\mathcal{P}_{n-1} \neq \mathcal{P}_n$ at most for the indices $n = n_1, n_2, \ldots$ or that the sequence $\mathcal{P}_n$ is nested satisfying $A_n(x) \in \{A_{n_k}(x), A_{n_{k+1}}(x)\}$ for $n \in \{n_k, n_k + 1, \ldots, n_{k+1}\}$ $(k = 1, 2, \ldots, x \in \mathcal{R}^d)$. Then m_n is strongly universally consistent.*

Let $\mathcal{P}_1, \ldots, \mathcal{P}_7$ be the partitions described in Figure 23.1. Then for $n \leq 7$ the first kind of assumptions of Theorem 23.4 is satisfied with $n_1 = 3$ and $n_2 = 7$, and the second kind of assumptions is satisfied with $n_1 = 1$ and $n_2 = 7$.

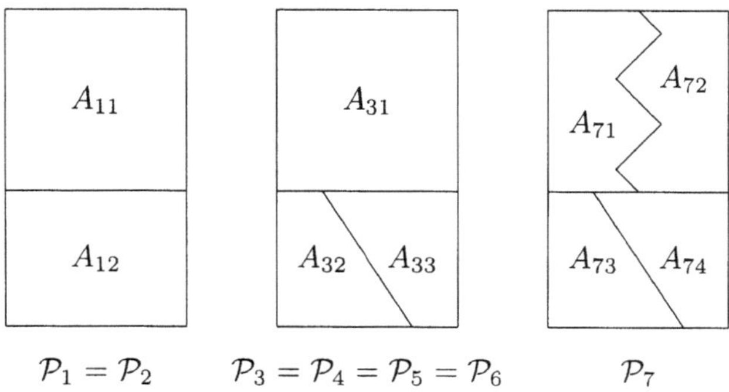

Figure 23.1. A sequence of nested partitions.

If Y is bounded then according to Theorem 23.1 conditions (4.1) and (4.2) imply strong consistency, even for nonnested $\mathcal{P}_n$.

In the following we give a simple example of a partition, which satisfies the second kind of assumptions of Theorem 23.4. Let $d = 1$. Via successive bisections we define partitions of $(0, 1)$: let $\mathcal{Q}_{2^k}$ be

$$\left\{ \left(0, \frac{1}{2^k}\right], \left(\frac{1}{2^k}, \frac{2}{2^k}\right], \ldots, \left(\frac{2^k - 1}{2^k}, 1\right) \right\}$$

and let $\mathcal{Q}_{2^k + j}$ be

$$\left\{ \left(0, \frac{1}{2^{k+1}}\right], \ldots, \left(\frac{2j - 1}{2^{k+1}}, \frac{2j}{2^{k+1}}\right], \left(\frac{2j}{2^{k+1}}, \frac{2j + 2}{2^{k+1}}\right], \ldots, \left(\frac{2^{k+1} - 2}{2^{k+1}}, 1\right) \right\}$$

for $j \in \{1, \ldots, 2^k - 1\}$, $k \in \{0, 1, \ldots\}$. By a fixed continuous bijective transformation of $(0, 1)$ to $\mathcal{R}$ (e.g., by $x \mapsto \tan(\pi(x - 1/2))$) let the partition $\mathcal{Q}_n$ of $(0, 1)$ be transformed into the partition $\tilde{\mathcal{P}}_n$ of $\mathcal{R}$. Then $\tilde{\mathcal{P}}_n$ fulfills the conditions of Theorem 23.4 with $n_k = 2^k$, besides (4.2), and with $\mathcal{P}_n = \tilde{\mathcal{P}}_{\lceil \sqrt{n} \rceil}$ all conditions of Theorem 23.4 are fulfilled.

We need some lemmas for the proof of Theorem 23.4. The following lemma is well-known from the classical Kolmogorov proof and from Etemadi's (1981) proof of the strong law of large numbers for independent and identically distributed integrable random variables.

Lemma 23.4. *For identically distributed random variables $Y_n \geq 0$ with $\mathbf{E}Y_n < \infty$ let Y_n^* be the truncation of Y_n at n, i.e., $Y_n^* := Y_n I_{\{Y_n \leq n\}} + n I_{\{Y_n > n\}}$. Then*

$$\frac{1}{n} \mathbf{E}Y_n^{*^2} \to 0 \ (n \to \infty), \tag{23.5}$$

$$\sum_{n=1}^{\infty} \frac{1}{n^2} \mathbf{E}Y_n^{*^2} < \infty, \tag{23.6}$$

and, for $n_{k+1} \geq Dn_k$ with $D > 1$,

$$\sum_{k=1}^{\infty} \frac{1}{n_k} \mathbf{E} Y_{n_k}^{*2} < \infty. \tag{23.7}$$

PROOF. Noticing that

$$\mathbf{E} Y_n^{*2} = \sum_{i=1}^{n} \int_{(i-1,i]} t^2 \mathbf{P}_Y(dt),$$

one obtains (23.5) from

$$\sum_{i=1}^{\infty} \frac{1}{i} \int_{(i-1,i]} t^2 \mathbf{P}_Y(dt) \leq \sum_{i=1}^{\infty} \int_{(i-1,i]} t \mathbf{P}_Y(dt) = \mathbf{E} Y < \infty$$

by the Kronecker lemma, and (23.6) follows from

$$
\begin{aligned}
\sum_{n=1}^{\infty} \frac{1}{n^2} \mathbf{E} Y_n^{*2} &= \sum_{n=1}^{\infty} \sum_{i=1}^{n} \frac{1}{n^2} \int_{(i-1,i]} t^2 \mathbf{P}_Y(dt) \\
&= \sum_{i=1}^{\infty} \int_{(i-1,i]} t^2 \mathbf{P}_Y(dt) \sum_{n=i}^{\infty} \frac{1}{n^2} \\
&\leq \sum_{i=1}^{\infty} \frac{2}{i} \int_{(i-1,i]} t^2 \mathbf{P}_Y(dt) \\
&\leq 2\mathbf{E} Y < \infty.
\end{aligned}
$$

In view of (23.7), one notices

$$\sum_{k=j}^{\infty} \frac{1}{n_k} \leq \sum_{k=j}^{\infty} \frac{1}{D^{k-j} \cdot n_j} = \frac{1}{n_j} \sum_{l=0}^{\infty} \left(\frac{1}{D}\right)^l = \frac{D}{(D-1)n_j}.$$

Now let m_i denote the minimal index m with $i \leq n_m$. Then

$$
\begin{aligned}
\sum_{k=1}^{\infty} \frac{1}{n_k} \mathbf{E} Y_{n_k}^{*2} &= \sum_{k=1}^{\infty} \frac{1}{n_k} \sum_{i=1}^{n_k} \int_{(i-1,i]} t^2 \mathbf{P}_Y(dt) \\
&= \sum_{i=1}^{\infty} \sum_{k=1}^{\infty} I_{\{i \leq n_k\}} \cdot \frac{1}{n_k} \int_{(i-1,i]} t^2 \mathbf{P}_Y(dt) \\
&= \sum_{i=1}^{\infty} \left(\sum_{k=m_i}^{\infty} \frac{1}{n_k} \right) \int_{(i-1,i]} t^2 \mathbf{P}_Y(dt) \\
&\leq \sum_{i=1}^{\infty} \frac{D}{(D-1) \cdot n_{m_i}} \int_{(i-1,i]} t^2 \mathbf{P}_Y(dt)
\end{aligned}
$$

$$\leq \; \frac{D}{D-1} \sum_{i=1}^{\infty} \frac{1}{i} \int_{(i-1,i]} t^2 \mathbf{P}_Y(dt)$$

$$\leq \; \frac{D}{D-1} \mathbf{E}Y < \infty.$$

$\square$

Lemma 23.5. *Let $K_n : \mathcal{R}^d \times \mathcal{R}^d \to \{0,1\}$ be a measurable function. Assume that a constant $\rho > 0$ exists with*

$$\int \frac{K_n(x,z)}{\int K_n(x,s)\mu(ds)} \mu(dx) \leq \rho \tag{23.8}$$

for all n, all z, and all distributions μ. If $K_{n-1} \neq K_n$ at most for the indices $n = n_1, n_2, \ldots$, where $n_{k+1} \geq D n_k$ for some fixed $D > 1$, then

$$\limsup_{n \to \infty} \int \sum_{i=1}^{n} \frac{Y_i K_n(x, X_i)}{1 + \sum\limits_{j \in \{1,\ldots,n\} \setminus \{i\}} K_n(x, X_j)} \mu(dx) \leq 2\rho \mathbf{E}Y \quad a.s., \tag{23.9}$$

for each integrable $Y \geq 0$.

PROOF. Without loss of generality we may assume $n_{k+1}/n_k \leq 2$ (as long as $n_{l+1}/n_l \geq 2$ for some l insert $\left\lceil \sqrt{n_l \cdot n_{l+1}} \right\rceil$ into the sequence n_k). Set

$$Y^{[N]} := Y I_{\{Y \leq N\}} + N \cdot I_{\{Y > N\}},$$

$$Y_i^{[N]} := Y_i I_{\{Y_i \leq N\}} + N \cdot I_{\{Y_i > N\}},$$

and, mimicking Lemma 23.4,

$$Y_n^* := Y_n I_{\{Y_n \leq n\}} + n I_{\{Y_n > n\}}.$$

Further, set

$$U_{n,i} := \int \frac{Y_i^* K_n(x, X_i)}{1 + \sum\limits_{j \in \{1,\ldots,n\} \setminus \{i\}} K_n(x, X_j)} \mu(dx)$$

and

$$V_{n,i} := U_{n,i} - \mathbf{E}U_{n,i}.$$

In the first step it will be shown

$$\mathbf{E}\left\{ \left(\sum_{i=1}^{N} V_{n,i} \right)^2 \right\} \leq \frac{c}{N} \mathbf{E}Y^{[N]^2} \tag{23.10}$$

for $4 \leq n \leq N \leq 2n$ with some suitable constant c.

Notice

$$\mathbf{E}\frac{1}{(1+B(n,p))^2} \le \frac{2}{(n+1)(n+2)p^2}, \tag{23.11}$$

$$\mathbf{E}\frac{1}{(1+B(n,p))^4} \le \frac{24}{(n+1)(n+2)(n+3)(n+4)p^4}, \tag{23.12}$$

for random variables $B(n,p)$ binominally distributed with parameters n and p (cf. Problem 23.1). Let

$$(X_1, Y_1), \ldots, (X_N, Y_N)$$

and

$$(\tilde{X}_1, \tilde{Y}_1), \ldots, (\tilde{X}_N, \tilde{Y}_N)$$

be i.i.d. copies of (X, Y) and let $U_{n,i,l}$ be obtained from $U_{n,i}$ via replacing (X_l, Y_l) by $(\tilde{X}_l, \tilde{Y}_l)$ ($l = 1, \ldots, N$). By Theorem A.3,

$$\mathbf{E}\left\{\left|\sum_{i=1}^{N} V_{n,i}\right|^2\right\}$$

$$= \mathbf{Var}\left\{\sum_{i=1}^{N} U_{n,i}\right\}$$

$$\le \frac{1}{2}\sum_{l=1}^{N}\mathbf{E}\left\{\left|\sum_{i=1}^{N} U_{n,i} - \sum_{i=1}^{N} U_{n,i,l}\right|^2\right\}$$

$$= \frac{1}{2}\sum_{l=1}^{N}\mathbf{E}\left\{\left|\sum_{i=1}^{N}\int\frac{Y_i^* K_n(x, X_i)}{1 + \sum_{j\in\{1,\ldots,n\}\setminus\{i\}} K_n(x, X_j)}\mu(dx)\right.\right.$$

$$- \sum_{i\in\{1,\ldots,N\}\setminus\{l\}}\int\frac{Y_i^* K_n(x, X_i)}{1 + \sum_{j\in\{1,\ldots,n\}\setminus\{i,l\}} K_n(x, X_j) + K_n(x, \tilde{X}_l)I_{\{l\le n\}}}\mu(dx)$$

$$\left.\left.- \int\frac{\tilde{Y}_l^* K_n(x, \tilde{X}_l)}{1 + \sum_{j\in\{1,\ldots,n\}\setminus\{l\}} K_n(x, X_j)}\mu(dx)\right|^2\right\}$$

$$=: \frac{1}{2}\sum_{l=1}^{N} A_l.$$

For $l \in \{1, \ldots, N\}$ we have

$$A_l$$

$$\leq 2\mathbf{E}\left\{\left|\int \frac{Y_l^* K_n(x, X_l) + \tilde{Y}_l^* K_n(x, \tilde{X}_l)}{1 + \sum\limits_{j \in \{1,\ldots,n\}\setminus\{l\}} K_n(x, X_j)} \mu(dx)\right|^2\right\}$$

$$+ 2\mathbf{E}\left\{\left|\sum\limits_{i \in \{1,\ldots,N\}\setminus\{l\}} \int \frac{Y_i^* K_n(x, X_i)[K_n(x, \tilde{X}_l) + K_n(x, X_l)]}{[1 + \sum\limits_{j \in \{1,\ldots,n\}\setminus\{i,l\}} K_n(x, X_j)]^2} \mu(dx)\right|^2\right\}$$

$$\leq 8\mathbf{E}\left\{\left|\int \frac{Y_l^* K_n(x, X_l)}{1 + \sum\limits_{j \in \{1,\ldots,n\}\setminus\{l\}} K_n(x, X_j)} \mu(dx)\right|^2\right\}$$

$$+ 8\mathbf{E}\left\{\left|\sum\limits_{i \in \{1,\ldots,N\}\setminus\{l\}} \int \frac{Y_i^* K_n(x, X_i) K_n(x, X_l)}{[1 + \sum\limits_{j \in \{1,\ldots,n\}\setminus\{i,l\}} K_n(x, X_j)]^2} \mu(dx)\right|^2\right\}$$

$$\text{(by exchangeability)}$$

$$\leq 8\mathbf{E}\left\{Y_1^{[N]^2} \left(\int \frac{K_n(x, X_1)}{1 + \sum\limits_{j \in \{2,\ldots,n\}} K_n(x, X_j)} \mu(dx)\right)^2\right\}$$

$$+ 8N\mathbf{E}\left\{Y_1^{[N]^2} \left(\int \frac{K_n(x, X_1) K_n(x, X_2)}{[1 + \sum\limits_{j \in \{3,\ldots,n\}} K_n(x, X_j)]^2} \mu(dx)\right)^2\right\}$$

$$+ 8N^2\mathbf{E}\left\{Y_1^{[N]} Y_2^{[N]} \int \frac{K_n(x, X_1) K_n(x, X_3)}{[1 + \sum\limits_{j \in \{4,\ldots,n\}} K_n(x, X_j)]^2} \mu(dx)\right.$$

$$\left. \times \int \frac{K_n(\tilde{x}, X_2) K_n(\tilde{x}, X_3)}{[1 + \sum\limits_{j \in \{4,\ldots,n\}} K_n(\tilde{x}, X_j)]^2} \mu(d\tilde{x})\right\}$$

$$\text{(by exchangeability and } 0 \leq Y_i^* \leq Y_i^{[N]})$$

$$=: \; 8B + 8C + 8D.$$

We have

$$
\begin{aligned}
B &= \mathbf{E}\left\{ \left|Y_1^{[N]}\right|^2 \int\!\!\int \right. \\
&\qquad \left. \frac{K_n(x,X_1)K_n(\tilde{x},X_1)}{[1+\sum\limits_{j\in\{2,\dots,n\}} K_n(x,X_j)][1+\sum\limits_{j\in\{2,\dots,n\}} K_n(\tilde{x},X_j)]}\,\mu(dx)\mu(d\tilde{x}) \right\} \\
&\leq \mathbf{E}\left\{ \left|Y_1^{[N]}\right|^2 \int\!\!\int K_n(x,X_1)\cdot K_n(\tilde{x},X_1) \right. \\
&\qquad \times \left(\mathbf{E}\left\{ \frac{1}{[1+\sum_{j=2}^n K_n(x,X_j)]^2} \right\} \right)^{1/2} \\
&\qquad \left. \times \left(\mathbf{E}\left\{ \frac{1}{[1+\sum_{j=2}^n K_n(\tilde{x},X_j)]^2} \right\} \right)^{1/2} \mu(dx)\mu(d\tilde{x}) \right\}
\end{aligned}
$$

(by independence and by the Cauchy–Schwarz inequality)

$$
\leq 2\mathbf{E}\left\{ \left|Y_1^{[N]}\right|^2 \frac{1}{n^2} \int \frac{K_n(x,X_1)}{\int K_n(x,s)\mu(ds)}\mu(dx) \int \frac{K_n(\tilde{x},X_1)}{\int K_n(\tilde{x},s)\mu(ds)}\mu(d\tilde{x}) \right\}
$$

(by (23.11))

$$
\leq \frac{2\rho^2}{n^2}\mathbf{E}Y^{[N]^2}
$$

(by (23.8)).

In a similar way we obtain

$$
C \leq \frac{17\rho^2 N}{n^3}\mathbf{E}Y^{[N]^2}
$$

by

$$
\frac{K_n(x,X_2)}{1+\sum_{j=3}^n K_n(x,X_j)} \leq \frac{1}{1+\sum_{j=3}^n K_n(x,X_j)} \leq 1
$$

and by the use of (23.11), (23.12), and (23.8), further, via exchangeability,

$$
D \leq 52\rho^2 \frac{N^2}{n^4}\mathbf{E}\left\{ \left|Y^{[N]}\right|^2 \right\}
$$

by the use of (23.12) and (23.8) (cf. Problem 23.2). These bounds yield (23.10).

In the second step a monotonicity argument will be used to show

$$\limsup_{n\to\infty} \sum_{i=1}^{n} \int \frac{Y_i^* K_n(x, X_i)}{1 + \sum_{j\in\{1,\dots,n\}\setminus\{i\}} K_n(x, X_j)}\mu(dx) \leq 2\rho\mathbf{E}Y \quad \text{a.s.} \qquad (23.13)$$

For

$$n_k \leq n \leq n_{k+1} - 1 \leq n_{k+1} \leq 2n_k$$

one has $K_n = K_{n_k}$, which implies

$$\sum_{i=1}^{n} \int \frac{Y_i^* K_n(x, X_i)}{1 + \sum_{j\in\{1,\dots,n\}\setminus\{i\}} K_n(x, X_j)}\mu(dx) \leq \sum_{i=1}^{n_{k+1}-1} U_{n_k,i}. \qquad (23.14)$$

Further

$$\mathbf{E}U_{n,i}$$

$$= \int \mathbf{E}\{Y_i^* K_n(x, X_i)\} \mathbf{E}\left\{\frac{1}{1 + \sum_{j\in\{1,\dots,n\}\setminus\{i\}} K_n(x, X_j)}\right\}\mu(dx)$$

$$\leq \int \mathbf{E}\{Y_i^* K_n(x, X_i)\} \frac{1}{n \cdot \int K_n(x, s)\mu(ds)}\mu(dx)$$

$$\text{(by independence and by Lemma 4.1)}$$

$$= \frac{1}{n}\mathbf{E}\left\{Y_i^* \int \frac{K_n(x, X_i)}{\int K_n(x, s)\mu(ds)}\mu(dx)\right\}$$

$$\leq \frac{\rho}{n}\mathbf{E}\{Y_i^*\}$$

$$\text{(by (23.8))},$$

which implies

$$\sum_{i=1}^{n_{k+1}-1} \mathbf{E}U_{n_k,i} \leq \rho\frac{n_{k+1}}{n_k}\mathbf{E}Y \leq 2\rho\mathbf{E}Y. \qquad (23.15)$$

Using (23.10), Lemma 23.4, and $n_{k+1} \leq 2n_k$ one obtains

$$\mathbf{E}\left\{\sum_{k=1}^{\infty}\left(\sum_{i=1}^{n_{k+1}-1} V_{n_k,i}\right)^2\right\} \leq \sum_{k=1}^{\infty}\frac{c}{n_{k+1}-1}\mathbf{E}\left\{\left|Y^{[n_{k+1}-1]}\right|^2\right\} < \infty,$$

which implies

$$\sum_{i=1}^{n_{k+1}-1} V_{n_k,i} \to 0 \quad (k \to \infty) \quad \text{a.s.} \qquad (23.16)$$

Now (23.16), (23.14), (23.15) yield (23.13).

In the third step we will prove the assertion (23.9). Because of

$$\sum_{i=1}^{\infty} \mathbf{P}\{Y_i \neq Y_i^*\} = \sum_{i=1}^{\infty} \mathbf{P}\{Y > i\} \leq \mathbf{E}Y < \infty$$

one has with probability one $Y_i = Y_i^*$ from a random index on. Then, because of (23.13), it suffices to show that, for each fixed l,

$$\int \frac{K_n(x, X_l)}{1 + \sum\limits_{j \in \{1,\dots,n\}\setminus\{l\}} K_n(x, X_j)} \mu(dx) \to 0 \text{ a.s.}$$

But this follows from

$$\sum_{n=1}^{\infty} \mathbf{E}\left\{\left| \int \frac{K_n(x, X_l)}{1 + \sum\limits_{j \in \{1,\dots,n\}\setminus\{l\}} K_n(x, X_j)} \mu(dx) \right|^2\right\}$$

$$\leq 2\rho^2 \sum_{n=1}^{\infty} \frac{1}{n^2}$$

(see above upper bound for B)

$$< \infty.$$

□

PROOF OF THEOREM 23.4. The assertion holds when $L > 0$ exists with $|Y| \leq L$ (see Theorem 23.1). According to Lemma 23.3 it suffices to show that, for some constant $c > 0$,

$$\limsup_{n \to \infty} \int \frac{\sum_{i=1}^{n} |Y_i| \cdot I_{A_n(x)}(X_i)}{\sum_{i=1}^{n} I_{A_n(x)}(X_i)} \mu(dx) \leq c \cdot \mathbf{E}|Y| \qquad (23.17)$$

for every distribution of (X, Y) with $\mathbf{E}|Y| < \infty$.

W.l.o.g. we can assume $Y \geq 0$ with probability one. Notice that the covering assumption (23.8) is fulfilled with

$$K_n(x, z) = I_{\{z \in A_n(x)\}}$$

and $\rho = 1$ (cf. (23.1)).

If $\mathcal{P}_{n-1} \neq \mathcal{P}_n$ at most for indices $n = n_1, n_2, \dots$, then (23.17) immediately follows from Lemma 23.5.

If $\mathcal{P}_n$ is nested with $A_n(x) \in \{A_{n_k}(x), A_{n_{k+1}}(x)\}$ for $n \in \{n_k, n_k + 1, \dots, n_{k+1}\}$ $(k = 1, 2, \dots, x \in \mathcal{R}^d)$, then let m_n^* and m_n^{**} be the sequences of estimates based on the sequences of partitions $\mathcal{P}_n^*$ and $\mathcal{P}_n^{**}$, respectively, where

$$\mathcal{P}_n^* = \mathcal{P}_{n_k} \quad \text{for } n_k \leq n < n_{k+1}$$

and

$$\mathcal{P}_n^{**} = \mathcal{P}_{n_{k+1}} \quad \text{for } n_k < n \leq n_{k+1}.$$

Then $m_n(x) \in \{m_n^*(x), m_n^{**}(x)\}$ for any $x \in \mathcal{R}^d$, $n > n_1$, which implies

$$m_n(x) \le m_n^*(x) + m_n^{**}(x).$$

By Lemma 23.5, (23.17) is fulfilled for $(m_n^*)_{n \ge n_1}$ and $(m_n^{**})_{n \ge n_1}$ and thus for m_n as well. $\qquad\square$

23.2 Kernel Estimates

For the strong consistency of kernel estimates, we consider a rather general class of kernels.

Definition 23.1. *The kernel K is called regular if it is nonnegative, and if there is a ball $S_{0,r}$ of radius $r > 0$ centered at the origin, and constant $b > 0$ such that*

$$1 \ge K(x) \ge bI_{\{x \in S_{0,r}\}}$$

and

$$\int \sup_{u \in x + S_{0,r}} K(u)\, dx < \infty. \tag{23.18}$$

Theorem 23.5. *Let m_n be the kernel estimate of the regression function m with a regular kernel K. Assume that there is an $L < \infty$ such that $\mathbf{P}\{|Y| \le L\} = 1$. If $h_n \to 0$ and $nh_n^d \to \infty$, then the kernel estimate is strongly consistent.*

For the proof we will need the following four lemmas. Put $K_h(x) = K(x/h)$.

Lemma 23.6. (COVERING LEMMA). *If the kernel is regular then there exists a finite constant $\rho = \rho(K)$ only depending upon K such that, for any $u \in \mathcal{R}^d$, $h > 0$, and probability measure μ,*

$$\int \frac{K_h(x - u)}{\int K_h(x - z)\mu(dz)} \mu(dx) \le \rho.$$

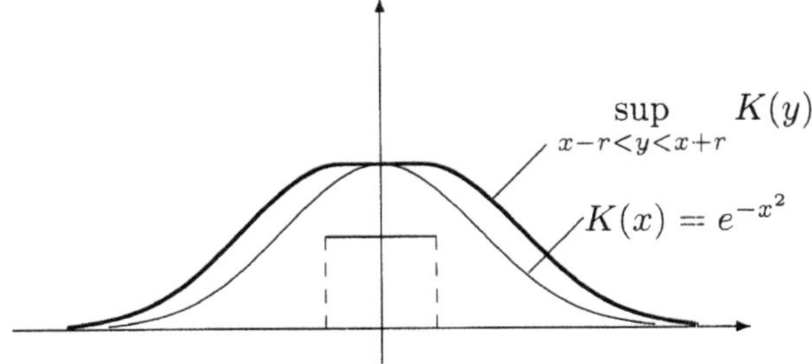

Figure 23.2. Regular kernel.

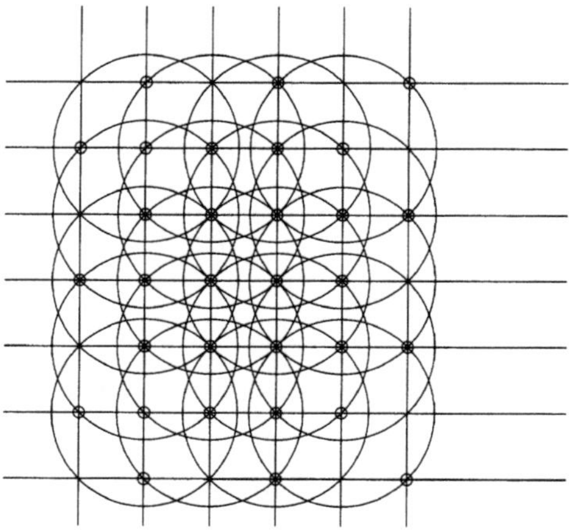

Figure 23.3. An example for a bounded overlap cover of $\mathcal{R}^2$.

Moreover, for any $\delta > 0$

$$\lim_{h \to 0} \sup_u \int \frac{K_h(x - u) I_{\{\|x-u\|>\delta\}}}{\int K_h(x - z)\mu(dz)} \mu(dx) = 0.$$

PROOF. First take a bounded overlap cover of $\mathcal{R}^d$ with translates of $S_{0,r/2}$, where $r > 0$ is the constant appearing in the definition of a regular kernel.

This cover has an infinite number of member balls, but every x gets covered at most k_1 times, where k_1 depends upon d only. The centers of the balls are called x_i $(i = 1, 2, \ldots)$. The integral condition (23.18) on K implies that

$$\sum_{i=1}^{\infty} \sup_{z \in x_i + S_{0,r/2}} K(z)$$

$$= \sum_{i=1}^{\infty} \frac{1}{\int_{S_{0,r/2}} dx} \int_{x \in S_{x_i,r/2}} \sup_{z \in x_i + S_{0,r/2}} K(z)\, dx$$

$$\leq \frac{1}{\int_{S_{0,r/2}} dx} \int \sum_{i=1}^{\infty} I_{\{x \in S_{x_i,r/2}\}} \sup_{z \in x + S_{0,r}} K(z)\, dx$$

(because $x \in S_{x_i,r/2}$ implies $x_i + S_{0,r/2} \subseteq x + S_{0,r}$ cf. Figure 5.8)

$$\leq \frac{k_1}{\int_{S_{0,r/2}} dx} \int \sup_{z \in x + S_{0,r}} K(z)\, dx \leq k_2$$

for another finite constant k_2. Furthermore,

$$K_h(x - u) \leq \sum_{i=1}^{\infty} \sup_{x \in u + hx_i + S_{0,rh/2}} K_h(x - u),$$

and, for $x \in u + hx_i + S_{0,rh/2}$,

$$\int K_h(x - z)\mu(dz) \geq b\mu(x + S_{0,rh}) \geq b\mu(u + hx_i + S_{0,rh/2}),$$

from which we conclude

$$\int \frac{K_h(x - u)}{\int K_h(x - z)\mu(dz)}\mu(dx)$$

$$\leq \sum_{i=1}^{\infty} \int_{x \in u + hx_i + S_{0,rh/2}} \frac{K_h(x - u)}{\int K_h(x - z)\mu(dz)}\mu(dx)$$

$$\leq \sum_{i=1}^{\infty} \int_{x \in u + hx_i + S_{0,rh/2}} \frac{\sup_{z \in hx_i + S_{0,rh/2}} K_h(z)}{b\mu(u + hx_i + S_{0,rh/2})}\mu(dx)$$

$$= \sum_{i=1}^{\infty} \frac{\mu(u + hx_i + S_{0,rh/2})\sup_{z \in hx_i + S_{0,rh/2}} K_h(z)}{b\mu(u + hx_i + S_{0,rh/2})}$$

$$\leq \frac{1}{b}\sum_{i=1}^{\infty} \sup_{z \in x_i + S_{0,r/2}} K(z) \leq \frac{k_2}{b},$$

where k_2 depends on K and d only. To obtain the second statement in the lemma, substitute $K_h(z)$ above by $K_h(z)I_{\{\|z\|>\delta\}}$ and notice that

$$\int \frac{K_h(x - u)I_{\{\|x-u\|>\delta\}}}{\int K_h(x - z)\mu(dz)}\mu(dx) \leq \frac{1}{b}\sum_{i=1}^{\infty} \sup_{z \in x_i + S_{0,r/2}} K(z)I_{\{\|z\|>\delta/h\}} \to 0$$

as $h \to 0$ by dominated convergence. $\qquad\square$

Lemma 23.7. *Let $0 < h \leq R < \infty$, and let $S \subset \mathcal{R}^d$ be a ball of radius R. Then, for every probability measure μ,*

$$\int_S \frac{1}{\sqrt{\mu(S_{x,h})}}\mu(dx) \leq \left(1 + \frac{R}{h}\right)^{d/2} c_d,$$

where c_d depends upon the dimension d only.

The proof of Lemma 23.7 is left to the reader (cf. Problem 23.3).
Define

$$m_n^*(x) = \frac{\sum_{i=1}^{n} Y_i K_{h_n}(x - X_i)}{n\mathbf{E}K_{h_n}(x - X)}.$$

Lemma 23.8. *Under the conditions of Theorem 23.5,*

$$\lim_{n \to \infty} \int \mathbf{E}|m(x) - m_n^*(x)|\mu(dx) = 0.$$

PROOF. By the triangle inequality

$$\int \mathbf{E}|m(x) - m_n^*(x)|\mu(dx)$$

$$\leq \int |m(x) - \mathbf{E}m_n^*(x)|\mu(dx) + \int \mathbf{E}|m_n^*(x) - \mathbf{E}m_n^*(x)|\mu(dx).$$

Concerning the first term on the right-hand side verify the conditions of Theorem 23.2 for

$$K_n(x, z) = \frac{K\left(\frac{x-z}{h_n}\right)}{\int K\left(\frac{x-u}{h_n}\right)\mu(du)}.$$

Part (i) follows from the covering lemma with $c = \rho$. Parts (ii) and (iv) are obvious. For (iii) note that again by the covering lemma,

$$\int \int K_n(x, z) I_{\{\|x-z\| > a\}} \mu(dz)\mu(dx)$$

$$= \int \frac{\int K\left(\frac{x-z}{h_n}\right) I_{\{\|x-z\| > a\}}\mu(dz)}{\int K\left(\frac{x-u}{h_n}\right)\mu(du)} \mu(dx) \to 0.$$

For the second term we have, with $h = h_n$,

$$\mathbf{E}\left\{|m_n^*(x) - \mathbf{E}m_n^*(x)|\right\}$$

$$\leq \sqrt{\mathbf{E}\left\{|m_n^*(x) - \mathbf{E}m_n^*(x)|^2\right\}}$$

$$= \sqrt{\frac{\mathbf{E}\left\{\left(\sum_{j=1}^n (Y_j K_h(x - X_j) - \mathbf{E}\{YK_h(x - X)\})\right)^2\right\}}{n^2(\mathbf{E}K_h(x - X))^2}}$$

$$= \sqrt{\frac{\mathbf{E}\left\{(YK_h(x - X) - \mathbf{E}\{YK_h(x - X)\})^2\right\}}{n(\mathbf{E}K_h(x - X))^2}}$$

$$\leq \sqrt{\frac{\mathbf{E}\left\{(YK_h(x - X))^2\right\}}{n(\mathbf{E}K_h(x - X))^2}}$$

$$\leq L\sqrt{\frac{\mathbf{E}\left\{(K_h(x - X))^2\right\}}{n(\mathbf{E}K_h(x - X))^2}}$$

$$\leq L\sqrt{K_{max}}\sqrt{\frac{\mathbf{E}K((x - X)/h)}{n(\mathbf{E}K((x - X)/h))^2}}$$

$$\leq L\sqrt{\frac{K_{max}}{b}}\sqrt{\frac{1}{n\mu(S_{x,h})}},$$

where we used the Cauchy-Schwarz inequality. Next we use the inequality above to show that the integral converges to zero. Divide the integral over $\mathcal{R}^d$ into two terms, namely, an integral over a large ball S centered at the origin, of radius $R > 0$, and an integral over S^c. For the integral outside the ball we have

$$\int_{S^c} \mathbf{E}\left\{|\mathbf{E}m_n^*(x) - m_n^*(x)|\right\}\mu(dx) \leq 2\int_{S^c} \mathbf{E}\left\{|m_n^*(x)|\right\}\mu(dx) \leq 2L\mu(S^c),$$

which can be small by the choice of the ball S. To bound the integral over S we employ Lemma 23.7:

$$
\begin{aligned}
\int_S \mathbf{E}\left\{|\mathbf{E}m_n^*(x) - m_n^*(x)|\right\}\mu(dx) \;\leq\;& L\sqrt{\frac{K_{max}}{b}}\frac{1}{\sqrt{n}}\int_S \frac{1}{\sqrt{\mu(S_{x,h})}}\mu(dx) \\
& \text{(by the inequality obtained above)} \\
\leq\;& L\sqrt{\frac{K_{max}}{b}}\frac{1}{\sqrt{n}}\left(1+\frac{R}{h}\right)^{d/2}c_d \\
\to\;& 0 \quad \text{(by assumption } nh^d \to \infty\text{).}
\end{aligned}
$$

Therefore,

$$\mathbf{E}\left\{\int |m(x) - m_n^*(x)|\mu(dx)\right\} \to 0.$$

$\square$

Lemma 23.9. *For n large enough*

$$\mathbf{P}\left\{\int |m(x) - m_n^*(x)|\mu(dx) > \epsilon\right\} \leq e^{-n\epsilon^2/(8L\rho)^2)}.$$

PROOF. We use a decomposition, as in the proof of strong consistency of the partitioning estimate,

$$
\begin{aligned}
& \int |m(x) - m_n^*(x)|\mu(dx) \\
=\;& \int \mathbf{E}|m(x) - m_n^*(x)|\mu(dx) \\
& + \int \Big(|m(x) - m_n^*(x)| - \mathbf{E}|m(x) - m_n^*(x)|\Big)\mu(dx).
\end{aligned}
$$

$$(23.19)$$

The first term on the right-hand side tends to 0 by Lemma 23.8. It remains to show that the second term on the right-hand side of (23.19) is small with large probability. To do this, we use McDiarmid's inequality (Theorem A.2) for

$$\int |m(x) - m_n^*(x)|\mu(dx) - \mathbf{E}\left\{\int |m(x) - m_n^*(x)|\mu(dx)\right\}.$$

Fix the training data at $((x_1, y_1), \ldots, (x_n, y_n))$ and replace the ith pair (x_i, y_i) by $(\widehat{x}_i, \widehat{y}_i)$, changing the value of $m_n^*(x)$ to $m_{ni}^*(x)$. Clearly, by the covering lemma (Lemma 23.6),

$$\int |m(x) - m_n^*(x)|\mu(dx) - \int |m(x) - m_{ni}^*(x)|\mu(dx)$$

$$\leq \int |m_n^*(x) - m_{ni}^*(x)|\mu(dx)$$

$$\leq \sup_{y \in \mathcal{R}^d} \int \frac{2LK_h(x-y)}{n\mathbf{E}K_h(x-X)}\mu(dx)$$

$$\leq \frac{2L\rho}{n}.$$

So by Theorem A.2, for n large enough,

$$\mathbf{P}\left\{\int |m(x) - m_n^*(x)|\mu(dx) > \epsilon\right\}$$

$$\leq \mathbf{P}\left\{\int |m(x) - m_n^*(x)|\mu(dx) - \mathbf{E}\left\{\int |m(x) - m_n^*(x)|\mu(dx)\right\} > \frac{\epsilon}{2}\right\}$$

$$\leq e^{-n\epsilon^2/(8L^2\rho^2)}.$$

The proof is now completed. $\qquad\qquad\square$

PROOF OF THEOREM 23.5.   As in the proof of Theorem 23.1, it suffices
to show that

$$\int |m_n(x) - m(x)|\mu(dx) \to 0 \text{ a.s.}$$

Obviously,

$$\int |m_n(x) - m(x)|\mu(dx)$$

$$\leq \int |m_n(x) - m_n^*(x)|\mu(dx) + \int |m_n^*(x) - m(x)|\mu(dx)$$

and, according to Lemma 23.9 ,

$$\int |m_n^*(x) - m(x)|\mu(dx) \to 0$$

with probability one. On the other hand,

$$|m_n^*(x) - m_n(x)|$$

$$= \left|\frac{\sum_{i=1}^n Y_i K_{h_n}(x-X_i)}{n\mathbf{E}K_{h_n}(x-X)} - \frac{\sum_{i=1}^n Y_i K_{h_n}(x-X_i)}{\sum_{i=1}^n K_{h_n}(x-X_i)}\right|$$

$$= \left|\sum_{i=1}^n Y_i K_{h_n}(x-X_i)\right| \left|\frac{1}{n\mathbf{E}K_{h_n}(x-X)} - \frac{1}{\sum_{i=1}^n K_{h_n}(x-X_i)}\right|$$

$$\leq\; L\left|\sum_{i=1}^{n} K_{h_n}(x - X_i)\right|\left|\frac{1}{n\mathbf{E}K_{h_n}(x - X)} - \frac{1}{\sum_{i=1}^{n} K_{h_n}(x - X_i)}\right|$$

$$=\; L\,|M_n^*(x) - 1|\,,$$

where $M_n^*(x)$ is the special form of $m_n^*(x)$ for $Y \equiv 1$. Therefore,

$$\int |m_n(x) - m_n^*(x)|\mu(dx) \leq L \int |M_n^*(x) - 1|\mu(dx) \to 0 \ \text{ a.s.},$$

which completes the proof. $\qquad\square$

The following theorem concerns the strong universal consistency of the kernel estimate with naive kernel and special sequences of bandwidths. As to more general kernels we refer to Walk (2002c).

Theorem 23.6. *Let $K = I_S$ and let h_n satisfy*

$$h_{n-1} \neq h_n \ \text{ at most for the indices } n = n_1, n_2, \ldots,$$

where $n_{k+1} \geq Dn_k$ for fixed $D > 1$ and

$$h_n \to 0, \quad nh_n^d \to \infty,$$

e.g., $h_n = ce^{-\gamma\lfloor q \log n\rfloor/q}$ with $c > 0$, $\quad 0 < \gamma d < 1$, $\quad q > 0$. Then m_n is strongly universally consistent.

PROOF. We argue as in the proof of Theorem 23.4. The assertion holds when $L > 0$ exists with $|Y| \leq L$ (cf. Theorem 23.5). According to Lemma 23.3 it suffices to show that, for some constant $c > 0$,

$$\limsup_{n\to\infty} \int \frac{\sum_{i=1}^{n} |Y_i| K_{h_n}(x - X_i)}{\sum_{i=1}^{n} K_{h_n}(x - X_i)}\mu(dx) \leq c\mathbf{E}|Y| \ \text{ a.s.}$$

for every distribution of (X, Y) with $\mathbf{E}|Y| < \infty$. But this follows from Lemma 23.5. $\qquad\square$

In Theorem 23.6 the statistician does not change the bandwidth at each change of n, as is done in the usual choice $h_n^* = cn^\gamma$, $c > 0$, $0 < \gamma d < 1$. But if the example choice in Theorem 23.6 is written in the form $h_n = c_n n^{-\gamma}$, one has $|c_n - c| \leq c(e^{\gamma/q} - 1)$ such that h_n and h_n^* are of the same order, and even the factor c in h_n^* can be arbitrarily well-approximated by use of a sufficiently large q in the definition of h_n. This is important in view of the rate of convergence under regularity assumptions.

The modification

$$m_n'(x) = \frac{\sum_{i=1}^{n} Y_i K_{h_n}(x - X_i)}{\max\{\delta, \sum_{i=1}^{n} K_{h_n}(x - X_i)\}}$$

of the classical kernel estimate, fixed $\delta > 0$ (see Spiegelman and Sacks (1980)), which for $1 \geq \delta > 0$ coincides with it for naive kernel, yields continuity of $m_n'(\cdot)$ if the kernel K is continuous. For K sufficiently smooth (e.g.,

Gaussian kernel $K(x) = e^{-\|x\|^2}$ or quartic kernel $K(x) = (1-\|x\|^2)^2 I_{\|x\|\le 1}$) Walk (2002c) showed strong universal consistency in the case $h_n = n^{-\gamma}$ ($0 < \gamma d < 1$). Here, for Gaussian kernel K, the proof of Theorem 23.6 can be modified majorizing K in the denominator by $K(\cdot/p)$, $1 < p < \sqrt{2}$, and noticing regularity of the kernel $K(\cdot/p)^2/K(\cdot)$. For a general smooth kernel one first shows strong consistency of $(m'_1 + \cdots + m'_n)/n$ by martingale theory (cf. Walk (2002a)) and then strong consistency of m'_n by a Tauberian argument of summability theory.

23.3   k-NN Estimates

As in Chapter 6, we shall assume that ties occur with probability 0.

Theorem 23.7. *Assume that* $\mathbf{P}\{|Y| \le L\} = 1$ *for some* $L < \infty$ *and that for each x the random variable $\|X - x\|$ is absolutely continuous. If $k_n \to \infty$ and $k_n/n \to 0$, then the k_n-NN regression function estimate is strongly consistent.*

PROOF. We show that, for sufficiently large n,

$$\mathbf{P}\left\{ \int |m(x) - m_n(x)|\mu(dx) > \epsilon \right\} \le 4e^{-n\epsilon^2/(18L^2\gamma_d^2)},$$

where γ_d has been defined in Chapter 6. Define $\rho_n(x)$ as the solution of the equation

$$\frac{k_n}{n} = \mu(S_{x,\rho_n(x)}).$$

Note that the condition that for each x the random variable $\|X - x\|$ is absolutely continuous implies that the solution always exists. (This is the only point in the proof where we use this assumption.) Also define

$$m_n^*(x) = \frac{1}{k_n} \sum_{j=1}^{n} Y_j I_{\{\|X_j - x\| < \rho_n(x)\}}.$$

The basis of the proof is the following decomposition:

$$|m(x) - m_n(x)| \le |m_n(x) - m_n^*(x)| + |m_n^*(x) - m(x)|.$$

For the first term on the right-hand side observe that, denoting $R_n(x) = \|X_{(k_n,n)}(x) - x\|$,

$$|m_n^*(x) - m_n(x)| \;=\; \frac{1}{k_n}\left| \sum_{j=1}^{n} Y_j I_{\{X_j \in S_{x,\rho_n(x)}\}} - \sum_{j=1}^{n} Y_j I_{\{X_j \in S_{x,R_n(x)}\}} \right|$$

$$\le \frac{L}{k_n} \sum_{j=1}^{n} \left| I_{\{X_j \in S_{x,\rho_n(x)}\}} - I_{\{X_j \in S_{x,R_n(x)}\}} \right|.$$

By considering the cases $\rho_n(x) \le R_n(x)$ and $\rho_n(x) > R_n(x)$ one gets that $I_{\{X_j \in S_{x,\rho_n(x)}\}} - I_{\{X_j \in S_{x,R_n(x)}\}}$ have the same sign for each j. It follows that

$$|m_n^*(x) - m_n(x)| \;\le\; L \left| \frac{1}{k_n} \sum_{j=1}^{n} I_{\{X_j \in S_{x,\rho_n(x)}\}} - 1 \right| = L|M_n^*(x) - M(x)|,$$

where M_n^* is defined as m_n^* with Y replaced by the constant random variable $Y = 1$, and $M \equiv 1$ is the corresponding regression function. Thus,

$$|m(x) - m_n(x)| \le L|M_n^*(x) - M(x)| + |m_n^*(x) - m(x)|. \qquad (23.20)$$

First we show that the expected values of the integrals of both terms on the right-hand side converge to zero. Then we use McDiarmid's inequality to prove that both terms are very close to their expected values with large probability. For the expected value of the first term on the right-hand side of (23.20), using the Cauchy-Schwarz inequality, we have

$$
\begin{aligned}
L\mathbf{E} \int |M_n^*(x) - M(x)|\mu(dx) \;&\le\; L \int \sqrt{\mathbf{E}\{|M_n^*(x) - M(x)|^2\}}\mu(dx) \\
&=\; L \int \sqrt{\mathbf{E}\{|M_n^*(x) - \mathbf{E}M_n^*(x)|^2\}}\mu(dx) \\
&=\; L \int \sqrt{\frac{1}{k_n^2} n \, \mathbf{Var}\{I_{\{X \in S_{x,\rho_n(x)}\}}\}}\mu(dx) \\
&\le\; L \int \sqrt{\frac{1}{k_n^2} n\mu(S_{x,\rho_n(x)})}\mu(dx) \\
&=\; L \int \sqrt{\frac{n}{k_n^2} \frac{k_n}{n}}\mu(dx) \\
&=\; \frac{L}{\sqrt{k_n}},
\end{aligned}
$$

which converges to zero. For the expected value of the second term on the right-hand side of (23.20), note that Theorem 6.1 implies that

$$\lim_{n \to \infty} \mathbf{E} \int |m(x) - m_n(x)|\mu(dx) = 0.$$

Therefore,

$$
\begin{aligned}
\mathbf{E} &\int |m_n^*(x) - m(x)|\mu(dx) \\
&\le\; \mathbf{E} \int |m_n^*(x) - m_n(x)|\mu(dx) + \mathbf{E} \int |m(x) - m_n(x)|\mu(dx) \\
&\le\; L\mathbf{E} \int |M_n^*(x) - M_n(x)|\mu(dx) + \mathbf{E} \int |m(x) - m_n(x)|\mu(dx) \to 0.
\end{aligned}
$$

Assume now that n is so large that

$$L\mathbf{E}\int |M_n^*(x) - M(x)|\mu(dx) + \mathbf{E}\int |m_n^*(x) - m(x)|\mu(dx) < \frac{\epsilon}{3}.$$

Then, by (23.20), we have

$$\mathbf{P}\left\{\int |m(x) - m_n(x)|\mu(dx) > \epsilon\right\} \qquad (23.21)$$

$$\leq \mathbf{P}\left\{\int |m_n^*(x) - m(x)|\mu(dx) - \mathbf{E}\int |m_n^*(x) - m(x)|\mu(dx) > \frac{\epsilon}{3}\right\}$$

$$+ \mathbf{P}\left\{L\int |M_n^*(x) - M(x)|\mu(dx) - \mathbf{E}L\int |M_n^*(x) - M(x)|\mu(dx) > \frac{\epsilon}{3}\right\}.$$

Next we get an exponential bound for the first probability on the right-hand side of (23.21) by McDiarmid's inequality (Theorem A.2). Fix an arbitrary realization of the data $D_n = \{(x_1, y_1), \ldots, (x_n, y_n)\}$, and replace (x_i, y_i) by $(\widehat{x}_i, \widehat{y}_i)$, changing the value of $m_n^*(x)$ to $m_{ni}^*(x)$. Then

$$\left|\int |m_n^*(x) - m(x)|\mu(dx) - \int |m_{ni}^*(x) - m(x)|\mu(dx)\right|$$

$$\leq \int |m_n^*(x) - m_{ni}^*(x)|\mu(dx).$$

But $|m_n^*(x) - m_{ni}^*(x)|$ is bounded by $2L/k_n$ and can differ from zero only if $\|x - x_i\| < \rho_n(x)$ or $\|x - \widehat{x}_i\| < \rho_n(x)$. Observe that $\|x - x_i\| < \rho_n(x)$ or $\|x - \widehat{x}_i\| < \rho_n(x)$ if and only if $\mu(S_{x,\|x-x_i\|}) < k_n/n$ or $\mu(S_{x,\|x-\hat{x}_i\|}) < k_n/n$. But the measure of such x's is bounded by $2 \cdot \gamma_d k_n/n$ by Lemma 6.2. Therefore,

$$\sup_{x_1,y_1,\ldots,x_n,y_n,\widehat{x}_i,\widehat{y}_i} \int |m_n^*(x) - m_{ni}^*(x)|\mu(dx) \leq \frac{2L}{k_n}\frac{2 \cdot \gamma_d k_n}{n} = \frac{4L\gamma_d}{n}$$

and, by Theorem A.2,

$$\mathbf{P}\left\{\left|\int |m(x) - m_n^*(x)|\mu(dx) - \mathbf{E}\int |m(x) - m_n^*(x)|\mu(dx)\right| > \frac{\epsilon}{3}\right\}$$

$$\leq 2e^{-n\epsilon^2/(72L^2\gamma_d^2)}.$$

Finally, we need a bound for the second term on the right-hand side of (23.21). This probability may be bounded by McDiarmid's inequality exactly in the same way as for the first term, obtaining

$$\mathbf{P}\left\{\left|L\int |M_n^*(x) - M(x)|\mu(dx) - \mathbf{E}L\int |M_n^*(x) - M(x)|\mu(dx)\right| > \frac{\epsilon}{3}\right\}$$

$$\leq 2e^{-n\epsilon^2/(72L^2\gamma_d^2)},$$

and the proof is completed. $\square$

Theorem 23.8. *Assume that for each x the random variable $\|X - x\|$ is absolutely continuous. If $k_n/\log n \to \infty$ and $k_n/n \to 0$ then the k_n-NN regression function estimate is strongly universally consistent.*

Before we prove Theorem 23.8 we will formulate and prove two lemmas.

Let A_i be the collection of all x that are such that X_i is one of its k_n nearest neighbors of x in $\{X_1, \ldots, X_n\}$. Here, we use some geometric arguments similar to those in the proof of Lemma 6.2. Similarly, let us define cones $x + C_1, \ldots, x + C_{\gamma_d}$, where x defines the top of the cones and the union of $C_1, \ldots, C_{\gamma_d}$ covers $\mathcal{R}^d$. Then

$$\bigcup_{j=1}^{\gamma_d} \{x + C_j\} = \mathcal{R}^d$$

regardless of how x is picked. According to the cone property, if $u, u' \in x + C_j$, and $\|x - u\| < \|x - u'\|$, then $\|u - u'\| < \|x - u'\|$. Furthermore, if $\|x - u\| \leq \|x - u'\|$, then $\|u - u'\| \leq \|x - u'\|$. In the space $\mathcal{R}^d$, define the sets

$$C_{i,j} = X_i + C_j \quad (1 \leq i \leq n, 1 \leq j \leq \gamma_d).$$

Let $B_{i,j}$ be the subset of $C_{i,j}$ consisting of all $x \in C_{i,j}$ that are among the k_n nearest neighbors of X_i in the set

$$\{X_1, \ldots, X_{i-1}, X_{i+1}, \ldots, X_n, x\} \cap C_{i,j}.$$

(If $C_{i,j}$ contains fewer than $k_n - 1$ of the X_l points $i \neq l$, then $B_{i,j} = C_{i,j}$.) Equivalently, $B_{i,j}$ is the subset of $C_{i,j}$ consisting of all x that are closer to X_i than the k_nth nearest neighbor of X_i in $\{X_1, \ldots, X_{i-1}, X_{i+1}, \ldots, X_n\} \cap C_{i,j}$.

Lemma 23.10. *Assume that for each x the random variable $\|X - x\|$ is absolutely continuous. Let $1 \leq i \leq n$. If $x \in A_i$, then $x \in \bigcup_{j=1}^{d} B_{i,j}$, and thus*

$$\mu(A_i) \leq \sum_{j=1}^{\gamma_d} \mu(B_{i,j}).$$

PROOF. To prove this claim, take $x \in A_i$. Then locate a j for which $x \in C_{i,j}$. We have to show that $x \in B_{i,j}$ to conclude the proof. Thus, we need to show that x is one of the k_n nearest neighbors of X_i in the set

$$\{X_1, \ldots, X_{i-1}, X_{i+1}, \ldots, X_n, x\} \cap C_{i,j}.$$

Take $X_l \in C_{i,j}$. If $\|X_l - X_i\| < \|x - X_i\|$, we recall that by the property of our cones that $\|x - X_l\| < \|x - X_i\|$, and thus X_l is one of the $k_n - 1$ nearest neighbors of x in $\{X_1, \ldots, X_n\}$ because of $x \in A_i$. This shows that in $C_{i,j}$ there are at most $k_n - 1$ points X_l closer to X_i than x. Thus x is one of the k_n nearest neighbors of X_i in the set

$$\{X_1 \ldots, X_{i-1}, X_{i+1}, \ldots, X_n, x\} \cap C_{i,j}.$$

This concludes the proof of the claim. $\square$

Lemma 23.11. *If $k_n/\log(n) \to \infty$ and $k_n/n \to 0$ then, for every $j \in \{1, \ldots, \gamma_d\}$,*

$$\limsup_{n \to \infty} \frac{n}{k_n} \max_{1 \le i \le n} \mu(B_{i,j}) \le 2 \ \ a.s.$$

PROOF. We prove that, for every j,

$$\sum_{n=1}^{\infty} \mathbf{P}\left\{ \frac{n}{k_n} \max_{1 \le i \le n} \mu(B_{i,j}) > 2 \right\} < \infty.$$

In order to do this we give a bound for

$$\mathbf{P}\{\mu(B_{i,j}) > p|X_i\}$$

for $0 < p < 1$. If $\mu(C_{i,j}) \le p$ then since $B_{i,j} \subseteq C_{i,j}$, we have $\mathbf{P}\{\mu(B_{i,j}) > p|X_i\} = 0$, therefore we assume that $\mu(C_{i,j}) > p$. Fix X_i. Define

$$G_{i,p} = C_{i,j} \cap S_{X_i, R_n(X_i)},$$

where $R_n(X_i) > 0$ is chosen such that $\mu(G_{i,p}) = p$. Observe that either $B_{i,j} \supseteq G_{i,p}$ or $B_{i,j} \subseteq G_{i,p}$, therefore we have the following dual relationship:

$$
\begin{aligned}
&\mathbf{P}\{\mu(B_{i,j}) > p|X_i\} \\
=\ &\mathbf{P}\{\mu(B_{i,j}) > \mu(G_{i,p})|X_i\} \\
=\ &\mathbf{P}\{B_{i,j} \supset G_{i,p}|X_i\} \\
=\ &\mathbf{P}\{G_{i,p} \text{ captures } < k_n \text{ of the points } X_l \in C_{i,j}, l \ne i|X_i\} \ .
\end{aligned}
$$

The number of points X_l ($l \ne i$) captured by $G_{i,p}$ given X_i is binomially distributed with parameters $(n-1, p)$, so by Lemma A.1, with $p = 2k_n/(n-1)$ and $\epsilon = p/2 = k_n/(n-1)$, we have that

$$
\begin{aligned}
&\mathbf{P}\left\{ \max_{1 \le i \le n} \mu(B_{i,j}) > p \right\} \\
\le\ &n\mathbf{P}\{\mu(B_{1,j}) > p\} \\
=\ &n\mathbf{E}\{\mathbf{P}\{\mu(B_{1,j}) > p|X_1\}\} \\
=\ &n\mathbf{E}\{\mathbf{P}\{G_{1,p} \text{ captures } < k_n \text{ of the points } X_l \in C_{1,j}, l \ne 1|X_1\}\} \\
\le\ &ne^{-(n-1)[p-\epsilon+\epsilon \log(\epsilon/p)]} \\
=\ &ne^{-2k_n+k_n+k_n \log 2} \\
\le\ &ne^{-k_n(1-\log 2)},
\end{aligned}
$$

which is summable because of $k_n/\log n \to \infty$. $\square$

PROOF OF THEOREM 23.8. By Lemma 23.3 and Theorem 23.7 it is enough to prove that there is a constant $c > 0$,

$$\limsup_{n \to \infty} \sum_{i=1}^{n} \int W_{ni}(x)\mu(dx)Y_i^2 \leq c\mathbf{E}Y^2 \quad \text{a.s.}$$

Observe that

$$\sum_{i=1}^{n} \int W_{ni}(x)\mu(dx)Y_i^2 = \frac{1}{k_n}\sum_{i=1}^{n}Y_i^2\mu(A_i) \leq \left(\frac{n}{k_n}\max_i \mu(A_i)\right)\frac{1}{n}\sum_{i=1}^{n}Y_i^2.$$

If we can show that

$$\limsup_{n \to \infty} \frac{n}{k_n}\max_i \mu(A_i) \leq c \quad \text{a.s.} \tag{23.22}$$

for some constant $c > 0$, then by the law of large numbers

$$\limsup_{n \to \infty}\left(\frac{n}{k_n}\max_i \mu(A_i)\right)\frac{1}{n}\sum_{i=1}^{n}Y_i^2 \leq \limsup_{n \to \infty} c\frac{1}{n}\sum_{i=1}^{n}Y_i^2 = c\mathbf{E}Y^2 \quad \text{a.s.,}$$

so we have to prove (23.22). But by Lemma 23.10,

$$\mu(A_i) \leq \sum_{j=1}^{\gamma_d}\mu(B_{i,j}),$$

therefore, Lemma 23.11 implies that (23.22) is satisfied with $c = 2\gamma_d$, so the proof of the theorem is completed. $\square$

23.4 Bibliographic Notes

Devroye and Györfi (1983) proved Theorem 23.1. Theorem 23.3 is due to Györfi (1991). Theorems 23.4 and 23.6 have been shown in Walk (2002a). Theorem 23.5 has been proved by Devroye and Krzyżak (1989). Lemma 23.7 is from Devroye, Györfi, and Lugosi (1996). Theorems 23.7 and 23.8 are due to Devroye et al. (1994). Strong consistency of the partitioning estimate with a cross-validated choice of partitions and also of the kernel estimate with a cross-validated bandwidth for bounded Y are in Kohler, Krzyżak, and Walk (2002).

Problems and Exercises

PROBLEM 23.1. Prove (23.11) and (23.12).
  HINT: Proceed as in the proof of Lemma 4.1.

PROBLEM 23.2. Show the bounds for B and C in the proof of Lemma 23.5.

PROBLEM 23.3. Prove Lemma 23.7.

HINT: Apply the covering in part (iv) of the proof of Theorem 5.1.

24

Semirecursive Estimates

24.1 A General Result

For a sequence of measurable real-valued functions $K_n(x, u)$ $(n = 1, 2, ...)$, $x, u \in \mathcal{R}^d$, we consider estimates of the form

$$m_n(x) = \frac{\sum_{i=1}^n Y_i K_i(x, X_i)}{\sum_{i=1}^n K_i(x, X_i)}, \qquad (24.1)$$

($0/0$ is 0 by definition). We call such regression function estimates semi-recursive since both the numerator and the denominator can be calculated recursively. Thus estimates can be updated sequentially when new observations became available. At the nth stage one has to store only the numerator and the denominator, not the whole set of observations $(X_1, Y_1), \ldots, (X_n, Y_n)$. Another simple interpretation of the estimate is that the denominator and the estimate are stored: put

$$
\begin{aligned}
m_1(x) &= Y_1, \\
f_1(x) &= K_1(x, X_1), \\
f_{n+1}(x) &= f_n(x) + K_{n+1}(x, X_{n+1}), \\
m_{n+1}(x) &= \left(1 - \frac{K_{n+1}(x, X_{n+1})}{f_{n+1}(x)}\right) m_n(x) + \frac{Y_{n+1} K_{n+1}(x, X_{n+1})}{f_{n+1}(x)},
\end{aligned}
$$

if $f_1(x) \neq 0$.

Consider estimates of the form (24.1) with general $K_n(x,z)$ satisfying

$$0 \leq K_n(x,z) \leq K_{max} \quad \text{for all } x,z \in \mathcal{R}^d \qquad (24.2)$$

for some $K_{max} \in \mathcal{R}$, $K_{max} \geq 1$.

As a byproduct we also consider pointwise consistency such that first we give sufficient conditions of strong pointwise consistency (Theorem 24.1), and then prove weak universal consistency (Lemma 24.1), and strong universal consistency (Lemma 24.2).

To simplify the notation we use the abbreviation mod μ to indicate that a relation holds for μ-almost all $x \in \mathcal{R}^d$. The Euclidean norm of a vector $x \in \mathcal{R}^d$ is denoted by $\|x\|$ and the Lebesgue measure by λ.

Theorem 24.1. *Assume that, for every distribution μ of X,*

$$\frac{\int K_n(x,z)f(z)\mu(dz)}{\int K_n(x,z)\mu(dz)} \rightarrow f(x) \ \ mod \ \mu \qquad (24.3)$$

for all square μ-integrable functions f on $\mathcal{R}^d$ and

$$\sum_{n=1}^{\infty} \mathbf{E}K_n(x,X_n) = \infty \ \ mod \ \mu. \qquad (24.4)$$

Then

$$m_n(x) \rightarrow m(x) \ \ a.s.$$

for μ-almost all x for all distributions of (X,Y) with $\mathbf{E}|Y|^2 < \infty$.

PROOF. We have to show that

$$\frac{\sum_{i=1}^n K_i(x,X_i)Y_i}{\sum_{i=1}^n K_i(x,X_i)} \rightarrow m(x) \ \text{a.s. mod } \mu. \qquad (24.5)$$

This follows from

$$\frac{\sum_{i=1}^n \mathbf{E}K_i(x,X_i)Y_i}{\sum_{i=1}^n \mathbf{E}K_i(x,X_i)} = \frac{\sum_{i=1}^n \int K_i(x,z)m(z)\mu(dz)}{\sum_{i=1}^n \int K_i(x,z)\mu(dz)} \rightarrow m(x) \ \text{mod } \mu \quad (24.6)$$

and

$$\frac{\sum_{i=1}^n (K_i(x,X_i)Y_i - \mathbf{E}K_i(x,X_i)Y_i)}{\sum_{i=1}^n \mathbf{E}K_i(x,X_i)} \rightarrow 0 \ \text{a.s. mod } \mu \qquad (24.7)$$

and

$$\frac{\sum_{i=1}^n K_i(x,X_i)}{\sum_{i=1}^n \mathbf{E}K_i(x,X_i)} \rightarrow 1 \ \text{a.s. mod } \mu. \qquad (24.8)$$

Convergence (24.6) is a consequence of the Toeplitz lemma (see Problem A.5), (24.3), and (24.4). In order to prove (24.7) we set

$$V_i = K_i(x,X_i)Y_i - \mathbf{E}K_i(x,X_i)Y_i$$

and

$$b_n = \sum_{i=1}^{n} \mathbf{E} K_i(x, X_i).$$

Put $M(x) = \mathbf{E}\{Y^2 | X = x\}$. Then

$$\mathbf{E} V_n^2 \leq \mathbf{E} K_n(x, X_n)^2 Y_n^2 \leq K_{max} \cdot \frac{\mathbf{E} K_n(x, X_n) M(X_n)}{\mathbf{E} K_n(x, X_n)} \cdot \mathbf{E} K_n(x, X_n),$$

and (24.3) implies

$$\sum_n \frac{\mathbf{E} V_n^2}{b_n^2} \leq d(x) \cdot \sum_n \frac{\mathbf{E} K_n(x, X_n)}{\{\sum_{i=1}^{n} \mathbf{E} K_i(x, X_i)\}^2}$$

for some $d(x) < \infty$ mod μ. By the Abel-Dini theorem (see Problem 24.1) one gets

$$\sum_n \frac{\mathbf{E} K_n(x, X_n)}{\{\sum_{i=1}^{n} \mathbf{E} K_i(x, X_i)\}^2} < \infty.$$

Therefore

$$\sum_n \frac{\mathbf{E} V_n^2}{b_n^2} < \infty \text{ mod } \mu,$$

and Theorem A.6 implies (24.7). Further (24.7) for $Y_i = 1$ implies (24.8). $\square$

Put

$$\overline{m}_n(x) = \frac{\mathbf{E} K_n(x, X) Y}{\mathbf{E} K_n(x, X)} = \frac{\int K_n(x, z) m(z) \mu(dz)}{\int K_n(x, z) \mu(dz)}.$$

Lemma 24.1. *Assume the conditions of Theorem 24.1. If, in addition, a constant $c > 0$ exists such that*

$$\sup_n \mathbf{E} \int \left(\frac{\sum_{i=1}^{n} (Y_i - \overline{m}_i(x)) K_i(x, X_i)}{\sum_{i=1}^{n} K_i(x, X_i)} \right)^2 \mu(dx) \leq c \mathbf{E} Y^2 \qquad (24.9)$$

and

$$\int \sup_n |\overline{m}_n(x)|^2 \mu(dx) \leq c \mathbf{E} Y^2, \qquad (24.10)$$

for every distribution of (X, Y) with square integrability of Y, then the sequence m_n is weakly universally consistent.

PROOF. Theorem 24.1 implies that

$$\frac{\sum_{i=1}^{n} Y_i K_i(x, X_i)}{\sum_{i=1}^{n} K_i(x, X_i)} \to m(x) \text{ a.s. mod } \mu \qquad (24.11)$$

especially for bounded Y. From (24.9) and (24.10) one obtains

$$\mathbf{E}\int\left(\frac{\sum_{i=1}^{n}Y_iK_i(x,X_i)}{\sum_{i=1}^{n}K_i(x,X_i)}\right)^2\mu(dx)\leq cEY^2$$

for each square integrable Y and $n\in\mathcal{N}$. From this result and relation (24.11) for bounded Y, which by Lebesgue's dominated convergence theorem yields weak (and strong) consistency in the boundedness case, one obtains the assertion in the general case by a truncation argument for the Y_i's (see the proof of Lemma 23.3).

Lemma 24.2. *(a) Assume the conditions of Lemma 24.1. If, in addition, a constant $c>0$ exists such that*

$$\int\sum_{n=1}^{\infty}\mathbf{E}\frac{Y_n^2K_n(x,X_n)^2}{\left(\sum_{i=1}^{n}K_i(x,X_i)\right)^2}\mu(dx)\leq c\mathbf{E}Y^2,\qquad(24.12)$$

and

$$\int\sum_{n=1}^{\infty}\mathbf{E}\frac{\overline{m}_n(x)^2K_n(x,X_n)^2}{\left(\sum_{i=1}^{n}K_i(x,X_i)\right)^2}\mu(dx)\leq c\mathbf{E}Y^2\qquad(24.13)$$

for every distribution of (X,Y) with square integrability of Y, and

$$K_1(x,z)=1\ \text{ for all }x,z\in\mathcal{R}^d\qquad(24.14)$$

or

$$K_n(x,z)\in\{0\}\cup[\alpha,\beta]\ \text{ for all }x,z\in\mathcal{R}^d\qquad(24.15)$$

for some $0<\alpha<\beta<\infty$, then the sequence m_n is strongly universally consistent.
(b) Conditions (24.13) and (24.12) imply (24.9).

PROOF. See Problem 24.2.

24.2 Semirecursive Kernel Estimate

The semirecursive kernel estimate is defined according to (24.1) by a kernel $K:\mathcal{R}^d\to\mathcal{R}_+$ and a sequence of bandwidths $h_n>0$ via

$$K_n(x,u)=K\left(\frac{x-u}{h_n}\right).\qquad(24.16)$$

Theorem 24.2. *Either let (24.16) hold for $n\geq2$ and let $K_1(x,u)=K(0)$, $x,u\in\mathcal{R}^d$, with symmetric Lebesgue-integrable kernel $K:\mathcal{R}^d\to\mathcal{R}_+$ satisfying*

$$\alpha H(\|x\|)\leq K(x)\leq\beta H(\|x\|),\ x\in\mathcal{R}^d,\qquad(24.17)$$

for some $0 < \alpha < \beta < \infty$ and nonincreasing $H : \mathcal{R}_+ \to \mathcal{R}_+$ with $H(+0) > 0$, or let (24.16) hold for $n \geq 1$ with $K : \mathcal{R}^d \to \mathcal{R}_+$ satisfying

$$\alpha I_{S_{0,R}} \leq K \leq \beta I_{S_{0,R}} \tag{24.18}$$

for some $0 < \alpha < \beta < \infty$, $0 < R < \infty$. Assume further

$$h_n \downarrow 0 \quad (n \to \infty), \quad \sum h_n^d = \infty. \tag{24.19}$$

Then the semirecursive kernel estimate is weakly and strongly universally consistent.

In both cases (trivially in the second case with $H = I_{[0,R]}$) the assumptions in Theorem 24.2 imply that

$$K \geq b I_{S_{0,R}} \text{ for some } b > 0, \ 0 < R < \infty,$$

and

$$r^d H(r) \to 0 \quad (r \to \infty).$$

The next covering lemma using balls plays an important role in the proof of Lemma 24.4.

Lemma 24.3. *Let A be a bounded set in R^d, $k \in \mathcal{N}$, $0 < r_1 < \cdots < r_k < \infty$. For each $x \in A$ let $S(x)$ be a closed ball centered at x with radius $r(x) \in \{r_1, \ldots, r_k\}$. Then there exists an $m \in \mathcal{N}$ depending only on d, but not on A or k with the following property: there exists a finite number of points $x_1, \ldots, x_l \in A$ such that $A \subseteq S(x_1) \cup \cdots \cup S(x_l)$ and each $v \in \mathcal{R}^d$ belongs to at most m of the sets $S(x_1), \ldots, S(x_l)$.*

PROOF. Choose x_1 such that $r(x_1)$ is the largest possible. Then choose $x_2 \in A - S(x_1)$ such that $r(x_2)$ is the largest possible, and choose $x_3 \in A - (S(x_1) \cup S(x_2))$ such that $r(x_3)$ is the largest possible, etc. The procedure terminates with the choice of x_l for some l, because $||x_i - x_j|| > r_1$ $(i \neq j)$ and, due to the boundedness of A, where $|| \cdot ||$ is the Euclidean norm. Thus $S(x_1), \ldots, S(x_l)$ cover A.

For each $v \in \mathcal{R}^d$ there exists a finite number m (depending only on d, but not on v) of congruent cones $C_1, \ldots, C_m$ with vertex v covering $\mathcal{R}^d$, such that two arbitrary rays in a cone starting at v form an angle less than or equal to $\pi/4$.

Then for all x, y in a cone with $||x - v|| \leq r, ||y - x|| > r$ for some $r \in (0, \infty)$ one has $||y - v|| > r$.

Consequently, there do not exist two points x_i, x_j in the same cone C_n $(n \in \{1, \ldots, m\})$ with $v \in S(x_i), v \in S(x_j)$, because $||x_i - x_j|| > \max\{r(x_i), r(x_j)\} = r(x_i)$ (the latter w.l.o.g.), $||x_i - v|| \leq r(x_i)$ imply $||x_j - v|| > r(x_i)$ in contrast to $||x_j - v|| \leq r(x_j)$. Therefore the number of x_i's with $v \in S(x_i)$ is at most m. This completes the proof. $\square$

In the following let μ, ν be measures on $\mathcal{B}_d$ assigning finite values to bounded sets. It is assumed that in the expressions

$$\sup_{h>0} \frac{\nu(S_{x,h})}{\mu(S_{x,h})},$$

$$\limsup_{h\to 0} \frac{\nu(S_{x,h})}{\mu(S_{x,h})},$$

$(x \in \mathcal{R}^d)$ h assumes countably many positive values thus making the expressions measurable functions of x. Here $0/0 = 0$.

Lemma 24.4. *There is a constant c depending on d such that*

$$\mu\left\{x \in \mathcal{R}^d : \sup_{h>0} \frac{\nu(S_{x,h})}{\mu(S_{x,h})} > \alpha\right\} \leq \frac{c}{\alpha}\nu(\mathcal{R}^d)$$

for any $\alpha > 0$.

PROOF. Let $H = \{h_1, h_2, \ldots\}$ be the countable set of positive h's. Let $\alpha > 0$ be fixed. Set

$$M = \left\{x \in \mathcal{R}^d : \sup_{h>0} \frac{\nu(S_{x,h})}{\mu(S_{x,h})} > \alpha\right\}$$

and let G be an arbitrary bounded Borel set. Define further

$$D_N = \left\{x \in G \cap M : \exists_{h\in\{h_1,\ldots,h_N\}} \frac{\nu(S_{x,h})}{\mu(S_{x,h})} > \alpha\right\}.$$

Then $D_N \uparrow G \cap M$. Let N be arbitrary. Choose $x_1, \ldots, x_l \in D_N$ according to Lemma 24.3 with corresponding $h(x_1), \ldots, h(x_l) \in \{h_1, \ldots, h_N\}$ and $m = c$ (depending only on d) such that, with notation $S_j = S_{x_j, h(x_j)}$,

$$\frac{\nu(S_j)}{\mu(S_j)} > \alpha \quad (j = 1, \ldots, l),$$

$$D_N \subseteq \bigcup_{j=1}^{l} S_j,$$

$$\sum_{j=1}^{l} I_{S_j} \leq c.$$

Then

$$\mu(D_N) \leq \sum_{j=1}^{l} \mu(S_j) < \frac{1}{\alpha}\sum_{j=1}^{l} \nu(S_j) = \frac{1}{\alpha}\sum_{j=1}^{l} \int_{\mathcal{R}^d} I_{S_j}\, d\nu \leq \frac{c}{\alpha}\nu(\mathcal{R}^d),$$

and, by $N \to \infty$,

$$\mu(G \cap M) \leq \frac{c}{\alpha}\nu(\mathcal{R}^d).$$

Letting $G \uparrow \mathcal{R}^d$, one obtains the assertion $\mu(M) \le \frac{c}{\alpha}\nu(\mathcal{R}^d)$. $\qquad\square$

Now we state the generalized pointwise Lebesgue density theorem:

Lemma 24.5. *Let f be a Borel measurable function integrable on $\mathcal{R}^d$. Then*

$$\lim_{h \to 0} \frac{\int_{S_{x,h}} |f(t) - f(x)|\mu(dt)}{\mu(S_{x,h})} = 0 \ \ a.s. \ mod \ \mu.$$

PROOF. For any $\epsilon > 0$, according to Theorem A.1 choose a continuous function g of compact support such that

$$\int_{\mathcal{R}^d} |f - g|\, d\mu < \frac{\epsilon^2}{2(c+1)}$$

with constant c from Lemma 24.4. We have

$$\frac{1}{\mu(S_{x,h})} \int_{S_{x,h}} |f(t) - f(x)|\mu(dt)$$

$$\le \frac{1}{\mu(S_{x,h})} \int_{S_{x,h}} |f - g|\, d\mu + |f(x) - g(x)|$$

$$+ \frac{1}{\mu(S_{x,h})} \int_{S_{x,h}} |g(t) - g(x)|\mu(dt).$$

Since g is continuous, the last term on the right-hand side converges to 0. Define the set

$$T_\epsilon = \left\{ x : \sup_{h>0} \frac{1}{\mu(S_{x,h})} \int_{S_{x,h}} |f - g|\, d\mu + |f(x) - g(x)| > \epsilon \right\}.$$

By Lemma 24.4 and the Markov inequality

$$\mu(T_\epsilon) \le \mu\left(\left\{ x : \sup_{h>0} \frac{1}{\mu(S_{x,h})} \int_{S_{x,h}} |f - g|\, d\mu > \epsilon/2 \right\} \right)$$

$$+ \mu(\{x : |f(x) - g(x)| > \epsilon/2\})$$

$$\le c(2/\epsilon) \int_{\mathcal{R}^d} |f - g|\, d\mu + (2/\epsilon) \int_{\mathcal{R}^d} |f - g|\, d\mu$$

$$= \frac{2(c+1)}{\epsilon} \int_{\mathcal{R}^d} |f - g|\, d\mu \le \epsilon,$$

where $\epsilon \to 0$ yields the assertion. In the proof we tacitly assumed that $\mu(S_{x,h}) > 0$ for all $x \in \mathcal{R}^d, h > 0$. We can show that $\mu(T) = 0$, where

$$T = \{x : \exists h_x > 0, \mu(S_{x,h_x}) = 0\}.$$

Let Q denote a countable dense set in $\mathcal{R}^d$. Then for each $x \in T$, there is $q_x \in Q$ with $\|x - q_x\| \le h_x/3$. This implies that $S_{q_x, h_x/2} \subset S_{x, h_x}$. Therefore

$\mu(S_{q_x,h_x/2}) = 0$, $x \in T$, and

$$S \subseteq \bigcup_{x \in T} S_{q_x,h_x/2}.$$

The right-hand side is a union of countably many sets of zero measure, and therefore $\mu(T) = 0$. $\qquad\square$

Lemma 24.6. *It holds that*

$$\limsup_{h \to 0} \frac{h^d}{\mu(S_{x,h})} = g(x) < \infty \ \ mod \ \mu.$$

PROOF. Let S_R be the open sphere centered at 0 with radius R and define the finite measure λ' by $\lambda'(B) = \lambda(B \cap S_R)$, $B \in \mathcal{B}_d$. We obtain

$$\mu\left(\left\{x \in S_R; \limsup_{h \to 0} \frac{\lambda(S_{x,h})}{\mu(S_{x,h})} = \infty\right\}\right)$$

$$= \ \mu\left(\left\{x \in S_R; \limsup_{h \to 0} \frac{\lambda'(S_{x,h})}{\mu(S_{x,h})} = \infty\right\}\right)$$

$$\leq \ \mu\left(\left\{x \in S_R; \sup_{h > 0} \frac{\lambda'(S_{x,h})}{\mu(S_{x,h})} = \infty\right\}\right)$$

$$= \ \lim_{s \to \infty} \mu\left(\left\{x \in S_R; \sup_{h > 0} \frac{\lambda'(S_{x,h})}{\mu(S_{x,h})} > s\right\}\right)$$

$$= \ 0$$

by Lemma 24.4 with $\nu = \lambda'$, then, by $R \to \infty$, the relation

$$\limsup_{h \to 0} \frac{\lambda(S_{x,h})}{\mu(S_{x,h})} < \infty \ \ mod \ \mu$$

and thus the assertion because $\lambda(S_{x,h}) = V_d h^d$, where V_d is the volume of the unit ball in $\mathcal{R}^d$. $\qquad\square$

Lemma 24.7. *Let $m \in L_2(\mu)$ and let m^* be the generalized Hardy-Littlewood maximal function of m defined by*

$$m^*(x) = \sup_{h > 0} \frac{1}{\mu(S_{x,h})} \int_{S_{x,h}} |m| \, d\mu, \quad x \in R^d.$$

Thus $m^* \in L_2(\mu)$ and

$$\int m^*(x)^2 \mu(dx) \leq c^* \int m(x)^2 \mu(dx),$$

where $c^* < \infty$ depends only on d.

PROOF. For arbitrary $\alpha > 0$ define $g_\alpha = m I_{[|m| \geq \alpha/2]}$ and let g_α^* be its generalized Hardy-Littlewood maximal function. One has

$$|m| \leq |g_\alpha| + \alpha/2,$$

$$m^* \leq g_\alpha^* + \alpha/2.$$

Thus, with the image measure μ_{m^*},

$$
\begin{aligned}
\mu_{m^*}((\alpha, \infty)) &= \mu(\{x \in \mathcal{R}^d : m^*(x) > \alpha\}) \\
&\leq \mu(\{x \in \mathcal{R}^d : g_\alpha^*(x) > \alpha/2\}) \\
&\leq \frac{2c}{\alpha} \int |g_\alpha| \, d\mu \\
&= \frac{2c}{\alpha} \int_{\{x \in \mathcal{R}^d : |m(x)| \geq \alpha/2\}} |m| \, d\mu
\end{aligned}
$$

with c depending only on d and the last inequality following from Lemma 24.4. Furthermore,

$$\int m^{*2} \, d\mu = \int_{R_+} s^2 \, d\mu_{m^*}(s) = 2 \int_{R_+} \alpha \mu_{m^*}((\alpha, \infty)) \, d\alpha$$

by using transformation of integrals and integration by parts. Thus

$$
\begin{aligned}
\int m^{*2} \, d\mu &\leq 4c \int_{R_+} \left(\int_{\{x \in \mathcal{R}^d : |m(x)| \geq \alpha/2\}} |m| \, d\mu \right) d\alpha \\
&= 4c \int_{\mathcal{R}^d} |m(x)| \left(\int_0^{2|m(x)|} d\alpha \right) \mu(dx) \\
&\qquad\qquad \text{(by Fubini's theorem)} \\
&= 8c \int_{\mathcal{R}^d} |m(x)|^2 \mu(dx).
\end{aligned}
$$

The proof is complete. $\qquad\square$

Lemma 24.8. *Assume*

$$
\begin{aligned}
c_1 H(\|x\|) &\leq K(x) \leq c_2 H(\|x\|), \quad c_1, c_2 > 0, \\
H(+0) &> 0, \tag{24.20} \\
t^d H(t) &\to 0 \quad as \quad t \to \infty, \tag{24.21}
\end{aligned}
$$

where H is a nonincreasing Borel function on $[0, \infty)$. Then, for all μ-integrable functions f,

$$\lim_{h \to 0} \frac{\int K((x-z)/h) f(z) \mu(dz)}{\int K((x-z)/h) \mu(dz)} = f(x) \mod \mu.$$

PROOF. Clearly

$$\left| \frac{\int K((x-z)/h) f(z) \mu(dz)}{\int K((x-z)/h) \mu(dz)} - f(x) \right|$$

$$\leq \frac{c_2}{c_1} \int H\left(\frac{\|x-y\|}{h} \right) |f(x) - f(y)| \mu(dy) \Big/ \int H\left(\frac{\|x-y\|}{h} \right) \mu(dy).$$

Observe

$$H(t) = \int_0^\infty I_{\{H(t)>s\}}(s)\,ds.$$

Thus

$$
\begin{aligned}
\int H\left(\frac{\|x-y\|}{h}\right)\mu(dy)
&= \int\left(\int_0^\infty I_{\{H\left(\frac{\|x-y\|}{h}\right)>s\}}\,ds\right)\mu(dy)\\
&= \int_0^\infty \mu\left\{y : H\left(\frac{\|x-y\|}{h}\right)>t\right\}dt\\
&= \int_0^\infty \mu(A_{t,h})\,dt,
\end{aligned}
$$

likewise,

$$\int H\left(\frac{\|x-y\|}{h}\right)|f(x)-f(y)|\mu(dy) = \int_0^\infty\left(\int_{A_{t,h}}|f(x)-f(y)|\mu(dy)\right)dt,$$

where $A_{t,h} = \{y : H(\|x-y\|/h) > t\}$.

Let $\delta = \epsilon h^d, \epsilon > 0$. Obviously,

$$\int_\delta^\infty\left(\int_{A_{t,h}}|f(x)-f(y)|\mu(dy)\right)dt\Big/\int_0^\infty\mu(A_{t,h})dt$$

$$\leq \sup_{t\geq\delta}\left(\int_{A_{t,h}}|f(x)-f(y)|\mu(dy)/\mu(A_{t,h})\right). \tag{24.22}$$

It is clear that the radii of sets $A_{t,h}$, $t \geq \delta$, do not exceed the radius of $A_{\delta,h}$, which in turn is h times that of the set $A_{\delta,1}$. The radius of $A_{\delta,1}$ does not exceed the length $H^+(\delta)$ of the interval $\{t : H(t) > \delta\}$. Thus the radius of $A_{\delta,h}$ is dominated by $hH^+(\delta)$. Now by (24.21) and the definition of δ, $hH^+(\delta) = hH^+(\epsilon h^d)$ converges to zero as $h \to 0$. Since $A_{t,h}$ is a ball, Lemma 24.5 implies that the right-hand side of (24.22) tends to zero at μ-almost all $x \in \mathcal{R}^d$. But

$$\int_0^\delta\left(\int_{A_{t,h}}|f(x)-f(y)|\mu(dy)\right)dt \leq (c_3 + |f(x)|)\delta$$

where $c_3 = \int|f(x)|\mu(dx)$. Using (24.20), and thus $cI_{\{\|x\|\leq r\}} \leq H(r)$ for suitable $c > 0$ and $r > 0$, we get

$$\int H\left(\frac{\|x-y\|}{h}\right)\mu(dy) \geq c\mu(S_{rh}) = \frac{c(rh)^d}{a_{rh}(x)}$$

where $a_h(x) = h^d/\mu(S_h)$. Using the above and the definition of δ we obtain

$$\int_0^\delta\left(\int_{A_{t,h}}|f(x)-f(y)|\mu(dy)\right)dt\Big/\int_0^\infty\mu(A_{t,h})\,dt$$

$$\leq \ \epsilon \left(\frac{c_3 + |f(x)|}{cr^d} \right) a_{rh}(x).$$

By Lemma 24.6 for μ-almost all x the right-hand side of the inequality above may be made arbitrarily small for ϵ small enough. The proof is completed. $\qquad\square$

Note that for $K(x) = I_{S_{0,1}}(x)$ Lemma 24.8 reduces to Lemma 24.5.

PROOF OF THEOREM 24.2. By Lemmas 24.1 and 24.2, this can be done by verifying (24.3), (24.4), (24.10), (24.12), and (24.13).

PROOF OF (24.3) AND (24.4). Under the assumptions of the theorem, (24.3) and (24.4) hold according to Lemmas 24.8 and 24.6.

PROOF OF (24.10). The function $x \to H(\|x\|)$ which is Riemann-integrable on compact spheres, can be approximated from below by a sequence of positive linear combinations

$$\sum_{k=1}^{N} c_{N,k} I_{S_{0,R_{N,k}}}$$

of indicator functions for spheres. Let m^* be the generalized Hardy-Littlewood maximal function for m, which is defined by

$$m^*(x) := \sup_{h>0} \frac{\int_{S_{x,h}} |m(t)|\mu(dt)}{\mu(S_{x,h})}, \quad x \in \mathcal{R}^d.$$

Because of Lemma 24.7, $m^* \in L^2(\mu)$ with

$$\int m^*(x)^2 \mu(dx) \leq c^* \int m(x)^2 \mu(dx) \leq c^* \mathbf{E} Y^2.$$

For each $h > 0$,

$$\int |m(t)| I_{S_{0,R_{N,k}}} \left(\frac{x-t}{h} \right) \mu(dt) \leq m^*(x) \int I_{S_{0,R_{N,k}}} \left(\frac{x-t}{h} \right) \mu(dt),$$

therefore, by the dominated convergence theorem,

$$\int |m(t)| H \left(\left\| \frac{x-t}{h} \right\| \right) \mu(dt)$$

$$= \lim_{N \to \infty} \int |m(t)| \sum_{k=1}^{N} c_{N,k} I_{S_{0,R_{N,k}}} \left(\frac{x-t}{h} \right) \mu(dt)$$

$$\leq m^*(x) \lim_{N \to \infty} \int \sum_{k=1}^{N} c_{N,k} I_{S_{0,R_{N,k}}} \left(\frac{x-t}{h} \right) \mu(dt)$$

$$= m^*(x) \int H \left(\left\| \frac{x-t}{h} \right\| \right) \mu(dt),$$

thus,

$$\int |m(t)| K\left(\frac{x-t}{h}\right) \mu(dt) \;\leq\; \beta \int |m(t)| H\left(\left\|\frac{x-t}{h}\right\|\right) \mu(dt)$$

$$\leq\; \beta m^*(x) \int H\left(\left\|\frac{x-t}{h}\right\|\right) \mu(dt)$$

$$\leq\; \frac{\beta}{\alpha} m^*(x) \int K\left(\frac{x-t}{h}\right) \mu(t).$$

Therefore,

$$\sup_{h>0} \frac{\int |m(t)| K(\frac{x-t}{h}) \mu(dt)}{\int K(\frac{x-t}{h}) \mu(dt)} \leq \frac{\beta}{\alpha} m^*(x),$$

and (24.10) is proved.

PROOF OF (24.12). We use the assumptions of the theorem with (24.16) for $n \geq 2$ and $K_1(x,u) = K(0) = 1$ (w.l.o.g.), $x, u \in \mathcal{R}^k$. For $n \geq 2$ one obtains

$$\mathbf{E}\frac{Y_n^2 K_n(x, X_n)^2}{\left(\sum\limits_{i=1}^{n} K_i(x, X_i)\right)^2} \leq \mathbf{E}\frac{1}{\left(1 + \sum\limits_{i=2}^{n-1} K(\frac{x-X_i}{h_i})\right)^2} \mathbf{E} Y^2 K\left(\frac{x-X}{h_n}\right)^2,$$

thus, by Fubini's theorem,

$$\int \sum_{n=1}^{\infty} \mathbf{E}\frac{Y_n^2 K_n(x, X_n)^2}{\left(\sum\limits_{i=1}^{n} K_i(x, X_i)\right)^2} \mu(dx)$$

$$\leq\; \mathbf{E}Y^2 + \int \mathbf{E}(Y^2 | X = z) \int \sum_{n=2}^{\infty} K\left(\frac{x-z}{h_n}\right)^2$$

$$\times \mathbf{E}\frac{1}{\left(1 + \sum\limits_{i=2}^{n-1} K(\frac{x-X_i}{h_i})\right)^2} \mu(dx)\mu(dz).$$

It suffices to show the existence of a constant $c_1 > 0$ such that

$$\sup_{z} \sum_{n=2}^{\infty} \mathbf{E} \int K\left(\frac{x-z}{h_n}\right)^2 \frac{1}{\left(1 + \sum\limits_{i=2}^{n-1} K(\frac{x-X_i}{h_i})\right)^2} \mu(dx) \leq c_1$$

for any distribution μ. A covering argument of Devroye and Krzyżak (1989) with a refined lower bound is used. Choose $R > 0$ such that $H(R) > 0$. Let $\mathcal{R}^d$ be covered by spheres $A_k = x_k + S_{0, R/2}$ such that every $x \in \mathcal{R}^d$ gets covered at most $k_1 = k_1(d)$ times. For each $n \geq 2$, $z \in \mathcal{R}^d$, we show that

$x \in z + h_n A_k$ implies

$$K\left(\frac{\cdot - x}{h_i}\right) \geq cK\left(\frac{\cdot - z}{h_i}\right) I_{A_k}\left(\frac{\cdot - z}{h_i}\right)$$

for all $i \in \{2, \ldots, n\}$ with $c_2 = \alpha H(R)/\beta H(0) \in (0,1]$. Without loss of generality let $z = 0$. With $x/h_n = \tilde{x}$ it suffices to show that

$$K\left(t - \tilde{x}\frac{h_n}{h_i}\right) \geq c_2 K(t)$$

for all $\tilde{x}, \; t \in A_k$ and all $n \geq 2, \; i \in \{2, \ldots, n\}$. Because of

$$\left\| t - \tilde{x}\frac{h_n}{h_i} \right\| \leq \max_{0 \leq r \leq 1} \| t - r\tilde{x} \|$$

$$= \max\{\|t\|, \|t - \tilde{x}\|\} \leq \max\{\|t\|, R\}$$

(since $h_n \leq h_i$) one has

$$H\left(\left\| t - \tilde{x}\frac{h_n}{h_i} \right\|\right) \geq H(R)H(\|t\|)/H(0)$$

in both cases $\|t - \tilde{x}h_n/h_i\| \leq \|t\|$, $\|t - \tilde{x}h_n/h_i\| \leq R$ by monotonicity of H, and thus the desired inequality.

Now for each $z \in \mathcal{R}^k$ one obtains

$$\sum_{n=2}^{\infty} \mathbf{E} \int_{\mathcal{R}^d} \frac{K(\frac{x-z}{h_n})^2}{\left(1 + \sum_{i=2}^{n-1} K(\frac{x-X_i}{h_i})\right)^2} \mu(dx)$$

$$= \sum_{k=1}^{\infty} \sum_{n=2}^{\infty} \mathbf{E} \int_{z+h_n A_k} \frac{K(\frac{x-z}{h_n})^2}{\left(1 + \sum_{i=2}^{n-1} K(\frac{x-X_i}{h_i})\right)^2} \mu(dx)$$

$$\leq \sum_{k=1}^{\infty} \sum_{n=2}^{\infty} \mathbf{E} \int_{z+h_n A_k} \frac{K(\frac{x-z}{h_n})^2}{\left(1 + c_2 \sum_{i=2}^{n-1} K(\frac{X_i-z}{h_i})I_{A_k}(\frac{X_i-z}{h_i})\right)^2} \mu(dx)$$

$$\leq \frac{1}{c_2^2} \sum_{k=1}^{\infty} \sum_{n=2}^{\infty} \mathbf{E} \int_{\mathcal{R}^d} \frac{K(\frac{x-z}{h_n})^2 I_{A_k}(\frac{x-z}{h_n})}{(1 + \sum_{i=2}^{n-1} K(\frac{X_i-z}{h_i})I_{A_k}(\frac{X_i-z}{h_i}))^2} \mu(dx)$$

$$\leq \frac{1}{c_2^2} \mathbf{E} \int \sum_{k=1}^{\infty} \sup_{s \in A_k} K(s) \sum_{n=2}^{\infty} \frac{K(\frac{x-z}{h_n})I_{A_k}(\frac{x-z}{h_n})}{\left(1 + \sum_{i=2}^{n-1} K(\frac{X_i-z}{h_i})I_{A_k}(\frac{X_i-z}{h_i})\right)^2} \mu(dx)$$

$$\leq \frac{1}{c_2^2} \sum_{k=1}^{\infty} \sup_{s \in A_k} K(s) \cdot \mathbf{E} \sum_{n=2}^{\infty} \frac{K(\frac{X_n-z}{h_n})I_{A_k}(\frac{X_n-z}{h_n})}{\left(1 + \sum_{i=2}^{n-1} K(\frac{X_i-z}{h_i})I_{A_k}(\frac{X_i-z}{h_i})\right)^2}$$

$$\leq \frac{4K_{max}^2}{c_2^2} \sum_{k=1}^{\infty} \sup_{s\in A_k} K(s) \cdot \mathbf{E} \sum_{n=2}^{\infty} \frac{K(\frac{X_n-z}{h_n})I_{A_k}(\frac{X_n-z}{h_n})}{\left(1 + \sum_{i=2}^{n} K(\frac{X_i-z}{h_i})I_{A_k}(\frac{X_i-z}{h_i})\right)^2}$$

$$\leq \frac{4K_{max}^2}{c_2^2} \sum_{k=1}^{\infty} \sup_{s\in A_k} K(s)$$

$$< \infty.$$

Here the formula

$$\sum_{n=2}^{N} \frac{a_n}{\left(1 + \sum_{i=2}^{n} a_i\right)^2} \leq 1 - \frac{1}{1 + \sum_{i=2}^{N} a_i},$$

valid for all sequences a_n with $a_n \geq 0$, $n \geq 2$ and obtainable by induction, is used, and also the properties of K.

The case $K_n(x,z) = K\left(\dfrac{x-z}{h_n}\right)$ with $\alpha I_{S_{0.R}} \leq K \leq \beta I_{S_{0.R}}$ $(0 < \alpha < \beta < \infty, 0 < R < \infty)$, hence w.l.o.g. $K = I_{S_{0.R}}$ is treated analogously, but in a slightly simpler way, using

$$\mathbf{E}\frac{Y_n^2 K(\frac{x-X_n}{h_n})^2}{\left(\sum_{i=1}^{n} K(\frac{x-X_i}{h_i})\right)^2} = \mathbf{E}\frac{Y_n^2 K(\frac{x-X_n}{h_n})}{\left(1 + \sum_{i=1}^{n-1} K(\frac{x-X_i}{h_i})\right)^2}$$

$$= \mathbf{E}\frac{1}{\left(1 + \sum_{i=1}^{n-1} K(\frac{x-X_i}{h_i})\right)^2}\mathbf{E}Y_n^2 K\left(\frac{x-X_n}{h_n}\right)$$

and

$$\sum_{n=1}^{\infty} \frac{I_{S_{0,R}\cap A_k}(\frac{X_n-z}{h_n})}{\left(\sum_{i=1}^{n} I_{S_{0,R}\cap A_k}(\frac{X_i-z}{h_i})\right)^2} \leq \sum_{n=1}^{\infty} \frac{1}{n^2} \leq 2$$

for all k and z.

PROOF OF (24.13). One notices

$$\int \sum_{n=1}^{\infty} \mathbf{E}\frac{\overline{m}_n(x)^2 K_n(x,X_n)^2}{\left(\sum_{i=1}^{n} K_i(x,X_i)\right)^2}\mu(dx)$$

$$\leq \int \sup_{n} \overline{m}_n(x)^2 \mathbf{E} \sum_{n=1}^{\infty} \frac{K_n(x,X_n)^2}{\left(\sum_{i=1}^{n} K_i(x,X_i)\right)^2}\mu(dx).$$

Under the assumptions of the theorem (w.l.o.g. $K(0) = 1$, in the case $\alpha I_{S_{0.R}} \leq K \leq \beta I_{S_{0.R}}$ even $K = I_{S_{0.R}}$) one has

$$\sum_{n=1}^{\infty} \frac{K_n(x,X_n)^2}{\left(\sum_{i=1}^{n} K_i(x,X_i)\right)^2} \leq K_{max} \cdot \sum_{n=1}^{\infty} \frac{K_n(x,X_n)}{\left(\sum_{i=1}^{n} K_i(x,X_i)\right)^2} \leq 2 \cdot K_{max}$$

for all x and all sequences X_n, according to the final argument in the proof of (24.12). This together with (24.10) yields the assertion. $\qquad\square$

24.3 Semirecursive Partitioning Estimate

For the semirecursive partitioning estimate we are given a sequence of (finite or countably infinite) partitions $\mathcal{P}_n = \{A_{n,1}, A_{n,2}, \ldots\}$ of $\mathcal{R}^d$, where $A_{n,1}, A_{n,2}, \ldots$ are Borel sets. Then put

$$K_n(x, u) = \sum_{j=1}^{\infty} I_{[x \in A_{n,j}, u \in A_{n,j}]}. \tag{24.23}$$

For $z \in \mathcal{R}^d$ set $A_n(z) = A_{n,j}$ if $z \in A_{n,j}$. Then (24.23) can be written in the form

$$K_n(x, u) = I_{A_n(x)}(u) = I_{A_n(u)}(x).$$

We call the sequence of partitions $\mathcal{P}_n$ nested if the sequence of generated σ-algebras $\mathcal{F}(\mathcal{P}_n)$ is increasing.

Theorem 24.3. *If*

$$\text{the sequence of partitions } \mathcal{P}_n \text{ is nested,} \tag{24.24}$$

$$diam\, A_n(z) := \sup_{u,v \in A_n(z)} \|u - v\| \to 0 \quad (n \to \infty) \tag{24.25}$$

for each $z \in \mathcal{R}^d$ and

$$\sum_{i=n}^{\infty} \lambda(A_i(z)) = \infty \tag{24.26}$$

for each z and n, then the semirecursive partitioning estimate is weakly and strongly universally consistent.

If the sequence of partitions $\mathcal{P}_n$ is nested, then the semirecursive partitioning estimator has the additional advantage that it is constant over each cell of $\mathcal{P}_n$. Such an estimator can be represented computationally by storing the constant numerator and denominator for each (nonempty) cell.

The proof of Theorem 24.3 applies the pointwise generalized Lebesgue density theorem for nested partitions (first part of Lemma 24.10).

In all lemmas of this section, μ and ν are assumed to be measures assigning finite values to bounded sets.

Lemma 24.9. *Assume (24.24). Then*

$$\mu \left\{ x \in \mathcal{R}^d : \sup_n \frac{\nu(A_n(x))}{\mu(A_n(x))} > \alpha \right\} \leq \frac{\nu(\mathcal{R}^d)}{\alpha}$$

for any $\alpha > 0$.

PROOF. Set

$$S := \left\{ x \in \mathcal{R}^d : \sup_{n>0} \frac{\nu(A_n(x))}{\mu(A_n(x))} > \alpha \right\}.$$

For $x \in S$ choose $n_x, i_x \in \mathcal{N}$ with $x \in A_{n_x, i_x}$ and

$$\frac{\nu(A_{n_x, i_x})}{\mu(A_{n_x, i_x})} > \alpha. \tag{24.27}$$

Clearly

$$S \subseteq \bigcup_{x \in S} A_{n_x, i_x}.$$

By the nestedness of P_n

$$\left\{ A_{n_j, i_j} : j \right\}$$
$$:= \left\{ A_{n_x, i_x} : x \in S \text{ and for every } y \in S : x \notin A_{n_y, i_y} \text{ or } n_y > n_x \right\}$$

is a disjoint finite- or infinite-countable cover of S. Using this and (24.27) we obtain

$$\begin{aligned}
\mu(S) &\leq \mu\left(\bigcup_j A_{n_j, i_j} \right) = \sum_j \mu(A_{n_j, i_j}) \\
&\leq \sum_j \frac{1}{\alpha} \nu(A_{n_j, i_j}) = \frac{1}{\alpha} \nu\left(\bigcup_j A_{n_j, i_j} \right) \\
&\leq \frac{1}{\alpha} \nu(\mathcal{R}^d)
\end{aligned}$$

and the assertion is proved. $\qquad\square$

Lemma 24.10. *Assume (24.24) and (24.25). Then*

$$\lim_{n \to \infty} \frac{\int_{A_n(x)} f(z)\mu(dz)}{\mu(A_n(x))} = f(x) \mod \mu,$$

and

$$\liminf_{n \to \infty} \frac{\mu(A_n(x))}{\lambda(A_n(x))} > 0 \mod \mu.$$

PROOF. See Problem 24.3.

Lemma 24.11. *Let the conditions of Lemma 24.9 be fulfilled, let $m \in L_2(\mu)$, and let m^* denote the generalized Hardy-Littlewood maximal function of m defined by*

$$m^*(x) = \sup_n \frac{1}{\mu(A_n(x))} \int_{A_n(x)} |m| \, d\mu, \quad x \in \mathcal{R}^d.$$

Thus $m^* \in L_2(\mu)$ and

$$\int m^*(x)^2 \mu(dx) \le c^* \int m(x)^2 \mu(dx),$$

where c^* depends only on d.

PROOF. See Problem 24.4.

PROOF OF THEOREM 24.3. By Lemmas 24.1 and 24.2, this can be done by verifying (24.3), (24.4), (24.10), (24.12), and (24.13).

PROOF OF (24.3). See Lemma 24.10.

PROOF OF (24.4). By Lemma 24.10 we have that

$$\liminf_{n \to \infty} \frac{\mu(A_i(x))}{\lambda(A_i(x))} > 0 \quad \mod \mu,$$

which together with (24.26) implies

$$\sum_{i=1}^n \mathbf{E} K_i(x, X_i) = \sum_{i=1}^n \frac{\mu(A_i(x))}{\lambda(A_i(x))} \cdot \lambda(A_i(x)) \to \infty \quad \mod \mu.$$

PROOF OF (24.10). See Lemma 24.11.

PROOF OF (24.12). It suffices to show the existence of a constant $c_1 > 0$ such that

$$\sup_z \mathbf{E} \sum_{n=1}^{\infty} \int K_n(x, z) \frac{1}{\left(1 + \sum_{i=1}^{n-1} K_i(x, X_i)\right)^2} \mu(dx) \le c_1$$

for any distribution μ. The sequence of partitions is nested, thus $x \in A_n(z)$ and $i \le n$ imply $z \in A_n(x) \subseteq A_i(x)$ which in turn implies $A_i(x) = A_i(z)$. Therefore,

$$\begin{aligned}
&\mathbf{E} \sum_{n=1}^{\infty} \int K_n(x, z) \frac{1}{\left(1 + \sum_{i=1}^{n-1} K_i(x, X_i)\right)^2} \mu(dx) \\
&= \mathbf{E} \sum_{n=1}^{\infty} \int I_{\{x \in A_n(z)\}} \frac{1}{\left(1 + \sum_{i=1}^{n-1} I_{\{X_i \in A_i(x)\}}\right)^2} \mu(dx) \\
&= \mathbf{E} \sum_{n=1}^{\infty} \int I_{\{x \in A_n(z)\}} \frac{1}{\left(1 + \sum_{i=1}^{n-1} I_{\{X_i \in A_i(z)\}}\right)^2} \mu(dx) \\
&= \mathbf{E} \sum_{n=1}^{\infty} \frac{\mu(A_n(z))}{\left(1 + \sum_{i=1}^{n-1} I_{\{X_i \in A_i(z)\}}\right)^2}
\end{aligned}$$

$$= \mathbf{E} \sum_{n=1}^{\infty} \frac{I_{\{X_n \in A_n(z)\}}}{\left(\sum_{i=1}^{n} I_{\{X_i \in A_i(z)\}}\right)^2}. \tag{24.28}$$

If for fixed $X_1, X_2, ...,$ n is the kth index with $I_{\{X_n \in A_n(z)\}} = 1$, then $\sum_{i=1}^{n} I_{\{X_i \in A_i(z)\}} = k$. Therefore,

$$\sum_{n=1}^{\infty} \frac{I_{\{X_n \in A_n(z)\}}}{\left(\sum_{i=1}^{n} I_{\{X_i \in A_i(z)\}}\right)^2} \le \sum_{k=1}^{\infty} \frac{1}{k^2} \le 2$$

which together with (24.28) yields the assertion.

PROOF OF (24.13). One notices

$$\int \sum_{n=1}^{\infty} \mathbf{E} \frac{\overline{m}_n(x)^2 K_n(x, X_n)^2}{\left(\sum_{i=1}^{n} K_i(x, X_i)\right)^2} \, \mu\,(dx)$$

$$\le \int \sup_n \overline{m}_n(x)^2 \mathbf{E} \sum_{n=1}^{\infty} \frac{K_n(x, X_n)}{\left(\sum_{i=1}^{n} K_i(x, X_i)\right)^2} \, \mu\,(dx).$$

One has

$$\sum_{n=1}^{\infty} \frac{K_n(x, X_n)}{\left(\sum_{i=1}^{n} K_i(x, X_i)\right)^2} \le 2$$

for all x and all sequences X_n, according to the final argument in the proof of (24.12). This together with (24.10) yields the assertion. $\square$

24.4   Bibliographic Notes

Theorems 24.1, 24.2, and 24.3 are due to Györfi, Kohler, and Walk (1998). In the literature, the semirecursive estimates have been considered with the general form

$$K_n(x, u) = \alpha_n K \left(\frac{x - u}{h_n} \right)$$

with a sequence of weights $\alpha_n > 0$. Motivated by a recursive kernel density estimate due to Wolverton and Wagner (1969b) and Yamato (1971), Greblicki (1974) and Ahmad and Lin (1976) proposed and studied semirecursive kernel estimates with $\alpha_n = 1/h_n^d$, see also Krzyżak and Pawlak (1983). The choice $\alpha_n = 1$ has been proposed and investigated by Devroye and Wagner (1980b). Consistency properties of this estimate were studied by Krzyżak and Pawlak (1984a), Krzyżak (1992), Greblicki and Pawlak (1987b), and Györfi, Kohler, and Walk (1998). Lemma 24.4 is related to Lemma 10.47 of Wheeden and Zygmund (1977). Lemma 24.6 is due to Devroye (1981). Lemma 24.7 deals with the Hardy-Littlewood maximal function (see Stein and Weiss (1971) and Wheeden and Zygmund (1977)). Concerning Lemma 24.8 see Greblicki, Krzyżak, and Pawlak (1984), Greblicki and Pawlak (1987a), and Krzyżak (1991). Semirecursive kernel estimates was applied to estimation of nonlinear, dynamic systems in Krzyżak (1993).

Problems and Exercises

PROBLEM 24.1. Prove the Abel-Dini theorem: for any sequence $a_n \geq 0$ with $a_1 > 0$,

$$\sum_{n=1}^{\infty} \frac{a_n}{\left(\sum_{i=1}^{n} a_i\right)^2} < \infty.$$

PROBLEM 24.2. Prove Lemma 24.2.

HINT: First show that

$$\int \left(\frac{\sum_{i=1}^{n} \overline{m}_i(x) K_i(x, X_i)}{\sum_{i=1}^{n} K_i(x, X_i)} - m(x)\right)^2 \mu(dx) \to 0 \text{ a.s.}$$

by the use of (24.3), (24.10), (24.4), (24.8), the Toeplitz lemma, and Lebesgue's dominated convergence theorem. Then formulate a recursion for $\{U_n\}$ with

$$U_n(x) = \frac{\sum_{i=1}^{n} (Y_i - \overline{m}_i(x)) K_i(x, X_i)}{\sum_{i=1}^{n} K_i(x, X_i)}$$

and show a.s. convergence of $\int U_n(x)^2 \mu(dx)$ by the use of (24.12), (24.13), and Theorem A.6 distinguishing cases (24.14) and (24.15). These results yield a.s. convergence of $\int (m_n(x) - m(x))^2 \mu(dx)$ and thus, by Lemma 24.1, the assertion of part (a). Prove part (b) analogously by taking expectations.

PROBLEM 24.3. Prove Lemma 24.10.

PROBLEM 24.4. Prove Lemma 24.11.

PROBLEM 24.5. Formulate and prove the variant of Lemma 23.3 by which the proof of Lemma 24.1 can be finished.

PROBLEM 24.6. For $d = 1$ and nonnested partitions where each partition consists of nonaccumulating intervals, prove the assertions of Lemma 24.9 with factor 2 on the right-hand side of the inequalities, and then prove Theorem 24.3 for $d = 1$ without condition (24.24).

HINT: Use arguments in the proof of Lemma 24.4.

PROBLEM 24.7. Prove both parts of Lemma 24.10 using a martingale convergence theorem (Theorem A.4).

HINT: Let $\mathcal{F}_n$ be the σ-algebra generated by the partition $\mathcal{P}_n$. Put

$$f_n(x) = \mathbf{E}\{f(X)|X \in A_n(x)\}.$$

Then $(f_n, \mathcal{F}_n)$ forms a convergent martingale on the probability space $(\mathcal{R}^d, \mathcal{B}_d, \mu)$.

25

Recursive Estimates

25.1 A General Result

Introduce a sequence of bounded, measurable, symmetric and nonnegative valued functions $K_n(x, z)$ on $\mathcal{R}^d \times \mathcal{R}^d$. Let $\{a_n\}$ be a sequence of positive numbers, then the estimator is defined by the following recursion:

$$m_1(x) = Y_1,$$

$$m_{n+1}(x) = m_n(x)(1 - a_{n+1}K_{n+1}(x, X_{n+1})) + a_{n+1}Y_{n+1}K_{n+1}(x, X_{n+1}).$$
$$(25.1)$$

By each new observation the estimator will be updated. At the nth stage one has only to store $m_n(x)$. The estimator is of a stochastic approximation type, especially of Robbins-Monro type (cf. Ljung, Pflug, and Walk (1992) with further references). $m_{n+1}(x)$ is obtained as a linear combination of the estimates $m_n(x)$ and Y_{n+1} with weights $1 - a_{n+1}K_{n+1}(x, X_{n+1})$ and $a_{n+1}K_{n+1}(x, X_{n+1})$, respectively.

Theorem 25.1. *Assume that there exists a sequence $\{h_n\}$ of positive numbers tending to 0 and a nonnegative nonincreasing function H on $[0, \infty)$ with $r^d H(r) \to 0$ $(r \to \infty)$ such that*

$$h_n^d K_n(x, z) \leq H(\|x - z\|/h_n), \qquad (25.2)$$

$$\sup_{x,z,n} a_n K_n(x, z) < 1, \qquad (25.3)$$

$$\liminf_n \int K_n(x,t)\mu(dt) > 0 \quad \text{for } \mu\text{-almost all } x, \tag{25.4}$$

$$\sum_n a_n = \infty, \tag{25.5}$$

and

$$\sum_n \frac{a_n^2}{h_n^{2d}} < \infty. \tag{25.6}$$

Then m_n is weakly and strongly universally consistent.

PROOF. The proof of the first statement can be done by the verification of the conditions of Stone's theorem (Theorem 4.1). If, by definition, a void product is 1, then by (25.1),

$$W_{n,i}(z) = \prod_{l=i+1}^{n} (1 - a_l K_l(z, X_l)) a_i K_i(z, X_i)$$

for $n \geq 2$, $i = 1, \dots, n$, and

$$W_{1,1}(z) = 1.$$

It is easy to check by induction that these are probability weights. To check condition (i) let b_n be defined by (25.1) if Y_i is replaced by $f(X_i)$ and $b_1(x) = f(X_1)$, where f is an arbitrary nonnegative Borel function. Then we prove that

$$\mathbf{E}b_n(X) = \mathbf{E}f(X), \tag{25.7}$$

which implies (i) with $c = 1$. Introduce the notation $\mathcal{F}_n$ for the σ-algebra generated by (X_i, Y_i) $(i = 1, 2, \dots, n)$,

$$
\begin{aligned}
&\mathbf{E}\{b_{n+1}(X)|\mathcal{F}_n\} \\
=\ &\mathbf{E}\{b_n(X)|\mathcal{F}_n\} + a_{n+1}\mathbf{E}\{(f(X_{n+1}) - b_n(X))K_{n+1}(X, X_{n+1})|\mathcal{F}_n\} \\
=\ &\mathbf{E}\{b_n(X)|\mathcal{F}_n\} + a_{n+1}\int\int (f(x) - b_n(z))K_{n+1}(z, x)\mu(dx)\mu(dz) \\
=\ &\mathbf{E}\{b_n(X)|\mathcal{F}_n\} + a_{n+1}\int\int (f(x) - b_n(x))K_{n+1}(x, z)\mu(dx)\mu(dz) \\
=\ &\mathbf{E}\{b_n(X)|\mathcal{F}_n\} + a_{n+1}\mathbf{E}\{(f(X) - b_n(X))K_{n+1}(X, X_{n+1})|\mathcal{F}_n\},
\end{aligned}
$$

where the symmetry of $K_n(x, z)$ was applied. Thus

$$\mathbf{E}b_{n+1}(X) = \mathbf{E}b_n(X) + a_{n+1}\mathbf{E}\{(f(X) - b_n(X))K_{n+1}(X, X_{n+1})\}.$$

Define another sequence

$$b_1^*(X) = f(X),$$

$$b_{n+1}^*(X) = b_n^*(X) + a_{n+1}(f(X) - b_n^*(X))K_{n+1}(X, X_{n+1}).$$

Then

$$b_n^*(X) = f(X)$$

and

$$\mathbf{E}b_{n+1}^*(X) = \mathbf{E}b_n^*(X) + a_{n+1}\mathbf{E}\{(f(X) - b_n^*(X))K_{n+1}(X, X_{n+1})\}.$$

For $\mathbf{E}\{b_n^*(X)\}$ and for $\mathbf{E}\{b_n(X)\}$ we have the same iterations and, thus,

$$\mathbf{E}\{b_n^*(X)\} = \mathbf{E}\{b_n(X)\},$$

so (25.7) and therefore (i) is proved. Set

$$p_n(x) = \int K_n(x, t)\mu(dt).$$

Then by condition (25.4) for each fixed i and for μ-almost all z,

$$\mathbf{E}W_{n,i}(z) \leq \frac{H(0)a_i}{h_i^d}e^{-\sum_{l=i+1}^{n} a_l\mathbf{E}K_l(z,X)} = \frac{H(0)a_i}{h_i^d}e^{-\sum_{l=i+1}^{n} a_l p_l(z)} \to 0.$$

$$(25.8)$$

Obviously,

$$\mathbf{E}\sum_{i=1}^{n} W_{n,i}^2(z) \to 0 \tag{25.9}$$

for μ-almost all z implies (v), so we prove (25.9) showing

$$\mathbf{E}\sum_{i=1}^{n} W_{n,i}^2(z) \leq \sum_{i=1}^{n} \mathbf{E}W_{n,i}(z)\frac{H(0)a_i}{h_i^d} \to 0$$

by the Toeplitz lemma and by (25.6) and (25.8). Concerning (iii) it is enough to show that for all $a > 0$ and for μ-almost all z,

$$\mathbf{E}\left\{\sum_{i=1}^{n} W_{n,i}(z)I_{[\|X_i-z\|>a]}\right\} \to 0.$$

By (25.4),

$$\liminf_n p_n(z) = 2p(z) > 0 \text{ for } \mu\text{-almost all } z,$$

so for such z there is an $n_0(z)$ such that, for $n > n_0(z)$,

$$p_n(z) \geq p(z).$$

Because of (25.8) it suffices to show, for these z,

$$\mathbf{E}\sum_{i=n_0(z)}^{n} W_{n,i}(z)I_{[\|X_i-z\|>a]} \to 0.$$

Because of the conditions

$$\mathbf{E} \sum_{i=n_0(z)}^{n} W_{n,i}(z) I_{[\|X_i - z\| > a]}$$

$$= \sum_{i=n_0(z)}^{n} \mathbf{E}\{W_{n,i}(z)\} \frac{\mathbf{E}K_i(z, X_i) I_{[\|X_i - z\| > a]}}{\mathbf{E}K_i(z, X_i)}$$

$$\leq \sum_{i=n_0(z)}^{n} \mathbf{E}\{W_{n,i}(z)\} \frac{h_i^{-d} H(a/h_i)}{p_i(z)}$$

$$\leq \sum_{i=n_0(z)}^{n} \mathbf{E}\{W_{n,i}(z)\} \frac{h_i^{-d} H(a/h_i)}{p(z)} \to 0$$

by the Toeplitz lemma. Thus the first statement is proved:

$$\mathbf{E}\|m_n - m\|^2 \to 0. \tag{25.10}$$

In order to prove the second statement note

$$m_{n+1}(x) - \mathbf{E}m_{n+1}(x)$$

$$= m_n(x) - \mathbf{E}m_n(x)$$

$$- a_{n+1}(m_n(x)K_{n+1}(x, X_{n+1}) - \mathbf{E}m_n(x)K_{n+1}(x, X_{n+1}))$$

$$+ a_{n+1}(Y_{n+1}K_{n+1}(x, X_{n+1}) - \mathbf{E}Y_{n+1}K_{n+1}(x, X_{n+1})),$$

therefore,

$$\mathbf{E}\{(m_{n+1}(x) - \mathbf{E}m_{n+1}(x))^2 | \mathcal{F}_n\} = I_1 + I_2 + I_3 + I_4 + I_5 + I_6,$$

where

$$I_1 = (m_n(x) - \mathbf{E}m_n(x))^2$$

and because of the independence of $m_n(x)$ and $K_{n+1}(x, X_{n+1})$,

$$I_2 = \mathbf{E}\{a_{n+1}^2(m_n(x)K_{n+1}(x, X_{n+1}) - \mathbf{E}\{m_n(x)K_{n+1}(x, X_{n+1})\})^2 | \mathcal{F}_n\}$$

$$= a_{n+1}^2(m_n(x)^2(\mathbf{E}K_{n+1}(x, X_{n+1})^2 - [\mathbf{E}K_{n+1}(x, X_{n+1})]^2)$$

$$+ (m_n(x) - \mathbf{E}m_n(x))^2[\mathbf{E}K_{n+1}(x, X_{n+1})]^2)$$

$$\leq \frac{H(0)^2 a_{n+1}^2}{h_{n+1}^{2d}} (m_n(x)^2 + (m_n(x) - \mathbf{E}m_n(x))^2)$$

and

$$I_3 = \mathbf{E}\{a_{n+1}^2(Y_{n+1}K_{n+1}(x, X_{n+1}) - \mathbf{E}Y_{n+1}K_{n+1}(x, X_{n+1}))^2 | \mathcal{F}_n\}$$

$$\leq a_{n+1}^2 \mathbf{E}Y_{n+1}^2 K_{n+1}(x, X_{n+1})^2$$

$$\leq \frac{H(0)^2 a_{n+1}^2}{h_{n+1}^{2d}} \mathbf{E}Y^2$$

and

$$
\begin{aligned}
I_4 \;&=\; -2a_{n+1}(m_n(x) - \mathbf{E}m_n(x))\mathbf{E}\{m_n(x)K_{n+1}(x, X_{n+1}) \\
&\qquad - \mathbf{E}\{m_n(x)K_{n+1}(x, X_{n+1})\}|\mathcal{F}_n\} \\
&=\; -2a_{n+1}(m_n(x) - \mathbf{E}m_n(x))^2 \mathbf{E}K_{n+1}(x, X_{n+1}) \\
&\leq\; 0
\end{aligned}
$$

and

$$
\begin{aligned}
I_5 \;&=\; 2a_{n+1}(m_n(x) - \mathbf{E}m_n(x)) \\
&\qquad \times \mathbf{E}\{Y_{n+1}K_{n+1}(x, X_{n+1}) - \mathbf{E}\{Y_{n+1}K_{n+1}(x, X_{n+1})\}|\mathcal{F}_n\} \\
&=\; 0
\end{aligned}
$$

and

$$
\begin{aligned}
I_6 \;& \\
=\;& -2a_{n+1}^2 \mathbf{E}\{(m_n(x)K_{n+1}(x, X_{n+1}) - \mathbf{E}m_n(x)K_{n+1}(x, X_{n+1})) \\
& \times (Y_{n+1}K_{n+1}(x, X_{n+1}) - \mathbf{E}Y_{n+1}K_{n+1}(x, X_{n+1}))|\mathcal{F}_n\} \\
=\;& -2a_{n+1}^2 m_n(x) \\
& \times \mathbf{E}\{m(X_{n+1})K_{n+1}(x, X_{n+1})^2 - K_{n+1}(x, X_{n+1})\mathbf{E}Y_{n+1}K_{n+1}(x, X_{n+1})\} \\
\leq\;& 2\frac{H(0)^2 a_{n+1}^2}{h_{n+1}^{2d}} |m_n(x)|(\mathbf{E}|m(X)| + \mathbf{E}|Y|) \\
\leq\;& 4\frac{H(0)^2 a_{n+1}^2}{h_{n+1}^{2d}} |m_n(x)|\mathbf{E}|Y| \\
\leq\;& 2\frac{H(0)^2 a_{n+1}^2}{h_{n+1}^{2d}} (m_n(x)^2 + \mathbf{E}Y^2).
\end{aligned}
$$

Thus summarizing

$$
\begin{aligned}
&\mathbf{E}\{(m_{n+1}(x) - \mathbf{E}m_{n+1}(x))^2|\mathcal{F}_n\} \\
\leq\;& \left(1 + \frac{H(0)^2 a_{n+1}^2}{h_{n+1}^{2d}}\right)(m_n(x) - \mathbf{E}m_n(x))^2 + 3\frac{H(0)^2 a_{n+1}^2}{h_{n+1}^{2d}}(m_n(x)^2 + \mathbf{E}Y^2).
\end{aligned}
$$

$$(25.11)$$

Analogously, by taking the integral with respect to μ, one obtains

$$
\begin{aligned}
&\mathbf{E}\{\|m_{n+1} - \mathbf{E}m_{n+1}\|^2|\mathcal{F}_n\} \\
\leq\;& \left(1 + \frac{H(0)^2 a_{n+1}^2}{h_{n+1}^{2d}}\right)\|m_n - \mathbf{E}m_n\|^2 + 3\frac{H(0)^2 a_{n+1}^2}{h_{n+1}^{2d}}(\|m_n\|^2 + \mathbf{E}Y^2).
\end{aligned}
$$

$$(25.12)$$

Relation (25.10) implies

$$\|\mathbf{E}m_n - m\|^2 \to 0 \tag{25.13}$$

and

$$\mathbf{E}\|m_n - \mathbf{E}m_n\|^2 \to 0 \tag{25.14}$$

and

$$\mathbf{E}\|m_n\|^2 = O(1). \tag{25.15}$$

Now according to Theorem A.5, because of (25.6) and (25.15), from (25.12) one obtains a.s. convergence of $\|m_n - \mathbf{E}m_n\|^2$. Because of (25.14),

$$\|m_n - \mathbf{E}m_n\|^2 \to 0$$

in probability, which together with the a.s. convergence of $\|m_n - \mathbf{E}m_n\|^2$ yields $\|m_n - \mathbf{E}m_n\|^2 \to 0$ a.s. This together with (25.13) yields the second statement.

25.2 Recursive Kernel Estimate

For a kernel $K : \mathcal{R}^d \to \mathcal{R}_+$ consider the recursive estimator (25.1) with

$$K_n(x, z) = \frac{1}{h_n^d} K\left(\frac{x - z}{h_n}\right), \quad x, z \in R^d. \tag{25.16}$$

Theorem 25.2. *Assume for the kernel K that there is a ball $S_{0,r}$ of radius $r > 0$ centered at the origin, and a constant $b > 0$ such that $K(x) \geq bI_{S_{0,r}}$ and that there is a nonnegative nonincreasing Borel function H on $[0, \infty)$ with $r^d H(r) \to 0$ $(r \to \infty)$ such that*

$$K(x) \leq H(\|x\|), \tag{25.17}$$

$$h_n > 0, \quad \lim_n h_n = 0, \tag{25.18}$$

$$a_n > 0, \quad \sum_n a_n = \infty, \tag{25.19}$$

$$\sup_x K(x) \sup_n \frac{a_n}{h_n^d} < 1, \tag{25.20}$$

and

$$\sum_n \frac{a_n^2}{h_n^{2d}} < \infty. \tag{25.21}$$

Then m_n is weakly and strongly universally consistent.

PROOF. See Problem 25.1.

25.3 Recursive Partitioning Estimate

For the recursive partitioning estimate we are given a sequence of (finite or countably infinite) partitions $\mathcal{P}_n = \{A_{n,1}, A_{n,2}, \ldots\}$ of $\mathcal{R}^d$, where $A_{n,1}, A_{n,2}, \ldots$ are Borel sets. For $z \in \mathcal{R}^d$ set $A_n(z) = A_{n,j}$ if $z \in A_{n,j}$, then consider the recursive estimator (25.1) with

$$K_n(x, z) = \frac{1}{h_n^d} I_{\{z \in A_n(x)\}}, \quad x, z \in R^d. \tag{25.22}$$

Theorem 25.3. *Assume that for a nested sequence of partitions with*

$$diam\, A_n(z) \leq h_n \to 0, \quad \liminf_{n \to \infty} \lambda(A_n(z))/h_n^d > 0, \tag{25.23}$$

for all $z \in \mathcal{R}^d$,

$$a_n > 0, \quad \sum_n a_n = \infty, \tag{25.24}$$

$$\sup_n \frac{a_n}{h_n^d} < 1, \tag{25.25}$$

and

$$\sum_n \frac{a_n^2}{h_n^{2d}} < \infty. \tag{25.26}$$

Then m_n is weakly and strongly universally consistent.

PROOF. See Problem 25.2.

25.4 Recursive NN Estimate

Introduce the recursive NN estimate such that it splits the data sequence $D_n = \{(X_1, Y_1), \ldots, (X_n, Y_n)\}$ into disjoint blocks of length $l_1, \ldots, l_N$, where $l_1, \ldots, l_N$ are positive integers. In each block find the nearest neighbor of x, and denote the nearest neighbor of x from the ith block by $X_i^*(x)$. Let $Y_i^*(x)$ be the corresponding label. Ties are broken by comparing indices as in Chapter 6. The recursive regression estimate is as follows: if $\sum_{i=1}^{N} l_i \leq n < \sum_{i=1}^{N+1} l_i$, then

$$m_n(x) = \frac{1}{N} \sum_{i=1}^{N} Y_i^*(x).$$

Theorem 25.4. *If $\lim_{N \to \infty} l_N = \infty$, then m_n is weakly and strongly universally consistent.*

PROOF. Put

$$\tilde{m}_N(x) = m_{\sum_{i=1}^N l_i}(x) = \frac{1}{N}\sum_{i=1}^N Y_i^*(x),$$

then it suffices to prove the weak and strong universal consistency of $\tilde{m}_N$. $\tilde{m}_N$ is an average of independent random variables taking values in L_2. First we show that concerning the bias

$$\|\mathbf{E}\tilde{m}_N - m\| \to 0,$$

which follows from the Toeplitz lemma if

$$\|\mathbf{E}Y_i^* - m\| \to 0.$$

Obviously,

$$
\begin{aligned}
\mathbf{E}Y_i^*(x) &= \mathbf{E}\sum_{j=\sum_{k=1}^{i-1} l_k+1}^{\sum_{k=1}^{i} l_k} Y_j I_{\{X_i^*(x)=X_j\}} \\
&= \mathbf{E}\sum_{j=1}^{l_i} Y_j I_{\{X^*(x)=X_j\}} \\
&= \mathbf{E}\sum_{j=1}^{l_i} m(X_j) I_{\{X^*(x)=X_j\}} \\
&= \mathbf{E}m(X^*(x)),
\end{aligned}
$$

where $X^*(x)$ is the nearest neighbor of x from $X_1,\ldots,X_{l_i}$. Because of Problem 6.3,

$$\mathbf{E}\{(m(X^*(X)) - m(X))^2\} \to 0,$$

therefore,

$$\mathbf{E}\{(\mathbf{E}\{m(X^*(X))|X\} - m(X))^2\} \to 0.$$

Turning next to the variation term we show that

$$C = \sup_i \mathbf{E}\|Y_i^* - \mathbf{E}Y_i^*\|^2 < \infty.$$

Put $\sigma^2(x) = \mathbf{E}\{Y^2|X=x\}$ and $C^* = \sup_{z,i}\int K_i(x,z)\mu(dx) \le \gamma_d$ then

$$
\begin{aligned}
\mathbf{E}\|Y_i^* - \mathbf{E}Y_i^*\|^2 &\le \mathbf{E}\|Y_i^*\|^2 \\
&= \mathbf{E}Y_i^*(X)^2 \\
&= \int\int \sigma^2(z) K_i(x,z)\mu(dx)\mu(dz) \\
&\le C^*\int \sigma^2(z)\mu(dz)
\end{aligned}
$$

$$= C^* \mathbf{E} Y^2$$
$$= C.$$

Thus

$$\mathbf{E}\|\tilde{m}_N - \mathbf{E}\tilde{m}_N\|^2 = \frac{1}{N^2} \sum_{i=1}^{N} \mathbf{E}\|Y_i^* - \mathbf{E}Y_i^*\|^2 \le C\frac{1}{N} \to 0,$$

so the weak universal consistency is proved. Let $\mathcal{F}_n$ denote the σ-algebra generated by $X_1, Y_1, \ldots, X_n, Y_n$. Concerning the strong universal consistency we can apply the almost supermartingale convergence theorem (Theorem A.5) since

$$\mathbf{E}\left\{ \|\tilde{m}_N - \mathbf{E}\tilde{m}_N\|^2 | \mathcal{F}_{\left\{\sum_{j=1}^{N-1} l_j\right\}} \right\}$$
$$= (1 - 1/N)^2 \|\tilde{m}_{N-1} - \mathbf{E}\tilde{m}_{N-1}\|^2 + 1/N^2 \mathbf{E}\|Y_N^* - \mathbf{E}Y_N^*\|^2$$
$$\le \|\tilde{m}_{N-1} - \mathbf{E}\tilde{m}_{N-1}\|^2 + C/N^2.$$

$\square$

25.5   Recursive Series Estimate

Introduce a sequence of functions on R^d: $\{\phi_1, \phi_2, \ldots\}$. Let $\{k_n\}$ be a nondecreasing sequence of positive integers and let

$$K_n(x, u) = \sum_{j=1}^{k_n} \phi_j(x)\phi_j(u), \ x, u \in R^d.$$

Then the recursive series estimate is defined by the following recursion:

$$m_1(x) = Y_1,$$

$$m_{n+1}(x) = m_n(x) - a_{n+1}(m_n(X_{n+1}) - Y_{n+1})K_{n+1}(x, X_{n+1}),$$

where $a_n > 0$.

Theorem 25.5. *Let the functions* $\{\phi_1, \phi_2, \ldots\}$ *be uniformly bounded by* 1 *and square integrable with respect to the Lebesgue measure* λ. *Assume, moreover, that they form an orthonormal system in* $L_2(\lambda)$ *and span* $L_2(\mu)$, *i.e., the set of finite linear combinations of the* ϕ_i*'s is dense in* $L_2(\mu)$. *If*

$$k_n \to \infty, \quad \sum_n a_n = \infty, \quad \sum_n a_n^2 k_n^2 < \infty,$$

then the recursive series estimate is weakly and strongly consistent.

Observe that Theorem 25.5 implies a strong universal consistency result if one has an example of ϕ_i's satisfying the conditions universally for all

μ. As a possible example consider $d = 1$ (it can be extended to $d > 1$). By Theorem A.1, for an arbitrary $f \in L_2(\mu)$, choose a continuous f^* of compact support with $\|f - f^*\|_\mu < \epsilon$. For instance, let (ϕ_i') be the Walsh system of support $(0, 1]$ and

$$\phi_{ij}(x) = \phi_i'(x - j), \; i = 0, 1, 2, \ldots, \; j = 0, \pm 1, \pm 2, \ldots.$$

Then $(\phi_{i,j})$ is an orthonormal system in $L_2(\lambda)$ and f^* can be approximated with respect to the supremum norm by linear combinations of ϕ_{ij}'s (see, e.g., Alexits (1961), Sections 1.6, 1.7, 4.2), which then approximate f, too, in $L_2(\mu)$. Instead of the Walsh system one can use the standard trigonometric system of support $(-\pi, \pi]$ or the system of Legendre polynomials of support $(-1, 1]$.

When k_n is fixed, say k, then the corresponding results hold with m replaced by the projection $P_k m$ of m onto $span\{\phi_1, \ldots, \phi_k\}$ with respect to $L_2(\mu)$. $P_k m$ is unique as an element in $L_2(\mu)$, but its representation as a linear combination of $\phi_1, \ldots, \phi_k$ is generally not unique. The a.s. convergence result for fixed k_n also follows from the general stochastic approximation results in Hilbert space (Fritz (1974), Györfi (1980; 1984), Walk (1985), Walk and Zsidó (1989)), where in the case of weights $a_n = 1/n$ only the stationarity and ergodicity of $\{(X_n, Y_n)\}$, together with the condition $\mathbf{E}Y^2 < \infty$, are assumed.

The proper choice of a_n and k_n can be made knowing the rate of the convergence, which is possible assuming regularity conditions on the distributions of (X, Y).

For the proof of Theorem 25.5 we use some lemmas and the following notations. Let $(\cdot, \cdot)$, $(\cdot, \cdot)_\lambda$, $\|\cdot\|$, $\|\cdot\|_\lambda$ denote the inner products and the norms in the spaces $L_2(\mu)$, $L_2(\lambda)$, respectively. Moreover, put

$$H_n = span\{\phi_1, \ldots, \phi_{k_n}\}.$$

Introduce the notation $\mathcal{F}_n$ for the σ-algebra generated by (X_i, Y_i) $(i = 1, 2, \ldots, n)$. Let, for $z \in L_2(\mu)$,

$$A_n z = z(X_n) \sum_{i=1}^{k_n} \phi_i(X_n)\phi_i, \quad \bar{A}_n z := \mathbf{E}A_n z,$$

and

$$b_n = Y_n \sum_{i=1}^{k_n} \phi_i(X_n)\phi_i, \quad \bar{b}_n := \mathbf{E}b_n.$$

Obviously,

$$\bar{A}_n z = \sum_{i=1}^{k_n} (z, \phi_i)\phi_i$$

and

$$\bar{b}_n = \sum_{i=1}^{k_n} (m, \phi_i)\phi_i$$

and

$$\bar{A}_n m = \bar{b}_n.$$

The above recursion can now be written as

$$m_{n+1} = m_n - a_n A_{n+1} m_n + a_n b_{n+1}. \qquad (25.27)$$

Lemma 25.1. *For each $z \in L_2(\mu)$ one has*

$$\begin{aligned}
\|\bar{A}_n z\| &\leq& k_n \|z\|, \\
\mathbf{E}\|A_n z\|^2 &\leq& k_n^2 \|z\|^2, \\
\mathbf{E}\|b_n\|^2 &\leq& k_n^2 \mathbf{E} Y_1^2, \\
\|\bar{A}_n z\|_\lambda^2 &\leq& k_n \|z\|^2, \\
\mathbf{E}\|A_n z\|_\lambda^2 &\leq& k_n \|z\|^2, \\
\mathbf{E}\|(A_{n+1} - \bar{A}_{n+1})m_n\|_\lambda^2 &\leq& k_{n+1}\mathbf{E}\|m_n\|^2,
\end{aligned}$$

and

$$\mathbf{E}\|b_n - \bar{b}_n\|_\lambda^2 \leq k_n \mathbf{E} Y_1^2.$$

PROOF. Using the boundedness assumption on the ϕ_i's, the proof is straightforward. In particular, one obtains

$$\|\bar{A}_n z\|_\lambda^2 = \sum_{i=1}^{k_n} (z, \phi_i)^2 \leq \sum_{i=1}^{k_n} \|z\|^2 \|\phi_i\|^2 \leq k_n \|z\|^2$$

and

$$\mathbf{E}\|A_n z\|_\lambda^2 = \mathbf{E}\left\{ z(X_n)^2 \sum_{i=1}^{k_n} \phi_i(X_n)^2 \right\} \leq k_n \mathbf{E} z(X_n)^2 = k_n \|z\|^2.$$

Moreover, by use of the independence assumption,

$$\begin{aligned}
\mathbf{E}\|(A_{n+1} - \bar{A}_{n+1})m_n\|_\lambda^2 &=& \mathbf{E}\{\mathbf{E}\{\|(A_{n+1} - \bar{A}_{n+1})m_n\|_\lambda^2 | \mathcal{F}_n\}\} \\
&=& \int \mathbf{E}\{\|(A_{n+1} - \bar{A}_{n+1})v\|_\lambda^2\} P_{m_n}(dv) \\
&\leq& \int \mathbf{E}\{\|A_{n+1}v\|_\lambda^2\} P_{m_n}(dv) \\
&\leq& k_{n+1} \int \mathbf{E}\{v(X_{n+1})^2\} P_{m_n}(dv) \\
&=& k_{n+1} \int \left[\int v(x)^2 \mu(dx)\right] P_{m_n}(dv)
\end{aligned}$$

$$= k_{n+1} \int \|v\|^2 P_{m_n}(dv)$$

$$= k_{n+1}\mathbf{E}\|m_n\|^2.$$

and

$$\mathbf{E}\|b_n - \bar{b}_n\|_\lambda^2 \le \mathbf{E}\|b_n\|_\lambda^2 \le k_n \mathbf{E} Y_1^2.$$

$\square$

A connection between the inner product in $L_2(\lambda)$ and the norm in $L_2(\mu)$ is given by the following lemma:

Lemma 25.2. *For each $z \in H_n$, one has*

$$(\bar{A}_n z, z)_\lambda = \|z\|^2.$$

PROOF.

$$(\bar{A}_n z, z)_\lambda \;=\; \sum_{i=1}^{k_n}(z, \phi_i)_\mu(z, \phi_i)_\lambda$$

$$=\; \int z(x) \sum_{i=1}^{k_n}(z, \phi_i)_\lambda \phi_i(x) \mu(dx)$$

$$=\; \int z(x)^2 \mu(dx)$$

$$=\; \|z\|^2.$$

$\square$

Lemma 25.3. *Let $z_1 \in L_2(\mu)$ and*

$$z_{n+1} = z_n - a_{n+1}\bar{A}_{n+1} z_n. \tag{25.28}$$

Then

$$\|z_n\| \to 0. \tag{25.29}$$

PROOF. In the first step we show that, for each starting point $z_1 \in L_2(\mu)$ the sequence $\|z_n\|$ is convergent. For this we write

$$\|z_{n+1}\|^2 = \|z_n\|^2 - 2a_{n+1}(\bar{A}_{n+1} z_n, z_n) + a_{n+1}^2\|\bar{A}_{n+1} z_n\|^2.$$

On the one hand, $(\bar{A}_{n+1} z_n, z_n) = \sum_{i=1}^{k_{n+1}}(z_n, \phi_i)^2 \ge 0$ and, on the other hand, $\|\bar{A}_{n+1} z_n\| \le k_{n+1}\|z_n\|$ because of Lemma 25.1. Therefore, by $\sum_n a_n^2 k_n^2 < \infty$, the assertion follows. In the second step we show that, for the sequence of bounded linear operators,

$$B_n = (I - a_{n+1}\bar{A}_{n+1}) \ldots (I - a_2 \bar{A}_2)$$

from $L_2(\mu)$ into $L_2(\mu)$, where I denotes the identity operator, and the sequence of norms $\|B_n\|$ is bounded. We notice that for each $z_1 \in L_2(\mu)$

the sequence $B_n z_1$ equals z_n given by recursion (25.28) and thus is bounded in $L_2(\mu)$ according to the first step. Now the uniform boundedness principle yields the assertion.

Our aim is to show (25.29), i.e., $\|B_n z_1\| \to 0$, for each starting point $z_1 \in L_2(\mu)$. Because $\bigcup_j H_j$ is dense in $L_2(\mu)$ and the sequence $\|B_n\|$ is bounded (according to the second step), so it suffices to show (25.29) for each starting point $z_1 \in \bigcup_j H_j$. This will be done in the third step. We notice

$$z_{n+1} = (I - a_{n+1}\bar{A}_{n+1}) \ldots (I - a_2\bar{A}_2)z_1.$$

Choose j such that $z_1 \in H_j$. Then $(I - a_j\bar{A}_j)\ldots(I - a_2\bar{A}_2)z_1 \in H_j$ for $j = 2, 3, \ldots$. Therefore, it suffices to prove that, for $z^* \in H_j$,

$$\|(I - a_{n+1}\bar{A}_{n+1})\ldots(I - a_{j+1}\bar{A}_{j+1})z^*\| \to 0.$$

Since $\sum_{n=j}^{\infty} a_n^2 k_n^2 < \infty$ (used below) it suffices to consider the case $j = 1$ with $z^* \in H_1$, i.e., one has to show that for $z_1 \in H_1$, $\|z_n\| \to 0$. In the recursion formula (25.28) take the norm square with respect to $L_2(\lambda)$ to obtain

$$\|z_{n+1}\|_\lambda^2 = \|z_n\|_\lambda^2 - 2a_{n+1}(\bar{A}_{n+1}z_n, z_n)_\lambda + a_{n+1}^2\|\bar{A}_{n+1}z_n\|_\lambda^2.$$

From Lemma 25.1,

$$\|\bar{A}_{n+1}z_n\|_\lambda^2 \le k_{n+1}\|z_n\|^2.$$

Further, noticing $z_n \in H_n$, we get, by Lemma 25.2,

$$(\bar{A}_{n+1}z_n, z_n)_\lambda = \|z_n\|^2.$$

Thus

$$\|z_{n+1}\|_\lambda^2 \le \|z_n\|_\lambda^2 - 2a_{n+1}\|z_n\|^2 + a_{n+1}^2 k_{n+1}\|z_n\|^2.$$

According to the first step, the sequence $\|z_n\|^2$ is bounded; thus, since $\sum_n a_n^2 k_n < \infty$, we have

$$\sum_n a_{n+1}^2 k_{n+1}\|z_n\|^2 < \infty.$$

Therefore

$$\sum_n a_{n+1}\|z_n\|^2 < \infty.$$

This, together with $\sum_n a_n = \infty$ and the convergence of $\|z_n\|^2$ (by the first step), yields $\|z_n\| \to 0$. $\qquad\square$

PROOF OF THEOREM 25.5. We use the notation at the beginning of this section. In the first step from the recursion formula (25.27) we obtain

$$m_{n+1} - m$$
$$= \ m_n - m - a_{n+1}\bar{A}_{n+1}(m_n - m) - a_{n+1}(A_{n+1} - \bar{A}_{n+1})(m_n - m)$$
$$- a_{n+1}A_{n+1}m + a_{n+1}b_{n+1}.$$

Now take the norm squared with respect to $L_2(\mu)$ and then conditional expectations using the independence assumption and the relation

$$-a_{n+1}A_{n+1}m + a_{n+1}b_{n+1} = -a_{n+1}(A_{n+1} - \bar{A}_{n+1})m + a_{n+1}(b_{n+1} - \bar{b}_{n+1}).$$

Thus

$$\mathbf{E}\{\|m_{n+1} - m\|^2 | \mathcal{F}_n\}$$
$$= \|m_n - m\|^2 - 2a_{n+1}(\bar{A}_{n+1}(m_n - m), m_n - m)$$
$$+ a_{n+1}^2 \|\bar{A}_{n+1}(m_n - m)\|^2$$
$$+ a_{n+1}^2 \mathbf{E}\{\|(A_{n+1} - \bar{A}_{n+1})(m_n - m) + A_{n+1}m - b_{n+1}\|^2 | \mathcal{F}_n\}.$$

Notice that, by Lemma 25.1,

$$(\bar{A}_{n+1}(m_n - m), m_n - m) = \sum_{i=1}^{k_{n+1}} (m_n - m, \phi_i)^2 \geq 0.$$

Application of Lemma 25.1 yields, by the independence assumption,

$$\mathbf{E}\{\|m_{n+1} - m\|^2 | \mathcal{F}_n\}$$
$$\leq \|m_n - m\|^2 + a_{n+1}^2 k_{n+1}^2 \|m_n - m\|^2$$
$$+ 3a_{n+1}^2 \mathbf{E}\{\|A_{n+1}(m_n - m)\|^2 | \mathcal{F}_n\}$$
$$+ 3a_{n+1}^2 \mathbf{E}\|A_{n+1}m\|^2 + 3a_{n+1}^2 \mathbf{E}\|b_{n+1}\|^2$$
$$\leq (1 + 4a_{n+1}^2 k_{n+1}^2)\|m_n - m\|^2 + 3a_{n+1}^2 k_{n+1}^2 (\|m\|^2 + \mathbf{E}Y_1^2).$$

Since $\sum_n a_n^2 k_n^2 < \infty$, by Theorem A.5 we obtain a.s. convergence of $\|m_n - m\|^2$ and convergence of $\mathbf{E}\|m_n - m\|^2$.

In the second step subtract the expectations in the recursion formula (25.27), where the independence assumption is used. We have

$$m_{n+1} - \mathbf{E}m_{n+1} = m_n - \mathbf{E}m_n - a_{n+1}\bar{A}_{n+1}(m_n - \mathbf{E}m_n)$$
$$- a_{n+1}(A_{n+1} - \bar{A}_{n+1})m_n + a_{n+1}(b_{n+1} - \bar{b}_{n+1}).$$

Take the norm squared with respect to $L_2(\lambda)$ and then the expectation using the independence assumption once more. This yields

$$\mathbf{E}\|m_{n+1} - \mathbf{E}m_{n+1}\|_\lambda^2$$
$$= \mathbf{E}\|m_n - \mathbf{E}m_n\|_\lambda^2 - 2a_{n+1}\mathbf{E}(\bar{A}_{n+1}(m_n - \mathbf{E}m_n), m_n - \mathbf{E}m_n)_\lambda$$
$$+ a_{n+1}^2 \mathbf{E}\|\bar{A}_{n+1}(m_n - \mathbf{E}m_n)\|_\lambda^2$$
$$+ a_{n+1}^2 \mathbf{E}\|(A_{n+1} - \bar{A}_{n+1})m_n - b_{n+1} + \bar{b}_{n+1}\|_\lambda^2.$$

Notice that $\mathbf{E}(\bar{A}_{n+1}(m_n - \mathbf{E}m_n), m_n - \mathbf{E}m_n)_\lambda = \mathbf{E}\|m_n - \mathbf{E}m_n\|^2$, by Lemma 25.2. Lemma 25.1 shows that

$$\mathbf{E}\|m_{n+1} - \mathbf{E}m_{n+1}\|_\lambda^2$$
$$\leq \mathbf{E}\|m_n - \mathbf{E}m_n\|_\lambda^2 - 2a_{n+1}\mathbf{E}\|m_n - \mathbf{E}m_n\|^2$$

$$+ a_{n+1}^2 k_{n+1} (\mathbf{E}\|m_n - \mathbf{E}m_n\|^2 + 2\mathbf{E}\|m_n\|^2 + 2\mathbf{E}Y_1^2).$$

Since $\mathbf{E}\|m_n\|^2 = O(1)$ and $\sum_n a_n^2 k_n < \infty$, we have

$$\sum_n a_{n+1}^2 k_{n+1} (\mathbf{E}\|m_n - \mathbf{E}m_n\|^2 + 2\mathbf{E}\|m_n\|^2 + 2\mathbf{E}Y_1^2) < \infty.$$

Thus

$$\sum_n a_{n+1} \mathbf{E}\|m_n - \mathbf{E}m_n\|^2 < \infty.$$

Since $\sum_n a_n = \infty$, there exists an index subsequence n' with

$$\mathbf{E}\|m_{n'} - \mathbf{E}m_{n'}\|^2 \to 0.$$

In the third step take the expectations in the recursion sequence to obtain

$$\mathbf{E}(m_{n+1} - m) = \mathbf{E}(m_n - m) - a_{n+1}\bar{A}_{n+1}\mathbf{E}(m_n - m)$$

by use of the independence assumption and of the relation $\bar{A}_{n+1}m = \bar{b}_{n+1}$.
By Lemma 25.3,

$$\|\mathbf{E}(m_n - m)\| \to 0.$$

Finally, for the index subsequence n' above, we obtain

$$\sqrt{\mathbf{E}\|m_{n'} - m\|^2} \leq \sqrt{\mathbf{E}\|m_{n'} - \mathbf{E}m_{n'}\|^2} + \|\mathbf{E}m_{n'} - m\| \to 0$$

by the second and third steps. Because, according to the first step, $\mathbf{E}\|m_n - m\|^2$ is convergent, we have

$$\mathbf{E}\|m_n - m\|^2 \to 0.$$

This, together with a.s. convergence of $\|m_n - m_n\|^2$, according to the first step, yields

$$\|m_n - m\|^2 \to 0 \quad \text{a.s.}$$

$\square$

25.6 Pointwise Universal Consistency

Let (X, Y) be as before and again our goal is to infer the regression function. Let $m_n(x)$ be an estimate of the regression function $m(x)$ based on the training sequence

$$(X_1, Y_1), \dots, (X_n, Y_n).$$

The main focus of this book is the L_2 error. As a slight detour, in this section we consider another error criterion.

Definition 25.1. *The sequence $m_n(x)$ is called* **strongly pointwise consistent** *if*

$$m_n(x) \to m(x) \quad a.s.$$

and for almost all x mod μ. The sequence $m_n(x)$ is called **strongly universally pointwise consistent (s.u.p.c.)** *if it is strongly pointwise consistent for all distributions of (X, Y) with $\mathbf{E}\,|Y| < \infty$.*

Thus s.u.p.c. is required that $m_n(x)$ converges to $m(x)$ with probability one and for μ-almost every x. This is equivalent to the statement, that for all distributions of (X, Y) with $\mathbf{E}\,|Y| < \infty$,

$$m_n(X) \to m(X) \quad \text{a.s.}$$

This notion of consistency is very strong, and it is not at all obvious how to construct such estimates, and how to prove their consistency, for example, it is still an open question whether the standard regression estimates are s.u.p.c. This problem occurs in function estimation for additive noise, where the noise distribution has a large tail. The difficulty is caused by the fact that in the neighborhood of x there are few observations in order to have strong consistency.

Consider some partitioning estimates.

Theorem 25.6. *In addition to the conditions of Theorem 24.3, assume that, for all x,*

$$n\lambda(A_n(x))/\log n \to \infty.$$

Then the partitioning estimate is strongly pointwise consistent for $|Y| < L$.

PROOF. See Problem 25.8.

The question whether standard partitioning estimates are s.u.p.c. is still open. For semirecursive and recursive partitioning estimates we have such consistency.

Theorem 25.7. *Under the conditions of Theorem 24.3 the semirecursive partitioning estimate is s.u.p.c.*

For the proof of Theorem 25.7 we establish two lemmas. The first is a variant of Lemma 23.3, and the second one is purely deterministic.

Lemma 25.4. *Let K_n be a sequence of measurable nonnegative functions with*

$$K_n(x, z) \le K_{\max}.$$

Assume that, for every distribution μ of X,

$$\frac{\int K_n(x, z) f(z) \mu(dz)}{\int K_n(x, z) \mu(dz)} \longrightarrow f(x) \quad mod\ \mu \tag{25.30}$$

for all μ-integrable functions f and

$$\sum_n \int K_n(x,z)\mu(dz) = \infty \quad mod \ \mu. \tag{25.31}$$

Assume further that a finite constant c^* exists such that

$$a.s. \quad \limsup_n \ \frac{\sum_{i=1}^{n} Y_i K_i(x, X_i)}{1 + \sum_{i=1}^{n} \int K_i(x,z)\mu(dz)} \le c^* m(x) \quad mod \ \mu \tag{25.32}$$

for all distributions $P_{(X,Y)}$ with $Y \ge 0$, $\mathbf{E}Y < \infty$. Let m_n be an estimate of the form

$$m_n(x) = \frac{\sum_{i=1}^{n} Y_i K_i(x, X_i)}{\sum_{i=1}^{n} K_i(x, X_i)},$$

where $\mathbf{E}|Y| < \infty$. Then

$$m_n(x) \longrightarrow m(x) \quad a.s. \ mod \ \mu.$$

PROOF. See Problem 25.4.

Lemma 25.5. *Let $0 \le r_n \le 1$, $R_n := r_1 + \cdots + r_n$, $R_0 := 0$. There is a sequence p_i of integers with $p_i \uparrow \infty$ and*

$$R_{p_i} \le i + 1, \tag{25.33}$$

$$\sum_{j=p_i}^{\infty} \frac{r_j}{(1 + R_j)^2} < \frac{1}{i}. \tag{25.34}$$

PROOF. Set $R_\infty := \lim R_n$ and $1/(1 + R_\infty) := 0$ if $R_\infty = \infty$. For $p \in \{2, 3, \ldots\}$ we have

$$\sum_{j=p}^{\infty} \frac{r_j}{(1 + R_j)^2} \le \sum_{j=p}^{\infty} \left(\frac{1}{1 + R_{j-1}} - \frac{1}{1 + R_j} \right) = \frac{1}{1 + R_{p-1}} - \frac{1}{1 + R_\infty}.$$

Choose $p_i \in \{2, 3, \ldots\}$ as the first index with

$$\frac{1}{1 + R_{p_i-1}} - \frac{1}{1 + R_\infty} < \frac{1}{i}.$$

Then (25.34) holds, and by definition of p_i,

$$R_{p_i-2} \le i - 1 \quad \text{if} \ \ p_i \ge 3,$$

thus (25.33). $\qquad\square$

PROOF OF THEOREM 25.7. We use Lemma 25.4 with

$$K_n(x,t) = I_{A_n(x)}(t).$$

Relations (25.30) and (25.31) follow from the assumptions by Lemma 24.10. It remains to verify (25.32) for $P_{(X,Y)}$ with $Y \geq 0$, $\mathbf{E}Y < \infty$. We shall use a suitable truncation. According to Lemma 25.5 with $r_n = \mu(A_n(t))$ we choose indices $p_i = p(t,i) \uparrow \infty$ $(i \to \infty)$ such that (25.33), (25.34) hold for all i. We define for $p(t,\cdot)$ an inverse function $q(t,\cdot)$ by

$$q(t,n) := \max\{i;\ p(t,i) \leq n\},$$

further the truncated random variables

$$Z_i := Y_i I_{[Y_i \leq q(X_i,i)]}.$$

It will be shown

$$\frac{\sum_{i=1}^n \left[Z_i I_{A_i(x)}(X_i) - \mathbf{E}Z_i I_{A_i(x)}(X_i) \right]}{1 + \sum_{i=1}^n \mu(A_i(x))} \longrightarrow 0 \ \text{ a.s. mod } \mu.$$

Because of (25.31) and Theorem A.6 it suffices to show

$$\sum_n \frac{\mathbf{E}Z_n^2 I_{A_n(x)}(X_n)}{\left(1 + \sum_{i=1}^n \mu(A_i(x))\right)^2} < \infty \ \text{ mod } \mu.$$

But this follows from

$$\int \sum_{n=1}^\infty \frac{\mathbf{E}Z_n^2 I_{A_n(x)}(X_n)}{\left(1 + \sum_{j=1}^n \mu(A_j(x))\right)^2} \mu(dx)$$

$$= \sum_{n=1}^\infty \int \left(\int \frac{\mathbf{E}\{Z_n^2 \mid X_n = t\} I_{A_n(x)}(t)}{\left(1 + \sum_{j=1}^n \mu(A_j(x))\right)^2} \mu(dx) \right) \mu(dt)$$

$$= \sum_{n=1}^\infty \int \int \sum_{i=1}^{q(t,n)} \int_{(i-1,i]} v^2 P_{Y|X=t}(dv) \frac{I_{A_n(t)}(x)}{\left(1 + \sum_{j=1}^n \mu(A_j(x))\right)^2} \mu(dx)\mu(dt)$$

$$= \int \sum_{i=1}^\infty \int_{(i-1,i]} v^2 P_{Y|X=t}(dv) \sum_{n=p(t,i)}^\infty \frac{\mu(A_n(t))}{\left(1 + \sum_{j=1}^n \mu(A_j(t))\right)^2} \mu(dt)$$

$$\leq \int \mathbf{E}\{Y|X = t\}\mu(dt)$$

$$= \mathbf{E}Y < \infty.$$

Here we obtain the third equality by noticing that for the nested sequence of partitions the relation $x \in A_n(t)$ and $j \leq n$ imply $A_j(x) = A_j(t)$ (as in the proof of Theorem 24.3), and the inequality is obtained by use of (25.34). Further

$$\limsup_n \frac{\sum_{i=1}^{n} \mathbf{E}Z_i I_{A_i(x)}(X_i)}{1 + \sum_{i=1}^{n} \mu(A_i(x))} \leq \lim_n \frac{\sum_{i=1}^{n} \int m(t) I_{A_i(x)}(t)\mu(dt)}{1 + \sum_{i=1}^{n} \mu(A_i(x))}$$

$$= m(x), \ \text{mod} \ \mu$$

because of (25.30), (25.31), and the Toeplitz lemma. Thus

$$\limsup_n \frac{\sum_{i=1}^{n} Z_i I_{A_i(x)}(X_i)}{1 + \sum_{i=1}^{n} \mu(A_i(x))} \leq m(x) \ \text{a.s. mod} \ \mu. \tag{25.35}$$

In the next step we show

$$\sum_n \mathbf{P}\{Z_n I_{A_n(x)}(X_n) \neq Y_n I_{A_n(x)}(X_n)\} < \infty \ \text{mod} \ \mu. \tag{25.36}$$

This follows from

$$\int \sum_{n=1}^{\infty} \mathbf{P}\{Y_n > q(X_n, n), X_n \in A_n(x)\}\mu(dx)$$

$$= \int \sum_{n=1}^{\infty} \int \mathbf{P}\{Y > q(t, n)|X = t\} I_{A_n(x)}(t)\mu(dt)\mu(dx)$$

$$= \sum_{n=1}^{\infty} \int \mathbf{P}\{Y > q(t, n)|X = t\}\mu(A_n(t))\mu(dt)$$

$$\leq \sum_{i=1}^{\infty} \int \mathbf{P}\{Y \in (i, i+1]|X = t\} \sum_{n=1}^{p(t,i+1)} \mu(A_n(t))\mu(dt)$$

$$\leq 3 \int \mathbf{E}\{Y|X = t\}\mu(dt)$$

$$= 3\mathbf{E}Y < \infty$$

by the use of (25.33). Because of (25.31),

$$1 + \sum_{i=1}^{n} \mu(A_i(x)) \to \infty \ \text{mod} \ \mu. \tag{25.37}$$

Relations (25.35), (25.36), and (25.37) yield (25.32). Now the assertion follows by Lemma 25.4. $\square$

Theorem 25.8. *Under the conditions of Theorem 25.3 together with $a_n/h_n^d = O(1/n)$ the recursive partitioning estimate is s.u.p.c.*

PROOF. Without loss of generality $Y_n \geq 0$ may be assumed. We use the notations

$$
\begin{aligned}
T_n(x) &:= a_n \overline{K}_n(x, X_n) := a_n \frac{1}{h_n^d} I_{A_n(x)}(X_n), \\
B_n(x) &:= \left[(1 - T_2(x))\ldots(1 - T_n(x))\right]^{-1}, \\
G_n(x) &:= T_n(x)B_n(x) \ (n = 2, 3, \ldots), \\
B_1(x) &:= 1, \quad G_1(x) := 1.
\end{aligned}
$$

Representations of the following kind are well-known (compare Ljung, Pflug, and Walk (1992), Part I, Lemma 1.1):

$$
B_n(x) = \sum_{i=1}^{n} G_i(x), \tag{25.38}
$$

$$
m_n(x) = B_n(x)^{-1} \sum_{i=1}^{n} G_i(x) Y_i. \tag{25.39}
$$

The assumption $a_n/h_n^d = O(1/n)$ yields $\sum \mathbf{E} T_n(x)^2 < \infty)$, thus by Theorem A.6 almost sure convergence of $\sum(T_n(x) - \mathbf{E} T_n(x))$. Relations (25.23) and (25.24) yield $\sum \mathbf{E} T_n(x) = \infty$ mod μ by the second part of Lemma 24.10. Therefore

$$
\sum T_n(x) = \infty \text{ a.s. mod } \mu
$$

and thus

$$
B_n(x) \uparrow \infty \text{ a.s. mod } \mu. \tag{25.40}
$$

Let $\widetilde{Y}_n := Y_n I_{[Y_n \leq n]}$. Then, by Lemma 23.4,

$$
\sum \frac{1}{n^2} \mathbf{E} \widetilde{Y}_n^2 < \infty, \tag{25.41}
$$

further,

$$
\sum \mathbf{P}\{Y_n \neq \widetilde{Y}_n\} = \sum \mathbf{P}\{Y > n\} \leq \mathbf{E} Y < \infty. \tag{25.42}
$$

By (25.39) we can use the representation

$$
m_n(x) = m_n^{(1)}(x) + m_n^{(2)}(x) + m_n^{(3)}(x)
$$

with

$$m_n^{(1)}(x) = \frac{1}{B_n(x)} \sum_{i=1}^{n} G_i(x) \left(\widetilde{Y}_i - \frac{\mathbf{E}\widetilde{Y}_i \overline{K}_i(x, X_i)}{\mathbf{E}\overline{K}_i(x, X_i)} \right),$$

$$m_n^{(2)}(x) = \frac{1}{B_n(x)} \sum_{i=1}^{n} G_i(x) \frac{\mathbf{E}\widetilde{Y}_i \overline{K}_i(x, X_i)}{\mathbf{E}\overline{K}_i(x, X_i)},$$

$$m_n^{(3)}(x) = \frac{1}{B_n(x)} \sum_{i=1}^{n} G_i(x)(Y_i - \widetilde{Y}_i).$$

In the first step we show

$$m_n^{(1)}(x) \to 0 \text{ a.s. mod } \mu. \tag{25.43}$$

By (25.40) and the Kronecker lemma it suffices to show a.s. convergence of

$$\sum T_n(x) \left(\widetilde{Y}_n - \frac{\mathbf{E}\widetilde{Y}_n \overline{K}_n(x, X_n)}{\mathbf{E}\overline{K}_n(x, X_n)} \right).$$

But this follows from

$$\sum \mathbf{E}\big(T_n(x)\widetilde{Y}_n\big)^2 < \infty,$$

which holds because of $a_n/h_n^d = O(1/n)$ and (25.41).

In the second step we show

$$m_n^{(2)}(x) \to m(x) \text{ a.s. mod } \mu. \tag{25.44}$$

Because of (25.38), (25.40), and the Toeplitz lemma it suffices to show

$$\frac{\mathbf{E}\widetilde{Y}_n \overline{K}_n(x, X_n)}{\mathbf{E}\overline{K}_n(x, X_n)} \to m(x) \text{ mod } \mu. \tag{25.45}$$

Because of (24.23), via the first part of Lemma 24.10, we have

$$\limsup_{n} \frac{\mathbf{E}\widetilde{Y}_n \overline{K}_n(x, X_n)}{\mathbf{E}\overline{K}_n(x, X_n)} \leq \lim_{n} \frac{\mathbf{E}Y \overline{K}_n(x, X)}{\mathbf{E}\overline{K}_n(x, X)} = m(x) \text{ mod } \mu,$$

on the other side, for each $c > 0$,

$$\liminf_{n} \frac{\mathbf{E}\widetilde{Y}_n \overline{K}_n(x, X_n)}{\mathbf{E}\overline{K}_n(x, X_n)} \geq \lim_{n} \frac{\mathbf{E}Y I_{[y \leq c]} \overline{K}_n(x, X)}{\mathbf{E}\overline{K}_n(x, X)}$$

$$= \mathbf{E}(Y I_{[Y \leq c]} | X = x) \text{ mod } \mu.$$

These relations with $c \to \infty$ yield (25.45).

In the third step we obtain

$$m_n^{(3)}(x) \to 0 \text{ a.s. mod } \mu \tag{25.46}$$

by (25.40) and (25.42). Now (25.43), (25.44), (25.46) yield the assertion. $\square$

For $k \geq 1$ let

$$\tau_{k,0}(x) = 0$$

and inductively define $\tau_{k,j}(x)$ as the jth recurrence time of $A_k(x)$:

$$\tau_{k,j}(x) = \inf\{t > \tau_{k,j-1}(x) : X_t \in A_k(x)\}.$$

We have that $\mu(A_k(x)) > 0 \mod \mu$, so that $\tau_{k,1}(x), \tau_{k,2}(x), \ldots$ are finite with probability one and for μ-almost every x. Clearly, $\{\tau_{k,j}(X)\}_{j \geq 0}$ is an increasing sequence of stopping times adapted to the filtration $\{\mathcal{F}_t\}_{t \geq 0}$, where $\mathcal{F}_t = \sigma(X, X_1, \ldots, X_t)$.

Given an integer sequence $\{J_k\}_{k \geq 1}$ such that $J_k \uparrow \infty$, we define the *modified estimates*

$$\hat{m}_k(x) = \frac{1}{J_k} \sum_{j=1}^{J_k} Y_{\tau_{k,j}(x)}.$$

This estimate is well-defined since $\tau_{k,j}(x)$ is finite with probability one and for μ-almost every x. Unfortunately, the sample size required for evaluating $\hat{m}_k(x)$ is random, it is $\tau_{k,J_k}(x)$.

To get a modified estimate of fixed sample size, we set

$$k_n(x) = \max\{k : \tau_{k,J_k}(x) \leq n\},$$

$$m_n(x) = \hat{m}_{k_n(x)}(x).$$

The strong universal pointwise consistency of $\hat{m}_k(x)$ implies that of $m_n(x)$, since $m_n(x)$ is a subsequence of $\hat{m}_k(x)$. The estimate $m_n(x)$ can be interpreted as a standard partitioning estimate with data-dependent partitioning.

The s.u.p.c. of modified partitioning estimates is open. Only their truncated versions are guaranteed to be s.u.p.c.

For any real number y and any truncation level $\mathcal{G} \geq 0$ we define the truncated value

$$y^{(\mathcal{G})} = y\, 1\{|y| \leq \mathcal{G}\}.$$

Given integers J_k such that $J_k \uparrow \infty$ and truncation levels $\mathcal{G}_j$ such that $\mathcal{G}_j \uparrow \infty$, we define the *modified truncated partitioning estimates*

$$\tilde{m}_k(x) = \frac{1}{J_k} \sum_{j=1}^{J_k} Y_{\tau_{k,j}(x)}^{(\mathcal{G}_j)}.$$

Let $j \mapsto K_j$ denote the inverse of the map $k \mapsto J_k$ (so that $j \leq J_k$ iff $K_k \leq k$), and let

$$\mathcal{R}_j = \sum_{k \geq K_j} \frac{1}{J_k^2}.$$

Assume that $J_k \uparrow \infty$ and $\mathcal{G}_j \uparrow \infty$ in such a way that

$$M = \sup_i \left[\mathcal{G}_{i+1} \left(\sum_{j \geq i+1} \mathcal{R}_j \right) \right] < \infty. \qquad (25.47)$$

Theorem 25.9. *Let $\{\mathcal{P}_k\}_{k \geq 1}$ be a nested sequence of partitions that asymptotically generate the Borel σ-field. If $J_k \uparrow \infty$ and $\mathcal{G}_j \uparrow \infty$ in such a way that (25.47) holds, then the modified truncated partitioning estimate $\tilde{m}_k^P(x)$ is s.u.p.c.*

Let

$$\tau_0^P(x) = 0$$

and inductively define

$$\tau_j^P(x) = \inf\{t > \tau_{j-1}^P(x) : X_t \in A_j(x)\}.$$

We may introduce the *modified recursive partitioning estimate*

$$m_k^{*P}(x) = \frac{1}{k} \sum_{j=1}^k Y_{\tau_j^P(x)}.$$

Theorem 25.10. *Suppose the partitions $\mathcal{P}_k$ are nested and asymptotically generate the Borel σ-field. Then the modified recursive partitioning estimate $m_k^{*P}(x)$ is s.u.p.c.*

Let us turn to the kernel estimates.

Theorem 25.11. *For the naive kernel assume that*

$$h_n \to 0 \quad and \quad nh_n^d / \log n \to \infty.$$

Then the kernel estimate is strongly pointwise consistent for $|Y| < L$.

PROOF. See Problem 25.9.

It is unknown whether the standard kernel estimate is s.u.p.c. However, it is known that truncation of the response variables yields estimates that are s.u.p.c.

Theorem 25.12. *Put*

$$h_n = Cn^{-\delta}, \quad 0 < \delta < 1/d,$$

and

$$\mathcal{G}_n = n^{1-\delta d}.$$

Then the truncated kernel estimate

$$m_n(x) = \frac{\sum_{i=1}^n Y_i^{(\mathcal{G}_n)} K\left(\frac{x - X_i}{h_n}\right)}{\sum_{i=1}^n K\left(\frac{x - X_i}{h_n}\right)}$$

with naive kernel is s.u.p.c.

Again the semirecursive and recursive estimates are easier.

Theorem 25.13. *Under the conditions concerning (24.18) of Theorem 24.2 the semirecursive kernel estimate is s.u.p.c.*

Theorem 25.14. *Under the conditions of Theorem 25.2 together with $K(x) \geq cH(\|x\|)$ for some $c > 0$ and $a_n/h_n^d = O(1/n)$ the recursive kernel estimate is s.u.p.c.*

Given $B = S_1(0)$ and a bandwidth sequence $\{h_k\}_{k \geq 1}$ such that $h_k \downarrow 0$, let

$$B_k(x) = x + h_k\, B.$$

Let

$$\tau_{k,0}^B(x) = 0$$

and inductively define

$$\tau_{k,j}^B(x) = \inf\{t > \tau_{k,j-1}^B(x) : X_t \in B_k(x)\}.$$

If $h_k > 0$ then $\mu(B_k(x)) > 0 \bmod \mu$, so all $\tau_{k,j}^B(x)$ are finite and $Y_{\tau_{k,1}^B(x)}$ is well-defined. We define the *modified truncated kernel estimate* as follows:

$$\hat{m}_k^B(x) = \frac{1}{J_k} \sum_{j=1}^{J_k} Y_{\tau_{k,j}^B(x)}^{(\mathcal{G}_j)}.$$

Yakowitz (1993) defined a modified kernel estimate of fixed sample size, which can be viewed as the modified kernel estimate without truncation and with $J_k = k$. Yakowitz called it the r-nearest-neighbor estimate, since it is closer in spirit to nearest neighbor estimates than to kernel estimates.

Theorem 25.15. *Let $\{h_k\}_{k \geq 1}$ be a bandwidth sequence such that $h_k \to 0$. If $J_k \uparrow \infty$ and $\mathcal{G}_j \uparrow \infty$ in such a way that (25.47) holds, then the modified truncated kernel estimate $\hat{m}_k^B(x)$ is s.u.p.c.*

Let

$$\tau_0^B(x) = 0$$

and inductively define

$$\tau_j^B(x) = \inf\{t > \tau_{j-1}^B(x) : X_t \in B_j(x)\}.$$

Then $\mu(B_k(x)) > 0 \bmod \mu$ and all $\tau_j^B(x)$ are finite with probability one by Poincaré's recurrence theorem. Thus one may define the *modified recursive kernel estimates*

$$m_k^{*B}(x) = \frac{1}{k} \sum_{j=1}^{k} Y_{\tau_j^B(x)}.$$

Theorem 25.16. *Suppose some bandwidth sequence $\{h_k\}_{k\geq 1}$ such that $h_k \to 0$ as $k \to \infty$. Then the modified recursive kernel estimate $m_k^{*B}(x)$ is s.u.p.c.*

Consider some nearest neighbor estimates.

Theorem 25.17. *If*

$$k_n/\log n \to \infty \quad and \quad k_n/n \to 0$$

then the nearest neighbor estimate is strongly pointwise consistent for $|Y| < L$.

PROOF. See Problem 25.10.

It is unknown whether the standard nearest neighbor estimate is s.u.p.c. Let $\{\ell_k\}_{k\geq 1}$ be a sequence of positive integers such that

$$\ell_k \uparrow \infty \ \text{ as } k \to \infty.$$

For each $k \geq 1$ one may subdivide the data into successive segments of length ℓ_k. Let $X_{k,j}(x)$ denote the nearest neighbor of x among the observations from the jth segment $X_{(j-1)\ell_k+1}, \ldots, X_{j\ell_k}$, and let $Y_{k,j}(x)$ denote the corresponding label of $X_{k,j}(x)$.

The *modified truncated nearest neighbor estimate* is defined by

$$\hat{m}_k^{NN}(x) = \frac{1}{J_k} \sum_{j=1}^{J_k} Y_{k,j}^{(\mathcal{G}_j)}(x).$$

Theorem 25.18. *Let $\{\ell_k\}_{k\geq 1}$ be a deterministic sequence of integers such that $\ell_k \to \infty$. If $J_k \uparrow \infty$ and $\mathcal{G}_j \uparrow \infty$ in such a way that (25.47) holds, then the modified truncated nearest neighbor estimate $\hat{m}_k^{NN}(x)$ is s.u.p.c.*

Given a sequence $\ell_1, \ell_2, \ldots$ of positive integers, we split the data sequence $(X_1, Y_1), (X_2, Y_2), \ldots$ into disjoint blocks with lengths $\ell_1, \ldots, \ell_k$ and find the nearest neighbor of x in each block. Let $X_j^*(x)$ denote the nearest neighbor of x from the jth block (ties are broken by selecting the nearest neighbor with the lowest index), and let $Y_j^*(x)$ denote the corresponding label. The *recursive nearest neighbor estimate* is defined as

$$\hat{m}_n^{NN}(x) = \frac{1}{k} \sum_{j=1}^{k} Y_j^*(x)$$

if $\sum_{j=1}^{k} \ell_j \leq n < \sum_{j=1}^{k+1} \ell_j$.

Theorem 25.19. *If $\ell_k \to \infty$, then the recursive nearest neighbor estimate is s.u.p.c.*

25.7 Bibliographic Notes

Theorems 25.1, 25.2, and 25.5 are due to Győrfi and Walk (1997; 1996). The estimator with (25.16) was introduced and investigated by Révész (1973). Under regularity conditions on m and under the condition that the real X has a density, Révész (1973) proved a large deviation theorem. Győrfi (1981) proved the weak universal consistency in the multidimensional case for a kernel of compact support. The weak universal consistency of the recursive NN estimate has been proved by Devroye and Wise (1980). Theorems 25.7, 25.8, 25.13, and 25.14 have been proved by Walk (2001). Stronger versions of Theorems 25.11 and 25.17 can be found in Devroye (1982b; 1981). Theorems 25.9, 25.15, 25.16, 25.18, and 25.19 are in Algoet and Győrfi (1999). Theorem 25.10 is due to Algoet (1999). Kozek, Leslie, and Schuster (1998) proved Theorem 25.12.

Problems and Exercises

PROBLEM 25.1. Prove Theorem 25.2.
  HINT: Apply Theorem 25.1 and Lemma 24.6.

PROBLEM 25.2. Prove Theorem 25.3.
  HINT: Apply Theorem 25.1 and Lemma 24.10.

PROBLEM 25.3. Prove Theorem 25.3 for $d = 1$, where each partition consists of nonaccumulating intervals, without the condition that the sequence of partitions is nested.
  HINT: Compare Problem 24.6.

PROBLEM 25.4. Prove Lemma 25.4.
  HINT: Use the truncation $Y_L = YI_{Y \leq L} + LI_{Y > L}$ for $Y \geq 0$ and notice for $m_L(x) = \mathbf{E}\{Y_L | X = x\}$ that for $\epsilon > 0$ an integer $L_0(x)$ exists with

$$|m_L(x) - m(x)| < \epsilon$$

for all $L > L_0(x)$ and for μ-almost all x.

PROBLEM 25.5. Prove Theorem 25.13 for a naive kernel.
  HINT: Argue according to the proof of Theorem 25.7, but use the covering argument of Theorem 24.2 and use Lemmas 24.5 and 24.6.

PROBLEM 25.6. Prove Theorem 25.14 for a naive kernel K.
  HINT: Argue according to the proof of Theorem 25.8, but with $\overline{K}_n(x, z) = \frac{1}{h_n^d} K\left(\frac{x-z}{h_n}\right)$, and use Lemmas 24.5 and 24.6.

PROBLEM 25.7. Prove Theorem 25.19.

PROBLEM 25.8. Prove Theorem 25.6.
  HINT: Put

$$\bar{m}_n(x) = \frac{\int_{A_n(x)} m(z)\mu(dz)}{\mu(A_n(x))},$$

then, by Lemma 24.10,

$$\bar{m}_n(x) \to m(x) \mod \mu.$$

Moreover,

$$m_n(x) = \frac{\dfrac{\sum_{i=1}^n Y_i I_{\{X_i \in A_n(x)\}}}{n\mu(A_n(x))}}{\dfrac{\sum_{i=1}^n I_{\{X_i \in A_n(x)\}}}{n\mu(A_n(x))}},$$

therefore, it is enough to show that

$$\frac{\sum_{i=1}^n Y_i I_{\{X_i \in A_n(x)\}}}{n\mu(A_n(x))} - \bar{m}_n(x) \to 0 \ \text{ a.s. mod } \mu$$

By Bernstein's inequality

$$\mathbf{P}\left\{ \left| \frac{\sum_{i=1}^n Y_i I_{\{X_i \in A_n(x)\}}}{n\mu(A_n(x))} - \bar{m}_n(x) \right| > \epsilon \right\}$$

$$\leq \ 2e^{-n\epsilon^2 \frac{\mu(A_n(x))^2}{2\,\mathbf{Var}(Y_1 I_{\{X_1 \in A_n(x)\}})+4L\epsilon\mu(A_n(x))/3}}$$

$$\leq \ 2e^{-n\epsilon^2 \frac{\mu(A_n(x))^2}{2L^2\mu(A_n(x))+4L\epsilon\mu(A_n(x))/3}}$$

$$\leq \ 2e^{-n\epsilon^2 \frac{\frac{\mu(A_n(x))}{\lambda(A_n(x))}\lambda(A_n(x))}{2L^2+4L\epsilon/3}},$$

which is summable because of the condition and because of

$$\liminf_{n\to\infty} \frac{\mu(A_n(x))}{\lambda(A_n(x))} > 0 \mod \mu$$

(cf. Lemma 24.10).

PROBLEM 25.9. Prove Theorem 25.11.

HINT: Proceed as for Problem 25.8 such that we refer to Lemmas 24.5 and 24.6.

PROBLEM 25.10. Prove Theorem 25.17.

HINT: Put

$$\bar{m}_R(x) = \frac{\int_{S_{x,R}} m(z)\mu(dz)}{\mu(S_{x,R})},$$

then, by Lemma 24.6,

$$\lim_{R\to 0} \bar{m}_R(x) = m(x) \mod \mu.$$

Given $\|x - X_{(k_n,n)}(x)\| = R$, the distribution of

$$(X_{(1,n)}(x), Y_{(1,n)}(x)), \ldots, (X_{(k_n,n)}(x), Y_{(k_n,n)}(x))$$

is the same as the distribution of the nearest neighbor permutation of the i.i.d.

$$(X_1^*(x), Y_1^*(x)), \ldots, (X_{k_n}^*(x), Y_{k_n}^*(x)),$$

where

$$\mathbf{P}\{X_i^*(x) \in A, Y_i^*(x) \in B\} = \mathbf{P}\{X_i \in A, Y_i \in B | X_i \in S_{x,R}\},$$

therefore,

$$\mathbf{E}\{Y_i^*(x)\} = \mathbf{E}\{Y_i | X_i \in S_{x,R}\} = \bar{m}_R(x).$$

Thus, by Hoeffding's inequality,

$$\mathbf{P}\left\{ \left| \frac{1}{k_n} \sum_{i=1}^{k_n} Y_{(i,n)}(x) - \bar{m}_R(x) \right| > \epsilon \Big| \|x - X_{(k_n,n)}(x)\| = R \right\}$$

$$= \mathbf{P}\left\{ \left| \frac{1}{k_n} \sum_{i=1}^{k_n} (Y_i^*(x) - \mathbf{E}Y_i^*(x)) \right| > \epsilon \right\}$$

$$\leq 2e^{-\frac{k_n \epsilon^2}{2L^2}},$$

which is summable because of $k_n / \log n \to \infty$.

26

Censored Observations

26.1 Right Censoring Regression Models

This chapter deals with nonparametric regression analysis in the presence of randomly right censored data.

Let Y be a nonnegative random variable representing the survival time of an individual or subject taking part in a medical or other experimental study, and let $X = (X^{(1)}, \ldots, X^{(d)})$ be a random vector of covariates, e.g., the medical file of a patient, jointly distributed with Y. In the model of right censoring the survival time Y_i is subject to right censoring so that the observable random variables are given by X_i, $Z_i = \min(Y_i, C_i)$, and $\delta_i = I_{\{Y_i \leq C_i\}}$. Here C_i is a nonnegative random variable, representing the censoring time, which could be the subject's time to withdrawal or the time until the end of the study.

It is well-known that in medical studies the observation on the survival time of a patient is often incomplete due to right censoring. Classical examples of the causes of this type of censoring are that the patient was alive at the termination of the study, that the patient withdrew alive during the study, or that the patient died from other causes than those under study. Another example of the same model we would like to offer here is in the world of medical insurance. Suppose an insurance company sets its various insurance rates on a basis of the lifetimes Y_i of its clients having a specific disorder (heart disease, diabetes, etc.). In addition, for each client a vector X_i of additional measurements is available from medical examinations. However, for several reasons, a patient may stop the contract with the insurance company at time C_i (due to lack of money or because of more favorable conditions elsewhere), so that the actual file of the patient with the insurance company is incomplete in the above sense: it does not contain Y_i, it contains only C_i with the indication that it is not the real life time.

The distribution of the observable random vector (X, Z, δ) does not identify the conditional distribution of Y given X. The problem of identifiability can be resolved by imposing additional conditions on the possible distributions of (X, Y, C).

Model A: Assume that C and (X, Y) are independent.
Model B: Assume that Y and C are conditionally independent given X.

Under these conditions we show consistent regression function estimates which imply that the distribution of (X, Z, δ) identifies the distribution of (X, Y).

Model A is plausible in all situations where the censoring is caused by extraneous circumstances, not related to any characteristics of the individual. This can clearly be argued to be the case in the medical insurance example described above, as well as in many classical survival studies. One could think, e.g., of volunteers participating in a medical study, but for reasons not related to the data contained in (X, Y) they withdraw prematurely from the study, such as because of lack of enthusiasm. Or the censoring may be caused by the (random) termination of the study, which can be assumed to occur independently of the persons participating in the study.

Obviously Model A is a special case of Model B, therefore the estimation problem for Model A is easier such that for Model A the estimates are simpler and the rate of convergence might be better. However, in most practical problems only the conditions of Model B are satisfied.

26.2 Survival Analysis, the Kaplan-Meier Estimate

In this section we present some results on distribution estimation for censored observation when there is no observation vector X. Thus let us first consider the case that there are no covariates.

Assume that Y and C are independent. We observe

$$(Z_1, \delta_1), \ldots, (Z_n, \delta_n)$$

where $\delta_i = I_{\{Y_i \leq C_i\}}$, and $Z_i = \min(Y_i, C_i)$.

Introduce

$$\begin{aligned}
F(t) &= \mathbf{P}\{Y > t\}, \\
G(t) &= \mathbf{P}\{C > t\}, \\
K(t) &= \mathbf{P}\{Z > t\} = F(t)G(t).
\end{aligned}$$

In survival analysis such tail distribution functions are called survival functions.

Define

$$T_F = \sup\{y : F(y) > 0\},$$
$$T_G = \sup\{y : G(y) > 0\},$$
$$T_K = \sup\{y : K(y) > 0\} = \min\{T_F, T_G\}.$$

The main subject of survival analysis is to estimate F and G. Kaplan and Meier (1958) invented such estimates, which are often called product-limit estimates. Let F_n and G_n be the Kaplan-Meier estimates of F and G, respectively, which are defined as

$$F_n(t) = \begin{cases} \prod_{i:Z_{(i)} \leq t} \left(\frac{n-i}{n-i+1}\right)^{\delta_{(i)}} & \text{if } t \leq Z_{(n)}, \\ 0 & \text{otherwise}, \end{cases} \tag{26.1}$$

and

$$G_n(t) = \begin{cases} \prod_{i:Z_{(i)} \leq t} \left(\frac{n-i}{n-i+1}\right)^{1-\delta_{(i)}} & \text{if } t \leq Z_{(n)}, \\ 0 & \text{otherwise}, \end{cases} \tag{26.2}$$

where $(Z_{(i)}, \delta_{(i)})$ $(i = 1, \ldots, n)$ are the n pairs of observed (Z_i, δ_i) ordered on the $Z_{(i)}$, i.e.,

$$Z_{(1)} \leq Z_{(2)} \leq \cdots \leq Z_{(n)} := T_{K_n}. \tag{26.3}$$

Note that since F is arbitrary, some of the Z_i may be identical. In this case the ordering of the Z_i's into $Z_{(i)}$'s is not unique. However, it is easy to see that the Kaplan-Meier estimator is unique. We can observe that $F_n(t)$ has jumps at uncensored sample points. Similarly, $G_n(t)$ has jumps at censored sample points. It is not at all obvious why they should be consistent. The first interpretation in this respect is due to Kaplan and Meier (1958) showing that the estimates are of maximum likelihood type.

Efron (1967) introduced another interpretation. He gave a computation-ally simple algorithm for calculating $F_n(t)$. It is a recursive rule working from the left to right for $Z_{(i)}$. Place probability mass $1/n$ at each of the points $Z_{(i)}$, that is, construct the conventional empirical distribution. If $Z_{(1)}$ is not censored $(\delta_{(1)} = 1)$ then keep its mass, otherwise remove its mass and redistribute it equally among the other points. Continue this procedure in a recursive way. Suppose that we have already considered the first $i - 1$ points. If $Z_{(i)}$ is not censored $(\delta_{(i)} = 1)$ then keep its current mass, other-wise remove its mass and redistribute it equally among the $n - i$ points to the right of it, $Z_{(i+1)}, \ldots, Z_{(n)}$. Make these steps $n - 1$ times, so $F_n(t) = 0$ for $t > Z_{(n)}$. It is easy to check that Efron's redistribution algorithm results in the same Kaplan-Meier estimate (cf. Problem 26.1).

Table 26.1. Efron's redistribution algorithm for $F_6(t)$.

$Z_{(1)}$	$Z_{(2)}$	$Z_{(3)}$	$Z_{(4)}$	$Z_{(5)}$	$Z_{(6)}$
1/6	1/6	1/6	1/6	1/6	1/6
1/6	1/6	1/6	1/6	1/6	1/6
1/6	0	5/24	5/24	5/24	5/24
1/6	0	0	5/18	5/18	5/18
1/6	0	0	5/18	5/18	5/18
1/6	0	0	5/18	5/18	5/18

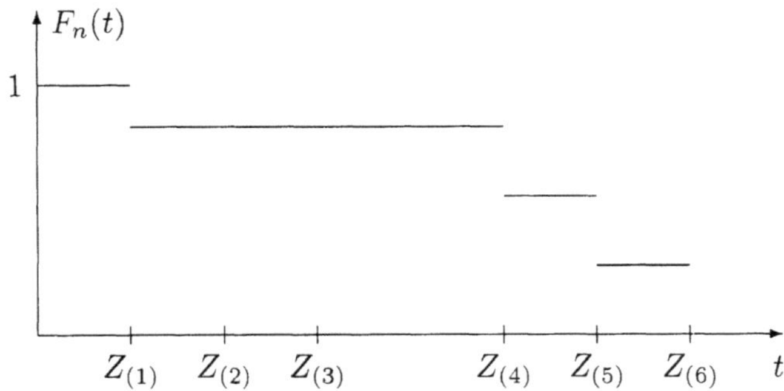

Figure 26.1. Illustration for $F_6(t)$.

EXAMPLE. For the sake of illustration consider the following example:

$$\{(Z_i, \delta_i)\}_{i=1}^6 = \{(5, 1), (2, 0), (6, 1), (1, 1), (3, 0), (7, 1)\}.$$

Then

$$\{(Z_{(i)}, \delta_{(i)})\}_{i=1}^6 = \{(1, 1), (2, 0), (3, 0), (5, 1), (6, 1), (7, 1)\}.$$

In Table 26.1 we describe the steps for calculating $F_6(t)$. Initially all sample points have mass $1/6$. $Z_{(1)}$ is not censored so there is no change in Step 1. $Z_{(2)}$ is censored, so its mass is redistributed among $Z_{(3)}, Z_{(4)}, Z_{(5)}, Z_{(6)}$. $Z_{(3)}$ is censored, too, this results in the next row. Since the last three sample points are not censored, there are no changes in the final three steps. From this final row we can generate $F_6(t)$ (Fig. 26.1).

Table 26.2 and Fig. 26.2 show the same for $G_6(t)$.

Peterson (1977) gave a third interpretation utilizing the fact that the survival functions $F(t)$ and $G(t)$ are the functions of the subsurvival functions

$$F^*(t) = \mathbf{P}\{Y > t, \delta = 1\}$$

and

$$G^*(t) = \mathbf{P}\{C > t, \delta = 0\},$$

Table 26.2. Efron's redistribution algorithm for $G_6(t)$.

$Z_{(1)}$	$Z_{(2)}$	$Z_{(3)}$	$Z_{(4)}$	$Z_{(5)}$	$Z_{(6)}$
1/6	1/6	1/6	1/6	1/6	1/6
0	1/5	1/5	1/5	1/5	1/5
0	1/5	1/5	1/5	1/5	1/5
0	1/5	1/5	1/5	1/5	1/5
0	1/5	1/5	0	3/10	3/10
0	1/5	1/5	0	0	3/5

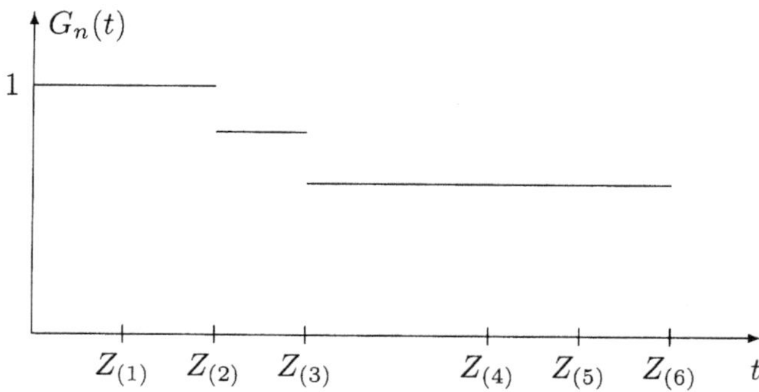

Figure 26.2. Illustration for $G_6(t)$.

and the subsurvival functions can be estimated by averages. This observation led to the following strong consistency theorem:

Theorem 26.1. *Assume that F and G have no common jumps. Then, for all $t < T_K$,*

$$F_n(t) \longrightarrow F(t) \quad a.s.$$

as $n \to \infty$ and

$$G_n(t) \to G(t) \quad a.s.$$

as $n \to \infty$.

PROOF. We prove pointwise convergence of $G_n(t)$ to $G(t)$ for all $t < T_K$. The case of $F_n(t)$ is similar. For a function S let $D(S)$ denote the set of the jump points of S. Then any tail distribution (survival) function can be expressed as

$$S(t) = \exp\left(\int_{[0,t]\setminus D(S)} \frac{dS(s)}{S(s)}\right) \cdot \prod_{\substack{s \in D(S) \\ s \leq t}} \frac{S(s)}{S(s^-)} \tag{26.4}$$

(see Cox (1972)), where $S(s^-) = \lim_{t \uparrow s} S(t)$. Thus

$$G(t) = \exp \left(\int_{[0,t] \setminus D(G)} \frac{dG(s)}{G(s)} \right) \cdot \prod_{\substack{s \in D(G) \\ s \leq t}} \frac{G(s)}{G(s^-)} \qquad (26.5)$$

and

$$G_n(t) = \prod_{\substack{s \in D(G_n) \\ s \leq t}} \frac{G_n(s)}{G_n(s^-)}$$

since $G_n(\cdot)$ is a discrete survival function. The subsurvival function $G^*(\cdot)$ can be expressed in terms of the survival functions $G(\cdot)$ and $F(\cdot)$

$$
\begin{aligned}
G^*(t) &= \mathbf{P}\{C > t, \delta = 0\} \\
&= \mathbf{P}\{C > t, Y > C\} \\
&= \mathbf{E}\{\mathbf{P}\{C > t, Y > C | C\}\} \\
&= \mathbf{E}\{I_{\{C > t\}} F(C)\} \\
&\quad \text{(because of independence of } Y \text{ and } C) \\
&= -\int_t^\infty F(s)\, dG(s).
\end{aligned}
\qquad (26.6)
$$

Note that

$$
\begin{aligned}
G^*(t) + F^*(t) &= \mathbf{P}\{Z > t\} = K(t) \\
&= \mathbf{P}\{\min(Y, C) > t\} = G(t) \cdot F(t).
\end{aligned}
\qquad (26.7)
$$

For real-valued functions H_1 and H_2 define a function ϕ by

$$
\begin{aligned}
&\phi(H_1, H_2, t) \\
&= \exp \left(\int_{[0,t] \setminus D(H_1)} \frac{dH_1(s)}{H_1(s) + H_2(s)} \right) \cdot \prod_{\substack{s \in D(H_1) \\ s \leq t}} \frac{H_1(s) + H_2(s)}{H_1(s^-) + H_2(s^-)}.
\end{aligned}
$$

We will show that

$$G(t) = \phi(G^*(\cdot), F^*(\cdot), t). \qquad (26.8)$$

Because of (26.6) G^* has the same jump points as G. Thus (26.8) follows from (26.5), since

$$
\begin{aligned}
\frac{dG^*(s)}{G^*(s) + F^*(s)} &= \frac{dG^*(s)}{G(s) \cdot F(s)} \qquad \text{(from (26.7))} \\[2mm]
&= \frac{F(s) \cdot dG(s)}{F(s) \cdot G(s)} \qquad \text{(from (26.6))}
\end{aligned}
$$

$$= \frac{dG(s)}{G(s)}$$

and

$$\frac{G^*(s) + F^*(s)}{G^*(s^-) + F^*(s^-)} = \frac{G(s) \cdot F(s)}{G(s^-) \cdot F(s^-)} \qquad \text{(from (26.7))}$$

$$= \frac{G(s)}{G(s^-)}$$

if s is a jump point of $G(\cdot)$, because by assumption $G(\cdot)$ and $F(\cdot)$ do not have common jumps, thus s cannot be a jump point of $F(\cdot)$, so $F(s) = F(s^-)$. Now we express the Kaplan-Meier estimator in terms of the empirical subsurvival functions

$$G_n^*(t) = \frac{1}{n} \sum_{i=1}^{n} I_{\{Z_i > t, \delta_i = 0\}}$$

and

$$F_n^*(t) = \frac{1}{n} \sum_{i=1}^{n} I_{\{Z_i > t, \delta_i = 1\}}.$$

First note that (26.7) is also true for the estimates, since

$$G_n^*(t) + F_n^*(t) = \frac{1}{n} \sum_{i=1}^{n} I_{\{Z_i > t, \delta_i = 0\}} + \frac{1}{n} \sum_{i=1}^{n} I_{\{Z_i > t, \delta_i = 1\}}$$

$$= \frac{1}{n} \sum_{i=1}^{n} I_{\{Z_i > t\}}$$

$$=: K_n(t)$$

and

$$G_n(t) \cdot F_n(t) = \prod_{Z_{(i)} \le t} \left(\frac{n-i}{n-i+1} \right)^{1 - \delta_{(i)}} \cdot \prod_{Z_{(i)} \le t} \left(\frac{n-i}{n-i+1} \right)^{\delta_{(i)}}$$

$$= \prod_{Z_{(i)} \le t} \left(\frac{n-i}{n-i+1} \right)$$

$$= \frac{n-1}{n} \cdot \frac{n-2}{n-1} \cdots \frac{n - \sum_{i=1}^{n} I_{\{Z_i \le t\}}}{n - \sum_{i=1}^{n} I_{\{Z_i \le t\}} + 1}$$

$$= \frac{1}{n} \sum_{i=1}^{n} I_{\{Z_i > t\}}$$

$$= K_n(t).$$

Therefore,

$$\frac{G_n^*(s) + F_n^*(s)}{G_n^*(s^-) + F_n^*(s^-)} = \frac{G_n(s) \cdot F_n(s)}{G_n(s^-) \cdot F_n(s^-)} = \frac{G_n(s)}{G_n(s^-)}$$

if s is a jump point of $G_n(\cdot)$, because by assumption $G_n(\cdot)$ and $F_n(\cdot)$ do not have common jumps a.s., thus $F_n(s) = F_n(s^-)$. Hence

$$G_n(t) = \phi(G_n^*(\cdot), F_n^*(\cdot), t)$$

for $t \le Z_{(n)}$. The empirical subsurvival functions $G_n^*(t)$ and $F_n^*(t)$ converge almost surely uniformly in t (Glivenko-Cantelli theorem). It is easy to show that

$$Z_{(n)} = T_{K_n} \to T_K \quad \text{a.s.} \quad \text{as } n \to \infty. \tag{26.9}$$

The almost sure convergence of $G_n(\cdot)$ to $G(\cdot)$ then follows from the fact that $\phi(H_1(\cdot), H_2(\cdot), t)$ is a continuous function of the argument $H_1(\cdot)$ and $H_2(\cdot)$, where the metric on the space of bounded functions is the supremum norm. $\quad\square$

The regression function is a conditional expectation, therefore we now consider the problem of estimating an expectation from censored observations. If the $\{Y_i\}$ are i.i.d. and available, then the arithmetic mean

$$\frac{1}{n} \sum_{i=1}^{n} Y_i$$

is the usual estimate of $\mathbf{E}\{Y\}$. If instead of the $\{Y_i\}$ only the $\{Z_i, \delta_i\}$ are available and, in addition, the censoring distribution G is known, then an unbiased estimate of $\mathbf{E}\{Y\}$ is

$$\frac{1}{n} \sum_{i=1}^{n} \frac{\delta_i Z_i}{G(Z_i)},$$

as can easily be verified by using properties of conditional expectation and the assumed independence of Y and C:

$$\mathbf{E}\left\{\frac{\delta Z}{G(Z)} \Big| X\right\} = \mathbf{E}\left\{\frac{I_{\{Y \le C\}} \min(Y, C)}{G(\min(Y, C))} \Big| X\right\}$$

$$= \mathbf{E}\left\{I_{\{Y \le C\}} \frac{Y}{G(Y)} \Big| X\right\}$$

$$= \mathbf{E}\left\{\mathbf{E}\left\{I_{\{Y \le C\}} \frac{Y}{G(Y)} \Big| X, Y\right\} \Big| X\right\}$$

$$= \mathbf{E}\left\{G(Y) \frac{Y}{G(Y)} \Big| X\right\}$$

$$= \mathbf{E}\{Y \mid X\}$$

$$= \ m(X). \tag{26.10}$$

If G is unknown then $\mathbf{E}\{Y\}$ can be estimated by the so-called Kaplan-Meier mean

$$M_n := \frac{1}{n}\sum_{j=1}^{n}\frac{\delta_j Z_j}{G_n(Z_j)} \ = \ \frac{1}{n}\sum_{j=1}^{n}\frac{\delta_j Y_j}{G_n(Y_j)}$$

$$= \ \int_0^{+\infty} F_n(y)\,dy \tag{26.11}$$

$$= \ \int_0^{T_F} F_n(y)\,dy,$$

where (26.11) follows from an identity in Susarla, Tsai, and Van Ryzin (1984) (cf. Problem 26.2). Theorem 26.1 already implies that, for fair censoring $(T_F \le T_G)$,

$$M_n \to \mathbf{E}\{Y\} \ \text{a.s.}$$

(cf. Problem 26.3).

26.3 Regression Estimation for Model A

The regression function $m(x)$ must be estimated from the censored data, which is an i.i.d. sequence of random variables:

$$\{(X_1, Z_1, \delta_1), \ldots, (X_n, Z_n, \delta_n)\},$$

where $\delta_i = I_{\{Y_i \le C_i\}}$, and $Z_i = \min(Y_i, C_i)$.

Throughout this section we rely on the following assumptions:

 (i) (X, Y) and C are independent;
 (ii) F and G have no common jumps;
 (iii) $T_F < \infty$; and
 (iv) G is continuous in T_K and $G(T_F) > 0$.

The last condition implies that $T_F < T_G$ and hence $T_K = T_F$ by definition of T_K.

We define **local averaging estimates for censored data** by

$$\tilde{m}_{n,1}(x) = \sum_{i=1}^{n} W_{ni}(x) \cdot \frac{\delta_i \cdot Z_i}{G_n(Z_i)}. \tag{26.12}$$

According to such an estimate we calculate first $G_n(t)$ from the sample $(Z_1, \delta_1), \ldots, (Z_n, \delta_n)$ and then proceed as in (26.12). The fact $0 \le Y \le T_F < \infty$ a.s. implies that $0 \le m(x) \le T_F$ $(x \in \mathcal{R}^d)$. In order to ensure that the estimates are bounded in the same way, set

$$m_n(x) = \min\left(T_{K_n}, \tilde{m}_{n,1}(x)\right) \tag{26.13}$$

with $T_{K_n} = Z_{(n)}$. Observe $0 \leq T_{K_n} \leq T_K = T_F < \infty$.

Our first result concerns the strong consistency of local averaging estimates for censored data.

In the censored case, the partitioning estimate is

$$m_n^{(part)}(x) = \min\left(T_{K_n}, \frac{\sum_{i=1}^n \frac{\delta_i Z_i}{G_n(Z_i)} I_{\{X_i \in A_n(x)\}}}{\sum_{i=1}^n I_{\{X_i \in A_n(x)\}}}\right). \tag{26.14}$$

Theorem 26.2. *Assume (i)–(iv). Then under the conditions (4.1) and (4.2), for the regression function estimate $m_n^{(part)}$, one has*

$$\int (m_n^{(part)}(x) - m(x))^2 \mu(dx) \to 0 \quad a.s. \qquad as \ n \to \infty.$$

The proof of this theorem uses the following lemma:

Lemma 26.1. *Assume that $T_F < \infty$, G is continuous, and $G(T_F) > 0$. Then*

$$\frac{1}{n} \sum_j \delta_j Z_j \left| \frac{1}{G(Z_j)} - \frac{1}{G_n(Z_j)} \right| \to 0 \quad a.s.$$

PROOF. Applying the identity

$$\begin{aligned}
|a - b| &= (a-b)^+ + (b-a)^+ \\
&= 2(a-b)^+ + b - a
\end{aligned}$$

we obtain

$$\begin{aligned}
\frac{1}{n} \sum_j \delta_j Z_j \left| \frac{1}{G(Z_j)} - \frac{1}{G_n(Z_j)} \right| &\leq 2\frac{1}{n} \sum_{j=1}^n \delta_j Z_j \left(\frac{1}{G(Z_j)} - \frac{1}{G_n(Z_j)} \right)^+ \\
&\quad + \frac{1}{n} \sum_{j=1}^n \frac{\delta_j Z_j}{G_n(Z_j)} - \frac{1}{n} \sum_{j=1}^n \frac{\delta_j Z_j}{G(Z_j)} \\
&:= I + II - III. \tag{26.15}
\end{aligned}$$

Next introduce

$$S_n(y) = \sup_{n \leq m} \left(1 - \frac{G(y)}{G_m(y)} \right)^+.$$

Since G is continuous, F and G have no common jumps and, therefore, because of Theorem 26.1,

$$G_n(y) \to G(y) \quad a.s.$$

as $n \to \infty$ for all $y < T_K = T_F$. Hence $S_n(y)$ has the following properties:

(a) $0 \leq S_n(y) \leq 1$; and

(b) $S_n(y) \to 0$ a.s. for all $y < T_K = T_F$ as $n \to \infty$.

By the dominated convergence theorem

$$\mathbf{E}S_n(Y) \to 0 \quad \text{as } n \to \infty,$$

and hence one can choose, for any $\epsilon > 0, n_0$ sufficiently large such that

$$0 \le \mathbf{E}S_{n_0}(Y) < \epsilon.$$

Now fix n_0 and take $n \ge n_0$. Then, since $\delta_j Z_j = \delta_j Y_j$ and from the definition of $S_n(y)$, expression I in (26.15) can be written as

$$
\begin{aligned}
I &= 2\frac{1}{n}\sum_{j=1}^{n} \frac{\delta_j Y_j}{G(Y_j)}\left(1 - \frac{G(Y_j)}{G_n(Y_j)}\right)^{+} \\
&\le 2\frac{1}{n}\sum_{j=1}^{n} \frac{\delta_j Y_j}{G(Y_j)} S_{n_0}(Y_j).
\end{aligned}
\tag{26.16}
$$

Taking $\lim\sup$ on both sides of (26.16) we obtain, by applying the strong law of large numbers to the sequence of random variables $V_j :=$ $\frac{\delta_j Y_j}{G(Y_j)}S_{n_0}(Y_j)$ and observing that $Y \le T_F < \infty$,

$$
\begin{aligned}
0 &\le \limsup_{n\to\infty} 2\frac{1}{n}\sum_{j=1}^{n} \delta_j Y_j \left(\frac{1}{G(Y_j)} - \frac{1}{G_n(Y_j)}\right)^{+} \\
&\le \limsup_{n\to\infty} 2\frac{1}{n}\sum_{j=1}^{n} \frac{\delta_j Y_j}{G(Y_j)} S_{n_0}(Y_j) \\
&= 2\mathbf{E}\left\{\frac{\delta Y}{G(Y)}S_{n_0}(Y)\right\} \quad \text{a.s.} \\
&\le 2T_F\mathbf{E}\left\{\frac{\delta}{G(Y)}S_{n_0}(Y)\right\} \\
&= 2T_F\mathbf{E}\{S_{n_0}(Y)\} < 2T_F\epsilon
\end{aligned}
$$

which can be made arbitrarily small by an appropriate choice of ϵ, hence $\lim_{n\to\infty} I = 0$ a.s. Using again $\delta_j Z_j = \delta_j Y_j$ and the strong law of large numbers, we obtain, for expression III, that

$$III = \frac{1}{n}\sum_{j=1}^{n} \frac{\delta_j Y_j}{G(Y_j)} \to \mathbf{E}\left\{\frac{\delta Y}{G(Y)}\right\} = \mathbf{E}\{Y\} \quad \text{a.s. as} \quad n \to \infty.$$

For the expression II we get

$$II = \frac{1}{n}\sum_{j=1}^{n} \frac{\delta_j Z_j}{G_n(Z_j)} = \int_0^{T_F} F_n(y)\,dy \to \int_0^{T_F} F(y)\,dy = \mathbf{E}Y \quad \text{a.s.}$$

This, combined with the result that $\lim_{n\to\infty} I = 0$ a.s., completes the proof of the lemma. $\qquad\square$

PROOF OF THEOREM 26.2. Because of

$$\int (m_n^{(part)}(x) - m(x))^2 \mu(dx) \le T_F \int |m_n^{(part)}(x) - m(x)|\mu(dx)$$

it is enough to show that

$$\int |m_n^{(part)}(x) - m(x)|\mu(dx) \to 0 \text{ a.s. as } n \to \infty.$$

Put

$$K_n(x,z) = I_{\{z \in A_n(x)\}},$$

then

$$m_n(x) = m_n^{(part)}(x) = \min\left(T_{K_n}, \frac{\sum_{i=1}^n \frac{\delta_i Z_i}{G_n(Z_i)} K_n(x, X_i)}{\sum_{i=1}^n K_n(x, X_i)}\right).$$

Introduce the following notations:

$$\bar{m}_n(x) = \min\left(T_{K_n}, \frac{\sum_{i=1}^n \frac{\delta_i Z_i}{G_n(Z_i)} K_n(x, X_i)}{n\mathbf{E}\{K_n(x, X)\}}\right)$$

and

$$\tilde{m}_n(x) = \min\left(T_{K_n}, \frac{\sum_{i=1}^n \frac{\delta_i Z_i}{G(Z_i)} K_n(x, X_i)}{n\mathbf{E}\{K_n(x, X)\}}\right).$$

Then

$$|m_n(x) - m(x)|$$
$$\le |m_n(x) - \bar{m}_n(x)| + |\bar{m}_n(x) - \tilde{m}_n(x)| + |\tilde{m}_n(x) - m(x)|$$
$$= I_{n,1}(x) + I_{n,2}(x) + I_{n,3}(x).$$

Because of Lemma 23.2 and (26.9),

$$\int I_{n,3}(x)\mu(dx) \to 0 \text{ a.s.}$$

We have that

$$\int \frac{K_n(x, X_i)}{\mathbf{E}K_n(x, X)}\mu(dx) = 1,$$

therefore,

$$\int I_{n,2}(x)\mu(dx) \le \frac{1}{n}\sum_{i=1}^n \left|\frac{\delta_i Z_i}{G(Z_i)} - \frac{\delta_i Z_i}{G_n(Z_i)}\right| \int \frac{K_n(x, X_i)}{\mathbf{E}K_n(x, X)}\mu(dx)$$

$$= \frac{1}{n}\sum_{i=1}^n \left|\frac{\delta_i Z_i}{G(Z_i)} - \frac{\delta_i Z_i}{G_n(Z_i)}\right|$$

$$\to 0 \text{ a.s.}$$

by Lemma 26.1. Moreover,

$$I_{n,1}(x) \leq m_n(x)\left|1 - \frac{\bar{m}_n(x)}{m_n(x)}\right|$$

$$\leq T_{K_n}\left|1 - \frac{\min\left(T_{K_n}, \frac{\sum_{i=1}^n \frac{\delta_i Z_i}{G_n(Z_i)} K_n(x,X_i)}{n\mathbf{E}\{K_n(x,X)\}}\right)}{\min\left(T_{K_n}, \frac{\sum_{i=1}^n \frac{\delta_i Z_i}{G_n(Z_i)} K_n(x,X_i)}{\sum_{i=1}^n K_n(x,X_i)}\right)}\right|.$$

If $\bar{m}_n(x) = T_{K_n}$, $m_n(x) = T_{K_n}$, then

$$\left|1 - \frac{\bar{m}_n(x)}{m_n(x)}\right| = 0.$$

If $\bar{m}_n(x) < T_{K_n}$, $m_n(x) < T_{K_n}$, then

$$\left|1 - \frac{\bar{m}_n(x)}{m_n(x)}\right| = \left|1 - \frac{\sum_{i=1}^n K_n(x,X_i)}{n\mathbf{E}\{K_n(x,X)\}}\right|.$$

If $\bar{m}_n(x) < T_{K_n}$, $m_n(x) = T_{K_n}$, then

$$\left|1 - \frac{\bar{m}_n(x)}{m_n(x)}\right| = 1 - \frac{\bar{m}_n(x)}{m_n(x)}$$

$$\leq 1 - \frac{\frac{\sum_{i=1}^n \frac{\delta_i Z_i}{G_n(Z_i)} K_n(x,X_i)}{n\mathbf{E}\{K_n(x,X)\}}}{\frac{\sum_{i=1}^n \frac{\delta_i Z_i}{G_n(Z_i)} K_n(x,X_i)}{\sum_{i=1}^n K_n(x,X_i)}}$$

$$= 1 - \frac{\sum_{i=1}^n K_n(x,X_i)}{n\mathbf{E}\{K_n(x,X)\}}$$

$$\leq \left|1 - \frac{\sum_{i=1}^n K_n(x,X_i)}{n\mathbf{E}\{K_n(x,X)\}}\right|.$$

If $\bar{m}_n(x) = T_{K_n}$, $m_n(x) < T_{K_n}$, then

$$\left|1 - \frac{\bar{m}_n(x)}{m_n(x)}\right| = \frac{\bar{m}_n(x)}{m_n(x)} - 1$$

$$\leq \frac{\frac{\sum_{i=1}^n \frac{\delta_i Z_i}{G_n(Z_i)} K_n(x,X_i)}{n\mathbf{E}\{K_n(x,X)\}}}{\frac{\sum_{i=1}^n \frac{\delta_i Z_i}{G_n(Z_i)} K_n(x,X_i)}{\sum_{i=1}^n K_n(x,X_i)}} - 1$$

$$= \frac{\sum_{i=1}^n K_n(x,X_i)}{n\mathbf{E}\{K_n(x,X)\}} - 1$$

$$\leq \left|1 - \frac{\sum_{i=1}^n K_n(x,X_i)}{n\mathbf{E}\{K_n(x,X)\}}\right|.$$

Thus, in general,

$$\int I_{n,1}(x)\mu(dx) \;\le\; T_F \int \left| 1 - \frac{\sum_{i=1}^{n} K_n(x, X_i)}{n\mathbf{E}\{K_n(x, X)\}} \right| \mu(dx)$$

$$\to \quad 0 \quad \text{a.s.}$$

by Lemma 23.2. $\qquad\qquad\qquad\qquad\qquad\qquad\qquad\qquad\qquad\qquad\qquad\qquad\square$

In the censored case, the kernel estimate is

$$m_n^{(kern)}(x) = \min\left(T_{K_n}, \frac{\sum_{i=1}^{n} \frac{\delta_i Z_i}{G_n(Z_i)} K_{h_n}(x - X_i)}{\sum_{i=1}^{n} K_{h_n}(x - X_i)} \right). \qquad (26.17)$$

Theorem 26.3. *Assume (i)–(iv). Then under the conditions of Theorem 23.5, for the regression function estimate* $m_n^{(kern)}$ *one has*

$$\int (m_n^{(kern)}(x) - m(x))^2 \mu(dx) \to 0 \quad \text{a.s. as } n \to \infty.$$

PROOF. See Problem 26.4.

In the censored case, the k-NN estimate is

$$m_n^{(k-NN)}(x) = \min\left(T_{K_n}, \sum_{i=1}^{n} \frac{1}{k} I_{\{X_i \text{ is among the k-NNs of } x\}} \frac{\delta_i Z_i}{G_n(Z_i)} \right).$$

$$(26.18)$$

Theorem 26.4. *Assume (i)-(iv). Then under the conditions of Theorem 23.7, for the regression function estimate* $m_n^{(k-NN)}$ *one has*

$$\int (m_n^{(k-NN)}(x) - m(x))^2 \mu(dx) \to 0 \quad \text{a.s.} \quad \text{as } n \to \infty.$$

The proof is a consequence of the following lemma:

Lemma 26.2. *Assume, that* $T_F < \infty, G$ *is continuous, and* $G(T_F) > 0$. *Moreover, assume that* $\{W_{n,i}\}$ *are subprobability weights such that, for all bounded* Y,

$$\lim_{n\to\infty} \int \left(\sum_{i=1}^{n} W_{n,i}(x) Y_i - m(x) \right)^2 \mu(dx) = 0 \quad \text{a.s.} \qquad (26.19)$$

and

$$\lim_{n\to\infty} \sum_{i=1}^{n} \left| \frac{\delta_i Z_i}{G(Z_i)} - \frac{\delta_i Z_i}{G_n(Z_i)} \right| \int W_{n,i}(x)\mu(dx) = 0 \quad \text{a.s.} \qquad (26.20)$$

Then for the regression function estimate $m_n(x)$, defined by (26.12) and (26.13), one has

$$\int (m_n(x) - m(x))^2 \mu(dx) \to 0 \quad a.s. \ as \ n \to \infty.$$

PROOF. Because of the conditions

$$\frac{\delta_i Z_i}{G(Z_i)} = \frac{\delta_i Y_i}{G(Y_i)} \leq \frac{T_F}{G(T_F)} < \infty,$$

therefore (26.19) and (26.10) imply that

$$\lim_{n\to\infty} \int \left(\sum_{i=1}^{n} W_{n,i}(x) \frac{\delta_i Z_i}{G(Z_i)} - m(x) \right)^2 \mu(dx) = 0 \ \text{a.s.}$$

Because $G(T_F) > 0$, we have $T_K = T_F$, and thus

$$T_{K_n} \to T_F \ \text{a.s. as } n \to \infty,$$

thus

$$\lim_{n\to\infty} \int \left(\min \left(T_{K_n}, \sum_{i=1}^{n} W_{n,i}(x) \frac{\delta_i Z_i}{G(Z_i)} \right) - m(x) \right)^2 \mu(dx) = 0,$$

and, therefore,

$$\int (m_n(x) - m(x))^2 \mu(dx)$$

$$\leq 2 \int \left(\min \left(T_{K_n}, \sum_{i=1}^{n} W_{n,i}(x) \frac{\delta_i Z_i}{G_n(Z_i)} \right) \right.$$

$$\left. - \min \left(T_{K_n}, \sum_{i=1}^{n} W_{n,i}(x) \frac{\delta_i Z_i}{G(Z_i)} \right) \right)^2 \mu(dx)$$

$$+ 2 \int \left(\min \left(T_{K_n}, \sum_{i=1}^{n} W_{n,i}(x) \frac{\delta_i Z_i}{G(Z_i)} \right) - m(x) \right)^2 \mu(dx)$$

$$\leq 2 T_K \sum_{i=1}^{n} \left| \frac{\delta_i Z_i}{G_n(Z_i)} - \frac{\delta_i Z_i}{G(Z_i)} \right| \int W_{n,i}(x) \mu(dx)$$

$$+ 2 \int \left(\min \left(T_{K_n}, \sum_{i=1}^{n} W_{n,i}(x) \frac{\delta_i Z_i}{G(Z_i)} \right) - m(x) \right)^2 \mu(dx)$$

$$\to \quad 0 \ \text{a.s.},$$

where we used (26.20).   □

PROOF OF THEOREM 26.4. We will verify the conditions of Lemma 26.2. Equation (26.19) follows from the fact that the estimate is consistent in the uncensored case. Concerning (26.20) we will need the following covering results (Lemmas 23.10 and 23.11):

$$\limsup_{n\to\infty} \frac{n}{k_n}\mu\left(\{x : X_i \text{ is among the } k_n\text{-NNs of } x\}\right) \leq const \text{ a.s.}$$

Now

$$\sum_{i=1}^{n}\left|\frac{\delta_i Z_i}{G(Z_i)} - \frac{\delta_i Z_i}{G_n(Z_i)}\right|\int W_{n,i}(x)\mu(dx)$$

$$= \sum_{i=1}^{n}\left|\frac{\delta_i Z_i}{G(Z_i)} - \frac{\delta_i Z_i}{G_n(Z_i)}\right|\int \frac{1}{k_n}I_{\{X_i \text{ is among the } k_n\text{-NNs of } x\}}\mu(dx)$$

$$= \sum_{i=1}^{n}\left|\frac{\delta_i Z_i}{G(Z_i)} - \frac{\delta_i Z_i}{G_n(Z_i)}\right|\frac{1}{k_n}\mu\left(\{x : X_i \text{ is among the } k_n\text{-NNs of } x\}\right)$$

$$\leq \frac{n}{k_n}\max_i\mu\left(\{x : X_i \text{ is among the } k_n\text{-NNs of } x\}\right)$$

$$\times\frac{1}{n}\sum_{i=1}^{n}\left|\frac{\delta_i Z_i}{G(Z_i)} - \frac{\delta_i Z_i}{G_n(Z_i)}\right|$$

$$\to 0$$

a.s. because of Lemma 26.1. □

26.4 Regression Estimation for Model B

Assume that Y and C are conditionally independent given X (Model B). Let

$$F(t|x) = \mathbf{P}\{Y > t|X = x\},$$

$$G(t|x) = \mathbf{P}\{C > t|X = x\},$$

$$K(t|x) = \mathbf{P}\{Z > t|X = x\} = F(t|x)G(t|x),$$

and

$$T_F(x) = \sup\{t : F(t|x) > 0\},$$

$$T_G(x) = \sup\{t : G(t|x) > 0\},$$

$$T_K(x) = \min\{T_F(x), T_G(x)\}.$$

We observe

$$(X_1, Z_1, \delta_1), \ldots, (X_n, Z_n, \delta_n),$$

where $\delta_i = I_{\{Y_i \leq C_i\}}$ and $Z_i = \min(Y_i, C_i)$. Two related problems are studied:

(1) estimating the conditional survival functions $F(t|x)$ and $G(t|x)$; and

(2) estimating the regression function $m(x) = \mathbf{E}\{Y|X = x\}$ from the censored data.

Define the local averaging estimates of $F(t|x)$ and $G(t|x)$ in the following way:

$$F_n(t|x) = \prod_{\substack{Z_{(i)} \leq t \\ \sum_{j=i}^{n} W_{n(j)}(x) > 0}} \left(\frac{\sum_{j=i+1}^{n} W_{n(j)}(x)}{\sum_{j=i}^{n} W_{n(j)}(x)} \right)^{\delta_{(i)}} I_{\{t \leq Z_{(n)}\}} \qquad (26.21)$$

and

$$G_n(t|x) = \prod_{\substack{Z_{(i)} \leq t \\ \sum_{j=i}^{n} W_{n(j)}(x) > 0}} \left(\frac{\sum_{j=i+1}^{n} W_{n(j)}(x)}{\sum_{j=i}^{n} W_{n(j)}(x)} \right)^{1-\delta_{(i)}} I_{\{t \leq Z_{(n)}\}} \qquad (26.22)$$

where $(Z_{(i)}, \delta_{(i)})$ $(i = 1, \ldots, n)$ are the n pairs of observed (Z_i, δ_i) ordered on the $Z_{(i)}$, i.e.,

$$Z_{(1)} \leq Z_{(2)} \leq \cdots \leq Z_{(n)},$$

and W_{ni} are subprobability weights depending on $X_1, \ldots, X_n$, i.e., $W_{ni}(x) \geq 0$ and $\sum_{j=1}^{n} W_{ni}(x) \leq 1$, and $W_{n(i)}$ are these weights ordered according to the ordering of $Z_{(i)}$.

Examples of local averaging-type estimates are the nearest neighbor estimates, the kernel estimates, and the partitioning estimates.

The estimates defined in (26.21) and (26.22) extend the Kaplan-Meier method to conditional survival functions. If $W_{ni}(x) = \frac{1}{n}$ for all i then

$$F_n(t|x) = \prod_{i:Z_{(i)} \leq t} \left(\frac{\sum_{j=i+1}^{n} W_{n(j)}(x)}{\sum_{j=i}^{n} W_{n(j)}(x)} \right)^{\delta_{(i)}} = \prod_{i:Z_{(i)} \leq t} \left(\frac{n-i}{n-i+1} \right)^{\delta_{(i)}}$$

for all x, which is the Kaplan-Meier estimator $F_n(t)$.

In this section we assume that:

(I) Y and C are conditionally independent given X;

(II) F and G have no common jumps; and

(III) $T_F(x) < T_G(x)$ for all x.

Now we will show that these estimates are consistent if the weights are the consistent k-NN, kernel, or partitioning weights.

Theorem 26.5. *Assume (I)–(III). If the weights $W_{ni}(x)$ are such that the local averaging regression estimate using probability weights is strongly pointwise consistent for bounded Y, then, for $t < T_K(x)$,*

(a) $F_n(t|x) \to F(t|x)$; and

(b) $G_n(t|x) \to G(t|x)$;

a.s. as $n \to \infty$ for μ-almost all x.

PROOF. The argument follows the reasoning of the proof of Theorem 26.1, here we prove the statement for F. For G the proof is similar. $F(\cdot|x)$ can be expressed as

$$F(t|x) = \exp\left(\int_{[0,t]\setminus D(F(\cdot|x))} \frac{dF(s|x)}{F(s|x)}\right) \cdot \prod_{\substack{s \in D(F(\cdot|x)) \\ s \leq t}} \frac{F(s|x)}{F(s^-|x)} \qquad (26.23)$$

and since $F_n(\cdot|x)$ is a discrete survival function

$$F_n(t|x) = \prod_{\substack{s \in D(F_n(\cdot|x)) \\ s \leq t}} \frac{F_n(s|x)}{F_n(s^-|x)},$$

where $D(S)$ is, as earlier, the set of jump points of function S. Define the conditional subsurvival functions

$$F^*(t|x) = \mathbf{P}\{Z > t, \delta = 1 | X = x\},$$

$$G^*(t|x) = \mathbf{P}\{Z > t, \delta = 0 | X = x\}.$$

$F^*(\cdot|x)$ can be expressed in terms of the survival functions $F(\cdot|x)$ and $G(\cdot|x)$,

$$F^*(t|x) = \int_t^\infty G(s|x)[-dF(s|x)], \qquad (26.24)$$

because the random variables Y and C are conditionally independent given X (cf. (26.11)). Note that

$$\begin{aligned}
F^*(t|x) + G^*(t|x) &= \mathbf{P}\{Z > t | X = x\} \\
&= K(t|x) \\
&= \mathbf{P}\{\min(Y, C) > t | X = x\} \\
&= F(t|x) \cdot G(t|x). \qquad (26.25)
\end{aligned}$$

For real-valued functions H_1 and H_2 define a function ϕ by

$$\phi(H_1, H_2, t)$$

$$= \exp\left(\int_{[0,t]\setminus D(H_1)} \frac{dH_1(s)}{H_1(s) + H_2(s)}\right) \cdot \prod_{\substack{s \in D(H_1) \\ s \leq t}} \frac{H_1(s) + H_2(s)}{H_1(s^-) + H_2(s^-)}.$$

We will show that

$$F(t|x) = \phi(F^*(\cdot|x), G^*(\cdot|x), t). \qquad (26.26)$$

Because of (26.24), F^* has the same jump points as F. Thus (26.26) follows from (26.23), since

$$\frac{dF^*(s|x)}{F^*(s|x) + G^*(s|x)} = \frac{dF^*(s|x)}{F(s|x) \cdot G(s|x)} \qquad \text{(from (26.25))}$$

$$= \frac{-G(s|x) \cdot dF(s|x)}{G(s|x) \cdot F(s|x)} \qquad \text{(from (26.24))}$$

$$= \frac{-dF(s|x)}{F(s|x)}$$

and

$$\frac{F^*(s|x) + G^*(s|x)}{F^*(s^-|x) + G^*(s^-|x)} = \frac{F(s|x) \cdot G(s|x)}{F(s^-|x) \cdot G(s^-|x)} \qquad \text{(from (26.25))}$$

$$= \frac{F(s|x)}{F(s^-|x)}$$

if s is a jump point of $F(\cdot|x)$ because, by assumption, $F(\cdot|x)$ and $G(\cdot|x)$ do not have common jumps, thus s cannot be a jump point of $G(\cdot|x)$, so $G(s|x) = G(s^-|x)$.

Now we express $F_n(\cdot|x)$ in terms of the local averaging estimates

$$F_n^*(t|x) = \sum_{i=1}^n W_{ni}(x) I_{\{Z_i > t, \delta_i = 1\}}$$

and

$$G_n^*(t|x) = \sum_{i=1}^n W_{ni}(x) I_{\{Z_i > t, \delta_i = 0\}}$$

of the conditional subsurvival functions.

First note that (26.25) is also true for the estimates, since

$$F_n^*(t|x) + G_n^*(t|x) = \sum_{i=1}^n W_{ni}(x) I_{\{Z_i > t, \delta_i = 1\}} + \sum_{i=1}^n W_{ni}(x) I_{\{Z_i > t, \delta_i = 0\}}$$

$$= \sum_{i=1}^n W_{ni}(x) I_{\{Z_i > t\}}$$

and

$$F_n(t|x) \cdot G_n(t|x) = \prod_{\substack{Z_{(i)} \le t \\ \sum_{j=i}^n W_{n(j)}(x) > 0}} \left(\frac{\sum_{j=i+1}^n W_{n(j)}(x)}{\sum_{j=i}^n W_{n(j)}(x)} \right)^{\delta_{(i)}}$$

$$\times \prod_{\substack{Z_{(i)} \leq t \\ \sum_{j=i}^{n} W_{n(j)}(x) > 0}} \left(\frac{\sum_{j=i+1}^{n} W_{n(j)}(x)}{\sum_{j=i}^{n} W_{n(j)}(x)} \right)^{1-\delta_{(i)}}$$

$$= \prod_{\substack{Z_{(i)} \leq t \\ \sum_{j=i}^{n} W_{n(j)}(x) > 0}} \left(\frac{\sum_{j=i+1}^{n} W_{n(j)}(x)}{\sum_{j=i}^{n} W_{n(j)}(x)} \right)$$

$$= \frac{\sum_{i=1}^{n} W_{ni}(x) I_{\{Z_i > t\}}}{\sum_{j=1}^{n} W_{nj}(x)}$$

$$= \sum_{i=1}^{n} W_{ni}(x) I_{\{Z_i > t\}}$$

because of $\sum_{j=1}^{n} W_{nj}(x) = 1$. Therefore

$$\frac{F_n^*(s|x) + G_n^*(s|x)}{F_n^*(s^-|x) + G_n^*(s^-|x)} = \frac{F_n(s|x) \cdot G_n(s|x)}{F_n(s^-|x) \cdot G_n(s^-|x)} = \frac{F_n(s|x)}{F_n(s^-|x)}$$

if s is a jump point of $F_n(\cdot|x)$ because, by assumption, $F_n(\cdot|x)$ and $G_n(\cdot|x)$ do not have common jumps a.s., thus $G_n(s|x) = G_n(s^-|x)$. Hence

$$F_n(t|x) = \phi(F_n^*(\cdot|x), G_n^*(\cdot|x), t)$$

for $t \leq Z_{(n)}$.

Function $\phi(H_1(\cdot), H_2(\cdot), t)$ is a continuous function of the argument $H_1(\cdot)$ and $H_2(\cdot)$ according to the sup norm topology. $F^*(\cdot|x)$ and $G^*(\cdot|x)$ are regression functions of indicator random variables, and $F_n(\cdot|x)$ and $G_n(\cdot|x)$ are regression estimates. They are pointwise consistent on $[0, T_K]$ because of the condition. The limits are nonincreasing, therefore the convergence is uniform, consequently, $F_n(\cdot|x)$ converges to $F(\cdot|x)$ almost surely. $\square$

From a consistent estimator $F_n(\cdot|x)$ of the conditional survival function $F(\cdot|x)$ we can obtain a consistent nonparametric estimator $m_n(x)$ of the regression function $m(x)$.

Let

$$m_n(x) = \int_0^\infty F_n(t|x)\, dt = \int_0^{T_{Kn}} F_n(t|x)\, dt. \tag{26.27}$$

Theorem 26.6. *Under the assumptions of Theorem 26.5*

$$\int_{\mathcal{R}^d} |m_n(x) - m(x)|^2 \mu(dx) \to 0 \quad a.s. \ as \ n \to \infty.$$

PROOF. Since Y is bounded by T_F, we have $m(x) \leq T_F$ and hence also

$$|m(x) - m_n(x)| \leq T_F.$$

$T_F(x) < T_G(x)$ implies that $T_F(x) = T_K(x)$ therefore

$$\begin{aligned}
\int |m(x) - m_n(x)|^2 \mu(dx) &\leq T_F \int |m(x) - m_n(x)| \mu(dx) \\
&= T_F \int \left| \int_0^\infty (F(t|x) - F_n(t|x)) dt \right| \mu(dx) \\
&\leq T_F \int \int_0^{T_F(x)} |F(t|x) - F_n(t|x)| \, dt \mu(dx),
\end{aligned}$$

which tends to zero a.s. by Theorem 26.5 and by the dominated convergence theorem. $\qquad\square$

To see that the integration in (26.27) can be carried out easily, we calculate the jump of the estimator $F_n(\cdot|x)$ defined by (26.21) in Z_i.

Lemma 26.3. *For probability weights*

$$m_n(x) = \sum_{i=1}^n W_{ni}(x) \frac{Z_i \delta_i}{G_n(Z_i|x)}. \tag{26.28}$$

PROOF. We show that the jump of the estimator $F_n(\cdot|x)$ in an observation Z_i is given by

$$dF_n(Z_i|x) = \frac{W_{ni}(x)\delta_i}{G_n(Z_i|x)},$$

where $G_n(\cdot|x)$ is the estimator of $G(\cdot|x)$ defined by (26.22). $F_n(\cdot|x)$ has no jump at a censored observation, therefore we consider only the jump in an uncensored observation, i.e., let $\delta_{(i)} = 1$,

$$\begin{aligned}
dF_n(Z_{(i)}|x) &= F_n(Z_{(i-1)}|x) - F_n(Z_{(i)}|x) \\
&= \prod_{\substack{k \leq i-1 \\ \sum_{j=k}^n W_{n(j)}(x) > 0}} \left(\frac{\sum_{j=k+1}^n W_{n(j)}(x)}{\sum_{j=k}^n W_{n(j)}(x)} \right)^{\delta_{(k)}} \\
&\quad - \prod_{\substack{l \leq i \\ \sum_{j=l}^n W_{n(j)}(x) > 0}} \left(\frac{\sum_{j=l+1}^n W_{n(j)}(x)}{\sum_{j=l}^n W_{n(j)}(x)} \right)^{\delta_{(l)}}
\end{aligned}$$

$$
= \prod_{\substack{k \le i-1 \\ \sum_{j=k}^{n} W_{n(j)}(x) > 0}} \left(\frac{\sum_{j=k+1}^{n} W_{n(j)}(x)}{\sum_{j=k}^{n} W_{n(j)}(x)} \right)^{\delta_{(k)}}
$$

$$
\times \left(1 - \left(\frac{\sum_{j=i+1}^{n} W_{n(j)}(x)}{\sum_{j=i}^{n} W_{n(j)}(x)} \right)^{\delta_{(i)}} \right)
$$

$$
= \prod_{\substack{k \le i-1 \\ \sum_{j=k}^{n} W_{n(j)}(x) > 0}} \left(\frac{\sum_{j=k+1}^{n} W_{n(j)}(x)}{\sum_{j=k}^{n} W_{n(j)}(x)} \right)^{\delta_{(k)}}
$$

$$
\times \frac{W_{n(i)}(x)}{\sum_{j=i}^{n} W_{n(j)}(x)} .
$$

If $W_{n(i)}(x) = 0$, then $dF_n(Z_{(i)}|x) = 0$, otherwise $\sum_{j=k}^{n} W_{n(j)}(x) > 0$ for all $k \le i$, thus

$$
dF_n(Z_{(i)}|x)
$$

$$
= \prod_{k \le i-1} \left(\frac{\sum_{j=k+1}^{n} W_{n(j)}(x)}{\sum_{j=k}^{n} W_{n(j)}(x)} \right)^{\delta_{(k)}} \frac{W_{n(i)}(x)}{\sum_{j=i}^{n} W_{n(j)}(x)}
$$

$$
= \prod_{k \le i} \left(\frac{\sum_{j=k+1}^{n} W_{n(j)}(x)}{\sum_{j=k}^{n} W_{n(j)}(x)} \right)^{\delta_{(k)}} \left(\frac{\sum_{j=i+1}^{n} W_{n(j)}(x)}{\sum_{j=i}^{n} W_{n(j)}(x)} \right)^{-1}
$$

$$
\times \frac{W_{n(i)}(x)}{\sum_{j=i}^{n} W_{n(j)}(x)}
$$

$$
(\text{since } \delta_{(i)} = 1)
$$

$$
= \prod_{k \le i} \left(\frac{\sum_{j=k+1}^{n} W_{n(j)}(x)}{\sum_{j=k}^{n} W_{n(j)}(x)} \right)^{\delta_{(k)}} W_{n(i)}(x) \frac{1}{\sum_{j=i+1}^{n} W_{n(j)}(x)} .
$$

Since $W_{n(j)}(x)$ are probability weights,

$$
\sum_{j=i+1}^{n} W_{n(j)}(x) = \frac{\sum_{j=i+1}^{n} W_{n(j)}(x)}{\sum_{j=1}^{n} W_{n(j)}(x)} = \prod_{k \le i} \left(\frac{\sum_{j=k+1}^{n} W_{n(j)}(x)}{\sum_{j=k}^{n} W_{n(j)}(x)} \right) .
$$

Thus, by the definition of $G_n(\cdot|x)$, and since $\delta_{(i)} = 1$,

$$
dF_n(Z_{(i)}|x) = \prod_{k \le i} \left(\frac{\sum_{j=k+1}^{n} W_{n(j)}(x)}{\sum_{j=k}^{n} W_{n(j)}(x)} \right)^{\delta_{(k)}-1} W_{n(i)}(x)
$$

$$= \frac{1}{\prod_{k \le i} \left(\frac{\sum_{j=k+1}^{n} W_{n(j)}(x)}{\sum_{j=k}^{n} W_{n(j)}(x)} \right)^{1-\delta_{(k)}}} W_{n(i)}(x)$$

$$= \frac{W_{n(i)}(x)\delta_{(i)}}{G_n(Z_{(i)}|x)}.$$

The integral of $F_n(t|x)$ can be expressed as the sum of Z_i times the jump of F_n in Z_i that is apparent once we calculate the area in question by decomposing it into horizontal stripes, rather than into vertical stripes, as is customary in the calculation of integrals. Thus

$$\int_0^\infty F_n(t|x)\,dt = \sum_{i=1}^{n} Z_i \frac{W_{ni}(x)\delta_i}{G_n(Z_i|x)}.$$

Hence the regression estimate defined by (26.27) can be written in the following form:

$$m_n(x) = \sum_{i=1}^{n} W_{ni}(x) \frac{Z_i \delta_i}{G_n(Z_i|x)}.$$

$\square$

We have consequences for local averaging estimates. Formally Theorem 26.6 assumed probability weights, which is satisfied for partitioning and kernel estimates if x is in the support set and n is large.

Corollary 26.1. *Under the conditions of Theorems 26.6 and 25.6, for the partitioning estimate defined by (26.28),*

$$\int_{\mathcal{R}^d} |m_n(x) - m(x)|^2 \mu(dx) \to 0 \quad (n \to \infty) \quad a.s.$$

Corollary 26.2. *Under the conditions of Theorems 26.6 and 25.11, for the kernel estimate defined by (26.28),*

$$\int_{\mathcal{R}^d} |m_n(x) - m(x)|^2 \mu(dx) \to 0 \quad (n \to \infty) \quad a.s.$$

Corollary 26.3. *Under the conditions of Theorems 26.6 and 25.17, for the nearest neighbor estimate defined by (26.28),*

$$\int_{\mathcal{R}^d} |m_n(x) - m(x)|^2 \mu(dx) \to 0 \quad (n \to \infty) \quad a.s.$$

26.5 Bibliographic Notes

Theorem 26.1 is due to Peterson (1977). A Glivenko-Cantelli type sharpening is due to Stute and Wang (1993), see also the survey article of Gill (1994). Carbonez, Györfi, and van der Meulen (1995) proved Theorem 26.2. Theorems 26.3 and 26.4 are due to Kohler, Máthé, and Pintér (2002). Under the assumption that Y and C are conditionally independent given X (Model B), regression estimation was considered by many authors, the first fully nonparametric approach was given by Beran (1981) who introduced a class of nonparametric regression estimates to estimate conditional survival functions in the presence of right censoring. Beran (1981) proved Theorem 26.5. Dabrowska (1987; 1989) proved some consistency results for this estimate. An alternative approach was taken by Horváth (1981) who proposed to estimate the conditional survival function by integrating an estimate of the conditional density. Lemmas 26.2, 26.3, and Theorem 26.6 are due to Pintér (2001).

Problems and Exercises

PROBLEM 26.1. Prove that the Kaplan-Meier and Efron estimates are the same.

HINT: Both estimates have jumps at sample points. By induction verify that the jumps are the same.

PROBLEM 26.2. Prove (26.11).

HINT: $\frac{\delta_j}{nG_n(Z_j)}$ is the jump of F_n in Z_j.

PROBLEM 26.3. Prove that the Kaplan-Meier mean is a consistent estimate of the expectation if $T_F < T_G$.

HINT: Apply Theorem 26.1 and the dominated convergence theorem.

PROBLEM 26.4. Prove Theorem 26.3.

HINT: With the notation

$$K_n(x, z) = K_{h_n}(x - z)$$

we copy the proof of Theorem 26.2 such that the only difference is the covering lemma (Lemma 23.6)

$$\int \frac{K_{h_n}(x - X_i)}{\mathbf{E}K_{h_n}(x - X)} \mu(dx) \leq \rho,$$

moreover, instead of Lemma 23.2, we refer to Lemma 23.9.

27
Dependent Observations

If the data are not i.i.d. then the regression problem has two versions.

For the **regression function estimation from dependent data** $D_n = \{(X_i, Y_i)\}_{i=1}^n$, the random vectors (X_i, Y_i) $(i = 1, 2, \ldots)$ are not i.i.d., but D_n and (X, Y) are independent. Because of this independence

$$\mathbf{E}\{Y|X, D_n\} = \mathbf{E}\{Y|X\},$$

therefore

$$\min_f \mathbf{E}\{(f(X, D_n) - Y)^2\} = \min_f \mathbf{E}\{(f(X) - Y)^2\},$$

so the best L_2 error for the regression problem is the same as the best L_2 error for i.i.d. data.

For the **time series problem** $(X, Y) = (X_{n+1}, Y_{n+1})$. Here

$$\mathbf{E}\{Y_{n+1}|X_{n+1}, D_n\} \text{ and } \mathbf{E}\{Y_{n+1}|X_{n+1}\}$$

are not identical in general, thus we may have that

$$\min_f \mathbf{E}\{(f(X_{n+1}, D_n) - Y_{n+1})^2\} < \min_f \mathbf{E}\{(f(X_{n+1}) - Y_{n+1})^2\},$$

thus in the time series problem the best possible L_2 error can be improved with respect to the i.i.d. case.

In this chapter we are interested only in the extreme dependence, when the data are stationary and ergodic.

27.1 Stationary and Ergodic Observations

Since our main interest is the ergodicity, we summarize first the basics of stationary and ergodic sequences.

Definition 27.1. *A sequence of random variables* $\mathbf{Z} = \{Z_i\}_{i=-\infty}^{\infty}$ *defined on a probability space* $(\Omega, \mathcal{F}, \mathbf{P})$ *is stationary if for any integers* k *and* n *the random vectors* $(Z_1, \ldots, Z_n)$ *and* $(Z_{k+1}, \ldots, Z_{k+n})$ *have the same distribution.*

A measurable-transformation T from Ω to Ω is called measure preserving if, for all $A \in \mathcal{F}$,

$$\mathbf{P}\{A\} = \mathbf{P}\{T^{-1}A\}.$$

It is easy to show that for any stationary sequence $\mathbf{Z}$ there are a random variable Z and a measure-preserving transformation T such that

$$Z_n(\omega) = Z(T^n\omega). \tag{27.1}$$

Definition 27.2. *A stationary sequence of random variables*

$$\mathbf{Z} = \{Z_i\}_{i=-\infty}^{\infty}$$

represented by a measure-preserving transformation T *is called ergodic if*

$$T^{-1}A = A$$

implies that $\mathbf{P}\{A\}$ *is either* 0 *or* 1.

The next lemma is the strong law of large numbers for ergodic sequences.

Lemma 27.1. (BIRKHOFF'S ERGODIC THEOREM). *Let* $\{Z_i\}_{-\infty}^{\infty}$ *be a stationary and ergodic process with* $\mathbf{E}|Z_1| < \infty$. *Then*

$$\lim_{n\to\infty} \frac{1}{n} \sum_{i=1}^{n} Z_i = \mathbf{E}Z_1 \quad a.s.$$

PROOF. First we prove the so-called maximal ergodic theorem: Let Z be an integrable real random variable and let T be a measure-preserving transformation, further

$$M_n = \max_{k=1,2,\ldots,n} \sum_{i=0}^{k-1} Z \circ T^i.$$

Then

$$\mathbf{E}\{ZI_{\{M_n>0\}}\} \geq 0.$$

In order to show this, notice that, for $k = 0, 1, \ldots, n$,

$$M_n^+ \geq \sum_{i=0}^{k-1} Z \circ T^i.$$

(The void sum is 0 by definition.) Thus

$$Z + M_n^+ \circ T \geq Z + \sum_{i=0}^{k-1} Z \circ T^{i+1} = \sum_{i=0}^{k} Z \circ T^i$$

for $k = 0, 1, \ldots, n$, and

$$Z \geq \max_{k=1,2,\ldots,n} \sum_{i=0}^{k} Z \circ T^i - M_n^+ \circ T \geq M_n - M_n^+ \circ T.$$

Because $M_n = M_n^+$ on $[M_n > 0] = [M_n^+ > 0]$, one obtains

$$\begin{aligned}
\mathbf{E}\{ZI_{\{M_n>0\}}\} &= \int_{M_n^+>0} Z \, d\mathbf{P} \\
&\geq \int_{M_n^+>0} (M_n^+ - M_n^+ \circ T) \, d\mathbf{P} \\
&= \int_\Omega M_n^+ \, d\mathbf{P} - \int_{M_n^+>0} M_n^+ \circ T \, d\mathbf{P} \\
&\geq \int_\Omega M_n^+ \, d\mathbf{P} - \int_\Omega M_n^+ \circ T \, d\mathbf{P} \\
&= \int_\Omega M_n^+ \, d\mathbf{P} - \int_{T^{-1}(\Omega)} M_n^+ \, d\mathbf{P}_T \\
&= \int_\Omega M_n^+ \, d\mathbf{P} - \int_\Omega M_n^+ \, d\mathbf{P} \\
&= 0,
\end{aligned}$$

since $\mathbf{P}\{T^{-1}(\Omega)\} = \mathbf{P}\{\Omega\} = 1$ and $\mathbf{P}_T = \mathbf{P}$ for measure-preserving T, and the proof of the maximal ergodic theorem is complete.

In order to prove Lemma 27.1 we use the representation of $\{Z_i\}_{i=-\infty}^{\infty}$ given after Definition 27.1. Without loss of generality assume $\mathbf{E}Z = 0$. It suffices to show

$$\bar{Z} := \limsup_{n \to \infty} \frac{1}{n} \sum_{i=0}^{n-1} Z_i \leq 0 \quad \text{a.s.},$$

because then by use of $\{-Z_i\}$ we get

$$\liminf_{n \to \infty} \frac{1}{n} \sum_{i=0}^{n-1} Z_i \geq 0 \quad \text{a.s.}$$

and thus

$$\frac{1}{n} \sum_{i=0}^{n-1} Z_i \to 0 \quad \text{a.s.}$$

Notice that $\bar{Z}$ is T-invariant, i.e., $\bar{Z} \circ T = \bar{Z}$. Because of

$$
\begin{aligned}
T^{-1}\{\bar{Z} > s\} &= T^{-1}\bar{Z}^{-1}((s, \infty)) \\
&= (\bar{Z}T)^{-1}((s, \infty)) \\
&= \bar{Z}^{-1}((s, \infty)) \\
&= \{\bar{Z} > s\}
\end{aligned}
$$

for each real s, by ergodicity of $\{Z_i\}$ the probability $\mathbf{P}\{\bar{Z} > s\}$ equals 0 cr 1 for each s. Thus a constant $\delta \in \mathcal{R} \cup \{-\infty, \infty\}$ exists such that

$$
\bar{Z} = \delta \quad \text{a.s.}
$$

We show that the assumption $\delta > 0$ leads to a contradiction. Set $\delta^* = \delta$ if $0 < \delta < \infty$ and $\delta^* = 1$ if $\delta = \infty$, further

$$
Z^* = Z - \delta^*
$$

and

$$
\tilde{M}_n = \max_{k=1,2,\ldots,n} \frac{1}{k} \sum_{i=0}^{k-1} Z^* \circ T^i.
$$

On the one hand, by the maximal ergodic theorem

$$
\mathbf{E}\{Z^* I_{\{\tilde{M}_n > 0\}}\} \geq 0.
$$

On the other hand,

$$
\{\tilde{M}_n > 0\} \uparrow \left\{ \sup_k \frac{1}{k} \sum_{i=0}^{k-1} Z^* \circ T^i > 0 \right\} = \left\{ \sup_k \frac{1}{k} \sum_{i=0}^{k-1} Z \circ T^i \geq \delta^* \right\}
$$

and

$$
\mathbf{P}\left\{ \sup_k \frac{1}{k} \sum_{i=0}^{k-1} Z \circ T^i \geq \delta^* \right\} = 1,
$$

since $\mathbf{P}\{\bar{Z} = \delta\} = 1$. Because of

$$
|Z^* I_{\{\tilde{M}_n > 0\}}| \leq |Z| + \delta^*
$$

Lebesgue's dominated convergence theorem yields

$$
\mathbf{E}\{Z^* I_{\{\tilde{M}_n > 0\}}\} \to \mathbf{E} Z^* = -\delta^* < 0
$$

for $n \to \infty$, which is a contradiction. $\qquad\square$

The main ingredient of the proofs in the sequel is known as Breiman's generalized ergodic theorem.

Lemma 27.2. *Let* $\mathbf{Z} = \{Z_i\}_{-\infty}^{\infty}$ *be a stationary and ergodic process. Let* T *denote the left shift operator. Let* f_i *be a sequence of real-valued functions such that for some function* f, $f_i(\mathbf{Z}) \to f(\mathbf{Z})$ *a.s. Assume that*

$\mathbf{E}\sup_i |f_i(\mathbf{Z})| < \infty$. *Then*

$$\lim_{n\to\infty} \frac{1}{n} \sum_{i=1}^{n} f_i(T^i\mathbf{Z}) = \mathbf{E}f(\mathbf{Z}) \quad a.s.$$

PROOF. Put

$$g_k = \inf_{k\le i} f_i$$

and

$$G_k = \sup_{k\le i} f_i,$$

then for any integer $k > 0$ Lemma 27.1 implies that

$$\limsup_{n\to\infty} \frac{1}{n} \sum_{i=1}^{n} f_i(T^i\mathbf{Z}) \le \limsup_{n\to\infty} \frac{1}{n} \sum_{i=1}^{n} G_k(T^i\mathbf{Z}) = \mathbf{E}G_k(\mathbf{Z}) \quad \text{a.s.}$$

since by our condition $\mathbf{E}G_k(\mathbf{Z})$ is finite. The monotone convergence theorem yields

$$\mathbf{E}G_k(\mathbf{Z}) \downarrow \mathbf{E}\left\{ \limsup_{i\to\infty} f_i(\mathbf{Z}) \right\} = \mathbf{E}f(\mathbf{Z}).$$

Similarly,

$$\liminf_{n\to\infty} \frac{1}{n} \sum_{i=1}^{n} f_i(T^i\mathbf{Z}) \ge \liminf_{n\to\infty} \frac{1}{n} \sum_{i=1}^{n} g_k(T^i\mathbf{Z}) = \mathbf{E}g_k(\mathbf{Z}) \quad \text{a.s.}$$

and

$$\mathbf{E}g_k(\mathbf{Z}) \uparrow \mathbf{E}\left\{ \liminf_{i\to\infty} f_i(\mathbf{Z}) \right\} = \mathbf{E}f(\mathbf{Z}).$$

$\square$

27.2 Dynamic Forecasting: Autoregression

The autoregression is a special time series problem where there is no X_i only Y_i. Here Cover (1975) formulated the following two problems:

Dynamic Forecasting. Find an estimator $\hat{E}(Y_0^{n-1})$ of the value $\mathbf{E}\{Y_n|Y_0^{n-1}\}$ such that for all stationary and ergodic sequences $\{Y_i\}$

$$\lim_{n\to\infty} |\hat{E}(Y_0^{n-1}) - \mathbf{E}\{Y_n|Y_0^{n-1}\}| = 0 \quad \text{a.s.}$$

Static Forecasting. Find an estimator $\hat{E}(Y_{-n}^{-1})$ of the value $\mathbf{E}\{Y_0|Y_{-\infty}^{-1}\}$ such that for all stationary and ergodic sequences $\{Y_i\}$

$$\lim_{n\to\infty} \hat{E}(Y_{-n}^{-1}) = \mathbf{E}\{Y_0|Y_{-\infty}^{-1}\} \quad \text{a.s.}$$

First we show that the dynamic forecasting is impossible.

Theorem 27.1. *For any estimator* $\{\hat{E}(Y_0^{n-1})\}$ *there is a stationary ergodic binary valued process* $\{Y_i\}$ *such that*

$$\mathbf{P}\left\{\limsup_{n\to\infty} |\hat{E}(Y_0^{n-1}) - \mathbf{E}\{Y_n|Y_0^{n-1}\}| \geq 1/4\right\} \geq \frac{1}{8}.$$

PROOF. We define a Markov process which serves as the technical tool for the construction of our counterexample. Let the state space S be the non-negative integers. From state 0 the process certainly passes to state 1 and then to state 2, at the following epoch. From each state $s \geq 2$, the Markov chain passes either to state 0 or to state $s+1$ with equal probabilities 0.5. This construction yields a stationary and ergodic Markov process $\{M_i\}$ with stationary distribution

$$\mathbf{P}\{M = 0\} = \mathbf{P}\{M = 1\} = \frac{1}{4}$$

and

$$\mathbf{P}\{M = i\} = \frac{1}{2^i} \quad \text{for } i \geq 2.$$

Let τ_k denote the first positive time of the occurrence of state $2k$:

$$\tau_k = \min\{i \geq 0 : M_i = 2k\}.$$

Note that if $M_0 = 0$ then $M_i \leq 2k$ for $0 \leq i \leq \tau_k$. Now we define the hidden Markov chain process $\{Y_i\}$, which we denote as $Y_i = f(M_i)$. It will serve as the stationary unpredictable time series. We will use the notation M_0^n to denote the sequence of states $M_0, \ldots, M_n$. Let $f(0) = 0$, $f(1) = 0$, and $f(s) = 1$ for all even states s. A feature of this definition of $f(\cdot)$ is that whenever $Y_n = 0, Y_{n+1} = 0, Y_{n+2} = 1$ we know that $M_n = 0$ and vice versa. Next we will define $f(s)$ for odd states s maliciously. We define $f(2k + 1)$ inductively for $k \geq 1$. Assume $f(2l+1)$ is defined for $l < k$. If $M_0 = 0$ (i.e., $f(M_0) = 0$, $f(M_1) = 0$, $f(M_2) = 1$), then $M_i \leq 2k$ for $0 \leq i \leq \tau_k$ and the mapping

$$M_0^{\tau_k} \to (f(M_0), \ldots, f(M_{\tau_k}))$$

is invertible. (It can be seen as follows: Given Y_0^n find $1 \leq l \leq n$ and positive integers $0 = r_0 < r_1 < \cdots < r_l = n + 1$ such that $Y_0^n = (Y_{r_0}^{r_1-1}, Y_{r_1}^{r_2-1}, \ldots, Y_{r_{l-1}}^{r_l-1})$, where $2 \leq r_{i+1} - 1 - r_i < 2k$ for $0 \leq i < l - 1$, $r_l - 1 - r_{l-1} = 2k$ and for $0 \leq i < l$, $Y_{r_i}^{r_{i+1}-1} = (f(0), f(1), \ldots, f(r_{i+1} - 1 - r_i))$. Now $\tau_k = n$ and $M_{r_i}^{r_{i+1}-1} = (0, 1, \ldots, r_{i+1} - 1 - r_i)$ for $0 \leq i < l$.

This construction is always possible under our postulates that $M_0 = 0$ and $\tau_k = n$.) Let

$$B_k^+ = \left\{ M_0 = 0, \hat{E}(f(M_0), \ldots, f(M_{\tau_k})) \geq \frac{1}{4} \right\}$$

and

$$B_k^- = \left\{ M_0 = 0, \hat{E}(f(M_0), \ldots, f(M_{\tau_k})) < \frac{1}{4} \right\}.$$

Now notice that the sets B_k^+ and B_k^- do not depend on the future values of $f(2r+1)$ for $r \geq k$. One of the two sets B_k^+, B_k^- has at least probability $1/8$. Now we specify $f(2k+1)$. Let $f(2k+1) = 1$, $I_k = B_k^-$ if $\mathbf{P}\{B_k^-\} \geq \mathbf{P}\{B_k^+\}$ and let $f(2k + 1) = 0$, $I_k = B_k^+$ if $\mathbf{P}\{B_k^-\} < \mathbf{P}\{B_k^+\}$. Because of the construction of $\{M_i\}$, on event I_k,

$$
\begin{aligned}
\mathbf{E}\{Y_{\tau_k+1}|Y_0^{\tau_k}\} &= f(2k+1)\mathbf{P}\{Y_{\tau_k+1} = f(2k+1)|Y_0^{\tau_k}\} \\
&= f(2k+1)\mathbf{P}\{M_{\tau_k+1} = 2k+1|M_0^{\tau_k}\} \\
&= 0.5f(2k+1).
\end{aligned}
$$

The difference of the estimate and the conditional expectation is at least $\frac{1}{4}$ on set I_k and this event occurs with probability not less than $1/8$. Finally, by Fatou's lemma,

$$
\begin{aligned}
&\mathbf{P}\left\{ \limsup_{n\to\infty}\{|\hat{E}(Y_0^{n-1}) - \mathbf{E}\{Y_n|Y_0^{n-1}\}| \geq 1/4\} \right\} \\
\geq\ &\mathbf{P}\left\{ \limsup_{n\to\infty}\{|\hat{E}(Y_0^{n-1}) - \mathbf{E}\{Y_n|Y_0^{n-1}\}| \geq 1/4, Y_0 = Y_1 = 0, Y_2 = 1\} \right\} \\
\geq\ &\mathbf{P}\left\{ \limsup_{k\to\infty}\{|\hat{E}(f(M_0), \ldots, f(M_{\tau_k})) \right. \\
&\quad\left. - \mathbf{E}\{f(M_{\tau_k+1})|f(M_0), \ldots, f(M_{\tau_k})\}| \geq 1/4, M_0 = 0\} \right\} \\
\geq\ &\mathbf{P}\left\{ \limsup_{k\to\infty} I_k \right\} \\
=\ &\mathbf{E}\left\{ \limsup_{k\to\infty} I_{I_k} \right\} \\
\geq\ &\limsup_{k\to\infty} \mathbf{E}\{I_{I_k}\} \\
=\ &\limsup_{k\to\infty} \mathbf{P}\{I_k\} \geq \frac{1}{8}.
\end{aligned}
$$

□

One-Step Dynamic Forecasting. Find an estimator $\hat{E}(Y_0^{n-1})$ of the value $\mathbf{E}\{Y_n|Y_{n-1}\}$ such that, for all stationary and ergodic sequences $\{Y_i\}$,

$$\lim_{n\to\infty} |\hat{E}(Y_0^{n-1}) - \mathbf{E}\{Y_n|Y_{n-1}\}| = 0 \quad \text{a.s.}$$

We show next that the one-step dynamic forecasting is impossible.

Theorem 27.2. *For any estimator $\{\hat{E}(Y_0^{n-1})\}$ there is a stationary ergodic real-valued process $\{Y_i\}$ such that*

$$\mathbf{P}\left\{\limsup_{n\to\infty}\{|\hat{E}(Y_0^{n-1}) - \mathbf{E}\{Y_n|Y_{n-1}\}| \geq 1/8\}\right\} \geq \frac{1}{8}.$$

PROOF. We will use the Markov process $\{M_i\}$ defined in the proof of Theorem 27.1. Note that one must pass through state s to get to any state $s' > s$ from 0. We construct a process $\{Y_i\}$ which is in fact just a relabeled version of $\{M_i\}$. This construct uses a different (invertible) function $f(\cdot)$, for $Y_i = f(M_i)$. Define $f(0)=0$, $f(s) = L_s + 2^{-s}$ if $s > 0$ where L_s is either 0 or 1 as specified later. In this way, knowing Y_i is equivalent to knowing M_i and *vice versa*. Thus $Y_i = f(M_i)$ where f is one-to-one. For $s \geq 2$ the conditional expectation is

$$\mathbf{E}\{Y_t|Y_{t-1} = L_s + 2^{-s}\} = \frac{L_{s+1} + 2^{-(s+1)}}{2}.$$

We complete the description of the function $f(\cdot)$ and thus the conditional expectation by defining L_{s+1} so as to confound any proposed predictor $\hat{E}(Y_0^{n-1})$. Let τ_s denote the time of the first occurrence of state s:

$$\tau_s = \min\{i \geq 0 : M_i = s\}.$$

Let $L_1 = L_2 = 0$. Suppose $s \geq 2$. Assume we specified L_i for $i \leq s$. Define

$$B_s^+ = \{Y_0 = 0, \hat{E}(Y_0^{\tau_s}) \geq 1/4\}$$

and

$$B_s^- = \{Y_0 = 0, \hat{E}(Y_0^{\tau_s}) < 1/4\}.$$

One of the two sets has at least probability 1/8. Take $L_{s+1} = 1$ and $I_s = B_s^-$ if $\mathbf{P}\{B_s^-\} \geq \mathbf{P}\{B_s^+\}$. Let $L_{s+1} = 0$ and $I_s = B_s^+$ if $\mathbf{P}\{B_s^-\} < \mathbf{P}\{B_s^+\}$. The difference of the estimation and the conditional expectation is at least 1/8 on set I_s and this event occurs with probability not less than 1/8. By Fatou's lemma,

$$\mathbf{P}\left\{\limsup_{n\to\infty}\{|\hat{E}(Y_0^{n-1}) - \mathbf{E}\{Y_n|Y_{n-1}\}| \geq 1/8\}\right\}$$

$$\geq \mathbf{P}\left\{\limsup_{s\to\infty}\{|\hat{E}(Y_0^{\tau_s}) - \mathbf{E}\{Y_{\tau_s+1}|Y_{\tau_s}\}| \geq 1/8. Y_0 = 0\}\right\}$$

$$\geq \mathbf{P}\left\{\limsup_{s\to\infty} I_s\right\}$$

$$\geq \limsup_{s\to\infty} \mathbf{P}\{I_s\} \geq 1/8.$$

$\square$

27.3   Static Forecasting: General Case

Consider the static forecasting problem. Assume, in general, that $\{(X_i, Y_i)\}$ is a stationary and ergodic time series. One wishes to infer the conditional expectation

$$\mathbf{E}\{Y_0 | X^0_{-\infty}, Y^{-1}_{-\infty}\}.$$

Let $\mathcal{P}_n = \{A_{n,j}\}$ and $\mathcal{Q}_n = \{B_{n,j}\}$ be sequences of nested partitions of R^d and R, respectively, and a_n and b_n the corresponding quantizers with reproduction points $a_{n,j} \in A_{n,j}$ and $b_{n,j} \in B_{n,j}$, respectively,

$$a_n(x) = a_{n,j} \ \text{ if } x \in A_{n,j}$$

and

$$b_n(y) = b_{n,j} \ \text{ if } y \in B_{n,j}.$$

The estimate is as follows: Set $\lambda_0 = 1$ and for $0 < k < \infty$ define recursively

$$\tau_k = \min\{t > 0 : a_k(X^{-t}_{-\lambda_{k-1}-t}) = a_k(X^0_{-\lambda_{k-1}}), b_k(Y^{-1-t}_{-\lambda_{k-1}-t}) = b_k(Y^{-1}_{-\lambda_{k-1}})\},$$

and

$$\lambda_k = \tau_k + \lambda_{k-1}.$$

The number τ_k is random, but is finite almost surely.

The kth estimate of $\mathbf{E}\{Y_0 | X^0_{-\infty}, Y^{-1}_{-\infty}\}$ is provided by

$$m_k = \frac{1}{k} \sum_{1 \le j \le k} Y_{-\tau_j}. \tag{27.2}$$

To obtain a fixed sample-size version we apply the same method as in Algoet (1992). For $0 < t < \infty$ let κ_t denote the maximum of the integers k such that $\lambda_k \le t$. Formally, let

$$\kappa_t = \max\{k \ge 0 : \lambda_k \le t\}.$$

Now put

$$\hat{m}_{-t} = m_{\kappa_t}. \tag{27.3}$$

Theorem 27.3. *Assume that $\{\mathcal{P}_k\}$ asymptotically generates the Borel σ-algebra, i.e., $\mathcal{F}(\mathcal{P}_k) \uparrow \mathcal{B}_d$ and*

$$\epsilon_k = \sup_j diam(B_{k,j}) \to 0.$$

Then

$$\lim_{k \to \infty} m_k = \mathbf{E}\{Y_0 | X^0_{-\infty}, Y^{-1}_{-\infty}\} \quad a.s.$$

and

$$\lim_{t \to \infty} \hat{m}_{-t} = \mathbf{E}\{Y_0 | X^0_{-\infty}, Y^{-1}_{-\infty}\} \quad a.s.$$

PROOF. Following the technique by Morvai (1995) we first show that for

$$\mathcal{F}_j = \mathcal{F}(a_j(X^0_{-\lambda_j}), b_j(Y^{-1}_{-\lambda_j}))$$

and Borel set C,

$$\mathbf{P}\{Y_{-\tau_j} \in C | \mathcal{F}_{j-1}\} = \mathbf{P}\{Y_0 \in C | \mathcal{F}_{j-1}\}. \tag{27.4}$$

For $0 < m < \infty$, $0 < l < \infty$, any sequences x^0_{-m} and y^{-1}_{-m}, and Borel set C we prove that

$$T^{-l}(\{a_{j-1}(X^0_{-m}) = a_{j-1}(x^0_{-m}),$$
$$b_{j-1}(Y^{-1}_{-m}) = b_{j-1}(y^{-1}_{-m}), \ \lambda_{j-1} = m, \ \tau_j = l, \ Y_{-l} \in C\})$$
$$= \{a_{j-1}(X^0_{-m}) = a_{j-1}(x^0_{-m}),$$
$$b_{j-1}(Y^{-1}_{-m}) = b_{j-1}(y^{-1}_{-m}), \ \lambda_{j-1} = m, \ \tilde{\tau}_j = l, \ Y_0 \in C\}$$

where T denotes the left shift operator and

$$\tilde{\tau}_k = \min\{t > 0 : a_k(X^t_{-\lambda_{k-1}+t}) = a_k(X^0_{-\lambda_{k-1}}), b_k(Y^{-1+t}_{-\lambda_{k-1}+t})$$
$$= b_k(Y^{-1}_{-\lambda_{k-1}})\}.$$

Since the event $\{\lambda_{j-1} = m\}$ is measurable with respect to $\mathcal{F}(a_{j-1}(X^0_{-m}), b_{j-1}(Y^{-1}_{-m}))$, either

$$\{a_{j-1}(X^0_{-m}) = a_{j-1}(x^0_{-m}), \ b_{j-1}(Y^{-1}_{-m}) = b_{j-1}(y^{-1}_{-m}), \ \lambda_{j-1} = m\} = \emptyset$$

and the statement is trivial, or

$$\{a_{j-1}(X^0_{-m}) = a_{j-1}(x^0_{-m}), \ b_{j-1}(Y^{-1}_{-m}) = b_{j-1}(y^{-1}_{-m}), \ \lambda_{j-1} = m\}$$
$$= \{a_{j-1}(X^0_{-m}) = a_{j-1}(x^0_{-m}), \ b_{j-1}(Y^{-1}_{-m}) = b_{j-1}(y^{-1}_{-m})\}.$$

Then

$$T^{-l}(\{a_{j-1}(X^0_{-m}) = a_{j-1}(x^0_{-m}),$$
$$b_{j-1}(Y^{-1}_{-m}) = b_{j-1}(y^{-1}_{-m}), \ \tau_j = l, \ Y_{-l} \in C\})$$
$$= T^{-l}(\{a_{j-1}(X^0_{-m}) = a_{j-1}(x^0_{-m}), \ a_j(X^{-l}_{-m-l}) = a_j(X^0_{-m}),$$
$$b_{j-1}(Y^{-1}_{-m}) = b_{j-1}(y^{-1}_{-m}), \ b_j(Y^{-1-l}_{-m-l}) = b_j(Y^{-1}_{-m}),$$
$$a_j(X^{-t}_{-m-t}) \neq a_j(X^0_{-m}) \text{ or } b_j(Y^{-1-t}_{-m-t}) = b_j(Y^{-1}_{-m}) \text{ for all } 0 < t < l,$$
$$Y_{-l} \in C\})$$
$$= \{a_{j-1}(X^l_{-m+l}) = a_{j-1}(x^0_{-m}), \ a_j(X^0_{-m}) = a_j(X^l_{-m+l}),$$
$$b_{j-1}(Y^{-1+l}_{-m+l}) = b_{j-1}(y^{-1}_{-m}), \ b_j(Y^{-1}_{-m}) = b_j(Y^{-1+l}_{-m+l}),$$
$$a_j(X^{-t+l}_{-m-t+l}) \neq a_j(X^l_{-m+l})$$
$$\text{ or } b_j(Y^{-1-t+l}_{-m-t+l}) = b_j(Y^{-1+l}_{-m+l}) \text{ for all } 0 < t < l,$$

$$Y_0 \in C\}$$
$$= \{a_{j-1}(X^0_{-m}) = a_{j-1}(x^0_{-m}), \ a_j(X^0_{-m}) = a_j(X^l_{-m+l}),$$
$$b_{j-1}(Y^{-1}_{-m}) = b_{j-1}(y^{-1}_{-m}), \ b_j(Y^{-1}_{-m}) = b_j(Y^{-1+l}_{-m+l}),$$
$$a_j(X^t_{-m+t}) \neq a_j(X^0_{-m}) \text{ or } b_j(Y^{-1+t}_{-m+t}) = b_j(Y^{-1}_{-m}) \text{ for all } 0 < t < l,$$
$$Y_0 \in C\}$$
$$= \{a_{j-1}(X^0_{-m}) = a_{j-1}(x^0_{-m}),$$
$$b_{j-1}(Y^{-1}_{-m}) = b_{j-1}(y^{-1}_{-m}), \ \tilde{\tau}_j = l, \ Y_0 \in C\},$$

and the equality of the events is proven. Now note that each generating atom of $\mathcal{F}_{j-1}$ (there are countably many of them) has the form

$$H = \{a_{j-1}(X^0_{-m}) = a_{j-1}(x^0_{-m}), \ b_{j-1}(Y^{-1}_{-m}) = b_{j-1}(y^{-1}_{-m}), \ \lambda_{j-1} = m\},$$

so we have to show that for any atom H the following equality holds:

$$\int_H \mathbf{P}\{Y_{-\tau_j} \in C | \mathcal{F}_{j-1}\} \, d\mathbf{P} = \int_H \mathbf{P}\{Y_0 \in C | \mathcal{F}_{j-1}\} \, d\mathbf{P},$$

which can be done using the properties of conditional probability and the previous identities:

$$\int_H \mathbf{P}\{Y_{-\tau_j} \in C | \mathcal{F}_{j-1}\} \, d\mathbf{P} = \mathbf{P}\{H \cap \{Y_{-\tau_j} \in C\}\}$$

$$= \sum_{1 \leq l < \infty} \mathbf{P}\{H \cap \{\tau_j = l, \ Y_{-l} \in C\}\}$$

$$= \sum_{1 \leq l < \infty} \mathbf{P}\{T^{-l}(H \cap \{\tau_j = l, \ Y_{-l} \in C\})\}$$

$$= \sum_{1 \leq l < \infty} \mathbf{P}\{H \cap \{\tilde{\tau}_j = l, \ Y_0 \in C\}\}$$

$$= \mathbf{P}\{H \cap \{Y_0 \in C\}\}$$

$$= \int_H \mathbf{P}\{Y_0 \in C | \mathcal{F}_{j-1}\} \, d\mathbf{P}.$$

Introduce another estimate:

$$\tilde{m}_k = \frac{1}{k} \sum_{1 \leq j \leq k} b_j(Y_{-\tau_j}).$$

By condition,

$$|m_k - \tilde{m}_k| \leq \frac{1}{k} \sum_{1 \leq j \leq k} |Y_{-\tau_j} - b_j(Y_{-\tau_j})| \leq \frac{1}{k} \sum_{1 \leq j \leq k} \epsilon_j \to 0,$$

thus it suffices to show that

$$\lim_{k \to \infty} \tilde{m}_k = \mathbf{E}\{Y_0 | X^0_{-\infty}, Y^{-1}_{-\infty}\} \quad \text{a.s.}$$

We have that

$$\tilde{m}_k - \mathbf{E}\{Y_0|X^0_{-\infty}, Y^{-1}_{-\infty}\}$$

$$= \frac{1}{k} \sum_{1 \le j \le k} (b_j(Y_{-\tau_j}) - \mathbf{E}\{b_j(Y_{-\tau_j})|\mathcal{F}_{j-1}\})$$

$$+ \frac{1}{k} \sum_{1 \le j \le k} (\mathbf{E}\{b_j(Y_{-\tau_j})|\mathcal{F}_{j-1}\} - \mathbf{E}\{b_j(Y_0)|\mathcal{F}_{j-1}\})$$

$$+ \frac{1}{k} \sum_{1 \le j \le k} (\mathbf{E}\{b_j(Y_0)|\mathcal{F}_{j-1}\} - \mathbf{E}\{Y_0|\mathcal{F}_{j-1}\})$$

$$+ \frac{1}{k} \sum_{1 \le j \le k} (\mathbf{E}\{Y_0|\mathcal{F}_{j-1}\} - \mathbf{E}\{Y_0|X^0_{-\infty}, Y^{-1}_{-\infty}\})$$

$$= I_1 + I_2 + I_3 + I_4.$$

By (27.4),

$$I_2 = 0,$$

moreover,

$$|I_3| \le \frac{1}{k} \sum_{1 \le j \le k} \epsilon_j \to 0.$$

Since $\mathcal{F}_j \uparrow \mathcal{F}(X^0_{-\infty}, Y^{-1}_{-\infty})$ the martingale convergence theorem (Theorem A.4) implies that

$$\mathbf{E}\{Y_0|\mathcal{F}_{j-1}\} \to \mathbf{E}\{Y_0|X^0_{-\infty}, Y^{-1}_{-\infty}\} \quad \text{a.s.,}$$

so

$$I_4 \to 0 \quad \text{a.s.}$$

Observe that I_1 is an average of martingale differences. Apply Theorem A.6 for $c_i = 1$, then we have to show that

$$\sup_j \mathbf{E}\{(b_j(Y_{-\tau_j}) - \mathbf{E}\{b_j(Y_{-\tau_j})|\mathcal{F}_{j-1}\})^2\} < \infty.$$

Use again (27.4) then

$$\begin{aligned}
\mathbf{E}\{(b_j(Y_{-\tau_j}) - \mathbf{E}\{b_j(Y_{-\tau_j})|\mathcal{F}_{j-1}\})^2\} &\le \mathbf{E}\{b_j(Y_{-\tau_j})^2\} \\
&= \mathbf{E}\{\mathbf{E}\{b_j(Y_{-\tau_j})^2|\mathcal{F}_{j-1}\}\} \\
&= \mathbf{E}\{\mathbf{E}\{b_j(Y_0)^2|\mathcal{F}_{j-1}\}\} \\
&= \mathbf{E}\{b_j(Y_0)^2\} \\
&\le \mathbf{E}\{2(Y_0^2 + \epsilon_j^2)\} \\
&\le 2\mathbf{E}\{Y_0^2\} + 2\max_k \epsilon_k^2 < \infty.
\end{aligned}$$

The second statement is an obvious consequence of the first one. $\square$

27.4   Time Series Problem: Cesàro Consistency

Introduce the notation

$$(\mathbf{X}, \mathbf{Y}) = \{(X_i, Y_i)\}_{i=-\infty}^{\infty},$$

then

$$\hat{m}_{-n}(X_{-n}^0, Y_{-n}^{-1}) = \hat{m}_{-n}(\mathbf{X}, \mathbf{Y}).$$

Define the estimate

$$m_n(X_0^n, Y_0^{n-1}) = \hat{m}_{-n}(T^n(\mathbf{X}, \mathbf{Y})).$$

We show that m_n is Cesàro consistent a.s.

Theorem 27.4. *Assume that* $|Y| \leq L$ *a.s. Then, under the conditions of Theorem 27.3,*

$$\lim_{n \to \infty} \frac{1}{n} \sum_{i=1}^n (m_i - \mathbf{E}\{Y_i | X_0^i, Y_0^{i-1}\})^2 = 0 \quad a.s.$$

PROOF. Put

$$\mathbf{Z} = (\mathbf{X}, \mathbf{Y})$$

and

$$f_n(\mathbf{Z}) = (\hat{m}_{-n}(\mathbf{X}, \mathbf{Y}) - \mathbf{E}\{Y_0 | X_{-n}^0, Y_{-n}^{-1}\})^2.$$

Then Theorem 27.3 and the martingale convergence theorem (Theorem A.4) imply that

$$f_n(\mathbf{Z}) \to f(\mathbf{Z}) = 0 \quad \text{a.s.}$$

Because of $|Y| \leq L$ we have that $|f_n(\mathbf{Z})| \leq L$, so by Lemma 27.2,

$$\lim_{n \to \infty} \frac{1}{n} \sum_{i=1}^n f_i(T^i \mathbf{Z}) = \lim_{n \to \infty} \frac{1}{n} \sum_{i=1}^n (m_i - \mathbf{E}\{Y_i | X_0^i, Y_0^{i-1}\})^2 = \mathbf{E}f(\mathbf{Z}) = 0 \quad \text{a.s.}$$

$\square$

27.5   Time Series Problem: Universal Prediction

Unfortunately, the Morvai, Yakowitz, and Györfi (1996) estimator consumes data very rapidly. Morvai, Yakowitz, and Algoet (1997) proposed a very simple estimator which seems to be more efficient and is weakly consistent. In this section we introduce another Cesàro consistent estimator, which is derived from a sequential prediction of a real-valued sequence. At each time instant $i = 1, 2, \ldots$, the predictor is asked to guess the value of the next outcome y_i of a sequence of real numbers $y_1, y_2, \ldots$ with knowledge of the past (x_1^i, y_1^{i-1}). Thus, the predictor's estimate, at time i, is based on the value of (x_1^i, y_1^{i-1}). Formally, the strategy of the predictor is a sequence $g = \{g_i\}_{i=1}^\infty$ of decision functions

$$g_i : (\mathcal{R}^d)^i \times \mathcal{R}^{i-1} \to \mathcal{R}$$

and the prediction formed at time i is $g_i(x_1^i, y_1^{i-1})$. After n rounds of play, the *normalized cumulative prediction error* on the string x_1^n, y_1^n is

$$L_n(g) = \frac{1}{n} \sum_{i=1}^n (g_i(x_1^i, y_1^{i-1}) - y_i)^2.$$

The fundamental limit for the predictability of the sequence can be determined based on a result of Algoet (1994) who showed that, for any prediction strategy g and stationary ergodic process $\{X_n, Y_n\}_{-\infty}^\infty$,

$$\liminf_{n \to \infty} L_n(g) \geq L^* \quad \text{a.s.,}$$

where

$$L^* = \mathbf{E}\left\{ \left(Y_0 - \mathbf{E}\{Y_0 | X_{-\infty}^0, Y_{-\infty}^{-1}\}\right)^2 \right\}$$

is the minimal mean square error of any prediction for the value of Y_0 based on the infinite past $X_{-\infty}^0, Y_{-\infty}^{-1}$. We prove this if both $|Y|$ and $|g_i(X_1^i, Y_1^{i-1})|$ are bounded by L. Then for $\mathcal{F}_{i-1} = \mathcal{F}(X_1^i, Y_1^{i-1})$ we get that

$$
\begin{aligned}
L_n(g) \;=\; & \frac{1}{n} \sum_{i=1}^n (g_i(X_1^i, Y_1^{i-1}) - Y_i)^2 \\[2mm]
=\; & \frac{1}{n} \sum_{i=1}^n \left((g_i(X_1^i, Y_1^{i-1}) - Y_i)^2 - \mathbf{E}\{(g_i(X_1^i, Y_1^{i-1}) - Y_i)^2 | \mathcal{F}_{i-1}\} \right) \\[2mm]
& + \frac{1}{n} \sum_{i=1}^n \mathbf{E}\{(g_i(X_1^i, Y_1^{i-1}) - Y_i)^2 | \mathcal{F}_{i-1}\} \\[2mm]
\geq\; & \frac{1}{n} \sum_{i=1}^n \left((g_i(X_1^i, Y_1^{i-1}) - Y_i)^2 - \mathbf{E}\{(g_i(X_1^i, Y_1^{i-1}) - Y_i)^2 | \mathcal{F}_{i-1}\} \right) \\[2mm]
& + \frac{1}{n} \sum_{i=1}^n \mathbf{E}\{(\mathbf{E}\{Y_i | \mathcal{F}_{i-1}\} - Y_i)^2 | \mathcal{F}_{i-1}\},
\end{aligned}
$$

where the first term on the right-hand side is an average of bounded martingale difference, so it tends to 0 a.s. (Theorem A.6), while the second term tends to L^* a.s. (Theorem A.4).

This lower bound gives sense to the following definition:

Definition 27.3. *A prediction strategy g is called* universal with respect to a class $\mathcal{C}$ of stationary and ergodic processes $\{(X_n, Y_n)\}_{-\infty}^{\infty}$, *if for each process in the class,*

$$\lim_{n \to \infty} L_n(g) = L^* \quad a.s.$$

We introduce our prediction strategy for bounded ergodic processes. We assume that $|Y_0|$ is bounded by a constant $B > 0$, with probability one, and the bound B is known.

The prediction strategy is defined, at each time instant, as a convex combination of *elementary predictors*, where the weighting coefficients depend on the past performance of each elementary predictor.

We define an infinite array of elementary predictors $h^{(k,\ell)}$ $(k, \ell = 1, 2, \dots)$ as follows. Let $\mathcal{P}_\ell = \{A_{\ell,j}\}$ and $\mathcal{Q}_\ell = \{B_{\ell,j}\}$ be sequences of finite partitions of $\mathcal{R}^d$ and $\mathcal{R}$, respectively, and let G_ℓ and H_ℓ be the corresponding quantizers:

$$G_\ell(x) = j \text{ if } x \in A_{\ell,j}$$

and

$$H_\ell(y) = j \text{ if } y \in B_{\ell,j}.$$

Fix positive integers k, ℓ, and for each $k + 1, k$-long strings s, z of positive integers, define the partitioning regression function estimate

$$\widehat{E}_n^{(k,\ell)}(x_1^n, y_1^{n-1}, s, z) = \frac{\sum_{\{k < i < n : G_\ell(x_{i-k}^i) = s, H_\ell(y_{i-k}^{i-1}) = z\}} y_i}{\left| \{k < i < n : G_\ell(x_{i-k}^i) = s, H_\ell(y_{i-k}^{i-1}) = z\} \right|},$$

$n > k + 1$, where $0/0$ is defined to be 0.

Now we define the elementary predictor $h^{(k,\ell)}$ by

$$h_n^{(k,\ell)}(x_1^n, y_1^{n-1}) = \widehat{E}_n^{(k,\ell)}(x_1^n, y_1^{n-1}, G_\ell(x_{n-k}^n), H_\ell(y_{n-k}^{n-1})), \ n = 1, 2, \dots.$$

That is, $h^{(k,\ell)}$ quantizes the sequence x_1^n, y_1^n according to the partitions $\mathcal{P}_\ell$ and $\mathcal{Q}_\ell$, and looks for all appearances of the last-seen quantized strings $G_\ell(x_{n-k}^n), H_\ell(y_{n-k}^{n-1})$ of length $k+1, k$ in the past. Then it predicts according to the average of the y_i's following the string.

The proposed prediction algorithm proceeds as follows: let $\{q_{k,\ell}\}$ be a probability distribution on the set of all pairs (k, ℓ) of positive integers such that for all $k, \ell, q_{k,\ell} > 0$. Put $c = 8B^2$, and define the weights

$$w_{t,k,\ell} = q_{k,\ell} e^{-(t-1)L_{t-1}(h^{(k,\ell)})/c}$$

and their normalized values

$$v_{t,k,\ell} = \frac{w_{t,k,\ell}}{\sum_{i,j=1}^{\infty} w_{t,i,j}}.$$

The prediction strategy g is defined by

$$g_t(x_1^t, y_1^{t-1}) = \sum_{k,\ell=1}^{\infty} v_{t,k,\ell} h^{(k,\ell)}(x_1^t, y_1^{t-1}), \ t = 1, 2, \ldots. \tag{27.5}$$

Theorem 27.5. *Assume that the sequences of partitions $\mathcal{P}_\ell$ and $\mathcal{Q}_\ell$ are nested and asymptotically generate the Borel σ-field. Then the prediction scheme g defined above is universal with respect to the class of all ergodic processes such that $\mathbf{P}\{Y_i \in [-B, B]\} = 1$.*

One of the main ingredients of the proof is the following lemma, whose proof is a straightforward extension of standard arguments in the prediction theory of individual sequences.

Lemma 27.3. *Let $\tilde{h}_1, \tilde{h}_2, \ldots$ be a sequence of prediction strategies (experts), and let $\{q_k\}$ be a probability distribution on the set of positive integers. Assume that $\tilde{h}_i(x_1^n, y_1^{n-1}) \in [-B, B]$ and $y_1^n \in [-B, B]^n$. Define*

$$w_{t,k} = q_k e^{-(t-1)L_{t-1}(\tilde{h}_k)/c}$$

with $c \geq 8B^2$, and

$$v_{t,k} = \frac{w_{t,k}}{\sum_{i=1}^{\infty} w_{t,i}}.$$

If the prediction strategy $\tilde{g}$ is defined by

$$\tilde{g}_t(x_1^t, y_1^{t-1}) = \sum_{k=1}^{\infty} v_{t,k} \tilde{h}_k(x_1^t, y_1^{t-1})$$

then, for every $n \geq 1$,

$$L_n(\tilde{g}) \leq \inf_k \left(L_n(\tilde{h}_k) - \frac{c \ln q_k}{n} \right).$$

Here $-\ln 0$ is treated as ∞.

PROOF. Introduce $W_1 = 1$ and $W_t = \sum_{k=1}^{\infty} w_{t,k}$ for $t > 1$. First we show that, for each $t > 1$,

$$\left[\sum_{k=1}^{\infty} v_{t,k} \left(y_t - \tilde{h}_k(x_1^t, y_1^{t-1}) \right) \right]^2 \leq -c \ln \frac{W_{t+1}}{W_t}. \tag{27.6}$$

Note that

$$W_{t+1} = \sum_{k=1}^{\infty} w_{t,k} e^{-\left(y_t - \tilde{h}_k(x_1^t, y_1^{t-1})\right)^2/c} = W_t \sum_{k=1}^{\infty} v_{t,k} e^{-\left(y_t - \tilde{h}_k(x_1^t, y_1^{t-1})\right)^2/c},$$

so that

$$-c \ln \frac{W_{t+1}}{W_t} = -c \ln \left(\sum_{k=1}^{\infty} v_{t,k} e^{-\left(y_t - \tilde{h}_k(x_1^t, y_1^{t-1})\right)^2/c} \right).$$

Therefore, (27.6) becomes

$$\exp\left(\frac{-1}{c}\left[\sum_{k=1}^{\infty} v_{t,k}\left(y_t - \tilde{h}_k(x_1^t, y_1^{t-1})\right)\right]^2\right) \geq \sum_{k=1}^{\infty} v_{t,k} e^{-\left(y_t - \tilde{h}_k(x_1^t, y_1^{t-1})\right)^2/c},$$

which is implied by Jensen's inequality and the concavity of the function $F_t(z) = e^{-(y_t-z)^2/c}$ for $c \geq 8B^2$. Thus, (27.6) implies that

$$\begin{aligned}
nL_n(\tilde{g}) &= \sum_{t=1}^{n}\left(y_t - \tilde{g}(x_1^t, y_1^{t-1})\right)^2 \\
&= \sum_{t=1}^{n}\left[\sum_{k=1}^{\infty} v_{t,k}\left(y_t - \tilde{h}_k(x_1^t, y_1^{t-1})\right)\right]^2 \\
&\leq -c\sum_{t=1}^{n}\ln\frac{W_{t+1}}{W_t} \\
&= -c\ln W_{n+1} \\
&= -c\ln\left(\sum_{k=1}^{\infty} w_{n+1,k}\right) \\
&= -c\ln\left(\sum_{k=1}^{\infty} q_k e^{-nL_n(\tilde{h}_k)/c}\right) \\
&\leq -c\ln\left(\sup_k q_k e^{-nL_n(\tilde{h}_k)/c}\right) \\
&= \inf_k\left(-c\ln q_k + nL_n(\tilde{h}_k)\right),
\end{aligned}$$

which concludes the proof. $\square$

PROOF OF THEOREM 27.5. By a double application of the ergodic theorem, as $n \to \infty$,

$$\begin{aligned}
&\widehat{E}_n^{(k,\ell)}(X_1^n, Y_1^{n-1}, s, z) \\
&= \frac{\frac{1}{n}\sum_{\{k<i<n:G_\ell(X_{i-k}^i)=s, H_\ell(Y_{i-k}^{i-1})=z\}} Y_i}{\frac{1}{n}\left|\{k<i<n:G_\ell(X_{i-k}^i)=s, H_\ell(Y_{i-k}^{i-1})=z\}\right|} \\
&\to \frac{\mathbf{E}\{Y_0 I_{\{G_\ell(X_{-k}^0)=s, H_\ell(Y_{-k}^{-1})=z\}}\}}{\mathbf{P}\{G_\ell(X_{-k}^0)=s, H_\ell(Y_{-k}^{-1})=z\}} \\
&= \mathbf{E}\{Y_0|G_\ell(X_{-k}^0)=s, H_\ell(Y_{-k}^{-1})=z\} \quad \text{a.s.}
\end{aligned}$$

s and z may have finitely many values, therefore,

$$\lim_{n\to\infty}\sup_{s,z}|\widehat{E}_n^{(k,\ell)}(X_1^n, Y_1^{n-1}, s, z) - \mathbf{E}\{Y_0|G_\ell(X_{-k}^0)=s, H_\ell(Y_{-k}^{-1})=z\}| = 0$$

a.s. Thus, by Lemma 27.2, as $n \to \infty$,

$$
\begin{aligned}
L_n(h^{(k,\ell)}) &= \frac{1}{n}\sum_{i=1}^{n}(h^{(k,\ell)}(X_1^i, Y_1^{i-1}) - Y_i)^2 \\
&= \frac{1}{n}\sum_{i=1}^{n}(\widehat{E}_n^{(k,\ell)}(X_1^i, Y_1^{i-1}, X_{i-k}^i, Y_{i-k}^{i-1}) - Y_i)^2 \\
&\to \mathbf{E}\left\{\left(Y_0 - \mathbf{E}\{Y_0|G_\ell(X_{-k}^0), H_\ell(Y_{-k}^{-1})\}\right)^2\right\} \\
&\overset{\text{def}}{=} \epsilon_{k,\ell} \quad \text{a.s.}
\end{aligned}
$$

Since the partitions $\mathcal{P}_\ell$ and $\mathcal{Q}_\ell$ are nested, $\mathbf{E}\left\{Y_0|G_\ell(X_{-k}^0), H_\ell(Y_{-k}^{-1})\right\}$ is a martingale indexed by the pair (k, ℓ). Thus, by the martingale convergence theorem (Theorem A.4) $\epsilon_{k,\ell}$ converges and, since the partitions asymptotically generate the Borel σ-field, the limit is L^*:

$$
\lim_{k,\ell \to \infty} \epsilon_{k,\ell} = \mathbf{E}\left\{\left(Y_0 - \mathbf{E}\{Y_0|X_{-\infty}^0, Y_{-\infty}^{-1}\}\right)^2\right\} = L^*.
$$

Now by Lemma 27.3,

$$
L_n(g) \le \inf_{k,\ell}\left(L_n(h^{(k,\ell)}) - \frac{c \ln q_{k,\ell}}{n}\right), \tag{27.7}
$$

and therefore,

$$
\begin{aligned}
\limsup_{n \to \infty} L_n(g) &\le \limsup_{n \to \infty} \inf_{k,\ell}\left(L_n(h^{(k,\ell)}) - \frac{c \ln q_{k,\ell}}{n}\right) \\
&\le \inf_{k,\ell} \limsup_{n \to \infty}\left(L_n(h^{(k,\ell)}) - \frac{c \ln q_{k,\ell}}{n}\right) \\
&\le \inf_{k,\ell} \limsup_{n \to \infty} L_n(h^{(k,\ell)}) \\
&= \inf_{k,\ell} \epsilon_{k,\ell} \\
&= \lim_{k,\ell \to \infty} \epsilon_{k,\ell} \\
&= L^* \quad \text{a.s.,}
\end{aligned}
$$

and the proof of the theorem is finished. $\square$

Theorem 27.5 shows that, asymptotically, the predictor g_t defined by (27.5) predicts as well as the optimal predictor given by the regression function $\mathbf{E}\{Y_t|X_{-\infty}^t, Y_{-\infty}^{t-1}\}$. In fact, g_t gives a Cesàro consistent estimate of the regression function in the following sense:

Corollary 27.1. *Under the conditions of Theorem 27.5,*

$$
\lim_{n \to \infty} \frac{1}{n}\sum_{i=1}^{n}\left(\mathbf{E}\{Y_i|X_{-\infty}^i, Y_{-\infty}^{i-1}\} - g_i(X_1^i, Y_1^{i-1})\right)^2 = 0 \quad \text{a.s.}
$$

PROOF. By Theorem 27.5,

$$\lim_{n \to \infty} \frac{1}{n} \sum_{i=1}^{n} \left(Y_i - g_i(X_1^i, Y_1^{i-1})\right)^2 = L^* \quad \text{a.s.}$$

Consider the following decomposition:

$$\left(Y_i - g_i(X_1^i, Y_1^{i-1})\right)^2$$
$$= \left(Y_i - \mathbf{E}\{Y_i|X_{-\infty}^i, Y_{-\infty}^{i-1}\}\right)^2$$
$$+ 2\left(Y_i - \mathbf{E}\{Y_i|X_{-\infty}^i, Y_{-\infty}^{i-1}\}\right)\left(\mathbf{E}\{Y_i|X_{-\infty}^i, Y_{-\infty}^{i-1}\} - g_i(X_1^i, Y_1^{i-1})\right)$$
$$+ \left(\mathbf{E}\{Y_i|X_{-\infty}^i, Y_{-\infty}^{i-1}\} - g_i(X_1^i, Y_1^{i-1})\right)^2.$$

Then the ergodic theorem implies that

$$\lim_{n \to \infty} \frac{1}{n} \sum_{i=1}^{n} \left(Y_i - \mathbf{E}\{Y_i|X_{-\infty}^i, Y_{-\infty}^{i-1}\}\right)^2 = L^* \quad \text{a.s.}$$

It remains to show that

$$\frac{1}{n} \sum_{i=1}^{n} \left(Y_i - \mathbf{E}\{Y_i|X_{-\infty}^i, Y_{-\infty}^{i-1}\}\right)\left(\mathbf{E}\{Y_i|X_{-\infty}^i, Y_{-\infty}^{i-1}\} - g_i(X_1^i, Y_1^{i-1})\right)$$

converges to 0 a.s., which is a consequence of Theorem A.6 with $c_i = 1$ since the martingale differences

$$Z_i = \left(Y_i - \mathbf{E}\{Y_i|X_{-\infty}^i, Y_{-\infty}^{i-1}\}\right)\left(\mathbf{E}\{Y_i|X_{-\infty}^i, Y_{-\infty}^{i-1}\} - g_i(X_1^i, Y_1^{i-1})\right)$$

are bounded by $4B^2$. $\quad\square$

27.6   Estimating Smooth Regression Functions

For regression function estimation from ergodic data there are similar negative findings:

Theorem 27.6. (NOBEL (1999)). *There is no weakly consistent regression estimate for all stationary and ergodic sequences $\{(X_i, Y_i)\}$ with $0 \leq X_i \leq 1$ and $Y_i \in \{0, 1\}$.*

Theorem 27.7. (YAKOWITZ AND HEYDE (1998)). *There is no weakly consistent autoregression estimate for all stationary and ergodic sequences $\{(X_i, Y_i)\}$, where $Y_i = X_{i+1}$.*

Because of these theorems we assume some smoothness on the regression function. Now we attack the problem of estimating the regression function $m(x)$ which combines partitioning estimation with a series expansion.

Let $\mathcal{P}_k = \{A_{k,i}\}$ be a nested cubic partition of $\mathcal{R}^d$ with volume $(2^{-k-2})^d$. Define $A_k(x)$ to be the partition cell of $\mathcal{P}_k$ in which x falls. Take

$$M_k(x) := \mathbf{E}\{Y|X \in A_k(x)\}. \tag{27.8}$$

By Lemma 24.10,

$$M_k(x) \to m(x) \tag{27.9}$$

for μ-almost all $x \in \mathcal{R}^d$. Motivated by (27.9) our analysis depends on the representation,

$$m(x) = M_1(x) + \sum_{k=2}^{\infty} \Delta_k(x) = \lim_k M_k(x)$$

for μ-almost all $x \in \mathcal{R}^d$, where

$$\Delta_k(x) = M_k(x) - M_{k-1}(x).$$

Now, for integer $k > 1$ and real $L > 0$, define

$$\Delta_{k,L}(x) = \text{sign}(M_k(x) - M_{k-1}(x)) \min(|M_k(x) - M_{k-1}(x)|, L2^{-k}).$$

Let

$$m_L(x) := M_1(x) + \sum_{i=1}^{\infty} \Delta_{i,L}(x).$$

Notice that $|\Delta_{i,L}| \le L2^{-i}$, hence $m_L(x)$ is well-defined for all $x \in \mathcal{R}^d$.

The crux of the truncated partitioning estimate is inference of the terms $\Delta_k(x) = M_k(x) - M_{k-1}(x)$. Define

$$\hat{M}_{k,n}(x) := \frac{\sum_{j=1}^{n} Y_j I_{\{X_j \in A_k(x)\}}}{\sum_{j=1}^{n} I_{\{X_j \in A_k(x)\}}}.$$

Now for $k > 1$, put

$$\hat{\Delta}_{k,n,L}(x) = \text{sign}(\hat{M}_{k,n}(x) - \hat{M}_{k-1,n}(x)) \min(|\hat{M}_{k,n}(x) - \hat{M}_{k-1,n}(x)|, L2^{-k})$$

and

$$\hat{m}_{n,L}(x) = \hat{M}_{1,n}(x) + \sum_{k=2}^{N_n} \hat{\Delta}_{k,n,L}(x).$$

Theorem 27.8. *Let $\{(X_i, Y_i)\}$ be a stationary ergodic time series. Assume $N_n \to \infty$. Then, for μ-almost all $x \in \mathcal{R}^d$,*

$$\hat{m}_{n,L}(x) \to m_L(x) \quad a.s. \tag{27.10}$$

If the support S of μ is a bounded subset of $\mathcal{R}^d$, then

$$\sup_{x \in S} |\hat{m}_{n,L}(x) - m_L(x)| \to 0 \quad a.s. \tag{27.11}$$

If in addition, either:

(i) $|Y| \leq D < \infty$ (D need not be known); or
(ii) μ is of bounded support,
then

$$\int (\hat{m}_{n,L}(x) - m_L(x))^2 \mu(dx) \to 0 \quad a.s. \tag{27.12}$$

PROOF. Define the support S of μ as

$$S := \{x \in \mathcal{R}^d : \mu(A_k(x)) > 0 \text{ for all } k \geq 1\}.$$

Then $\mu(S) = 1$. Assume $\mu(A_{k,i}) > 0$. Then by the ergodic theorem, as $n \to \infty$,

$$\frac{\sum_{j=1}^{n} I_{\{X_j \in A_{k,i}\}}}{n} \to \mathbf{P}\{X \in A_{k,i}\} = \mu(A_{k,i}) \quad \text{a.s.}$$

Similarly,

$$\frac{\sum_{j=1}^{n} I_{\{X_j \in A_{k,i}\}} Y_j}{n} \to \mathbf{E}\{Y I_{\{X \in A_{k,i}\}}\} = \int_{A_{k,i}} m(z) \mu(dz) \quad \text{a.s.},$$

which is finite since $\mathbf{E}Y$ is finite. Now apply this for $A_k(x) = A_{k,i}$. One can write $\hat{M}_{k,n}(x)$ as the ratio of these two almost surely convergent sequences. Thus, for all $x \in S$,

$$\hat{M}_{k,n}(x) \to M_k(x) \quad \text{a.s.}$$

and so

$$\hat{\Delta}_{k,n,L}(x) \to \Delta_{k,L}(x) \quad \text{a.s.}$$

Let integer $R > 1$ be arbitrary. Let n be so large that $N_n > R$. For all $x \in S$,

$$|\hat{m}_{n,L}(x) - m_L(x)|$$

$$\leq \quad |\hat{M}_{1,n}(x) - M_1(x)| + \sum_{k=2}^{N_n} |\hat{\Delta}_{k,n,L}(x) - \Delta_{k,L}(x)| + \sum_{k=N_n+1}^{\infty} |\Delta_{k,L}(x)|$$

$$\leq \quad |\hat{M}_{1,n}(x) - M_1(x)| + \sum_{k=2}^{R} |\hat{\Delta}_{k,n,L}(x) - \Delta_{k,L}(x)|$$

$$+ \sum_{k=R+1}^{\infty} (|\hat{\Delta}_{k,n,L}(x)| + |\Delta_{k,L}(x)|)$$

$$\leq \quad |\hat{M}_{1,n}(x) - M_1(x)| + \sum_{k=2}^{R} |\hat{\Delta}_{k,n,L}(x) - \Delta_{k,L}(x)| + 2L \sum_{k=R+1}^{\infty} 2^{-k}$$

$$\leq \quad |\hat{M}_{1,n}(x) - M_1(x)| + \sum_{k=2}^{R} |\hat{\Delta}_{k,n,L}(x) - \Delta_{k,L}(x)| + L2^{-(R-1)}.$$

Now, for all $x \in S$,

$$|\hat{M}_{1,n}(x) - M_1(x)| + \sum_{k=2}^{R} |\hat{\Delta}_{k,n,L}(x) - \Delta_{k,L}(x)| \to 0 \quad \text{a.s.}$$

Then

$$\limsup_{n\to\infty} |\hat{m}_{n,L}(x) - m_L(x)| \leq L2^{-(R-1)} \quad \text{a.s.}$$

Since R was arbitrary, (27.10) is proved. Now we prove (27.11). Assume the support S of μ is bounded. Let $\mathcal{A}_k$ denote the set of cells from partition $\mathcal{P}_k$ with nonempty intersection with S. That is, define

$$\mathcal{A}_k = \{A \in \mathcal{P}_k : A \cap S \neq \emptyset\}.$$

Since S is bounded, $\mathcal{A}_k$ is a finite set. For $A \in \mathcal{P}_k$, let $a(A)$ be the center of A. Then

$$\sup_{x\in S} \left(|\hat{M}_{1,n}(x) - M_1(x)| + \sum_{k=2}^{R} |\hat{\Delta}_{k,n,L}(x) - \Delta_{k,L}(x)| \right)$$

$$\leq \quad \max_{A\in\mathcal{A}_1} |\hat{M}_{1,n}(a(A)) - M_1(a(A))|$$

$$+ \sum_{k=2}^{R} \max_{A\in\mathcal{A}_k} |\hat{\Delta}_{k,n,L}(a(A)) - \Delta_{k,L}(a(A))|$$

$$\to \quad 0$$

keeping in mind that only finitely many terms are involved in the maximization operation. The rest of the proof goes virtually as before. Now we prove (27.12),

$$|\hat{m}_{n,L}(x) - m_L(x)|^2$$

$$\leq \quad 2\left(|\hat{M}_{1,n}(x) - M_1(x)|^2 + \left(M_1(x) + \sum_{k=2}^{N_n} \hat{\Delta}_{k,n,L}(x) - m_L(x) \right)^2 \right).$$

If condition (i) holds, then for the first term we have dominated convergence

$$|\hat{M}_{1,n}(x) - M_1(x)|^2 \leq (2D)^2,$$

and for the second one, too,

$$\left| M_1(x) + \sum_{k=2}^{N_n} \hat{\Delta}_{k,n}(x) - m_L(x) \right| \leq \sum_{k=2}^{\infty} (|\hat{\Delta}_{k,n}| + |\Delta_{k,L}|)$$

$$\leq \quad 2L,$$

and thus (27.12) follows by Lebesgue's dominated convergence theorem,

$$0 = \int \lim_{n \to \infty} |\hat{m}_n(x) - m_L(x)|^2 \mu(dx) = \lim_{n \to \infty} \int |\hat{m}_n(x) - m_L(x)|^2 \mu(dx)$$

a.s. If condition (ii) holds then (27.12) follows from (27.11).   □

Corollary 27.2. *Assume $m(x)$ is Lipschitz continuous with Lipschitz constant $L/\sqrt{d}$. Then Theorem 27.7 holds with $m_L(x) = m(x)$ for $x \in S$.*

PROOF. Since $m(x)$ is Lipschitz with constant $L/\sqrt{d}$, for $x \in S$,

$$
\begin{aligned}
|M_k(x) - m(x)| &\leq \left| \frac{\int_{A_k(x)} m(y)\mu(dy)}{\mu(A_k(x))} - m(x) \right| \\
&\leq \frac{1}{\mu(A_k(x))} \int_{A_k(x)} |m(y) - m(x)|\mu(dy) \\
&\leq \frac{1}{\mu(A_k(x))} \int_{A_k(x)} (L/\sqrt{d})(2^{-k-2}\sqrt{d})\mu(dy) \\
&= L2^{-k-2}.
\end{aligned}
$$

For $x \in S$ we get

$$
\begin{aligned}
|M_k(x) - M_{k-1}(x)| &\leq |M_k(x) - m(x)| + |m(x) - M_{k-1}(x)| \\
&\leq L2^{-k-2} + L2^{-k-1} \\
&< L2^{-k}.
\end{aligned}
$$

Hence for μ-almost all $x \in \mathcal{R}^d$,

$$m_L(x) = M_1(x) + \sum_{k=2}^{\infty} \Delta_k(x) = m(x),$$

and Corollary 27.2 is proved.   □

If there is no truncation, that is, $L = \infty$, then $\hat{m}_n = \hat{M}_{N_n.n}$; that is, $\hat{m}_n$ is the standard partitioning estimate.

Our consistency is not universal, however, since m should be Lipschitz continuous.

N_n can be data dependent, provided $N_n \to \infty$ a.s.

In order to introduce the truncated kernel estimate let $K(x)$ be a non-negative kernel function with

$$bI_{\{x \in S_{O,r}\}} \leq K(x) \leq I_{\{x \in S_{O,1}\}},$$

where $0 < b \leq 1$ and $0 < r \leq 1$. ($S_{z,r}$ denotes the ball around z with radius r.) Choose

$$h_k = 2^{-k-2}$$

and

$$M_k^*(x) = \frac{\mathbf{E}\{YK(\frac{X-x}{h_k})\}}{\mathbf{E}\{K(\frac{X-x}{h_k})\}} = \frac{\int m(z)K(\frac{z-x}{h_k})\mu(dz)}{\int K(\frac{z-x}{h_k})\mu(dz)}.$$

Let

$$\Delta_k^*(x) = M_k^*(x) - M_{k-1}^*(x).$$

By Lemma 24.8 we have (27.9). Now for $k > 1$, define

$$\Delta_{k.L}^*(x) = \text{sign}(M_k^*(x) - M_{k-1}^*(x))\min(|M_k^*(x) - M_{k-1}^*(x)|, L2^{-k}).$$

Let

$$m_L^*(x) := M_1^*(x) + \sum_{i=1}^{\infty} \Delta_{i,L}^*(x).$$

Put

$$\hat{M}_{k,n}^*(x) := \frac{\sum_{j=1}^{n} Y_j K(\frac{X_j-x}{h_k})}{\sum_{j=1}^{n} K(\frac{X_j-x}{h_k})}.$$

For $k > 1$, put

$$\hat{\Delta}_{k,n,L}^*(x) = \text{sign}(\hat{M}_{k,n}^*(x) - \hat{M}_{k-1,n}^*(x))\min(|\hat{M}_{k,n}^*(x) - \hat{M}_{k-1,n}^*(x)|, L2^{-k})$$

and

$$\hat{m}_{n,L}^*(x) = \hat{M}_{1,n}^*(x) + \sum_{k=2}^{N_n} \hat{\Delta}_{k,n,L}^*(x).$$

Theorem 27.9. *Assume $N_n \to \infty$. Then, for μ-almost all $x \in \mathcal{R}^d$,*

$$\hat{m}_n^*(x) \to m_L^*(x) \quad a.s.$$

If, in addition, $|Y| \leq D < \infty$ (D need not be known), then

$$\int (\hat{m}_n^*(x) - m_L^*(x))^2 \mu(dx) \to 0 \quad a.s.$$

Corollary 27.3. *Assume $m(x)$ is Lipschitz continuous with Lipschitz constant $L/2$. Then Theorem 27.9 holds with $m_L^*(x) = m(x)$.*

27.7 Bibliographic Notes

Under various mixing conditions on the data D_n the consistency of the conventional regression estimates has been summarized in Györfi et al. (1989). Lemma 27.1 is due to Birkhoff (1931). Versions of Lemma 27.2 have been proved by Algoet (1994), Barron (1985), Breiman (1957) and Maker (1940).

Bailey (1976) showed that the problem of dynamic forecasting cannot be solved. Bailey's counterexample uses the technique of cutting and stacking developed by Ornstein (1974) (see also Shields (1991)). Bailey hasn't published his result, so Ryabko (1988) rediscovered this result with a much simpler proof. Theorem 27.1 is due to Bailey (1976) and Ryabko (1988), while the current proof together with Theorem 27.2 is from Győrfi, Morvai, and Yakowitz (1998) based on a clever counterexample of Ryabko (1988). For the static forecasting binary sequences Ornstein (1978) gave an estimator, which has been extended to real-valued data by Algoet (1992). A much simpler estimate can be found in Morvai, Yakowitz, and Győrfi (1996) (cf. Theorem 27.3). Theorem 27.5 has been proved in Győrfi, and Lugosi (2002). Lemma 27.3 is due to Kivinen and Warmuth (1999) and Singer and Feder (1999). Concerning the prediction of individual sequences we refer to Haussler, Kivinen, and Warmuth (1998). Theorems 27.8 and 27.9 and their corollaries are from Yakowitz et al. (1999).

Problems and Exercises

PROBLEM 27.1. Prove Theorem 27.9.

Appendix A
Tools

A.1   A Denseness Result

The universal consistency of some nonparametric regression estimates is shown such that first the consistency is proved for continuous regression functions, and then it is extended to arbitrary regression functions. This extension is possible because of the following theorem:

Theorem A.1. *For any $p \geq 1$ and any probability measure μ, the set of continuous functions of bounded support is dense in $L_p(\mu)$, i.e. for any $f \in L_p(\mu)$ and $\epsilon > 0$ there is a continuous function g with compact support such that*

$$\int |f(x) - g(x)|^p \, \mu(dx) \leq \epsilon.$$

PROOF. Without loss of generality assume that $f \geq 0$. The case of arbitrary f can be handled similarily as the case $f \geq 0$ by using the decomposition $f = f^+ - f^-$ where $f^+ = \max\{f, 0\}$ and $f^- = -\min\{f, 0\}$. $f \in L_p(\mu)$ implies that there is an open sphere S centered at the origin with radius R and an integer $K > 0$ such that

$$\int \left| f(x) - \min\{f(x), K\} I_{\{x \in S\}} \right|^p \mu(dx) \leq \epsilon/2^p.$$

Because of $|a + b|^p \leq 2^{p-1}(|a|^p + |b|^p)$, we have that

$$\int |f(x) - g(x)|^p \mu(dx) \leq 2^{p-1} \int \left| f(x) - \min\{f(x), K\} I_{\{x \in S\}} \right|^p \mu(dx)$$

$$+ 2^{p-1} \int \left| \min\{f(x), K\} I_{\{x \in S\}} - g(x) \right|^p \mu(dx),$$

and it suffices to show that there is a continuous function g with support in the closure $\bar{S}$ of S such that

$$\int_S \left| \min\{f(x), K\} I_{\{x \in S\}} - g(x) \right|^p \mu(dx) \leq \epsilon/2^p.$$

Without loss of generality, assume that $K = 1$. Put

$$f^*(x) = \min\{f(x), 1\} I_{\{x \in S\}}.$$

Using the technique of Lusin's theorem we show that there is a continuous function g with support in $\bar{S}$ such that $0 \leq g \leq 1$ and

$$\mu\left(\{x \in \bar{S} : f^*(x) \neq g(x)\}\right) \leq \epsilon/2^p. \tag{A.1}$$

Obviously, this will imply the theorem since

$$\int_{\bar{S}} |f^*(x) - g(x)|^p \mu(dx) \leq \mu\left(\{x \in \bar{S}; f^*(x) \neq g(x)\}\right) \leq \epsilon/2^p.$$

We have now that $0 \leq f^* \leq 1$. A function s is called a *simple function* if its range consists of finitely many points in $[0, \infty)$. We can construct a monotonically increasing sequence of simple functions $s_1 \leq s_2 \leq \cdots \leq f^*$ such that $s_n(x) \to f^*(x)$ as $n \to \infty$ for every $x \in S$. Indeed, for $n = 1, 2, \ldots$, define the sets

$$E_{n,i} = \left\{ x \in \bar{S} : \frac{i-1}{2^n} \leq f^*(x) < \frac{i}{2^n} \right\} \quad (1 \leq i < 2^n)$$

and

$$E_{n,2^n} = \left\{ x \in \bar{S} : \frac{2^n - 1}{2^n} \leq f^*(x) \leq 1 \right\}$$

and the simple function

$$s_n = \sum_{i=1}^{2^n} \frac{i-1}{2^n} I_{E_{n,i}}.$$

Sets $E_{n,i}$ are inverse images of intervals through a measurable function and thus are measurable. Clearly the sequence s_n is monotonically increasing, $s_n \leq f^*$, and $|s_n(x) - f^*(x)| \leq 2^{-n}$ for all $x \in S$.

Now define $t_1 = s_1$ and $t_n = s_n - s_{n-1}$ $(n = 2, 3, \ldots)$. By definition of s_n, $s_n(x) \neq s_{n-1}(x)$ implies $s_n(x) - s_{n-1}(x) = 2^{-n}$, therefore $2^n t_n$ is the

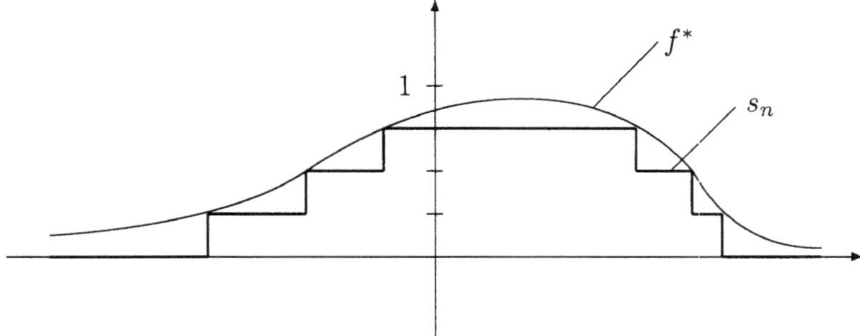

Figure A.1. Approximation by step function.

indicator function of a set $T_n \subset S$ and

$$f^*(x) = \sum_{n=1}^{\infty} t_n(x).$$

Then there exist compact sets U_n and open sets V_n such that $U_n \subset T_n \subset V_n \subset S \subset \bar{S}$ and $\mu(V_n - U_n) < 2^{-n}\epsilon/2^p$. Next we need a special version of Urysohn's lemma which states that for any compact set U and bounded open set V such that $U \subset V$ there exists a continuous function $h : \mathcal{R}^d \to [0,1]$ such that $h(x) = 1$ for $x \in U$ and $h(x) = 0$ for $x \in \mathcal{R}^d \setminus V$. In order to show this special case of Urysohn's lemma for any set A introduce the function

$$d(x, A) = \inf_{z \in A} \|x - z\|.$$

Then $d(x, A)$ (as a function of x) is continuous, and $d(x, A) = 0$ iff x lies in the closure of A. Such a function h can be defined as

$$h(x) = \frac{d(x, \mathcal{R}^d \setminus V)}{d(x, U) + d(x, \mathcal{R}^d \setminus V)}.$$

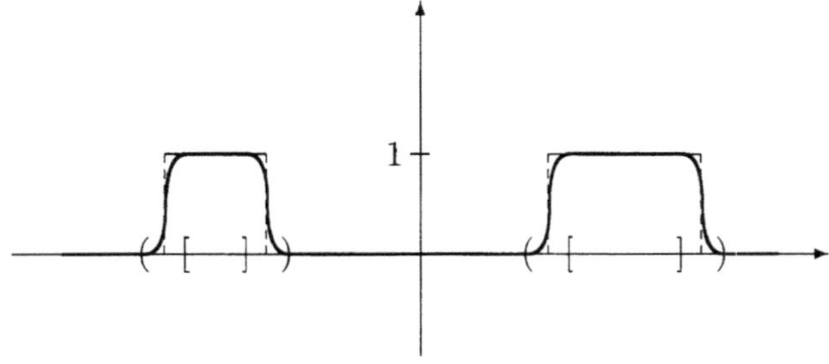

Figure A.2. Approximation of a step function by a continuous function.

Thus by Urysohn's lemma there are continuous functions $h_n : \mathcal{R}^d \to [0,1]$ that assume value 1 on U_n and are zero outside V_n. Let

$$g(x) = \sum_{n=1}^{\infty} 2^{-n} h_n(x).$$

Since the functions h_n are bounded by 1 this series converges uniformly on S and thus g is continuous. Also its support lies in $\bar{S}$. Since $2^{-n} h_n(x) = t_n(x)$ except in $V_n - U_n$, we obtain $g(x) = f^*(x)$ except in $\bigcup_{n=1}^{\infty}(V_n - U_n)$, but

$$\mu\left(\bigcup_{n=1}^{\infty}(V_n - U_n)\right) \leq \sum_{n=1}^{\infty}\mu(V_n - U_n) \leq \epsilon/(2^p)\sum_{n=1}^{\infty}2^{-n} = \epsilon/2^p,$$

which proves (A.1).   □

Corollary A.1. *For any $p \geq 1$ and any probability measure μ, the set of infinitely often differentiable functions of bounded support is dense in $L_p(\mu)$.*

PROOF. We notice that the function g in the proof of Theorem A.1 can be approximated arbitrarily well by an infinitely often differentiable function having bounded support with respect to the supremum norm and thus also with respect to $\|\cdot\|_{L_p(\mu)}$. In fact, choose $K \in C^{\infty}(\mathcal{R}^d)$ with $K \geq 0$, $\int K(x)\,dx = 1$, $K(x) = 0$ for $\|x\| > 1$, e.g.,

$$K(x) = \left\{ \begin{array}{ll} c \cdot \exp\left(-1/(1 - \|x\|)\right) & \text{if } \|x\| < 1, \\ 0 & \text{else,} \end{array}\right.$$

with suitable $c > 0$, and set

$$g_h(x) = \frac{1}{h^d}\int_{\mathcal{R}^d} K\left(\frac{x - z}{h}\right) g(z)\,dz, \quad h > 0.$$

Then g_h is infinitely often differentiable, has bounded support, and satisfies

$$\sup_{x \in \mathcal{R}^d} |g(x) - g_h(x)| \leq \sup_{x,z \in \mathcal{R}^d, \|x-z\| \leq h} |g(x) - g(z)| \to 0 \quad (h \to 0).$$

□

A.2   Inequalities for Independent Random Variables

Lemma A.1. (CHERNOFF (1952)). *Let B be a binomial random variable with parameters n and p. Then, for $1 > \epsilon > p > 0$,*

$$\mathbf{P}\{B > n\epsilon\} \leq e^{-n\left[\epsilon \log \frac{\epsilon}{p} + (1-\epsilon)\log\frac{1-\epsilon}{1-p}\right]} \leq e^{-n[p - \epsilon + \epsilon\log(\epsilon/p)]}$$

and, for $0 < \epsilon < p < 1$,

$$\mathbf{P}\{B < n\epsilon\} \le e^{-n\left[\epsilon \log \frac{\epsilon}{p} + (1-\epsilon) \log \frac{1-\epsilon}{1-p}\right]} \le e^{-n[p-\epsilon+\epsilon \log(\epsilon/p)]}.$$

PROOF. We proceed by Chernoff's exponential bounding method. In particular, for arbitrary $s > 0$,

$$
\begin{aligned}
\mathbf{P}\{B > n\epsilon\} &= \mathbf{P}\{sB > sn\epsilon\} \\
&= \mathbf{P}\{e^{sB} > e^{sn\epsilon}\} \\
&\le e^{-sn\epsilon}\mathbf{E}\{e^{sB}\} \\
&\qquad \text{(by the Markov inequality)} \\
&= e^{-sn\epsilon} \sum_{k=0}^{n} e^{sk} \binom{n}{k} p^k (1-p)^{n-k} \\
&= e^{-sn\epsilon}(e^s p + 1 - p)^n \\
&= [e^{-s\epsilon}(e^s p + 1 - p)]^n.
\end{aligned}
$$

Next choose s such that

$$e^s = \frac{\epsilon}{1-\epsilon}\frac{1-p}{p}.$$

With this value we get

$$
\begin{aligned}
e^{-s\epsilon}(e^s p + 1 - p) &= e^{-\epsilon \cdot \log\left(\frac{\epsilon}{1-\epsilon}\frac{1-p}{p}\right)} \cdot \left(\frac{\epsilon}{1-\epsilon}\frac{1-p}{p} \cdot p + 1 - p\right) \\
&= e^{-\epsilon \cdot \log\left(\frac{\epsilon}{p}\frac{1-p}{1-\epsilon}\right)} \cdot \left(\epsilon \cdot \frac{1-p}{1-\epsilon} + 1 - p\right) \\
&= e^{-\epsilon \cdot \log \frac{\epsilon}{p} - \epsilon \cdot \log \frac{1-p}{1-\epsilon} + \log \frac{1-p}{1-\epsilon}} \\
&= e^{-\epsilon \cdot \log \frac{\epsilon}{p} + (1-\epsilon) \cdot \log \frac{1-p}{1-\epsilon}},
\end{aligned}
$$

which implies the first inequality.

The second inequality follows from

$$
\begin{aligned}
(1-\epsilon)\log \frac{1-\epsilon}{1-p} &= -(1-\epsilon)\log \frac{1-p}{1-\epsilon} \\
&= -(1-\epsilon)\log\left(1 + \frac{\epsilon - p}{1-\epsilon}\right) \\
&\ge -(1-\epsilon) \cdot \frac{\epsilon - p}{1-\epsilon} \\
&\qquad \text{(by } \log(1+x) \le x) \\
&= p - \epsilon.
\end{aligned}
$$

To prove the second half of the lemma, observe that $n - B$ is a binomial random variable with parameters n and $1 - p$. Hence for $\epsilon < p$ the results of the first step imply that

$$
\begin{aligned}
\mathbf{P}\{B < n\epsilon\} &= \mathbf{P}\{n - B > n(1 - \epsilon)\} \\
&\leq e^{-n\left[(1-\epsilon)\log\frac{1-\epsilon}{1-p} + \epsilon\log\frac{\epsilon}{p}\right]} \\
&= e^{-n\left[\epsilon\log\frac{\epsilon}{p} + (1-\epsilon)\log\frac{1-\epsilon}{1-p}\right]} \\
&\leq e^{-n[p - \epsilon + \epsilon\log(\epsilon/p)]}.
\end{aligned}
$$

$\square$

Lemma A.2. (BERNSTEIN (1946)). *Let* $X_1, \ldots, X_n$ *be independent real-valued random variables, let* $a, b \in \mathcal{R}$ *with* $a < b$, *and assume that* $X_i \in [a, b]$ *with probability one* $(i = 1, \ldots, n)$. *Let*

$$
\sigma^2 = \frac{1}{n}\sum_{i=1}^{n}\mathbf{Var}\{X_i\} > 0.
$$

Then, for all $\epsilon > 0$,

$$
\mathbf{P}\left\{\left|\frac{1}{n}\sum_{i=1}^{n}(X_i - \mathbf{E}\{X_i\})\right| > \epsilon\right\} \leq 2e^{-\frac{n\epsilon^2}{2\sigma^2 + 2\epsilon(b-a)/3}}.
$$

PROOF. Set $Y_i = X_i - \mathbf{E}\{X_i\}$ $(i = 1, \ldots, n)$. Then we have, with probability one,

$$
|Y_i| \leq b - a \quad \text{and} \quad \mathbf{E}\{Y_i^2\} = \mathbf{Var}\{X_i\} \quad (i = 1, \ldots, n).
$$

By Chernoff's exponential bounding method we get, for arbitrary $s > 0$,

$$
\begin{aligned}
\mathbf{P}\left\{\frac{1}{n}\sum_{i=1}^{n}(X_i - \mathbf{E}\{X_i\}) > \epsilon\right\} &= \mathbf{P}\left\{\frac{1}{n}\sum_{i=1}^{n}Y_i > \epsilon\right\} \\
&= \mathbf{P}\left\{s\sum_{i=1}^{n}Y_i - sn\epsilon > 0\right\} \\
&\leq \mathbf{E}\left\{e^{s\sum_{i=1}^{n}Y_i - sn\epsilon}\right\} \\
&= e^{-sn\epsilon}\prod_{i=1}^{n}\mathbf{E}\{e^{sY_i}\},
\end{aligned}
$$

by the independence of Y_i's. Because of $|Y_i| \leq b - a$ a.s.

$$
e^{sY_i} = 1 + sY_i + \sum_{j=2}^{\infty}\frac{(sY_i)^j}{j!}
$$

$$\leq \; 1 + sY_i + \sum_{j=2}^{\infty} \frac{s^j Y_i^2 (b-a)^{j-2}}{2 \cdot 3^{j-2}}$$

$$= \; 1 + sY_i + \frac{s^2 Y_i^2}{2} \sum_{j=2}^{\infty} \left(\frac{s\,(b-a)}{3} \right)^{j-2}$$

$$= \; 1 + sY_i + \frac{s^2 Y_i^2}{2} \frac{1}{1 - s(b-a)/3}$$

if $|s(b-a)/3| < 1$. This, together with $\mathbf{E}\{Y_i\} = 0$ $(i = 1,\ldots,n)$ and $1 + x \leq e^x$ $(x \in \mathcal{R})$, implies

$$\mathbf{P}\left\{ \frac{1}{n} \sum_{i=1}^{n} (X_i - \mathbf{E}\{X_i\}) > \epsilon \right\}$$

$$\leq \; e^{-sn\epsilon} \prod_{i=1}^{n} \left(1 + \frac{s^2 \, \mathbf{Var}\{X_i\}}{2} \frac{1}{1 - s(b-a)/3} \right)$$

$$\leq \; e^{-sn\epsilon} \prod_{i=1}^{n} \exp\left(\frac{s^2 \, \mathbf{Var}\{X_i\}}{2} \frac{1}{1 - s(b-a)/3} \right)$$

$$= \; \exp\left(-sn\epsilon + \frac{s^2 n \sigma^2}{2(1 - s(b-a)/3)} \right).$$

Set

$$s = \frac{\epsilon}{\epsilon(b-a)/3 + \sigma^2}.$$

Then

$$\left| \frac{s(b-a)}{3} \right| < 1$$

and

$$-sn\epsilon + \frac{s^2 n \sigma^2}{2(1 - s(b-a)/3)}$$

$$= \frac{-n\epsilon^2}{\epsilon(b-a)/3 + \sigma^2} + \frac{\epsilon^2}{(\epsilon(b-a)/3 + \sigma^2)^2} \cdot \frac{n\sigma^2}{2\left(1 - \frac{\epsilon(b-a)/3}{\epsilon(b-a)/3+\sigma^2}\right)}$$

$$= \frac{-n\epsilon^2}{\epsilon(b-a)/3 + \sigma^2} + \frac{\epsilon^2}{\epsilon(b-a)/3 + \sigma^2} \cdot \frac{n\sigma^2}{2\left(\epsilon(b-a)/3 + \sigma^2 - \epsilon(b-a)/3\right)}$$

$$= \frac{-n\epsilon^2}{2\epsilon(b-a)/3 + 2\sigma^2},$$

hence

$$\mathbf{P}\left\{ \frac{1}{n} \sum_{i=1}^{n} (X_i - \mathbf{E}X_i) > \epsilon \right\} \leq \exp\left(\frac{-n\epsilon^2}{2\epsilon(b-a)/3 + 2\sigma^2} \right).$$

Similarly,

$$\mathbf{P}\left\{\frac{1}{n}\sum_{i=1}^{n}(X_i - \mathbf{E}X_i) < -\epsilon\right\} = \mathbf{P}\left\{\frac{1}{n}\sum_{i=1}^{n}(-X_i - \mathbf{E}\{-X_i\}) > \epsilon\right\}$$

$$\leq \exp\left(\frac{-n\epsilon^2}{2\epsilon(b-a)/3 + 2\sigma^2}\right),$$

which implies the assertion. $\qquad\square$

Lemma A.3. (HOEFFDING (1963)). *Let $X_1, \ldots, X_n$ be independent real-valued random variables, let $a_1, b_1, \ldots, a_n, b_n \in \mathcal{R}$, and assume that $X_i \in [a_i, b_i]$ with probability one ($i = 1, \ldots, n$). Then, for all $\epsilon > 0$,*

$$\mathbf{P}\left\{\left|\frac{1}{n}\sum_{i=1}^{n}(X_i - \mathbf{E}\{X_i\})\right| > \epsilon\right\} \leq 2e^{-\frac{2n\epsilon^2}{\frac{1}{n}\sum_{i=1}^{n}|b_i - a_i|^2}}.$$

PROOF. Let $s > 0$ be arbitrary. Similarly to the proof of Lemma A.2 we get

$$\mathbf{P}\left\{\frac{1}{n}\sum_{i=1}^{n}(X_i - \mathbf{E}X_i) > \epsilon\right\}$$

$$\leq \exp(-sn\epsilon) \cdot \prod_{i=1}^{n}\mathbf{E}\left\{\exp\left(s \cdot (X_i - \mathbf{E}X_i)\right)\right\}.$$

We will show momentarily

$$\mathbf{E}\left\{\exp\left(s \cdot (X_i - \mathbf{E}X_i)\right)\right\} \leq \exp\left(\frac{s^2(b_i - a_i)^2}{8}\right) \quad (i = 1, \ldots, n), \quad (A.2)$$

from which we can conclude

$$\mathbf{P}\left\{\frac{1}{n}\sum_{i=1}^{n}(X_i - \mathbf{E}X_i) > \epsilon\right\} \leq \exp\left(-sn\epsilon + \frac{s^2}{8}\sum_{i=1}^{n}(b_i - a_i)^2\right).$$

The right-hand side is minimal for

$$s = \frac{4n\epsilon}{\sum_{i=1}^{n}(b_i - a_i)^2}.$$

With this value we get

$$\mathbf{P}\left\{\frac{1}{n}\sum_{i=1}^{n}(X_i - \mathbf{E}X_i) > \epsilon\right\}$$

$$\leq \exp\left(-\frac{4n\epsilon^2}{\frac{1}{n}\sum_{i=1}^{n}(b_i - a_i)^2} + \frac{2n\epsilon^2}{\frac{1}{n}\sum_{i=1}^{n}(b_i - a_i)^2}\right)$$

$$= \exp\left(-\frac{2n\epsilon^2}{\frac{1}{n}\sum_{i=1}^{n}(b_i - a_i)^2}\right).$$

This implies that

$$\mathbf{P}\left\{\left|\frac{1}{n}\sum_{i=1}^{n}(X_i - \mathbf{E}X_i)\right| > \epsilon\right\}$$

$$= \mathbf{P}\left\{\frac{1}{n}\sum_{i=1}^{n}(X_i - \mathbf{E}X_i) > \epsilon\right\} + \mathbf{P}\left\{\frac{1}{n}\sum_{i=1}^{n}(-X_i - \mathbf{E}\{-X_i\}) > \epsilon\right\}$$

$$\leq 2\exp\left(-\frac{2n\epsilon^2}{\frac{1}{n}\sum_{i=1}^{n}(b_i - a_i)^2}\right).$$

So it remains to show (A.2). Fix $i \in \{1,\ldots,n\}$ and set

$$Y = X_i - \mathbf{E}X_i.$$

Then $Y \in [a_i - \mathbf{E}X_i, b_i - \mathbf{E}X_i] =: [a, b]$ with probability one, $a - b = a_i - b_i$, and $\mathbf{E}Y = 0$. We have to show

$$\mathbf{E}\left\{\exp(sY)\right\} \leq \exp\left(\frac{s^2(b-a)^2}{8}\right). \tag{A.3}$$

Because of e^{sx} convex we have

$$e^{sx} \leq \frac{x-a}{b-a}e^{sb} + \frac{b-x}{b-a}e^{sa} \quad \text{for all } a \leq x \leq b,$$

thus

$$\mathbf{E}\{\exp(sY)\} \quad \leq \quad \frac{\mathbf{E}\{Y\} - a}{b-a}e^{sb} + \frac{b - \mathbf{E}\{Y\}}{b-a}e^{sa}$$

$$= \quad e^{sa}\left(1 + \frac{a}{b-a} - \frac{a}{b-a}e^{s(b-a)}\right)$$

$$\text{(because of } \mathbf{E}\{Y\} = 0\text{)}.$$

Setting

$$p = -\frac{a}{b-a}$$

we get

$$\mathbf{E}\{\exp(sY)\} \quad \leq \quad (1 - p + p \cdot e^{s(b-a)})e^{-sp(b-a)} = e^{\Phi(s(b-a))},$$

where

$$\Phi(u) = \ln\left((1 - p + pe^u)e^{-pu}\right) = \ln\left(1 - p + pe^u\right) - pu.$$

Next we make a Taylor expansion of Φ. Because of

$$\Phi(0) = 0,$$

$$\Phi'(u) = \frac{pe^u}{1 - p + pe^u} - p, \quad \text{hence } \Phi'(0) = 0$$

and

$$\Phi''(u) = \frac{(1 - p + pe^u)pe^u - pe^u pe^u}{(1 - p + pe^u)^2} = \frac{(1-p)pe^u}{(1 - p + pe^u)^2}$$

$$\leq \frac{(1-p)pe^u}{4(1-p)pe^u} = \frac{1}{4}$$

we get, for any $u > 0$,

$$\Phi(u) = \Phi(0) + \Phi'(0)u + \frac{1}{2}\Phi''(\eta)u^2 \leq \frac{1}{8}u^2$$

for some $\eta \in [0, u]$. We conclude

$$\mathbf{E}\{\exp(sY)\} \leq e^{\Phi(s(b-a))} \leq \exp\left(\frac{1}{8}s^2(b-a)^2\right),$$

which proves (A.3). $\qquad\square$

A.3 Inequalities for Martingales

Let us first recall the notion of martingales. Consider a probability space $(\Omega, \mathcal{F}, \mathbf{P})$.

Definition A.1. *A sequence of integrable random variables $Z_1, Z_2, \ldots$ is called a* martingale *if*

$$\mathbf{E}\{Z_{i+1}|Z_1, \ldots, Z_i\} = Z_i \quad \text{with probability one}$$

for each $i > 0$.

Let $X_1, X_2, \ldots$ be an arbitrary sequence of random variables. The sequence $Z_1, Z_2, \ldots$ of integrable random variables is called a martingale *with respect to the sequence $X_1, X_2, \ldots$ if for every $i > 0$, Z_i is a measurable function of $X_1, \ldots, X_i$, and*

$$\mathbf{E}\{Z_{i+1}|X_1, \ldots, X_i\} = Z_i \quad \text{with probability one.}$$

Obviously, if $Z_1, Z_2, \ldots$ is a martingale with respect to $X_1, X_2, \ldots$, then $Z_1, Z_2, \ldots$ is a martingale, since

$$\mathbf{E}\{Z_{i+1}|Z_1, \ldots, Z_i\} = \mathbf{E}\{\mathbf{E}\{Z_{i+1}|X_1, \ldots, X_i\}|Z_1, \ldots, Z_i\}$$

$$= \mathbf{E}\{Z_i|Z_1, \ldots, Z_i\}$$

$$= Z_i.$$

The most important examples of martingales are sums of independent zero-mean random variables. Let $U_1, U_2, \ldots$ be independent random variables with zero mean. Then the random variables

$$S_i = \sum_{j=1}^{i} U_j, \quad i > 0,$$

form a martingale. Martingales share many properties of sums of independent variables. The role of the independent random variables is played here by a so-called martingale difference sequence.

Definition A.2. *A sequence of integrable random variables $V_1, V_2, \ldots$ is a* martingale difference sequence *if*

$$\mathbf{E}\{V_{i+1}|V_1, \ldots, V_i\} = 0 \quad \text{with probability one}$$

for every $i > 0$.

A sequence of integrable random variables $V_1, V_2, \ldots$ is called a martingale difference sequence *with respect to a sequence of random variables $X_1, X_2, \ldots$ if for every $i > 0$, V_i is a measurable function of $X_1, \ldots, X_i$, and*

$$\mathbf{E}\{V_{i+1}|X_1, \ldots, X_i\} = 0 \quad \text{with probability one.}$$

Again, it is easily seen that if $V_1, V_2, \ldots$ is a martingale difference sequence with respect to a sequence $X_1, X_2, \ldots$ of random variables, then it is a martingale difference sequence. Also, any martingale $Z_1, Z_2, \ldots$ leads naturally to a martingale difference sequence by defining

$$V_i = Z_i - Z_{i-1}$$

for $i > 1$ and $V_1 = Z_1$.

In Problem A.1 we will show the following extension of the Hoeffding inequality:

Lemma A.4. (HOEFFDING (1963), AZUMA (1967)). *Let $X_1, X_2, \ldots$ be a sequence of random variables, and assume that $V_1, V_2, \ldots$ is a martingale difference sequence with respect to $X_1, X_2, \ldots$. Assume, furthermore, that there exist random variables $Z_1, Z_2, \ldots$ and nonnegative constants $c_1, c_2, \ldots$ such that for every $i > 0$, Z_i is a function of $X_1, \ldots, X_{i-1}$, and*

$$Z_i \leq V_i \leq Z_i + c_i \quad a.s.$$

Then, for any $\epsilon > 0$ and n,

$$\mathbf{P}\left\{\sum_{i=1}^{n} V_i \geq \epsilon\right\} \leq e^{-2\epsilon^2/\sum_{i=1}^{n} c_i^2}$$

and

$$\mathbf{P}\left\{\sum_{i=1}^{n} V_i \leq -\epsilon\right\} \leq e^{-2\epsilon^2/\sum_{i=1}^{n} c_i^2}.$$

Next we present a generalization of Hoeffding's inequality, due to Mc-Diarmid (1989). The result will equip us with a powerful tool to handle complicated functions of independent random variables. The inequality has found many applications in combinatorics, as well as in nonparametric statistics (see McDiarmid (1989) and Devroye (1991) for surveys, see also Devroye, Györfi and Lugosi (1996)).

Theorem A.2. (MCDIARMID (1989)). *Let* $Z_1, \ldots, Z_n$ *be independent random variables taking values in a set A and assume that $f : A^n \to \mathcal{R}$ satisfies*

$$\sup_{\substack{z_1,\ldots,z_n, \\ z_i' \in A}} |f(z_1,\ldots,z_n) - f(z_1,\ldots,z_{i-1},z_i',z_{i+1},\ldots,z_n)| \leq c_i, \ 1 \leq i \leq n.$$

Then, for all $\epsilon > 0$,

$$\mathbf{P}\left\{f(Z_1,\ldots,Z_n) - \mathbf{E}f(Z_1,\ldots,Z_n) \geq \epsilon\right\} \leq e^{-2\epsilon^2 / \sum_{i=1}^n c_i^2},$$

and

$$\mathbf{P}\left\{\mathbf{E}f(Z_1,\ldots,Z_n) - f(Z_1,\ldots,Z_n) \geq \epsilon\right\} \leq e^{-2\epsilon^2 / \sum_{i=1}^n c_i^2}.$$

PROOF. See Problem A.2. $\qquad\qquad\square$

The following theorem concerns the variance of a function of independent random variables:

Theorem A.3. (EFRON AND STEIN (1981), STEELE (1986), DEVROYE (1991)). *Let $Z_1,\ldots,Z_n,\tilde{Z}_1,\ldots,\tilde{Z}_n$ be independent m-dimensional random vectors where the two random vectors Z_k and $\tilde{Z}_k$ have the same distribution $(k = 1,\ldots,n)$. For measurable $f : \mathcal{R}^{m\cdot n} \to \mathcal{R}$ assume that $f(Z_1,\ldots,Z_n)$ is square integrable. Then*

$$\mathbf{Var}\{f(Z_1,\ldots,Z_n)\}$$
$$\leq \ \frac{1}{2}\sum_{k=1}^n \mathbf{E}|f(Z_1,\ldots,Z_k,\ldots,Z_n) - f(Z_1,\ldots,\tilde{Z}_k,\ldots,Z_n)|^2.$$

PROOF. Versions of the following simple proof together with applications can be found in Devroye, Györfi, and Lugosi (1996), Lugosi (2002), and Walk (2002a). Set

$$\begin{aligned}
V &= f(Z_1,\ldots,Z_n) - \mathbf{E}f(Z_1,\ldots,Z_n), \\
V_1 &= \mathbf{E}\{V|Z_1\}, \\
V_k &= \mathbf{E}\{V|Z_1,\ldots,Z_k\} - \mathbf{E}\{V|Z_1,\ldots,Z_{k-1}\}, \ k \in \{2,\ldots,n\}.
\end{aligned}$$

$V_1, \ldots, V_n$ form a martingale difference sequence with respect to $Z_1, \ldots, Z_n$, hence, for $k < l$,

$$
\begin{aligned}
\mathbf{E}\{V_k V_l\} &= \mathbf{E}\{\mathbf{E}\{V_k V_l | Z_1, \ldots, Z_{l-1}\} \\
&= \mathbf{E}\{V_k \mathbf{E}\{V_l | Z_1, \ldots, Z_{l-1}\} \\
&= \mathbf{E}\{V_k \cdot 0\} \\
&= 0.
\end{aligned}
$$

We have that

$$
V = \sum_{k=1}^{n} V_k
$$

and

$$
\mathbf{E}V^2 = \sum_{k=1}^{n} \sum_{l=1}^{n} \mathbf{E}\{V_k V_l\} = \sum_{k=1}^{n} \mathbf{E}V_k^2 .
$$

Put $f = f(Z_1, \ldots, Z_n)$. To bound $\mathbf{E}V_k^2$, note that, by Jensen's inequality,

$$
\begin{aligned}
V_k^2 &= \left(\mathbf{E}\{f | Z_1, \ldots, Z_k\} - \mathbf{E}\{f | Z_1, \ldots, Z_{k-1}\} \right)^2 \\
&= \left(\mathbf{E}\left\{ \mathbf{E}\{f | Z_1, \ldots, Z_n\} \right. \right. \\
&\qquad \left. \left. - \mathbf{E}\{f | Z_1, \ldots, Z_{k-1}, Z_{k+1}, \ldots, Z_n\} \middle| Z_1, \ldots, Z_k \right\} \right)^2 \\
&\leq \mathbf{E}\left\{ (f - \mathbf{E}\{f | Z_1, \ldots, Z_{k-1}, Z_{k+1}, \ldots, Z_n\})^2 \middle| Z_1, \ldots, Z_k \right\} ,
\end{aligned}
$$

and, therefore,

$$
\begin{aligned}
\mathbf{E}V_k^2 &\leq \mathbf{E}\left\{ (f - \mathbf{E}\{f | Z_1, \ldots, Z_{k-1}, Z_{k+1}, \ldots, Z_n\})^2 \right\} \\
&= \frac{1}{2} \mathbf{E}\left\{ \left(f(Z_1, \ldots, Z_n) - f(Z_1, \ldots, \tilde{Z}_k, \ldots, Z_n) \right)^2 \right\} ,
\end{aligned}
$$

where at the last step we used the elementary fact that on a conditional probability space given $Z_1, \ldots, Z_{k-1}, Z_{k+1}, \ldots, Z_n$, if U and V are independent and identically distributed random variables, then $\mathbf{Var}(U) = (1/2)\mathbf{E}\{(U - V)^2\}$. $\qquad \square$

A.4   Martingale Convergences

The following results from martingale theory will be used in Chapters 24 and 25:

Definition A.3. *A sequence of integrable random variables $V_1, V_2, \ldots$ is called a* supermartingale *with respect to a sequence of random variables*

$X_1, X_2, \ldots$ *if for every* $i > 0$ *the random variable* V_i *is a measurable function of* $X_1, \ldots, X_i$, *and*

$$\mathbf{E}\{V_{i+1}|X_1, \ldots, X_i\} \leq V_i \quad \textit{with probability one.}$$

In the next theorem we will use the abbreviation $V^- = \max\{0, -V\}$ for the negative part of a random variable V.

Theorem A.4. (SUPERMARTINGALE CONVERGENCE THEOREM; DOOB (1940)). *A supermartingale* $V_1, V_2, \ldots$ *with* $\sup_{i>0} \mathbf{E}\{V_i^-\} < \infty$ *converges with probability one to an integrable random variable.*

For the standard proof see Doob (1953) or Bauer (1996). A simple proof can be found in Johansen and Karush (1966). In the following, another simple proof from Nevelson and Khasminskii (1973) is given, for which the most important tool is the Upcrossing lemma.

To formulate this lemma, we need the following definition: Let $a < b$ be real numbers and let $x_1, \ldots, x_n \in \mathcal{R}$. The *number of upcrossings* of $[a, b]$ by $x_1, \ldots, x_n$ is the number of times the sequence $x_1, \ldots, x_n$ passes from $\leq a$ to $\geq b$ (in one or several steps).

Lemma A.5. (UPCROSSING LEMMA). *Let* $V_1, V_2, \ldots$ *be a supermartingale. For an interval* $[a, b]$, $a < b$, *and* $T \in \mathcal{N}$ *let* $H = H_{a,b}^T$ *be the random number of upcrossings of* $[a, b]$ *by* $V_1, V_2, \ldots, V_T$. *Then*

$$\mathbf{E}H_{a,b}^T \leq \frac{\mathbf{E}\{(a - V_T)^+\}}{b - a}.$$

PROOF. We show first that there is a sequence of binary valued random variables $\{I_k\}$ such that I_k is measurable with respect to $X_1, \ldots, X_k$ and

$$(b - a)H \leq \sum_{t=1}^{T-1} I_t(V_{t+1} - V_t) + (a - V_T)^+.$$

Let t_1 be the first (random) time moment for which $V_t \leq a$ and let t_2 be the first (random) moment after t_1 for which $V_t \geq b$, and let t_3 be the first (random) moment after t_2 for which $V_t \leq a$, etc.

Then during $[t_{2H}, T]$ the process does not have an upcrossing of the interval $[a, b]$ and

$$(b - a)H \leq \sum_{i=1}^{H}(V_{t_{2i}} - V_{t_{2i-1}}).$$

Set $I_1 = 1$ if $t_1 = 1$ (i.e., if $V_1 \leq a$) and $I_1 = 0$ otherwise. Choose $I_2, I_3, \ldots, I_T \in \{0, 1\}$ such that $I_1, I_2, \ldots, I_T$ changes its binary values at the moments (indices) t_k. We distinguish two cases.

CASE 1: $t_{2H+1} \geq T$

This means

$$V_j > a \text{ for } j \in \{t_{2H} + 1, \ldots, T - 1\}$$

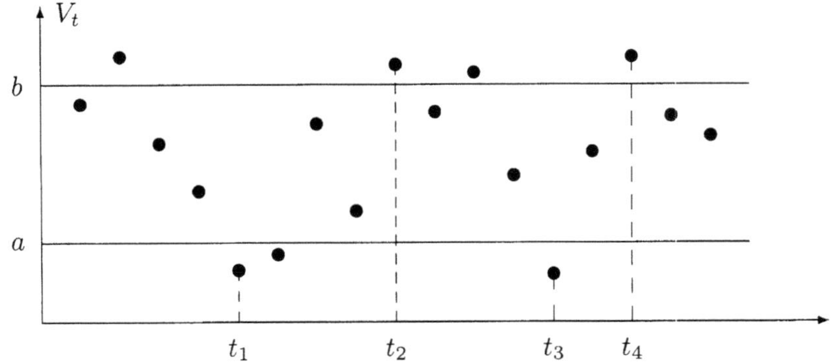

Figure A.3. Crossings of $[a, b]$.

which implies

$$I_{t_{2H}} = I_{t_{2H}+1} = \cdots = I_{T-1} = 0,$$

therefore,

$$\sum_{t=1}^{T-1} I_t(V_{t+1} - V_t) = \sum_{i=1}^{H}(V_{t_{2i}} - V_{t_{2i-1}}) \geq (b-a)H.$$

CASE 2: $t_{2H+1} < T$
This means

$$V_j > a \text{ for } j \in \{t_{2H}+1, \ldots, t_{2H+1}-1\}, \ V_{t_{2H-1}} \leq a,$$

and

$$V_j < b \text{ for } j \in \{t_{2H+1}, \ldots, T\},$$

therefore,

$$I_{t_{2H}} = I_{t_{2H}+1} = \cdots = I_{t_{2H+1}-1} = 0$$

and

$$I_{t_{2H+1}} = I_{t_{2H+1}+1} = \cdots = I_{T-1} = 1,$$

which implies

$$\sum_{t=1}^{T-1} I_t(V_{t+1} - V_t) = \sum_{i=1}^{H}(V_{t_{2i}} - V_{t_{2i-1}}) + V_T - V_{t_{2H+1}}$$

$$\geq \sum_{i=1}^{H}(V_{t_{2i}} - V_{t_{2i-1}}) + (V_T - a)$$

$$\geq (b-a)H - (a - V_T)^+.$$

Hence, in both cases,

$$(b-a)H \le \sum_{t=1}^{T-1} I_t(V_{t+1} - V_t) + (a - V_T)^+$$

and I_k is measurable with respect to $X_1, \ldots, X_k$. Thus

$$
\begin{aligned}
(b-a)\mathbf{E}H_{a,b}^T &\le \sum_{t=1}^{T-1} \mathbf{E}\{I_t(V_{t+1} - V_t)\} + \mathbf{E}\{(a-V_T)^+\} \\
&= \sum_{t=1}^{T-1} \mathbf{E}\{\mathbf{E}\{I_t(V_{t+1} - V_t)|X_1, \ldots X_t\}\} + \mathbf{E}\{(a-V_T)^+\} \\
&= \sum_{t=1}^{T-1} \mathbf{E}\{I_t(\mathbf{E}\{V_{t+1}|X_1, \ldots X_t\} - V_t)\} + \mathbf{E}\{(a-V_T)^+\} \\
&\le \mathbf{E}\{(a-V_T)^+\}.
\end{aligned}
$$

□

PROOF OF THEOREM A.4. Let B be the event that V_t has no limit, that is,

$$B = \left\{ \liminf_{t\to\infty} V_t < \limsup_{t\to\infty} V_t \right\}.$$

Thus

$$B = \bigcup_{r_1 < r_2} \left\{ \liminf_{t\to\infty} V_t < r_1 < r_2 < \limsup_{t\to\infty} V_t \right\},$$

where the union is taken over all rational numbers $r_1 < r_2$, therefore,

$$B = \bigcup_{r_1 < r_2} \{H_{r_1,r_2}^{(\infty)} = \infty\}.$$

By the monotone convergence theorem and the Upcrossing lemma

$$
\begin{aligned}
\mathbf{E}\{H_{r_1,r_2}^{(\infty)}\} &= \mathbf{E}\left\{ \lim_{T\to\infty} H_{r_1,r_2}^T \right\} \\
&= \lim_{T\to\infty} \mathbf{E}\{H_{r_1,r_2}^T\} \\
&\le \limsup_{T\to\infty} \frac{\mathbf{E}\{(r_1 - V_T)^+\}}{r_2 - r_1} \\
&\le \limsup_{T\to\infty} \frac{|r_1| + \mathbf{E}V_T^-}{r_2 - r_1} < \infty,
\end{aligned}
$$

therefore,

$$\mathbf{P}\{H_{r_1,r_2}^{(\infty)} = \infty\} = 0,$$

so

$$\mathbf{P}\{B\} = 0.$$

This proves that $\{V_t\}$ converges with probability one. What is left to show is that

$$\mathbf{E}\left\{\left|\lim_{t\to\infty} V_t\right|\right\} < \infty.$$

By Fatou's lemma

$$
\begin{aligned}
\mathbf{E}\left\{\left|\lim_{t\to\infty} V_t\right|\right\} &= \mathbf{E}\left\{\liminf_{t\to\infty} |V_t|\right\} \\
&\leq \liminf_{t\to\infty} \mathbf{E}\{|V_t|\} \\
&= \liminf_{t\to\infty}(\mathbf{E}\{V_t\} + 2\mathbf{E}\{V_t^-\}) \\
&\leq \liminf_{t\to\infty}(\mathbf{E}\{V_1\} + 2\mathbf{E}\{V_t^-\}) \\
&\qquad \text{(because } V_1, V_2, \ldots \text{ is a supermartingale)} \\
&\leq \mathbf{E}\{V_1\} + 2\sup_t \mathbf{E}\{V_t^-\} < \infty.
\end{aligned}
$$

$\square$

Theorem A.5. (A.S. CONVERGENCE OF ALMOST SUPERMARTINGALES; ROBBINS AND SIEGMUND (1971)). *Let* $V_1, V_2, \ldots$, *also* $\alpha_1, \alpha_2, \ldots$ *and* $\beta_1, \beta_2, \ldots$, *be sequences of integrable nonnegative random variables with* $\sum_{i=1}^{\infty} \mathbf{E}\alpha_i < \infty$, $\sum_{i=1}^{\infty} \mathbf{E}\beta_i < \infty$, *and let* $X_1, X_2, \ldots$ *be a sequence of random variables such that for every* i, V_i, α_i, β_i *are measurable functions of* $X_1, \ldots, X_i$ *and*

$$\mathbf{E}\{V_{i+1}|X_1, \ldots, X_i\} \leq (1 + \alpha_i)V_i + \beta_i.$$

Then the sequence $V_1, V_2, \ldots$, *which is called a nonnegative almost supermartingale, converges with probability one.*

For related results compare also Aizerman, Braverman, and Rozonoer (1964), Gladyshev (1965), MacQueen (1967), and Van Ryzin (1969).

PROOF. The random variables

$$Z_i = \left(\prod_{j=1}^{i-1}(1 + \alpha_j)\right)^{-1} V_i - \sum_{k=1}^{i-1}\left(\prod_{j=1}^{k}(1 + \alpha_j)\right)^{-1} \beta_k, \quad i > 0,$$

form a supermartingale with respect to $X_1, X_2, \ldots$. The relation $\sup_i \mathbf{E}Z_i^- < \infty$ immediately follows from the nonnegativity of α_i, β_i, V_i and $\sum_{k=1}^{\infty} \mathbf{E}\beta_k < \infty$. By Theorem A.4 the sequence $Z_1, Z_2 \ldots$ is convergent with probability one. This, together with the almost sure convergence of the series $\sum_{k=1}^{\infty} \alpha_k$

and that of the sequence $\prod_{j=1}^{i} (1 + \alpha_j)$ (which follows from

$$\prod_{j=1}^{i}(1 + \alpha_j) \le \prod_{j=1}^{i+1}(1 + \alpha_j) \le \prod_{j=1}^{i+1} e^{\alpha_j} \le e^{\sum_{j=1}^{\infty} \alpha_j} < \infty \quad \text{a.s.})$$

and with almost sure convergence of the series $\sum_{k=1}^{\infty} \beta_k$, yields the assertion. $\qquad \square$

The following theorem contains, in the case of independence random variables, a classical theorem of Kolmogorov:

Theorem A.6. (CHOW (1965); CF. STOUT (1974)). *Let $\{Z_i\}$ be a martingale difference sequence with* $\mathbf{Var}(Z_i) = \sigma_i^2$ *and* $c_i > 0$. *If*

$$\sum_{i=1}^{\infty} c_i = \infty \quad \text{and} \quad \sum_{n=1}^{\infty} \frac{\sigma_n^2}{(\sum_{i=1}^{n} c_i)^2} < \infty,$$

then

$$\lim_{n \to \infty} \frac{\sum_{i=1}^{n} Z_i}{\sum_{i=1}^{n} c_i} = 0 \quad a.s.$$

PROOF. Put

$$V_n = \frac{\sum_{i=1}^{n} Z_i}{\sum_{i=1}^{n} c_i}.$$

Then

$$\mathbf{E}\{V_{n+1}^2 | Z_1, \ldots, Z_n\} = \mathbf{E}\left\{ \left| \frac{\sum_{i=1}^{n} Z_i}{\sum_{i=1}^{n+1} c_i} + \frac{Z_{n+1}}{\sum_{i=1}^{n+1} c_i} \right|^2 \Big| Z_1, \ldots, Z_n \right\}$$

$$= \left| \frac{\sum_{i=1}^{n} Z_i}{\sum_{i=1}^{n+1} c_i} \right|^2 + \mathbf{E}\left\{ \left| \frac{Z_{n+1}}{\sum_{i=1}^{n+1} c_i} \right|^2 \Big| Z_1, \ldots, Z_n \right\}$$

$$= \left(\frac{\sum_{i=1}^{n} c_i}{\sum_{i=1}^{n+1} c_i} \right)^2 V_n^2 + \frac{\mathbf{E}\{Z_{n+1}^2 | Z_1, \ldots, Z_n\}}{(\sum_{i=1}^{n+1} c_i)^2}$$

$$\le V_n^2 + \frac{\mathbf{E}\{Z_{n+1}^2 | Z_1, \ldots, Z_n\}}{(\sum_{i=1}^{n+1} c_i)^2},$$

thus by Theorem A.5, V_n^2 is convergent a.s. By the Kronecker lemma (cf. Problem A.6)

$$\lim_{n \to \infty} \frac{\sum_{i=1}^{n} \sigma_i^2}{(\sum_{i=1}^{n} c_i)^2} = \lim_{n \to \infty} \frac{1}{(\sum_{i=1}^{n} c_i)^2} \sum_{k=1}^{n} \left(\sum_{i=1}^{k} c_i \right)^2 \cdot \frac{\sigma_k^2}{\left(\sum_{i=1}^{k} c_i \right)^2} = 0$$

because of $\sum_{i=1}^{\infty} c_i = \infty$, and thus

$$
\begin{aligned}
\lim_{n \to \infty} \mathbf{E}\left\{ \left(\frac{\sum_{i=1}^{n} Z_i}{\sum_{i=1}^{n} c_i} \right)^2 \right\}
&= \lim_{n \to \infty} \frac{1}{\left(\sum_{i=1}^{n} c_i\right)^2} \mathbf{E}\left\{ \left(\sum_{i=1}^{n} Z_i \right)^2 \right\} \\
&= \lim_{n \to \infty} \frac{1}{\left(\sum_{i=1}^{n} c_i\right)^2} \sum_{i=1}^{n} \mathbf{E}\left\{ Z_i^2 \right\} \\
&= \lim_{n \to \infty} \frac{\sum_{i=1}^{n} \sigma_i^2}{\left(\sum_{i=1}^{n} c_i\right)^2} \\
&= 0,
\end{aligned}
$$

therefore the limit of V_n is 0 a.s. $\qquad\square$

Problems and Exercises

PROBLEM A.1. Prove Lemma A.4.

HINT: Show that

$$
\mathbf{E}\left\{ e^{sV_i} | X_1, \ldots, X_{i-1} \right\} \le e^{s^2 c_i^2 / 8},
$$

and, therefore,

$$
\begin{aligned}
\mathbf{P}\left\{ \sum_{i=1}^{n} V_i > \epsilon \right\}
&\le \frac{\mathbf{E}\left\{ \prod_{i=1}^{n} e^{sV_i} \right\}}{e^{s\epsilon}} \\
&= \frac{\mathbf{E}\left\{ \mathbf{E}\left\{ \prod_{i=1}^{n} e^{sV_i} | X_1, \ldots, X_{n-1} \right\} \right\}}{e^{s\epsilon}} \\
&= \frac{\mathbf{E}\left\{ \prod_{i=1}^{n-1} e^{sV_i} \mathbf{E}\left\{ e^{sV_n} | X_1, \ldots, X_{n-1} \right\} \right\}}{e^{s\epsilon}} \\
&\le \frac{\mathbf{E}\left\{ \prod_{i=1}^{n-1} e^{sV_i} e^{s^2 c_n^2 / 8} \right\}}{e^{s\epsilon}} \\
&\le \frac{\prod_{i=1}^{n} e^{s^2 c_i^2 / 8}}{e^{s\epsilon}}.
\end{aligned}
$$

Finish as in the proof of Lemma A.3.

PROBLEM A.2. Prove Theorem A.2.

HINT: Use the decomposition

$$
f(Z_1, \ldots, Z_n) - \mathbf{E}f(Z_1, \ldots, Z_n) = \sum_{i=1}^{n} V_i
$$

of the proof of Theorem A.3 and apply Lemma A.4 for the martingale difference sequence $\{V_n\}$.

PROBLEM A.3. Prove Bernstein's inequality for martingale differences: let $X_1, \ldots, X_n$ be martingale differences such that $X_i \in [a, b]$ with probability one $(i = 1, \ldots, n)$. Assume that, for all i,

$$\mathbf{E}\{X_i^2 | X_1, \ldots, X_{i-1}\} \le \sigma^2 \text{ a.s.}$$

Then, for all $\epsilon > 0$,

$$\mathbf{P}\left\{\left|\frac{1}{n}\sum_{i=1}^{n} X_i\right| > \epsilon\right\} \le 2e^{-\frac{n\epsilon^2}{2\sigma^2 + 2\epsilon(b-a)/3}}.$$

PROBLEM A.4. Show Theorem A.6 by use of Theorem A.4, but without use of Theorem A.5.

PROBLEM A.5. (TOEPLITZ LEMMA). Assume that, for the double array a_{ni},

$$\sup_n \sum_{i=1}^{\infty} |a_{ni}| < \infty$$

and

$$\lim_{n\to\infty} \sum_{i=1}^{\infty} a_{ni} = a$$

with $|a| < \infty$ and, for each fixed i,

$$\lim_{n\to\infty} a_{ni} = 0,$$

moreover, for the sequence b_n,

$$\lim_{n\to\infty} b_n = b.$$

Then

$$\lim_{n\to\infty} \sum_{i=1}^{\infty} a_{ni} b_i = ab.$$

PROBLEM A.6. (KRONECKER LEMMA). Let $x_1, x_2, \ldots$ be a sequence of real numbers such that

$$\sum_{k=1}^{\infty} x_k$$

converges and let $\sigma_1, \sigma_2, \ldots$ be a sequence of positive real numbers which tends monotonically to infinity. Then

$$\frac{1}{\sigma_n} \sum_{k=1}^{n} \sigma_k x_k \to 0 \quad (n \to \infty).$$

HINT: Set $v_n = \sum_{k=n+1}^{\infty} x_k$. Then $v_0 = s$, $x_n = v_{n-1} - v_n$, and

$$\sum_{k=1}^{n} \sigma_k x_k = \sum_{k=1}^{n} \sigma_k(v_{k-1} - v_k) = \sum_{k=1}^{n-1}(\sigma_{k+1} - \sigma_k)v_k + \sigma_1 s - \sigma_n v_n.$$

Apply the Toeplitz lemma.

Notation

- s.u.p.c. strongly universally pointwise consistent.
- RBF radial basis function.
- ANN artificial neural network.
- NN-clustering scheme nearest neighbor-clustering scheme.
- ULLN uniform law of large numbers.
- k-NN k-nearest neighbor.
- a.s. almost surely.
- w.r.t. with respect to.
- w.l.o.g. without loss of generality.
- i.i.d. independent and identically distributed.
- mod μ assertion holds for μ-almost all arguments.
- $\mathcal{Z}$ set of integers $0, \pm 1, \pm 2, \pm 3, \ldots$
- $\mathcal{N}$ set of natural numbers $1, 2, 3, \ldots$.
- $\mathcal{N}_0$ set of nonnegative integers.
- $\mathcal{R}$ set of real numbers.
- $\mathcal{R}_+$ set of nonnegative real numbers.
- $\mathcal{R}^d$ set of d-dimensional real vectors.
- I_A indicator of an event A.
- $I_B(x) = I_{\{x \in B\}}$ indicator function of a set B.
- $|A|$ cardinality of a finite set A.
- A^c complement of a set A.
- $A \triangle B$ symmetric difference of sets A, B.
- $f \circ g$ composition of functions f, g.
- log natural logarithm (base e).
- $\lfloor x \rfloor$ integer part of the real number x.

- $\lceil x \rceil$ upper integer part of the real number x.
- $x^+ = x_+ = \max\{x, 0\}$ positive part of the real number x.
- $x^- = x_- = \max\{-x, 0\}$ negative part of the real number x.
- $X \in \mathcal{R}^d$ observation vector, vector-valued random variable.
- $Y \in \mathcal{R}$ response, real random variable.
- $D_n = \{(X_1, Y_1), \ldots, (X_n, Y_n)\}$ training data, sequence of i.i.d. pairs that are independent of (X, Y), and have the same distribution as that of (X, Y).
- $m(x) = \mathbf{E}\{Y | X = x\}$ regression function.
- $m_n : \mathcal{R}^d \times \{\mathcal{R}^d \times \mathcal{R}\}^n \to \mathcal{R}$ regression estimate. The short notation $m_n(x) = m_n(x, D_n)$ is also used.
- $\mu(A) = \mathbf{P}\{X \in A\}$ probability measure of X.
- $\mu_n(A) = \frac{1}{n} \sum_{i=1}^{n} I_{\{X_i \in A\}}$ empirical measure corresponding to $X_1, \ldots, X_n$.
- λ Lebesgue measure on $\mathcal{R}^d$.
- $x^{(1)}, \ldots, x^{(d)}$ components of the d-dimensional column vector x.
- $(x_1, x_2) = \sum_{j=1}^{d} x_1^{(j)} x_2^{(j)}$ inner product of $x_1, x_2 \in \mathcal{R}^d$.
- $\|x\| = \sqrt{\sum_{i=1}^{d} \left(x^{(i)}\right)^2}$ Euclidean norm of $x \in \mathcal{R}^d$.
- $\|f\| = \left\{\int f(x)^2 \mu(dx)\right\}^{1/2}$ $L_2(\mu)$ norm of $f : \mathcal{R}^d \to \mathcal{R}$.
- $\|f\|_\infty = \sup_{x \in \mathcal{R}^d} |f(x)|$ supremum norm of $f : \mathcal{R}^d \to \mathcal{R}$.
- $\|f\|_{\infty, A} = \sup_{x \in A} |f(x)|$ supremum norm of $f : \mathcal{R}^d \to \mathcal{R}$ restricted to $A \subseteq \mathcal{R}^d$.
- $\mathcal{P}$ partition of $\mathcal{R}^d$.
- $X_{(k,n)}(x), X_{(k,n)}, X_{(k)}(x)$ kth nearest neighbor of x among $X_1, \ldots, X_n$.
- $K : \mathcal{R}^d \to \mathcal{R}$ kernel function.
- $h, h_n > 0$ smoothing factor for a kernel rule.
- $K_h(x) = K(x/h)$ scaled kernel function.
- $\mathcal{F}$ class of functions $f : \mathcal{R}^d \to \mathcal{R}$.
- $\mathcal{F}^+$ class of all subgraphs of functions $f \in \mathcal{F}$.
- $\mathcal{D}$ class of distributions of (X, Y).
- $s(\mathcal{A}, n)$ n-th shatter coefficient of the class $\mathcal{A}$ of sets.
- $V_\mathcal{A}$ Vapnik–Chervonenkis dimension of the class $\mathcal{A}$ of sets.
- $\mathcal{N}\left(\epsilon, \mathcal{G}, \|\cdot\|_{L_p(\nu)}\right)$ ϵ-covering number of $\mathcal{G}$ w.r.t. $\|\cdot\|_{L_p(\nu)}$.
- $\mathcal{N}_p\left(\epsilon, \mathcal{G}, z_1^n\right)$ L_p ϵ-covering number of $\mathcal{G}$ on z_1^n.
- $\mathcal{M}\left(\epsilon, \mathcal{G}, \|\cdot\|_{L_p(\nu)}\right)$ ϵ-packing number of $\mathcal{G}$ w.r.t. $\|\cdot\|_{L_p(\nu)}$.
- $\mathcal{M}_p\left(\epsilon, \mathcal{G}, z_1^n\right)$ L_p ϵ-packing number of $\mathcal{G}$ on z_1^n.
- $S_{x,r} = \{y \in \mathcal{R}^d : \|y - x\| \leq r\}$ closed Euclidean ball in $\mathcal{R}^d$ centered at $x \in \mathcal{R}^d$, with radius $r > 0$.
- $C(\mathcal{R}^d)$ set of all continuous functions $f : \mathcal{R}^d \to \mathcal{R}$.
- $C^l(A)$ set of all l times continuously differentiable functions $f : A \to \mathcal{R}$, $A \subseteq \mathcal{R}^d$.
- $C_0^\infty(\mathcal{R}^d)$ set of all infinitely often continuously differentiable functions $f : \mathcal{R}^d \to \mathcal{R}$ with compact support.

- $z = \arg\min_{x \in D} f(x)$ abbreviation for $z \in D$ and $f(z) = \min_{x \in D} f(x)$.
- $\delta_{j,k}$ Kronecker symbol, equals one if $j = k$ and zero otherwise.
- $\operatorname{diam}(A) = \sup_{x,z \in A} ||x - z||$ diameter of a set $A \subset \mathcal{R}^d$.

Bibliography

Agarwal, G. G. (1989). Splines in statistics. *Bulletin of the Allahabad Mathematical Society*, 4:1–55.

Agarwal, G. G. and Studden, W. J. (1980). Asymptotic integrated mean square error using least squares and bias minimizing splines. *Annals of Statistics*, 8:1307–1325.

Ahmad, I. A. and Lin, P. E. (1976). Nonparametric sequential estimation of a multiple regression function. *Bulletin of Mathematical Statistics*, 17:63–75.

Aizerman, M. A., Braverman, E. M., and Rozonoer, L. I. (1964). The probability problem of pattern recognition learning and the method of potential functions. *Automation and Remote Control*, 25:1175–1190.

Akaike, H. (1954). An approximation to the density function. *Annals of the Institute of Statistical Mathematics*, 6:127–132.

Akaike, H. (1974). A new look at the statistical model identification. *IEEE Transactions on Automatic Control*, 19:716–723.

Alesker, S. (1997). A remark on the Szarek-Talagrand theorem. *Combinatorics, Probability, and Computing*, 6:139–144.

Alexander, K. (1984). Probability inequalities for empirical processes and a law of the iterated logarithm. *Annals of Probability*, 4:1041–1067.

Alexits, G. (1961). *Convergence Problems of Orthogonal Series*. Pergamon Press, Oxford, UK.

Algoet, P. (1992). Universal schemes for prediction, gambling and portfolio selection. *Annals of Probability*, 20:901–941.

Algoet, P. (1994). The strong law of large numbers for sequential decisions under uncertainty. *IEEE Transactions on Information Theory*, 40:609–633.

Algoet, P. (1999). Universal schemes for learning the best nonlinear predictor given the infinite past and side information. *IEEE Transactions on Information Theory*, 45:1165–1185.

Algoet, P. and Györfi, L. (1999). Strong universal pointwise consistency of some regression function estimation. *Journal of Multivariate Analysis*, 71:125–144.

Allen, D. M. (1974). The relationship between variable selection and data augmentation and a method for prediction. *Technometrics*, 16:125–127.

Alon, N., Ben-David, S., and Haussler, D. (1997). Scale-sensitive dimensions, uniform convergence, and learnability. *Journal of the ACM*, 44:615–631.

Amemiya, T. (1985). *Advanced Econometrics*. Harvard University Press, Cambridge, MA.

Andrews, D. W. K. and Whang, Y. J. (1990). Additive interactive regression models: Circumvention of the curse of dimensionality. *Econometric Theory*, 6:466–479.

Antoniadis, A., Grégoire, G., and Vial, P. (1997). Random design wavelet curve smoothing. *Statistics and Probability Letters*, 35:225–232.

Antos, A., Györfi, L., and Kohler, M. (2000). Lower bounds on the rate of convergence of nonparametric regression estimates. *Journal of Statistical Planning and Inference*, 83:91–100.

Antos, A. and Lugosi, G. (1998). Strong minimax lower bound for learning. *Machine Learning*, 30:31–56.

Asmus, V. V., Vadász, V., Karasev, A. B., and Ketskeméty, L. (1987). An adaptive and automatic method for estimating the crops prospects from satellite images (in Russian). *Researches on the Earth from Space (Issledovanie Zemli iz Kosmosa)*, 6:79–88.

Azuma, K. (1967). Weighted sums of certain dependent random variables. *Tohoku Mathematical Journal*, 68:357–367.

Bailey, D. (1976). *Sequential schemes for classifying and predicting ergodic processes*. PhD Thesis, Stanford University.

Baraud, Y., Comte, F., and Viennet, G. (2001). Adaptive estimation in autoregression or β-mixing regression via model selection. *Annals of Statistics*, 29:839–875.

Barron, A. R. (1985). The strong ergodic theorem for densities: generalized Shannon-McMillan-Breiman theorem. *Annals of Probability*, 13:1292–1303.

Barron, A. R. (1989). Statistical properties of artificial neural networks. In *Proceedings of the 28th Conference on Decision and Control*, pages 280–285. Tampa, FL.

Barron, A. R. (1991). Complexity regularization with application to artificial neural networks. In *Nonparametric Functional Estimation and Related Topics*, Roussas, G., editor, pages 561–576. NATO ASI Series, Kluwer Academic Publishers, Dordrecht.

Barron, A. R. (1993). Universal approximation bounds for superpositions of a sigmoidal function. *IEEE Transactions on Information Theory*, 39:930–944.

Barron, A. R. (1994). Approximation and estimation bounds for artificial neural networks. *Machine Learning*, 14:115–133.

Barron, A. R., Birgé, L., and Massart, P. (1999). Risk bounds for model selection via penalization. *Probability Theory and Related Fields*, 113:301–413.

Bartlett, P. L. and Anthony, M. (1999). *Neural Network Learning: Theoretical Foundations*. Cambridge University Press, Cambridge.

Bauer, H. (1996). *Probability Theory*. de Gruyter, New York.

Beck, J. (1979). The exponential rate of convergence of error for k_n-NN nonparametric regression and decision. *Problems of Control and Information Theory*, 8:303–311.

Beirlant, J. and Györfi, L. (1998). On the asymptotic L_2-error in partitioning regression estimation. *Journal of Statistical Planning and Inference*, 71:93–107.

Bellman, R. E. (1961). *Adaptive Control Processes*. Princeton University Press, Princeton, NJ.

Beran, R. (1981). Nonparametric regression with randomly censored survival data. Technical Report, University of California, Berkeley.

Bernstein, S. N. (1946). *The Theory of Probabilities*. Gastehizdat Publishing House, Moscow.

Bhattacharya, P. K. and Mack, Y. P. (1987). Weak convergence of k–NN density and regression estimators with varying k and applications. *Annals of Statistics*, 15:976–994.

Bickel, P. J. and Breiman, L. (1983). Sums of functions of nearest neighbor distances, moment bounds, limit theorems and a goodness of fit test. *Annals of Probability*, 11:185–214.

Bickel, P. J., Klaasen, C. A. J., Ritov, Y., and Wellner, J. A. (1993). *Efficient and Adaptive Estimation for Semiparametric Models*. The Johns Hopkins University Press, Baltimore, MD.

Birgé, L. (1983). Approximation dans les espaces métriques et théorie de l'estimation. *Zeitschrift für Wahrscheinlichkeitstheorie und verwandte Gebiete*, 65:181–237.

Birgé, L. (1986). On estimating a density using Hellinger distance and some other strange facts. *Probability Theory and Related Fields*, 71:271–291.

Birgé, L. and Massart, P. (1993). Rates of convergence for minimum contrast estimators. *Probability Theory and Related Fields*, 97:113–150.

Birgé, L. and Massart, P. (1998). Minimum contrast estimators on sieves: exponential bounds and rate of convergence. *Bernoulli*, 4:329–375.

Birkhoff, G. D. (1931). Proof of the ergodic theorem. *Proceedings of the National Academy of Sciences U.S.A.*, 17:656–660.

Bosq, D. (1996). *Nonparametric Statistics for Stochastic Processes*. Springer-Verlag, New York.

Bosq, D. and Lecoutre, J. P. (1987). *Théorie de l' Estimation Fonctionnelle*. Economica, Paris.

Breiman, L. (1957). The individual ergodic theorem of information theory. *Annals of Mathematical Statistics*, 28:809–811.

Breiman, L. (1993). Fitting additive models to regression data. *Computational Statistics and Data Analysis*, 15:13–46.

Breiman, L. and Friedman, J. H. (1985). Estimating optimal transformations for multiple regression and correlation. *Journal of the American Statistical Association*, 80:580–598.

Breiman, L., Friedman, J. H., Olshen, R. A., and Stone, C. J. (1984). *Classification and Regression Trees*. Wadsworth International, Belmont, CA.

Bretagnolle, J. and Huber, C. (1979). Estimation des densités: risque minimax. *Zeitschrift für Wahrscheinlichkeitstheorie und verwandte Gebiete*, 47:119–137.

Broomhead, D. S. and Lowe, D. (1988). Multivariable functional interpolation and adaptive networks. *Complex Systems*, 2:321–323.

Burman, P. (1990). Estimation of generalized additive models. *Journal of Multivariate Analysis*, 32:230–255.

Cacoullos, T. (1965). Estimation of a multivariate density. *Annals of the Institute of Statistical Mathematics*, 18:179–190.

Carbonez, A., Györfi, L., and van der Meulen, E. C. (1995). Partitioning-estimates of a regression function under random censoring. *Statistics and Decisions*, 13:21–37.

Chen, S., Cowan, C. F. N., and Grant, P. M. (1991). Orthogonal least squares learning algorithm for radial basis networks. *IEEE Transactions on Neural Networks*, 2:302–309.

Chen, Z. (1991). Interaction spline models and their convergence rates. *Annals of Statistics*, 19:1855–1868.

Cheng, P. E. (1995). A note on strong convergence rates in nonparametric regression. *Statistics and Probability Letters*, 24:357–364.

Chernoff, H. (1952). A measure of asymptotic efficiency of tests of a hypothesis based on the sum of observations. *Annals of Mathematical Statistics*, 23:493–507.

Chiu, S. T. (1991). Some stabilized bandwidth selectors for nonparametric regression. *Annals of Statistics*, 19:1528–1546.

Chow, Y. S. (1965). Local convergence of martingales and the law of large numbers. *Annals of Mathematical Statistics*, 36:552–558.

Chui, C. (1992). *Wavelets: A Tutorial in Theory and Applications*. Academic Press, Boston, MA.

Clark, R. M. (1975). A calibration curve for radiocarbon dates. *Antiquity*, 49:251–266.

Cleveland, W. S. (1979). Robust locally weighted regression and smoothing scatterplots. *Journal of the American Statistical Association*, 74:829–836.

Collomb, G. (1977). Quelques proprietes de la méthode du noyau pour l'estimation nonparametrique de la regression en un point fixe. *Comptes Rendus de l'Académie des Sciences de Paris*, 285:28–292.

Collomb, G. (1979). Estimation de la regression par la méthode des k points les plus proches: propriétés de convergence ponctuelle. *Comptes Rendus de l'Académie des Sciences de Paris*, 289:245–247.

Collomb, G. (1980). *Estimation de la regression par la méthode des k points les plus proches avec noyau*. Lecture Notes in Mathematics #821, Springer-Verlag, Berlin. 159–175.

Collomb, G. (1981). Estimation non parametrique de la regression: revue bibliographique. *International Statistical Review*, 49:75–93.

Collomb, G. (1985). Nonparametric regression: an up-to-date bibliography. *Statistics*, 16:300–324.

Cover, T. M. (1968a). Estimation by the nearest neighbor rule. *IEEE Transactions on Information Theory*, 14:50–55.

Cover, T. M. (1968b). Rates of convergence for nearest neighbor procedures. In *Proceedings of the Hawaii International Conference on Systems Sciences*, pages 413–415. Honolulu, HI.

Cover, T. M. (1975). Open problems in information theory. In *1975 IEEE-USSR Joint Workshop on Information Theory*, Forney, G. D., editor, pages 35–36. IEEE Press, New York.

Cover, T. M. and Hart, P. E. (1967). Nearest neighbor pattern classification. *IEEE Transactions on Information Theory*, 13:21–27.

Cox, D. D. (1984). Multivariate smoothing spline functions. *SIAM Journal on Numerical Analysis*, 21:789–813.

Cox, D. D. (1988). Approximation of least squares regression on nested subspaces. *Annals of Statistics*, 16:713–732.

Cox, D. R. (1972). Regression models and life tables. *Journal of the Royal Statistical Society Series B*, 34:187–202.

Csibi, S. (1971). Simple and compound processes in iterative machine learning. Technical Report, CISM Summer Course, Udine, Italy.

Cybenko, G. (1989). Approximations by superpositions of sigmoidal functions. *Mathematics of Control, Signals, and Systems*, 2:303–314.

Dabrowska, D. M. (1987). Nonparametric regression with censored data. *Scandinavian J. Statistics*, 14:181–197.

Dabrowska, D. M. (1989). Uniform consistency of the kernel conditional Kaplan-Meier estimate. *Annals of Statistics*, 17:1157–1167.

Daniel, C. and Wood, F. S. (1980). *Fitting Equations to Data*. Wiley, New York, 2nd edition.

Daubechies, I. (1992). *Ten Lectures on Wavelets*. SIAM, Philadelphia, PA.

Davidson, R. and MacKinnon, J. G. (1993). *Estimation and Inference in Econometrics*. Oxford University Press, New York.

de Boor, C. (1978). *A Practical Guide to Splines*. Springer-Verlag, New York.

Devroye, L. (1978a). The uniform convergence of nearest neighbor regression function estimators and their application in optimization. *IEEE Transactions on Information Theory*, 24:142–151.

Devroye, L. (1978b). The uniform convergence of the Narayada-Watson regression function estimate. *The Canadian Journal of Statistics*, 6:179–191.

Devroye, L. (1981). On the almost everywhere convergence of nonparametric regression function estimates. *Annals of Statistics*, 9:1310–1319.

Devroye, L. (1982a). Bounds on the uniform deviation of empirical measures. *Journal of Multivariate Analysis*, 12:72–79.

Devroye, L. (1982b). Necessary and sufficient conditions for the almost everywhere convergence of nearest neighbor regression function estimates. *Zeitschrift für Wahrscheinlichkeitstheorie und verwandte Gebiete*, 61:467–481.

Devroye, L. (1988). Automatic pattern recognition: A study of the probability of error. *IEEE Transactions on Pattern Analysis and Machine Intelligence*, 10:530–543.

Devroye, L. (1991). Exponential inequalities in nonparametric estimation. In *Nonparametric Functional Estimation and Related Topics*, Roussas, G., editor, pages 31–44. NATO ASI Series, Kluwer Academic Publishers, Dordrecht.

Devroye, L. and Györfi, L. (1983). Distribution-free exponential bound on the L_1 error of partitioning estimates of a regression function. In *Proceedings of the Fourth Pannonian Symposium on Mathematical Statistics*, Konecny, F., Mogyoródi, J., and Wertz, W., editors, pages 67–76. Akadémiai Kiadó, Budapest, Hungary.

Devroye, L. and Györfi, L. (1985). *Nonparametric Density Estimation: The L_1 View*. Wiley, New York.

Devroye, L., Györfi, L., and Krzyżak, A. (1998). The Hilbert kernel regression estimate. *Journal of Multivariate Analysis*, 65:209–227.

Devroye, L., Györfi, L., Krzyżak, A., and Lugosi, G. (1994). On the strong universal consistency of nearest neighbor regression function estimates. *Annals of Statistics*, 22:1371–1385.

Devroye, L., Györfi, L., and Lugosi, G. (1996). *Probabilistic Theory of Pattern Recognition*. Springer-Verlag, New York.

Devroye, L. and Krzyżak, A. (1989). An equivalence theorem for L_1 convergence of the kernel regression estimate. *Journal of Statistical Planning and Inference*, 23:71–82.

Devroye, L. and Lugosi, G. (2001). *Combinatorial Methods in Density Estimation*. Springer-Verlag, New York.

Devroye, L. and Wagner, T. J. (1976). Nonparametric discrimination and density estimation. Technical Report 183, Electronics Research Center, University of Texas.

Devroye, L. and Wagner, T. J. (1979). Distribution-free inequalities for the deleted and holdout error estimates. *IEEE Transactions on Information Theory*, 25:202–207.

Devroye, L. and Wagner, T. J. (1980a). Distribution-free consistency results in nonparametric discrimination and regression function estimation. *Annals of Statistics*, 8:231–239.

Devroye, L. and Wagner, T. J. (1980b). On the L_1 convergence of kernel estimators of regression functions with applications in discrimination. *Zeitschrift für Wahrscheinlichkeitstheorie und verwandte Gebiete*, 51:15–21.

Devroye, L. and Wise, G. L. (1980). Consistency of a recursive nearest neighbor regression function estimate. *Journal of Multivariate Analysis*, 10:539–550.

Dippon, J., Fritz, P., and Kohler, M. (2002). A statistical approach to case based reasoning, with application to breast cancer data. *Computational Statistics and Data Analysis*, (to appear).

Donoho, D. (1997). Cart and best–ortho–basis: a connection. *Annals of Statistics*, 25:1870–1911.

Donoho, D. and Johnstone, I. M. (1994). Ideal spatial adaptation by wavelet shrinkage. *Biometrika*, 81:425–455.

Donoho, D. and Johnstone, I. M. (1998). Minimax estimation via wavelet shrinkage. *Annals of Statistics*, 26:879–921.

Donoho, D., Johnstone, I. M., Kerkyacharian, G., and Picard, D. (1995). Wavelet shrinkage: Asymptopia? *Journal of the Royal Statistical Society, Series B*, 57:301–369.

Doob, J. L. (1940). Regularity properties of certain families of chance variables. *Transactions of the American Mathematical Society*, 47:455–486.

Doob, J. L. (1953). *Stochastic Processes*. Wiley, New York.

Draper, N. R. and Smith, H. (1981). *Applied Regression Analysis, 2nd ed.* Wiley, New York.

Duchon, J. (1976). Interpolation des fonctions de deux variables suivant le principe de la flexion des plaques minces. *RAIRO Analyse Numérique*, 10:5–12.

Dudley, R. M. (1978). Central limit theorems for empirical measures. *Annals of Probability*, 6:899–929.

Dudley, R. M. (1984). A course on empirical processes. In *Ecole de Probabilité de St. Flour 1982, pages 1–142*. Lecture Notes in Mathematics #1097, Springer-Verlag, New York.

Dudley, R. M. (1987). Universal Donsker classes and metric entropy. *Annals of Probability*, 15:1306–1326.

Dyn, N. (1989). Interpolation and approximation by radial basis functions. In *Approximation Theory VI*, Chui, C. K., Schumaker, L. L., and Ward, J. D., editors, pages 211–234. Academic Press, New York.

Efromovich, S. (1999). *Nonparametric Curve Estimation: Methods, Theory, and Applications*. Springer-Verlag, New York.

Efron, B. (1967). The two-sample problem with censored data. In *Proceedings of the Fifth Berkeley Symposium on Mathematical Statistics and Probability*, pages 831–853. University of California Press, Berkeley.

Efron, B. and Stein, C. (1981). The jackknife estimate of variance. *Annals of Statistics*, 9:586–596.

Eggermont, P. P. B. and LaRiccia, V. N. (2001). *Maximum Penalized Likelihood Estimation. Vol. I: Density Estimation*. Springer-Verlag, New York.

Emery, M., Nemirovsky, A. S., and Voiculescu, D. (2000). *Lectures on Probability Theory and Statistics*. Springer-Verlag, New York.

Engel, J. (1994). A simple wavelet approach to nonparametric regression from recursive partitioning schemes. *Journal of Multivariate Analysis*, 49:242–254.

Etemadi, N. (1981). An elementary proof of the strong law of large numbers. *Zeitschrift für Wahrscheinlichkeitstheorie und Verwandte Gebiete*, 55:119–122.

Eubank, R. L. (1999). *Nonparametric Regression and Spline Smoothing*. Marcel Dekker, New York.

Fan, J. (1993). Local linear regression smoothers and their minimax efficiencies. *Annals of Statistics*, 21:196–216.

Fan, J. and Gijbels, I. (1992). Variable bandwidth and local linear regression smoothers. *Annals of Statistics*, 20:2008–2036.

Fan, J. and Gijbels, I. (1995). *Local Polynomial Modeling and its Applications*. Chapman and Hall, London.

Fan, J., Hu, T. C., and Truong, Y. K. (1994). Robust nonparametric function estimation. *Scandinavian Journal of Statistics*, 21:433–446.

Faragó, A. and Lugosi, G. (1993). Strong universal consistency of neural network classifiers. *IEEE Transactions on Information Theory*, 39:1146–1151.

Farebrother, R. W. (1988). *Linear Least Squares Computations*. Marcel Dekker, New York.

Farebrother, R. W. (1999). *Fitting Linear Relationships: A History of the Calculus of Observations 1750–1900*. Springer-Verlag, New York.

Fix, E. and Hodges, J. L. (1951). Discriminatory analysis. Nonparametric discrimination: Consistency properties. Technical Report 4, Project Number 21-49-004, USAF School of Aviation Medicine, Randolph Field, TX.

Fix, E. and Hodges, J. L. (1952). Discriminatory analysis: small sample performance. Technical Report 21-49-004, USAF School of Aviation Medicine, Randolph Field, TX.

Friedman, J. H. (1977). A recursive partitioning decision rule for nonparametric classification. *IEEE Transactions on Computers*, 26:404–408.

Friedman, J. H. (1991). Multivariate adaptive regression splines (with discussion). *Annals of Statistics*, 19:1–141.

Friedman, J. H. and Stuetzle, W. (1981). Projection pursuit regression. *Journal of the American Statistical Association*, 76:817–823.

Friedman, J. H. and Tukey, J. W. (1974). A projection pursuit algorithm for exploratory data analysis. *IEEE Transactions on Computers*, 23:881–889.

Fritz, J. (1974). Learning from ergodic training sequence. In *Limit Theorems of Probability Theory*, Révész, P., editor, pages 79–91. North-Holland, Amsterdam.

Funahashi, K. (1989). On the approximate realization of continuous mappings by neural networks. *Neural Networks*, 2:183–192.

Gaenssler, P. and Rost, D. (1999). *Empirical and Partial-sum Processes; Revisited as Random Measure Processes*. Centre for Mathematical Physics and Stochastics, University of Aarhus, Denmark.

Gallant, A. R. (1987). *Nonlinear Statistical Models*. Wiley, New York.

Gasser, T. and Müller, H.-G. (1979). Kernel estimation of regression functions. In *Smoothing Techniques for Curve Estimation*, Gasser, T. and Rosenblatt, M., editors, pages 23–68. Lecture Notes in Mathematics #757, Springer-Verlag, Heidelberg.

Geman, S. and Hwang, C.-R. (1982). Nonparametric maximum likelihood estimation by the method of sieves. *Annals of Statistics*, 10:401–414.

Gessaman, M. P. (1970). A consistent nonparametric multivariate density estimator based on statistically equivalent blocks. *Annals of Mathematical Statistics*, 41:1344–1346.

Gill, R. D. (1994). Lectures on survival analysis. In *Lectures on Probability Theory*, Bernard, P., editor, pages 115–241. Springer-Verlag.

Giné, E. (1996). Empirical processes and applications: An overview. *Bernoulli*, 2:1–28.

Girosi, F. (1994). Regularization theory, radial basis functions and networks. In *From Statistics to Neural Networks. Theory and Pattern Recognition Applications*, Cherkassky, V., Friedman, J. H., and Wechsler, H., editors, pages 166–187. Springer-Verlag, Berlin.

Girosi, F. and Anzellotti, G. (1992). Convergence rates of approximation by translates. *M.I.T. AI Memo. No.1288*, MIT, Cambridge, MA.

Girosi, F. and Anzellotti, G. (1993). Rates of convergence for radial basis functions and neural networks. In *Artificial Neural Networks for Speech and Vision*, Mammone, R. J., editor, pages 97–113. Chapman and Hall, London.

Gladyshev, E. G. (1965). On stochastic approximation. *Theory of Probability and its Applications*, 10:275–278.

Glick, N. (1973). Sample-based multinomial classification. *Biometrics*, 29:241–256.

Gordon, L. and Olshen, R. A. (1980). Consistent nonparametric regression from recursive partitioning schemes. *Journal of Multivariate Analysis*, 10:611–627.

Gordon, L. and Olshen, R. A. (1984). Almost surely consistent nonparametric regression from recursive partitioning schemes. *Journal of Multivariate Analysis*, 15:147–163.

Greblicki, W. (1974). Asymptotically optimal probabilistic algorithms for pattern recognition and identification. Technical Report, Monografie No. 3, Prace Naukowe Instytutu Cybernetyki Technicznej Politechniki Wroclawskiej No. 18, Wroclaw, Poland.

Greblicki, W. (1978a). Asymptotically optimal pattern recognition procedures with density estimates. *IEEE Transactions on Information Theory*, 24:250–251.

Greblicki, W. (1978b). Pattern recognition procedures with nonparametric density estimates. *IEEE Transactions on Systems, Man and Cybernetics*, 8:809–812.

Greblicki, W., Krzyżak, A., and Pawlak, M. (1984). Distribution-free pointwise consistency of kernel regression estimate. *Annals of Statistics*, 12:1570–1575.

Greblicki, W. and Pawlak, M. (1987a). Necessary and sufficient conditions for Bayes risk consistency of a recursive kernel classification rule. *IEEE Transactions on Information Theory*, 33:408–412.

Greblicki, W. and Pawlak, M. (1987b). Necessary and sufficient consistency conditions for a recursive kernel regression estimate. *Journal of Multivariate Analysis*, 23:67–76.

Green, P. J. and Silverman, B. W. (1994). *Nonparametric Regression and Generalized Linear Models: a Roughness Penalty Approach*. Chapman and Hall, London.

Grenander, U. (1981). *Abstract Inference*. Wiley, New York.

Guerre, E. (2000). Design adaptive nearest neighbor regression estimation. *Journal of Multivariate Analysis*, 75:219–255.

Györfi, L. (1975). An upper bound of error probabilities for multihypothesis testing and its application in adaptive pattern recognition. *Problems of Control and Information Theory*, 5:449–457.

Györfi, L. (1978). On the rate of convergence of nearest neighbor rules. *IEEE Transactions on Information Theory*, 29:509–512.

Györfi, L. (1980). Stochastic approximation from ergodic sample for linear regression. *Zeitschrift für Wahrscheinlichkeitstheorie und Verwandte Gebiete*, 54:47–55.

Györfi, L. (1981). Recent results on nonparametric regression estimate and multiple classification. *Problems of Control and Information Theory*, 10:43–52.

Györfi, L. (1984). Adaptive linear procedures under general conditions. *IEEE Transactions on Information Theory*, 30:262–267.

Györfi, L. (1991). Universal consistencies of a regression estimate for unbounded regression functions. In *Nonparametric Functional Estimation and Related Topics*, Roussas, G., editor, pages 329–338. NATO ASI Series, Kluwer Academic Publishers, Dordrecht.

Györfi, L. and Györfi, Z. (1975). On the nonparametric estimate of a posteriori probabilities of simple statistical hypotheses. In *Colloquia Mathematica Societatis János Bolyai: Topics in Information Theory*, pages 299–308. Keszthely, Hungary.

Györfi, L., Härdle, W., Sarda, P., and Vieu, P. (1989). *Nonparametric Curve Estimation from Time Series*. Springer Verlag, Berlin.

Györfi, L., Kohler, M., and Walk, H. (1998). Weak and strong universal consistency of semi-recursive partitioning and kernel regression estimate. *Statistics and Decisions*, 16:1–18.

Györfi, L. and Lugosi, G. (2002). Strategies for sequential prediction of stationary time series. In *Modeling Uncertainity: An Examination of its Theory, Methods and Applications*, Dror, M., L'Ecuyer, P., and Szidarovszky, F., editors, pages 225–248. Kluwer Academic Publishers, Dordrecht.

Györfi, L., Morvai, G., and Yakowitz, S. (1998). Limits to consistent on-line forecasting for ergodic time series. *IEEE Transactions on Information Theory*, 44:886–892.

Györfi, L., Schäfer, D., and Walk, H. (2002). Relative stability of global errors in nonparametric function estimations. *IEEE Transactions on Information Theory*, (to appear).

Györfi, L. and Walk, H. (1996). On strong universal consistency of a series type regression estimate. *Mathematical Methods of Statistics*, 5:332–342.

Györfi, L. and Walk, H. (1997). Consistency of a recursive regression estimate by Pál Révész. *Statistics and Probability Letters*, 31:177–183.

Hald, A. (1998). *A History of Mathematical Statistics from 1750 to 1930*. Wiley, New York.

Hall, P. (1984). Asymptotic properties of integrated square error and cross-validation for kernel estimation of a regression function. *Zeitschrift für Wahrscheinlichkeitstheorie und verwandte Gebiete*, 67:175–196.

Hall, P. (1988). Estimating the direction in which a data set is most interesting. *Probability Theory and Related Fields*, 80:51–77.

Hall, P. and Turlach, B. A. (1997). Interpolation methods for nonlinear wavelet regression with irregular spaced design. *Annals of Statistics*, 25:1912–1925.

Hamers, M. and Kohler, M. (2001). A bound on the expected maximal deviations of sample averages from their means. *Preprint 2001-9, Mathematical Institute A, University of Stuttgart*.

Härdle, W. (1990). *Applied Nonparametric Regression*. Cambridge University Press, Cambridge, UK.

Härdle, W., Hall, P., and Marron, J. S. (1988). How far are automatically chosen regression smoothing parameter selectors from their optimum? *Journal of the American Statistical Association*, 83:86–101.

Härdle, W. and Kelly, G. (1987). Nonparametric kernel regression estimation: optimal choice of bandwidth. *Statistics*, 1:21–35.

Härdle, W., Kerkyacharian, G., Picard, D., and Tsybakov, A. B. (1998). *Wavelets, Approximation, and Statistical Applications*. Springer-Verlag, New York.

Härdle, W. and Marron, J. S. (1985). Optimal bandwidth selection in nonparametric regression function estimation. *Annals of Statistics*, 13:1465–1481.

Härdle, W. and Marron, J. S. (1986). Random approximations to an error criterion of nonparametric statistics. *Journal of Multivariate Analysis*, 20:91–113.

Härdle, W. and Stoker, T. M. (1989). Investigating smooth multiple regression by the method of average derivatives. *Journal of the American Statistical Association*, 84:986–995.

Harrison, D. and Rubinfeld, D. L. (1978). Hedonic prices and the demand for clean air. *Journal of Environmental Economics and Management*, 5:81–102.

Hart, D. J. (1997). *Nonparametric Smoothing and Lack-of-Fit Tests*. Springer-Verlag, New York.

Hastie, T. and Tibshirani, R. J. (1990). *Generalized Additive Models*. Chapman and Hall, London, U. K.

Hastie, T., Tibshirani, R. J., and Friedman, J. H. (2001). *The Elements of Statistical Learning*. Springer-Verlag, New York.

Haussler, D. (1992). Decision theoretic generalizations of the PAC model for neural net and other learning applications. *Information and Computation*, 100:78–150.

Haussler, D., Kivinen, J., and Warmuth, M. K. (1998). Sequential prediction of individual sequences under general loss functions. *IEEE Transactions on Information Theory*, 44:1906–1925.

Hertz, J., Krogh, A., and Palmer, R. G. (1991). *Introduction to the Theory of Neural Computation*. Addison-Wesley, Redwood City, CA.

Hewitt, E. and Ross, K. A. (1970). *Abstract Harmonic Analysis II*. Springer-Verlag, New York.

Hoeffding, W. (1963). Probability inequalities for sums of bounded random variables. *Journal of the American Statistical Association*, 58:13–30.

Höllig, K. (1998). *Grundlagen der Numerik*. MathText, Zavelstein, Germany.

Hornik, K. (1991). Approximation capabilities of multilayer feedforward networks. *Neural Networks*, 4:251–257.

Hornik, K. (1993). Some new results on neural network approximation. *Neural Networks*, 6:1069–1072.

Hornik, K., Stinchcombe, M., and White, H. (1989). Multi-layer feedforward networks are universal approximators. *Neural Networks*, 2:359–366.

Horowitz, J. L. (1998). *Semiparametric Methods in Econometrics*. Springer-Verlag, New York.

Horváth, L. (1981). On nonparametric regression with randomly censored data. In *Proc. Third Pannonian Symp. on Math. Statist. Visegrad, Hungary*, Mogyoródi, J., Vincze, I., and Wertz, W., editors, pages 105–112. Akadémiai Kiadó, Budapest, Hungary.

Hristache, M., Juditsky, A., Polzehl, J., and Spokoiny, V. G. (2001). Structure adaptive approach for dimension reduction. *Annals of Statistics*, 29:1537–1566.

Huang, J. (1998). Projection esrtimation in multiple regression with applications to functional ANOVA models. *Annals of Statistics*, 26:242–272.

Ibragimov, I. A. and Khasminskii, R. Z. (1980). On nonparametric estimation of regression. *Doklady Akademii Nauk SSSR*, 252:780–784.

Ibragimov, I. A. and Khasminskii, R. Z. (1981). *Statistical Estimation: Asymptotic Theory*. Springer-Verlag, New York.

Ibragimov, I. A. and Khasminskii, R. Z. (1982). On the bounds for quality of nonparametric regression function estimation. *Theory of Probability and its Applications*, 27:81–94.

Ichimura, H. (1993). Semiparametric least squares (sls) and weighted sls estimation of single-index models. *Journal of Econometrics*, 58:71–120.

Johansen, S. and Karush, J. (1966). On the semimartingale convergence theorem. *Annals of Mathematical Statistics*, 37:690–694.

Kaplan, E. L. and Meier, P. (1958). Nonparametric estimation from incomplete observations. *J. Amer. Statist. Assoc.*, 53:457–481.

Katkovnik, V. Y. (1979). Linear and nonlinear methods for nonparametric regression analysis. *Avtomatika*, 5:35–46.

Katkovnik, V. Y. (1983). Convergence of linear and nonlinear nonparametric estimates of "kernel" type. *Automation and Remote Control*, 44:495–506.

Katkovnik, V. Y. (1985). *Nonparametric Identification and Data Smoothing: Local Approximation Approach (in Russian)*. Nauka, Moscow.

Kivinen, J. and Warmuth, M. K. (1999). Averaging expert predictions. In *Computational Learning Theory: Proceedings of the Fourth European Conference, Eurocolt'99*, Simon, H. U. and Fischer, P., editors, pages 153–167. Springer.

Kohler, M. (1998). Nonparametric regression function estimation using interaction least squares splines and complexity regularization. *Metrika*, 47:147–163.

Kohler, M. (1999). Universally consistent regression function estimation using hierarchical B-splines. *Journal of Multivariate Analysis*, 68:138–164.

Kohler, M. (2000a). *Analyse von nichtparametrischen Regressionsschätzern unter minimalen Voraussetzungen. Habilitationsschrift*. Shaker Verlag, Aachen.

Kohler, M. (2000b). Inequalities for uniform deviations of averages from expectations, with applications to nonparametric regression. *Journal of Statistical Planning and Inference*, 89:1–23.

Kohler, M. (2002a). Nonlinear orthogonal series estimates for random design regression. *To appear in Journal of Statistical Planning and Inference*.

Kohler, M. (2002b). Universal consistency of local polynomial kernel estimates. *To appear in Annals of the Institute of Statistical Mathematics*.

Kohler, M. and Krzyżak, A. (2001). Nonparametric regression estimation using penalized least squares. *IEEE Transaction on Information Theory*, 47:3054–3058.

Kohler, M., Krzyżak, A., and Schäfer, D. (2002). Application of structural risk minimization to multivariate smoothing spline regression estimates. *Bernoulli*, 8:1–15.

Kohler, M., Krzyżak, A., and Walk, H. (2002). Strong universal consistency of automatic kernel regression estimates. *Submitted*.

Kohler, M., Máthé, K., and Pintér, M. (2002). Prediction from randomly right censored data. *Journal of Multivariate Analysis*, 80:73–100.

Korostelev, A. P. and Tsybakov, A. B. (1993). *Minimax Theory of Image Reconstruction*. Springer-Verlag, Berlin.

Kovac, A. and Silverman, B. W. (2000). Extending the scope of wavelet regression methods by coefficient-dependent thresholding. *Journal of the American Statistical Association*, 95:172–183.

Kozek, A. S., Leslie, J. R., and Schuster, E. F. (1998). On a universal strong law of large numbers for conditional expectations. *Bernoulli*, 4:143–165.

Krahl, D., Windheuser, U., and Zick, F.-K. (1998). *Data Mining. Einsatz in der Praxis*. Addison–Wesley, Bonn.

Krzyżak, A. (1986). The rates of convergence of kernel regression estimates and classification rules. *IEEE Transactions on Information Theory*, 32:668–679.

Krzyżak, A. (1990). On estimation of a class of nonlinear systems by the kernel regression estimate. *IEEE Transactions on Information Theory*, 36:141–152.

Krzyżak, A. (1991). On exponential bounds on the Bayes risk of the kernel classification rule. *IEEE Transactions on Information Theory*, 37:490–499.

Krzyżak, A. (1992). Global convergence of the recursive kernel regression estimates with applications in classification and nonlinear system estimation. *IEEE Transactions on Information Theory*, IT-38:1323–1338.

Krzyżak, A. (1993). Identification of nonlinear block-oriented systems by the recursive kernel regression estimate. *Journal of the Franklin Institute*, 330(3):605–627.

Krzyżak, A. (2001). Nonlinear function learning using optimal radial basis function networks. *Journal on Nonlinear Analysis*, 47:293–302.

Krzyżak, A. and Linder, T. (1998). Radial basis function networks and complexity regularization in function learning. *IEEE Transactions on Neural Networks*, 9(2):247–256.

Krzyżak, A., Linder, T., and Lugosi, G. (1996). Nonparametric estimation and classification using radial basis function nets and empirical risk minimization. *IEEE Transactions on Neural Networks*, 7(2):475–487.

Krzyżak, A. and Niemann, H. (2001). Convergence and rates of convergence of radial basis functions networks in function learning. *Journal on Nonlinear Analysis*, 47:281–292.

Krzyżak, A. and Pawlak, M. (1983). Universal consistency results for Wolverton-Wagner regression function estimate with application in discrimination. *Problems of Control and Information Theory*, 12:33–42.

Krzyżak, A. and Pawlak, M. (1984a). Almost everywhere convergence of recursive kernel regression function estimates. *IEEE Transactions on Information Theory*, 30:91–93.

Krzyżak, A. and Pawlak, M. (1984b). Distribution-free consistency of a nonparametric kernel regression estimate and classification. *IEEE Transactions on Information Theory*, 30:78–81.

Krzyżak, A. and Schäfer, D. (2002). Nonparametric regression estimation by normalizeed radial basis function networks. *Preprint, Mathematisches Institut A, Universität Stuttgart.*

Kulkarni, S. R. and Posner, S. E. (1995). Rates of convergence of nearest neighbor estimation under arbitrary sampling. *IEEE Transactions on Information Theory*, 41:1028–1039.

Lecoutre, J. P. (1980). Estimation d'une fonction de régression pour par la méthode du regressogramme á blocks équilibrés. *Comptes Rendus de l'Académie des Sciences de Paris*, 291:355–358.

Ledoux, M. (1996). On Talagrand's deviation inequalities for product measures. *ESAIM: Probability and Statistics*, 1:63–87.

Ledoux, M. and Talagrand, M. (1991). *Probability in Banach Space.* Springer-Verlag, New York.

Lee, W. S., Bartlett, P. L., and Williamson, R. C. (1996). Efficient agnostic learning of neural networks with bounded fan-in. *IEEE Transactions on Information Theory*, 42(6):2118–2132.

Li, K. C. (1984). Consistency for cross-validated nearest neighbor estimates in nonparametric regression. *Annals of Statistics*, 12:230–240.

Li, K. C. (1986). Asymptotic optimality of c_l and generalized cross-validation in ridge regression with application to spline smoothing. *Annals of Statistics*, 14:1101–1112.

Li, K. C. (1987). Asymptotic optimality for c_p, c_l cross-validation and generalized cross-validation: discrete index set. *Annals of Statistics*, 15:958–975.

Light, W. A. (1992). Some aspects of radial basis function approximation. In *Approximation Theory, Spline Functions and Applications*, Singh, S. P., editor, pages 163–190. NATO ASI Series, Kluwer Academic Publishers, Dordrecht.

Linton, O. B. (1997). Efficient estimation of additive nonparametric regression models. *Biometrika*, 84:469–474.

Linton, O. B. and Härdle, W. (1996). Estimating additive regression models with known links. *Biometrika*, 83:529–540.

Linton, O. B. and Nielsen, J. B. (1995). A kernel method of estimating structured nonparametric regression based on marginal integration. *Biometrika*, 82:93–100.

Ljung, L., Pflug, G., and Walk, H. (1992). *Stochastic Approximation and Optimization of Random Systems.* Birkhäuser, Basel, Boston, Berlin.

Loftsgaarden, D. O. and Quesenberry, C. P. (1965). A nonparametric estimate of a multivariate density function. *Annals of Mathematical Statistics*, 36:1049–1051.

Lugosi, G. (2002). Pattern classification and learning theory. In *Principles of Nonparametric Learning*, Györfi, L., editor, pages 5–62. Springer-Verlag, Wien NewYork.

Lugosi, G. and Nobel, A. (1996). Consistency of data-driven histogram methods for density estimation and classification. *Annals of Statistics*, 24:687–706.

Lugosi, G. and Nobel, A. (1999). Adaptive model selection using empirical complexities. *Annals of Statistics*, 27:1830–1864.

Lugosi, G. and Zeger, K. (1995). Nonparametric estimation via empirical risk minimization. *IEEE Transactions on Information Theory*, 41:677–678.

Lunts, A. L. and Brailovsky, V. L. (1967). Evaluation of attributes obtained in statistical decision rules. *Engineering Cybernetics*, 3:98–109.

Mack, Y. P. (1981). Local properties of k–nearest neighbor regression estimates. *SIAM Journal on Algebraic and Discrete Methods*, 2:311–323.

MacQueen, J. (1967). Some methods for classification and analysis of multivariate observations. In *Proceedings of the Fifth Berkeley Symposium on Mathematical Statistics and Probability*, Neyman, J., editor, pages 281–297. University of California Press, Berkeley and Los Angeles, California.

Maindonald, J. H. (1984). *Statistical Computation*. Wiley, New York.

Maiorov, V. E. and Meir, R. (2000). On the near optimality of the stochastic approximation of smooth functions by neural networks. *Advances in Computational Mathematics*, 13:79–103.

Maker, P. T. (1940). The ergodic theorem for a sequence of functions. *Duke Math J.*, 6:27–30.

Mallows, C. L. (1973). Some comments on C_p. *Technometrics*, 15:661–675.

Mammen, E. and van de Geer, S. (1997). Locally adaptive regression splines. *Annals of Statistics*, 25:387–413.

Manski, C. F. (1988). Identification of binary response models. *Journal of the American Statistical Association*, 83:729–738.

Massart, P. (1990). The tight constant in the Dvoretzky-Kiefer-Wolfowitz inequality. *Annals of Probability*, 18:1269–1283.

McCaffrey, D. F. and Gallant, A. R. (1994). Convergence rates for single hidden layer feedforward networks. *Neural Networks*, 7(1):147–158.

McCulloch, W. and Pitts, W. (1943). A logical calculus of ideas immanent in neural activity. *Bulletin of Mathematical Biophysics*, 5:115–133.

McDiarmid, C. (1989). On the method of bounded differences. In *Surveys in Combinatorics 1989*, pages 148–188. Cambridge University Press, Cambridge, UK.

Meyer, Y. (1993). *Wavelets: Algorithms and Applications.* SIAM, Philadelphia, PA.

Mhaskar, H. N. (1996). Neural networks for optimal approximation of smooth and analytic functions. *Neural Computation,* 8:164–177.

Mielniczuk, J. and Tyrcha, J. (1993). Consistency of multilayer perceptron regression estimators. *Neural Networks,* 6:1019–1022.

Minsky, M. L. and Papert, S. (1969). *Perceptrons: An Introduction to Computational Geometry.* MIT Press, Cambridge, MA.

Moody, J. and Darken, J. (1989). Fast learning in networks of locally-tuned processing units. *Neural Computation,* 1:281–294.

Morvai, G. (1995). *Estimation of conditional distributions for stationary time series.* PhD Thesis, Technical University of Budapest.

Morvai, G., Yakowitz, S., and Algoet, P. (1997). Weakly convergent nonparametric forecasting of stationary time series. *IEEE Transactions on Information Theory,* 43:483–498.

Morvai, G., Yakowitz, S., and Györfi, L. (1996). Nonparametric inference for ergodic, stationary time series. *Annals of Statistics,* 24:370–379.

Nadaraya, E. A. (1964). On estimating regression. *Theory of Probability and its Applications,* 9:141–142.

Nadaraya, E. A. (1970). Remarks on nonparametric estimates for density functions and regression curves. *Theory of Probability and its Applications,* 15:134–137.

Nadaraya, E. A. (1989). *Nonparametric Estimation of Probability Densities and Regression Curves.* Kluwer Academic Publishers, Dordrecht.

Nemirovsky, A. S., Polyak, B. T., and Tsybakov, A. B. (1983). Estimators of maximum likelihood type for nonparametric regression. *Soviet Mathematics Doklady,* 28:788–792.

Nemirovsky, A. S., Polyak, B. T., and Tsybakov, A. B. (1984). Signal processing by the nonparametric maximum likelihood method. *Problems of Information Transmssion,* 20:177–192.

Nemirovsky, A. S., Polyak, B. T., and Tsybakov, A. B. (1985). Rate of convergence of nonparametric estimators of maximum-likelihood type. *Problems of Information Transmission,* 21:258–272.

Neumann, M. H. and Spokoiny, V. G. (1995). On the efficiency of wavelet estimators under arbitrary error distributions. *Mathematical Methods of Statistics,* 4:137–166.

Nevelson, M. B. and Khasminskii, R. Z. (1973). *Stochastic Approximation and Recursive Estimation.* American Mathematical Society, Providence, R.I.

Newey, W. K. (1994). Kernel estimation of partial means and a general variance estimator. *Econometric Theory*, 10:233–253.

Nicoleris, T. and Yatracos, Y. G. (1997). Rates of convergence of estimates, Kolmogorov's entropy and the dimensionality reduction principle in regression. *Annals of Statistics*, 25:2493– 2511.

Nilsson, N. J. (1965). *Learning Machines: Foundations of Trainable Pattern Classifying Systems*. McGraw-Hill, New York.

Niyogi, P. and Girosi, F. (1996). On the relationship between generalization error, hypothesis complexity, and sample complexity for radial basis functions. *Neural Computation*, 8:819–842.

Nobel, A. (1996). Histogram regression estimation using data dependent partitions. *Annals of Statistics*, 24:1084–1105.

Nobel, A. (1999). Limits to classification and regression estimation from ergodic processes. *Annals of Statistics*, 27:262–273.

Nolan, D. and Pollard, D. (1987). U-processes: Rates of convergence. *Annals of Statistics*, 15:780–799.

Ornstein, D. S. (1974). *Ergodic Theory, Randomness and Dynamical Systems*. Yale University Press, New Haven.

Ornstein, D. S. (1978). Guessing the next output of a stationary process. *Israel Journal of Mathematics*, 30:292–296.

Park, J. and Sandberg, I. W. (1991). Universal approximation using radial-basis-function networks. *Neural Computation*, 3:246–257.

Park, J. and Sandberg, I. W. (1993). Approximation and radial-basis-function networks. *Neural Computation*, 5:305–316.

Parzen, E. (1962). On the estimation of a probability density function and the mode. *Annals of Mathematical Statistics*, 33:1065–1076.

Pawlak, M. (1991). On the almost everywhere properties of the kernel regression estimate. *Annals of the Institute of Statistical Mathematics*, 43:311–326.

Peterson, A. V. (1977). Expressing the Kaplan-Meier estimator as a function of empirical subsurvival functions. *J. Amer. Statist. Assoc.*, 72:854–858.

Pintér, M. (2001). *Consistency results in nonparametric regression estimation and classification*. PhD Thesis, Technical University of Budapest.

Poggio, T. and Girosi, F. (1990). A theory of networks for approximation and learning. *Proceedings of the IEEE*, 78:1481–1497.

Pollard, D. (1984). *Convergence of Stochastic Processes*. Springer-Verlag, New York.

Pollard, D. (1986). Rates of uniform almost sure convergence for empirical processes indexed by unbounded classes of functions. Manuscript.

Pollard, D. (1989). Asymptotics via empirical processes. *Statistical Science*, 4:341–366.

Pollard, D. (1990). *Empirical Processes: Theory and Applications*. NSF-CBMS Regional Conference Series in Probability and Statistics, Institute of Mathematical Statistics, Hayward, CA.

Polyak, B. T. and Tsybakov, A. B. (1990). Asymptotic optimality of the c_p-test for the orthogonal series estimation of regression. *Theory of Probability and its Applications*, 35:293–306.

Powel, J. L. (1994). Estimation of semiparametric models. In *Handbook of Econometrics*, Engle, R. F. and McFadden, D. F., editors, pages 79–91. Elsevier, Amsterdam.

Powel, J. L., Stock, J. H., and Stoker, T. M. (1989). Semiparametric estimation of index coefficients. *Econometrica*, 51:1403–1430.

Powell, M. J. D. (1987). Radial basis functions for multivariable interpolation: a review. In *Algorithms for Approximation*, Mason, J. C. and Cox, M. G., editors, pages 143–167. Oxford University Press, Oxford, UK.

Powell, M. J. D. (1992). The theory of radial basis functions approximation. In *Advances in Numerical Analysis III, Wavelets, Subdivision Algorithms and Radial Basis Functions*, Light, W. A., editor, pages 105–210. Clarendon Press, Oxford, UK.

Prakasa Rao, B. L. S. (1983). *Nonparametric Functional Estimation*. Academic Press, New York.

Rafajlowicz, E. (1987). Nonparametric orthogonal series estimators of regression: a class attaining the optimal convergence rate in L_2. *Statistics and Probability Letters*, 5:213–224.

Rao, R. C. (1973). *Linear Statistical Inference and Its Applications*. Wiley, New York, 2nd edition.

Reinsch, C. (1967). Smoothing by spline functions. *Numerische Mathematik*, 10:177–183.

Rejtő, L. and Révész, P. (1973). Density estimation and pattern classification. *Problems of Control and Information Theory*, 2:67–80.

Révész, P. (1973). Robbins-Monro procedure in a Hilbert space and its application in the theory of learning processes. *Studia Scientiarium Mathematicarum Hungarica*, 8:391–398.

Rice, J. and Rosenblatt, M. (1983). Smoothing splines: regression, derivatives and deconvolution. *Annals of Statistics*, 11:141–156.

Ripley, B. D. (1996). *Pattern Recognition and Neural Networks*. Cambridge University Press, Cambridge, UK.

Robbins, H. and Siegmund, D. (1971). A convergence theorem for nonnegative almost supermartingales and some applications. In *Optimizing Methods in Statistics*, Rustagi, J., editor, pages 233–257. Academic Press, New York.

Robinson, P. M. (1987). Asymptotically efficient estimation in the presence of heteroscedasticity of unknown form. *Econometrica*, 55:167–182.

Robinson, P. M. (1988). Root-n-consistent semiparametric regression. *Econometrica*, 56:931–954.

Rosenblatt, F. (1958). The perceptron: A probabilistic model for information storage and organization in the brain. *Psychological Review*, 65:386–408.

Rosenblatt, F. (1962). *Principles of Neurodynamics: Perceptrons and the Theory of Brain Mechanisms*. Spartan Books, Washington, DC.

Rosenblatt, M. (1956). Remarks on some nonparametric estimates of a density function. *Annals of Mathematical Statistics*, 27:832–837.

Royall, R. M. (1966). *A class of nonparametric estimators of a smooth regression function*. PhD Thesis, Stanford University, Stanford, CA.

Rudin, W. (1964). *Principles of Mathematical Analysis*. McGraw-Hill, New York.

Rudin, W. (1966). *Real and Complex Analysis*. McGraw-Hill, New York.

Rumelhart, D. E., Hinton, G. E., and Williams, R. J. (1986). Learning internal representations by error propagation. In *Parallel Distributed Processing Vol. I*, Rumelhart, D. E., McClelland, J. L., and the PDP Research Group, editors. MIT Press, Cambridge, MA. Reprinted in: J. A. Anderson and E. Rosenfeld, *Neurocomputing—Foundations of Research*, MIT Press, Cambridge, MA, pp. 673–695, 1988.

Rumelhart, D. E. and McClelland, J. L. (1986). *Parallel Distributed Processing: Explorations in the Microstructure of Cognition. Volume 1. Foundations*. MIT Press, Cambridge, MA.

Ryabko, B. Y. (1988). Prediction of random sequences and universal coding. *Problems of Information Transmission*, 24:87–96.

Ryzin, J. V. (1966). Bayes risk consistency of classification procedures using density estimation. *Sankhya Series A*, 28:161–170.

Ryzin, J. V. (1969). On strong consistency of density estimates. *Annals of Mathematical Statistics*, 40:1765–1772.

Samarov, A. M. (1993). Exploring regression structure using nonparametric functional estimation. *Journal of the American Statistical Association*, 88:836–847.

Sauer, N. (1972). On the density of families of sets. *Journal of Combinatorial Theory Series A*, 13:145–147.

Schoenberg, I. J. (1964). Spline functions and the problem of graduation. *Proceedings of the National Academy of Sciences U.S.A.*, 52:947–950.

Schumaker, L. (1981). *Spline Functions: Basic Theory.* Wiley, New York.

Seber, G. A. F. (1977). *Linear Regression Analysis.* Wiley, New York.

Shelah, S. (1972). A combinatorial problem: Stability and order for models and theories in infinity languages. *Pacific Journal of Mathematics*, 41:247–261.

Shen, X. (1998). On the method of penalization. *Statistica Sinica*, 8:337–357.

Shen, X. and Wong, W. H. (1994). Convergence rate of sieve estimates. *Annals of Statistics*, 22:580–615.

Shibata, R. (1976). Selection of the order of an autoregressive model by Akaike's information criterion. *Biometrica*, 63:117–126.

Shibata, R. (1981). An optimal selection of regression variables. *Biometrika*, 68:45–54.

Shields, P. C. (1991). Cutting and stacking: a method for constructing stationary processes. *IEEE Transactions on Information Theory*, 37:1605–1614.

Shorack, G. R. and Wellner, J. A. (1986). *Empirical Processes with Applications.* Wiley, New York.

Simonoff, J. S. (1996). *Smoothing Methods in Statistics.* Springer-Verlag, New York.

Singer, A. and Feder, M. (1999). Universal linear prediction by model order weighting. *IEEE Transactions on Signal Processing*, 47:2685–2699.

Speckman, P. (1985). Spline smoothing and optimal rates of convergence in nonparametric regression models. *Annals of Statistics*, 13:970–983.

Spiegelman, C. and Sacks, J. (1980). Consistent window estimation in nonparametric regression. *Annals of Statistics*, 8:240–246.

Steele, J. M. (1975). *Combinatorial entropy and uniform limit laws.* PhD Thesis, Stanford University, Stanford, CA.

Steele, J. M. (1986). An Efron-Stein inequality for nonsymmetric statistics. *Annals of Statistics*, 14:753–758.

Stein, E. M. (1970). *Singular Integrals and Differentiability Properties of Functions.* Princeton University Press, Princeton, New Jersey, NJ.

Stein, E. M. and Weiss, G. (1971). *Introduction to Fourier Analysis on Euclidean Spaces.* Princeton University Press, Princeton, NJ.

Stigler, S. M. (1999). *Statistics on the Table: The History of Statistical Concepts and Methods.* Harvard University Press, Cambridge, MA.

Stoer, J. (1993). *Numerische Mathematik, Vol. 1.* Springer-Verlag, Berlin.

Stoker, T. M. (1991). Equivalence of direct, indirect and slope estimators of average derivatives. In *Economics and Statistics*, Barnett, W. A., Powel, J., and Tauchen, G., editors, pages 79–91. Cambridge University Press, Cambridge, UK.

Stone, C. J. (1977). Consistent nonparametric regression. *Annals of Statistics*, 5:595–645.

Stone, C. J. (1980). Optimal rates of convergence for nonparametric estimators. *Annals of Statistics*, 8:1348–1360.

Stone, C. J. (1982). Optimal global rates of convergence for nonparametric regression. *Annals of Statistics*, 10:1040–1053.

Stone, C. J. (1985). Additive regression and other nonparametric models. *Annals of Statistics*, 13:689–705.

Stone, C. J. (1994). The use of polynomial splines and their tensor products in multivariate function estimation. *Annals of Statistics*, 22:118–184.

Stone, C. J., Hansen, M. H., Kooperberg, C., and Truong, Y. K. (1997). Polynomial splines and their tensor product in extended linear modeling. *Annals of Statistics*, 25:1371–1410.

Stone, M. (1974). Cross-validatory choice and assessment of statistical predictions. *Journal of the Royal Statistical Society*, 36:111–147.

Stout, W. F. (1974). *Almost Sure Convergence*. Academic Press, New York.

Stute, W. (1984). Asymptotic normality of nearest neighbor regression function estimates. *Annals of Statistics*, 12:917–926.

Stute, W. and Wang, J. L. (1993). The strong law under random censoring. *Annals of Statistics*, 21:1591–1607.

Susarla, V., Tsai, W. Y., and Ryzin, J. V. (1984). A Buckley-James-type estimator for the mean with censored data. *Biometrika*, 71:624–625.

Szarek, S. J. and Talagrand, M. (1997). On the convexified Sauer-Shelah theorem. *Journal of Combinatorial Theory, Series B*, 69:183–192.

Talagrand, M. (1994). Sharper bounds for Gaussian and empirical processes. *Annals of Probability*, 22:28–76.

Thompson, J. W. and Tapia, R. A. (1990). *Nonparametric Function Estimation, Modeling, and Simulation*. SIAM, Philadelphia.

Tsybakov, A. B. (1986). Robust reconstruction of functions by the local-approximation method. *Problems of Information Transmission*, 22:133–146.

Tukey, J. W. (1947). Nonparametric estimation II. Statistically equivalent blocks and tolerance regions. *Annals of Mathematical Statistics*. 18:529–539.

Tukey, J. W. (1961). Curves as parameters and touch estimation. *Proceedings of the Fourth Berkeley Symposium*, pages 681–694.

van de Geer, S. (1987). A new approach to least squares estimation, with applications. *Annals of Statistics*, 15:587–602.

van de Geer, S. (1990). Estimating a regression function. *Annals of Statistics*, 18:907–924.

van de Geer, S. (2000). *Empirical Process in M-Estimation*. Cambridge University Press, New York.

van de Geer, S. (2001). Least squares estimation with complexity penalties. *Report MI 2001-04, Mathematical Institute, Leiden University*.

van de Geer, S. and Wegkamp, M. (1996). Consistency for the least squares estimator in nonparametric regression. *Annals of Statistics*, 24:2513–2523.

van der Vaart, A. W. and Wellner, J. A. (1996). *Weak Convergence and Empirical Processes, with Applications to Statistics*. Springer-Verlag, New York.

Vapnik, V. N. (1982). *Estimation of Dependencies Based on Empirical Data*. Springer-Verlag, New York.

Vapnik, V. N. (1998). *Statistical Learning Theory*. Wiley, New York.

Vapnik, V. N. and Chervonenkis, A. Y. (1971). On the uniform convergence of relative frequencies of events to their probabilities. *Theory of Probability and its Applications*, 16:264–280.

Vapnik, V. N. and Chervonenkis, A. Y. (1974). *Theory of Pattern Recognition*. Nauka, Moscow. (in Russian); German translation: *Theorie der Zeichenerkennung*, Akademie Verlag, Berlin, 1979.

Wahba, G. (1975). Smoothing noisy data with spline functions. *Numerische Mathematik*, 24:383–393.

Wahba, G. (1990). *Spline Models for Observational Data*. SIAM, Philadelphia, PA.

Wahba, G., Wang, Y., Gu, C., Klein, R., and Klein, B. (1995). Smoothing spline ANOVA for exponential families, with application to the Wisconsin epidemiological study of diabetic retinopathy. *Annals of Statistics*, 23:1865–1895.

Wahba, G. and Wold, S. (1975). A completely automatic French curve: fitting spline functions by cross-validation. *Communications in Statistics – Theory and Methods*, 4:1–17.

Walk, H. (1985). Almost sure convergence of stochastic approximation processes. *Statistics and Decisions*, Supplement Issue No. 2:137–141.

Walk, H. (2001). Strong universal pointwise consistency of recursive regression estimates. *Annals of the Institute of Statistical Mathematics*, 53:691–707.

Walk, H. (2002a). Almost sure convergence properties of Nadaraya-Watson regression estimates. In *Modeling Uncertainty: An Examination of its Theory, Methods and Applications*, Dror, M., L'Ecuyer, P., and Szidarovszky, F., editors, pages 201–223. Kluwer Academic Publishers, Dordrecht.

Walk, H. (2002b). On cross–validation in kernel and partitioning regression estimation. *Statistics and Probability Letters*, (to appear).

Walk, H. (2002c). Strong universal consistency of smooth kernel regression estimates. *Preprint 2002-8, Math. Inst. A, Universität Stuttgart*.

Walk, H. and Zsidó, L. (1989). Convergence of the Robbins-Monro method for linear problems in a Banach space. *Journal of Mathematical Analysis and Applications*, 139:152–177.

Wand, M. P. and Jones, M. C. (1995). *Kernel Smoothing*. Chapman and Hall, London.

Watson, G. S. (1964). Smooth regression analysis. *Sankhya Series A*, 26:359–372.

Wegman, E. J. and Wright, I. W. (1983). Splines in statistics. *Journal of the American Statistical Association*, 78:351–365.

Wheeden, R. L. and Zygmund, A. (1977). *Measure and Integral*. Marcel Dekker, New York.

White, H. (1990). Connectionist nonparametric regression: multilayer feedforward networks can learn arbitrary mappings. *Neural Networks*, 3:535–549.

White, H. (1991). Nonparametric estimation of conditional quantiles using neural networks. In *Proceedings of the 23rd Symposium of the Interface: Computing Science and Statistics*, pages 190–199. American Statistical Association, Alexandria, VA.

Whittaker, E. (1923). On a new method of graduation. *Proceedings of the Edinburgh Mathematical Society*, 41:63–75.

Wolverton, C. T. and Wagner, T. J. (1969a). Asymptotically optimal discriminant functions for pattern classification. *IEEE Transactions on Information Theory*, 15:258–265.

Wolverton, C. T. and Wagner, T. J. (1969b). Recursive estimates of probability densities. *IEEE Transactions on Systems, Science and Cybernetics*, 5:246–247.

Wong, W. H. (1983). On the consistency of cross-validation in kernel nonparametric regression. *Annals of Statistics*, 11:1136–1141.

Xu, L., Krzyżak, A., and Oja, E. (1993). Rival penalized competitive learning for clustering analysis, RBF net and curve detection. *IEEE Transactions on Neural Networks*, 4:636–649.

Xu, L., Krzyżak, A., and Yuille, A. L. (1994). On radial basis function nets and kernel regression: approximation ability, convergence rate and receptive field size. *Neural Networks*, 7:609–628.

Yakowitz, S. (1993). Nearest neighbor regression estimation for null-recurrent Markov time series. *Stochastic Processes and their Applications*, 48:311–318.

Yakowitz, S., Györfi, L., Kieffer, J., and Morvai, G. (1999). Strongly consistent nonparametric estimation of smooth regression functions for stationary ergodic sequences. *Journal of Multivariate Analysis*, 71:24–41.

Yakowitz, S. and Heyde, C. (1998). Stationary Markov time series with long range dependence and long, flat segments with application to nonparametric estimation. *Submitted*.

Yamato, H. (1971). Sequential estimation of a continuous probability density function and the mode. *Bulletin of Mathematical Statistics*, 14:1–12.

Zhang, P. (1991). Variable selection in nonparametric regression with continuous covarates. *Annals of Statistics*, 19:1869–1882.

Zhao, L. C. (1987). Exponential bounds of mean error for the nearest neighbor estimates of regression functions. *Journal of Multivariate Analysis*, 21:168–178.

Zhu, L. X. (1992). A note on the consistent estimator of nonparametric regression constructed by splines. *Computers and Mathematics with Applications*, 24:65–70.

Author Index

Subject Index

Springer Series in Statistics

(continued from p. ii)

Kotz/Johnson (Eds.): Breakthroughs in Statistics Volume II.
Kotz/Johnson (Eds.): Breakthroughs in Statistics Volume III.
Küchler/Sørensen: Exponential Families of Stochastic Processes.
Le Cam: Asymptotic Methods in Statistical Decision Theory.
Le Cam/Yang: Asymptotics in Statistics: Some Basic Concepts, 2nd edition.
Liu: Monte Carlo Strategies in Scientific Computing.
Longford: Models for Uncertainty in Educational Testing.
Mielke, Jr./Berry: Permutation Methods: A Distance Function Approach.
Pan/Fang: Growth Curve Models and Statistical Diagnostics.
Parzen/Tanabe/Kitagawa: Selected Papers of Hirotugu Akaike.
Politis/Romano/Wolf: Subsampling.
Ramsay/Silverman: Applied Functional Data Analysis: Methods and Case Studies.
Ramsay/Silverman: Functional Data Analysis.
Rao/Toutenburg: Linear Models: Least Squares and Alternatives.
Reinsel: Elements of Multivariate Time Series Analysis, 2nd edition.
Rosenbaum: Observational Studies, 2nd edition.
Rosenblatt: Gaussian and Non-Gaussian Linear Time Series and Random Fields.
Särndal/Swensson/Wretman: Model Assisted Survey Sampling.
Schervish: Theory of Statistics.
Shao/Tu: The Jackknife and Bootstrap.
Simonoff: Smoothing Methods in Statistics.
Singpurwalla and Wilson: Statistical Methods in Software Engineering:
    Reliability and Risk.
Small: The Statistical Theory of Shape.
Sprott: Statistical Inference in Science.
Stein: Interpolation of Spatial Data: Some Theory for Kriging.
Taniguchi/Kakizawa: Asymptotic Theory of Statistical Inference for Time Series.
Tanner: Tools for Statistical Inference: Methods for the Exploration of Posterior
    Distributions and Likelihood Functions, 3rd edition.
van der Vaart/Wellner: Weak Convergence and Empirical Processes: With
    Applications to Statistics.
Verbeke/Molenberghs: Linear Mixed Models for Longitudinal Data.
Weerahandi: Exact Statistical Methods for Data Analysis.
West/Harrison: Bayesian Forecasting and Dynamic Models, 2nd edition.

GPSR Compliance
The European Union's (EU) General Product Safety Regulation (GPSR) is a set
of rules that requires consumer products to be safe and our obligations to
ensure this.

If you have any concerns about our products, you can contact us on

ProductSafety@springernature.com

In case Publisher is established outside the EU, the EU authorized
representative is:

Springer Nature Customer Service Center GmbH
Europaplatz 3
69115 Heidelberg, Germany

www.ingramcontent.com/pod-product-compliance
Ingram Content Group UK Ltd.
Pitfield, Milton Keynes, MK11 3LW, UK
UKHW020643070726
13597UKWH00018B/128